Springer Series in Materials Science

Volume 215

The Springer Series in Materials Science covers the complete spectrum of materials physics, including fundamental principles, physical properties, materials theory and design. Recognizing the increasing importance of materials science in future device technologies, the book titles in this series reflect the state-of-the-art in understanding and controlling the structure and properties of all important classes of materials.

More information about this series athttp://www.springer.com/series/856

Carlo Massobrio · Jincheng Du
Marco Bernasconi · Philip S. Salmon
Editors

Molecular Dynamics Simulations of Disordered Materials

From Network Glasses to Phase-Change Memory Alloys

Springer

Editors
Carlo Massobrio
IPCMS
Strasbourg University
Strasbourg
France

Marco Bernasconi
Department of Materials Science
University of Milan Bicocca
Milan
Italy

Jincheng Du
Department of Materials Science
and Engineering
University of North Texas
Denton, TX
USA

Philip S. Salmon
Department of Physics
University of Bath
Bath
UK

ISSN 0933-033X ISSN 2196-2812 (electronic)
Springer Series in Materials Science
ISBN 978-3-319-15674-3 ISBN 978-3-319-15675-0 (eBook)
DOI 10.1007/978-3-319-15675-0

Library of Congress Control Number: 2015933149

Springer Cham Heidelberg New York Dordrecht London

Printed on acid-free paper

Springer International Publishing AG Switzerland is part of Springer Science+Business Media (www.springer.com)

Preface

The purpose of this book is to identify current achievements and properly assess the state of the art in the atomic scale modelling of structurally disordered (glassy) materials. More precisely, we intend to bring to the attention of the readership representative examples of systems for which the structural information provided by molecular dynamics has been instrumental in bringing significant progresses in the area of glass science. The underlying motivation of this collection of contributions rests on the notion that glassy materials are intrinsically devoid of regular structural organization.

Early attempts to extract information on glass structure were based on a combination of indirect experimental evidence (quite often obtained by associating measured spectral features with specific crystalline-like motifs) and phenomenological models. The resulting descriptions of the glass structures were highly qualitative and unable to account for the role of chemical bonding in determining the nature of the structural units, their connectivity as well as the extent of their correlation and order. Advances in algorithms and high performing computer facilities capable of handling realistic models and to extend the size and timescale of dynamical simulations have represented a major step forward in promoting this class of simulations to reliable *virtual* experiments. Indeed, recent years have witnessed the advent of atomic scale modelling as a new approach for understanding the properties of glass. This approach is characterized by a clear distinction between the notion of "glasses" as ideal "statistical mechanics" models and their treatment as real materials of interest in material science and technology. By focusing on real glasses a computational material scientist seeks a precise knowledge of structural properties for a given system by using quantitative tools. This strategy is radically different from qualitative assessments that are equally valid and applicable to any disordered system but do not target any correlation between atomic structure and bonding properties. Investigating glasses in the framework of computational material science is a theoretical strategy legitimated by the increased reliability of both classical molecular dynamics (CMD) and first-principles molecular dynamics (FPMD). This will be exemplified in this book and it is fully substantiated by the observation that CMD and FPMD are able to produce models more and more realistic, since their predictive power increases at a very fast pace.

To set the scene for a proper account of relevant issues in the area of disordered network, this volume opens with a contribution (by Philip S. Salmon and Anita Zeidler) having a predominant experimental character and yet containing several useful considerations on the role played by atomic scale modelling in the understanding of short and intermediate range order. While the essence of classical molecular dynamics is intuitively accessible to any practitioner willing to model a system by employing a suitable interatomic potential, the concepts inherent in first-principles molecular dynamics are less straightforward to grasp. This is because FPMD requires the knowledge and the control of a specific methodology combining electronic structure concepts and newtonian dynamics. For this reason, a chapter written by Mauro Boero and co. is devoted to this issue. Moving a further step into the methodology to tackle problems related to the glassy state organization, the contribution by the team of Riccardo Mazzarello focuses on metadynamics as a tool to understand nucleation and phase changes involving the disordered state. Moving into actual modelling of glassy materials, the proper description of iono-covalent bonding is extremely challenging within an effective interatomic potential framework. In the first contribution devoted to modelling of glasses within classical (and yet realistic) molecular dynamics, Liping Huang and John Kieffer are able to describe under which conditions potential models can be used to study archetypical, iono-covalent glass formers. Along the same lines, Pedone and Menziani address the issue of the development of reliable and transferable empirical potentials, optimization of the glass forming procedures and experimental validation of the resulting structures. At the crossroad between simulation methodology (applied to amorphous recrystallization) and realistic modelling of interface phenomena, the classical modelling developed by Evelyne Lampin is able to account for the morphology and the dynamics of a crystal/amorphous interface. In his contribution, Jincheng Du addresses the issue of atomic-scale modelling of multicomponent oxide glasses. Once again, the focus is on the capabilities of classical molecular dynamics to model (with no explicit account of the electronic structure) interactions requiring the account of polarizability for systems that can contain several hundred thousand atoms. The team of Monia Montorsi is also very much concerned by this issue, as shown by the quite realistic modelling of complex transition metal oxides. The section of the book devoted to the applications of classical molecular dynamics models and methods ends with a large series of examples (by Mark Wilson) for silica and carbon system, based on highly refined interatomic potentials containing n-body and/or polarization effects. Interestingly, these models turn out to be quite realistic for systems having different dimensionalities. At the crossroad between classical and first-principles molecular dynamics, Antonio Tilocca addresses a very important issue of glass science, namely the role of these materials in determining and regulating biological functions such as biodegradation. Dynamical effects are nicely highlighted, within the framework of surface reactivity and ion migration. First-principles molecular dynamics (FPMD) is the common ingredient of the last set of contributions, all inspired and nurtured by the predictive power of electronic structure calculations for the potential energy surfaces, combined with newtonian dynamics. For instance, Matthieu Micoulaut establishes the link between the connectivity of such realistic models and the topological constraint theory. The contribution by Assil Bouzid and co. (C. Massobrio/M. Boero team) traces back the modelling of Ge_xSe_{1-x} chalcogenides from the early stages until the last realizations, with a focus on the comparison between $GeSe_4$ and GeS_4 glasses. The peculiar properties of glass surfaces (for silica and chalcogenides) are addressed by Guido Ori and co., with implications for the development of classic force fields based on a consistent definition of charges depending on the local environment. The case of a prototypical network-forming systems based on trigonal units is presented in great detail by Guillaume Ferlat, focussing on the ring structure of glassy B_2O_3. First-principles molecular dynamics approaches have been widely employed in recent years to gain valuable insight into the properties of a very important class of disordered networks, the so-called phase change materials. These are of great interest for optical recording and memory devices. This books ends with five contributions related to the structural and bonding properties of specific chalcogenide alloys very much employed in this context. As shown by Caravati and co., FPMD can also be used as an input for the creation of smart interatomic potentials (the so-called Neural Network ones) enabling realistic crystallization studies on quite large samples (4000 atoms). Structure and crystallization dynamics are also tackled by Jaakko Akola and co. on the prototypical phase change material (PCM hereafter and in the remainder of the book) $Ge_2Sb_2Te_5$, while a large variety of structural behaviours common to sub-systems inherent in the PCMs are reviewed in the contribution by Jean-Yves Raty and co., with a special emphasis for the criterion of structural stability. Finally, the effect of doping on phase change materials is considered in the contribution of the teams headed by David Drabold and Stephen Elliott, respectively. While both contain information on transition metal doping, the second paper also provides information on carbon and nitrogen doping.

Overall, we are convinced that the research efforts presented in this volume are higly representative of the impact of atomic-scale molecular dynamics modelling towards the understanding of structural and topological features of glass. Whenever it appears possible, sufficiently realistic and convenient, glass can be simulated by using interatomic potential, these tools becoming more and more refined, especially when they are derived from electronic structure potential energy surfaces. For situations where the accuracy of first-principle molecular dynamics is required, glasses are studied via a quantitative account of chemical bonding, through first-principles molecular dynamics, yielding trajectories that evolve self-consistently as a function of the network topology and of the changes induced by temperature.

Based on the above assertions, we conclude that molecular dynamics applied to glass has evolved from a computer-based tool complementary to experiments to a reliable and authoritative source of atomic-scale information on its own.

Strasbourg

Carlo Massobrio
Jincheng Du
Marco Bernasconi
Philip S. Salmon

Contents

Contributors

Jaakko Akola Department of Physics, Tampere University of Technology, Tampere, Finland

M. Bauchy Department of Civil and Environmental Engineering, University of California, Los Angeles, CA, USA

Marco Bernasconi Department of Materials Science, University of Milano-Bicocca, Milano, Italy

Christophe Bichara CINAM—CNRS Université Aix-Marseille, Marseille Cedex 9, France

Mauro Boero Institut de Physique et Chimie des Materiaux de Strasbourg (IPCMS), University of Strasbourg—CNRS UMR 7504, Strasbourg, France

Assil Bouzid Institut de Physique et Chimie des Materiaux de Strasbourg (IPCMS), University of Strasbourg—CNRS UMR 7504, Strasbourg, France

Giulia Broglia Department of Science and Methods for Engineering, University of Modena and Reggio Emilia, Reggio Emilia, Italy

Sebastiano Caravati Department of Materials Science, University of Milano-Bicocca, Milano, Italy

B. Coasne Multiscale Materials Science for Energy and Environment, CNRS-MIT (UMI 3466), Cambridge, MA, USA; Institut Charles Gerhard Montpellier, CNRS (UMR 5253), ENSCM, Université Montpellier 2, Montpellier Cedex 5, France; Department of Civil and Environmental Engineering, Massachusetts Institute of Technology, Cambridge, MA, USA

D.A. Drabold Department of Physics and Astronomy, Ohio University, Athens, Ohio, USA

Jincheng Du Department of Materials Science and Engineering, University of North Texas, Denton, TX, USA

S.R. Elliott Department of Chemistry, University of Cambridge, Bath, UK

Guillaume Ferlat IMPMC, Université Pierre et Marie Curie, Paris, France

H. Flores-Ruiz Laboratoire de Physique Thorique de la Matire Condense, Paris Cedex 05, France

Liping Huang Department of Materials Science and Engineering, Rensselaer Polytechnic Institute, Troy, NY, USA

Robert O. Jones Peter Grünberg Institut PGI-1 and JARA/HPC, Forschungszentrum Jülich, Jülich, Germany

Janne Kalikka Singapore University of Technology and Design, Singapore, Singapore

John Kieffer Department of Materials Science and Engineering, University of Michigan, Ann Arbor, MI, USA

Evelyne Lampin Institute of Electronics, Microelectronics and Nanotechnology, Villeneuve d'Ascq Cedex, France

Sébastien Le Roux Institut de Physique et Chimie des Materiaux de Strasbourg (IPCMS), University of Strasbourg—CNRS UMR 7504, Strasbourg, France

T.H. Lee Department of Chemistry, University of Cambridge, Bath, UK

Vincenzo Lordi Lawrence Livermore National Laboratory, Livermore, CA, USA

Carlo Massobrio Institut de Physique et Chimie des Materiaux de Strasbourg (IPCMS), University of Strasbourg—CNRS UMR 7504, Strasbourg, France

Riccardo Mazzarello Institute for Theoretical Solid State Physics, RWTH Aachen, Aachen, Germany

Maria Cristina Menziani Dipartimento di Scienze Chimiche e Geologiche, Università degli Studi di Modena e Reggio Emilia, Modena, Italy

M. Micoulaut Laboratoire de Physique Thorique de la Matire Condense, Paris Cedex 05, France

Monia Montorsi Department of Science and Methods for Engineering, University of Modena and Reggio Emilia, Reggio Emilia, Italy

Consuelo Mugoni Department of Engineering "Enzo Ferrari", University of Modena and Reggio Emilia, Modena, Italy

Guido Ori Multiscale Materials Science for Energy and Environment, CNRS-MIT (UMI 3466), Cambridge, MA, USA; Institut Charles Gerhardt Montpellier, CNRS UMR 5253, University of Montpellier II, ENSCM, Montpellier, France; Institut de Physique et Chimie des Matériaux de Strasbourg, University of Strasbourg—CNRS UMR 7504, Strasbourg, France

Céline Otjacques Physics of Solids, Interfaces and Nanostructures, University of Liége B5, Liége, Belgium

Burak Ozdamar Institut de Physique et Chimie des Materiaux de Strasbourg (IPCMS), University of Strasbourg—CNRS UMR 7504, Strasbourg, France

Alfonso Pedone Dipartimento di Scienze Chimiche e Geologiche, Università degli Studi di Modena e Reggio Emilia, Modena, Italy

Rengin Peköz Max Planck Institute for Polymer Research, Mainz, Germany

Binay Prasai Department of Physics and Astronomy, Ohio University, Athens, Ohio, USA

Jean-Yves Raty Physics of Solids, Interfaces and Nanostructures, University of Liége B5, Liége, Belgium

Ider Ronneberger Institute for Theoretical Solid State Physics, RWTH Aachen, Aachen, Germany

Philip S. Salmon Department of Physics, University of Bath, Bath, UK

J.M. Skelton Department of Chemistry, University of Bath, Bath, UK

Gabriele C. Sosso Faculty of Maths and Physical Sciences, London Centre for Nanotechnology, University College London, London, UK

Antonio Tilocca Department of Chemistry, University College London, London, UK

Christine Tugène Institut de Physique et Chimie des Matériaux de Strasbourg, University of Strasbourg—CNRS UMR 7504, Strasbourg, France

Mark Wilson Physical and Theoretical Chemistry Laboratory, Department of Chemistry, University of Oxford, Oxford, UK

Anita Zeidler Department of Physics, University of Bath, Bath, UK

Chapter 1
The Atomic-Scale Structure of Network Glass-Forming Materials

Philip S. Salmon and Anita Zeidler

Abstract A prerequisite for understanding the physico-chemical properties of network glass-forming materials is knowledge about their atomic-scale structure. The desired information is not, however, easy to obtain because structural disorder in a liquid or glass leads to complexity. It is therefore important to design experiments to give site-specific information on the structure of a given material in order to test the validity of different molecular dynamics models. In turn, once a molecular dynamics scheme contains the correct theoretical ingredients, it can be used both to enrich the information obtained from experiment and to predict the composition and temperature/pressure dependence of a material's properties, a first step in using the principles of rational design to prepare glasses with novel functional properties. In this chapter the symbiotic relationship between experiment and simulation is explored by focussing on the structures of liquid and glassy $ZnCl_2$ and $GeSe_2$, and on the structure of glassy GeO_2 under pressure. Issues to be addressed include extended range ordering on a nanometre scale, the formation of homopolar (like-atom) bonds, and the density-driven mechanisms of network collapse.

1.1 Introduction

Network glass-forming materials are important in a broad range of scientific and technological disciplines, ranging from photonics [1] to magmas in planetary science [2]. It is therefore desirable to have realistic microscopic models of these materials in order to predict their behaviour when different chemical components are added, and when the state conditions are changed. A prerequisite for guiding in the development of a model is unambiguous information from experiment on the atomic-scale structure and dynamics in order to provide a critical test of its predictions.

P.S. Salmon (✉) · A. Zeidler
Department of Physics, University of Bath, Bath BA2 7AY, UK
e-mail: p.s.salmon@bath.ac.uk

A. Zeidler
e-mail: a.zeidler@bath.ac.uk

C. Massobrio et al. (eds.), *Molecular Dynamics Simulations of Disordered Materials*, Springer Series in Materials Science 215,
DOI 10.1007/978-3-319-15675-0_1

Structure refinement methods such as Reverse Monte Carlo (RMC) [3, 4] and Empirical Potential Structure Refinement (EPSR) [5, 6] are widely used by experimentalists to model measured diffraction data. In these methods, the atoms in a 3-dimensional starting model are moved in order to give configurations with diffraction patterns that are in agreement with experiment, subject to imposed constraints such as the measured number density, the inability of neighbouring particles to overlap, and the type and quantity of local structural units as provided by e.g. nuclear magnetic resonance (NMR) and/or extended x-ray absorption fine structure (EXAFS) spectroscopy experiments. The structural models therefore have the benefit of being consistent with the experimental data used in their construction[1] and, since they are based on 3-dimensional particle configurations, information can be obtained on three- and higher-body correlations. The reliability of the structural features in a given model will, however, depend on the sensitivity of the experimental data to the relevant correlations, the results for higher body correlations need to be treated with caution because diffraction data provides information only at the pair-correlation function level, and the final configurations can be sensitive to the choice of starting model as shown by work on SiO_2 glass [7, 8] and water [9–11]. For this reason, it is usually best to construct a realistic starting model so that the use of RMC or EPSR amounts to a refinement of that model using the experimental results as a reference.[2] Owing to the nature of their construction, RMC or EPSR models do not provide information on the particle dynamics, and since the modelling procedures are driven by experimental data they cannot be used if this information is unavailable i.e. the refinement methods have in this sense no predictive power.

Molecular dynamics methods, which are extensively used to model the structure of network glass-forming materials, also provide the atomic-scale dynamics (e.g. the vibrational density of states and self-diffusion coefficients), thus enriching the information made available on a given material. A comparison of this dynamical information with experiment can provide a particularly severe test for the validity of a model for a particular material. Furthermore, if the theory underlying the calculations has the correct ingredients then the simulations can be used to predict the composition and temperature/pressure dependence of a material's properties. Oftentimes, the search for the correct theoretical ingredients is not, however, trivial and different approaches involve trade-offs between e.g. the accuracy in describing a particular bonding scheme versus the number of atoms that can be dealt with on a realistic computational timescale.

For example, ionic interaction models can give an accurate description of the measured structure of glass-forming systems such as $ZnCl_2$ [14–17], provided that anion polarisation effects are taken into account [18–20], and the relative simplicity of these models allows for the coverage of relatively long length and time scales. Such models are, however, inappropriate for glass-forming materials such as $GeSe_2$ where

[1] In the literature, the results from RMC or EPSR models are sometimes erroneously referred to as 'experimental results' when comparisons are made with molecular dynamics simulations.

[2] Increasingly, molecular dynamics is being used to provide the starting models for refinement procedures, see e.g. [12, 13].

the electronegativity difference between the atomic species is small and homopolar (like-atom) bonds are prevalent [21–23]. These features necessitate a first-principles density-functional based approach in which the electronic structure is taken into explicit account but where the simulation results can be sensitive to the choice of density functional [24–40]. These methods allow only for the investigation of relatively small systems for short times, although this may not be such an important issue when investigating e.g. the operation of phase-change memory alloys where the pertinent length and time scales are small relative to those associated with glass formation. In the investigation of glass-forming materials, there is also the question as how best to prepare accurate molecular dynamics models given the use of fast simulated quench-rates [37, 41–43].

In the following, the role of experiment in guiding molecular dynamics simulations of network glass-forming systems will be illustrated by considering a small set of materials with the MX_2 stoichiometry. Particular attention will be paid to the results obtained from the method of neutron diffraction with isotope substitution (NDIS) since it has been extensively used to obtain information at the partial structure factor level. An excellent starting point is provided by molten MX_2 salts where NDIS results have helped in the development of a reliable ionic interaction model for glass-forming materials like $ZnCl_2$. Next, the $GeSe_2$ system is considered where NDIS results have played a major role in the continuing development of first-principles molecular dynamics methods for describing the structure and properties of this and other chalcogenide glass-formers.[3] Finally, GeO_2 glass is considered where the results from recent in situ high-pressure NDIS experiments are helping to arbitrate between competing molecular dynamics models for the density-driven network collapse.

1.2 Outline Diffraction Theory

In a neutron diffraction experiment on a liquid or glassy MX_2 system, the coherent scattered intensity measured with respect to the magnitude of the scattering vector k can be represented by the total structure factor [44]

$$F(k) = c_M^2 b_M^2 [S_{MM}(k) - 1] + 2c_M c_X b_M b_X [S_{MX}(k) - 1] + c_X^2 b_X^2 [S_{XX}(k) - 1] \quad (1.1)$$

where c_α and b_α denote the atomic fraction and bound coherent scattering length of chemical species α, respectively. $S_{\alpha\beta}(k)$ is a so-called Faber-Ziman [45] partial structure factor which is related to the partial pair-distribution function $g_{\alpha\beta}(r)$ by the Fourier transform relation

[3] Chalcogenide glass-forming materials are those containing one or more of the chalcogen elements S, Se and Te.

$$g_{\alpha\beta}(r) = 1 + \frac{1}{2\pi^2 \rho r} \int_0^\infty \mathrm{d}k\, k \left[S_{\alpha\beta}(k) - 1 \right] \sin(kr), \tag{1.2}$$

where ρ is the atomic number density of the system and r is a distance in real space. The mean coordination number of atoms of type β, contained in a volume defined by two concentric spheres of radii r_1 and r_2 centred on an atom of type α, is given by

$$\bar{n}_\alpha^\beta = 4\pi\, \rho\, c_\beta \int_{r_1}^{r_2} \mathrm{d}r\, r^2 g_{\alpha\beta}(r). \tag{1.3}$$

The full set of $S_{\alpha\beta}(k)$ functions for an MX_2 system can be extracted from the measured diffraction patterns by applying the NDIS method, provided that isotopes are available with a sufficiently large neutron scattering length contrast [44, 46, 47].

The total structure factor can also be expressed in terms of the Bhatia-Thornton [48] number-number, concentration-concentration and number-concentration partial structure factors denoted by $S_{NN}(k)$, $S_{CC}(k)$ and $S_{NC}(k)$, respectively. These partial structure factors are related to fluctuations (in the liquid or glass) of the number density, concentration and their cross-correlation, respectively. Equation (1.1) can be re-written as

$$F(k) = \langle b \rangle^2 \left[S_{NN}(k) - 1 \right] + c_M c_X (b_M - b_X)^2 \left\{ \left[S_{CC}(k)/c_M c_X \right] - 1 \right\} + 2 \langle b \rangle (b_M - b_X) S_{NC}(k) \tag{1.4}$$

where $\langle b \rangle = c_M b_M + c_X b_X$ is the average coherent neutron scattering length. The relationships between the two sets of partial structure factors are given by

$$S_{NN}(k) = c_M^2 S_{MM}(k) + c_X^2 S_{XX}(k) + 2 c_M c_X S_{MX}(k), \tag{1.5}$$

$$S_{CC}(k) = c_M c_X \left\{ 1 + c_M c_X \left[S_{MM}(k) + S_{XX}(k) - 2 S_{MX}(k) \right] \right\}, \tag{1.6}$$

$$S_{NC}(k) = c_M c_X \left\{ c_M \left[S_{MM}(k) - S_{MX}(k) \right] - c_X \left[S_{XX}(k) - S_{MX}(k) \right] \right\}. \tag{1.7}$$

The Fourier transforms of $S_{NN}(k)$, $S_{CC}(k)$ and $S_{NC}(k)$ are the partial pair-distribution functions $g_{NN}(r)$, $g_{CC}(r)$ and $g_{NC}(r)$, respectively. The relationships between the $g_{IJ}(r)$ (I, J = N, C) and $g_{\alpha\beta}(r)$ (α, β = M, X) functions are given by

$$g_{NN}(r) = c_M^2 g_{MM}(r) + c_X^2 g_{XX}(r) + 2 c_M c_X g_{MX}(r), \tag{1.8}$$

$$g_{CC}(r) = c_M c_X \left[g_{MM}(r) + g_{XX}(r) - 2 g_{MX}(r) \right], \tag{1.9}$$

$$g_{NC}(r) = c_M \left[g_{MM}(r) - g_{MX}(r) \right] - c_X \left[g_{XX}(r) - g_{MX}(r) \right]. \tag{1.10}$$

If $b_M = b_X$ the incident neutrons in a diffraction experiment cannot distinguish between the different scattering nuclei and the measured total structure factor gives $S_{NN}(k)$ directly (see (1.4)). The corresponding Fourier transform $g_{NN}(r)$ therefore describes the sites of the scattering nuclei and, since it cannot distinguish between the chemical species that decorate those sites, it gives information on the topological ordering. If $\langle b \rangle = 0$, however, the measured total structure factor gives $S_{CC}(k)$ directly and its Fourier transform $g_{CC}(r)$ describes the chemical ordering of the M and X atomic species. The $g_{CC}(r)$ function will have a positive or negative peak at a given distance when there is a preference for like or unlike neighbours, respectively (see (1.9)). The $g_{NC}(r)$ function describes the correlation between the sites described by $g_{NN}(r)$ and their occupancy by a given chemical species.

In practice, a diffractometer can only access a finite k-space range with a maximum cutoff value k_{max}. Provided that sufficiently small k-values can be accessed, a reciprocal-space function such as $F(k)$ will therefore be truncated by a modification function given by $M(k) = 1$ for $k \leq k_{max}$ and $M(k) = 0$ for $k > k_{max}$. In consequence, the real-space information corresponding to $F(k)$ is obtained by the Fourier transform relation

$$G(r) = \frac{1}{2\pi^2 \rho r} \int_0^\infty dk\, k\, F(k) M(k) \sin(kr). \tag{1.11}$$

The desired r-space information is therefore convoluted with the Fourier transform of $M(k)$, the effect of which becomes negligible if k_{max} is sufficiently large that $F(k)$ is featureless at higher k-values. To give smoother r-space functions, other expressions for $M(k)$ are used such as the Lorch [49] modification function where $M(k) = \sin(\pi k/k_{max})/(\pi k/k_{max})$ for $k \leq k_{max}$ and $M(k) = 0$ for $k > k_{max}$.[4] To facilitate a like-for-like comparison between measured and molecular dynamics results, the reciprocal-space functions constructed from simulations are often Fourier transformed according to (1.11) with k_{max} set at the experimental value.

1.3 Ionic Interaction Models for MX_2 Glass-Forming Materials

The NDIS method has been used to measure the full set of partial structure factors for molten salts with the MX_2 stoichiometry. The effect on the structure of varying the cation to anion size ratio was thereby investigated for liquid $BaCl_2$ [51], $SrCl_2$ [52], $CaCl_2$ [53], $MgCl_2$ [54], $NiCl_2$ [55] and $ZnCl_2$ [14] where the radius of Cl^- is 1.81 Å and the cation radii are 1.35 Å (Ba^{2+}), 1.18 Å (Sr^{2+}), 1.00 Å (Ca^{2+}), 0.72 Å (Mg^{2+}), 0.69 Å (Ni^{2+}) and 0.74 Å (Zn^{2+}) [56].[5] Of these liquids, only $ZnCl_2$ readily forms a glass by bulk-quenching methods, and corner-sharing $ZnCl_4$ tetrahedra are the predominant structural motifs.

---

[4] A rigorous derivation of the Lorch modification function and its corresponding real-space representation is given in [50].

[5] The radii correspond to six-fold coordinated ions.

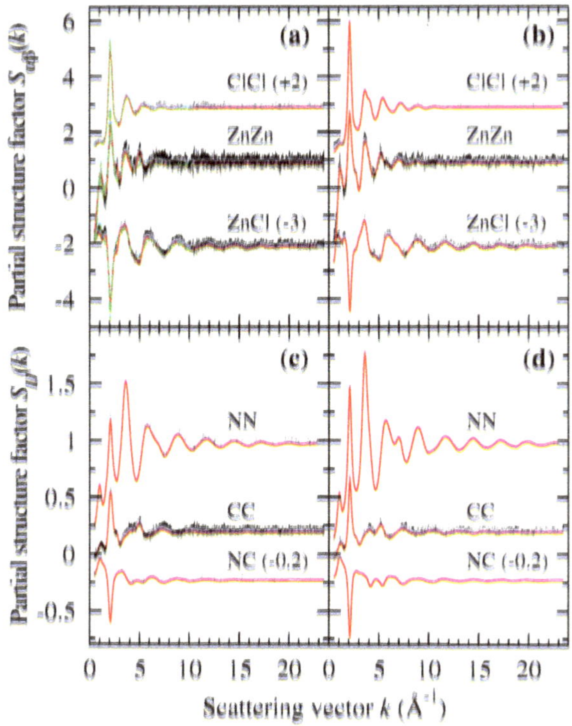

Fig. 1.1 The Faber-Ziman $S_{\alpha\beta}(k)$ (α, β = M, X) and Bhatia-Thornton $S_{IJ}(k)$ (I, J = N, C) partial structure factors for liquid and glassy $ZnCl_2$. The points with vertical (*black*) error bars are the measured functions in (**a**) and (**c**) for the liquid at 332(5) °C [16] and in (**b**) and (**d**) for the glass at 25(1) °C [15, 16]. The solid (*red*) curves are the Fourier back transforms of the corresponding partial pair-distribution functions after the unphysical oscillations at r-values smaller than the distance of closest approach between the centres of two atoms are set to the calculated limit at $r = 0$. The broken (*green*) curves in (**a**) are from the polarisable ion model of Sharma and Wilson [63] for the liquid at 327 °C

The full set of partial structure factors recently measured for liquid and glassy $ZnCl_2$ are shown in Fig. 1.1 [15, 16]. The prominent first sharp diffraction peak (FSDP) in $S_{ZnZn}(k)$ at a scattering vector $k_{FSDP} \simeq 1$ Å^{-1} is a signature of structural complexity on an intermediate length scale with a periodicity given by $2\pi/k_{FSDP}$ and with a correlation length given by $2\pi/\Delta k_{FSDP}$ where Δk_{FSDP} is the full-width at half-maximum of the FSDP [57]. As shown in Fig. 1.1, the principal peaks[6] in the Faber-Ziman partial structure factors align at a common scattering vector $k_{PP} \simeq 2.1$ Å^{-1} and it follows from (1.5)–(1.7) that the principal peaks in the Bhatia-Thornton [48] partial structure factors $S_{IJ}(k)$ also align at this common position. The measured $S_{NN}(k)$ function for the liquid shows a clear 'three-peak' character that is not shared

[6] A so-called principal peak or trough at $k_{PP} \simeq 2$–3 Å^{-1} is a common feature in the partial structure factors for liquid and glassy materials [47].

with the other molten salts listed above, and all of the partial structure factors $S_{IJ}(k)$ (I, J = N, C) for both the liquid and glass display an FSDP [58, 59] e.g. there are concentration fluctuations on an intermediate length scale that will be discussed further in Sect. 1.4.3.

The experimental results for molten $ZnCl_2$ feature a nearest-neighbour Zn–Zn distance that is comparable to the nearest-neighbour Cl–Cl distance. This observation is not expected on the basis of a rigid ion model (RIM) for the interionic interactions in which the ions are non-deformable and the Coulomb repulsion between divalent cations is large. The experimental results for molten $ZnCl_2$ have therefore been attributed to a manifestation of 'covalent' effects in the bonding [60]. As shown by Wilson and Madden [18], however, it is possible to describe the structure of $ZnCl_2$ within the framework of an ionic interaction model, provided that account is taken of the anion polarisability α_X. The effect of this polarisability is shown in Fig. 1.2 where two simulations are made on an MX_2 system in which the M^{2+} and X^- ions take full formal charges but α_X is either set to zero, corresponding to a RIM, or set to 20 au, corresponding to a polarisable ion model (PIM) [61]. An FSDP develops in $S_{MM}(k)$ at $k_{FSDP} \simeq 1.2$ Å^{-1} as the anion polarisability is increased to $\alpha_X = 20$ au and the principal peaks in all three of the Faber-Ziman partial structure factors align at a common value $k_{PP} \simeq 2$ Å^{-1}. The anion polarisation shields the Coulomb repulsion between divalent cations which reduces the mean M-X-M bond angle between MX_4 tetrahedra, leading to a shortening of the mean M-M distance relative to the RIM. This shielding leads to regions in which there is either an enhanced or diminished

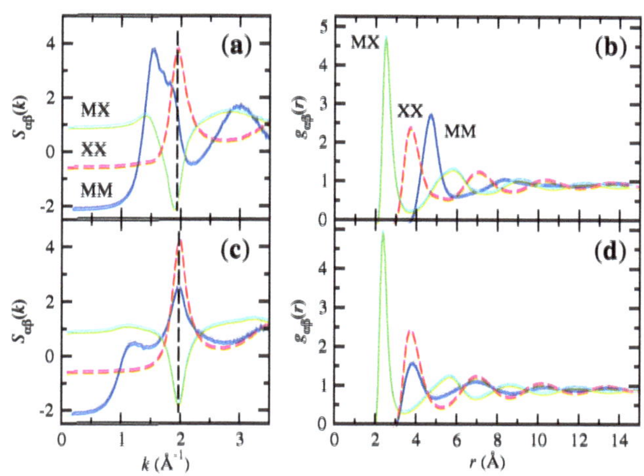

Fig. 1.2 The Faber-Ziman partial structure factors $S_{\alpha\beta}(k)$ and partial pair-distribution functions $g_{\alpha\beta}(r)$ (α, β = M, X) as calculated for models using two different values for the anion polarisability α_X [61]. The curves in (**a**) and (**b**) correspond to a rigid ion model (RIM) with $\alpha_X = 0$, while the curves in (**c**) and (**d**) correspond to a polarisable ion model (PIM) with $\alpha_X = 20$ au. The introduction of anion polarisability leads to the appearance of an FSDP in $S_{MM}(k)$ at $k_{FSDP} \simeq 1.2$ Å^{-1} and to an alignment of the principal peaks in all three $S_{\alpha\beta}(k)$ functions at $k_{PP} \simeq 2$ Å^{-1}. The alignment of the principal peaks in (**c**) arises from in-phase large-r oscillations in the $g_{\alpha\beta}(r)$ functions shown in (**d**)

cation density relative to a RIM [62] i.e. there is a modulation of the cation-cation correlations on an intermediate length scale that gives rise to the FSDP in $S_{MM}(k)$. The $S_{\alpha\beta}(k)$ functions predicted for liquid $ZnCl_2$ by using a PIM with $\alpha_X = 20$ au [63] are shown in Fig. 1.1a.

On cooling a liquid to form a glass, there is a sharpening of the peaks in the measured partial structure factors in accordance with a loss of thermal disorder (Fig. 1.1). Since the FSDP is already a sharp feature and is the peak that occurs at the smallest k-value, it might be expected to dominate the large-r behaviour of the partial pair-distribution functions. This is not, however, the case as can be shown by investigating the Bhatia-Thornton partial pair-correlation functions $rh_{NN}(r) \equiv r\,[g_{NN}(r) - 1]$, $rh_{CC}(r) \equiv rg_{CC}(r)$ and $rh_{NC}(r) \equiv rg_{NC}(r)$, which enable a separation of the contributions to the structure from topological versus chemical ordering [15, 50, 64, 65]. As shown in Fig. 1.3, the measured $rh_{IJ}(r)$ functions for $ZnCl_2$ glass show

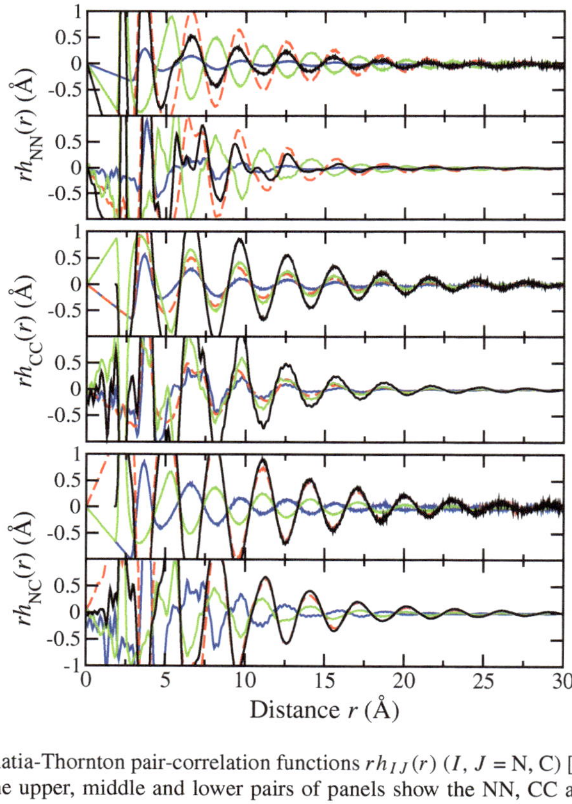

Fig. 1.3 The Bhatia-Thornton pair-correlation functions $rh_{IJ}(r)$ $(I, J = \text{N, C})$ [solid dark (*black*) curves] where the upper, middle and lower pairs of panels show the NN, CC and NC functions, respectively. For each pair, the upper panel gives the function obtained for a polarisable ion model (PIM) with $\alpha_X = 20$ au [20] and the lower panel gives the measured function for glassy $ZnCl_2$ [15, 16]. Each function is broken down into its contributions from $rh_{XX}(r)$ [broken (*red*) curves], $rh_{MX}(r)$ [light solid (*green*) curves] and $rh_{MM}(r)$ [solid (*blue*) curves]. The abscissa for the simulated functions are scaled by 1.98/2.09 to reflect the relative positions of the principal peak in the simulated and measured $S_{NN}(k)$ partial structure factors

large-r oscillations that extend to distances of several nanometres, well beyond the regime associated with the FSDP, with a common periodicity given by $2\pi/k_{PP}$ and a common decay length that is related to $2\pi/\Delta k_{PP}$ where Δk_{PP} is the full-width at half-maximum of a principal peak. The number of correlated ions is therefore large e.g. 4060 for a sphere of radius 30 Å in glassy $ZnCl_2$ where $\rho = 0.0359$ Å^{-3} [16]. The character of this extended range ordering is captured by the PIM with α_X = 20 au as indicated in Fig. 1.2d by the in-phase oscillations in the $g_{\alpha\beta}(r)$ functions at large r-values, and by the $rh_{IJ}(r)$ functions illustrated in Fig. 1.3. A PIM therefore reproduces all of the main features in the structure of $ZnCl_2$ that are observed by experiment.

1.3.1 Simple Theory for Extended Range Ordering

The character of the extended range ordering in network glass-forming materials such as $ZnCl_2$ can be addressed by using simple theory. Let the pair-potential describing the interactions between two ions labelled by i and j separated by a distance r be represented by a RIM given by the expression [66]

$$\phi_{ij}(r) = \phi_{ij}^{\text{sr}}(r) + \frac{Z_i Z_j e^2}{\epsilon\, r} - \frac{A_{ij}}{r^6} \tag{1.12}$$

where $Z_i e$ is the charge on the ith ion, e is the elementary charge, $\varepsilon \equiv 4\pi\varepsilon_r\varepsilon_0$, ε_r is the dimensionless relative dielectric constant of the medium in which the ions are embedded, and ε_0 is the vacuum permittivity. In this equation, $\phi_{ij}^{\text{sr}}(r)$ describes the short-ranged repulsive interactions, $\phi_{ij}^{\text{Coul}}(r) \propto r^{-1}$ describes the Coulomb interactions, and $\phi_{ij}^{\text{disp}}(r) = -A_{ij}r^{-6}$ describes the dispersion interactions where the parameter A_{ij} ($\geq$0) depends on the ion polarisability [67].

For this RIM, a simple power-law dependence for the ultimate decay of the pair correlation functions is expected i.e. $rh_{NN}(r) \to r^{-5}$, $rh_{CC}(r) \to r^{-9}$ and $rh_{NC}(r) \to r^{-7}$ [50, 68]. However, if the dispersion terms are absent in (1.12), then a pole analysis of the k-space solutions to the Ornstein-Zernike equations following the method of Evans and co-workers [69, 70] leads, in the case when the system density is sufficiently high, to the following expressions for the asymptotic decay of the partial pair-correlation functions [50]

$$rh_{NN}(r) \to 2|\mathscr{A}_{NN}| \exp(-a_0 r) \cos(a_1 r - \theta_{NN}), \tag{1.13}$$

$$rh_{CC}(r) \to 2c_M c_X |\mathscr{A}_{CC}| \exp(-a_0 r) \cos(a_1 r - \theta_{CC}), \tag{1.14}$$

$$rh_{NC}(r) \to 2|\mathscr{A}_{NC}| \exp(-a_0 r) \cos(a_1 r - \theta_{NC}). \tag{1.15}$$

The $rh_{IJ}(r)$ are therefore exponentially damped oscillatory functions with a common decay length given by a_0^{-1} and a common wavelength for the oscillations given by

$2\pi/a_1$. The $\mathscr{A}_{IJ}$ are complex numbers with amplitudes related by $|\mathscr{A}_{NN}||\mathscr{A}_{CC}| = |\mathscr{A}_{NC}|^2$ and the phases are related by $\theta_{NN} + \theta_{CC} = 2\theta_{NC}$. Equations (1.13)–(1.15) also hold for binary mixtures of hard-spheres having different diameters, i.e. when both the Coulomb and dispersion terms are absent from (1.12), where the common wavelength of oscillation is set by one or other of the hard sphere sizes depending on the thermodynamic conditions [71]. The effect on (1.13)–(1.15) of introducing anion polarisability has yet to be fully explored.

1.3.2 Relative Fragility of Tetrahedral Glass-Forming MX_2 Liquids

A systematic variation of the anion polarisability α_X within a PIM has been used to investigate the relative "fragility" of network glass-forming MX_2 liquids in which the predominant structural motifs are MX_4 tetrahedra [61]. The fragility is a measure of the rate at which the dynamical properties of a liquid change on approaching the glass transition temperature T_g and can be quantified in terms of a fragility index $m = \mathrm{d}\log_{10}\eta/\mathrm{d}\left(T_g/T\right)|_{T=T_g}$ where η is the liquid viscosity and T is the absolute temperature [72, 73]. Figure 1.4a shows the measured relation between

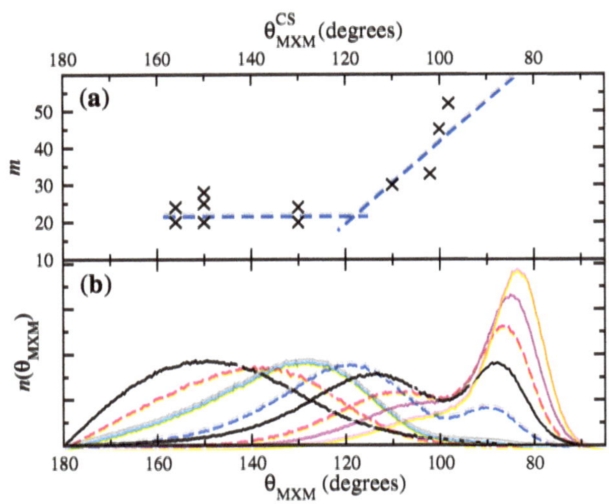

Fig. 1.4 **a** The dependence of the measured fragility index m on the M–X–M bond angle for corner-sharing tetrahedra θ^{CS}_{MXM} for a series of MX_2 network glass-forming materials. The measured θ^{CS}_{MXM} values correspond, from left to right, to BeF_2 [77], SiO_2 [78], GeO_2 [78], $ZnCl_2$ [77], GeS_2 [79], $ZnBr_2$ (estimated) and $GeSe_2$ [59]. The fragility values are taken from [73–76]. **b** The M–X–M bond angle distribution $n(\theta_{MXM})$ as calculated using a polarisable ion model (PIM) where the curves, appearing from left to right, correspond to anion polarisability α_X values of 0, 5, 10, 15, 17.5, 20, 22.5 and 25 au, respectively. The figure is taken from Wilson and Salmon [61]

m [73–76] and the mean M–X–M bond angle for corner-sharing MX_4 tetrahedra θ^{CS}_{MXM} [59, 77–79]. The fragility is small and approximately invariant for large θ^{CS}_{MXM} values, characteristic of networks dominated by corner-sharing units in systems like BeF_2, SiO_2 and GeO_2 [77, 78], but increases when θ^{CS}_{MXM} reduces below $\simeq 120°$ and edge-sharing units become numerous in systems like GeS_2 and $GeSe_2$ [59, 79]. The molecular dynamics simulations also show this trend, where the M–X–M bond angle distribution $n(\theta_{MXM})$ for different α_X values is shown in Fig. 1.4(b) and the associated fragility was assessed from the temperature dependence of the cation self-diffusion coefficient [61]. As α_X is increased above $\sim$15 au, the fragility increases as a second peak due to edge-sharing units appears in n (θ_{MXM}) at an angle smaller than the peak associated with θ^{CS}_{MXM}. This trend towards increasing fragility with increasing fraction of edge-sharing motifs is also anticipated for other glass-forming liquids [61].

The relative fragility of tetrahedral MX_2 network glass-forming materials manifests itself in the relative importance of the FSDP versus the principal peak in the measured $S_{NN}(k)$ functions [59, 65]. For example, in the relatively fragile glass-forming system $ZnCl_2$ the anion packing fraction in the glass is large at 0.647(9) and the mean inter-tetrahedral M–X–M bond angle is 111°, whereas in the strong glass-forming system GeO_2 the anion packing fraction in the glass is much smaller at 0.495(5) and the mean M–X–M bond angle is larger at 132(2)°. These differences between dense and more open networks of tetrahedra lead to a principal peak in $S_{NN}(k)$ that is a more prominent feature than the FSDP for more fragile glass-formers, with the converse relation holding for strong glass-formers. Hence, there is competition between the intermediate and extended range ordering in these MX_2 materials that is won by the latter with increasing density.

1.4 Covalent Effects in MX_2 Glass-Forming Materials: Structure of Liquid and Glassy $GeSe_2$

In Sect. 1.3 'ionic' network glass-forming systems were considered, where the properties of materials like $ZnCl_2$ can be reproduced by using an interaction model based on discrete closed-shell ions with integer charges i.e. there was no need to consider 'covalent' interactions that arise from the formation of chemical bonds in which pairs of electrons are shared between atoms. Indeed, many material properties that have been attributed to 'covalency' may in fact be explained in terms of 'ionic' interactions provided that effects such as polarisation, compression and deformation are taken into explicit account [19]. We now consider the prototypical network glass-forming material $GeSe_2$ where the small electronegativity difference between Ge and Se and the observation of broken chemical order precludes the successful employment of an ionic interaction model, necessitating a first-principles molecular dynamics approach [35, 37].

1.4.1 Diffraction Results for Liquid and Glassy $GeSe_2$

The NDIS method has been used to measure the full set of partial structure factors for liquid $GeSe_2$ at 784(3) °C and for glassy $GeSe_2$ at 26(1) °C [21–23]. Several of the main results are presented in the following, while a more complete discussion of these results in the context of other experimental work on liquids and glasses in the binary Ge-Se system is given elsewhere [47, 80].

The Faber-Ziman $S_{\alpha\beta}(k)$ and corresponding $g_{\alpha\beta}(r)$ functions are shown in Figs. 1.5, 1.6 and 1.7, and the Bhatia-Thornton $S_{IJ}(k)$ functions are shown in

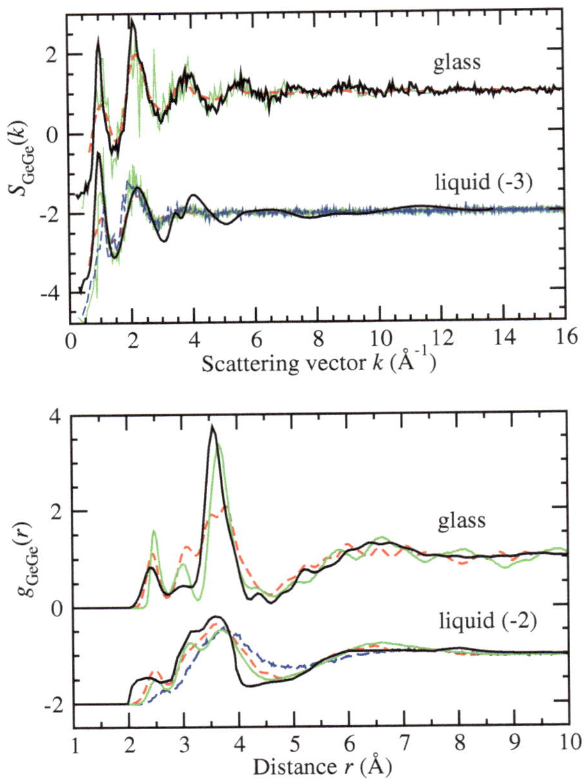

Fig. 1.5 The measured and simulated $S_{GeGe}(k)$ and $g_{GeGe}(r)$ functions for liquid and glassy $GeSe_2$. The dark solid (*black*) curves give the measured functions for the liquid [21] and glass [22], where a spline fit to the liquid state $S_{GeGe}(k)$ function is shown for clarity of presentation. For the liquid, the light broken (*red*) curves show the LDA results of Cobb and Drabold [25], the dark broken (*blue*) curves show the LDA results of Massobrio et al. [30], and the light solid (*green*) curves show the BLYP results of Micoulaut et al. [39]. For the glass, the broken (*red*) curves show the LDA results of Zhang and Drabold [29] and the light solid (*green*) curves show the BLYP results of Massobrio and co-workers [87]

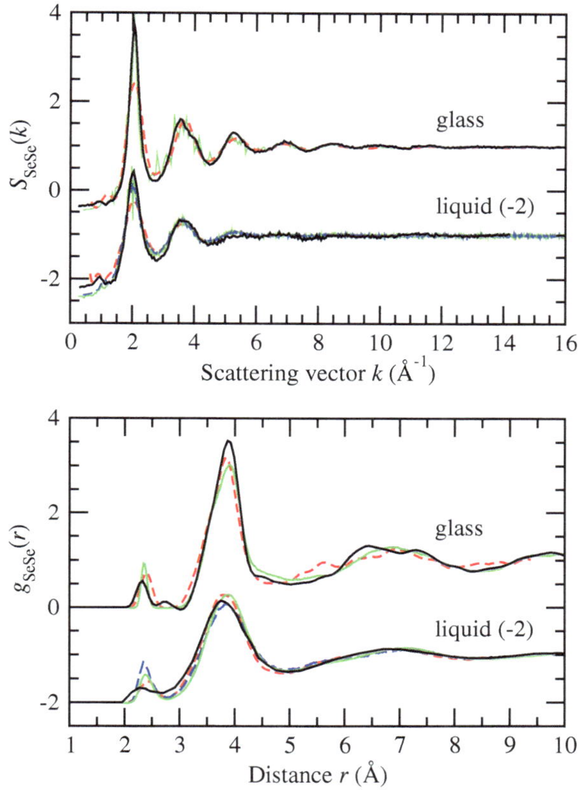

Fig. 1.6 The measured and simulated $S_{\mathrm{SeSe}}(k)$ and $g_{\mathrm{SeSe}}(r)$ functions for liquid and glassy $GeSe_2$. The dark solid (*black*) curves give the measured functions for the liquid [21] and glass [22]. For the liquid, the light broken (*red*) curves show the LDA results of Cobb and Drabold [25], the dark broken (*blue*) curves show the LDA results of Massobrio et al. [30], and the light solid (*green*) curves show the BLYP results of Micoulaut et al. [39]. For the glass, the broken (*red*) curves show the LDA results of Zhang and Drabold [29] and the light solid (*green*) curves show the BLYP results of Massobrio and co-workers [87]

Figs. 1.8, 1.9 and 1.10.[7] The overall features in the measured functions are similar to those observed for liquid and glassy $ZnCl_2$ as befits a material with a structure that is also based predominantly on MX_4 tetrahedra. For example, a prominent FSDP is observed in the partial structure factor describing the pair-correlations between the more electropositive chemical species i.e. in $S_{\mathrm{GeGe}}(k)$ at $k_{\mathrm{FSDP}} \simeq 1$ Å^{-1}; an FSDP also manifests itself as a feature in all three of the $S_{IJ}(k)$ functions; the principal peaks in the $S_{\alpha\beta}(k)$ and $S_{IJ}(k)$ functions share a common position which leads in the case of the glass to prominent extended range ordering [15, 50]; and $S_{\mathrm{NN}}(k)$ has the same 'three-peak' character as shown for $ZnCl_2$ (Fig. 1.1). There are, however, subtle but important differences in structure, including clear evidence for homopolar bonds.

For the liquid, the main peaks in $g_{\mathrm{GeSe}}(r)$ and $g_{\mathrm{SeSe}}(r)$ occur at 2.42(2) and 3.75(2) Å, respectively, giving a Ge–Se:Se–Se distance ratio of 0.645(6) as compared

[7] The r-space functions for the liquid were obtained from a maximum entropy analysis in which homopolar bonds were not excluded. Those for the glass were obtained from a procedure aimed at removing the effect in r-space of the modification function $M(k)$. A more complete discussion is given in [21, 23, 80].

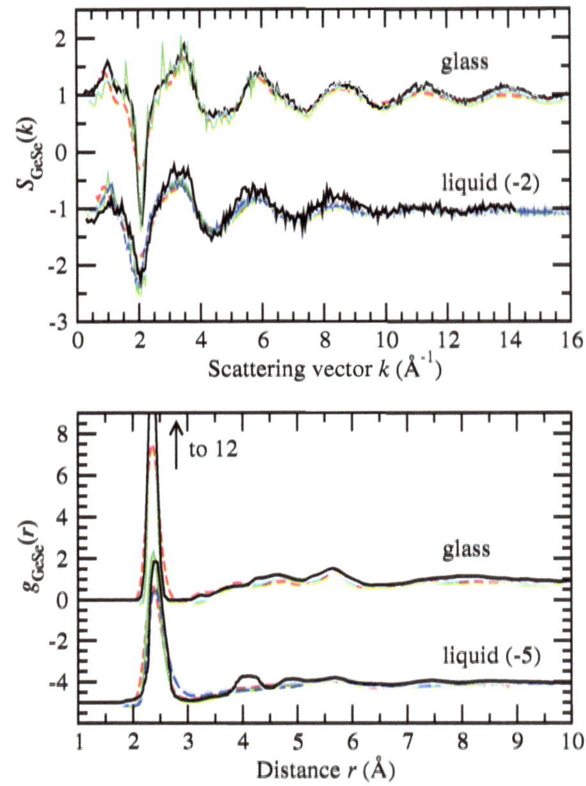

Fig. 1.7 The measured and simulated $S_{GeSe}(k)$ and $g_{GeSe}(r)$ functions for liquid and glassy $GeSe_2$. The dark solid (*black*) curves give the measured functions for the liquid [21] and glass [22]. For the liquid, the light broken (*red*) curves show the LDA results of Cobb and Drabold [25], the dark broken (*blue*) curves show the LDA results of Massobrio et al. [30], and the light solid (*green*) curves show the BLYP results of Micoulaut et al. [39]. For the glass, the broken (*red*) curves show the LDA results of Zhang and Drabold [29] and the light solid (*green*) curves show the BLYP results of Massobrio and co-workers [87]

to a ratio of $\sqrt{8/3} = 0.612$ for regular $GeSe_4$ tetrahedra. This ratio and the Ge–Se coordination number $\bar{n}_{Ge}^{Se} = 3.5(2)$ are consistent with the presence in the melt of a large number of distorted tetrahedral $GeSe_4$ motifs. In comparison, for the glass the main peaks in $g_{GeSe}(r)$ and $g_{SeSe}(r)$ occur at 2.36(2) and 3.89(2) Å, respectively, giving a Ge-Se:Se-Se distance ratio of 0.607(6), and $\bar{n}_{Ge}^{Se} = 3.7(1)$ i.e. the $GeSe_4$ motifs in the glass appear to be more regular than in the liquid.

In the high-temperature crystalline phase of $GeSe_2$ [81] there are equal numbers of both corner-sharing (CS) and edge-sharing (ES) tetrahedra and each Ge atom in a CS or ES tetrahedron has three or four nearest-neighbouring Ge atoms giving a Ge–Ge coordination number $\bar{n}_{Ge}^{Ge} = 3.5$. The Ge–Ge distance for ES tetrahedra is the shortest at 3.049 Å and the next largest distance is 3.508 Å. The measured $g_{GeGe}(r)$ functions for both the liquid and glass also support the existence of substantial fractions of both CS and ES tetrahedra. For the liquid, ES motifs manifest themselves by a short low-r cutoff value of $\simeq$2.8 Å for the first main peak in $g_{GeGe}(r)$ and by the relatively small coordination number for this peak of $\bar{n}_{Ge}^{Ge} = 2.9(3)$ [21, 80]. For the glass, ES motifs show themselves by a peak in $g_{GeGe}(r)$ at 3.02(2) Å with a coordination number $\bar{n}_{Ge}^{Ge} = 0.34(5)$. If the glass does not contain extended chains of ES units then this

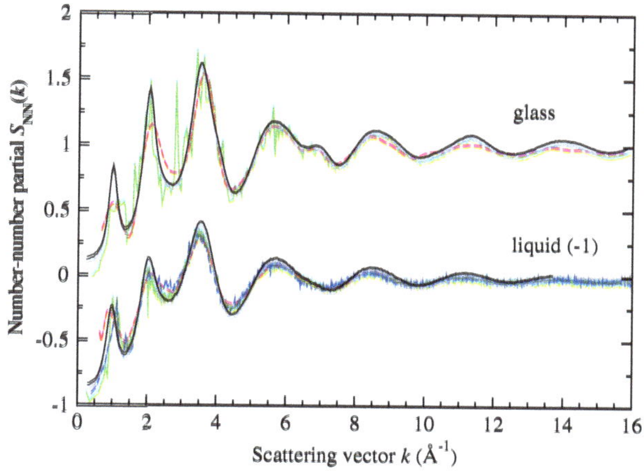

Fig. 1.8 The measured and simulated $S_{NN}(k)$ functions for liquid and glassy $GeSe_2$. The dark solid (*black*) curves give the measured functions for the liquid [21] and glass [22]. For the liquid, the light broken (*red*) curves show the LDA results of Cobb and Drabold [25], the dark broken (*blue*) curves show the LDA results of Massobrio et al. [30], and the light solid (*green*) curves show the BLYP results of Micoulaut et al. [39]. For the glass, the broken (*red*) curves show the LDA results of Zhang and Drabold [29] and the light solid (*green*) curves show the BLYP results of Massobrio and co-workers [87]

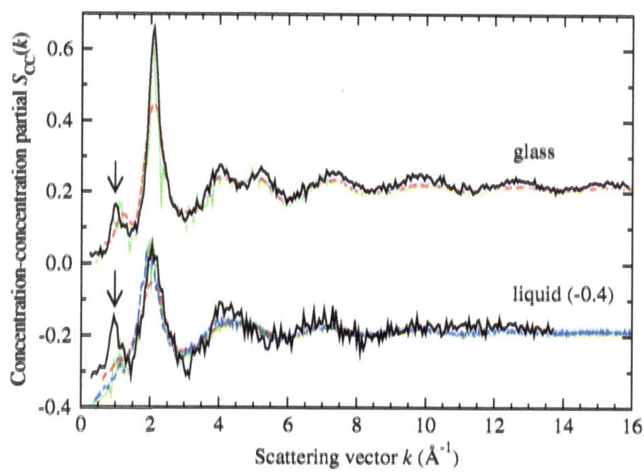

Fig. 1.9 The measured and simulated $S_{CC}(k)$ functions for liquid and glassy $GeSe_2$. The dark solid (*black*) curves give the measured functions for the liquid [21] and glass [22]. For the liquid, the light broken (*red*) curves show the LDA results of Cobb and Drabold [25], the dark broken (*blue*) curves show the LDA results of Massobrio et al. [30], and the light solid (*green*) curves show the BLYP results of Micoulaut et al. [39]. For the glass, the broken (*red*) curves show the LDA results of Zhang and Drabold [29] and the light solid (*green*) curves show the BLYP results of Massobrio and co-workers [87]. The vertical arrows point to the FSDP

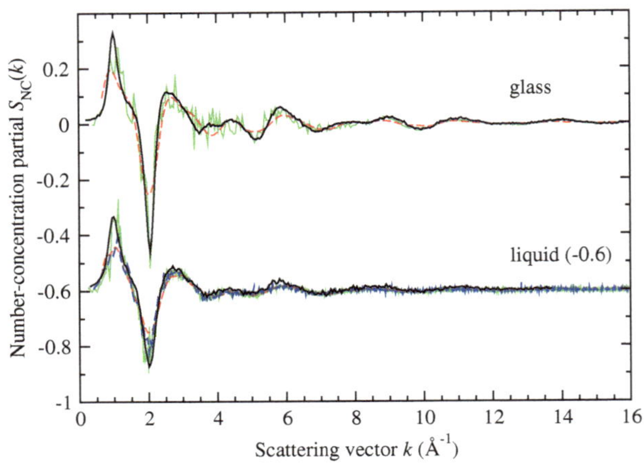

Fig. 1.10 The measured and simulated $S_{NC}(k)$ functions for liquid and glassy $GeSe_2$. The dark solid (*black*) curves give the measured functions for the liquid [21] and glass [22]. For the liquid, the light broken (*red*) curves show the LDA results of Cobb and Drabold [25], the dark broken (*blue*) curves show the LDA results of Massobrio et al. [30], and the light solid (*green*) curves show the BLYP results of Micoulaut et al. [39]. For the glass, the broken (*red*) curves show the LDA results of Zhang and Drabold [29] and the light solid (*green*) curves show the BLYP results of Massobrio and co-workers [87]

coordination number can be written as $\bar{n}_{\rm Ge}^{\rm Ge} = \left(N_{\rm Ge,ES} \times 1\right)/N_{\rm Ge}$ where $N_{\rm Ge,ES}$ is the number of Ge atoms in ES units and $N_{\rm Ge}$ is the total number of Ge atoms. Hence 34(5) % of the Ge atoms in the glass are involved in ES motifs.

As shown by the $g_{\alpha\beta}(r)$ functions of Figs. 1.5 and 1.6, there is clear evidence for a substantial number of Ge–Ge and Se–Se homopolar bonds in both the liquid and glass, with measured Ge–Ge distances of 2.33(3) Å (liquid) and 2.42(2) Å (glass) and with measured Se–Se distances of 2.30(2) Å (liquid) and 2.32(2) Å (glass). These distances are comparable to the Ge–Ge contact distances in liquid GeSe and amorphous Ge (2.36–2.47 Å) and to the Se–Se contact distances in liquid GeSe and Se (2.34–2.35 Å) [80]. For the liquid, the corresponding Ge–Ge and Se–Se coordination numbers are 0.25(10) and 0.23(5), respectively, and since $\bar{n}_{\rm Ge}^{\rm Se} = 3.5(2)$ the total Ge and Se coordination numbers are $\bar{n}_{\rm Ge} \equiv \bar{n}_{\rm Ge}^{\rm Se} + \bar{n}_{\rm Ge}^{\rm Ge} = 3.8(2)$ and $\bar{n}_{\rm Se} \equiv \bar{n}_{\rm Se}^{\rm Ge} + \bar{n}_{\rm Se}^{\rm Se} = 2.0(1)$. For the glass, the corresponding Ge–Ge and Se–Se coordination numbers are 0.25(5) and 0.20(5), respectively, and since $\bar{n}_{\rm Ge}^{\rm Se} = 3.7(1)$ the total Ge and Se coordination numbers are $\bar{n}_{\rm Ge} = 4.0(1)$ and $\bar{n}_{\rm Se} = 2.05(7)$.[8] These $\bar{n}_{\rm Ge}$ and $\bar{n}_{\rm Se}$ values imply that Ge and Se are, within the experimental error, fourfold and twofold coordinated

[8] For the glass, an estimate of the number of Ge atoms in CS tetrahedra $N_{\rm Ge,CS}$ can be obtained by taking $N_{\rm Ge} = N_{\rm Ge,ES} + N_{\rm Ge,CS} + N_{\rm Ge,homo}$ where $N_{\rm Ge,homo}$ is the number of Ge atoms in homopolar Ge–Ge bonds (see Appendix). If there are no extended chains of ES units then the corresponding coordination number $\bar{n}_{\rm Ge}^{\rm Ge} = \left(N_{\rm Ge,ES} \times 1\right)/N_{\rm Ge} = 0.34(5)$ and if homopolar bonds form only in pairs then the corresponding coordination number $\bar{n}_{\rm Ge}^{\rm Ge} = \left(N_{\rm Ge,homo} \times 1\right)/N_{\rm Ge} = 0.25(5)$. Hence $N_{\rm Ge,CS}/N_{\rm Ge} = 1 - 0.34(5) - 0.25(5) = 0.41(7)$ such that $N_{\rm Ge,ES}/N_{\rm Ge,CS} = 0.34(5)/0.41(7) = 0.83(16)$ [23].

in both the liquid and glass i.e. both chemical species have a full outer shell of eight electrons. The observation by the NDIS method of homopolar bonds in $GeSe_2$ glass is consistent with the findings from Raman, Mössbauer and x-ray emission spectroscopy experiments [23, 82, 83], and the fraction of homopolar bonds is in agreement with an estimate based on the law of mass action (see Appendix).

As shown in Figs. 1.8, 1.9 and 1.10, the overall features in a given Bhatia-Thornton $S_{IJ}(k)$ function are similar for both liquid and glassy $GeSe_2$, but the peaks for the glass are generally sharper than for the liquid in accordance with a loss of thermal disorder. A notable feature is the appearance of an FSDP in $S_{CC}(k)$ at $k_{FSDP} \simeq 1$ Å^{-1} which, from inspection of (1.6) and Figs. 1.5, 1.6 and 1.7, arises predominantly from the Ge–Ge correlations. The significance of an FSDP in $S_{CC}(k)$ will be discussed in Sect. 1.4.3.

1.4.2 First-Principles Molecular Dynamics Simulations of Liquid and Glassy $GeSe_2$

The $GeSe_2$ system has been the subject of extensive first-principles molecular dynamics simulations in which the electronic structure is taken into explicit account, as befits a material in which the electronegativities of the different chemical species are similar and the bonding takes an iono-covalent character [24–40].

Drabold and co-workers used an electronic-structure scheme within the local density approximation (LDA) for the exchange and correlation energy which does not evolve self-consistently with the atomic motion, together with a minimal basis set [24, 25, 29]. As illustrated in Figs. 1.5, 1.6, 1.7, 1.8, 1.9 and 1.10 the models do, however, reproduce many of the features in the NDIS results for the liquid and glass such as homopolar Ge–Ge and Se–Se bonds and the appearance of an FSDP in $S_{CC}(k)$.

Massobrio and co-workers first investigated liquid $GeSe_2$ by using fully self-consistent LDA calculations, but it was found that this approach led to structures that were too disordered (Figs. 1.5, 1.6, 1.7, 1.8, 1.9 and 1.10). This limitation was attributed to an underestimation of the ionic contribution to the bonding [26–28, 30] which led inter alia to use of the Perdew and Wang [84] generalised gradient approximation for the exchange and correlation energy, and to improved agreement with experiment [26–28, 30, 37]. Nevertheless, discrepancies remained that were particularly noticeable with regards to the Ge–Ge and concentration-concentration partial pair-correlation functions. These limitations led to an employment of the Becke, Lee, Yang and Parr (BLYP) generalised gradient approximation for the exchange and correlation energy [85, 86] to further enhance a localised distribution of the valence electrons. The results from this approach are leading to good agreement with the NDIS results for the liquid (Figs. 1.5, 1.6, 1.7, 1.8, 1.9 and 1.10) and to diffusion coefficients that are in better agreement with those expected from experiment [36, 39].

The procedure used to quench a liquid to the glass is expected to affect the resultant structure, especially since the simulated quench rates are many orders of magnitude faster than experimental ones. Massobrio and Pasquarello [34] devised a protocol that leads to marked differences between the glass and liquid structures, but it transpired that use of the NVT ensemble with a number density set at the measured value $\rho = 0.034$ Å^{-3} led to a marked overpressure of ~1 GPa [40]. This problem was addressed by increasing the size of the simulation box to reduce the number density to $\rho = 0.0326$ Å^{-3}, essentially eliminating the overpressure and leading to a structure that is in better agreement with experiment [40]. It was later found that this revised density is in excellent agreement with the value $\rho = 0.0324(1)$ Å^{-3} obtained from recent measurements on $GeSe_2$ glass [87]. As shown in Figs. 1.5, 1.6, 1.7, 1.8, 1.9 and 1.10, the combined use of the quench-rate protocol, revised number density and BLYP functional is now capturing all of the main features found from the NDIS experiments on $GeSe_2$ glass under ambient conditions [87]. It will be interesting to see whether this latest model yields a vibrational density of states for the ambient pressure glass that is in good agreement with experiment [88, 89].

1.4.3 Concentration Fluctuations on an Intermediate Length Scale

An FSDP in $S_{CC}(k)$ is observed for the liquid and glassy forms of $GeSe_2$ (Fig. 1.9) and $ZnCl_2$ (Fig. 1.1) and for several other MX_2 network glass-forming materials [31, 58, 59]. This peak has been a source of controversy because it was not predicted from previous investigations of these materials by using classical molecular dynamics simulations or integral-equation calculations [90–93]. Also, if these systems can be treated as purely ionic materials containing point-like cations and anions, then $S_{CC}(k)$ is related to the charge-charge partial structure factor $S_{ZZ}(k)$ by the equation $S_{CC}(k) = c_M c_X S_{ZZ}(k)$ such that an FSDP in $S_{CC}(k)$ implies a non-uniformity in the charge distribution on an intermediate length scale [58].

First-principles molecular dynamics simulations of liquid or amorphous $GeSe_2$, $SiSe_2$ and SiO_2 have been undertaken to examine the issue of concentration versus charge fluctuations on an intermediate length scale [31, 37]. In this work $S_{CC}(k)$, which depends on the atomic positions, was calculated separately from $S_{ZZ}(k)$, which depends on the valence-electron density. No FSDP was found for $S_{ZZ}(k)$ i.e. no evidence was found for charge fluctuations on an intermediate length scale. This led to a proposal for three classes of network-forming materials. Class I systems have perfect chemical order and no FSDP in $S_{CC}(k)$, class II systems have a moderate number of defects in an otherwise chemically ordered network and an FSDP in $S_{CC}(k)$, and class III systems have a large degree of chemical disorder, feature a rich variety of structural motifs, and show no FSDP in $S_{CC}(k)$ [31].

The appearance of charge neutrality on an intermediate length scale provides an important constraint on the network connectivity leading to these different network types. For example, in chemically ordered class I systems like SiO_2 the network is made from the same type of charge-neutral motif and concentration fluctuations need not occur on an intermediate range. In class II materials like $GeSe_2$ and $SiSe_2$, however, there is a moderate number of defects leading to a variability of M-centred structural motifs with different charges. These motifs must form a network in which there is charge neutrality on the length scale of a few structural motifs, leading to an arrangement with a non-uniform distribution of M-atoms i.e. to the appearance of concentration fluctuations on an intermediate length scale. By comparison, in more chemically disordered class III systems, the network structure is broken-up and the intermediate range order disappears such that the FSDP becomes a less prominent feature in $F(k)$ and disappears from $S_{CC}(k)$.[9]

Liquid and glassy $ZnCl_2$ are chemically ordered materials [14–16] and, according to the above, should therefore be categorised as class I network-forming materials. NDIS experiments show, however, that there is an FSDP in $S_{CC}(k)$ (Fig. 1.1), and edge-sharing tetrahedra are indicated by molecular dynamics [94] and RMC [16, 17] models and by Raman spectroscopy [95]. Edge-sharing motifs containing mis-coordinated atoms (i.e. those not satisfying the '8-N' rule) are primarily responsible for the FSDP in $S_{CC}(k)$ for $GeSe_2$ [33], and the majority of Si atoms in $SiSe_2$ are involved in edge-sharing conformations [37, 96]. In contrast, the measured partial structure factors for glassy SiO_2 do not show an FSDP in $S_{CC}^{BT}(k)$ [97] and the network, based on corner-sharing SiO_4 tetrahedra, is chemically-ordered [98]. On the other hand, glassy GeO_2 also forms a chemically-ordered network based on corner-sharing GeO_4 tetrahedra, but in this case the measured partial structure factors *do* show a small FSDP in $S_{CC}^{BT}(k)$ [65, 99]. For this material, a first-principles molecular dynamics model also shows a small FSDP in $S_{CC}^{BT}(k)$ and, although its origin is unknown, it is not related to coordination defects since they were not present in the model [100].

This evidence suggests a revised definition for class I and II network-forming MX_2 materials along the lines suggested in [101]. Class I systems form chemically ordered corner-sharing networks, class II systems form networks that incorporate both corner- and edge-sharing motifs, and class III systems form networks that are chemically disordered. Class I and II networks may both exhibit an FSDP in $S_{CC}^{BT}(k)$ where, in the case of class II systems, this feature originates primarily from edge-sharing motifs that may contain structural defects. In the case of class I systems, however, the FSDP does not originate from the fourfold rings associated with these edge-sharing conformations. Thus, the concentration-concentration partial structure factor is proving to be a sensitive probe of the ordering in network glass-forming materials [17].

[9]In [31] a first-principles molecular dynamics model for liquid $GeSe_2$ using the Perdew and Wang generalised gradient approximation was given as an example of a class III system. More recent models of this material using the BLYP generalised gradient approximation reduce the chemical disorder and produce a more pronounced FSDP in $S_{CC}(k)$, in better accord with experiment (Fig. 1.9). The measured FSDP in $S_{CC}(k)$ for glassy $GeSe_2$ is accurately reproduced by first-principles molecular dynamics simulations using the BLYP generalised gradient approximation (Fig. 1.9).

1.5 Density-Driven Mechanisms of Network Collapse in MX_2 Glasses: Structure of GeO_2 Under Pressure

The structural changes in glasses and liquids induced by high-pressure and/or high-temperature conditions can have a profound effect on their physico-chemical properties [72, 98, 102, 103]. A notable example is provided by so-called polyamorphic transitions, where the variation of pressure and/or temperature leads to an abrupt transformation between two phases having identical compositions but different densities [72, 98, 102–107]. It is therefore important to unravel the mechanisms by which these transformations occur in order to establish the underlying relationships to the network structure. This is not, however, a straightforward task as competing processes are often at work. For example, the compression of a network formed by corner-sharing tetrahedral MO_4 motifs could lead to a retention of these motifs but to a change in the distribution of primitive ring sizes,[10] as indicated by the ring statistics for different density polymorphs of crystalline SiO_2 [108]. Compression may, however, also lead to an alteration in character of the network-forming motifs as they transform to higher-coordinated polyhedra such as MO_5 square pyramids or MO_6 octahedra. In the following, the case example of GeO_2 glass under cold-compression (i.e. pressurisation at constant temperature) will be considered. The results highlight the usefulness of NDIS in helping to test the efficacy of the different molecular dynamics models that have been proposed for this material, and demonstrate the need for atomic interaction models that can be reliably transferred to high-pressure conditions.

Figure 1.11 shows the difference functions $\Delta F_{\rm Ge}(k)$ and $\Delta F_{\rm O}(k)$ as measured for GeO_2 glass at pressures up to 8 GPa by employing the in situ high-pressure NDIS method [109]. In these experiments, the total structure factors $F(k)$ are measured for two samples that are identical in every respect, except for the isotopic enrichment of the Ge atoms. These $F(k)$ functions are then subtracted in order to simplify the complexity of correlations associated with a single measurement [110, 111]. For example, $\Delta F_{\rm Ge}(k)$ is given by

$$\Delta F_{\rm Ge}(k)/\text{barn} = 0.124(3)\,[S_{\rm GeO}(k) - 1] + 0.081(2)\,[S_{\rm GeGe}(k) - 1] \qquad (1.16)$$

and, because the contribution from $S_{\rm OO}(k)$ has been eliminated, it gives site-specific information on the germanium atom correlations. Similarly, $\Delta F_{\rm O}(k)$ is given by

$$\Delta F_{\rm O}(k)/\text{barn} = 0.0875(5)\,[S_{\rm GeO}(k) - 1] + 0.1497(2)\,[S_{\rm OO}(k) - 1] \qquad (1.17)$$

and, because the contribution from $S_{\rm GeGe}(k)$ has been eliminated, it gives complementary site-specific information on the oxygen atom correlations. The corresponding

[10] A ring is a measure of the network topology and is a closed path usually chosen to pass along the bonds which connect nearest-neighbour atoms. A ring is primitive if it cannot be decomposed into smaller rings [108].

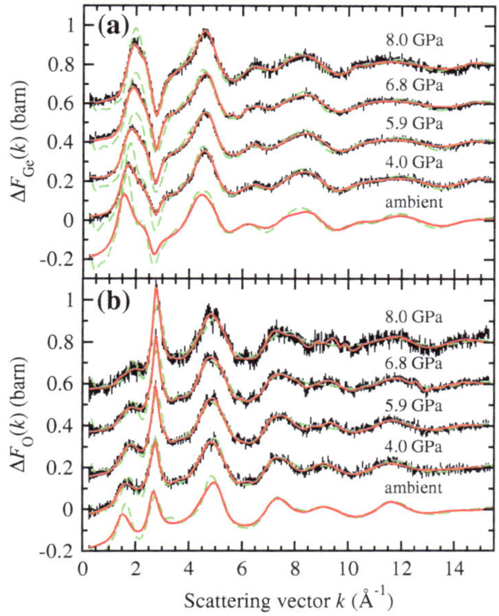

Fig. 1.11 The pressure dependence of the difference functions **a** $\Delta F_{\mathrm{Ge}}(k)$ and **b** $\Delta F_{\mathrm{O}}(k)$ for GeO_2 glass at ambient temperature [109]. The vertical bars give the statistical errors on the measured data points, the solid (*red*) curves give the Fourier transforms of the corresponding real-space functions shown in Fig. 1.12, and the broken (*green*) curves give the molecular dynamics results obtained by using the DIPPIM. The high-pressure data sets have been shifted vertically for clarity of presentation. The figure is adapted from Wezka et al. [109]

real-space functions $\Delta G_{\mathrm{Ge}}(r)$ and $\Delta G_{\mathrm{O}}(r)$ are obtained by replacing $S_{\alpha\beta}(k)$ by $g_{\alpha\beta}(r)$ in (1.16) and (1.17), respectively, and are shown in Fig. 1.12.

The first peak in both $\Delta G_{\mathrm{Ge}}(r)$ and $\Delta G_{\mathrm{O}}(r)$ originates from the Ge–O correlations, and the dependence on pressure of the corresponding Ge–O bond distance r_{GeO} and coordination number $\bar{n}_{\mathrm{Ge}}^{\mathrm{O}}$ is shown in Fig. 1.13. The second peaks in these functions originate from nearest-neighbour Ge–Ge and O–O correlations, respectively. The dependence on pressure of the corresponding peak positions r_{GeGe} and r_{OO} is also shown in Fig. 1.13, along with the O–O coordination number $\bar{n}_{\mathrm{O}}^{\mathrm{O}}$ which was obtained by assuming minimal overlap between the O–O and Ge–O correlations as observed under ambient conditions [99]. The resolution of these peaks, which is made possible by using the difference function method, also enables an estimate of the mean O–Ge–O and Ge–O–Ge bond angles as deduced from the measured nearest-neighbour distances according to $\cos(\theta_{\mathrm{OGeO}}) = 1 - r_{\mathrm{OO}}^2/2r_{\mathrm{GeO}}^2$ and $\cos(\theta_{\mathrm{GeOGe}}) = 1 - r_{\mathrm{GeGe}}^2/2r_{\mathrm{GeO}}^2$. The results are plotted in Fig. 1.14 as a function of the reduced density ρ/ρ_0 where ρ_0 is the value of the number density ρ at ambient pressure. This reduced density facilitates a ready comparison with the bond angles measured by diffraction experiments on the α-quartz polymorph of crystalline GeO_2 [112, 113]. The density-to-pressure conversion was made using the equation of state measured by Hong et al. [114].

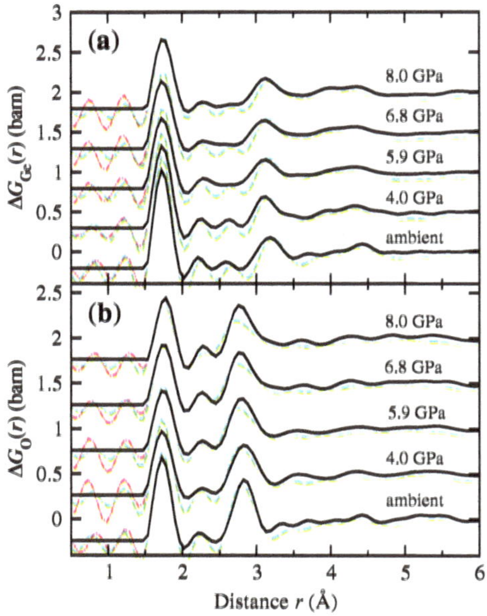

Fig. 1.12 The solid (*black*) curves show the difference functions **a** $\Delta G_{Ge}(r)$ and **b** $\Delta G_{O}(r)$ obtained by spline fitting and Fourier transforming the measured reciprocal-space functions shown in Fig. 1.11 at pressures ranging from ambient to 8 GPa. The chained (*red*) curves show the oscillations at r-values smaller than the distance of closest approach between two atoms. The broken (*green*) curves give the molecular dynamics results obtained by Fourier transforming the simulated functions shown in Fig. 1.11 after applying the same maximum cutoff k_{max} as for the neutron diffraction data. The high-pressure data sets have been shifted vertically for clarity of presentation. The figure is adapted from Wezka et al. [109]

In Figs. 1.11, 1.12, 1.13 and 1.14, the diffraction results are compared to molecular dynamics simulations made using the so-called DIPole-Polarisable Ion Model (DIPPIM) where the atomic interaction potentials include dipole-polarisation effects [43, 109, 115]. The potentials were parameterised using ab initio simulations as opposed to experimental results and should therefore be largely unbiased in their predictions of the glass structure. The DIPPIM is the only model currently available that gives, for a single set of parameters, a good account of both the structural and vibrational properties of glassy GeO_2 at ambient pressure along with the dynamical properties of liquid GeO_2 at elevated temperatures [115]. The ambient pressure glass was obtained by a quench-from-the-melt procedure, and the high-pressure configurations were generated by a cold-compression procedure [41] in which the cell lengths and particle positions were rescaled for each new density. The methodology did not reproduce the measured equation of state [43], so the simulations were made with the glass density set at measured values [109].

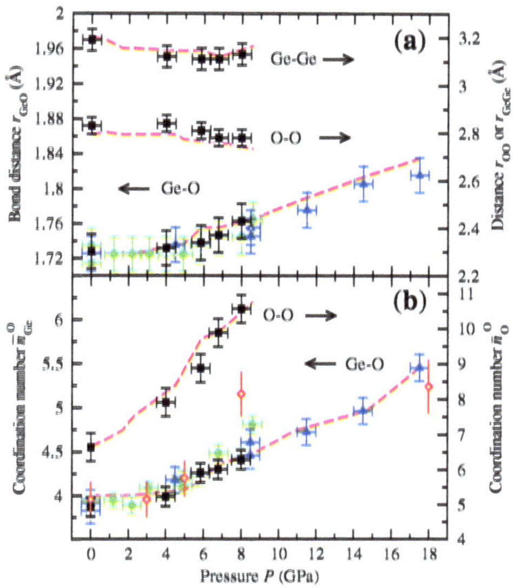

Fig. 1.13 The pressure dependence of **a** the nearest-neighbour Ge–O, O–O and Ge–Ge distances and **b** the Ge–O and O–O coordination numbers. The results from NDIS experiments (■) and molecular dynamics simulations using the DIPPIM (broken (*red*) curves) [109] are compared to those obtained from the neutron diffraction work of Drewitt et al. [110] [(*green*) •] and Salmon et al. [111] [(*blue*) ▲]. In (**b**) the Ge–O coordination numbers from IXS experiments [116] are also shown [(*red*) ◇]. The figure is taken from Wezka et al. [109]. © IOP Publishing. Reproduced by permission of IOP Publishing. All rights reserved

The DIPPIM molecular dynamics results are in good accord with the measured pressure dependence of the difference functions in both reciprocal and real space (Figs. 1.11 and 1.12), an agreement that also extends to the mean nearest-neighbour distances, coordination numbers, and O–Ge–O and Ge–O–Ge bond angles (Figs. 1.13 and 1.14). The model's predictions for the reduced density dependence of the fractions of four-, five- and six-fold coordinated Ge atoms, and of twofold and threefold coordinated oxygen atoms, are shown in Fig. 1.14. This dependence for the fractions of GeO_4, GeO_5 and GeO_6 units is not in agreement with estimates based on inelastic x-ray scattering (IXS) experiments [116], a discrepancy that may originate from the use of data from crystalline standards containing trigonal bipyramidal GeO_5 units to analyse the measured IXS spectra: the predominant GeO_5 units found from the molecular dynamics results are distorted square pyramids.[11] In contrast, other

[11] As shown in Fig. 1.13b, the Ge–O coordination number obtained at ~8 GPa ($\rho/\rho_0 \sim 1.4$) from the IXS experiments is large relative to the value obtained from neutron diffraction experiments in a regime for which the IXS data give, relative to molecular dynamics, a much greater fraction of GeO_6 units relative to GeO_4 and GeO_5 units (Fig. 1.14a).

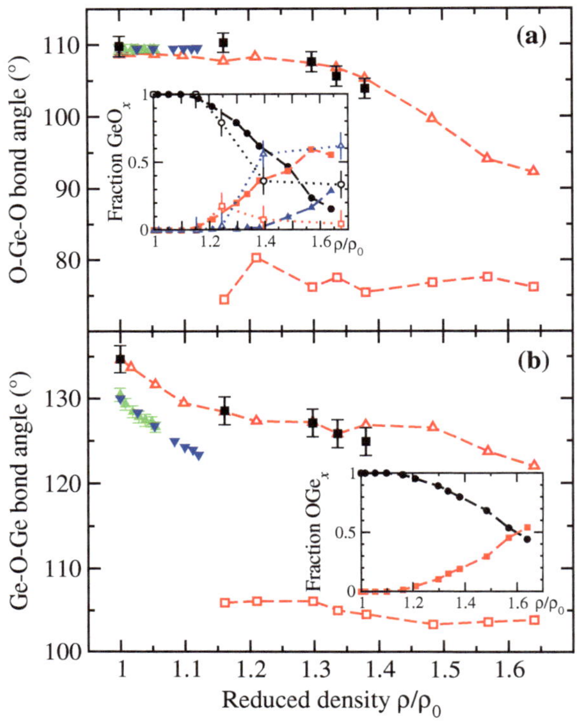

Fig. 1.14 The reduced density ρ/ρ_0 dependence of the **a** O–Ge–O and **b** Ge–O–Ge bond angles as measured for GeO_2 glass [109] (■) and for the α-quartz polymorph of crystalline GeO_2 in the work of Jorgensen [112] [(*green*) ▲] and Glinnemann et al. [113] [(*blue*) ▼]. Also shown are the peak positions in the DIPPIM bond angle distributions for the glass, where the appearance of a second branch at higher reduced densities corresponds to the development of an additional peak or shoulder in these bond-angle distributions. The first branch originates at ambient density from tetrahedral GeO_4 motifs [(*red*) △)], and the second branch appears at higher densities as these motifs are replaced by GeO_5 and GeO_6 units [(*red*) □]. The insets show the DIPPIM results for the density dependence of (**a**) the fraction of GeO_x species, where $x = 4$ (•), 5 [(*red*) ■] or 6 [(*blue*) ▲], and (**b**) the fraction of OGe_x species, where $x = 2$ (•) or 3 [(*red*) ■]. In (**a**), the inset also shows the fraction of GeO_x species from IXS experiments [116] where $x = 4$ (○), 5 [(*red*) □] or 6 [(*blue*) △]. The figure is taken from Wezka et al. [109].

models for the pressure-induced structural changes in GeO_2 glass, as obtained by using the Oeffner-Elliott two-body potentials [117] in classical molecular dynamics simulations [118–122] or first-principles molecular dynamics simulations using a generalised gradient approximation [123], do not reproduce basic features such as the pressure dependence of the measured Ge–O bond lengths and coordination numbers (Fig. 1.15). The Oeffner-Elliott potentials were initially employed to model the α-quartz and rutile-like phases of GeO_2 and the phase transition between the α- and β-quartz phases of this material [117].

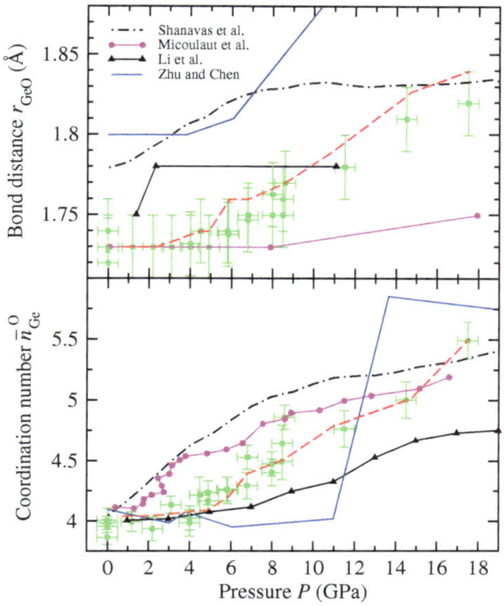

Fig. 1.15 The pressure dependence of the Ge–O bond distance r_{GeO} and coordination number $\bar{n}_{Ge}^{O}$ for GeO_2 glass. The results from neutron diffraction are shown by the (*green*) squares with error bars [109–111] and the results from different molecular dynamics simulations are given by the curves. The results from the DIPPIM interaction potentials [broken light (*red*) curves] are in agreement with the experimental data. In contrast, the results of Micoulaut et al. [118, 119] [solid (*magenta*) curves with circles], Shanavas et al. [120] [chained dark (*black*) curves] and Li et al. [122] [solid curves with triangles] obtained by using the Oeffner-Elliott interaction potentials [117], and the results of Zhu and Chen [123] [solid dark (*blue*) curves] obtained by using first-principles methods, are not consistent with the measured data. The figure is adapted from Wezka et al. [109]

1.6 Conclusions and Future Perspectives

This chapter has focussed on several prototypical MX_2 network glass-forming materials in order to illustrate the benefits of having detailed structural information from experiment to guide in the development of realistic molecular dynamics models. Many of the pertinent experimental results have originated from the NDIS method because this can be used to provide information at the partial structure factor level.

In the case of $ZnCl_2$ it was found that the main structural features can be accounted for within the framework of an ionic interaction model, provided that anion polarisation effects are taken into account. This led to a systematic investigation of the relationship between the structure and fragility of tetrahedral glass forming liquids where the anion polarisability was used as an adjustable parameter in order to change the network connectivity. The model reproduces the measured trends, and correlates increased fragility with an increase in number of edge-sharing units, thus emphasising the importance of these configurations on the dynamics of tetrahedral glass-forming liquids. In the case of $GeSe_2$ an ionic interaction model does not reproduce measured features such as homopolar bonds, and first-principles molecular dynamics need to be employed. The progression to the use in self-consistent calculations of the Becke-Lee-Yang-Parr (BLYP) functional has led to models that are in better agreement with the NDIS and other experimental results. Finally, in the case of GeO_2 under cold-compression, the parameterisation of a polarisable ion model using ab initio results has led to predictions for the pressure-induced structural changes that are in accord with experiment, and to insight into the mechanisms of density-driven network collapse.

When assessing the results obtained from a molecular dynamics model, it is valuable to make comparisons with diffraction data in both real and reciprocal space. For example, the short-range order will manifest itself by the appearance of a peak in a partial pair-distribution function at a small r-value, whereas the intermediate range order leads to a more subtle r-space modulation that will manifest itself by the appearance of an FSDP in the corresponding reciprocal space function e.g. $S_{\mathrm{MM}}(k)$ [57]. It is therefore convenient to examine details of the short-range order of a model in real space, and details of the intermediate-range order of this model in reciprocal space. It can also be valuable to make a like-for-like comparison of a modelled r-space function with experiment by following the experimental procedure i.e. by taking a simulated reciprocal space function, truncating it at the value for k_{max} set by a diffractometer, and Fourier transforming (Sect. 1.2). Interestingly, although the quench rates used in simulation are fast, the simulated pair-correlation functions can be in good accord with experiment i.e. modelled glass structures are not necessarily those of the corresponding high-temperature liquids caught in time.

Once the correct theoretical scheme has been established, molecular dynamics models can be used to enrich the information about a material that can be extracted, and they can also be used to predict the changes in material properties that occur when e.g. new chemical species are added or the state conditions are varied. Indeed, one of the ultimate aims of making realistic atomistic models for network glass-forming systems is the development of new materials following the principles of rational design [124–126] i.e. the strategy of creating new glasses with the desired functionality, based on an ability to predict the structure of a glass and how this will affect its physical properties. The rational design approach stands to gain more ground as network glass-forming materials continue to reveal more and more of their structural secrets.

Acknowledgments It is a pleasure to thank everyone who has contributed towards the experimental programme of research at Bath and UEA into the nature of network glass-forming materials, including Ian Penfold, Chris Benmore, Paul Lond, Erol Okan, Jian Liu, Shuqin Xin, Jonathan Wasse, Takeshi Usuki, Ingrid Petri, Richard Martin, James Drewitt, Prae Chirawatkul, Dean Whittaker, Kamil Wezka, Keiron Pizzey, Ruth Rowlands, Annalisa Polidori and Harry Bone. Special thanks also go to Adrian Barnes (Bristol), Pierre Chieux (ILL), Wilson Crichton (ESRF), Gabriel Cuello (ILL), Henry Fischer (ILL) and Stefan Klotz (Paris) for their contributions to the experimental work; and to Mauro Boero (Strasbourg), Assil Bouzid (Strasbourg), Sébastien Le Roux (Strasbourg), Dario Marrocchelli (MIT), Carlo Massobrio (Strasbourg), Matthieu Micoulaut (Paris), Alfredo Pasquarello (Lausanne) and Mark Wilson (Oxford) for all their contributions on the molecular dynamics front. The latter are also thanked for agreeing to a close dialogue with the experimental teams, where the feedback has been mutually beneficial in helping to decode the complexity of network glass-forming materials, and has also led to a fuller appreciation of both the advantages and limitations of experimental versus molecular dynamics methods. The support of the EPSRC (Grant: EP/J009741/1) and Institut Laue-Langevin (ILL) is gratefully acknowledged.

Appendix: Concentration of Defects in $GeSe_2$ Glass from the Law of Mass Action

Following Feltz [127, 128], consider the reversible reaction

$$2\text{Ge}-\text{Se} \rightleftharpoons \text{Ge}-\text{Ge} + \text{Se}-\text{Se} \tag{1.18}$$

where homopolar or defect bonds are formed in pairs, and for which the law of mass action gives an equilibrium constant

$$K = \frac{[\text{Ge}-\text{Ge}]\,[\text{Se}-\text{Se}]}{[\text{Ge}-\text{Se}]^2} = \exp\left(-\frac{\Delta G}{\text{R}T}\right) \tag{1.19}$$

where [A−B] represents the concentration of A−B bonds, ΔG is the standard reaction Gibbs energy, R is the molar gas constant, and T is the absolute temperature. From (1.18) it follows that the concentration of Ge−Ge or Se−Se defect bonds n_d = [Ge−Ge] = [Se−Se] where the Ge−Ge homopolar bonds might be in ethane-like $Se_{3/2}Ge-GeSe_{3/2}$ units as suggested by ^{119}Sn Mössbauer spectroscopy experiments [83, 129] and the Se−Se homopolar bonds might be in dimers linking Ge-centred tetrahedra. Equation (1.19) can therefore be re-written as

$$\frac{n_d}{n_0} = \exp\left(-\frac{\Delta G}{2\text{R}T}\right) \tag{1.20}$$

where $n_0 \equiv$ [Ge−Se]. If the concentration of defects is small such that $n_d \ll n_0$ then the latter is approximately equal to the concentration of Ge−Se bonds in a non-defected system.

ΔG can be estimated from the difference between the Ge−Se, Ge−Ge and Se−Se bond enthalpies which take values of 225, 188 and 227 kJ mol^{-1}, respectively, at 298 K i.e. $\Delta G \simeq \Delta H = 2\times225 - 188 - 227 = 35$ kJ mol^{-1} [127]. Hence, an estimate for the fraction of defects in the melt at the glass transition temperature (T_g = 665 K) is given by $n_d/n_0 \simeq 0.042$. Alternatively, if $n_d \equiv N_d/V$ and $n_0 \equiv N_{\text{bond}}/V$, where N_d is the number of Ge−Ge or Se−Se homopolar bonds and N_{bond} is the total number of bonds, it follows that $N_d/N_{\text{bond}} \simeq 0.042$. This ratio is probably a lower limit because the value of ΔG used in the calculation is likely to decrease when the entropy term ΔS is taken into account ($\Delta G = \Delta H - T\Delta S$ if the absolute temperature T is constant), and the reaction enthalpy ΔH is likely to be smaller at T_g as compared to room temperature [127, 128].

Let the total number of atoms in the system be denoted by $N = N_{\rm Ge} + N_{\rm Se}$ where $N_{\rm Ge}$ and $N_{\rm Se}$ are the numbers of Ge and Se atoms, respectively, such that the atomic fractions are given by $c_{\rm Ge} = N_{\rm Ge}/N$ and $c_{\rm Se} = N_{\rm Se}/N$. From the NDIS results on $GeSe_2$ glass [22, 23], the coordination number for Ge–Ge homopolar bonds $\bar{n}_{\rm Ge}^{\rm Ge} = 0.25(5)$. If these bonds form only in pairs then $\bar{n}_{\rm Ge}^{\rm Ge} = \left(N_{\rm Ge,homo} \times 1\right)/N_{\rm Ge}$ such that the number of Ge–Ge bonds is given by $N_{\rm Ge-Ge} = \left(\bar{n}_{\rm Ge}^{\rm Ge} \times N_{\rm Ge}\right)/2$ where the factor of two avoids double counting and $N_{\rm Ge} = N/3$. It follows that $N_{\rm Ge-Ge} = 0.042(8)N$. Similarly, from the NDIS results the coordination number for Se–Se homopolar bonds $\bar{n}_{\rm Se}^{\rm Se} = 0.20(5)$. If these bonds form only in pairs then $\bar{n}_{\rm Se}^{\rm Se} = \left(N_{\rm Se,homo} \times 1\right)/N_{\rm Se}$ such that the number of Se–Se bonds is given by $N_{\rm Se-Se} = \left(\bar{n}_{\rm Se}^{\rm Se} \times N_{\rm Se}\right)/2$ where the factor of two avoids double counting and $N_{\rm Se} = 2N/3$. It follows that $N_{\rm Se-Se} = 0.067(17)N$. Thus, within the experimental error, $N_{\rm Ge-Ge} \sim N_{\rm Se-Se}$ as in the model of Feltz [127] such that $N_d \simeq \left(N_{\rm Ge-Ge} + N_{\rm Se-Se}\right)/2 = 0.05(2)N$.

For $GeSe_2$, the number of Ge–Se bonds in a non-defected system $N_{\rm bond} = (N_{\rm Ge}Z_{\rm Ge} + N_{\rm Se}Z_{\rm Se})/2 = (c_{\rm Ge}Z_{\rm Ge} + c_{\rm Se}Z_{\rm Se})\,N/2$ where Z_α is the number of bonds formed by chemical species α. Since $Z_{\rm Ge} = 4$, $Z_{\rm Se} = 2$, $c_{\rm Ge} = 1/3$, $c_{\rm Se} = 2/3$ it follows that $N_{\rm bond} = 4N/3$. Thus $N_d/N_{\rm bond} \simeq 0.04(2)$ for the NDIS results, which is in agreement with the value $N_d/N_{\rm bond} \simeq 0.042$ estimated by using the law of mass action.

References

1. M. Yamane, Y. Asahara, *Glasses for Photonics* (Cambridge University Press, Cambridge, 2000)
2. S.Y. Park, S.K. Lee, Geochim. Cosmochim. Acta **80**, 125 (2012)
3. R.L. McGreevy, L. Pusztai, Mol. Simul. **1**, 359 (1988)
4. R.L. McGreevy, J. Phys.: Condens. Matter **13**, R877 (2001)
5. A.K. Soper, Chem. Phys. **202**, 295 (1996)
6. A.K. Soper, Phys. Rev. B **72**, 104204 (2005)
7. D.A. Keen, Phase Transitions **61**, 109 (1997)
8. M.G. Tucker, D.A. Keen, M.T. Dove, K. Trachenko, J. Phys.: Condens. Matter **17**, S67 (2005)
9. A.K. Soper, J. Phys.: Condens. Matter **19**, 335206 (2007)
10. A.K. Soper, C.J. Benmore, Phys. Rev. Lett. **108**, 259603 (2012)
11. A. Zeidler, P.S. Salmon, H.E. Fischer, J.C. Neuefeind, J.M. Simonson, H. Lemmel, H. Rauch, T.E. Markland, Phys. Rev. Lett. **108**, 259604 (2012)
12. L.B. Skinner, A.C. Barnes, P.S. Salmon, H.E. Fischer, J.W.E. Drewitt, V. Honkimäki, Phys. Rev. B **85**, 064201 (2012)
13. L.B. Skinner, A.C. Barnes, P.S. Salmon, L. Hennet, H.E. Fischer, C.J. Benmore, S. Kohara, J.K.R. Weber, A. Bytchkov, M.C. Wilding, J.B. Parise, T.O. Farmer, I. Pozdnyakova, S.K. Tumber, K. Ohara, Phys. Rev. B **87**, 024201 (2013)
14. S. Biggin, J.E. Enderby, J. Phys. C: Solid State Phys. **14**, 3129 (1981)
15. P.S. Salmon, R.A. Martin, P.E. Mason, G.J. Cuello, Nature **435**, 75 (2005)
16. A. Zeidler, P.S. Salmon, R.A. Martin, T. Usuki, P.E. Mason, G.J. Cuello, S. Kohara, H.E. Fischer, Phys. Rev. B **82**, 104208 (2010)
17. A. Zeidler, P. Chirawatkul, P.S. Salmon, T. Usuki, S. Kohara, H.E. Fischer, W.S. Howells, J. Non-Cryst. Solids **407**, 235 (2015)

18. M. Wilson, P.A. Madden, J. Phys.: Condens. Matter **5**, 6833 (1993)
19. P.A. Madden, M. Wilson, Chem. Soc. Rev. **25**, 339 (1996)
20. M. Wilson, Phys. Chem. Chem. Phys. **14**, 12701 (2012)
21. I.T. Penfold, P.S. Salmon, Phys. Rev. Lett. **67**, 97 (1991)
22. I. Petri, P.S. Salmon, H.E. Fischer, Phys. Rev. Lett. **84**, 2413 (2000)
23. P.S. Salmon, I. Petri, J. Phys.: Condens. Matter **15**, S1509 (2003)
24. M. Cobb, D.A. Drabold, R.L. Cappelletti, Phys. Rev. B **54**, 12162 (1996)
25. M. Cobb, D.A. Drabold, Phys. Rev. B **56**, 3054 (1997)
26. C. Massobrio, A. Pasquarello, R. Car, Phys. Rev. Lett. **80**, 2342 (1998)
27. C. Massobrio, A. Pasquarello, R. Car, J. Am. Chem. Soc. **121**, 2943 (1999)
28. C. Massobrio, A. Pasquarello, R. Car, Comput. Mater. Sci. **17**, 115 (2000)
29. X. Zhang, D.A. Drabold, Phys. Rev. B **62**, 15695 (2000)
30. C. Massobrio, A. Pasquarello, R. Car, Phys. Rev. B **64**, 144205 (2001)
31. C. Massobrio, M. Celino, A. Pasquarello, Phys. Rev. B **70**, 174202 (2004)
32. C. Massobrio, A. Pasquarello, J. Phys.: Condens. Mater **19**, 415111 (2007)
33. C. Massobrio, A. Pasquarello, Phys. Rev. B **75**, 014206 (2007)
34. C. Massobrio, A. Pasquarello, Phys. Rev. B **77**, 144207 (2008)
35. M. Wilson, B.K. Sharma, C. Massobrio, J. Chem. Phys. **128**, 244505 (2008)
36. M. Micoulaut, R. Vuilleumier, C. Massobrio, Phys. Rev. B **79**, 214205 (2009)
37. C. Massobrio, Lect. Notes Phys. **795**, 343 (2010)
38. C. Massobrio, M. Micoulaut, P.S. Salmon, Solid State Sci. **12**, 199 (2010)
39. M. Micoulaut, S. Le Roux, C. Massobrio, J. Chem. Phys. **136**, 224504 (2012)
40. A. Bouzid, C. Massobrio, J. Chem. Phys. **137**, 046101 (2012)
41. Y. Liang, C.R. Miranda, S. Scandolo, High Press. Res. **28**, 35 (2008)
42. C. Massobrio, M. Celino, P.S. Salmon, R.A. Martin, M. Micoulaut, A. Pasquarello, Phys. Rev. B **79**, 174201 (2009)
43. D. Marrocchelli, M. Salanne, P.A. Madden, J. Phys.: Condens. Matter **22**, 152102 (2010)
44. H.E. Fischer, A.C. Barnes, P.S. Salmon, Rep. Prog. Phys. **69**, 233 (2006)
45. T.E. Faber, J.M. Ziman, Philos. Mag. **11**, 153 (1965)
46. J.E. Enderby, D.M. North, P.A. Egelstaff, Philos. Mag. **14**, 961 (1966)
47. P.S. Salmon, A. Zeidler, Phys. Chem. Chem. Phys. **15**, 15286 (2013)
48. A.B. Bhatia, D.E. Thornton, Phys. Rev. B **2**, 3004 (1970)
49. E. Lorch, J. Phys. C: Solid State Phys. **2**, 229 (1969)
50. P.S. Salmon, J. Phys.: Condens. Matter **18**, 11443 (2006)
51. F.G. Edwards, R.A. Howe, J.E. Enderby, D.I. Page, J. Phys. C: Solid State Phys. **11**, 1053 (1978)
52. R.L. McGreevy, E.W.J. Mitchell, J. Phys. C: Solid State Phys. **15**, 5537 (1982)
53. S. Biggin, J.E. Enderby, J. Phys. C: Solid State Phys. **14**, 3577 (1981)
54. S. Biggin, M. Gay, J.E. Enderby, J. Phys. C: Solid State Phys. **17**, 977 (1984)
55. R.J. Newport, R.A. Howe, N.D. Wood, J. Phys. C: Solid State Phys. **18**, 5249 (1985)
56. R.D. Shannon, Acta Crystallogr. A **32**, 751 (1976)
57. P.S. Salmon, Proc. R. Soc. Lond. A **445**, 351 (1994)
58. P.S. Salmon, Proc. R. Soc. Lond. A **437**, 591 (1992)
59. P.S. Salmon, J. Phys.: Condens. Matter **19**, 455208 (2007)
60. J.E. Enderby, A.C. Barnes, Rep. Prog. Phys. **53**, 85 (1990)
61. M. Wilson, P.S. Salmon, Phys. Rev. Lett. **103**, 157801 (2009)
62. M. Wilson, P.A. Madden, Phys. Rev. Lett. **80**, 532 (1998)
63. B.K. Sharma, M. Wilson, J. Phys.: Condens. Matter **20**, 244123 (2008)
64. P.S. Salmon, J. Phys.: Condens. Matter **17**, S3537 (2005)
65. P.S. Salmon, A.C. Barnes, R.A. Martin, G.J. Cuello, Phys. Rev. Lett. **96**, 235502 (2006)
66. M.J.L. Sangster, M. Dixon, Adv. Phys. **25**, 247 (1976)
67. N.C. Pyper, in *Advances in Solid State Chemistry*, vol. 2, ed. by C.R.A. Catlow (JAI Press, London, 1991), p. 223
68. R. Kjellander, B. Forsberg, J. Phys. A: Math. Gen. **38**, 5405 (2005)

69. R. Evans, R.J.F. Leote de Carvalho, J.R. Henderson, D.C. Hoyle, J. Chem. Phys. **100**, 591 (1994)
70. R.J.F. Leote de Carvalho, R. Evans, Mol. Phys. **83**, 619 (1994)
71. C. Grodon, M. Dijkstra, R. Evans, R. Roth, J. Chem. Phys. **121**, 7869 (2004)
72. C.A. Angell, Science **267**, 1924 (1995)
73. V.N. Novikov, Y. Ding, A.P. Sokolov, Phys. Rev. E **71**, 061501 (2005)
74. J. Málek, J. Shánělová, J. Non-Cryst. Solids **243**, 116 (1999)
75. E.A. Pavlatou, S.N. Yannopoulos, G.N. Papatheodorou, G. Fytas, J. Phys. Chem. B **101**, 8748 (1997)
76. J. Ruska, H. Thurn, J. Non-Cryst. Solids **22**, 277 (1976)
77. J.A.E. Desa, A.C. Wright, J. Wong, R.N. Sinclair, J. Non-Cryst. Solids **51**, 57 (1982)
78. J. Neuefeind, K.-D. Liss, Ber. Bunsenges. Phys. Chem. **100**, 1341 (1996)
79. A. Zeidler, J.W.E. Drewitt, P.S. Salmon, A.C. Barnes, W.A. Crichton, S. Klotz, H.E. Fischer, C.J. Benmore, S. Ramos, A.C. Hannon, J. Phys.: Condens. Matter **21**, 474217 (2009)
80. P.S. Salmon, J. Non-Cryst. Solids **353**, 2959 (2007)
81. G. Dittmar, H. Schäfer, Acta Crystallogr. B **32**, 2726 (1976)
82. S.B. Mamedov, N.D. Aksenov, L.L. Makarov, Yu.F. Batrakov, J. Non-Cryst. Solids **195**, 272 (1996)
83. P. Boolchand, W.J. Bresser, Philos. Mag. B **80**, 1757 (2000)
84. J.P. Perdew, Y. Wang, Phys. Rev. B **45**, 13244 (1992)
85. A.D. Becke, Phys. Rev. A **38**, 3098 (1988)
86. C. Lee, W. Yang, R.G. Parr, Phys. Rev. B **37**, 785 (1988)
87. K. Wezka, A. Bouzid, K.J. Pizzey, P.S. Salmon, A. Zeidler, S. Klotz, H.E. Fischer, C.L. Bull, M.G. Tucker, M. Boero, S. Le Roux, C. Tugène, C. Massobrio, Phys. Rev. B **90**, 054206 (2014)
88. L. Giacomazzi, C. Massobrio, A. Pasquarello, Phys. Rev. B **75**, 174207 (2007)
89. L. Giacomazzi, C. Massobrio, A. Pasquarello, J. Phys.: Condens. Matter **23**, 295401 (2011)
90. P. Vashishta, R.K. Kalia, G.A. Antonio, I. Ebbsjö, Phys. Rev. Lett. **62**, 1651 (1989)
91. P. Vashishta, R.K. Kalia, I. Ebbsjö, Phys. Rev. B **39**, 6034 (1989)
92. H. Iyetomi, P. Vashishta, R.K. Kalia, Phys. Rev. B **43**, 1726 (1991)
93. P. Vashishta, R.K. Kalia, J.P. Rino, I. Ebbsjö, Phys. Rev. B **41**, 12197 (1990)
94. B.K. Sharma, M. Wilson, Phys. Rev. B **73**, 060201(R) (2006)
95. S.N. Yannopoulos, A.G. Kalampounias, A. Chrissanthopoulos, G.N. Papatheodorou, J. Chem. Phys. **118**, 3197 (2003)
96. M. Celino, C. Massobrio, Phys. Rev. Lett. **90**, 125502 (2003)
97. Q. Mei, C.J. Benmore, S. Sen, R. Sharma, J.L. Yarger, Phys. Rev. B. **78**, 144204 (2008)
98. G.N. Greaves, S. Sen, Adv. Phys. **56**, 1 (2007)
99. P.S. Salmon, A.C. Barnes, R.A. Martin, G.J. Cuello, J. Phys.: Condens. Matter **19**, 415110 (2007)
100. L. Giacomazzi, P. Umari, A. Pasquarello, Phys. Rev. B **74**, 155208 (2006)
101. M. Wilson, B.K. Sharma, J. Chem. Phys. **128**, 214507 (2008)
102. V.V. Brazhkin, A.G. Lyapin, J. Phys.: Condens. Matter **15**, 6059 (2003)
103. M.C. Wilding, M. Wilson, P.F. McMillan, Chem. Soc. Rev. **35**, 964 (2006)
104. G.N. Greaves, M.C. Wilding, S. Fearn, D. Langstaff, F. Kargl, S. Cox, Q. Vu Van, O. Majérus, C.J. Benmore, R. Weber, C.M. Martin, L. Hennet, Science **322**, 566 (2008)
105. A.C. Barnes, L.B. Skinner, P.S. Salmon, A. Bytchkov, I. Pozdnyakova, T.O. Farmer, H.E. Fischer, Phys. Rev. Lett. **103**, 225702 (2009)
106. A.C. Barnes, L.B. Skinner, P.S. Salmon, A. Bytchkov, I. Pozdnyakova, T.O. Farmer, H.E. Fischer, Phys. Rev. Lett. **106**, 119602 (2011)
107. D. Daisenberger, T. Deschamps, B. Champagnon, M. Mezouar, R.Q. Cabrera, M. Wilson, P.F. McMillan, J. Phys. Chem. B **115**, 14246 (2011)
108. C.S. Marians, L.W. Hobbs, J. Non-Cryst, Solids **124**, 242 (1990)
109. K. Wezka, P.S. Salmon, A. Zeidler, D.A.J. Whittaker, J.W.E. Drewitt, S. Klotz, H.E. Fischer, D. Marrocchelli, J. Phys.: Condens. Matter **24**, 502101 (2012)

110. J.W.E. Drewitt, P.S. Salmon, A.C. Barnes, S. Klotz, H.E. Fischer, W.A. Crichton, Phys. Rev. B **81**, 014202 (2010)
111. P.S. Salmon, J.W.E Drewitt, D.A.J. Whittaker, A. Zeidler, K. Wezka, C.L. Bull, M.G. Tucker, M.C. Wilding, M. Guthrie, D. Marrocchelli, J. Phys.: Condens. Matter **24**, 415102 (2012)
112. J.D. Jorgensen, J. Appl. Phys. **49**, 5473 (1978)
113. J. Glinnemann, H.E. King Jr, H. Schulz, Th. Hahn, S.J. La Placa, F. Dacol, Z. Kristallogr. **198**, 177 (1992)
114. X. Hong, G. Shen, V.B. Prakapenka, M. Newville, M.L. Rivers, S.R. Sutton, Phys. Rev. B **75**, 104201 (2007)
115. D. Marrocchelli, M. Salanne, P.A. Madden, C. Simon, P. Turq, Mol. Phys. **107**, 443 (2009)
116. G. Lelong, L. Cormier, G. Ferlat, V. Giordano, G.S. Henderson, A. Shukla, G. Calas, Phys. Rev. B **85**, 134202 (2012)
117. R.D. Oeffner, S.R. Elliott, Phys. Rev. B **58**, 14791 (1998)
118. M. Micoulaut, J. Phys.: Condens. Matter **16**, L131 (2004)
119. M. Micoulaut, Y. Guissani, B. Guillot, Phys. Rev. E **73**, 031504 (2006)
120. K.V. Shanavas, N. Garg, S.M. Sharma, Phys. Rev. B **73**, 094120 (2006)
121. M. Micoulaut, X. Yuan, L.W. Hobbs, J. Non-Cryst. Solids **353**, 1961 (2007)
122. T. Li, S. Huang, J. Zhu, Chem. Phys. Lett. **471**, 253 (2009)
123. X.F. Zhu, L.F. Chen, Phys. B **404**, 4178 (2009)
124. J.D. Martin, S.J. Goettler, N. Fossé, L. Iton, Nature **419**, 381 (2002)
125. P.S. Salmon, Nat. Mater. **1**, 87 (2002)
126. M.M. Smedskjaer, J.C. Mauro, S. Sen, Y. Yue, Chem. Mater. **22**, 5358 (2010)
127. A. Feltz, in *Physics of Disordered Materials*, ed. by D. Adler, H. Fritzche, S.R. Ovshinsky (Plenum, New York, 1985), p. 203
128. A. Feltz, *Amorphous Inorganic Materials and Glasses* (VCH, Weinheim, 1993), p. 97
129. P. Boolchand, J. Grothaus, W.J. Bresser, P. Suranyi, Phys. Rev. B **25**, 2975 (1982)

Chapter 2
First-Principles Molecular Dynamics Methods: An Overview

Mauro Boero, Assil Bouzid, Sebastien Le Roux, Burak Ozdamar and Carlo Massobrio

Abstract This chapter proposes an overview of computational approaches used nowadays in the field of first-principles simulations to model amorphous and liquid materials. The scope is to bring to the attention of the readership advances and (still existing) limitations in the description of the interactions among atoms which, starting in general from an ordered crystallographic structure, undergo significant modifications in the underlying electronic structure for the disordered phases. These subtle details are difficult to capture by resorting on classical model potentials and call for an accurate description of the quantum mechanical description of the intimate constituent of a glassy compound. The heavy computational workload associated can be nowadays overcome in virtue of the increasing computing power of last-generation high performance computers. Also of paramount importance are advances in algorithms and methods capable of providing the required speed-up in terms of both performances and accuracy.

M. Boero (✉) · A. Bouzid · S. Le Roux · B. Ozdamar · C. Massobrio
Institut de Physique et Chimie des Materiaux de Strasbourg (IPCMS),
University of Strasbourg and CNRS, UMR 7504,
23 rue du Loess, 67034 Strasbourg, France
e-mail: mauro.boero@ipcms.unistra.fr

A. Bouzid
e-mail: assil.bouzid@ipcms.unistra.fr

S. Le Roux
e-mail: Sebastien.leroux@ipcms.unistra.fr

B. Ozdamar
e-mail: Burak.Ozdamar@ipcms.unistra.fr

C. Massobrio
e-mail: Carlo.Massobrio@ipcms.unistra.fr

C. Massobrio et al. (eds.), *Molecular Dynamics Simulations of Disordered Materials*, Springer Series in Materials Science 215,
DOI 10.1007/978-3-319-15675-0_2

2.1 Introduction

A major target in atomic-scale simulations, not only limited to the amorphous state, is to reproduce in a realistic way physical and chemical events occurring in materials. The formation and finite temperature evolution of a glass, in this respect, represent one of the processes that have long challenged first principles methods and to many extent still are. The inherent complexity of a disordered material clearly implies something that goes beyond the simple calculation of the electronic structure of a given set of coordinates $\mathbf{R}_I$ representing, for instance, the positions of atoms in a bulk solid. This is particularly true when the solid is not a crystal, because a single configuration would correspond just to a specific topology of a number of atoms from a few hundreds to a few thousands, but still far from what a glass should be. It is exactly this idea that started the entire field called Molecular Dynamics (MD). Basically, the scope of MD is to study a system of *interacting* particles by recreating it on a computer in a way as close as possible to nature and by simulating its dynamics over a *physical length of time* relevant to the properties of interest.

From an historical point of view, the MD approach was introduced by Alder and Wainwright [1, 2] in the late 1950s to study the interactions of hard spheres. Many important insights concerning the behavior of simple liquids emerged from their studies, but due to the limitations of the computational facilities and the pioneering stage of the MD, we had to wait until 1964 to see the first simulation using a realistic potential, or more precisely a truncated Lennard-Jones type potential, for liquid Ar by Rahman [3]. And only 10 years later, the first molecular dynamics simulation of a realistic liquid water system was done by Rahman and Stillinger [4]. In a nutshell, any MD method is an iterative numerical scheme for solving some equations of motion (EOM), which represent the physical evolution of the system under study. Unfortunately, modeling the interaction of atoms with an analytic potential $V(\mathbf{R}_I)$, especially when chemical bonds evolve in time and/or are broken and formed is a hard task solved only for a very limited class of chemical species and very specific chemical processes.

On the other hand, the electronic structure for a general many-body system can be determined with a computationally reasonable workload by means of the density functional theory (DFT), originally proposed in the early 60s by Kohn, Hohenberg and Sham [5, 6], with important contributions also from the group of Pople [7–9]. Its importance in the advancement of computational quantum chemistry and related fields was acknowledged by the Nobel Prize in Chemistry in 1998 awarded jointly to Kohn and Pople.

2.1.1 A Brief Overview of Density Functional Theory

The DFT is a formulation of the many-body quantum mechanics in terms of an electron density distribution, $\rho(\mathbf{x})$, which describes the ground state of a system composed of interacting electrons and point-like nuclei having positions $\{\mathbf{R}_I\}$. Several excellent books and review articles have been published on the fundamentals of DFT [10, 11]. For this reason, we shall limit here the discussion to the basic details necessary to the ongoing discussion. The first step in DFT consists in giving an explicit form for the electron density distribution. In practice, single-particle wavefunctions $\psi_i(\mathbf{x})$ are used to express the many-body mathematical function $\rho(\mathbf{x})$. The dramatic simplification is the fact that the specific analytic form of the complex function $\psi_i(\mathbf{x})$ does not matter, but only its square modulus, so that the electron density reads

$$\rho(\mathbf{x}) = \sum_{i=1}^{N^{occ}} f_i |\psi_i(\mathbf{x})|^2 \tag{2.1}$$

This expression is clearly a single Slater determinant constructed from wavefunctions representing all the N^{occ} occupied orbitals. The coefficients f_i are the (integer) occupation numbers, and they are equal to 1 in the case in which the spin is explicitly considered (spin-unrestricted) or equal to 2 if the spin is neglected and energy levels are considered as doubly-occupied (spin-restricted). Furthermore, the wavefunctions $\psi_i(\mathbf{x})$ are subject to the orthonormality constraint

$$\int \psi_i^*(\mathbf{x})\psi_j(\mathbf{x})d^3x = \delta_{ij} \tag{2.2}$$

as in any quantum mechanics approach. The Kohn-Sham (KS) DFT total energy of the system in its ground state is then written as

$$E^{\mathrm{KS}}[\{\psi_i\}] = E_k[\{\psi_i\}] + E_H[\rho] + E_{xc}[\rho] + E_{eI}[\rho] + E_{II} \tag{2.3}$$

In (2.3) the first three terms on the right-hand side (E_k, E_H, E_{xc}) describe all the electron-electron interactions, the fourth term (E_{eI}) refers to the electron-nucleus interaction and the fifth one (E_{II}) corresponds to the nucleus-nucleus interaction. More explicitly, E_k is the Schrödinger-like kinetic energy expressed in terms of the single-particle wavefunctions $\psi_i(\mathbf{x})$ as

$$E_k[\{\psi_i\}] = \sum_{i=1}^{N^{occ}} f_i \int d^3x \psi_i^*(\mathbf{x}) \left(-\frac{1}{2}\nabla^2\right) \psi_i(\mathbf{x}) \tag{2.4}$$

We can remark that this expression for the kinetic energy does not depend on the density $\rho(\mathbf{x})$ but directly on the wavefunctions. The second term, E_H, is the Hartree energy, i.e. the Coulomb electrostatic interaction between two charge distributions

$$E_H[\rho] = \int\int d^3x d^3y \frac{\rho(\mathbf{x})\rho(\mathbf{y})}{|\mathbf{x}-\mathbf{y}|} \tag{2.5}$$

The exchange interaction and the electron correlations due to many-body effects are represented by the term $E_{xc}[\rho]$, whose exact analytical expression is unknown.

There are good approximations derived from the homogeneous electron gas limit for the exchange interaction [12], the so-called local density approximation (LDA), whose name comes from the fact that an interacting but homogeneous electron distribution is assumed, in which $\rho(\mathbf{x})$ depends just on the local point $\mathbf{x}$. Similarly, in the LDA version of the correlation energy [13], the explicit analytic form of the functional comes from a parameterization of the results of random phase approximation calculations. Due to the insufficiency of a simple LDA approximation for many real systems, non-local approximations including the gradient of the density, are often adopted and the exchange-correlation functional becomes $E_{xc}[\rho, \nabla\rho]$. In practical applications, however, the gradient enters only with its modulus, thus adding only a modest computational cost. These generalized gradient corrections [14–16] are indeed a bit arbitrary, in the sense that they do not represent a regular perturbation expansion. Nonetheless, they are generally based on solid physical and mathematical argumentations and can be assessed a posteriori by test calculations and comparisons with both exact results and experiments [9]. The electrostatic interaction between electrons and nuclei, is given by

$$E_{eI}[\rho] = -\int d^3x \sum_{I=1}^{M} \frac{Z_I \rho(x)}{|\mathbf{x} - \mathbf{R}_I|} \tag{2.6}$$

where Z_I is the charge of the Ith nucleus. However, in practice, this expression "as is" is computationally expensive. In fact, two different length scales come into play: a small one for the core electrons, characterized by rapidly varying wavefunctions, especially in the region very close to the nucleus, and a longer one for the valence electrons that form chemical bonds and vary more smoothly. Clearly, the first one would dominate and add a computational workload that would make impractical simulations of large systems. To overcome this problem, one can observe that core electrons are generally inert and do not participate to chemical bonds. This crucial observation led to the use of pseudopotentials [17–20]. Namely, core electrons are eliminated and a potential describing the core-valence interaction is built by fitting to the all-electron solutions of the Schrödinger or Dirac equation for the single atom of the chemical species considered [17]. In a pseudopotential (PP) approach, the electron-nucleus interaction is rewritten as

$$E_{eI}[\rho] = \int d^3x V_{ps}(\mathbf{x} - \mathbf{R}_I) \cdot \rho(\mathbf{x}) \tag{2.7}$$

Finally, the fifth and last term in right-hand side of (2.3) is simply the Coulomb interaction between two classical nuclei I and J and is written as

$$E_{II} = \sum_{I<J}^{M} \frac{Z_I Z_J}{|\mathbf{R}_I - \mathbf{R}_J|} \tag{2.8}$$

where Z_I and Z_J are the net valence charge in a PP approach.

2.1.2 The Basis Set Issue

The total energy E^{tot} of the ground state of such a system of interacting electrons and nuclei can then be obtained by minimizing the KS functional with respect to the single-particle orbitals $\psi_i(\mathbf{x})$, which, in practice, means solving the KS Schrödinger-like equations given by the variational derivative of the KS functional with respect to the single-particle wavefunctions $\psi_i(\mathbf{x})$. A question still unanswered at this point is "what is $\psi_i(\mathbf{x})$?". The answer is the selection of a proper basis set on which orbitals can be expanded. One possible choice is a localized basis set expressing the one-electron wavefunctions as

$$\psi_i(\mathbf{x}) = \sum_{k=1}^{M} c_i^k \phi_k(\mathbf{x}; \{\mathbf{R}_I\}) \tag{2.9}$$

and the number of analytic functions used, M, is also an indicator of the computational cost of the quantum calculation, in the obvious sense that the larger the basis set, the higher the computational workload. In most of the HF applications, a very popular, somehow minimal, basis set is represented by Slater-type orbitals (STO)

$$\phi_k(\mathbf{x}, \{\mathbf{R}_I\}) \equiv \phi_{\mathbf{k}}^{\mathrm{STO}}(\mathbf{x} - \mathbf{R}_I) = \phi_{\mathbf{k}}^{\mathrm{STO}}(\mathbf{r}) = N_{\mathbf{k}} \cdot r_x^{k_x} r_y^{k_y} r_z^{k_z} \exp\left(-\zeta_{\mathbf{k}} |\mathbf{r}|\right)$$

or Gaussian-type orbitals (GTO)

$$\phi_k(\mathbf{x}, \{\mathbf{R}_I\}) \equiv \phi_{\mathbf{k}}^{\mathrm{GTO}}(\mathbf{x} - \mathbf{R}_I) = \phi_{\mathbf{k}}^{\mathrm{GTO}}(\mathbf{r}) = N_{\mathbf{k}} \cdot r_x^{k_x} r_y^{k_y} r_z^{k_z} \exp\left(-\alpha_{\mathbf{k}} \cdot r^2\right)$$

where $\mathbf{r} = \mathbf{x} - \mathbf{R}_I$. When basis sets of these kinds are used, the constants $N_{\mathbf{k}}$, $\zeta_{\mathbf{k}}$ and $\alpha_{\mathbf{k}}$ are kept fixed during the electronic structure calculation, whereas the coefficients c_i^k are allowed to vary until they are fully optimized [21]. It must be remarked that orbitals expanded in a localized basis set depend on the atomic positions $\mathbf{R}_I$. As a consequence, in any calculation in which the forces acting on the ions are required, the explicit derivatives of these wavefunctions with respect to $\mathbf{R}_I$ must be computed, leading to non-Hellmann-Feynman force components known in the literature as Pulay forces [22]. An alternative rather popular in physics is represented by plane waves (PW)

$$\psi_i(\mathbf{x}) = \sum_{\mathbf{G}=0}^{\mathbf{G}^{\max}} c_i(\mathbf{G}) e^{i\mathbf{G}\cdot\mathbf{x}} \tag{2.10}$$

where the sum is truncated at a suitable cut-off $\mathbf{G}^{\max}$.

2.2 First Principles Molecular Dynamics

Until the early 1980s, few applications of DFT went beyond the static calculations of the electronic structure. Nonetheless, finite temperature and entropy effects are two of the dominant features in matter and their role is often far from negligible. In this respect, the so called first principles molecular dynamics (FPMD) has represented a huge step forward in quantum simulations of condensed phases. In this particular combination of DFT and MD, the interactions among atoms, instead of being described by an analytical function $V(\{\mathbf{R}_I\})$ of the atomic coordinates $\mathbf{R}_I$, is directly computed from the DFT total energy E^{tot}, which is simultaneously a function of the electron wavefunctions and the atomic coordinates. More specifically, the Born-Oppenheimer (BO) approximation [23] allows to disentangle the motions of the electrons and the nuclei, and each time the nuclei $\mathbf{R}_I(t)$ are displaced from given positions at time t to new positions $\mathbf{R}_I(t+\delta t)$ at a subsequent time $t+\delta t$, an optimization of the electronic structure has to be performed. Then the forces acting on the nuclei are estimated via direct calculations of the gradient of the total energy with respect to the ionic position and the variables $\mathbf{R}_I(t)$ updated to $\mathbf{R}_I(t+\delta t)$. The set of equations that one has to solve iteratively reads

$$M_I \ddot{\mathbf{R}}_I = -\nabla_{\mathbf{R}_I} \min_{\{\psi_i\}} E^{\text{tot}}\left[\{\psi_i\}, \{\mathbf{R}_I\}\right] \tag{2.11}$$

$$\frac{\delta E^{\text{tot}}}{\delta \psi_i^*} \equiv H^{\text{tot}} \psi_i(\mathbf{x}) = \varepsilon_i \psi_i(\mathbf{x}) \tag{2.12}$$

This iterative procedure assumes that the electronic structure is recomputed and the full diagonalization of the Hamiltonian is performed at each time step t along the discrete trajectory $\{\mathbf{R}_I(t)\}$.

2.2.1 Car-Parrinello Molecular Dynamics

An alternative to this scheme, which has represented a real breakthrough in first principles dynamical simulations was proposed in 1985 by Car and Parrinello [24]. The scope (and driving force) was to overcome the two major efforts arising in FPMD: on one hand one has to integrate the EOM for the nuclear positions as in (2.11), which represent the long time scale part to the problem. On the other hand, one has to propagate dynamically the smooth time-evolving (ground state) electronic subsystem. The Car-Parrinello molecular dynamics (CPMD) is able to satisfy this second requirement in a numerically stable way and makes an acceptable compromise for the time step length of the nuclear motion. The formulation is an extension of a classical molecular dynamics Lagrangean in which the wavefunctions representing the electronic degrees of freedom are added to the system, along with any other dynamical variable $q_\alpha(t)$, such as a thermostat [25, 26], a barostat [27–29], etc.

$$\begin{aligned} \mathcal{L}^{\text{CP}} &= \frac{1}{2}\sum_I M_I \dot{\mathbf{R}}_I^2 + \sum_i \mu \int d^3x \left|\dot{\psi}_i(\mathbf{x})\right|^2 + \frac{1}{2}\sum_\alpha \eta_\alpha \dot{q}_\alpha^2 - E^{tot}\left[\rho, \{\mathbf{R}_I\}, q_\alpha\right] \\ &+ \sum_{ij} \lambda_{ij} \left(\int d^3x\, \psi_i^*(\mathbf{x}) \psi_j(\mathbf{x}) - \delta_{ij} \right) \end{aligned} \tag{2.13}$$

The first three terms on the right-hand side of (2.13) are the kinetic energies of the nuclei, the electrons and the additional dynamical variables, the fourth one is the total energy, i.e. the DFT functional discussed above, and the last addendum is the orthonormality constraint for the wavefunctions. The kinetic energy for the electronic degrees of freedom is the novelty of the CPMD approach. This is an original strategy to update on-the-fly the wavefunctions when ions undergo a dynamical displacement, avoiding expensive diagonalizations required by the solution of (2.11) at each time step. In fact, the Euler-Lagrange EOM read

$$\mu\ddot{\psi}_i(\mathbf{x}) = -\frac{\delta E^{\mathrm{tot}}}{\delta\psi_i^*} + \sum_j \lambda_{ij}\psi_j(\mathbf{x}) \tag{2.14}$$

$$M_I\ddot{\mathbf{R}}_I = -\nabla_{\mathbf{R}_I} E^{\mathrm{tot}} \tag{2.15}$$

$$\eta_\alpha\ddot{q}_\alpha = -\frac{\partial E^{\mathrm{tot}}}{\partial q_\alpha} \tag{2.16}$$

The fictitious mass μ assigned to the orbitals $\psi_i(\mathbf{x})$ is the parameter that controls the speed of the updating of the wavefunctions with respect to the nuclear positions and, for this reason, it determines the degree of adiabaticity of the two subsystems, electrons and nuclei. A rigorous mathematical proof has been given by Bornemann and Schütte [30], showing that the CPMD trajectory $\{\mathbf{R}^{\mathrm{CP}}(t)\}$ remains close to the BO one $\{\mathbf{R}^{\mathrm{BO}}(t)\}$ and the upper bound is proportional to the square root of the fictitious mass μ

$$\left|\mathbf{R}^{CP}(t) - \mathbf{R}^{BO}(t)\right| < C \cdot \sqrt{\mu} \tag{2.17}$$

where C is a positive constant. Numerical simulations to probe of this theorem have been performed by Grossman et al. [31] and Schwegeler et al. [32], showing that an accurate choice of the input parameter μ provides a good control of the adiabaticity.

The fact that the CPMD was a milestone step forward in realistic simulations of materials, at various thermodynamics conditions, can be easily seen by the number of publications in first principles molecular dynamics (FPMD) before and after 1985, i.e. after the original CPMD publication [21]. Indeed, the original Car-Parrinello publication has more than 6500 citations in 2014 (source: ISI Web of Science), and, to acknowledge the importance of the method, the international PACS (Physics and Astronomy Classification Scheme) introduced in 1996 a new identification number, 71.15.Pd, to classify Car-Parrinello related publications. Since then, the method has been applied to a wide variety of materials, ranging from solids, to liquids and to biological systems [33].

Although it is by no means a restriction or a requirement of the method [34], PWs are often used as a convenient basis set for the coding of the CPMD EOM, since they have several good properties: (i) accuracy can be systematically improved in a fully variational way, (ii) PWs are independent from atomic positions (i.e. no Pulay forces [22] involved), (iii) they can be easily distributed in parallel processing. However, we must observe that the fact that PWs they are not localized can lead to

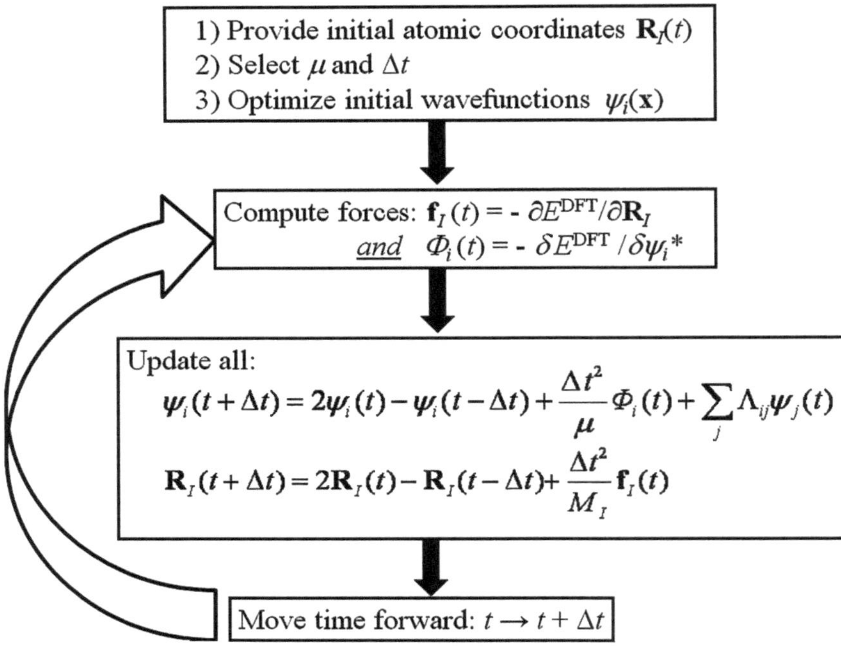

Fig. 2.1 Schematic blocks diagram for the practical implementation in a computer code of the Car-Parrinello method. The various *symbols* in the blocks are given in the text

inefficiencies for small clusters or surfaces placed in a large simulation cell. The EOM are implemented in a discrete finite differences way, via a Verlet, or velocity-Verlet, algorithm [35]. The general scheme is summarized in Fig. 2.1.

Namely, the dynamical variables $\mathbf{R}_I$(t) are updated at a rate Δt, while the electronic degrees of freedom are updated at a rate $\Delta t/\mu^{1/2}$. For most of the applications published so far [31, 32], Δt and μ are taken to be ~3–5 au and ~300–600 au, respectively. Of course, the (quantum) time scale of electrons is dominating in this kind of approaches and simulations times are of the order of few tens or, at very best, hundreds of ps. As far as the tractable system size is concerned, for a system with N electrons and $\mathbf{G}^{max}$ PWs, the scaling of the various parts composing the CPMD algorithm is $O(N\,\mathbf{G}^{max})$ for the kinetic term, $O(N\,\mathbf{G}^{max}\log\mathbf{G}^{max})$ for the local potential and $O(N^2\,\mathbf{G}^{max})$ for both the non-local term and wavefunctions orthogonalization procedure.

This approach has been used to simulate amorphous materials [36] for the first time just a few years after its introduction. In particular, in a pioneering work, a bulk amorphous silicon (a-Si) with realistic properties was obtained through a rapid quenching from a melted liquid-phase of silicon, followed by a thermal annealing on a time scale of a few picoseconds. This was on the verge of what could be afforded with the computational resources available at that time. The resulting structure was only affected by one issue in disagreement with experiments: a high bond-angle deviation.

The origin of this disagreement arises from the insufficient thermal annealing for structural relaxation and could be cured only upon advances in HPC architectures, computer power and algorithms.

In fact, it was only 10 years after the introduction of the CPMD method that amorphous silica (a-SiO_2), i.e. the first glass, could eventually be simulated [37].

Later [38, 39], oxygen vacancies (Fig. 2.2) and E' point defects present in glassy SiO_2 could be studied in great detail, including also full ab-initio calculations of the hyperfine parameters experimentally detected by electron-nuclear double resonance (ENDOR) experiments. Indeed, these types of measurements are nowadays routinely done to identify this class of paramagnetic defects. In the ENDOR technique, some Si atoms are substituted with their isotopes ^{29}Si. This confers a non-zero nuclear spin $\mathbf{I}$ to the atomic nucleus that couples to the electron spin $\mathbf{S}$ via a tensor $\mathbf{A}$. On the theoretical front, the calculation from first principles DFT approaches does not pose particular problems since the hyperfine interaction is still a ground state property which can be expressed in terms of the electronic density $\rho(\mathbf{x})$. The interaction between an electron spin ($\mathbf{S}$) and a nuclear ($\mathbf{I}$) spin is in fact described by the Hamiltonian

$$H^{hf} = \mu_0 g_e \mu_e g_I \mu_I \left[\frac{2}{3} \mathbf{I} \cdot \mathbf{S} \cdot \delta(x - \mathbf{R}_I) + \frac{1}{4\pi} \frac{3(\mathbf{I} \cdot \mathbf{u})(\mathbf{S} \cdot \mathbf{u}) - \mathbf{I} \cdot \mathbf{S}}{r^3} \right] \quad (2.18)$$

where μ_0 is the vacuum permeability, g_e the free-electron g-factor, μ_e the Bohr magneton, g_I the nuclear gyromagnetic ratio, μ_I the nuclear magneton of the atom located at $\mathbf{R}_I$ and $\mathbf{u} = \mathbf{r}/|\mathbf{r}|$. In this expression, the nucleus is assumed to be a point-like object with a magnetic dipole moment $\mu = \mu_I g_I \mathbf{I}$. This expression can be easily

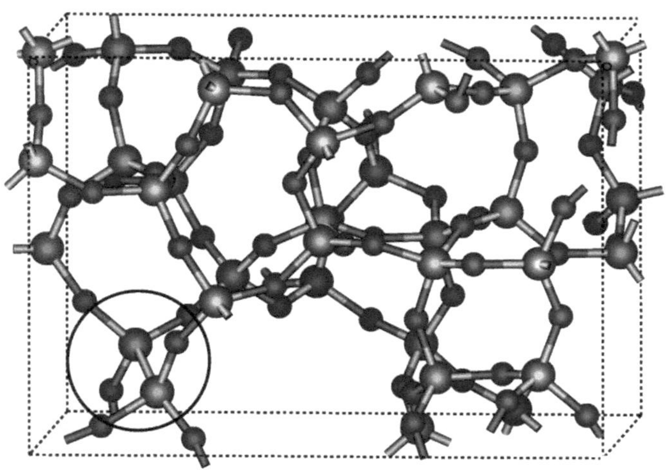

Fig. 2.2 Details of the amorphous silica model used in [38, 39] with an oxygen vacancy present as an Si–Si bond enclosed by a *circle*. Cell dimensions are 10.08 × 10.08 × 14.26 Å^3. The model contains 95 atoms with periodic boundary conditions. The *small dark spheres* are O atoms and the *large light-gray spheres* are Si atoms

integrated over the electronic degrees of freedom

$$\sum_i \langle \psi_i | H^{hf} | \psi_i \rangle = \mathbf{I} \cdot \mathbf{A} \cdot \mathbf{S} \tag{2.19}$$

to give the 3 × 3 matrix **A** which is the so-called hyperfine coupling tensor $A_{ij} = a\delta_{ij} + b_{ij}$. In (2.19) the first term is the Fermi contact interaction, that has no classical analog, while the second non-isotropic term is the dipole interaction. These are explicitly written as

$$a = \frac{2\mu_0}{3} g_e \mu_e g_I \mu_I \rho_{spin}(\mathbf{R}_I) \tag{2.20}$$

$$b_{ij} = \frac{\mu_0}{4\pi} g_e \mu_e g_I \mu_I \int \frac{3 r_i r_j - \delta_{ij} r^2}{r^5} \rho_{spin}(\mathbf{r})\, d^3 r \tag{2.21}$$

being $\rho_{spin}(\mathbf{r}) = \rho_\uparrow(\mathbf{r}) - \rho_\downarrow(\mathbf{r})$ the spin density of the system. The importance of these type of defects stems from the the fact that they originate very easily in a-SiO_2 as a consequence of radiation damage. Once generated, it is very hard to get rid of them and they are associated to the so-called color centers affecting seriously the properties and performances of (e.g.) optical fibers.

For analogous reasons, which again translate into the intrinsic computational cost, extensive and accurate first-principles investigations of chalcogenide glasses could be successfully afforded only recently [40–42]. These systems represent a hot topic and constitute a major target in the field of optoelectronics and optical support devices (CD, DVD, optical memory, etc.).

More specifically, upon structural changes, these materials are able to realize different network topologies, as in the simple example provided in Fig. 2.3. These different phases, and the possibility of their mutual interconversion, can be associated to a specific *bit* of information, hence becoming appealing for nanoelectronics applications as storage supports.

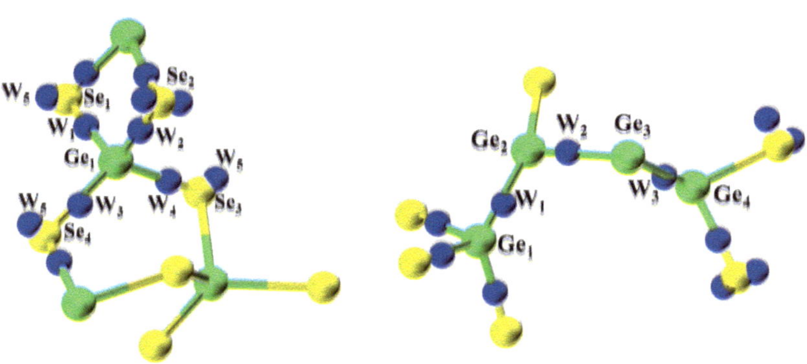

Fig. 2.3 Details of a dynamically evolving Ge_2Se_3 amorphous structure. Beside tetrahedral configurations in which a Ge atom (*green*) is surrounded by four Se atoms (*yellow*), ring structures can be realized, including a number of atoms ranging from four (*top left panel*) to six (*bottom left panel*) or more. These, eventually evolve in chains including only atoms of the same chemical species (homopolar bonds) as in the *right panel*. The *blue* balls are a short-hand visualization of the centers of the wavefunctions [43] contributing to the chemical bonds

2.3 Advanced Methods

2.3.1 Second Generation Car-Parrinello Dynamics

One of the major stumbling blocks in first-principles dynamical approaches is the inherent computational cost, associated to the electronic structure calculations all along the dynamics. This has limited the affordable length and time scales for several years and, in fact, only the most advanced high performing computer platforms recently available have allowed to increase the system size to about a thousand of atoms and simulation times towards hundreds of picosecond. Yet, many phenomena still call for a substantial boost. These are, for instance, diffusion in solids or, in the case of glasses generation from the melt, a less rapid cooling rate suitable to avoid numerically induced structural problems [36]. While linear scaling methods can be a viable way to access larger system sizes, they still have to face the problem of the simulation time scale. Moreover, the crossover point at which linear scaling methods become advantageous has remained fairly large, especially if high accuracy is needed. An interesting attempt at overcoming these limitations has been proposed in 2007 by Parrinello and coworkers [44]. The basic idea is to join the advantages of the BO approach and those of the CPMD.

These two approaches have, indeed, nearly complementary features as sketched in Table 2.1. Following the CPMD formulation as described in Sect. 2.2.1, it can be remarked that the Lagrangean mathematical formulation for the propagation of the wavefunctions is stable by construction, hence providing a reliable integration. This *stability* feature must then be preserved. Concerning the *efficiency*, large integration steps Δt are required and possibly a small, or better zero, deviation from the BO surface should be kept all along the dynamics to get a high *accuracy*. The mathematical result of this list of requirement resumes into a modified EOM for the ions, which now read

$$M_I \ddot{\mathbf{R}}_I = -\frac{\partial E_{NSC}}{\partial \mathbf{R}_I} + \sum_{i,j} \Lambda_{ij} \frac{\partial}{\partial \mathbf{R}_I} \langle \psi_i | \psi_j \rangle - 2 \frac{\partial \langle \psi_i |}{\partial \mathbf{R}_I} \left(\frac{\delta E_{NSC}}{\delta \langle \psi_i |} - \sum_j \Lambda_{ij} | \psi_j \rangle \right) \tag{2.22}$$

While the first term in the right hand of the equation is clear, the rest seems a bit puzzling at a first glance. Indeed, in the original formulation [44], the basis set adopted

Table 2.1 Comparison of the main features of the CPMD and the BO methods

	CPMD	BO
Conservation of constants of motion	**Good**	Convergence dependent
Electronic optimization	**Not needed**	Needed
Hamiltonian diagonalization	**Not needed**	Needed
Integration step Δt	Small	**Large**
Minimum of the BO surface	Approximate	**Exact**

is not a plane-wave expansion, but rather a localized basis set as the ones discussed in Sect. 2.1.2. Thus, the basis functions, hence the electronic wavefunctions, depend now also on the atomic coordinates $\mathbf{R}_I$ and the request of orthogonality at each step is released to save time, meaning that the scalar product $\langle\psi_i|\psi_j\rangle$ is no longer vanishing. Analogously, the total energy E^{tot} is not re-optimized as in full self-consistent BO procedures and for this reason is indicated as non-self-consistent energy E_{NSC} in (2.22). Wavefunctions are propagated now according to an algorithm which resembles the original CPMD formulation in the sense that second order EOM are used, however a damping term (first-order derivative) is present which reminds a sort of steepest-descent algorithm typical of the BO dynamics. The net result is a Langevin-type dynamics for the electron wavefunctions of the type,

$$\mu\frac{d^2}{dt^2}|\psi_i\rangle + \gamma\frac{d}{dt}|\psi_i\rangle = -\frac{\delta E_{NSC}}{\delta\langle\psi_i|} + \sum_j \Lambda_{ij}|\psi_i\rangle \tag{2.23}$$

which is then solved by resorting on a predictor-corrector scheme. With no pretention of completeness, we can briefly summarize the procedure for solving these EOM numerically as follows. On a first instance, in a localized basis set $\{|q\rangle\}$, the electronic wavefunctions are expanded as

$$|\psi_i\rangle = \sum_{q=1}^{M} C_{iq}|q\rangle \tag{2.24}$$

on the M functions composing the localized basis. Then the $N \times M$ matrix of the expansion coefficient is written as

$$\mathbf{C} = \left[C_{iq}\right] = \begin{pmatrix} C_{11} & \cdot & \cdot & C_{1M} \\ \cdot & \cdot & \cdot & \cdot \\ \cdot & \cdot & \cdot & \cdot \\ C_{N1} & \cdot & \cdot & C_{NM} \end{pmatrix} \qquad \mathbf{C}^T = \begin{pmatrix} C_{11} & \cdot & \cdot & C_{N1} \\ \cdot & \cdot & \cdot & \cdot \\ \cdot & \cdot & \cdot & \cdot \\ C_{1M} & \cdot & \cdot & C_{MN} \end{pmatrix} \tag{2.25}$$

and the density matrix becomes $\mathbf{P} = \mathbf{C}\mathbf{C}^T = \mathbf{PSP}$. The $M \times M$ matrix indicated as $\mathbf{S}$ has the usual form

$$\mathbf{S} = \left[S_{qq'}\right] = \begin{pmatrix} S_{11} & \cdot & \cdot & S_{1M} \\ \cdot & \cdot & \cdot & \cdot \\ \cdot & \cdot & \cdot & \cdot \\ S_{M1} & \cdot & \cdot & S_{MM} \end{pmatrix} \qquad S_{qq'} = \sum_{i=1}^{N} C^*_{iq}C_{iq} \tag{2.26}$$

Hence, the (DFT) total energy can be rewritten as $E^{\text{tot}}[\mathbf{C}, \mathbf{R}_I]$ which can be used in a straightforward way to write a BO dynamics

$$M\ddot{\mathbf{R}}_I = -\nabla_{\mathbf{R}_I}\min\left\{E^{\text{tot}}\left[\mathbf{C}, \mathbf{R}_I\right]\right\} \quad \forall \mathbf{C}^T \cdot \mathbf{S} \cdot \mathbf{C} = \mathbf{1} \tag{2.27}$$

under the given constraint on C which resumes in an implicit orthogonality condition. However we have to remark that: (i) diagonalization and minimization of E^{tot} are required in BO; (ii) Hellman-Feynman forces are just one component since Pulay forces due to the local basis set are present [22]; residual force components appear due to non-self-consistency (NSC) of the approach. To take into account all the points above, the basic strategy can be summarized in five major points:

1. Do not use brutally BO dynamics
2. Propagate the electronic variables in time according to the CP original idea of updating on-the-fly to avoid expensive full diagonalization operations
3. Use a (good) propagation algorithm $\mathbf{C}(t_n) = f(\mathbf{C}(t_{n-1}),\ldots,\mathrm{C}(t_{n-m}))$ depending on previous $m \in [1, K]$ time steps
4. Select the appropriate number of steps K to keep $\mathbf{C}(t_n)$ as close as possible to the (electronic) ground state
5. Enforce convergence on the BO surface, correct this propagation CPMD-like afterwards

Point 3 corresponds to the first move in the numerical integration procedure and we can identify it as the "predictor" part directly deriving from a standard numerical integration of the Car-Parrinello type EOM. Point 5, instead, is the "corrector" needed afterwards to better converge the wavefunctions and to restore the neglected self-consistent loop. The explicit algorithms used to fulfill all the points listed above are, for the predictor, the propagator of Kofala [45]

$$\mathbf{C}^{predictor}(t_n) = \sum_{m=1}^{K} (-1)^{m+1} \frac{\binom{2K}{K-m}}{\binom{2K-2}{K-1}} \cdot m \cdot \mathbf{C}(t_{n-m})\,\mathbf{C}^T(t_{n-m}) \times \mathbf{S}(t_{n-m})\,\mathbf{C}(t_{n-1}) \tag{2.28}$$

and the corrector eventually giving the true $\mathbf{C}(t_n)$ coefficients reads

$$\mathbf{C}(t_n) = \frac{K}{2K-1}\mathrm{MIN}\left[\mathbf{C}^{predictor}(t_n)\right] + \left(1 - \frac{K}{2K-1}\right)\mathbf{C}^{predictor}(t_n) \tag{2.29}$$

We have to remark that in the algorithm above, MIN indicates a single minimization step, not an iterative procedure as in standard BO approaches. This allows to save a lot of computational time and the use of a not necessarily fully converged wavefunctions at the predictor propagation stage allows for large integration steps, thus resulting in a remarkable boost in the dynamics. Indeed, it was only after the introduction of this approach that it was possible to simulate ternary glasses, such as In_3SbTe_2 [46], which represent a challenge in the field of phase changing materials for next-generation non-volatile memory devices and also a major computational task from the point of view of materials modeling. More precisely, a realistic amorphous state could be obtained for the first time, showing that the dominant chemical bonding is In–Sb and In–Te and that, contrary to the pristine crystal, defective octahedral structures with p-type bonding represent the main feature of the vitreous phase of such a material.

2.3.2 First Principles Molecular Dynamics with Hot Electrons

All the methods discussed so far assume that the wavefunctions of the system, namely the valence electrons, are on their ground state (exactly or approximately) all along the dynamics. This is however not always the case in nature, as well as in experiments. A simple example is represented by zero-gap materials, such as metals, in which electrons can at least partially escape from their ground state and go populating temporarily states that, from a ground state point of view, should be occupied. Electrons can be displaced away from their ground state also in insulating amorphous materials upon laser irradiation. This induces, in turn, weakening of the chemical bonds and structural reorganization of the atomic network. Such a statement is far from being a mere speculation: recent experiments have shown the practical possibility to create Si nano-structures in a matrix of SiO_2 without requiring expensive procedures to reach the silica melting point. By either immersing a SiO_2 glass in a molten salt at about 1000 K and injecting an electric current through a metal wire [47] or irradiating silica with a femtosecond laser [48, 49] Si–Si seeds and pure Si nano-structures have been found. Although at the pioneering stage, these techniques are a promising tool to create Si nano-dots, nanowires or nano-arrays of different patterns in a matrix of glass in a controlled way.

On the theoretical front, the need for treating "hot" electrons has given rise to what is nowadays called Free Energy Molecular Dynamics (FEMD) [50]. This approach is a generalization of the Kohn-Sham DFT formulation in the case in which electrons are allowed to assume a finite temperature T_e not necessarily identical to the ionic temperature T. Starting with a system of N objects, either electrons (i.e. occupied states) or states (i.e. occupied and empty states), its general quantum mechanical description is given by its Hamiltonian $\hat{H}$, and the Schrödinger or, better, KS equation reads

$$\hat{H}|\psi_i\rangle = E_i|\psi_i\rangle \tag{2.30}$$

Now, if the system is not at $T_e = 0$ K, but at a finite temperature $T_e > 0$, then each state E_i has, in the simplest Boltzmann case, a finite probability $\exp(-E_i/k_B T_e)$ of being occupied. By summing up on all states, we arrive at the expression

$$e^{-\beta F} = \frac{1}{N}\sum_{i=0}^{N} e^{-\beta E_i} = \left\langle e^{-\beta\hat{H}}\right\rangle = Z \quad \Rightarrow \quad F = -\frac{1}{\beta}\log Z \tag{2.31}$$

for the partition function Z and the free energy F, respectively, and using the standard notation $\beta = 1/(k_B T_e)$. Then, as found in any statistical mechanics textbook, a better and more comprehensive description of such a system is given by its free energy, rather than the Hamiltonian. This basic idea of the FEMD is then to replace the KS functional by the corresponding free energy one in which the statistics is not the simple Boltzmann one sketched here, but the Fermi-Dirac statistics, since electrons are fermionic particles. Thus, the partition function and the electronic free energy become

$$Z \equiv \det\left[1 + e^{-\beta\left(\hat{H}-\mu\right)}\right] \quad \Rightarrow \quad \Omega = -\frac{1}{\beta}\log Z \tag{2.32}$$

respectively. In the equation above, however, the number of states is somehow arbitrary and this simple expression would resemble to a grand canonical expression, difficult to handle and even more difficult to translate into a computer code, where the number of particle must be finite for obvious technical reasons. The microcanonical ensemble is restored by imposing that the total number of electrons N is constant and this constraint is added via a Lagrange multiplier μ as a new addend $+\mu\ N$ to (2.32). It is then easy to identify μ as the chemical potential of the system or, in solid state physics, as the Fermi level of the system. By adding also the ion-ion interaction E_{II}, the FEMD functional becomes

$$F\left[\rho, \{\mathbf{R}_I\}\right] = \Omega\left[\rho\right] + \mu N + E_{II} \tag{2.33}$$

As in standard DFT, we have now a functional dependent on both the electronic density ρ and the ionic coordinates $\mathbf{R}_I$. The Hamiltonian implicitly contained in Ω is still the KS one, i.e.

$$\hat{H} = -\frac{1}{2}\nabla^2 + V_H\left(\mathbf{x}\right) + \sum_I V_{eI}\left(\mathbf{x} - \mathbf{R}_I\right) + \frac{\delta E_{xc}}{\delta \rho\left(\mathbf{x}\right)} \tag{2.34}$$

with obvious notations. Nonetheless, the fact that the electronic temperature is finite changes deeply the electron density, which is still formally written as in any KS approach, i.e.

$$\rho(\mathbf{x}) = \sum_i f_i |\psi_i\left(\mathbf{x}\right)|^2 \tag{2.35}$$

but the occupation numbers f_i are no longer 1 (occupied) or 0 (unoccupied). Instead, they are given by the Fermi-Dirac distribution

$$f_i = \left[1 + \exp\left(\frac{E_i - \mu}{k_B T_e}\right)\right]^{-1} \tag{2.36}$$

Hence, also non-integer occupation numbers are kept into account and the number of states to be included in the sum of (2.35) increases upon increasing of T_e. In practical applications, the sum has to be performed at least until $f_i < 10^{-5}$, compatible with the typical DFT accuracy and precision allowed by the computer architectures and compilers.

All the ingredients necessary to perform first principles molecular dynamics are now available, since the electronic structure can be updated as in BO-like dynamics by computing the variational derivative

$$\frac{\delta F}{\delta \psi_i^* (\mathbf{x})} = \frac{\delta F}{\delta \rho (\mathbf{x})} \psi_\iota (\mathbf{x}) \tag{2.37}$$

and, analogously, forces acting on the ions are just gradients of F

$$M_I \ddot{\mathbf{R}}_I = -\nabla F \tag{2.38}$$

Since the relaxation time of the electrons τ is of the order of femtoseconds, it is always orders of magnitude smaller than the relaxation time of the ions [51], $\tau << t^{\text{relax}}{}_{\text{Ions}}$, thus the adiabaticity still holds also in FEMD.

One of the first applications of this approach to amorphous materials was performed by Silvestrelli and coworkers [52]. By focusing on the laser melting of silicon, they found that a high concentration of excited electrons dramatically weakens the covalent bond, with the net result of inducing their disruption. This process brings then the system to a liquid and eventually to an amorphous phase. What the authors observed is that the system undergoes a melting transition to a metallic state. In contrast to ordinary liquid silicon, the new liquid is characterized by a high coordination number and a strong reduction of covalent bonding effects.

Coming back to the technique mentioned at the beginning of this paragraph, the observed formation of pure Si structures in a SiO_2 glass matrix via laser excitations represents an appealing tool for applications in nanotechnology. The FEMD approach allowed for the first time to inspect at an atomic-scale level the underlying mechanism [53, 54]. More precisely, it was shown how electron excitations can weaken Si–O bonds in a bulk SiO_2, dislodging O atoms and thus allowing the two Si dangling bonds to reconstruct in stable Si–Si structures (Fig. 2.4).

Such a process is favored by the different diffusivity of O (fast) and Si (slow) which play a decisive role in the process. In fact, once O is dislodged, it can diffuse very easily away from its original site and this departure leaves behind one or more oxygen vacancies similar to the one shown in Fig. 2.2. These vacancies then reconstruct giving rise to the formation of Si–Si structures more or less extended as shown in Fig. 2.4. The major advantage of this laser-induced Si islands formation is the fact that it occurs at temperatures below the melting point of silica (Fig. 2.5), with clear economic advantages. Moreover it provides a practical tool to draw pure Si nanostructures (dots, wires, arrays, etc.) on a simple glass substrate.

2.3.3 Beyond the Local Minimum Exploration: Free Energy Sampling Techniques

Beside inspection of the behavior of matter at finite temperature, one of the main reasons that justify the use of first principles dynamical approaches is the possibility of studying cleavage and formation of bonds in chemical reactions. As discussed in previous paragraphs, this still represents a challenge, because long time scales are required whenever relatively high energy barriers separating reactants and products must be overcome. In the absence of ultra-fast phenomena of the type discussed in

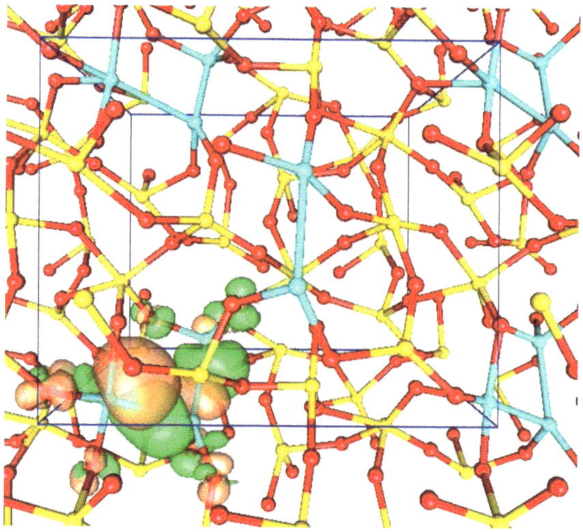

Fig. 2.4 Simulated laser-induced formation of Si–Si structures (*cyan*) in a matrix of silica glass (a-SiO_2) where Si atoms are shown are *yellow* balls and O atoms as *red* balls. The electronic excitation is $\hbar\,\omega = 2.16\,\text{eV}$, far below the energy gap of the material and corresponding to an electronic temperature of 25,000 K. On the *lower left* corner, two Si dangling bonds nearby are in the process of joining to reconstruct a Si–Si bond. Isosurfaces for these two dangling bonds in *green* and *brown* are shown at values of $|\psi_i|^2 = 5 \times 10^{-3}$ Å^{-3}

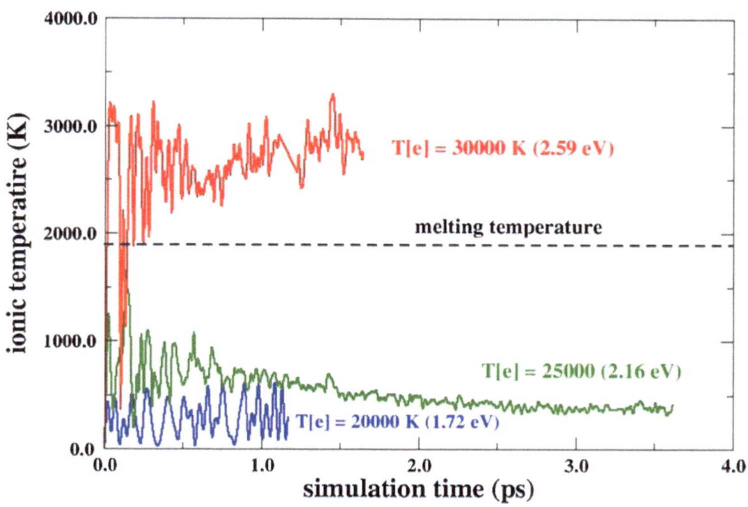

Fig. 2.5 Evolution of the ionic temperature during the FEMD simulations at different electronic temperatures. For low electron excitations below about 2 eV, no chemical bond disruption was observed and the net effect is only the creation of phonons (*blue line*). On the contrary, increasing excitations (laser frequency) induce chemical and topological modifications on the system and the realization of pure Si structures in silica

Sect. 2.3.2, these long time scales can range from μs to hours. These time scales are clearly beyond the reach of any classical or quantum molecular dynamics approach. Several methods have been proposed over the years to overcome this difficulty. Here we limit our discussion to a couple of them, since they are by far the most successful and widely applied ones in the field of disordered materials and condensed phases.

One of the first approaches introduced and used in condensed phases, also in the sense of the historical development, was a biasing technique known as the Blue Moon [55]. This approach was better refined over the years [56] to become more user-friendly and easily implementable in a computer code. For all the details, we refer the reader to the cited original publications. Just to summarize the essential points, let us recall that in the case of first principles dynamical simulation, the method relies on the identification of a reaction coordinate, or order parameter, $\xi = \xi(\mathbf{R}_I)$ of a given subset of atomic coordinates (Lagrangean variables) $\mathbf{R}_I$ able to track the activated process or chemical reaction on which one wants to focus. The simplest example is represented by the distance $\xi = |\mathbf{R}_I - \mathbf{R}_J|$ between two atoms that are expected to form or break a chemical bond. This analytical function is added to, e.g., a Car-Parrinello Lagrangean as a holonomic constraint,

$$\mathcal{L} = \mathcal{L}^{CP} + \lambda_k \cdot [\xi(\mathbf{R}_I) - \xi_k] \tag{2.39}$$

where λ_k is a Lagrange multiplier and ξ_k is a given assigned value of $\xi(\mathbf{R}_I)$ that one wants to sample through molecular dynamics. For each sampled value, a constrained dynamics is performed and then the average constraint force f_ξ computed as the time average

$$f_\xi = \langle \lambda_k \rangle_t = \frac{\partial F}{\partial \xi} \tag{2.40}$$

will be the gradient of the free energy F along ξ. The free energy profile of the reaction is then computed as

$$\Delta F = \int_{\xi_0}^{\xi_f} f_\xi d\xi \tag{2.41}$$

where ξ_0 is the initial value of the reaction coordinate and ξ_f the final one. As an example of application of this method, we can cite the discovery in silica of E' centers, analogous to the one discussed in Sect. 2.2.1, in the absence of oxygen vacancies [57]. During a local structural change in which an O atom is moved away in a Si–O–Si structure a new stable energy minimum can be reached in which a threefold oxygen, a fivefold Si (known as "floating bond") and a dangling bond can co-exist. More precisely, according on the spin state and net charge Q of the system, such a minimum can give rise to either a threefold oxygen in the case of a singlet ground state and a charge state Q = +1. Conversely, an unsaturated Si atom carrying a dangling bond can survive in the triplet state originating a paramagnetic defect detectable by spin-resonance or ENDOR techniques. The energy barrier, obtained via Blue Moon ensemble dynamics, depend on the spin state of the system as well

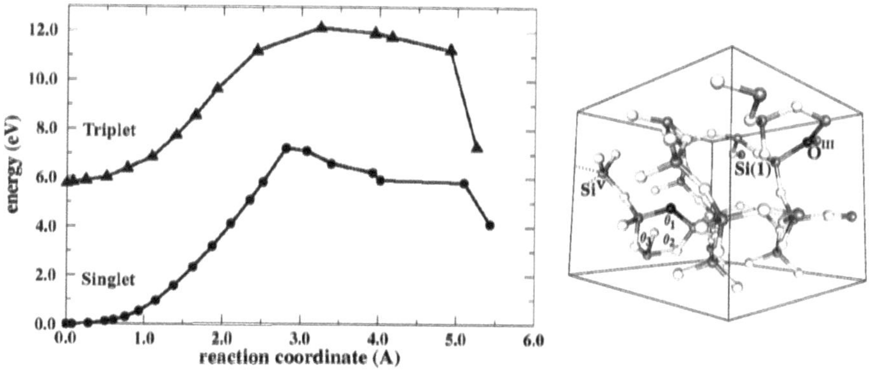

Fig. 2.6 Free energy profile for the O displacement in bulk silica as a function of the reaction coordinate intended as the distance of the selected O site with respect to its initial position. The *lower curve* (*filled circles*) refers to the singlet, while the *upper curve* (*triangles*) to the triplet state. The *right panel* show the final structure in which a fivefold Si site (Si^V), a threefold oxygen (O^{III}) and a dangling bond on Si(1) are present. This local disordered glassy structure gives rise to ring structures as the one indicated by the angles θ_1, θ_2, and θ_3, typical of a-SiO_2

and is shown in the left panel of Fig. 2.6, whereas the local topology of such a defect is reported in the right panel of the same figure.

In the CPMD approach, a useful and recently very popular method is to extend the Lagrangean to include as dynamical variables also the reaction coordinates that drive a slowly occurring process (on the time scale of the simulation), such as a chemical reaction, and that can discriminate between reactants and products. Such an approach is known as metadynamics [58, 59] and makes use of a biasing potential technique to force the system to escape from an initial local minimum. The biasing potential is generally a function of one or more reaction coordinates driving the activated process and these coordinates crucially depend on the system under study and on the reaction pathway that has to be simulated. Different formulations of metadynamics are possible, as extensively discussed elsewhere [60]. The metadynamics formulation that we revise here is the variational version allowing for a straightforward extension of the Lagrangean formulation of (2.13). Namely, the Car-Parrinello Lagrangean is extended to include the specific reaction coordinates, or "collective variables" (CVs) as additional dynamical variables which will be indicated as $s_\alpha(t)$ hereafter. These are selected (by the user) to describe the reaction mechanism on which the simulation is supposed to focus. In order to make the system evolve from an initial local initial minimum to another one, a history dependent potential $V(s_\alpha, t)$ is added. Such a penalty potential is generally expressed as a superposition of small analytic functions, namely Gaussians, which have the scope of preventing the system from visiting a region of the phase space already explored. The reactive system is thus described by a Lagrangean that reads

$$\mathcal{L} = \mathcal{L}^{\text{CPMD}} + \sum_{\alpha} \frac{1}{2} M_{\alpha} \dot{s}_{\alpha}^{2}(t) - \sum_{\alpha} \frac{1}{2} k_{\alpha} \left[s_{\alpha}(q) - s_{\alpha}(t) \right]^{2} - V(s_{\alpha}, t) \tag{2.42}$$

where the argument q of $s_{\alpha}(q)$ can be any function of an arbitrary set of atomic coordinates, such as a distance $|\mathbf{R}_I(t) - \mathbf{R}_J(t)|$, or of the electronic wavefunctions, or any other physical/chemical/geometrical quantity needed to drive the system from reactants to products. As mentioned above, the history dependent potential has the explicit Gaussian form

$$V(s_{\alpha}, t) = \int_{0}^{t} dt' |\dot{\mathbf{s}}(t')| \delta \left[\frac{\dot{\mathbf{s}}(t')}{|\dot{\mathbf{s}}(t')|} \left(\mathbf{s} - \mathbf{s}(t') \right) \right] A(t') \exp \left[- \frac{\left(\mathbf{s} - \mathbf{s}(t') \right)^{2}}{2(\Delta s)^{2}} \right] \tag{2.43}$$

where $\mathbf{s} = (s_1, \ldots, s_{\alpha}, \ldots)$ is the vector formed by the set of reaction coordinates and the Gaussian amplitude $A(t')$ has the dimensions of an energy. In this definition, the two crucial parameters are $A(t')$ and Δs. The first one, the amplitude of the Gaussian functions, should be small enough to give a good resolution in the free energy surface (FES) reconstruction, compatible with the accuracy of the underlying theory (DFT in the ongoing discussion). The second one, Δs, is the spread of these Gaussians and regulates practically how fine is the sampling of the CVs. Both of these parameters are system dependent in the sense that their choice depends on at least a rough knowledge of about how deep or shallow the minima are and how extended they are. If no information is available, rough exploratory metadynamics simulations with different parameters can be done to recover the missing pieces of information. It must also be remarked that, at variance with reaction path optimization or sampling methods, no previous knowledge of the final products is required.

In a typical first principles molecular dynamics simulation, a new Gaussian contribution is added to the potential $V(s_{\alpha}, t)$ every given time interval τ amounting to few hundreds of "ordinary" molecular dynamics steps. In fact, in between a Gaussian and the next one, the system must be allowed to fully re-equilibrate and to explore exhaustively the available portion of phase space before blocking one part of it as the introduction of a new penalty function is designed to do. The schematic view of this approach is summarized by Fig. 2.7.

In the original formulation of metadynamics, the FES $F(s_{\alpha})$, was assumed to be

$$F(s_{\alpha}) = - \lim_{t \to \infty} V(s_{\alpha}, t) + const. \tag{2.44}$$

where the limit has to be intended in the sense that the metadynamics is continued and additional Gaussians are added until the selected portion of the phase space, spanned by the CVs s_{α}, is saturated. This qualitative criterion has been shown to be insufficient to assess whether or not the convergence has been attained [61].

A better estimation of the FES can be achieved by continuing the simulation beyond the simple one-shot filling of all the minima. More precisely, one should allow the system to pursue its (meta)dynamics for at least a couple of passages back

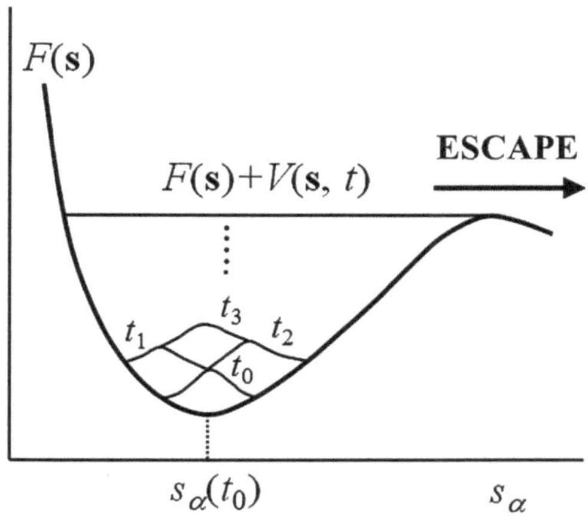

Fig. 2.7 Schematic picture of the metadynamics approach. Starting with a system in an initial local minimum, small Gaussian functions of suitable height and width are added during the Car-Parrinello dynamics until the system reaches the first maximum and escapes toward the next local minimum, where the procedure is reiterated until saturation of the phase space

and forth from the reactants side to the products side in the phase space spanned by the CVs. In fact, when all the minima of the FES are saturated, the system can diffuse in a nearly free manner from reactants to product. When this occurs, the arithmetic average of the penalty potential can be computed to estimate the FES as

$$F(s_\alpha) = -\frac{1}{t_{simul} - t_{diff}} \int_{t_{diff}}^{t_{simul}} V(s_\alpha, t)\, dt \tag{2.45}$$

where t_{simul} is the total simulation time and t_{diff} is the time at which the CVs start diffusing.

Early examples of application of this reactive CPMD approach were the chemical rearrangement of azulene into naphthalene [62], where new mechanisms and alternative possible pathways were evidenced for the first time, the synthetic organic reactions in supercritical water for the production of nylon synthetic fibers [63], and phase transitions in various materials from zeolites to graphene [64].

Coming to the materials targeted in the present book, specifically silica and related glasses, a recent application [65] of metadynamics was performed to inspect the initial stages of the mechanism of oxidation of an exposed surface of silicon resulting in the formation of SiO_2. In semiconductors and glass technology, this is a fundamental process of primary importance, since it implies a dramatic change in the nature of the material and, hence, in its performances in practical applications. The use of metadynamics in the inspection of the initial oxidation of Si(100) surfaces allowed to identify various competing reaction paths having comparable free energy barriers. The overall picture is summarized in Fig. 2.8. A crucial stable geometry, precursor of all the alternative pathways, was found to be ubiquitous during the oxidation and links the dissociation of O_2 and the oxidation of sub-surfaces.

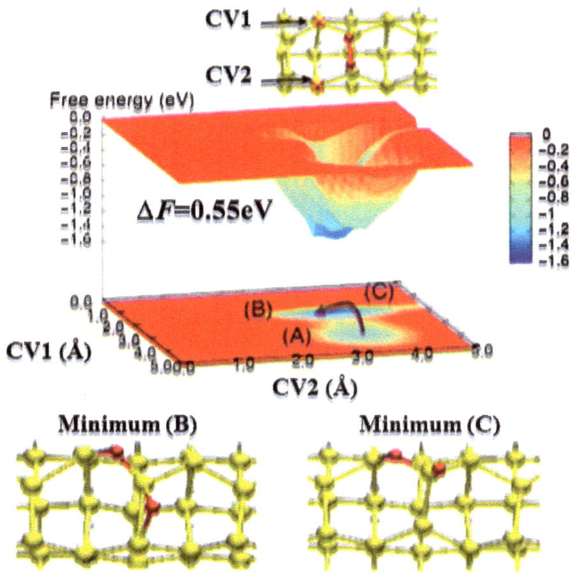

Fig. 2.8 Free energy landscape and main configurations of the O atoms adsorbed on an exposed Si(100) surface as provided by metadynamics in which two collective variables are used. These are the distances of the two atoms of an approaching O_2 molecule from Si sites. The initial position of the non-dissociated oxygen dimer shown on *top of the figure* can evolve into two possible dissociated forms corresponding to the two minima B and C on the free energy surface. The associated barrier is about 0.55 eV

References

1. B.J. Alder, T.E.J. Wainwright, Chem. Phys. **27**, 1208 (1957)
2. B.J. Alder, T.E.J. Wainwright, J. Chem. Phys. **31**, 459 (1959)
3. A. Rahman, Phys. Rev. **136**, A405 (1964)
4. F.H. Stillinger, A. Rahman, J. Chem. Phys. **60**, 1545 (1974)
5. P. Hohenberg, W. Kohn, Phys. Rev. **136**, B864 (1964)
6. W. Kohn, L.J. Sham, J. Phys. Rev. **140**, A1133 (1965)
7. D.J. Defeers, B.A. Levi, S.K. Pollack, W.J. Hehre, J.S. Binkley, J.A. Pople, J. Am. Chem. Soc. **101**, 4085 (1979)
8. J.A. Pople, M. Head-Gordon, D.J. Fox, K. Raghavachari, L.A. Curtiss, J. Chem. Phys. **90**, 5622 (1989)
9. B.G. Johnson, P.M.W. Gill, J.A. Pople, J. Chem. Phys. **98**, 5612 (1993)
10. R.G. Parr, W. Yang, *Density-Functional Theory of Atoms and Molecules* (Oxford University Press, New York, 1989). ISBN 0-19-504279-4
11. D. Marx, J. Hutter, *Ab Initio Molecular Dynamics: Basic Theory and Advanced Methods* (Cambridge University Press, New York, 2009)
12. A.D. Becke, Phys. Rev. **38**, 3098 (1988)
13. J.P. Perdew, A. Zunger, Phys. Rev. B **23**, 5048 (1981)
14. J.P. Perdew, K. Burke, M. Ernzerhof, Phys. Rev. Lett. **77**, 3865 (1996)
15. F.A. Hamprecht, A.J. Cohen, D.J. Tozer, N.C. Handy, J. Chem. Phys. **109**, 6264 (1998)
16. C. Lee, W. Yang, R.G. Parr, Phys. Rev. B **37**, 785 (1988)
17. D.R. Hamann, M. Schluter, C. Chang, Phys. Rev. Lett. **43**, 1494 (1979)
18. G.B. Bachelet, D.R. Hamann, M. Schluter, Phys. Rev. B **26**, 4199 (1982)
19. N. Troullier, J.L. Martins, Phys. Rev. B **43**, 1993 (1991)
20. D. Vanderbilt, Phys. Rev. B **41**, 7892 (1990)
21. W.J. Hehre, L. Radom, P.v.R. Schleyer, J.A. Pople, Ab Initio Molecular Orbital Theory (Wiley, New York, 1986). ISBN 978-0-471-81241-8

22. P. Pulay, in *Calculation of Forces by Non-Hellmann-Feynman Methods, Chapter 9 of "The Force Concept in Chemistry"*, ed. by M. Deb (Van Nostrand, New York, 1981)
23. M. Born, J.R. Oppenheimer, Annalen der Physik **84**, 457 (1927)
24. R. Car, M. Parrinello, Phys. Rev. Lett. **55**, 2471 (1985)
25. S. Nosé, J. Chem. Phys. **81**, 511 (1984)
26. W.G. Hoover, Phys. Rev. A **31**, 1695 (1985)
27. H.C. Andersen, J. Chem. Phys. **72**, 2384 (1980)
28. M. Parrinello, A. Rahman, Phys. Rev. Lett. **45**, 1196 (1980)
29. M. Parrinello, A. Rahman, J. Appl. Phys. **52**, 7182 (1981)
30. F.A. Bornemann, C. Schütte, Numerische Matematik **78**, 359 (1998)
31. J.C. Grossman, E. Schwegler, E.W. Draeger, F. Gygi, G. Galli, J. Chem. Phys. **120**, 300 (2004)
32. E. Schwegler, J.C. Grossman, F. Gygi, G. Galli, J. Chem. Phys. **121**, 5400 (2004)
33. CHEMPHYSCHEM (Parrinello Festschrift) **6**, 2005
34. R. Car, M. Parrinello, Solid State Comm. **62**, 403 (1987)
35. L. Verlet, Phys. Rev. **159**, 98 (1967)
36. I. Stich, R. Car, M. Parrinello, Phys. Rev. B **44**, 11092 (1991)
37. J. Sarnthein, A. Pasquarello, R. Car, Phys. Rev. Lett. **74**, 4682 (1995)
38. M. Boero, A. Pasquarello, J. Sarnthein, R. Car, Phys. Rev. Lett. **78**, 887 (1997)
39. D. Donadio, M. Bernasconi, M. Boero, Phys. Rev. Lett. **87**, 195504 (2001)
40. S. Le Roux, A. Bouzid, M. Boero, C. Massobrio, Phys. Rev. B **86**, 224201 (2012)
41. M. Bauchy, M. Micoulaut, M. Boero, C. Massobrio, Phys. Rev. Lett. **110**, 165501 (2013)
42. M. Celino, S. Le Roux, G. Ori, B. Coasne, A. Bouzid, M. Boero, C. Massobrio, Phys. Rev. B **88**, 174201 (2013)
43. N. Marzari, D. Vanderbilt, Phys. Rev. B **56**, 12847 (1997)
44. T.D. Kühne, M. Krack, F.R. Mohamed, M. Parrinello, Phys. Rev. Lett. **98**, 066401 (2007)
45. J. Kofala, J. Comput. Chem. **25**, 335 (2004)
46. J.H. Los, T.D. Kühne, S. Gabardi, M. Bernasconi, Phys. Rev. B **88**, 174203 (2013)
47. T. Nohira, K. Yasuda, Y. Ito, Nat. Mater. **2**, 397 (2003)
48. N. Fukata, Y. Yamamoto, K. Murakami, M. Hase, M. Kitajima, Appl. Phys. Lett. **87**, 3495 (2003)
49. K. Miura, K. Hirao, Y. Shimotsuma, M. Sakakura, S. Kanehira, Appl. Phys. A **93**, 183 (2008)
50. A. Alavi, J. Kohanoff, M. Parrinello, D. Frenkel, Phys. Rev. Lett. **73**, 2599 (1994)
51. K. Seibert, G.C. Cho, W. Kütt, H. Kurz, D.H. Reitze, J.I. Dadap, H. Ahn, M.C. Downer, A.M. Malvezzi, Phys. Rev. B **42**, 2842 (1990)
52. P.L. Silvestrelli, A. Alavi, M. Parrinello, D. Frenkel, Phys. Rev. Lett. **77**, 3149 (1996)
53. M. Boero, A. Oshiyama, P.L. Silvestrelli, K. Murakami, Appl. Phys. Lett. **86**, 201910 (2005)
54. M. Boero, A. Oshiyama, P.L. Silvestrelli, K. Murakami, Physica B **376–377**, 945 (2006)
55. J.P. Ryckaert and Ciccotti, J. Chem. Phys. **78**, 7368 (1983)
56. M. Sprik, G. Ciccotti, J. Chem. Phys. **109**, 7737 (1998)
57. M. Boero, A. Oshiyama, P.L. Silvestrelli, Phys. Rev. Lett. **91**, 206401 (2003)
58. A. Laio, M. Parrinello, Proc. Nat. Acad. Sci. USA **99**, 12562 (2002)
59. M. Iannuzzi, A. Laio, M. Parrinello, Phys. Rev. Lett. **90**, 238302 (2003)
60. A. Barducci, M. Bonomi, M. Parrinello, WIREs Comput. Mol. Sci. **1**, 826 (2011)
61. A. Laio, F.L. Gervasio, Rep. Prog. Phys. **71**, 126601 (2008)
62. A. Stirling, M. Iannuzzi, A. Laio, M. Parrinello, Chem. Phys. Chem. **5**, 1558 (2004)
63. M. Boero, T. Ikeshoji, K. Teralura, C.C. Liew, M. Parrinello, J. Am. Chem. Soc. **126**, 6280 (2004)
64. R. Martonak, A. Laio, M. Bernasconi, C. Ceriani, P. Raiteri, F. Zipoli, M. Parrinello, Zeitschrift für Kristall **220**, 489 (2005)
65. K. Kolizumi, M. Boero, Y. Shigeta, A. Oshiyama, Phys. Rev. B **85**, 205314 (2012)

Chapter 3
Metadynamics Simulations of Nucleation

Ider Ronneberger and Riccardo Mazzarello

Abstract This chapter offers an overview of recent applications of the metadynamics method to the study of nucleation and related phenomena. In the first section, the classical nucleation theory and the metadynamics method are introduced. The second section is devoted to applications, including computational studies of the surface tension, which affects the size and energy of criticial nuclei, and investigation of crystal nucleation from the amorphous and supercooled liquid state.

3.1 Introduction

In this introductory subsection, we discuss the classical theory of homogeneous nucleation and of crystal growth and the metadynamics method, including a mention of recent advances aiming at increasing the efficiency and the flexibility of this technique.

3.1.1 Classical Nucleation Theory

A first order phase transition from a metastable to a stable phase can be divided in two major processes, namely the nucleation and the growth of the stable phase. The formation of nuclei in the interior of a uniform substance without impurities is called homogeneous nucleation. If container walls or impurities like dust are present, they

I. Ronneberger · R. Mazzarello (✉)
Institute for Theoretical Solid State Physics, RWTH Aachen,
Otto-Blumenthal-Strasse, 52074 Aachen, Germany
e-mail: mazzarello@physik.rwth-aachen.de

I. Ronneberger
e-mail: ronneberger@physik.rwth-aachen.de

C. Massobrio et al. (eds.), *Molecular Dynamics Simulations of Disordered Materials*, Springer Series in Materials Science 215,
DOI 10.1007/978-3-319-15675-0_3

generally act as preferential nucleation sites, due to the lower formation energies of nuclei. This phenomenon is called heterogeneous nucleation. The classical nucleation theory (CNT) as developed in the early 20th century describes these processes based on thermodynamic and kinetic arguments.

In the following, we discuss homogeneous nucleation only. The starting point of the CNT is to consider the reversible work of formation W required to form a cluster of the stable phase of a certain size N (here N being the number of atoms in the cluster) within the metastable phase [1]:

$$\begin{aligned} W(N) &= N\Delta\mu + \gamma A \\ &= V\Delta g + \gamma A, \end{aligned}$$

where $\Delta\mu$ is the difference in the chemical potential between the stable and the metastable phase and $\Delta g = \frac{N\Delta\mu}{V}$ is the corresponding free energy difference per unit volume, often called the driving force. $\Delta\mu$ and Δg are negative quantities. The second term γA is a positive interfacial free energy contribution due to the formation of an interface of area A and surface tension γ. By assuming a spherical cluster shape with radius r, so that $A = 4\pi r^2 = (36\pi)^{1/3}\,(\bar{v}N)^{2/3}$ ($\bar{v}$ denotes the atomic volume V/N), one obtains the following dependence of W on N or r:

$$\begin{aligned} W(N) &= N\Delta\mu + (36\pi)^{1/3}\,(\bar{v}N)^{2/3}\gamma, \\ W(r) &= \frac{4\pi}{3}r^3\Delta g + 4\pi r^2\gamma. \end{aligned}$$

For small N the interfacial term dominates, whereas for large N the driving force takes over. Therefore there exists a maximum for $W(N)$ at a certain cluster size N_c, called critical cluster size, with the corresponding critical radius r_c (see Fig. 3.1).

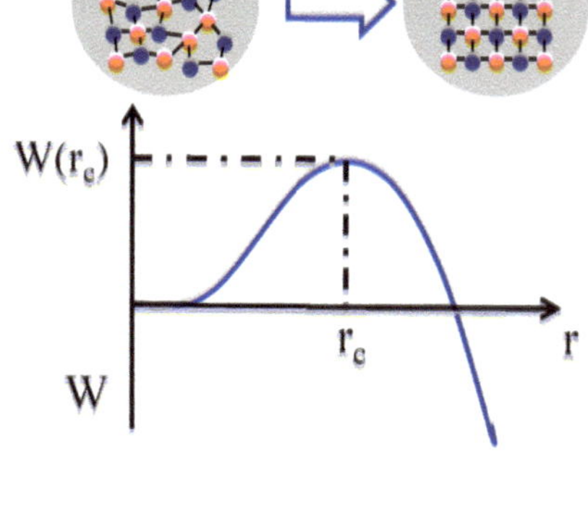

Fig. 3.1 Nucleation barrier separating the metastable disordered phase from the crystalline phase

The critical size and the maximum of $W(N)$ are easily computed by setting $dW/dN = 0$:

$$N_c = \frac{32\pi}{3\bar{v}} \frac{\gamma^3}{|\Delta g|^3}, \tag{3.1}$$

$$W(N_c) = \frac{16\pi}{3} \frac{\gamma^3}{(\Delta g)^2}. \tag{3.2}$$

Statistical fluctuations (the strength of which is determined by the thermal energy) lead to the formation of clusters of the new stable phase, which tend to dissolve if they are smaller than the critical size (precritical), whereas they grow on average if they exceed N_c (postcritical). Hence, nucleation can be related to the formation of postcritical clusters.

The experimentally accessible quantity in crystallization processes is the nucleation rate. The first proposed kinetic model of nucleation by Volmer and Weber [2] estimated it to be the product of the equilibrium cluster size distribution, $F^{eq}(N)$, and the rate of single-atom addition, $k^{+}(N)$ (forward rate), taken at the critical size N_c:

$$I^{VW} = F^{eq}(N_c) \cdot k^{+}(N_c). \tag{3.3}$$

Here the equilibrium cluster size distribution is

$$F^{eq}(N) = N_A \cdot \exp\left(-W(N)/k_{\mathrm{B}}T\right) \tag{3.4}$$

for $N < N_c$ (N_A is the Avogadro constant) and 0 elsewhere, i.e. clusters are assumed to grow to macroscopic sizes once they have reached the critical size. However, this model does not describe the cluster size distribution correctly, as argued by Becker and Döring [3], because it does not take into account the fact that critical and postcritical nuclei have a finite probability to lose atoms. Becker and Döring derived an expression for the steady-state nucleation rate, I^{st}, by including also the effects of the backward rate $k^{-}(N+1)$. The result is

$$I^{st} = F^{eq}(N_c) \cdot k^{+}(N_c) \cdot Z, \tag{3.5}$$

which differs from the I^{VW} by the Zeldovich factor,

$$Z = \left(\frac{|\Delta\mu|}{6\pi k_{\mathrm{B}} T N_c}\right)^{1/2}. \tag{3.6}$$

The expression for the forward rate $k^{+}(N_c)$ depends on the nature of the local rearrangements occurring at the interface. In the case of diffusion-limited crystallization, which involves significant changes of coordination numbers and local order at the interface, k^{+} is given by $N_c^s 6D/\lambda^2$, where N_c^s is the number of surface atoms of the critical cluster, D denotes the bulk diffusion coefficient of the metastable phase and λ the average interatomic distance.

In the case of collision-limited crystallization, on the other hand, attachment of atoms to the crystalline interface is driven by atomic collisions and no large rearrangement occurs. In this case, $k^+ = N_c^s v_s/\lambda$, where v_s is the sound velocity in the metastable phase. Assuming that crystallization is limited by diffusion, expression (3.5) can be rewritten as

$$I^{st} = \underbrace{\frac{24DN_c^{2/3}N_A}{\lambda^2}\left(\frac{|\Delta\mu|}{6\pi k_B T N_c}\right)^{1/2}\cdot}_{\text{dynamical factor}} \quad \underbrace{\exp\left(-\frac{W(N_c)}{k_B T}\right)}_{\text{critical cluster formation probability}} \cdot \tag{3.7}$$

Hence, the steady-state nucleation rate is determined by the product of thermodynamic and dynamical factors, which are governed by the height of the free energy barrier for nucleation and the diffusivity, respectively.

The dynamical prefactor of formula (3.7) can be orders of magnitude lower than those determined in experiments. Also, the expression for the free energy barrier $W(N_c)$ has come under intense scrutiny, due to the simplifying assumptions made in CNT. In particular, CNT employs macroscopic thermodynamic quantities (e.g. the surface tension γ) to describe the behaviour of microscopic objects, namely the crystalline clusters. The assumption of spherical clusters is also oversimplified. Hence, accurate estimations of the critical cluster size N_c and of the corresponding work of formation $W(N_c)$ are of major interest and several computational approaches have been developed to derive these quantities from molecular dynamics (MD) or Monte-Carlo simulations. In Sect. 3.1.3 we will introduce metadynamics, a recently developed enhanced sampling approach, which appears to be promising in this respect.

3.1.2 Crystal Growth

The crystal growth processes at crystal-amorphous or crystal-liquid interfaces, including the growth of postcritical crystalline clusters in the relevant metastable phase, can be controlled by long range diffusion of atoms (diffusion-controlled growth) or by small rearrangements at the interface (interface-controlled growth). Long range diffusion is important if a change in composition occurs during crystallization. This process should not be confused with the diffusion-limited crystallization introduced in the previous section, which occurs in the case of interface-controlled growth only.

In the following we consider the latter process. The crystal growth velocity v_g can be obtained by calculating the difference between the attachment and detachment rate:

$$v_g = \lambda k^+ \left[1 - \exp\left(-\frac{\Delta\mu}{k_B T}\right)\right], \tag{3.8}$$

The expression for k^+ depends on whether crystallization is diffusion-limited or collision-limited, in full analogy with the nucleation case. In the diffusion-limited case, v_g reads:

$$v_g = f\frac{6D}{\lambda}\left[1 - \exp\left(-\frac{\Delta\mu}{k_B T}\right)\right], \tag{3.9}$$

where f denotes the fraction of available sites for attachment at the crystalline interface.

3.1.3 Metadynamics

An inherent problem of MD simulations is their inability to investigate the transitions between phases of a physical system separated by high energy barriers. More precisely, since the overcoming of such barriers due to thermal fluctuations is a rare event, extremely long simulation times (compared to the typical timesteps of MD simulations, of the order of fs) would be required to observe these transitions. To solve this problem, several enhanced sampling methods [4–6] have been developed, which, in combination with MD or Monte Carlo simulations, enable one to study these processes, as well as to calculate the free energy barriers separating the relevant phases. Metadynamics (MTD) [7–9] belongs to the class of adaptive nonequilibrium methods [10, 11], in which the simulation history is stored and exploited to explore the phase space. This method is enjoying ever growing popularity, which stems from its simplicity and flexibility.

MTD is based on the assumption that the relevant processes can be described by a small set of coarse-grained variables, called collective variables (CVs), describing the essential degrees of freedom of the system. The CVs $\xi = \xi(\{\boldsymbol{r}\})$ themselves are functions of the atomic coordinates $\{\boldsymbol{r}\} = \boldsymbol{r}_1, \boldsymbol{r}_2, ..., \boldsymbol{r}_N$. In general, ξ is a vector comprising of d CVs as components. To drive the system out of the local free energy minimum in which it is trapped, a history-dependent biasing potential is added in the space of the selected CVs.

The Helmholtz free energy of the system is given by

$$F = -\frac{1}{\beta}\ln Z, \tag{3.10}$$

where $\beta = (k_B T)^{-1}$ and Z is the canonical partition function,

$$Z = \frac{1}{N!h^{3N}}\int \exp\left(-\beta\left(\sum_{i=1}^{N}\frac{p_i^2}{2m_i} + U(\{\boldsymbol{r}\})\right)\right)dp^N dr^N = \frac{1}{N!\Lambda^{3N}}Z_c, \tag{3.11}$$

where $\boldsymbol{p}_i$ is the momentum of particle i, $\Lambda = \sqrt{h^2\beta/2\pi m}$ is the thermal de Broglie wavelength and Z_c is the configurational partition function:

$$Z_c = \int \exp\left(-\beta U(\{\boldsymbol{r}\})\right) d\boldsymbol{r}^N. \tag{3.12}$$

The description of the free energy surface (FES) in the coarse-grained d-dimensional space of the CVs is obtained by integrating out the irrelevant degrees of freedom of the system (and neglecting the constant contribution due to the kinetic energy):

$$F(\xi) = -k_B T \ln\left(\int \exp\left(-\beta U(\{\boldsymbol{r}\})\right) \delta\left(\xi - \xi(\{\boldsymbol{r}\})\right) d\boldsymbol{r}^N\right). \tag{3.13}$$

The probability distribution for the CVs is easily expressed in terms of $F(\xi)$:

$$P(\xi) = \frac{\exp(-F(\xi)/(k_B T))}{\int \exp(-F(\xi)/(k_B T)) d\xi}. \tag{3.14}$$

The history-dependent bias potential V_b added to the potential energy consists of a sum of energy hills centered at the already visited points in the CV space. Without MTD, the system may become trapped in some local minimum of the FES, if the surrounding barriers are high compared to the thermal energy. In the presence of the potential V_b, the frequently visited regions of the CV space will accumulate more and more energy hills as penalties. As a result, the minimum of the FES will be gradually filled, until the system will finally escape it (see Fig. 3.2).

The hills are usually chosen to be repulsive Gaussian functions. In this case, the bias potential V_b can be written as (for $d = 1$):

$$V_b(\xi(\{\boldsymbol{r}\}), t) = w \sum_{t'=\tau, 2\tau, \ldots < t} \exp\left(-\frac{(\xi(\{\boldsymbol{r}\}) - \xi_{t'})^2}{2\sigma_\xi^2}\right), \tag{3.15}$$

where $\xi_{t'}$ is the value of the CV of the system taken at time t', σ_ξ denotes the width of the Gaussians and τ^{-1} is their deposition frequency.

Assuming that the size of the hills is small compared to the size of the free energy basins of the system, after a sufficiently long time the high barriers of the FES will

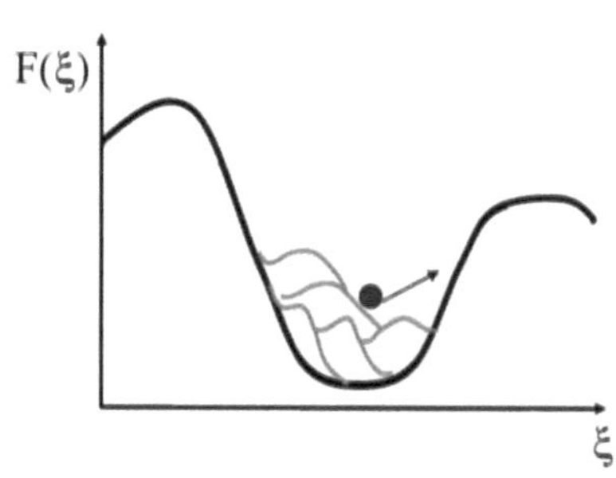

Fig. 3.2 The energy hills added during the simulation fill up the free energy well in the space of the collective variable ξ

be flattened and the evolution of the CVs will eventually become diffusive in the CV space. In this limit, the biasing potential yields an estimate of the underlying FES:

$$\lim_{t \to \infty} -V_b(\xi, t) \sim F(\xi). \tag{3.16}$$

Intuitively, $-V_b$ is expected to approach the exact FES in the long time limit, if the Gaussians are sufficiently small. Perhaps less intuitively, it can be shown that a relation similar to (3.16) holds exactly for the average of V_b over independent MTD runs, $\langle V_b \rangle_{MTD}$, when the evolution of the system in the CV space is of the Langevin type [12]:

$$\lim_{t \to \infty} -\langle V_b(\xi, t) \rangle_{MTD} = F(\xi). \tag{3.17}$$

In a single run, however, the estimated FES does not converge to a definite value but fluctuates around the exact value. The accuracy and efficiency of the MTD algorithm depend on the size and shape of the energy hills, that is the height w and the width σ_ξ of the Gaussians, and on the deposition interval τ. The exact expression for the error as a function of these quantities has been provided in [12]. A simpler, empirical expression was also derived [13], which reads

$$\langle \epsilon^2 \rangle_{MTD} \propto \frac{\Xi \sigma_\xi w}{\beta D_\xi \tau}, \tag{3.18}$$

where $\langle \epsilon^2 \rangle_{MTD}$ is the average quadratic deviation of a MTD run from the exact FES, D_ξ is the diffusion coefficient of the system in the coarse-grained space and Ξ is the linear size of the domain (assumed to be cubic) of the CVs. The latter expression was shown to compare well with the exact one [12]. In particular, the linear dependence of $\langle \epsilon^2 \rangle_{MTD}$ on the ratio w/τ (deposition rate) holds rigorously. Obviously, reducing the size of the Gaussians leads to smaller errors, however it also increases the simulation time needed to fill the FES. Recent works have focused on the possibility of adapting the Gaussian height [14] and/or width [15] during the run, so as to increase the accuracy of the FES reconstruction, without sacrificing the efficiency of the method. These developments are discussed briefly in the following two subsections.

3.1.3.1 Well-Tempered Metadynamics

The use of a constant Gaussian height w causes the bias potential to fluctuate around the exact FES in the diffusive limit, which can make the assessment of the convergence problematic. On the other hand, if the height of the hills decreased down to zero with time, the fluctuations would vanish in the long time limit. Therefore, a natural extension of the method is to enable one to reduce w during the simulation. This is done in the so called well-tempered metadynamics scheme [14]. Within this scheme, the Gaussian height is rescaled as $w = w_0 \cdot \exp[-V_b(\xi, t)/\Delta T]$.

This rescaling leads to a smooth convergence of the biasing potential. Furthermore, it can be shown that both the rate dV_b/dt and w tend to zero like $\sim 1/t$ as $t \rightarrow \infty$ [14]. The free energy surface can be recovered as:

$$\lim_{t \rightarrow \infty} V_b\left(\xi, t\right) = -\frac{\Delta T}{T + \Delta T} \cdot F\left(\xi\right). \tag{3.19}$$

The parameter ΔT controls the speed of exploration of the CV space: in the limiting cases of $\Delta T \rightarrow \infty$ and $\Delta T = 0$, the well-tempered MTD is equivalent to standard MTD and non-biased MD, respectively. For intermediate cases, the calculated FES from (3.19) coincides with the exact one at temperature T.

3.1.3.2 Adaptive Gaussians

The Gaussian width σ_ξ is typically chosen to be of the order of the standard deviation of the corresponding CV, computed in a preliminary (unbiased) MD run of the initial state. However, the standard deviation depends on the local FES and it would thus be desirable to adapt σ_ξ to the FES during the simulation. More generally, in the case of a multidimensional CV space ($d > 1$), where the hills are given by multivariate Gaussians,

$$\exp\left(-\frac{1}{2}\left(\boldsymbol{\xi}\left(\boldsymbol{r}\right) - \boldsymbol{\xi}_{t'}\right)^T \Sigma \left(\boldsymbol{\xi}\left(\boldsymbol{r}\right) - \boldsymbol{\xi}_{t'}\right)\right), \tag{3.20}$$

it would be advantageous to vary the covariance matrix $\Sigma = \left(\sigma_{ij}^{-2}\right)$, which determines the shape and width of the hills, during the run. Branduardi et al. [15] have introduced a scheme to control the efficiency of MTD using adaptive Gaussians, wherein the optimal value of the covariance matrix is estimated on the fly either dynamically, i.e. by calculating the evolution of the correlation functions $\langle \xi_i \xi_j \rangle$ along the trajectory, or geometrically, by calculating the metrics of the microscopic coordinates.

3.1.3.3 Lagrangian Metadynamics

The use of MTD in combination with ab initio methods poses a challenge in the presence of high free energy barriers, even if MTD is used for the sole purpose of accelerating the occurrence of a rare event. The reason for this is that, to overcome high barriers, a large amount of energy must be transferred to the CVs on the short time scales of ab initio simulations (typically, tens or hundreds of picoseconds), which might induce instabilities in the dynamics. To cure this problem and, at the same time, increase the efficiency of MTD, an extended Lagrangian method has been proposed, in which the CVs are treated as dynamical variables [16]. The Lagrangian that describes the ionic and electronic dynamics, $\mathscr{L}_{i,e}$, is supplemented with additional terms driving the dynamics of the CVs:

$$\mathscr{L}_{tot} = \mathscr{L}_{i,e} + \sum_{i=1}^{d} \left[\frac{1}{2} M_i \left(\frac{d\tilde{\xi}_i}{dt} \right)^2 - \frac{1}{2} k_i \left(\xi_i(\boldsymbol{r}) - \tilde{\xi}_i \right)^2 \right] + V_b \left(\tilde{\xi}, t \right). \quad (3.21)$$

In this formula, the second term is a fictitious kinetic energy for the dynamical CVs $\tilde{\xi}$, the third term is a potential energy consisting of harmonic potentials centered at $\xi_i(\boldsymbol{r})$ (i.e. the value of the CVs corresponding to the instantaneous configuration of the ions) and V_b is the usual biasing potential calculated at the instantaneous values of $\tilde{\xi}_i$ (which generally differs from $\xi_i(\boldsymbol{r})$). The CVs are subject to the force due to the biasing potential and to the elastic force. By tuning the coupling constants k_i and the fictitious masses M_i, one can make the typical frequencies $\omega_i = \sqrt{k_i/M_i}$ describing the elastic motion of the CVs so small so as to decouple the latter from the ionic and electronic dynamics. The adiabatic separation of these motions ensures that there is no spurious effect on the evolution of the ions due to the fictitious dynamics of the CVs. However, when tuning k_i and M_i, one must keep in mind that (a) k_i should not be very small, in order to prevent the typical values of $\xi_i(\boldsymbol{r}) - \tilde{\xi}_i$ from becoming too large, and (b) M_i should not be extremely large, otherwise the CVs will relax to the equilibrium distribution too slowly, thus spoiling the efficiency of the method [8, 17].

3.1.3.4 Choice of the Collective Variables

The choice of a proper set of CVs is crucial but sometimes very difficult. The set of CVs should be able to discriminate between initial, intermediate and final states, and should describe all the relevant events driving the process under study. On the other hand, their number should be small (maximum 3–4), for the time it takes to fill the FES depends exponentially on the dimensionality of the coarse-grained space of CVs. Neglecting a relevant CV in a MTD simulation will typically lead to hysteresis phenomena [8] and, needless to say, to completely wrong FESs. To alleviate these problems, several approaches have been developed. In [18], a new method was introduced, which is based on multiple, parallel MTD simulations employing different sets of CVs and on periodic exchanges between their bias potentials, according to a replica exchange scheme. Tribello et al. [19] instead proposed a method to identify the minima of complex, multidimensional FESs, which enables one to use one-dimensional CVs to describe the regions near the basins and to bias the system locally.

3.2 Applications

Since the beginning of the development of MTD, many successful applications have been reported in various disciplines, ranging from molecular biochemisty and physical chemistry to materials science. The application of MTD to biomolecular systems has been particularly fruitful, especially in studies of the conformational transitions of macromolecules, such as folding and unfolding of proteins [18, 20, 21]. Investigations of chemical reactions, as well as of phase transitions in crystalline systems, are numerous too. So far only a limited number of MTD studies of crystalline nucleation in liquid and amorphous systems is available in the literature. The reason for this lies in the complexity of the nucleation processes, which involve a large number of degrees of freedom and require the accumulation of a huge amount of statistical data for an adequate sampling of the CV space. In spite of these difficulties, the latest advances in MTD discussed in the previous section make it a promising tool for nucleation studies as well.

In the following subsections, we give an overview of recent MTD works about nucleation processes and related phenomena. Most of the studies discussed here employed classical MD simulations, because of their moderate computational cost. On the other side, ab initio techniques are generally needed to make reliable predictions for systems exhibiting complex chemical bonding mechanisms. Phase change materials (PCMs) belong to this category, in that their crystalline and amorphous phases exhibit different local order and chemical bonding. In the last part of this section, we present our preliminary work in modelling nucleation in PCMs by virtue of ab initio MTD.

The first subsection is about MTD calculations of the solid-liquid interfacial free energy. Accurate knowledge of this quantity is important to estimate the size of the critical nuclei and of the nucleation barriers using CNT. In the second subsection, we focus on homogeneous nucleation in several systems, including Lennard-Jones systems, ice, NaCl and PCMs.

3.2.1 Solid-Liquid Interfacial Free Energy

The surface tension γ is a key thermodynamic quantity which plays a central role in a wide range of physical and chemical processes, in particular in first-order phase trasitions induced by nucleation and growth (see also Sect. 3.1.1). It is defined as the reversible work to form an interface of unit area between two different phases. The surface tension makes the existence of metastable phases possible in the first place, for it is a positive energy contribution hindering the transition to the stable phase.

In spite of its importance, available data on the solid-liquid surface tension γ_{sl} are scarce. Its experimental measurement is highly challenging and usually afflicted with large uncertainties. The computer simulations offer an alternative route to estimate γ_{sl} and to interpret experimental results Several techniques have been developed for this purpose, including the cleavage potential method [22] and the capillary fluctuation method [23].

Recently, Angioletti-Uberti et al. [24, 25] have used a novel approach based on MTD to calculate γ_{sl} at the melting temperature T_m. Since the chemical potentials $\mu(P, T)$ (P denotes the pressure) of the solid and liquid bulk phases coincide at T_m, the Gibbs free energy $G(P, T_m)$ of a system containing a solid-liquid interface can be written as:

$$G(P, T_m) = \mu(P, T_m) N + G_{ex}. \tag{3.22}$$

The presence of an interface gives an excess free energy term $G_{ex} = A\gamma_{ls}$ (A is the area of the interface). Thus, γ_{ls} can be readily obtained as the change in G upon going from the pure bulk phases to a state with an interface, divided by the area A. Angioletti-Uberti et al. used MTD to calculate the changes in G. In the following two subsections we outline their approach and review their results about Argon and Lead.

3.2.1.1 Lennard-Jones Model of Argon

In [24], Angioletti-Uberti et al. computed γ_{ls} for Argon at T_m, using Lennard-Jones potentials and Well-Tempered MTD [14]. For this purpose, they considered an elongated unit cell, say in z direction, and divided it into two equally sized regions A and B along that direction. They chose the center of the unit cell as the origin, so that the two regions corresponded to $|z| \leqq \frac{c}{4}$ and $|z| > \frac{c}{4}$, here c being the length of the unit cell in z direction (see Fig. 3.3).

Due to the periodic boundary conditions, the models contained two interfaces. Then they defined two CVs in such a way that the presence of the two interfaces was explicitly taken into account. The two CVs, s_A and s_B, were constructed as

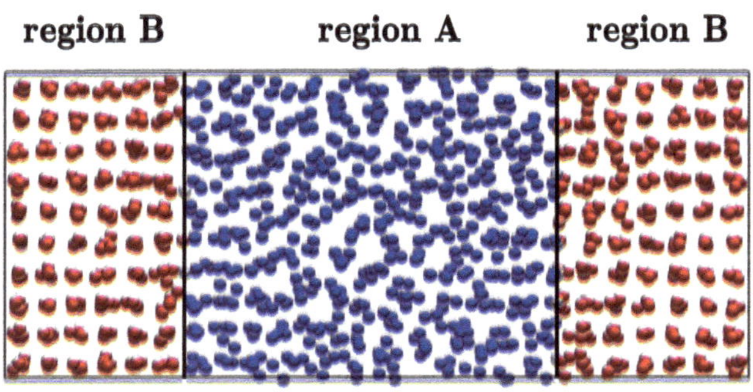

Fig. 3.3 Reproduction of the model considered in [24]. The cell was divided into two equally sized regions (**a**) and (**b**), in order to define an appropriate CV describing the solid-liquid interfaces

orientation-dependent (i.e. not rotationally invariant) order parameters ϕ, averaged over the corresponding regions A and B:

$$s_A = \frac{\sum_i \phi(\mathbf{r}_i) \cdot c_z(\mathbf{r}_i)}{\sum_i c_z(\mathbf{r}_i)}, \tag{3.23}$$

$$s_B = \frac{\sum_i \phi(\mathbf{r}_i) \cdot (1 - c_z(\mathbf{r}_i))}{\sum_i (1 - c_z(\mathbf{r}_i))}. \tag{3.24}$$

The interface cutoff function $c_z(\mathbf{r})$ takes the constant value 1 in the region A and 0 in the region B, except for the narrow interface regions around $\pm\frac{c}{4}$, where it changes smoothly from 1 to 0:

$$c_z(\mathbf{r}) = \begin{cases} 1 & |z| \leq c/4 - \Delta z/2 \\ 0 & |z| \geq c/4 + \Delta z/2 \\ \left[(u-1)^2 (1+2u)\right] & c/4 - \Delta z/2 < |z| < c/4 + \Delta z/2, \end{cases} \tag{3.25}$$

where $u = (|z| - c/4)/\Delta z$ and Δz is the width of the transition domain.

A suitable order parameter $\phi(\mathbf{r})$ should distinguish locally between the ordered (solid-like) and disordered (liquid-like) environment of an atom. Usually it is taken to be a function adapted to the symmetry of the solid phase, so that it takes nonzero values in the crystal-like configurations and zero in the liquid-like ones (ideal liquids are isotropic). One possible choice is a linear combination of spherical harmonics in a similar fashion as the popular Steinhardt order parameter Q_l [26] defined in the Sect. 3.2.1 Angioletti-Uberti et al. instead chose polynomials adapted to the face-centered cubic (fcc) symmetry of Ar, since they are cheaper to compute than spherical harmonics. Their polynomials have the form (using the same notation as in [27]):

$$c_\alpha(\mathbf{r}) = \left[\frac{x^4y^4}{|\mathbf{r}|^8}\left(1 - \frac{z^4}{|\mathbf{r}|^4}\right) + \frac{y^4z^4}{|\mathbf{r}|^8}\left(1 - \frac{x^4}{|\mathbf{r}|^4}\right) + \frac{z^4x^4}{|\mathbf{r}|^8}\left(1 - \frac{y^4}{|\mathbf{r}|^4}\right)\right]. \tag{3.26}$$

The order parameter $\phi(\mathbf{r})$ was computed as a weighted average of $c_\alpha(\mathbf{r})$ over all the atoms,

$$\phi(\mathbf{r}_i) = \frac{\sum_{j \neq i} c_r\left(|\mathbf{r}_j - \mathbf{r}_i|\right) c_\alpha\left(\mathbf{r}_j - \mathbf{r}_i\right)}{\sum_{j \neq i} c_r\left(|\mathbf{r}_j - \mathbf{r}_i|\right)}, \tag{3.27}$$

where the weight was given by a smooth, radial cutoff function $c_r(r)$. For $c_r(r)$, a similar functional form was used as for the interface cutoff function, c_z, with $|z|$ replaced with r and suitably defined inner and outer cutoff radii. Furthermore, ϕ was rescaled, such that it was equal to 1 for the perfect fcc solid and 0 for the ideal liquid.

The converged FES at $T = T_m$ of a $(7 \times 7 \times 12)$-supercell model containing 2352 atoms is shown in Fig. 3.4. The two basins associated with the bulk solid and bulk liquid phases have equal minima, as is expected to occur at T_m. In contrast, the

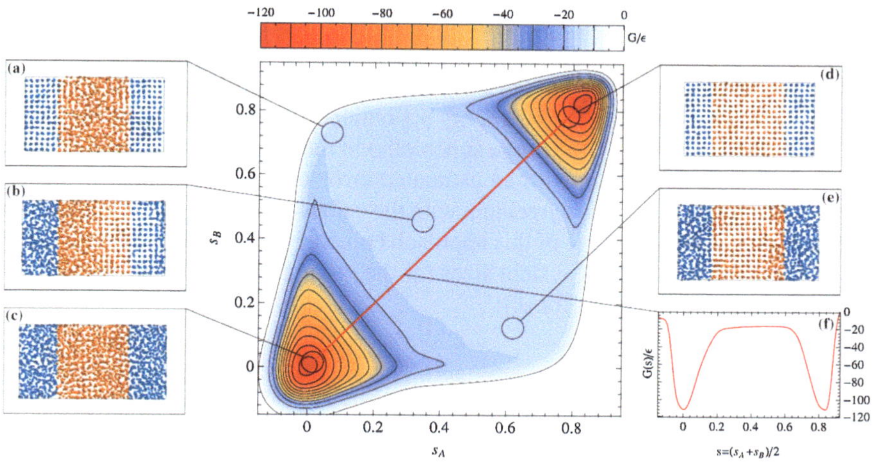

Fig. 3.4 The 2D FES obtained from the well-tempered MTD simulations. The energy is in reduced units. Selected snapshots of the system are also shown. Reprinted with permission from [24].

configurations on the flat transition barrier contain solid-liquid interfaces, which give rise to a nonzero G_{ex}. Depending on where the interfaces are located with respect to the boundaries between the regions A and B, these configurations correspond to different points on the line $s_B = 1 - s_A$. Three representative configurations are depicted in Fig. 3.4, together with the bulk solid and the bulk liquid structures.

Because of the symmetry of the two regions A and B, the FES has a mirror symmetry with respect to the line $s_A = s_B$. Moreover the transition region is quasi-planar. In fact, the domains of the two-dimensional FES needed to be explored for an adequate estimation of G_{ex} are only a small portion of the full FES. Therefore, one of the CVs was then restricted into a small range of values by adding reflective walls (i.e. large energy hills) to the potential energy, so that only the other CV was allowed to vary by a notable amount, thus making the MTD virtually one-dimensional. By doing so, the needed computational cost was lowered substantially.

Finite size effects were analyzed by varying the size of the supercell either in the z direction or in the xy-plane. The lattice constants along the x and y directions were fixed during the simulations, so that the supercell changed its size only in the z direction to adjust to the density change due to partial solidification or melting. The strain-related free energy contribution arising from this contraint is a finite size effect, which vanishes in the thermodynamic limit.

The most important finite size error comes from the interaction of the two interfaces and the expected plateau of the FES appears gradually by increasing the cell dimension along z and hence the distance between the interfaces. The minimum length of the supercell along z yielding a reasonable plateau was determined to be around 8 cell units. The variation of the width of the xy-plane showed that the residual slope of the plateau vanished if the lattice constants along x and y exceeded some correlation length, $2 \cdot L_{corr}$. The value of L_{corr} was found to be comparable to the distance at which the pair-correlation function $g(r)$ of the liquid approaches 1. In the case of Lennard-Jones argon, L_{corr} turned out to be equal to 3 cell constants.

The values for the surface free energy γ_{sl} obtained by this approach range from 0.39 (small supercells) to 0.36 (large supercells) in reduced units (i.e. in units of the Lennard-Jones parameters), with an estimated error on the latter value of roughly 2–3 %. This result is in close agreement with the values calculated by other methods. The computational advantage of this approach compared to the other methods lies in the fact that relatively small system sizes (of the order of 1000 atoms) were needed to achieve reliable estimates for γ_{sl}.

3.2.1.2 Semi-empirical Model of Pb

Angioletti-Uberti [25] used the same approach as described in Sect. 3.2.1.1 for the study of the liquid-solid interface of Pb at coexistence.

Semi-empirical many-body potentials of the Gupta-type (GU) [28, 29] and of the Finnis-Sinclair-type (FS) [30, 31] were chosen to describe Pb. The GU potential shows better agreement with experiments in terms of the pair-correlation function of the bulk liquid phase, whereas the FS potential yields values of several thermodynamics quantities which are closer to experimental data. In particular, the inherent melting temperature of the FS potential, T_m^{FS}, turns out to be 630 K, in fair agreement with the experimental value of 600.61 K, while $T_m^{GU} = 502$ K.

The simulation model consisted of a conventional fcc supercell of size $(5 \times 5 \times 12)$ with 1200 atoms, for which finite size effects were found to be negligible from comparisons with trial simulations employing larger supercells. The MTD simulations were essentially one-dimensional, owing to the use of a potential wall, as explained in Sect. 3.2.1.1.

The FES for both potential types are depicted in Fig. 3.5.

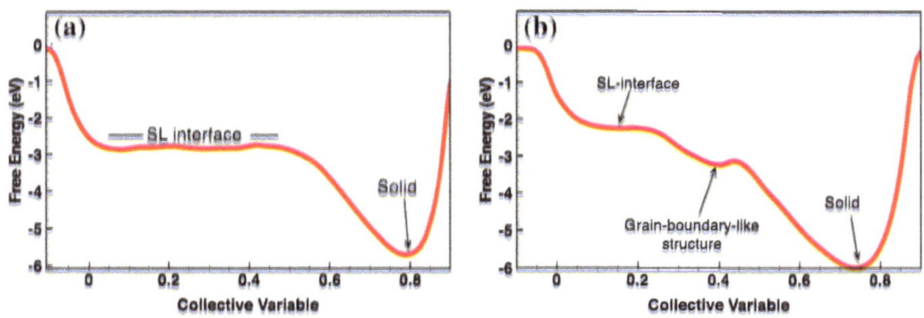

Fig. 3.5 The FES for the solid-liquid transition in Pb, obtained using GU (**a**) and FS (**b**) potentials. Reproduced by permission of IOP Publishing from [25]. Copyright 2011 IOP Publishing. All rights reserved

The one dimensional FES for the GU potential displays the expected flat plateau in the region where the solid-liquid interface is present. On the other hand, in the case of the FS potential, a small shoulder in the FES appears at values of the CV s around 0.4. According to the author, the feature can be attributed to the occurrence of additional "grain-boundary-like" configurations, namely solid-solid interfaces with different orientations. Close inspection of the evolution of the FS model indeed showed that the disordering process of the solid sometimes ceased and turned back into a solidification of the liquid. The recrystallized parts had different orientations than the original solid phase. These configurations resembled then a grain boundary. However, quenching of these configurations to 0 K resulted in a number of interstitial defects, due to structural relaxations, which rather suggests that they were not real grain boundaries. The reason for the development of these reorientations could not be clarified. The MTD bias alone could be ruled out as the origin for their appearance, owing to the absence of "grain-boundary-like" configurations in the case of the GU potential. Surely the characteristic structure of the FS potential seems to have a crucial influence on the probability of these configurations.

Here we speculate about the roots and the potential solutions of this problem. We think that both the FS potential and the precise definition of the CVs are relevant. A conceivable explanation could lie in the explicit form of the CV constructed from the local order parameter c_α given in (3.26), in combination with the energy barriers determined by the FS potential. c_α is defined as a polynomial which yields high values for certain preferential directions (see also Fig. 2 in [24]). By this choice, the solid configurations with non-preferential orientations take lower values of c_α and thus make the global CV s_A, which is an average of the local parameter c_α, less distinguishable from the values taken by the liquid phase or the configurations with an interface. The strategy of using orientation-dependent local order parameters should work well only if the expected configurations (liquid or mixed liquid-solid) in the corresponding range of the CV values are more easily accessible than the reoriented solid configurations. This might be not be satisfied in the case of the FS potential, leading to comparable free energy barriers for the latter phenomena. Appropriate modifications of the CV or the local order parameter could probably help to resolve the problem.

3.2.2 Nucleation in Liquids and Amorphous Materials

One of the primary goals for the understanding of crystallization of liquids and amorphous materials is to gain a microscopic picture of the nucleation processes. MTD, which accelerates the occurrence of rare events, appears to be well suited for this purpose.

Lennard-Jones systems are ideal starting models for nucleation studies. Since they have been extensively studied by a number of different approaches, they also serve as an excellent testing ground for MTD. In Sect. 3.2.2.1, we present the first study of nucleation in Lennard-Jones Argon using MTD.

In the subsequent subsections, we review recent results about MTD simulations of nucleation employing more sophisticated classical potentials (ice and NaCl) and ab initio methods (PCMs).

3.2.2.1 Lennard-Jones Argon

Trudu el al. [32] performed both transition path sampling (TPS) [33] and MTD simulations of the Lennard-Jones model of Argon at pressure $P = 0.25$ kbar and temperatures ranging from 0.7 T_m to 0.8 T_m. The system size of 6912 atoms was considered to be sufficiently large to avoid significant periodic boundary effects (the expected nucleus size is $N_c \sim 200$).

In the preliminary TPS simulations performed at $T = 0.8\ T_m$, the authors generated a large number of statistically independent trajectories from a crystallization trajectory by altering the velocities randomly at selected points along the path (keeping the total kinetic energy constant) and by propagating the system forward and backward in time. The new crystallization trajectories thus obtained were used as starting points to generate new sets of trajectories using the same procedure. In total, the authors obtained ten statistically independent crystallization trajectories.

Then the authors performed a commitment probability analysis (CPA) [33], which consists in assessing the probability for the system to go to the liquid or the crystalline state while changing randomly the velocities. This analysis enabled them to identify the transition state ensemble, i.e. the set of configurations from which the system evolves to the two phases with equal probability. This ensemble turned out to contain crystalline nuclei with a broad size distribution and far from spherical shapes. By averaging over the distribution, a critical size of $N_c = 240 \pm 34$ atoms was obtained.

This value is in stark disagreement with CNT, which yields a critical nucleus size of ~80 atoms (and a free energy barrier ΔG_c of 18.5 $k_B T$). The authors showed that a better agreement with their TPS simulations could be obtained by relaxing the assumption of spherical clusters and instead adopting an ellipsoidal shape for the nuclei. The resulting more general model, which was dubbed "extended CNT" model, provided values of N_c and ΔG_c of about 150 particles and 35 $k_B T$ respectively.

Subsequently, Trudu et al. used MTD to study nucleation as a function of the undercooling T/T_m. The employed CVs were the potential energy U and the Steinhardt bond order parameter Q_l defined as

$$Q_l = \sqrt{\frac{4\pi}{2l+1} \sum_{m=-l}^{l} \left| \frac{1}{N} \sum_i N_i q_{lm}(i) \right|^2} \tag{3.28}$$

$$q_{lm}(i) = \frac{1}{N_i} \sum_{j \in \Omega_i}^{N_i} Y_{lm}(\hat{r}_{ij}), \tag{3.29}$$

where i is an atomic index, $\mathbf{r}_{ij}$ is the vector connecting atom i and j (with $r_{ij} = |\mathbf{r}_{ij}|$ and $\hat{r}_{ij} = \mathbf{r}_{ij}/r_{ij}$) and the sum is taken over the set Ω_i (with N_i elements) of the

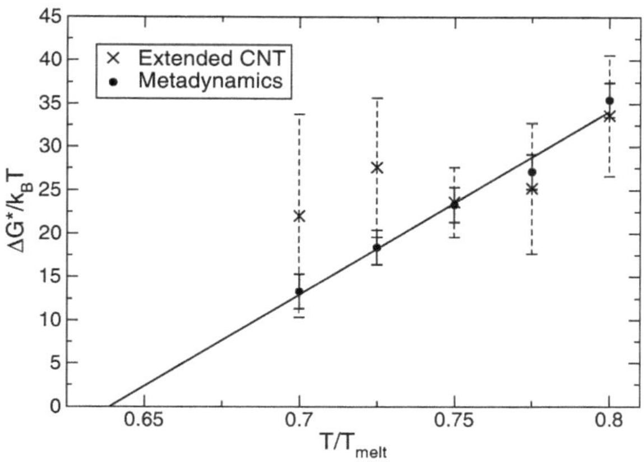

Fig. 3.6 Free energy barrier at different temperatures, as computed from MTD and the extended CNT model. The *continuous line* indicates the linear fit to the MTD data. Reprinted with permission from [32]. Copyright 2006 American Physical Society. All rights reserved

nearest neighbour indices of atom i. $Y_{lm}(\hat{r}_{ij})$ are spherical harmonics. In this work the parameter l was set to 6 and computed locally on a subset of selected atoms ($N_{sub} = 350$). The authors decided to use the local form of Q_6 in order to circumvent multiple nucleation at different sites in the simulation box, which was observed in previous studies employing global variables. A more thorough discussion of this issue will be given in Sect. 3.2.2.2.

Several MTD simulations were carried out in the undercooling range between 0.7 and 0.8 T/T_m. At $T = 0.8\ T_m$, the calculated ΔG_c value of 35.4 $k_B T$ agrees well with the one attained by the extendend CNT model, thus providing evidence for the ability of MTD to describe nucleation processes.

In the next step, the transition states of the MTD trajectories were identified by the CPA and the values of ΔG_c were re-estimated from the extended CNT model applied to these states. The results are reported in Fig. 3.6 as a function of T, together with the ΔG_c values obtained by MTD. A nice linear behavior of the MTD values of ΔG_c down to 0.7 T/T_m can be seen. The extended CNT model follows this line down to 0.75 T/T_m and then starts to deviate from it below this temperature, due to the large fluctuations in the cluster size at the transition state.

At very deep undercooling, ΔG_c is expected to vanish and the system should enter a regime of spinodal decomposition, rather than nucleation. The solid-liquid interface becomes diffuse and the crystallization proceeds by a collective mechanism. Trudu et al. performed standard (i.e. without biasing potential) MD runs at $T/T_m = 0.65$ and $T/T_m = 0.7$, which confirmed the existence of the spinodal regime. Their model crystallized in less than 200 ps at $T = 0.65\ T_m$, suggesting very low ΔG. The corresponding crystalline clusters displayed a diffuse branched structure. MTD runs at these temperatures provided FESs without liquid basins, implying the instability (instead of metastability) of the liquid phase, and the crystallization occurred similarly in a collective way.

In summary, this work validated the MTD method and the choice of the CVs and paved the way for further MTD studies of crystallization.

3.2.2.2 Ice

The next example for a successful application of MTD is the work of Quigley and Rodger [34] on ice nucleation in supercooled water. It is the first reported isothermal-isobaric (NPT) simulation where ice nucleation was realized. The authors aimed to achieve a polymorph selection from dynamical nucleation trajectories: MTD offers the possibility to generate such trajectories, in contrast to techniques like umbrella sampling. A proper choice of the CVs is crucial, for unsuitable CVs can restrict the available pathways and, consequently, the polymorph selection. Quigley and Rodger employed the global Steinhardt parameters Q_4 and Q_6, the tetrahedral parameter ζ and the potential energy U, and investigated the corresponding 4-dimensional FES. In previous works, these CVs had been shown to be capable of connecting the liquid phase to the cubic and hexagonal polymporphs of ice (denoted as I_c and I_h, respectively). The use of 4 CVs is quite remarkable: as discussed in Sect. 3.1.3, the computational cost of MTD simulations increases exponentially with the number of CVs. The exploration of a four-dimensional FES thus Should have required a substantial amount of computational resources.

The Steinhardt parameter has been introduced in Sect. 3.2.2.1. For water the sum of the spherical harmonics q_{lm} runs over all oxygen-oxygen pairs and was smoothed with a radial cutoff function $f(r_{ij})$. The tetrahedral parameter is defined as

$$\zeta = \frac{1}{4N} \sum_{i=1}^{N} \sum_{j \neq i}^{N_i} \sum_{k>j}^{N_i} f(r_{ij}) f(r_{ik}) \left(\cos \theta_{jik} + \frac{1}{3} \right)^2 , \tag{3.30}$$

where θ_{jik} is the angle formed by atoms j, i and k. The same radial cutoff function f as for q_{lm} in Q_l was used for ζ. The value of the parameters U and ζ were scaled so as to have equally large equilibrium fluctuations at the thermodynamics conditions of the simulations ($T = 180$ K and $P = 1$ atm).

The MTD simulations were initialized in the supercooled liquid state. The models contained 512 and 576 molecules. TIP4P force fields [35] were used to model the interactions between water molecules. The authors used different supercells, including orthorhombic and hexagonal unit cells commensurate with the I_h symmetry and cubic cells commensurate with I_c. However, regardless of the choice of the unit cell, only the nucleation of I_c ice was observed. The authors concluded that this form of ice is more stable than the hexagonal one, at least within the limits of the force field.

The calculated four-dimensional FES indicate that the liquid and solid I_c basins are separated by a barrier with a height of approximately 79 $k_B T$. However, the system size turned out to be too small for an accurate estimate of the barrier. In fact, due to the self-interaction between the periodic images, crystallizations occurred in an anisotropic way and elongated clusters formed, which eventually generated thin slabs of ice.

Interestingly, Quigley and Rodger observed the formation of single crystallites in their simulations. On the other hand, it is known that finite size effects can lead to nucleation in multiple regions of the simulation box. The cause for this effect was analyzed in detail by ten Wolde et al. in [27]. The nucleation of multiple crystallites is favorable if the translational entropy gain of creating such clusters overweighs the free energy cost associated with the formation of larger solid-liquid interface areas. The tendency of a system with a given size to have single or multiple crystallites depends on the solid fraction χ in it, and the free energy F varies as $\sim\chi^2$ for the former case and as $\sim\chi^{2/3}$ if multiple clusters are present. Hence, for small values of χ below some critical χ_c, multiple crystallites are favoured, but will become unstable above it. The critical fraction χ_c itself depends on the system size as $\sim V^{-1/4}$. Therefore, in the infinite size limit, $\chi_c = 0$, which shows that the appearance of multiple crystalline clusters is a finite size effect. Although in [27] an estimation of χ_c was given, based on the approximation that the crystallites behave like an ideal gas, its accurate evaluation remains problematic, in general.

The use of local order parameters biasing only a relatively small portion of the system, as described in Sect. 3.2.2.1, can circumvent this problem. Quigley and Rodger didn't observe multiple crystallites, in spite of the fact that they used global order parameters. This behaviour might be due to a large value of the surface tension γ. In [27], it was shown that χ_c depends on the surface tension γ as $\sim (\gamma / \exp(\text{const.} \cdot \gamma))^{1/4}$. Hence, for large γ, the formation of multiple nuclei becomes too energetically costly, even for relatively small cell sizes.

To shed light on this point, Quigley and Rodger also investigated a model of TIP4P water containing 2496 molecules, using the same 4 CVs introduced above [36]. They showed that water crystallizes from a single I_c nucleus in this model as well. We refer the reader to Fig. 4 in [36] which depicts the evolution of the crystalline nucleus.

3.2.2.3 NaCl in Water

In a recent paper by Giberti et al. [37], the nucleation of NaCl from its solution in water was investigated by MTD. The authors introduced a new CV, based on the solute density $\rho(\mathbf{r})$. Since the latter quantity is homogeneous in the absence of crystalline nuclei, whereas it varies abruptly across solid-liquid interfaces, the authors employed the integral of the squared gradient of $\rho(\mathbf{r})$ as CV:

$$S = \frac{1}{2} \int d\mathbf{r} \left(\nabla\rho(\mathbf{r})\right)^2 . \tag{3.31}$$

In order to get a smooth, coarse-grained distribution of $\rho(\mathbf{r})$, Giberti et al. approximated it as a sum of Gaussians centered at the atomic positions $\mathbf{r}_i$:

$$\rho(\mathbf{r}) = \sum_{i=1}^{N} \frac{1}{\sigma^3 (2\pi)^{3/2}} \exp\left(-\frac{(\mathbf{r} - \mathbf{r}_i)}{2\sigma^2}\right). \tag{3.32}$$

The authors used the gromos53a6 [38] and SPC/E [39] force fields to describe the ions and water, respectively. A model of 6.15 M solution of NaCl in water (containing

360 NaCl units and 2531 water molecules) was equilibrated in an NPT ensemble at $T = 300$ K and $P = 1$ bar. Then the biasing potential was turned on and a 50 ns canonical (NVT) MTD run was performed. By probing high values of S during the simulation, strongly hydrated amorphous NaCl structures were mostly found, which however tended to dissolve in unbiased MD test runs.

Ordered configurations of NaCl occurred as well, albeit much less frequently, showing two types of arrangements, namely the usual rocksalt structure and, surprisingly, the wurtzite structure. To assess their stability and lifetime, the authors took the observed structures out of the solution, embedded them in water and conducted unbiased MD simulations. Both types of crystallites exhibited long lifetimes of the order of 25–30 ns, much longer than those of the amorphous aggregates. Moreover, some of the rocksalt nuclei displayed a spontaneous transition to the wurtzite structure before dissolution.

In order to rule out the possibility that the wurtzite clusters were an artifact of the employed classical force fields, the authors performed an ab initio structural optimization of bulk NaCl in the wurtzite configuration and found that this is indeed a metastable phase. The energy of the structure was found to be only 11.3 kJ mol^{-1} higher than that of the rocksalt arrangement.

In addition, the difference between the free energy of formation of wurtzite ($\Delta G_{wurtzite}(r)$) and rocksalt ($\Delta G_{rocksalt}(r)$) spherical nuclei was estimated using the CNT (Formula 3.2). The driving forces Δg of the bulk phases were calculated from the average energies of the NVT simulations, in combination with the estimations of the entropy in the quasi-harmonic approximation. The surface tensions γ were evaluated using the Kirkwood-Buff formula, which relates the mechanical stress and γ itself. For this purpose, NVT simulations of the interface between a saturated solution and (a) the $(10\bar{1}0)$ surface of the wurtzite structure and (b) the (100) surface of the rocksalt phase were performed.

The results for the difference $\Delta G_{wurtzite} - \Delta G_{rocksalt}$ are shown in Fig. 3.7.

They indicate that the wurtzite structure is more stable than the rocksalt one for small nuclei. This property was shown to stem from the fact that γ (the dominant term for small nuclei) is lower in the wurtzite structure by 10 %, although Δg is larger by 20 %. The authors explained the lower values of γ for the wurtzite structure by its stronger ability to solvate its surface, indicated by the higher number of hydrogen bonds formed at the interface. The difference in γ turned out to be even more pronounced (70 %) in the case of an interface with pure water, instead of the solution. The CNT provides an estimated "critical" radius of about 5.7 nm, below which the wurtzite nucleus is more stable. The conversion to the rocksalt phase should thereby occur when the nucleus exceeds this size.

In conclusion, the insensitivity of the CV S to the symmetry of the emerging phase offered the possibility to explore several pathways during nucleation and find different polymorphs, including metastables ones. However, this CV probably needs to be coupled to other more selective CVs, if the evaluation of the free energy basins is required.

We expect that CVs similar to S, such as the integral of the squared gradient of a spatially averaged Q_6, could prove to be effective in finding intermediate phases during nucleation of amorphous materials as well.

3.2.2.4 Phase Change Materials

In the examples discussed in the previous sections, only classical MD potentials were utilized, because of their moderate computational cost. However, accurate MD simulations of complex systems, such as PCMs, require a more realistic, quantum mechanical modelling of the chemical bonds between the elements. In principle, ab initio MD (AIMD) simulations [40] based on density functional theory [41, 42] are a more suitable tool to describe such systems, nevertheless they require a large computational effort, rooted in the nature of the problem to solve. For instance, in the case of Born-Oppenheimer dynamics [43], a self-consistent diagonalization of the Kohn-Sham Hamiltonian must be carried out at each MD step. Although one can avoid the self-consistent iterative process by using state-of-the-art, Car-Parrinello like techniques [44, 45], the maximum system size and time scale that can currently be investigated by AIMD simulations are of the order of a thousand atoms and a few nanoseconds, respectively. Generally, these limitations pose a big obstacle for the study of rare events such as nucleation, even when AIMD is combined with enhanced sampling methods.

PCMs [46–48] undergo fast and reversible transitions between a crystalline and an amorphous phase as a function of temperature. Furthermore, the two phases are very stable at room temperature and exhibit a pronounced optical and electronic constrast. These properties have led to important applications in optical devices and non-volatile

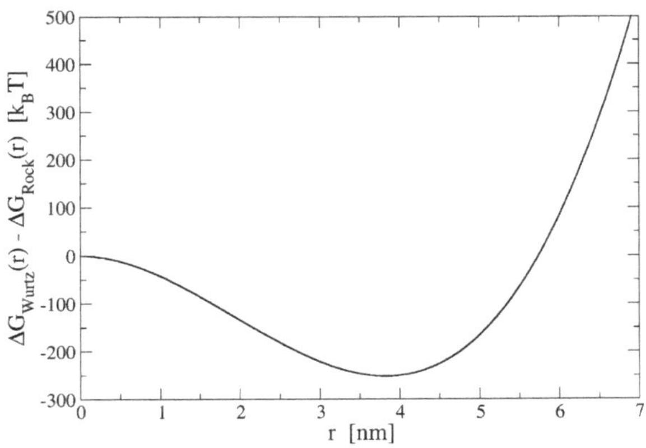

Fig. 3.7 The difference between the free energy of formation of wurtzite and rocksalt nuclei as a function of the radius of the nuclei. Reprinted with permission from [37]. Copyright 2013 American Chemical Society. All rights reserved

memories. Remarkably, PCMs display extremely fast crystallization at temperatures close the crystallization temperature, thereby providing the possibility to investigate them by AIMD. In fact, several AIMD simulations of PCMs were reported in recent years, focusing on the crystallization process [49–51], as well as on the structural and electronic properties of the amorphous [52–54] and crystalline phases [55–58]. Systems containing a few hundreds of atoms were successfully crystallized within a few hundreds of picoseconds by several authors. However, these studies were not able to provide accurate information about the crystal growth velocity, nor about typical nucleation barriers and nucleation rates. MTD simulations, in combination with AIMD, seem a promising path to study the nucleation mechanisms in this specific material class. Guided by these perspectives, we have started to tackle this problem by combining MTD with the method by Kühne et al. [45], as implemented in the cp2k package [59].

In the following, we summarize the results of our preliminary calculations for the GeTe alloy, a prototypical PCM. Here we should mention that important progress has been made recently in the construction of a novel, classical neural-network potential for GeTe [60], fitted against ab initio data, which shows an accuracy comparable to that of AIMD and is 4 orders of magnitude faster than the AIMD method by Kühne et al. [45]. However, since the development of neural-network potentials for 3- and 4-component PCMs remains a challenge, AIMD will likely be the only viable method for a systematic study of PCMs in the years to come.

The stable crystalline phase of GeTe at room temperature is a distorted rocksalt phase. Starting from a cubic supercell of this phase containing 216 and 512 atoms (corresponding to $3 \times 3 \times 3$ and $4 \times 4 \times 4$ supercells of the conventional unit cell), amorphous models of GeTe were generated by melting the system and subsequent quenching from the melt, with a quenching rate of 25 K/ps. The volume of the cell was fixed to have the experimental density of amorphous GeTe during the melt-quenching process. The structural properties of the so obtained amorphous models were validated by comparing to available theoretical and experimental data. For the study of crystallization using MTD, the volume of the cell and the atomic coordinates of amorphous GeTe were rescaled to yield an intermediate density of 5.85 g/cm^3 between the amorphous 5.56 g/cm^3 [61] and the crystalline value 6.13 g/cm^3 [62]. Temperature was set to 600 K, to be compared with the melting temperature T_m of 1000 K measured experimentally (a very similar value of T_m was obtained by MD simulations employing the neural-network potential mentioned above [63]).

The first CV was constructed from the sum q_{lm} defined in (3.29). Defining the $(2l + 1)$-dimensional vector $\mathbf{q}_l$, whose components are the $q_{lm}(i)$ (as in [27]),

$$\mathbf{q}_l(i) = \begin{pmatrix} q_{l,l} \\ q_{l,l-1} \\ \cdots \\ q_{l,-l+1} \\ q_{l,-l} \end{pmatrix} = (q_{lm}(i))_{m=-l,l}\,, \tag{3.33}$$

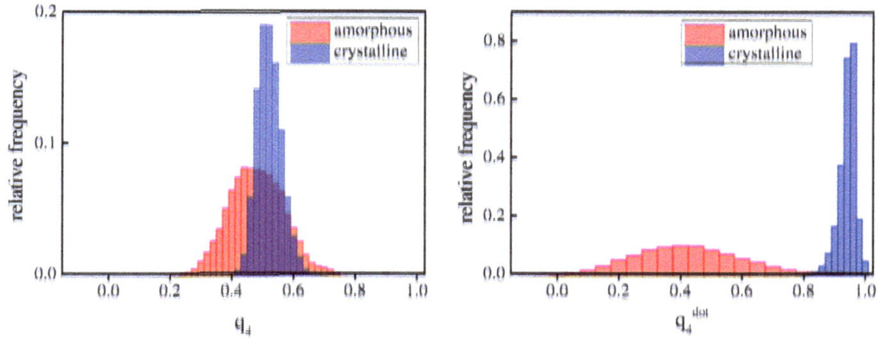

Fig. 3.8 Distribution of the bond order variables q_l and q_l^{dot} in the amorphous and the crystalline phase of GeTe at T = 600 K

the CV q_l^{dot} was constructed as a sum of bond order correlations C_{ij} between neighboring atoms i and j,

$$q_l^{dot}(i) = \frac{1}{N_i} \sum_{j \in \Omega_i} C_{ij}, \tag{3.34}$$

where C_{ij} is given by a dot product of the form

$$C_{ij} = \frac{\mathbf{q}_l(i) \cdot \mathbf{q}_l^*(j)}{\|\mathbf{q}_l(i)\| \cdot \|\mathbf{q}_l(i)\|}. \tag{3.35}$$

The C_{ij} are normalized by dividing by the norm of the vectors $\|\mathbf{q}_l(i)\|$, so as to get a value of 1 for the perfect crystalline configuration with completely correlated bonds. The C_{ij} are usually used to define a connection between solid-like particles and, thus, to analyze the solid clusters in nucleation studies, whereas the global Steinhardt parameter Q_l is commonly employed to bias the system. However, Q_l is expected to be an effective CV only if its local form $q_l(i) = \sqrt{\frac{4\pi}{2l+1} \sum_{m=-l}^{l} |q_{lm}(i)|^2}$ is capable to discriminate between the local structure of the liquid/amorphous and solid phases in a satisfactory way. This requirement can be checked by calculating the distribution function of $q_l(i)$ in the two phases. In Fig. 3.8, the distributions of $q_{l=4}(i)$ are compared with those of $q_{l=4}^{dot}(i)$, for the case of GeTe. It turns out that the overlap between the distributions of the crystalline and amorphous phase is significantly smaller in the case q_4^{dot}, thus indicating that it is more appropriate to consider a global CV constructed from $q_4^{dot}(i)$, rather than from $q_4(i)$ (though the global Q_l is not constructed as a direct sum of $q_l(i)$, but as a sum of $q_{lm}(i)$ before taking the absolute value). Therefore, we considered the global CV defined as $Q_4^{dot} = \frac{1}{N} \sum_i q_4^{dot}(i)$.

Since the coordination numbers in GeTe are known to change upon crystallization [52], we also used a continuous form of this quantity, n_{coord}, as CV, given by

Table 3.1 MTD parameters

N_{atoms}	512
ω in meV/ps/atom	5.4
σ_1, σ_2	0.02
ΔT in K	20,000
$N_{Walkers}$	50

$$n_{coord} = \frac{1 - \left(r_{ij}/r_{cut}\right)^p}{1 - \left(r_{ij}/r_{cut}\right)^q}, \tag{3.36}$$

with $p = 6$, $q = 12$ and $r_{cut} = 3.2$ Å.

Some of the parameters used for the well-tempered MTD are given in Table 3.1. The parameters were tuned during several trial calculations with high deposition rates.

For the sake of computational efficiency, the multiple walker (MW) scheme was applied, in which several independent simulations, i.e. walkers, ran simultaneously and deposited hills to a common shared biasing potential [64]. The efficiency of this scheme scales almost linearly with the number of walkers.

The initialization of the MW MTD plays an important role for an efficient filling of the basins. An optimal spread of the walkers at the starting time generally depends on the FES, which, however, is not known a priori. Furthermore, the walkers should lose memory of their initial positions, before the biasing potential converges. The corresponding relaxation time of the walkers was derived heuristically in [64].

The walkers of this study were initialized by performing short preliminary MTD runs with a small number of walkers and low deposition rate. After achieving sufficiently large separations of the walkers within the amorphous basin in the CV space, the deposition rate was raised gradually and new walkers were added. We employed 50 walkers.

In the following, we restrict the discussion to the 512-atom model. The primary goal of the preliminary simulations presented was to realize fast crystallization of GeTe, not to determine the FES accurately. Hence, the MTD simulation employing the global Q_4^{dot} and n_{coord} as CVs was terminated after the walkers ran for an average time of 200 ps, despite the fact that the FES was not yet fully converged. Some of the walkers led to a full crystallization of the model. The crystallized walkers showed different orientations, not necessarily parallel to the cell axes, such as the structure shown in Fig. 3.9. This structure contains compositional and interstitial defects: hence, its total energy is significantly higher than that of the perfect solid.

Although the FES is not filled completely, its main features can be recognized from the biasing potential collected (Fig. 3.10). An extensive basin associated with the amorphous configurations is readily identified, which is connected with the smaller crystalline basin over a transition region. When the MTD run was terminated, only a fraction of the walkers had crystallized, in the sense that the Q_4^{dot} CV had reached values higher than the selected threshold value of 0.7 (see Figs. 3.8 and 3.10). This fact indicates that the FES is still far from convergence. In fact, to obtain converged values of the FES, all of the walkers must sample the basins of interest sufficiently well and display diffusive motion in the CV space.

The major reason for terminating the MTD was the presence of multiple crystallites during the crystallization of some walkers. This behaviour is in contrast to the case of ice nucleation, where Quigley and Rodger observed the formation of single nuclei when biasing the global order parameters Q_6 (see the Sect. 3.2.2.2). Instead, the use of the global form of Q_4^{dot} seems to be inappropriate for GeTe, at least for models containing up to 512 atoms. As discussed in Sect. 3.2.2.2, the critical solid fraction χ_c, below which multiple crystallites are favoured, increases for decreasing values of γ. Hence, small values of γ make the formation of multiple nuclei favourable (for sufficiently small cell sizes). Apparently, this is the case for GeTe, although the precise value of γ is unknown.

To resolve this problem, we changed the CV Q_4^{dot} to be a local variable, which was computed for a selected subset of atoms in a similar way as in the work by Trudu et al. [32] (see also Sect. 3.2.2.1). The subset was chosen to consist of atoms inscribed in a sphere comprising approximately ~100 atoms. This number is expected to be larger than the critical size at the selected temperature of 600 K. To test this variable, we ran new one-dimensional MTD simulations. Indeed, single nuclei were generated during these simulations (see Fig. 3.11). These models can also be used as starting points for unbiased MD simulations to determine the growth velocity of postcritical crystalline nuclei. Furthermore, a rough estimation of the critical size of the nuclei should be possible, even without filling the FES completely, by using transition path sampling or similar methods.

In summary, we realized the crystallization of models of GeTe via ab initio MTD, starting from a fully amorphous phase. The formation of multiple crystallites was observed if a global Q_4^{dot} CV was used; by restricting the CV to act locally on selected atoms, this problem was overcome. However quantitative results about the FES cannot be extracted from the presented preliminary simulations. Though an acceleration of nucleation events was achieved, more sampling is needed to obtain

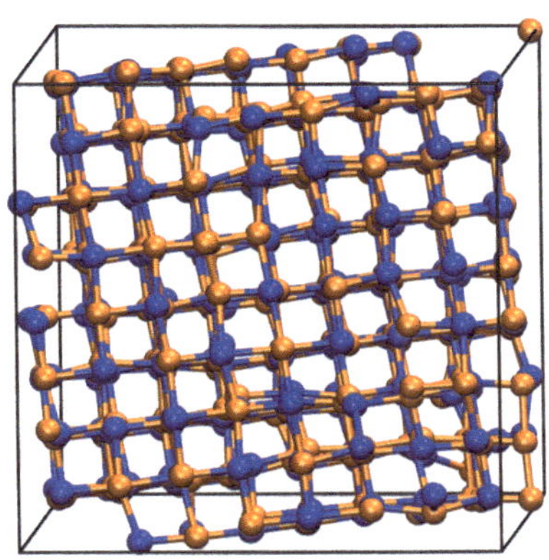

Fig. 3.9 An example of a crystallized walker. Ge (*orange*) and Te (*blue*)

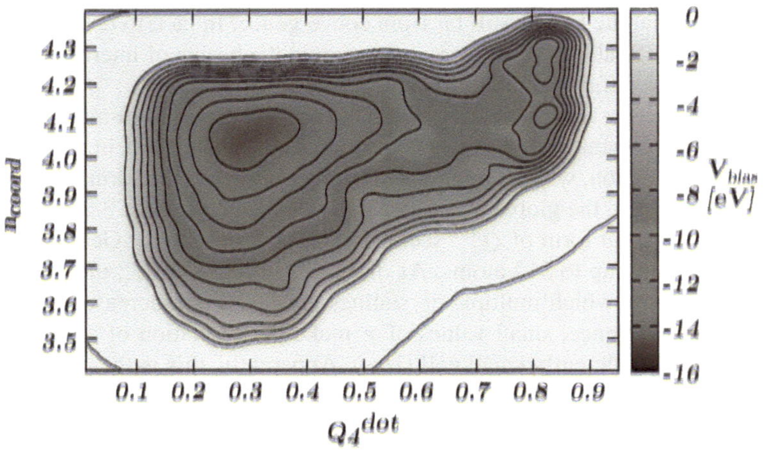

Fig. 3.10 FES in the CV space of Q_l^{dot} and n_{coord} for the GeTe model containing 512 atoms. As discussed in the text, the FES is not converged (in particular, the basin corresponding to the crystalline state is not fully explored) and provides only qualitative information about the two basins corresponding to amorphous and crystalline GeTe

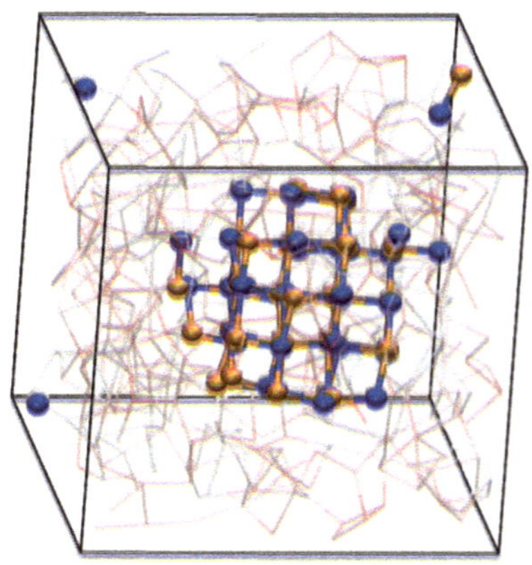

Fig. 3.11 Generation of a single crystalline nucleus by using a local Q_4^{dot} variable

a converged FES. An optimal initialization of the MW simulations requires some foreknowledge about the system and the expected energy scales of the FES. In particular, the starting points of the walkers play an important role, considering the fact that only a small fraction of the walkers explored the basin of the crystalline phase in the FES presented in Fig. 3.10.

The recently developed neural-network potential for GeTe should allow one to reconstruct the FES of GeTe as a function of temperature, at an affordable computational cost. It should also enable a systematic optimization of the CVs and of the parameters of MTD, which could then be employed for ab initio MTD investigations of nucleation in chemically more complex PCMs.

References

1. K. Kelton, A.L. Greer, *Nucleation in Condensed Matter: Applications in Materials and Biology* (Elsevier, Amsterdam, 2010)
2. M. Volmer, A. Weber, Zeitschrift fúr Physikalische Chemie **119**(277), 1 (1926)
3. R. Becker, W. Döring, Annalen der Physik **24**(2), 719 (1935)
4. M. Sprik, G. Ciccotti, J. Chem. Phys. **109**(18), 7737 (1998). http://dx.doi.org/10.1063/1.477419. http://scitation.aip.org/content/aip/journal/jcp/109/18/10.1063/1.477419
5. P. Bash, U. Singh, F. Brown, R. Langridge, P. Kollman, Science **235**(4788), 574 (1987). doi:10.1126/science.3810157. http://www.sciencemag.org/content/235/4788/574.abstract
6. G.N. Patey, J.P. Valleau, J. Chem. Phys. **63**(6), 2334 (1975). http://dx.doi.org/10.1063/1.431685. http://scitation.aip.org/content/aip/journal/jcp/63/6/10.1063/1.431685
7. A. Laio, M. Parrinello, Proc. Natl. Acad. Sci. **99**(20), 12562 (2002). doi:10.1073/pnas.202427399. http://www.pnas.org/content/99/20/12562.abstract
8. A. Laio, F.L. Gervasio, Rep. Progr. Phys. **71**(12), 126601 (2008). http://stacks.iop.org/0034-4885/71/i=12/a=126601
9. A. Barducci, M. Bonomi, M. Parrinello, Wiley Interdiscip. Rev. Comput. Mol. Sci. **1**(5), 826 (2011). doi:10.1002/wcms.31. http://dx.doi.org/10.1002/wcms.31
10. F. Wang, D.P. Landau, Phys. Rev. Lett. **86**, 2050 (2001). doi:10.1103/PhysRevLett.86.2050. http://link.aps.org/doi/10.1103/PhysRevLett.86.2050
11. E. Darve, A. Pohorille, J. Chem. Phys. **115**(20), 9169(2001). http://dx.doi.org/10.1063/1.1410978. http://scitation.aip.org/content/aip/journal/jcp/115/20/10.1063/1.1410978
12. G. Bussi, A. Laio, M. Parrinello, Phys. Rev. Lett. **96**, 090601 (2006). doi:10.1103/PhysRevLett.96.090601. http://link.aps.org/doi/10.1103/PhysRevLett.96.090601
13. A. Laio, A. Rodriguez-Fortea, F.L. Gervasio, M. Ceccarelli, M. Parrinello, J. Phys. Chem. B **109**(14), 6714 (2005). doi:10.1021/jp045424k. http://pubs.acs.org/doi/abs/10.1021/jp045424k
14. A. Barducci, G. Bussi, M. Parrinello, Phys. Rev. Lett. **100**, 020603 (2008). doi:10.1103/PhysRevLett.100.020603. http://link.aps.org/doi/10.1103/PhysRevLett.100.020603
15. D. Branduardi, G. Bussi, M. Parrinello, J. Chem. Theory Comput. **8**(7), 2247 (2012). doi:10.1021/ct3002464. http://pubs.acs.org/doi/abs/10.1021/ct3002464
16. M. Iannuzzi, A. Laio, M. Parrinello, Phys. Rev. Lett. **90**, 238302 (2003). doi:10.1103/PhysRevLett.90.238302. http://link.aps.org/doi/10.1103/PhysRevLett.90.238302
17. B. Ensing, A. Laio, M. Parrinello, M.L. Klein, J. Phys. Chem. B **109**(14), 6676 (2005). doi:10.1021/jp045571i. http://pubs.acs.org/doi/abs/10.1021/jp045571i
18. S. Piana, A. Laio, J. Phys. Chem. B **111**(17), 4553 (2007). doi:10.1021/jp067873l. http://pubs.acs.org/doi/abs/10.1021/jp067873l
19. G.A. Tribello, M. Ceriotti, M. Parrinello, Proc. Natl. Acad. Sci. **107**(41), 17509 (2010). doi:10.1073/pnas.1011511107. http://www.pnas.org/content/107/41/17509.abstract
20. G. Bussi, F.L. Gervasio, A. Laio, M. Parrinello, J. Am. Chem. Soc. **128**(41), 13435 (2006). doi:10.1021/ja062463w. http://pubs.acs.org/doi/abs/10.1021/ja062463w
21. M. Bonomi, D. Branduardi, F.L. Gervasio, M. Parrinello, J. Am. Chem. Soc. **130**(42), 13938 (2008). doi:10.1021/ja803652f. http://pubs.acs.org/doi/abs/10.1021/ja803652f
22. J.J. Hoyt, M. Asta, A. Karma, Phys. Rev. Lett. **86**, 5530 (2001). doi:10.1103/PhysRevLett.86.5530. http://link.aps.org/doi/10.1103/PhysRevLett.86.5530
23. J.Q. Broughton, G.H. Gilmer, J. Chem. Phys. **84**(10), 5759 (1986). http://dx.doi.org/10.1063/1.449884. http://scitation.aip.org/content/aip/journal/jcp/84/10/10.1063/1.449884
24. S. Angioletti-Uberti, M. Ceriotti, P.D. Lee, M.W. Finnis, Phys. Rev. B **81**, 125416 (2010). doi:10.1103/PhysRevB.81.125416. http://link.aps.org/doi/10.1103/PhysRevB.81.125416
25. S. Angioletti-Uberti, J. Phys. Condens. Matter **23**(43), 435008 (2011). http://stacks.iop.org/0953-8984/23/i=43/a=435008
26. P.J. Steinhardt, D.R. Nelson, M. Ronchetti, Phys. Rev. B **28**, 784 (1983). doi:10.1103/PhysRevB.28.784. http://link.aps.org/doi/10.1103/PhysRevB.28.784

27. P.R. ten Wolde, M.J. Ruiz-Montero, D. Frenkel, Faraday Discuss **104**, 93 (1996). doi:10.1039/FD9960400093. http://dx.doi.org/10.1039/FD9960400093
28. R.P. Gupta, Phys. Rev. B **23**, 6265 (1981). doi:10.1103/PhysRevB.23.6265. http://link.aps.org/doi/10.1103/PhysRevB.23.6265
29. F. Cleri, V. Rosato, Phys. Rev. B **48**, 22 (1993). doi:10.1103/PhysRevB.48.22. http://link.aps.org/doi/10.1103/PhysRevB.48.22
30. M.W. Finnis, J.E. Sinclair, Philos. Mag. A **50**(1), 45 (1984). doi:10.1080/01418618408244210. http://www.tandfonline.com/doi/abs/10.1080/01418618408244210
31. A. Landa, P. Wynblatt, A. Girshick, V. Vitek, A. Ruban, H. Skriver, Acta Materialia **46**(9), 3027 (1998). http://dx.doi.org/10.1016/S1359-6454(97)00496-5. http://www.sciencedirect.com/science/article/pii/S1359645497004965
32. F. Trudu, D. Donadio, M. Parrinello, Phys. Rev. Lett. **97**, 105701 (2006). doi:10.1103/PhysRevLett.97.105701. http://link.aps.org/doi/10.1103/PhysRevLett.97.105701
33. P.G. Bolhuis, C. Dellago, D. Chandler, Faraday Discuss **110**, 421 (1998). doi:10.1039/A801266K. http://dx.doi.org/10.1039/A801266K
34. D. Quigley, P.M. Rodger, J. Chem. Phys. **128**(15),154518 (2008). http://dx.doi.org/10.1063/1.2888999. http://scitation.aip.org/content/aip/journal/jcp/128/15/10.1063/1.2888999
35. W.L. Jorgensen, J. Chandrasekhar, J.D. Madura, R.W. Impey, M.L. Klein, J. Chem. Phys. **79**(2), 926 (1983). http://dx.doi.org/10.1063/1.445869. http://scitation.aip.org/content/aip/journal/jcp/79/2/10.1063/1.445869
36. D. Quigley, P. Rodger, Mol. Simul. **35**(7), 613 (2009). doi:10.1080/08927020802647280. http://www.tandfonline.com/doi/abs/10.1080/08927020802647280
37. F. Giberti, G.A. Tribello, M. Parrinello, J. Chem. Theory Comput. **9**(6), 2526 (2013). doi:10.1021/ct4002027. http://pubs.acs.org/doi/abs/10.1021/ct4002027
38. C. Oostenbrink, A. Villa, A.E. Mark, W.F. Van Gunsteren, J. Comput. Chem. **25**(13), 1656 (2004). doi:10.1002/jcc.20090. http://dx.doi.org/10.1002/jcc.20090
39. H.J.C. Berendsen, J.R. Grigera, T.P. Straatsma, J. Phys. Chem. **91**(24), 6269 (1987). doi:10.1021/j100308a038. http://pubs.acs.org/doi/abs/10.1021/j100308a038
40. R.M. Martin, *Electronic Structure: Basic Theory and Practical Methods* (Cambridge University Press, Cambridge, 2004)
41. P. Hohenberg, W. Kohn, Phys. Rev. **136**, B864 (1964). doi:10.1103/PhysRev.136.B864. http://link.aps.org/doi/10.1103/PhysRev.136.B864
42. W. Kohn, L.J. Sham, Phys. Rev. **140**, A1133 (1965). doi:10.1103/PhysRev.140.A1133. http://link.aps.org/doi/10.1103/PhysRev.140.A1133
43. G. Kresse, J. Hafner, Phys. Rev. B **47**, 558 (1993). doi:10.1103/PhysRevB.47.558. http://link.aps.org/doi/10.1103/PhysRevB.47.558
44. R. Car, M. Parrinello, Phys. Rev. Lett. **55**, 2471 (1985). doi:10.1103/PhysRevLett.55.2471. http://link.aps.org/doi/10.1103/PhysRevLett.55.2471
45. T.D. Kühne, M. Krack, F.R. Mohamed, M. Parrinello, Phys. Rev. Lett. **98**, 066401 (2007). doi:10.1103/PhysRevLett.98.066401. http://link.aps.org/doi/10.1103/PhysRevLett.98.066401
46. M. Wuttig, N. Yamada, Nat. Mater. **6**(11), 824 (2007). http://dx.doi.org/10.1038/nmat2009
47. S. Raoux, W. Wełnic, D. Ielmini. Chem. Rev. **110**(1), 240 (2010). doi:10.1021/cr900040x. http://pubs.acs.org/doi/abs/10.1021/cr900040x
48. D. Lencer, M. Salinga, M. Wuttig, Adv. Mater. **23**(18), 2030 (2011). doi:10.1002/adma.201004255. http://dx.doi.org/10.1002/adma.201004255
49. J. Hegedüs, S.R. Elliott, Nat. Mater. **7**(5), 399 (2008). http://dx.doi.org/10.1038/nmat2157
50. T.H. Lee, S.R. Elliott, Phys. Rev. Lett. **107**, 145702 (2011). doi:10.1103/PhysRevLett.107.145702. http://link.aps.org/doi/10.1103/PhysRevLett.107.145702
51. J. Kalikka, J. Akola, J. Larrucea, R.O. Jones, Phys. Rev. B **86**, 144113 (2012). doi:10.1103/PhysRevB.86.144113. http://link.aps.org/doi/10.1103/PhysRevB.86.144113
52. J. Akola, R.O. Jones, Phys. Rev. B **76**, 235201 (2007). doi:10.1103/PhysRevB.76.235201. http://link.aps.org/doi/10.1103/PhysRevB.76.235201

53. S. Caravati, M. Bernasconi, T.D. Kühne, M. Krack, M. Parrinello, Appl. Phys. Lett. **91**(17), 171906 (2007). http://dx.doi.org/10.1063/1.2801626. http://scitation.aip.org/content/aip/journal/apl/91/17/10.1063/1.2801626
54. W. Zhang, I. Ronneberger, Y. Li, R. Mazzarello, Adv. Mater. **24**(32), 4387 (2012). doi:10.1002/adma.201201507. http://dx.doi.org/10.1002/adma.201201507
55. J.L.F. Da Silva, A. Walsh, H. Lee, Phys. Rev. B **78**, 224111 (2008). doi:10.1103/PhysRevB.78.224111. http://link.aps.org/doi/10.1103/PhysRevB.78.224111
56. D. Lencer, M. Salinga, B. Grabowski, T. Hickel, J. Neugebauer, M. Wuttig, Nat. Mater. **7**(12), 972 (2008). http://dx.doi.org/10.1038/nmat2330
57. K. Shportko, S. Kremers, M. Woda, D. Lencer, J. Robertson, M. Wuttig, Nat. Mater. **7**(8), 653 (2008). http://dx.doi.org/10.1038/nmat2226
58. W. Zhang, A. Thiess, P. Zalden, R. Zeller, P.H. Dederichs, J.Y. Raty, M. Wuttig, S. Blügel, R. Mazzarello, Nat. Mater. **11**(11), 952 (2012). http://dx.doi.org/10.1038/nmat3456
59. J. VandeVondele, M. Krack, F. Mohamed, M. Parrinello, T. Chassaing, J. Hutter, Comput. Phys. Commun. **167**(2), 103 (2005). http://dx.doi.org/10.1016/j.cpc.2004.12.014. http://www.sciencedirect.com/science/article/pii/S0010465505000615
60. G.C. Sosso, G. Miceli, S. Caravati, J. Behler, M. Bernasconi, Phys. Rev. B **85**, 174103 (2012). doi:10.1103/PhysRevB.85.174103. http://link.aps.org/doi/10.1103/PhysRevB.85.174103
61. D.B. Dove, M.B. Heritage, K.L. Chopra, S.K. Bahl, Appl. Phys. Lett. **16**(3), 138 (1970). http://dx.doi.org/10.1063/1.1653128. http://scitation.aip.org/content/aip/journal/apl/16/3/10.1063/1.1653128
62. T. Chattopadhyay, J.X. Boucherle, H.G. vonSchnering, J. Phys. C Solid State Phys. **20**(10), 1431 (1987). http://stacks.iop.org/0022-3719/20/i=10/a=012
63. G.C. Sosso, J. Behler, M. Bernasconi, Physica Status Solidi (B) **249**(10), 1880 (2012). doi:10.1002/pssb.201200355. http://dx.doi.org/10.1002/pssb.201200355
64. P. Raiteri, A. Laio, F.L. Gervasio, C. Micheletti, M. Parrinello, J. Phys. Chem. B **110**(8), 3533 (2006). doi:10.1021/jp054359r. http://pubs.acs.org/doi/abs/10.1021/jp054359r

Chapter 4
Challenges in Modeling Mixed Ionic-Covalent Glass Formers

Liping Huang and John Kieffer

Abstract Archetypical glass formers such as SiO_2, GeO_2 and B_2O_3 pose an especial challenge for atomistic level modeling due to the mixed ionic-covalent bonding and the highly polarizable oxygen ion. Though significant improvements have been made in the past few decades in developing potential models for such systems, mostly based on pair-wise potentials, with or without taking into account of three-body or many-body effects, there is still much room for further advancement in the development of reliable, effective, and transferable potential models for mixed ionic-covalent glass formers.

4.1 Introduction

Atomistic simulation and modeling of glass is a challenging but effective method to investigate its structure and properties. Among various challenges we are facing, developing accurate potential models to describe the interactions in glass is one of the most difficult tasks. The first step to develop such a potential is to construct a functional form for modeling the interactions between ions without taking explicitly into account the electronic degrees of freedom. Archetypical glass formers such as SiO_2, GeO_2 and B_2O_3 are of mixed ionic-covalent nature in bonding. Furthermore, the oxygen ion with a lone pair of electrons is highly polarizable [1]. Therefore, realistic potential models for such systems may often need to have: (1) two-body terms, to describe the ionic character of the bond, including the long-range electrostatic interactions, the short-range exchange-repulsion as a consequence of the Pauli principle, and the dispersion due to correlated fluctuations of electrons; (2) three-body

L. Huang (✉)
Department of Materials Science and Engineering, Rensselaer Polytechnic Institute, MRC 202,110 8th Street, Troy, NY 12180, USA
e-mail: huangL5@rpi.edu

J. Kieffer
Department of Materials Science and Engineering, University of Michigan, 2018 H.H. Dow Building, 2300 Hayward Street, Ann Arbor, MI 48109, USA
e-mail: kieffer@umich.edu

C. Massobrio et al. (eds.), *Molecular Dynamics Simulations of Disordered Materials*, Springer Series in Materials Science 215,
DOI 10.1007/978-3-319-15675-0_4

terms, mimicking covalent interactions to constrain O–M–O and M–O–M bond angle distribution (where M is a network former ion); (3) many-body terms, arising from the polarizability of the oxygen ion in response to the local electrostatic field. Many attempts have been made to develop effective potentials with acceptable accuracy and computational efficiency by selectively including some of the above mentioned terms. Along the line of reducing computational cost, various short-ranged versions of potential models with screened and/or truncated interactions have been developed [2–6], which require modest computational resources without compromising the force field accuracy. Balancing model accuracy and computational efficiency is one of the key elements to developing force fields. Another important aspect of the potential development is how to parameterize the model. Most of the potential models used for the simulations of glass systems were parameterized from fitting to relevant crystalline or mineral systems, and/or potential energy surface of small clusters from quantum mechanical calculations. Recently, matching force, stress and energy of condensed systems from first principles calculations [7] or fitting structures from Car-Parrinello molecular-dynamics (CPMD) simulations [8] have been used for potential parameterization. In the following, we will briefly review the efforts in the past few decades that have been devoted to the development of force fields for mixed ionic-covalent glass formers, with a focus on the functional form and the parameterization procedure for systems containing SiO_2, GeO_2 and B_2O_3. As a touchstone for the veracity of a potential model is its ability to reproduce the vibrational dynamics of the physical system, we will compare the performance of force fields based the vibrational density of states (VDOS), infrared (IR) or Raman spectra, if available.

4.2 Functional Form

4.2.1 Two-Body Interaction

In atomistic simulations, the ionic limit is normally represented by a pair-wise potential consisting of a Coulombic term to describe the long-range electrostatic interaction between point charges, a short-range Born–Mayer–Huggins (BMH) exponential term to approximate the overlap repulsion between localized orbitals on nearby atoms, and an attractive dispersion term. For mixed ionic-covalent glass formers such as SiO_2, GeO_2 and B_2O_3, the following form is commonly used:

$$\phi_{ij} = \frac{q_i q_j}{r_{ij}} + A_{ij} e^{-r_{ij}/\rho_{ij}} - \frac{C_{ij}}{r_{ij}^6} \tag{4.1}$$

where ϕ_{ij} is the interaction energy between ions i and j, r_{ij} is the distance between ions i and j, q_i is the full or partial charge on ion i that plays the role of a parameter, and A_{ij}, ρ_{ij}, and C_{ij} are free parameters. Among others, the TTAM [9], BKS [10] and CHIK [8] potentials for SiO_2, Tsuchiya's [11] and Oeffner and Elliott's (OE) potential [12] for GeO_2, as well as Takada's potential [13] for B_2O_3 are of the general form of (4.1). Much work has been done to improve upon this simple BMH form of the potential. Other analytical functions such as Morse potentials were often used to replace (such as Takada [14], Stixrude [15], Demiralp [16], Tangney [7] for SiO_2, Hoang [17] for GeO_2 and Takada [13] for B_2O_3) or supplement (Yamahara [18] for SiO_2) the short-range repulsion in (4.1). In some cases, the dispersion term was not included, such as the potential used by Woodcock [19] for SiO_2, Karthikeyan [20] for GeO_2, Verhoef [21] and Soules [22] for B_2O_3. For silica, a variety of potential models have been proposed in the literature. The original pair-wise transferable rigid ion model (TRIM) potential [19, 23] and its modifications [24–27] are based on full charges, and generally give elastic moduli significantly larger than the experimental values. This is also shared by other pair-wise potentials using formal charges [28–30]. Later on pair-wise additive models, such as the TTAM [9, 31], BKS [10], CHIK [8] and Takada's [14] potentials for SiO_2, OE potential [12] for GeO_2 and some of Verhoef's [21] and Takada's [13] models for B_2O_3, all used partial charges, which significantly improved the accuracy of these models in terms of reproducing their structure and properties. This was based on the idea that the reorganization of the electronic density via charge transfer and polarization effects can be modeled by using effective charges instead of formal charges. The performance of TTAM and BKS potentials has been widely tested and critically evaluated by many authors [32–44]. Both TTAM and BKS potentials can reproduce the structure and many thermomechanical properties of various silica polymorphs, silica liquid and glass fairly well. It seems that both potentials overestimate the inter-tetrahedral angle, indicating that the Si–O–Si bending interaction is too weak. This can be easily understood because there is no term in these potential models to account for the lone pair of electrons on the oxygen, which is responsible for the bent Si–O–Si configuration. Both potentials also give a poor representation of the vibrational density of states (VDOS). As seen in Fig. 4.1, the calculated stretching frequencies in TTAM are too low, which are improved significantly in BKS (Fig. 4.2). The VDOS calculated from TTAM and BKS model is not very reliable at intermediate frequencies, compared to experiments and ab initio calculations. A systematic study by Hemmati and Angell [45, 46] has shown that none of the pair-wise potentials available at the time was capable of satisfactorily reproducing the short-time dynamic properties (e.g., IR spectra) [45] and long-time dynamic properties (e.g., diffusivities) [46] of SiO_2. Soules et al. [41] extensively tested a number of pair-wise force fields (such as BKS [10], CHIK [8], Soules' [6] and Takada's [14]) and found in all cases the activation energies for diffusion are lower than the experimental activation energies for viscosity. In short, pair-wise potentials can give very good equilibrium structures. However, it appears that dynamical properties, such as VDOS and IR spectra, can only be accurately reproduced by using more sophisticated models that include the covalent bonding and/or polarization effects more precisely.

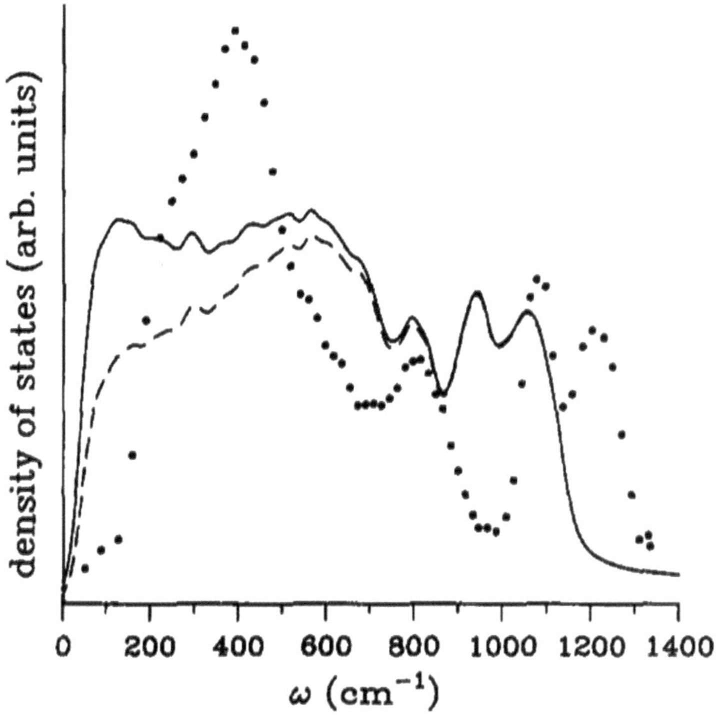

Fig. 4.1 Velocity autocorrelation function (calculation from TTAM: *continuous line*) and VDOS (experiments: points; calculation from TTAM: *broken line*). Figure taken from [34]

4.2.2 Three-Body Interaction

Simulations based on simple pair-wise potentials, especially the more recent approaches that assume partial charges for individual species, are quite successful in reproducing structural characteristics. The absence of directional forces tends to give higher atomic mobility and therefore a more relaxed glass structure. For the same reason pair-wise models often predict premature transitions between low- and higher-coordinated structural units upon compression. Later three-body terms were added and parameterized to constrain O–M–O and M–O–M bond angle distributions so that these angles agree more closely with the experimental values. A variety of three-body potentials have been described in the literature, such as Sanders [48], Catlow [49], Lasaga and Gibbs [30], Stixrude [15], Vashishta [50], Feuston and Garofalini [4], Huang and Kieffer's [51] potentials for SiO_2, Nanba's [52] potential for GeO_2, Verhoef [21], Takada [13], Huang and Kieffer's [53] potentials for B_2O_3. Huang and Kieffer's three-body potentials [51, 53] for SiO_2 and B_2O_3 are based on the same general formula, differ from others mostly in the functional forms chosen to describe the covalent attractive interactions

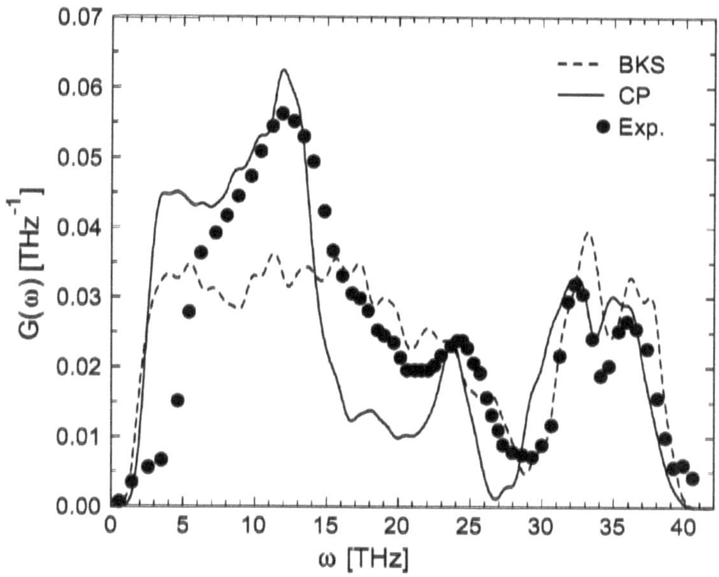

Fig. 4.2 VDOS calculated from ab initio (*solid line*) and BKS (*dashed line*), compared to neutron scattering experiments (*filled circles*). Figure taken from [47]

and orientational constraints, e.g., the angular restoring forces within M–O–M and O–M–O groups are symmetric with respect to the equilibrium bond angles. Generally speaking, three-body potentials improve the structural, elastic and dielectric properties over pair-wise potentials. However, Taraskin and Elliott [54] used Feuston and Garofalini's [4] three-body potential for SiO_2 and found out that it gives relatively poor agreement with the experimental dynamical structure factor. There is no obvious double-peak feature in the VDOS calculated from Vashishta's three-body potential [50] (Fig. 4.3) and Sander's shell model with three-body interactions [48] (Fig. 4.4c). Based on the above observations, it seems rather disappointing that more complicated potentials including three-body interactions and/or polarizability effects do not necessarily provide a more realistic description of a physical system. It is not clear this is a consequence of the parameterization or due to the form of the force field.

4.2.3 Dynamic Charge Transfer

Electronic densities of anions, in particular, are strongly affected by their interactions with their environment. So charges on ions, and change of the character of the bonding from covalent to ionic need to be dynamically determined based on the immediate atomic environment during MD simulations. Demiralp, Cagin, and Goddard (DCG) developed a fluctuating-charge potential in which charges are allowed to re-adjust

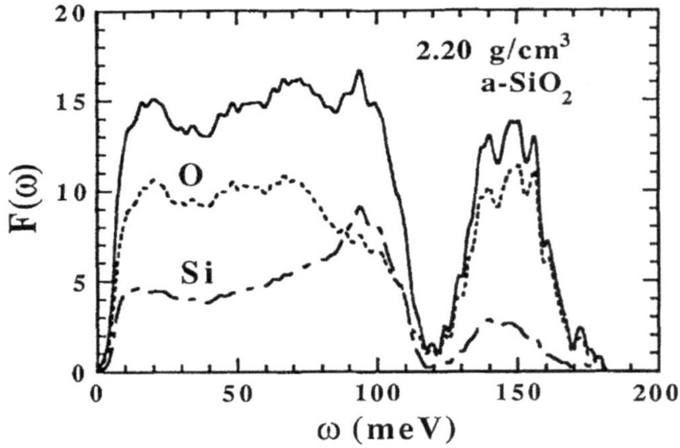

Fig. 4.3 Partial and total VDOS for silica glass calculated from Vashishta's three-body potential. Figure taken from [50]

instantaneously to the atomic configurations [16]. These charges are calculated using the charge equilibration (QEq) procedure developed by Rappé and Goddard [57]. In addition to electrostatics, a two-body Morse stretch potential was included to account for short-range non-electrostatic interactions. This MS-Q potential was applied to SiO_2, where it can describe well the fourfold coordinated and sixfold coordinated systems (such as quartz and stishovite), silica glass, and the pressure-induced phase transition from quartz to stishovite. Herzbach et al. [42] systemically studied the structural, thermomechanical, and dynamic properties of SiO_2 by using the DCG fluctuating-charge potential [16], the polarizable force field proposed by Tangney and Scandolo (TS) [7], the pair-wise BKS potential [10] and ab initio calculations. As seen in Fig. 4.5, the fluctuating-charge model coincides slightly better than the BKS model with ab initio data at the lower frequencies of VDOS of α-quartz. The peaks at 11 and 22 THz are reasonably well reproduced and the overall VDOS is similar to the curve from ab initio calculations. But the two peaks of the intra-tetrahedral modes collapse on just a single peak though exist at roughly the right frequency [42].

By treating charge as a variable that changes with time (like atomic coordinates do) during MD simulations, Yasukawa extended the bond-order method of Tersoff to the Si/SiO_2 system [58] by including self-consistent charge determination and an electrostatic term, following the Rappé and Goddard approach [57]. Built upon the Yasukawa's potential, the first and second generation of charge-optimized many-body (COMB) potential for Si/SiO_2 system were developed [59, 60], which take into account the effect of charge transfer using both the electronegativity equalization principle and many-body interactions. The second generation of COMB potential can describe very well the structure and mechanical properties of several crystalline silica polymorphs and amorphous silica [59].

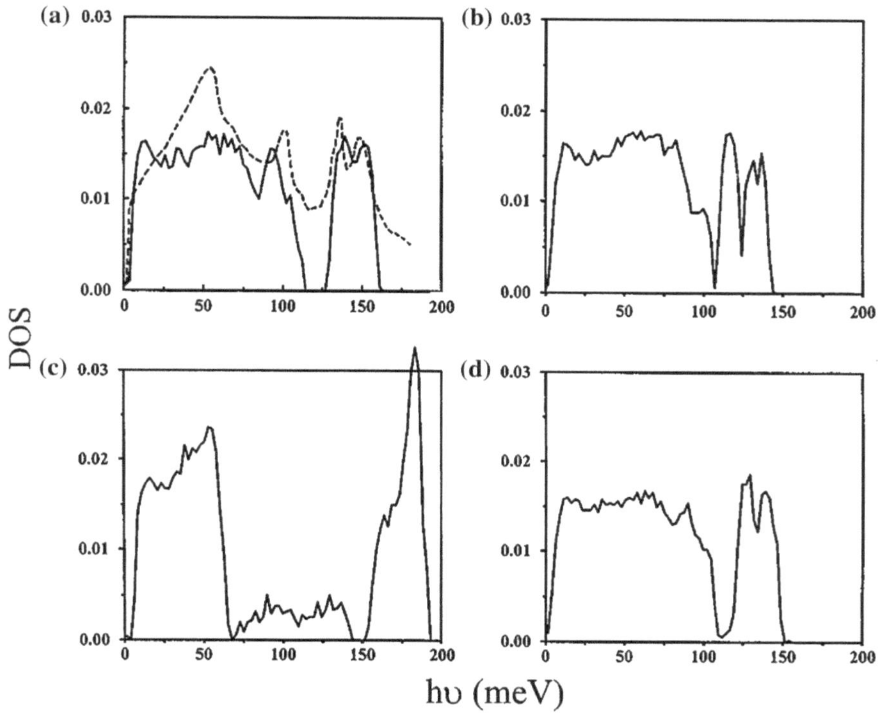

Fig. 4.4 VDOS calculated from **a** a two-body potential (VB) derived by Kramer et al. on a quantum-chemical calculation of an H_4SiO_4 cluster [55], **b** a two-body potential (TS) derived by Tsuneyuki et al. using a Hartree-Fock calculation on SiO_4^{-4} cluster [9], **c** a three-body (3B) potential by Sander, Leslie, and Catlow with a shell-model description and with three-body interactions [48]; and **d** a two-body potential (KR) proposed by Kramer et al. using a mixed self-consistent field and empirical procedures [10]. *Note* the KR potential here is the BKS potential. Figure taken from [56]

In our charge-transfer three-body potentials [51, 53], the effective charges on ions are dynamically determined based on the immediate atomic environment during MD simulations. This feature can accommodate the drastic charge redistribution associated with the breaking and formation of bonds. But even within an intact bond, subtle charge fluctuations resulting from bond stretching vibrations can be accounted for with this functional. The refinement of this feature constitutes the principal improvement over the initial form of this interaction model [61, 62], and is key to a better description of the properties of silica. Using this potential, we can simulate the α-to-β transformations in various crystalline forms of silica without need to change potential parameters [51, 63]. Our study also showed that the calculated IR spectra across the α-to-β cristobalite transformation are in good agreement with the experimental ones, both in terms of spectra peak positions and their relative intensities [51]. Comparing the VDOS of silica glass calculated from our charge-transfer three-body potential (Fig. 4.6a) with that from experiments [64, 65] (Fig. 4.6b), an excellent agreement in the peak position for all three bands (around 400, 800 and 1100 cm^{-1})

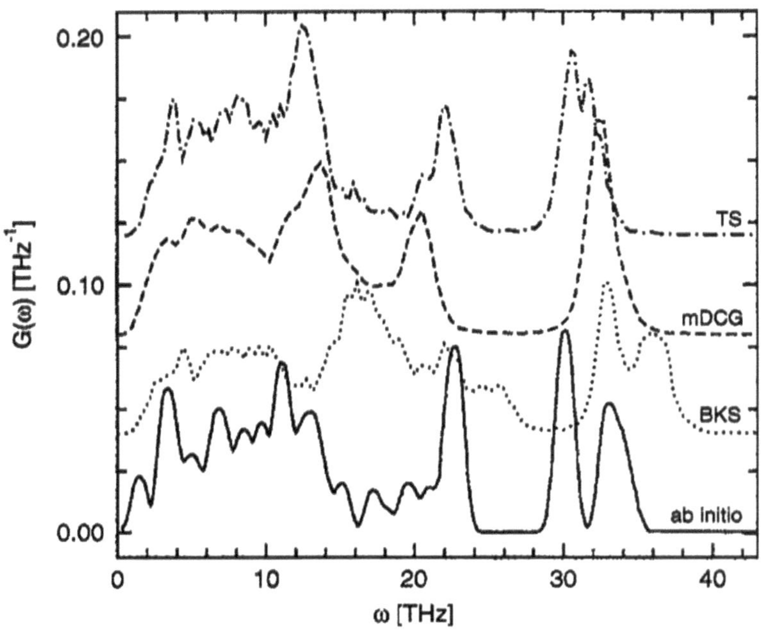

Fig. 4.5 VDOS of α-quartz at 300 K calculated from BKS, mDCG, TS and ab initio. For clarity the curves for BKS, mDCG and TS have been shifted upwards. *Note* mDCG is a modified version of the energy surface proposed by Demiralp, Cagin, and Goddard (DCG). Figure taken from [42]

is found. Most importantly, in agreement with experiments, we see more features in the low and intermediate range of frequencies. Most other models for silica glass only give the high frequency peak and a broad, relatively featureless spectral band in the intermediate and low range of frequencies.

4.2.4 *Polarizability Effect*

In earlier studies, the polarizability effect was described by using the shell model in which an ion is described as comprising a massless shell of certain charge and a core in which the mass is concentrated, a harmonic spring connects the shell with the core [48, 49]. Recently, many-body potentials based on the polarizable ion model have been developed for SiO_2, GeO_2 and B_2O_3 [1, 7, 66, 67]. In all of these models, the oxygen atoms are treated as polarizable ions. The polarizable ion model allows for a more realistic representation of dipole induction by both Coulombic and short-range interactions than the shell model so that the polarization effects may be parameterized unambiguously by ab initio electronic structure calculations, rather than empirically. As a result, numerous unusual structural properties, which

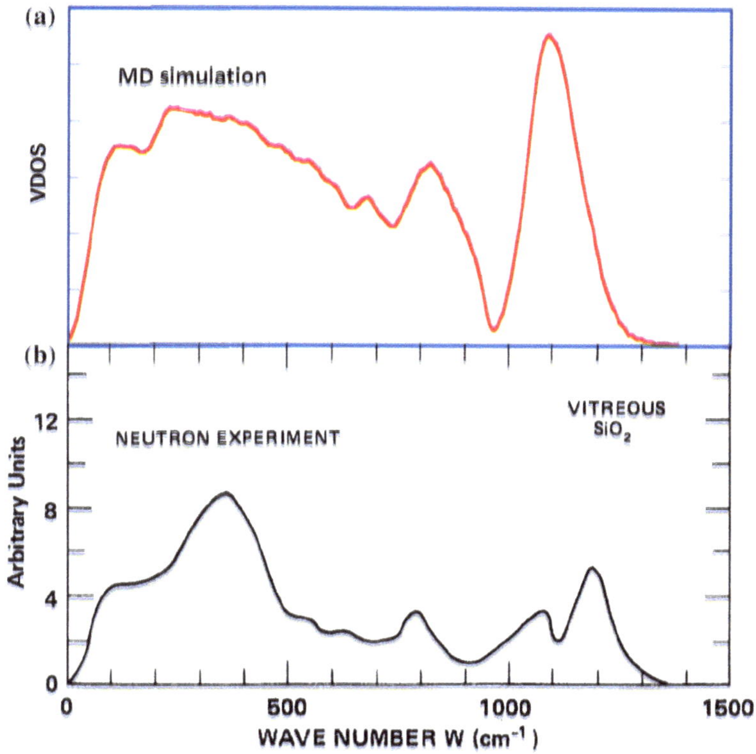

Fig. 4.6 VDOS of silica glass from **a** Huang and Kieffer's charge-transfer three-body potential, **b** Neutron scattering experiments [65]

had previously been attributed to "covalent effects", have been shown to be a consequence of polarization effects within a wholly ionic model [1]. It has been shown that the inclusion of a realistic description of polarization effects in essentially ionic models (with full formal charges) of several network-forming materials can correct the M–O–M bond-angle distribution [68, 69]. The importance of polarization for the relative energies of different crystalline polymorphs of SiO_2 and for their elastic constants has already been demonstrated [70, 71]. The addition of the many-body polarizability effects was found necessary to obtain a correct description of IR spectra of amorphous SiO_2 and GeO_2 as seen in Fig. 4.7. The many-body polarization effects have also been shown to be essential for stabilizing boron atoms in boroxol rings in B_2O_3 glass [66], producing a maximum ring fraction of ~33 %, better than other models without the polarization effect, although more than 60 % boroxol rings have been observed by various experimental techniques [72–77].

Following Wilson and Madden's approach [1], Tangney and Scandolo (TS) used a more flexible form for the potential which incorporates the effects of dipole polarization of the oxygen ions including the effective dipoles induced by the short-range

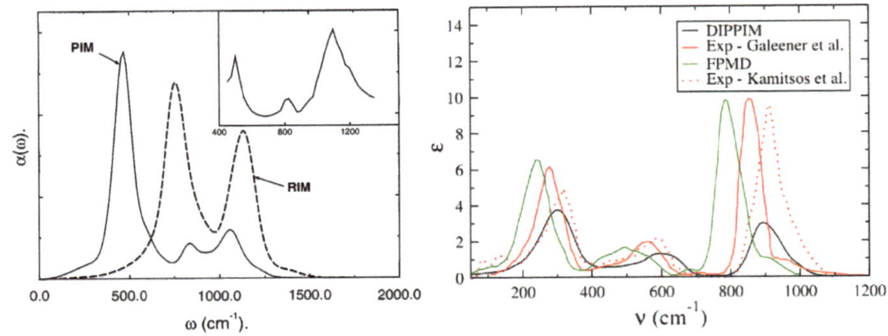

Fig. 4.7 *Left* The IR absorption spectrum (arbitrary units) calculated for models of amorphous SiO_2 in which a simple pair-wise potential is used (RIM—*dashed line*) and when this potential is supplemented with an account of polarization effects (PIM—*solid line*) [1]. *Right* Imaginary part of the dielectric function calculated from the DIPole-Polarizable Ion Model (DIPPIM) potential [67], compared to first principles molecular dynamics (FPMD) [78] and experimental results [65, 79]. Figures taken from [1, 67]

forces between ions [7]. The TS potential has proven very successful in reproducing many of the structural and dynamical properties (IR and Raman spectra) of several crystalline phases of silica [80]. As seen in Fig. 4.5, VDOS of α–quartz silica from the TS model almost coincides with the ab initio result, only the tetrahedral breathing mode is slightly shifted in frequency. It has also been shown that the TS potential remedies all of the qualitative failures of the BKS potential, including the c/a ratio anomaly at the α–β transition in quartz, the stability of cristobalite and tridymite [42]. However, when the TS model was used to construct amorphous silica structures via simulated annealing, it produced a large number of anomalous two-membered rings [81]. Two-membered rings are formed by edge-sharing tetrahedra and are often observed as defect sites on silica surfaces [82]. However, there is no experimental evidence and no other reported simulations that support their existence in the bulk silica at room temperature. Failure of the TS model to anneal properly could indicate that the polarizable model may need to be augmented with more complex features, such as the ability to undergo charge transfer. Another disadvantage of the TS model is its low efficiency. Due to the expensive self-consistent computations of the dipole moments on the oxygen atoms at every time step, up to two orders of magnitude more computer time is needed in MD simulations with the TS model than with a simple pair-wise potential such as the BKS model [42].

4.2.5 Reactive Force Field

Usually reactive events such as bond breaking and formation can be only described by quantum mechanical (QM) calculations. To circumvent the limited size ($\sim 10^2$ atoms) accessible by QM, the hybridization of different methodologies, such as quantum mechanical/molecular mechanics (QM/MM) schemes, have been employed to do QM calculations near the reactive sites, and MM calculations in the far field [83]. However, they require adaptive selection of reactive sites during simulations and sophisticated treatment of QM/MM boundaries, and neither of them is trivial. An alternative approach is to incorporate chemical reactions into MD simulations by describing essential chemical reactions using semi-empirical force fields, in which interatomic potentials are parameterized to reproduce datasets obtained with accurate QM calculations. Recent advances in reactive force fields like ReaxFF, REBO and COMB potentials have opened up a possibility to study reactive processes in a wide range of materials such as hydrocarbons, oxides, nitramines and polymers on very large systems (million-to-billion atom MD simulations) [59, 60, 84–90]. The ReaxFFSiO reactive force field for Si/SiO_2 systems [84] have been developed based on the ReaxFFCH reactive force field for hydrocarbons. ReaxFFSiO has been tested against a substantial data set derived from quantum chemical calculations on small clusters and on condensed systems covering both reactive and nonreactive aspects of silicon oxides. The COMB potentials for Si/SiO_2 system were developed by fitting to the structural, mechanical and energetic information of α-quartz obtained from experiments and first principles calculations [59, 60].

Potential models containing three-body terms usually discriminate in favor of a pair of bonds with desired geometry emanating from the vertex, which makes bond breaking and formation difficult to model in MD simulations. To simulate systems like SiO_2, GeO_2 and B_2O_3 with mixed ionic-covalent interactions, in which the structural building block can exhibit multiple coordination states, it is essential to use a potential model that allows for smooth coordination changes during simulations. Nanba et al. developed a multi-body potential function for germanate glasses, in which two energy minima in the angular term for a given three-atom unit were introduced to permit a transformation between different coordination states, such as four- and sixfold coordinated Ge sites [52]. To model mixed ionic-covalent glass formers, we have developed coordination-dependent charge-transfer three-body potentials [53] that have the following features: (1) *dynamic charge redistribution*—a charge transfer term controls the extent of charge polarization in a covalent bond, as well as the amount of charge transferred between atoms upon rupture or formation of such a bond; (2) *conditional three-body interactions*—the directional character of the covalent bonding is coupled to the degree of covalency in atomic interactions and vice versa; (3) *variable coordination number*—both two- and three-body interactions depend on the effective number of nearest neighbors of an atom, which is evaluated dynamically based on the local environment of this atom. To help understand the physics behind this potential model, we will briefly re-iterate its functional formula here:

$$\phi_i = q_i \sum_{j=1}^{N} \frac{q_j}{4\pi\varepsilon_0 r_{ij}} + \sum_{j=1}^{NC} C_{ij} e^{(\sigma_i+\sigma_j-r_{ij})\cdot\rho_{ij}} + \sum_{j=1}^{NC-1} \sum_{k=j+1}^{NC} (\varphi_{ij} + \varphi_{ik})\Lambda_{ijk} \quad (4.2)$$

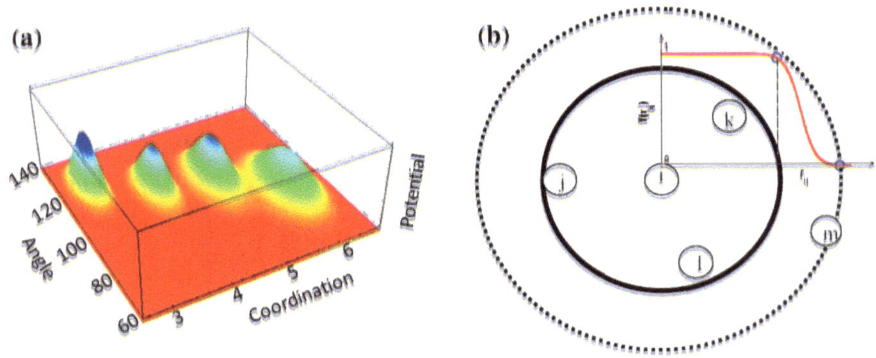

Fig. 4.8 **a** Graphical representation of the three-body potential term as a function of the angle between adjacent bonds and the effective coordination number. Negative energies are plotted upwards for better visualization; **b** sketch of the neighbor-cutoff spheres and the function to determine the effective coordination number Z_i of the central particle i

where ε_0 is the dielectric constant of vacuum, r_{ij} is the interatomic distance, q_i is the charge on atom i. σ_i is a measure of the size of the atom i, and ρ_{ij} describes the hardness in the repulsion between atoms with overlapping electron orbitals. The three-body term accounts for the directionality in covalent bonds. It comprises purely attractive terms $\varphi_{ij} = -C_{ij}\frac{\kappa_{ij}}{\eta_{ij}}\zeta_{ij}e^{(\lambda_{ij}-r_{ij})\eta_{ij}}$ and $\varphi_{ik} = -C_{ik}\frac{\kappa_{ik}}{\eta_{ik}}\zeta_{ik}e^{(\lambda_{ik}-r_{ik})\eta_{ik}}$, that act in radial direction between pairs of bonded atoms, and an angular constraint term $\Lambda_{ijk} = e^{-\gamma_{ijk}(\bar{\theta}-\theta_{ijk})^2}$, whose magnitude and equilibrium angle depend on the effective coordination number of the central atom (Fig. 4.8a). To determine the effective coordination number we define two concentric spherical regions around each particle (Fig. 4.8b). Atoms within the inside sphere are considered full neighbors of the central atom; those located in between the inner and outer sphere are considered partial neighbors, and those outside of the outer sphere are not included as neighbors. The effective coordination number Z_i of particle i is then calculated according to

$$Z_i = \sum_{j \neq i}^{NC} f\left(\left|r_{ij}\right|\right), \tag{4.3}$$

where NC is the total number of atoms contained within the outers cutoff sphere, and $f\left(\left|r_{ij}\right|\right) = \left(1 + e^{\left(\left|r_{ij}\right| - a_Z\right)b_Z}\right)^{-1}$ is a function that varies continuously between 1 at short distance and 0 at large distance and describes the contribution that each particle makes towards the effective coordination number. The parameters a_Z and b_Z are adjusted so that this function reaches the two limiting values at the radius of the inner and outer coordination region, respectively. Accordingly, the effective coordination number may be non-integer. When an atom m is approaching the central atom i, which is already surround by three other atoms j, k, and l, either a bond exchange or a coordination change may take place. During such changeovers, the energy of the configuration must vary continuously, and we accomplish this by constructing the energy term for the angular part of the potential to not only depend on the angle formed by a given the triplet $\langle ijk \rangle$, but also on the effective coordination number Z_i of the central atom, i.e.,

$$\Lambda_{ijk} = \sum_{Z_0=3}^{6} C_{Z_0} e^{-A_{Z_0}(Z_0-Z_i)^2} e^{-\gamma_{Z_0}(\bar{\theta}_{Z_0}-\theta_{ijk})^2}, \tag{4.4}$$

where θ_{Z_0} is the equilibrium angle of the triplet in one of the fundamental coordination environments, for example, three- and four-coordinated species will tend to form sp^2 and sp^3 hybridized bonds with equilibrium angle $\bar{\theta}_3 = 120°$ and $\bar{\theta}_3 = 109.5°$, respectively. C_{Z_0}, A_{Z_0} and γ_{Z_0} control the depth of the potential well, the coordination-dependent width, and the angle-dependent width of the three-body potential, respectively. These three parameters can be adjusted so that the depth and widths of the potential well are the same for all possible coordination states, or certain coordination states could be favored over others. Λ_{ijk} and its first and second derivatives are continuous, which allow atoms to smoothly make a transition from one coordination environment to another.

As a result of breaking or forming covalent bonds, charge transfer takes place between the central atom and all NC atoms within the outer radius of the coordination environment. The net charge associated with atom i is calculated according to

$$q_i = q_i^0 - 2\sum_{j=1}^{NC} \Delta_{ij}\zeta_{ij}, \tag{4.5}$$

where q_i^0 is the charge of the isolated atom and $\zeta_{ij} = \left(1 + e^{b_c(r_{ij}-a_c)}\right)^{-1}$ is the charge transfer function. a_c and b_c are empirical parameters. Electroneutrality is assured by requiring that $\Delta_{ij} = -\Delta_{ji}$. The net charge on the atoms will decrease with the increase of coordination number (for example, the charges on B ions change from +1.7 to +1.48 in their threefold and fourfold coordinated states, respectively), which in turn will reduce the strength of the atomic interactions in the direction of the bond, or vice versa. This dependence is consistent with theoretical calculations which have demonstrated that bond strengths decrease and bond lengths increase with increasing coordination number.

This potential model can accurately describe the structural and thermo-mechanical properties of crystalline and amorphous SiO_2 and B_2O_3 within a wide pressure and temperature regime [51, 53, 63, 91–95], and successfully capture the 4- to 6-coordination change in SiO_2 [95] and the 3- to 4-coordination change in B_2O_3 [53] under pressure.

4.2.6 Screened and/or Truncated Force Field

In potential models for mixed ionic-covalent glass formers, there is always the Coulombic part, describing electrostatic interactions between charges with 1/r long-range spatial dependence. Ewald summation is most frequently used to account for the Coulombic interactions in systems with periodic boundary conditions, however, it is computationally very expensive. By a careful choice of the separation into the real space and the reciprocal space contributions, $O(N^{3/2})$ scaling with respect to the number of particles can be achieved. The computational cost of long-range electrostatics is of particular concern for potentials including the polarizability effects since several electric field evaluations are required to self-consistently determine the dipole moments at each time step. There has been considerable effort in the literature to develop screened potentials [2–6]. Soules proposed a screened two-body potential in which the Coulomb force is replaced by the force field of a charge at the center of a sphere containing uniform charge density of opposite sign [6]. A similar treatment was adopted in Feuston and Garofalini's three-body potential [4]. Using Soules' simple screening, the same structures and dynamical properties as with the Ewald summation can be obtained with a cutoff of 11.0 Å, but ~20 times faster. If a shorter cutoff of 5.5 Å is used, the calculations can run another factor of 20 times faster [41]. Other techniques, such as the Yukawa screening and the Wolf summation method, have been used for screening the long-range part of the BKS potential [96]. Various static and dynamic quantities were computed and compared to results from simulations using the Ewald summation. Very good agreement was obtained at $r_c \approx 10$ Å for the Wolf method, and slightly larger cutoffs have to be used in the Yukawa method in order to obtain the same accuracy with respect to static and dynamic quantities as for the Wolf method, as seen in Fig. 4.9.

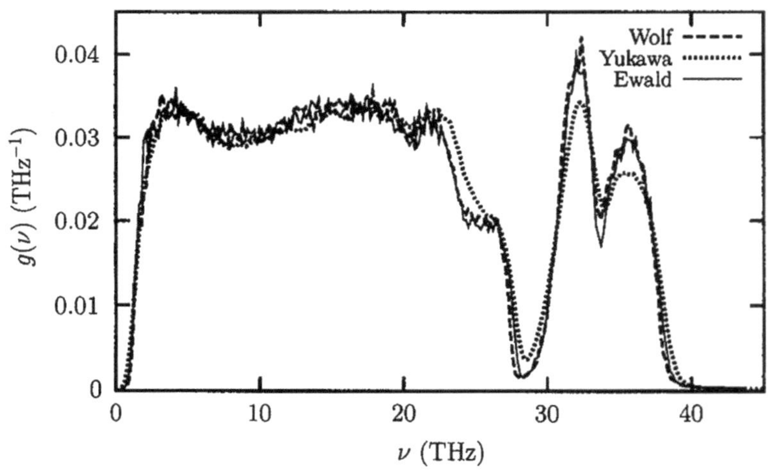

Fig. 4.9 VDOS from the BKS potential using Ewald, Yukawa and Wolf method. Figure taken from [96]

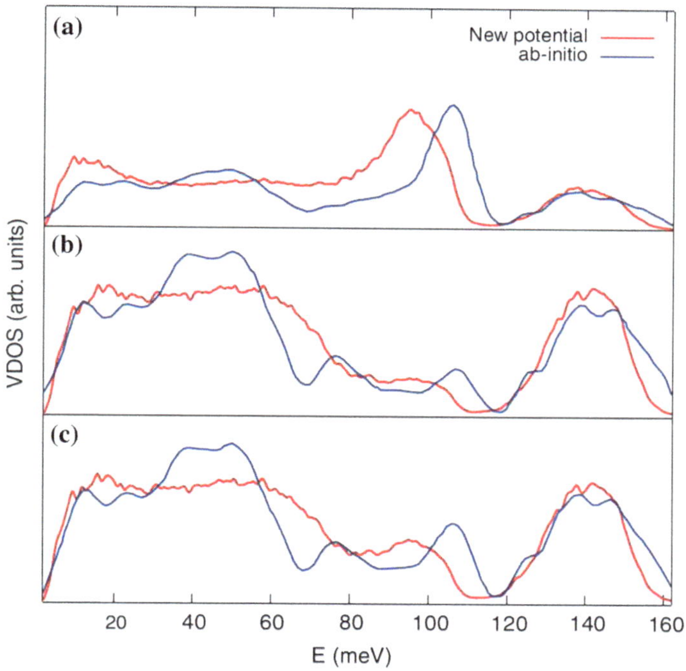

Fig. 4.10 VDOS of amorphous silica at 300 K calculated with the new short-ranged TS potential using the Wolf summation method, compared with an initio MD study. **a** Partial VDOS for silicon atoms, **b** partial VDOS for oxygen atoms, and **c** generalized VDOS. Figure taken from [2]

Recently, linear scaling with the number of particles has been achieved for the TS polarizable potential by using the Wolf summation method [2, 3] or the Yukawa screening method [5] without significant loss of accuracy. By optimizing the damping and truncation of the long-ranged potential while maintaining energy conservation, simulations can be performed at a comparatively small cutoff of 8 Å using the Wolf summation method [2]. The new potential reproduces the key features of the VDOS of amorphous silica from ab initio MD simulation as seen in Fig. 4.10, although it was not optimized for amorphous states at 300 K. By using an effective cutoff of 10 Å in the Yukawa screening method, the short-ranged potential reproduces ab initio and experimental structural, elastic, and vibrational properties of both α-quartz and amorphous silica to a high degree of accuracy [5].

It should be pointed out that non-Coulombic contributions to potentials are also often truncated and shifted to save computational time. Sometimes, the truncation has significant influence on the physical properties of the system under study. For example, in the BKS potential used by Vollmayr et al. the non-Coulombic contribution to the potential was truncated and shifted at a distance of 5.5 Å to correct the room temperature density for amorphous silica [38]. Our previous study showed that if the short-range cutoff is increased to 10 Å, a higher density of $\sim$2.38g/cm^3 for silica glass is obtained, compared with a density of 2.29 g/cm^3 using the cutoff of 5.5 Å, which in turn has a significant effect on the elastic moduli of silica glass [43]. It is noteworthy that the CHIK potential reproduces the equation of state of liquid silica with accuracy because the short-range part of the potential is truncated and shifted at 6.5 Å [8]. If not truncated, the density of the simulated melt is too high, e.g., at 2600 K, density of 2.34 g/cm^3 is obtained, higher than the experimental value of 2.2 g/cm^3 [97]. Although the truncation has no theoretical background, it is an efficient way to fit the pressure of a simulated silica melt. Therefore, the cutoff is an important parameter, which should be reported together with other parameters to complete the potential model.

4.3 Potential Parameterization

Once a functional form is decided to capture different aspects of the potential curve: (1) zero order derivative, (2) first order derivative, (3) second order derivative, and (4) anharmonicity. Free parameters of the potential model are obtained either through fitting them to ab initio data for small clusters or condensed systems, or adjusting the parameters to reproduce known experimental results, or some combination of these methods. In the following, we will selectively choose a few examples to illustrate the commonly used parameterization procedures.

4.3.1 Fitting to Experimental Data

Because of its empirical nature, a potential model must be validated by optimizing its parameters to reproduce the physical properties of known structures, e.g., static properties such as cohesive energy, density, bond lengths, bond angles, elastic moduli, dynamic properties such as VDOS, IR and Raman spectra, as well as transport properties such as diffusion coefficients. Reproduction of these quantities ascertains the right position and depth, slope and curvature of the minimum and tail of the potential function. The position and depth of the potential well largely determines bond length, density and defect formation energy, while the shape of the minimum affects atomic vibrations and elastic moduli. It is the shape of the tail that mainly tells how easily atoms can migrate out the potential well. Earlier potential models were usually parameterized by fitting to experimental data. For example, Tsuchiya empirically optimized a pair-wise potential to reproduce the structures, bulk moduli and thermal expansivities of GeO_2 polymorphs [11], but this potential fails to predict the dynamical properties properly. This motivated Oeffner and Elliott to re-parameterize the same functional form for GeO_2 based on ab initio calculations [12].

4.3.2 Fitting to Ab Initio Small Clusters

A new source of data for fitting potentials was made possible by the development of ab initio techniques. Fitting to ab initio calculations of small clusters were widely used to develop potentials for SiO_2 [9, 28], GeO_2 [12] and B_2O_3 [66]. For example, Tsuneyuki et al. have derived a simple two-body potential (TTAM) for silica based on ab initio calculations of small clusters, where the use of fractional charges seems to be one of the main ingredients for the success of the TTAM potential model [9]. The effective charges for Si and O were determined to be +2.4 and −1.2. Dynamically stable structures were obtained for a variety of silica polymorphs and liquid SiO_2 under ambient pressure using the TTAM potential. However, the predicted Si–O–Si angle is consistently too small for the structure of α-quartz under pressure [98]. This potential model was used in the lattice dynamics calculations of the vibrational frequencies of α-quartz, α-cristobalite, coesite, and stishovite [99]. The comparison with experiments is reasonable but, in general, the calculated stretching frequencies are somewhat low and the bending frequencies are too high.

Oeffner and Elliott (OE) [12] followed the Tsuneyuki's approach to derive an interatomic potential for GeO_2 by fitting the potential parameters to a potential energy surface of a $Ge(OH)_4$ cluster calculated with a Hartree-Fock program. The OE potential reproduces well both structural and dynamical properties of the two most common crystalline phases of GeO_2 (e.g., quartz with coordination of 4, and rutile with coordination of 6). Oeffner and Elliot presented two set of parameters, one corresponding to the so-called "original potential", fitted from ab initio calculations, and the other one corresponding to the "rescaled potential" to better reproduce the vibrational properties of α-quartz GeO_2. Peralta et al. used the "rescale potential" to study amorphous GeO_2 [100] and found good agreement with the experimental [101] and first principles results [78, 102] in the structural properties, except the Ge–Ge distance is slightly overestimated (3.26 vs. 3.16 Å of experimental value). A reasonable agreement with the experimental [103] and first principles [102, 104] results in the VDOS (Fig. 4.11) was obtained by Peralta et al. [100]. However, the "rescaled potential" yields too fluid a melt if compared with the "original potential" and the DIPPIM potential [67] and experimental data. The "rescaled potential" was used by Micoulaut et al. [105] and gave diffusion coefficients more than one order of magnitude larger than the values obtained using the "original potential" by Hawlitzky et al. [106].

4.3.3 Fitting to Ab Initio Small Clusters and Experiments

It was explicitly shown by van Beest et al. [10] that, as the range of the interatomic forces goes beyond the nearest neighbors, the need arises for complementing microscopic information with macroscopic information to optimize the interaction potentials. These authors obtained a two-body potential (BKS) for silica by iteratively

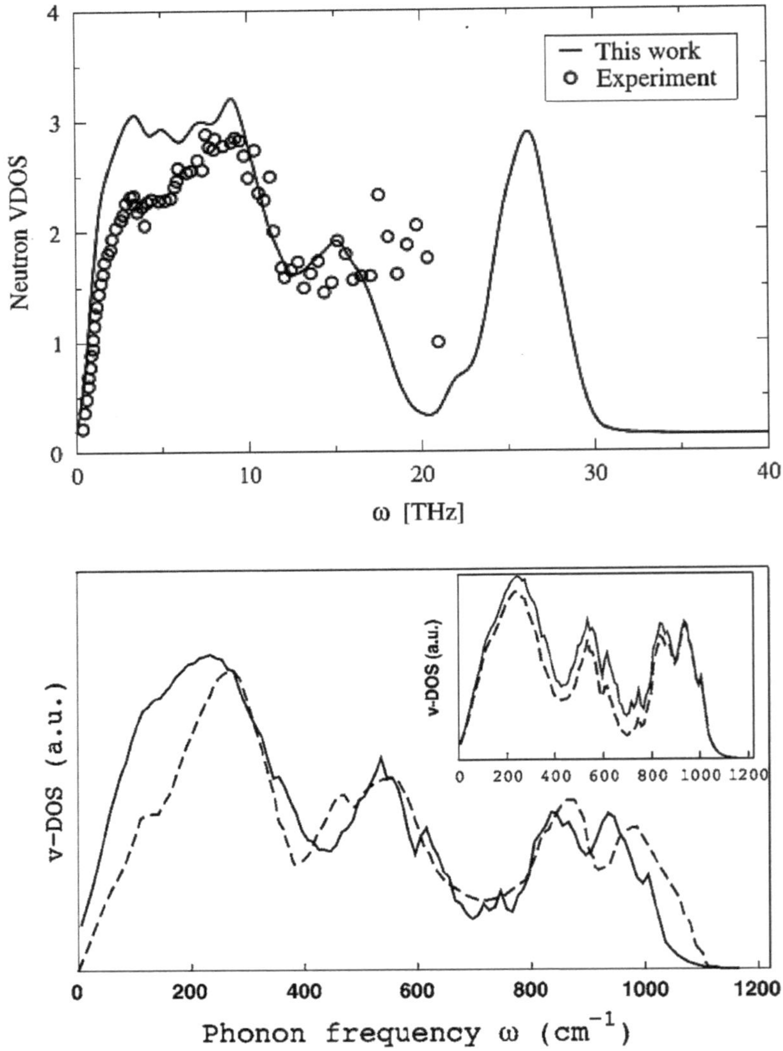

Fig. 4.11 *Top* Neutron VDOS for amorphous GeO_2 obtained from MD simulation using the rescaled OE potential (*solid line*) compared with experimental results [103] (*open circles*). *Bottom* VDOS of amorphous GeO_2 from first principles calculations [102] (*solid line*) and experimental inelastic neutron spectra (INS) [65] (*dashed line*). *Inset* partial VDOS of amorphous GeO_2 for O atoms with (*solid line*) and without (*dashed line*) the contribution of threefold coordinated O atoms. Figures taken from [100, 101]

fitting to ab initio potential energy surface of a H_4SiO_4 cluster and optimizing the effective charges on Si and O atoms to the bulk properties of α-quartz [10]. In this model, only two different short-range interactions, namely those describing the Si–O bond and the O–O non-bonded interactions are considered. These modify the Coulomb repulsion and ensure the tetrahedral arrangement of oxygen atoms around the silicon atom. The resulting potential can reproduce a large range of structural, mechanical, thermodynamic and dynamic properties of silica in many different phases and under many different thermodynamic conditions with reasonable accuracy [8, 10, 38, 41–44, 47, 96]. This is quite surprising for a simple pair-wise potential. However, there are several deficiencies in the BKS potential that should be pointed out. For instance, although the O–Si–O angles are calculated correctly, the Si–O–Si angles are overestimated, indicating a weaker Si–O–Si bending interaction. This can be easily understood because there is no inter-tetrahedral interaction in the BKS potential to account for the lone pair of electrons on oxygen responsible for the bending in the Si–O–Si arrangement. It was found that surfaces generated by the BKS potential have a higher concentration of defects (e.g., concentration of two-membered rings) than those generated with CPMD [107]. The equation of state, as far as it is known experimentally, is not reproduced well [7]. Herzbach et al. showed that the BKS model of cristobalite and tridymite silica are not thermodynamically stable [42]. In order for these phases to appear stable in MD simulations, appropriate periodic boundary conditions must be chosen, i.e., boundary condition which is incommensurate with the modes leading into the energetically favored structures. We found out that the stability of β-cristobalite silica strongly depends on the system size. As shown in Fig. 4.12, for a cubic system with 648 atoms, a first-order phase

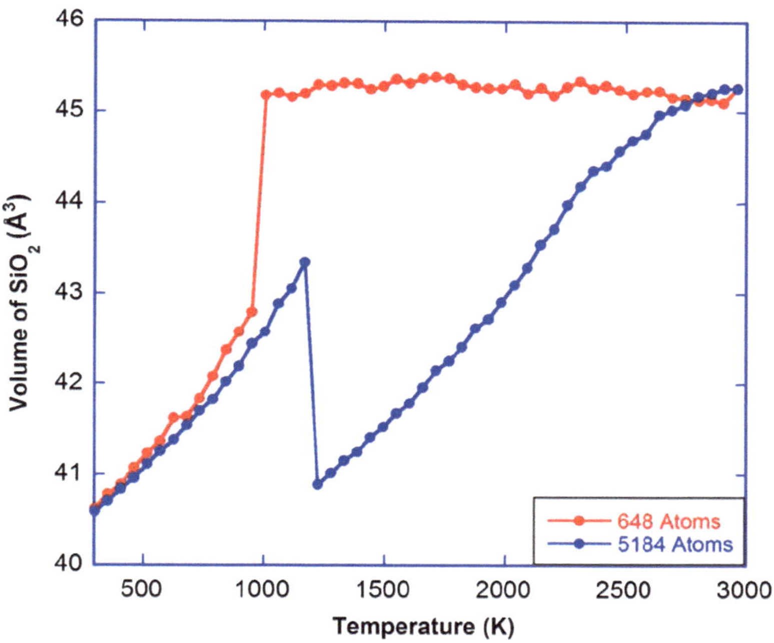

Fig. 4.12 Volume of SiO_2 as a function of temperature from MD simulations using the BKS potential

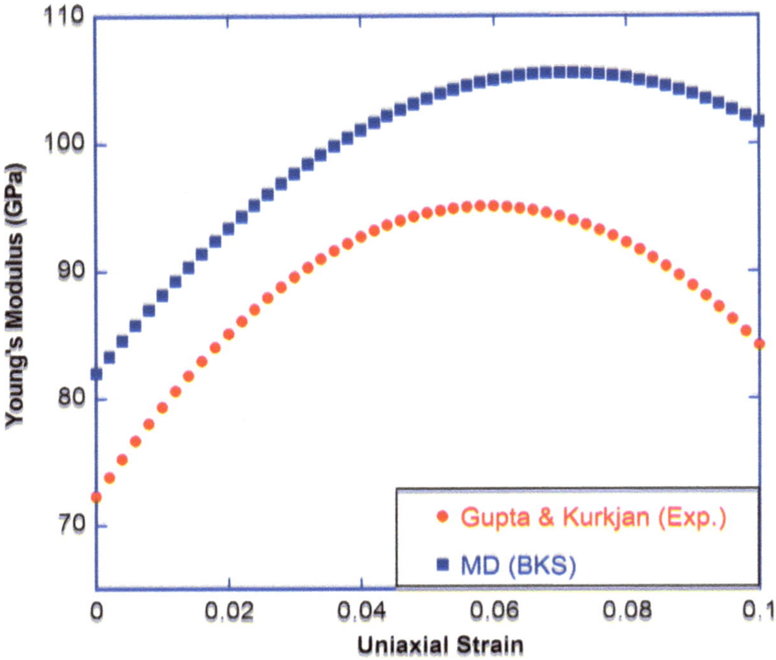

Fig. 4.13 Comparison of the non-linear elasticity of silica glass in MD simulations using the BKS potential and in Gupta and Kurkjian's experiments [107]

transformation from α- to β-cristobalite is observed; α-cristobalite has a positive thermal expansion coefficient, while β-cristobalite has zero or slightly negative thermal expansion coefficient, all in good agreement with experimental observations, even though the transformation temperature is overestimated [51]. However, when a cubic system with 5184 atoms is used, β-cristobalite cannot form properly. More importantly, the ground state of silica is not reproduced correctly: BKS stishovite is found to be more stable than α-quartz at zero pressure and temperature [42]. This is in strong contrast to experimental findings and to density-functional theory (DFT) calculations. In terms of elastic properties, BKS can reproduce the elastic constants of α-quartz very well as they were used as inputs for parameterization. Our study showed that the BKS model can describe the trend of the non-linear elasticity of silica glass very well (Fig. 4.13), but it overestimates the Young's modulus at ambient conditions, so is the Poisson's ratio [43, 44]. As shown in Fig. 4.2, the lack of the three-body interactions/polarizability effects in the BKS potential makes it less satisfactory in calculating the vibrational properties.

4.3.4 Fitting to CPMD Simulations

The parameterization of the TTAM and BKS potential was based on ab initio calculations of small clusters. In contrast to that, the development of the CHIK potential of the same functional formula was based on a CPMD simulation on a bulk system of 114 particles [8]. A structural fitting procedure was used for potential parameterization. The idea was to match the partial pair correlation functions, as obtained with the new effective potential, with those obtained from CPMD. The CHIK potential was shown to be superior to the BKS model with respect to various properties of amorphous silica (in particular the density at low pressures). It also reproduces very accurately the experimental equation of state, various structural properties and the anomalous diffusion dynamics of silica under pressure. Figure 4.14 shows the VDOS calculated from the CHIK potential, compared to that from the BKS model and CPMD simulations. Note that the BKS potential has been fitted to reproduce the high-frequency band of the vibrational spectrum. Thus, it gives a better agreement with CPMD than the CHIK model, since no vibrational properties were used in the fitting procedure of the CHIK potential. The CHIK model seems to be better than the BKS model in the intermediate frequency band around 20–30 THz, in contrast to the BKS case, a single peak is observed, albeit at a slightly lower frequency than in the CPMD results. Below 20 THz, the BKS and CHIK results are very similar and do not agree well with the CPMD results. In order to significantly improve the description of vibrational properties, it might be necessary to account for polarization effects in the model potential, as shown in studies by Wilson et al. [1], and Tangney and Scandolo [7].

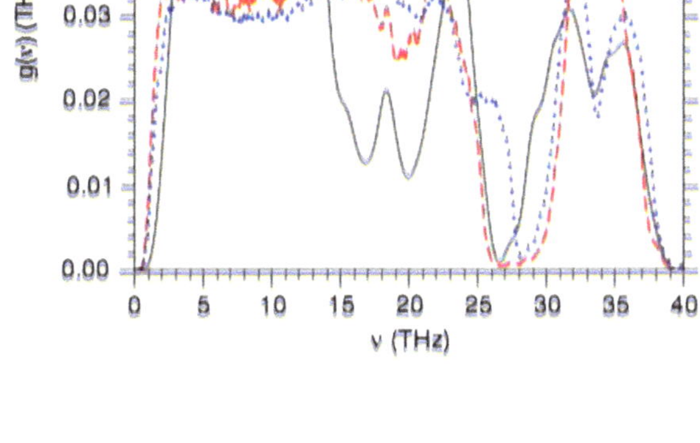

Fig. 4.14 VDOS from classical MD simulations using the BKS and CHIK potentials and from CPMD simulations. Figure taken from [8]

4.3.5 Fitting to Ab Initio Condensed Systems

Tangney and Scandolo (TS) potential explicitly incorporates many-body effects by treating the oxygen atoms as polarizable ions, with dipole moments determined self-consistently in response to the local electrostatic field [7]. The TS potential was parameterized by using a nonlinear force matching procedure based on forces, energies, and stresses for an extended database of configurations of 3000 K liquid silica evaluated from ab intitio calculations without any reliance on experimental data. Ab initio calculations were performed on a relatively large unit cell with 72 atoms under periodic boundary conditions to minimize the finite size and surface effects which may be present in cluster calculations. Such ab initio parameterized force field has been shown to achieve good transferability and provide an accurate description of a wide variety of properties of silica crystals, liquid and glass [7, 42, 80, 81]. A similar parameterization procedure was adopted in the short-ranged versions of the TS potential for silica by using the Wolf summation [2, 3] or the Yukawa's screening method [5]. The DIPPIM potential model for GeO_2 developed by Marrocchelli et al. includes a pair-wise potential, together with an account of the polarization effects [67]. The parameters in the DIPPIM potential were determined by matching the dipoles and forces on the ions calculated from ab initio on configurations of condensed systems. Formal ionic charges (Ge^{4+}, O^{2-}) were used in the DIPPIM potential with a hope to ensure a better transferability. This potential can reproduce the structural, dynamical and vibrational properties of GeO_2 liquid and glass to a high degree of precision [67].

4.4 Concluding Remarks

As shown in this review, great efforts have been made in the past few decades to develop potential models for mixed ionic-covalent glass formers. Significant advancements in potential development have taken place in terms of the functional form, the improved efficiency and the parameterization procedure. It remains a great challenge and there is still much room for further improvement in the construction of reliable, effective, and transferable potential models for such systems. The following questions need to be addressed for future potential development for mixed ionic-covalent glass formers.

(1) *What are the most important interactions to be included in the potential model for the systems or conditions under study?* Adding interaction (and parameters) to the model inevitably improves agreement with experimental or first principles results. However, the reason for the improvement may only be the greater flexibility allowed by the additional parameters rather than the physical significance of the additional interaction.
(2) *What parameterization procedure to use and what data set (from experiments and/or first principles calculations) should be used for fitting the potential*

parameters? In order to directly compare force fields and determine whether certain interactions are necessary, the force field construction needs to be systematically derived from a common data set using the same parameterization procedure. So far, matching force, stress, energy and structure from first principles calculations have been used for potential parameterization, but there is no reason other data such as VDOS cannot be included in the fitting procedure to improve the potential models. For systems with scarce and/or unreliable experimental data, first principles calculations have shown to be able to provide a viable data set.

(3) *From what phases/thermodynamic conditions the data set should be taken for fitting parameters to ensure transferability of a potential model?* A potential model usually performs well in the physical situation where it is parameterized, but often loses its predictive capability once moving away from it. Although the transferability of a potential model largely depends on the flexibility in its functional form, the phases/thermodynamic conditions from which the data are taken for fitting parameters can also play an important role. A good example is the better performance of the CHIK potential than the BKS model with the same functional form.

(4) *Can potential models be developed for individual elements like the pseudopotentials used in first principles calculations?* So the same potential can be used for a common species when changing from one material to another. So far, most of potentials developed for mixed ionic-covalent glass formers are based on individual systems. For example, even though both the BKS and the OE potentials for SiO_2 and GeO_2 have the same functional form, but the charge on oxygen ion and the parameters for the O–O interaction are different. A combination of the BKS and the OE potentials would not be suitable for the description of oxide melts containing both GeO_2 and SiO_2. Recent progress in the COMB potentials shows that it might be possible to develop element-based potentials that can be transferable from one material to another. For example, in the COMB potentials for Si/SiO_2 and Hf/HfO_2, the same parameters are used for oxygen atoms in both systems so it is possible to study $Hf_xSi_{1-x}O_y$ films or the growth of hafnia films on Si or SiO_2 without additional effort to develop a new potential.

(5) *How to improve the computational efficiency of complex and accurate potential models?* Generally speaking, more complex potentials with appropriate parameterization can better reproduce a wider range of physical behaviors of a system under study, albeit often with an undesirable increase in the computational cost. Short-ranged versions of the TS potentials prove that balance between computational efficiency and model accuracy can be achieved, which is a crucial requirement for large-scale atomistic simulations.

Potential development in the past for mixed ionic-covalent glass formers was mainly based on independent efforts. Construction of the functional form and selection of the input data set and fitting procedure were strongly influenced by the developer's personal choices. Developing potential seemed more like art than science, sometimes hit-or-miss for no obvious reason. Such a process for potential development is of very low efficiency, and cannot catch up with the much need for accurate potential models for many glasses of technological importance. In the future, a constant dialogue and a strong collaboration between computational physicists, chemists and materials scientists, computer scientists and mathematicians would be much beneficial for the endeavor of potential development for mixed ionic-covalent glass formers.

Acknowledgments L. Huang would like to acknowledge the financial support from the National Science Foundation under Grant No. DMR-1105238 and DMR-1255378.

References

1. M. Wilson et al., Phys. Rev. Lett. **77**(19), 4023 (1996)
2. P. Beck et al., J. Chem. Phys. **23**, 135 (2011)
3. P. Brommer et al., J. Chem. Phys. **19**, 132 (2010)
4. B.P. Feuston et al., J. Chem. Phys. **89**(9), 5818–5824 (1988)
5. J.R. Kermode et al., J. Chem. Phys. **9**, 133 (2010)
6. T.F. Soules, J. Non-Cryst. Solids **123**(1–3), 48–70 (1990)
7. P. Tangney et al., J. Chem. Phys. **117**(19), 8898–8904 (2002)
8. A. Carre et al., EPL **82**(1) (2008)
9. S. Tsuneyuki et al., Phys. Rev. Lett. **61**, 869 (1988)
10. B.W.H. van Beest et al., Phys. Rev. Lett. **54**, 1955 (1990)
11. T. Tsuchiya et al., Phys. Chem. Miner. **25**(2), 94–100 (1998)
12. R.D. Oeffner et al., Phys. Rev. B **58**(22), 14791–14803 (1998)
13. A. Takada et al., J. Phys.: Condens. Matter **7**, 8659 (1995)
14. A. Takada et al., J. Non-Cryst. Solids **345**, 224–229 (2004)
15. L. Stixrude et al., Phys. Chem. Miner. **16**(2), 199–206 (1988)
16. E. Demiralp et al., Phys. Rev. Lett. **82**(8), 1708–1711 (1999)
17. V.V. Hoang, J. Phys.-Condens. Matter **18**(3), 777–786 (2006)
18. K. Yamahara et al., J. Non-Cryst. Solids **291**(1–2), 32–42 (2001)
19. L.V. Woodcock et al., J. Chem. Phys. **65**, 1565 (1976)
20. A. Karthikeyan et al., J. Non-Cryst. Solids **281**(1–3), 152–161 (2001)
21. A.H. Verhoef et al., J. Non-Cryst. Solids **146**, 267 (1992)
22. T.F. Soules, J. Chem. Phys. **73**(8), 4032–4036 (1980)
23. C.A. Angell et al., Science **218**, 885 (1982)
24. T.F. Soules, J. Chem. Phys. **71**, 4570 (1979)
25. S.K. Mitra et al., Philos. Mag. B **43**, 365 (1981)
26. S.H. Garofalini, J. Chem. Phys. **76**, 3189 (1982)
27. N.V. Doan, Philos. Mag. A **49**(5), 683 (1984)
28. A.C. Lasaga et al., Phys. Chem. Miner. **14**, 107 (1987)
29. R.L. Erikson et al., J. Geochim. Cosmochim. Acta. **51**, 1209 (1987)
30. J.D. Kubicki et al., Am. Miner. **73**, 941 (1988)
31. S. Tsuneyuki et al., Phys. Rev. Lett. **74**(16), 3197 (1995)
32. J.R. Chelikowsky et al., Phys. Rev. Lett. **65**, 3309 (1990)
33. R.G. Della Valle et al., J. Chem. Phys. **94**(7), 5056 (1991)
34. R.G. Della Valle et al., Chem. Phys. **179**, 411 (1994)
35. R.G. Della Valle et al., Phys. Rev. B **54**, 3809 (1996)
36. J.S. Tse et al., J. Chem. Phys. **95**(12), 9176 (1991)
37. J.S. Tse et al., Phys. Rev. B **46**, 5933 (1992)
38. K. Vollmayr et al., Phys. Rev. B **54**(22), 15808 (1996)

39. I. Saika-Voivod et al., Nature **412**, 514 (2001)
40. I. Saika-Voivod et al., Phys. Rev. E **63**(1), 011202 (2001)
41. T.F. Soules et al., J. Non-Cryst. Solids **357**(6–7), 1564–1573 (2011)
42. D. Herzbach et al., J. Chem. Phys. **12**, 123 (2005)
43. F.L. Yuan et al., J. Non-Cryst. Solids **358**(24), 3481–3487 (2012)
44. F.L. Yuan et al., Sci. Rep. **4** (2014)
45. M. Hemmati et al., J. Non-Cryst. Solids **217**(2–3), 236 (1997)
46. M. Hemmati et al., in *Comparison of Pair-Potential Models for the Simulation of Liquid $Si\,O_2$: Thermodynamic, Angular-Distribution, and Diffusional Properties, in Physics Meets Mineralogy: Condensed-Matter Physics in Geosciences*, ed. by H. Aoki, Y. Syono, R.J. Hemley (Cambridge University Press, Cambridge, 2000), p. 325
47. M. Benoit et al., Europhys. Lett. **60**(2), 269–275 (2002)
48. M.J. Sanders et al., J. Chem. Soc. Chem. Commun. **19**, 1271 (1984)
49. C.R.A. Catlow et al., Philos. Mag. A-Phys. Condens. Matter Struct. Defects Mech. Prop. **58**(1), 123–141 (1988)
50. P. Vashishta et al., Phys. Rev. B **41**(17), 12197–12209 (1990)
51. L.P. Huang et al., J. Chem. Phys. **118**(3), 1487 (2003)
52. T. Nanba et al., J. Non-Cryst. Solids **277**(2–3), 188–206 (2000)
53. L.P. Huang et al., Phys. Rev. B **74**, 224107 (2006)
54. S.N. Taraskin et al., Phys. Rev. B **55**(1), 117–123 (1997)
55. G.J. Kramer et al., Phys. Rev. B **43**, 5068 (1991)
56. R.A. Murray et al., Phys. Rev. B **39**, 2 (1989)
57. A.K. Rappe et al., J. Phys. Chem. **95**, 3358 (1991)
58. A. Yasukawa, JSME Int. J. Ser. A-Mech. Mater. Eng. **39**(3), 313–320 (1996)
59. T.-R. Shan et al., Phys. Rev. B **82**(23) (2010)
60. J. Yu et al., Phys. Rev. B **75**(8) (2007)
61. D.C. Anderson et al., J. Chem. Phys. **98**, 8978 (1993)
62. L. Duffrène et al., J. Phys. Chem. Solids **59**, 1025 (1998)
63. L.P. Huang et al., Phys. Rev. Lett. **95**, 215901 (2005)
64. J.M. Carpenter et al., Phys. Rev. Lett. **54**(5), 441 (1985)
65. F.L. Galeener, Phys. Rev. B **27**, 1052 (1983)
66. J.K. Maranas et al., J. Chem. Phys. **115**(14), 6578 (2001)
67. D. Marrocchelli et al., Mol. Phys. **107**(4–6), 443–452 (2009)
68. M. Wilson et al., J. Phys.-Condens. Matter **5**(37), 6833–6844 (1993)
69. M. Wilson et al., J. Phys.-Condens. Matter **6**, A151–A155 (1994)
70. D.J. Lacks et al., Phys. Rev. B **48**(5), 2889–2908 (1993)
71. D.J. Lacks et al., J. Geophys. Res.-Solid Earth **98**(B12), 22147–22155 (1993)
72. A.C. Hannon et al., J. Non-Cryst. Solids **177**, 299 (1994)
73. P.A.V. Johnson et al., J. Non-Cryst. Solids **50**, 281 (1982)
74. C. Joo et al., J. Non-Cryst. Solids **261**, 282 (2000)
75. R.L. Mozzi et al., J. Appl. Cryst. **3**, 251 (1970)
76. R.N. Sinclair et al., Phys. Chem. Glasses **41**(5), 286 (2000)
77. G.E. Walrafen et al., J. Chem. Phys. **92**, 6987 (1990)
78. L. Giacomazzi et al., Phys. Rev. B **74**(15) (2006)
79. E.I. Kamitsos et al., J. Phys. Chem. **100**(28), 11755–11765 (1996)
80. Y.F. Liang et al., J. Chem. Phys. **125**(19) (2006)
81. S. Paramore et al., J. Chem. Theory Comput. **4**(10), 1698–1708 (2008)
82. S.M. Levine et al., J. Chem. Phys. **86**(5), 2997–3002 (1987)
83. M.H. Du et al., J. Chem. Phys. **119**(13), 6418–6422 (2003)
84. A.C.T. van Duin et al., J. Phys. Chem. A **107**(19), 3803–3811 (2003)
85. A.C.T. van Duin et al., J. Phys. Chem. A **105**(41), 9396–9409 (2001)
86. A. Strachan et al., Phys. Rev. Lett. **91**(9) (2003)
87. Q. Zhang et al., Phys. Rev. B **69**(4) (2004)
88. P. Vashishta et al., J. Propuls. Power **23**(4), 688–692 (2007)

89. D.W. Brenner, Phys. Rev. B **42**(15), 9458–9471 (1990)
90. D.W. Brenner et al., J. Phys. Condens. Matter **14**(4), 783–802 (2002)
91. L.P. Huang et al., Phys. Rev. B **69**, 224203 (2004)
92. L.P. Huang et al., Glass Sci. Technol. **77**, 124 (2004)
93. F.L. Yuan et al., Phys. Rev. B **85**(13) (2012)
94. L.P. Huang et al., J. Non-Cryst. Solids **349**, 1 (2004)
95. L.P. Huang et al., Nat. Mater. **5**(12), 977–981 (2006)
96. A. Carre et al., J. Chem. Phys. **127**(11) (2007)
97. B. Guillot et al., Geochimica Et Cosmochimica Acta **80**, 51–69 (2012)
98. J.R. Chelikowsky et al., Phys. Rev. B **15**, 10866 (1990)
99. R.G. Guido et al., J. Chem. Phys. **94**, 5056 (1991)
100. J. Peralta et al., J. Phys.-Condens. Matter **14**(20) (2008)
101. P.S. Salmon et al., J. Phys.-Condens. Matter **41**(19) (2007)
102. E. Scalise et al., Appl. Phys. Lett. **98**(20) (2011)
103. O. Pilla et al., J. Non-Cryst. Solids **322**(1–3), 53–57 (2003)
104. L. Giacomazzi et al., Phys. Rev. Lett. **95**(7) (2005)
105. M. Micoulaut et al., Phys. Rev. E **73**(3) (2006)
106. M. Hawlitzky et al., J. Phys.-Condens. Matter **20**(28) (2008)
107. C. Mischler et al., Comput. Phys. Commun. **147**(1–2), 222–225 (2002)
108. P.K. Gupta et al., J. Non-Cryst. Solids **351**(27–29), 2324–2328 (2005)

Chapter 5
Computational Modeling of Silicate Glasses: A Quantitative Structure-Property Relationship Perspective

Alfonso Pedone and Maria Cristina Menziani

Abstract This article reviews the present state of Quantitative Structure-Property Relationships (QSPR) in glass design and gives an outlook into future developments. First an overview is given of the statistical methodology, with particular emphasis to the integration of QSPR with molecular dynamics simulations to derive informative structural descriptors. Then, the potentiality of this approach as a tool for interpretative and predictive purposes is highlighted by a number of recent inspiring applications.

5.1 Introduction

Major global challenges in strategic fields such as chemistry, pharmaceutics, medicine, photonics, optics, electronics, clean energy and waste management can be addressed by the development of advanced technologies based on glassy materials. To this goal the correct understanding of the glass structure-property relationships is mandatory, since this is a prerequisite for improving specific properties and achieve greater focus on end-user application requirements, designing glass compositions for new applications, developing environmentally friendly processes and product, reducing development costs and speed time to market [1–3].

Notwithstanding the huge improvement in experimental methodologies, like X-ray Absorption Fine Structure, Neutron Diffraction, Nuclear Magnetic Resonance, Infrared and Raman spectroscopy, the elucidation of the glass structure still remain a difficult task [4]. In fact, quite often, difficulties in data interpretation of multi-component glasses and apparent contradictory structural evidences from different techniques have to be faced.

A. Pedone · M.C. Menziani (✉)
Dipartimento di Scienze Chimiche e Geologiche, Università degli Studi di Modena e Reggio Emilia, Via Campi 183, 41125 Modena, Italy
e-mail: menziani@unimore.it

A. Pedone
e-mail: alfonso.pedone@unimore.it

C. Massobrio et al. (eds.), *Molecular Dynamics Simulations of Disordered Materials*, Springer Series in Materials Science 215,
DOI 10.1007/978-3-319-15675-0_5

In this context, the use of large databases of experimentally measured glass properties [5, 6] has facilitated the systematic modelling of glasses and prediction of their properties by statistical analyses of composition-property relationships. However, these tools suffer from important drawbacks: (i) the range of compositions studied is determined by presence or absence of the experimental data required; (ii) the treatment of glasses of complex compositions involving multiple network formers and modifiers is unpractical; and (iii) their objective is exclusively predictive, thus they do not allow a detailed physical understanding for the observed property-composition dependence at the atomic scale [7, 8].

The advent of computational simulation techniques as an accepted component of material development is one of the most important advances in material research. Molecular Dynamics (MD) is nowadays well established as a powerful tool to provide an atomic scale picture of the structure and insight into the behavior of complex glasses in different environments and under different conditions. Recent advances in the construction of interatomic potential allow the correct quantitative evaluation of the numerical values of structural, mechanical, thermal, electrical and transport properties for simple glasses [9–15]. However, accurate and reliable evaluation of the same properties for multicomponent glasses has proved far more difficult.

In these cases, i.e. when a direct comparison with experimental observables is not possible, the results of Molecular Dynamics simulations can be used to provide the numerical representation of structure (codified by structural descriptors) to be related with the experimental properties of interest through mathematical models. This implies a shift from empirical composition-property relationships to computational structure-property relationships, thus acquiring an immense practical importance in the development of predictive and interpretative models [16]. This approach, called Quantitative Structure-Property Relationships (QSPR), is well known and extensively applied in the area of drug discovery, and chemical toxicology modeling. However, its application in the field of material design is only recently being explored [17–19].

In the following, a brief overview of the methodology used in QSPR of glasses is given. The mathematical method of choice in relation to the dataset under study is discussed together with the critical role of informative computational-derived descriptors and of appropriate model building and validation. Then, the potentiality of this approach as a tool for interpretative and predictive purposes is highlighted by a number of recent applications concerning the modeling of density, glass transition temperature, crystallization temperature, leaching, chemical durability, elastic properties, and NMR features. Finally, we discuss the future developments that will hopefully improve the QSPR approach described and overcome some current limitations.

5.2 Quantitative Structure-Property Relationship Analysis

QSPR methods are based on the hypothesis that changes in the structure are reflected in changes in observed macroscopic properties of materials. The basic strategy of QSPR analysis is to find optimum statistical correlations, which can then be used for

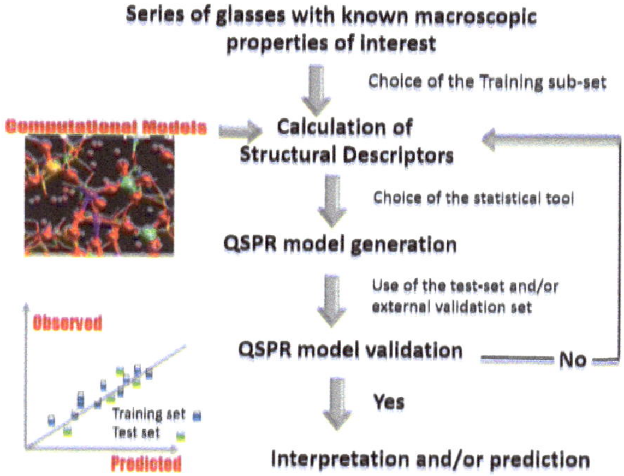

Fig. 5.1 Workflow of QSPR modeling

(a) the prediction of the properties for compounds as yet unmeasured or even not yet synthesized; (b) the detailed analysis of structural characteristics that can give rise to the property of interest.

Two very recent reviews provide an in-depth description of the main concepts involved in QSPR modeling of discrete molecules [20], and materials [21]. Therefore, only a summary of the important elements of the QSPR modeling process in the context of glass design is provided here, underlying the basic character of statistical analysis that has been ignored for too long in glass science.

The process of constructing a QSPR model includes the following steps, summarized in Fig. 5.1: (1) selection of a data set; (2) generation of various structural descriptors by means of MD simulations; application of variable selection or/and data reduction methods on the calculated descriptors in order to identify a small subset of these descriptors that are relevant to the macroscopic material properties being modeled (in some cases this step may not be required); (3) generation of linear/multilinear or non-linear relationship between the descriptors and the global material property (4) validation of the model to assess its reliability, robustness, predictivity, and domain of applicability.

5.2.1 Data Set

The key requirement for QSPR modeling is a reliable data set of glasses whose macroscopic properties of interest are known and microscopic structures can be reasonably well-defined by computational simulations. This is termed training data.

The use of heterogeneous experimental data from different sources and laboratories can affect the quality of QSPR models, by increasing the noise in the modeled response, thus affecting the stability and predictivity of models. Other potential obstacle in the development of robust, predictive, and reliable models is the insufficient data size (the range of composition is limited by the occurrence of crystallization, phase separation or other phenomena, or the range of measured property values is too small) and the dependency of the macroscopic property from additional factors besides microscopic structure, such as the history of the material: how it was synthesized, processed, and prepared for testing.

5.2.2 Structural Descriptors

The formulation of informative QSPR models adequate for multicomponent disordered systems is anything but trivial and their predictive and interpretative power depends critically on the information content of the descriptors utilized [22]. The selection of descriptors for meaningful QSPR models implies the knowledge of what features of the structure are measured by a given descriptor and of how the microscopic properties influence the macroscopic (measured) properties in a mechanistic way. Without this knowledge it is hard to apply a "reverse QSPR approach" to optimize materials directly.

To this regard, MD simulations can provide a plethora of promptly available descriptors among which to select the most informative ones. The linear correlation matrix, made up of the correlation coefficients "r" between couples of descriptors, gives an overview of the collinearities existing between them and help in their selections in relation to the statistical model (simple or multilinear regression) of choice and to the interpretation of the properties of interest. The minimal number of orthogonal (not correlated) descriptors of possible relevance to important physicochemical parameters relating to the series of compounds under discussion must be selected for multilinear equations, so that the overall relationships are highly significant by standard statistical criteria.

For oxide glasses, simple descriptors such as average bond lengths, bond angles, coordination numbers, percentage of bridging oxygen (BO) or non-bridging oxygen (NBO) attached to different cations, etc... can be derived from simple statistical averaging or from radial and pair distribution functions and their deconvolution, once the appropriate cut-off distances are defined [23–25]. Others descriptors can be defined as a combination of these ingredients. Finally, useful descriptors of the mid-range structure of the glasses are derived from the Q^n species (Q designates 'quaternary' and n the number of BO oxygens connected to other network former cations), ring size distribution, void size distribution and free volume [23–27].

5.2.3 Regression Analysis

There is no particular method that is ideal for all problems. The choice of an algorithm should be based on the nature of the data, and also whether the final goal is to build a predictive or interpretative model.

Various statistical methods are nowadays available to build models that describe the empirical relationship between the structure and property of interest. Classical methods, such as single or multiple linear regression (MLR), partial least squares (PLS), neural networks (NN), and support vector machine (SVM), are being upgraded by improving the kernel algorithms or by combining them with other methods, including novel approaches, such as gene expression programming (GEP), project pursuit regression (PPR), and local lazy regression (LLR) [28].

To avoid the risk of "by chance" correlation, statistical models requires significantly more data points than descriptors, since any data set can be fitted to a regression line given enough parameters. For example, in MLR analysis a useful rule of thumb is that the ratio between the number of objects in the data set and the number of descriptors should be at least five to one. Moreover, statistical modelling techniques follow the principle of parsimony postulated by William of Occam and called Occam's Razor (i.e. among a set of equally good explanations for a given phenomenon, the simplest one is the most probable) which means that the models should have as few parameters as possible (i.e. a variable is retained in the model only if its removal causes a significant decrease of the statistical parameters compared to those of the current model) and simple explanations have to be preferred than those more complex.

Therefore, according with the number of data point available in the data set, the simple or multiple linear regressions remain as popular choices for QSPR studies of glasses, since they allow an easier interpretation of the phenomena that determine the variation in the observable properties.

The final model built from the optimal parameters will then undergoes validation with a testing set of glasses to ensure that the model is appropriate and useful for prediction and/or interpretation.

5.2.4 Model Validation

Several procedures are available to determine the reliability and statistical significance of the model. The performance of regression models is commonly measured by the "explained variance" for the response variable y, denoted R^2, and the residual standard deviation (S^2), calculated using the following equations:

$$R^2 = 1 - \frac{\Sigma(\Delta^2)}{\Sigma[(\text{Observation} - \text{Average observation})^2]} \tag{5.1}$$

$$S^2 = \Sigma(\Delta^2)/DF \tag{5.2}$$

where, Δ are the model residuals (differences between the experimentally observed and the calculated property values), and DF the degrees of freedom (difference between the number of independent experimental data points and the number of variables including the intercept).

Both statistical parameters provide a measure of how well the model can predict new outcomes, however S^2 is a more robust estimates of the predictive ability of models because, unlike R^2 does not depend on the number of data points in the training set or on the number of descriptors in the model. Good QSPR models have R^2 values close to 1.0 and S^2 values small.

Cross-validation methods are often also applied. This involves omitting in turn one (leave-one-out) or more (leave-many-out) data points from the training set, generating a QSPR model using the remaining data points, and then predicting the properties of the data point(s) omitted. However, it has often been shown that the use of only this criterion gives an overly optimistic estimate of the predictivity of models [29].

The statistics of prediction of an independent external test set provide the best estimate of the performance of a model. However, the splitting of the data set in training set (used to develop the model) and the test set (used to estimate how well the model predicts unseen data) is not a suitable solution for small-sized data sets and an extensive use of internal validation procedures is recommend.

5.2.4.1 Outliers

For unimodal and symmetrical distributions, data point with deviations at least twice greater than the standard deviation of the data are usually considered outliers. Outliers that cause a poor fit degrade the predictive value of the model; however, care must be taken when excluding these outliers. They can be a clue in incorrectly measured experimental property or in the inadequacy of the model in capturing some important attribute of the material. Indeed, important microscopic properties of the material have not been accounted for in the model and/or the outlier represents an extreme point for this property.

5.3 Applications of QSPR Analysis

Among the examples reported in the literature we focus here on three cases instrumental to demonstrate the achievements of this approach in: (a) gaining insight into the physical processes determining the properties of interest (interpretative role of QSPR analysis); (b) predicting missing data and optimize property for intended application (predictive role of QSPR analysis); (c) assisting in experimental data collection and rationalization and support the design or assessment of foreseen experiments (assisting role of QSPR analysis). These are illustrated in relation to their performances on different properties.

5.3.1 QSPR Models for Density

Density, one of the most important properties in industrial glass production, is perhaps the simplest physical property that can be measured; nevertheless its dependence upon composition is not straightforward. A number of linear expressions, empirically derived by assuming additivity upon components, are available in the literature to predict this property [30]. However, the underlying additivity assumption limited the validity of these equations to narrow ranges of concentration [31]. Moreover, success in the estimation of the density from the chemical composition has been demonstrated only for glasses containing one network former cation (for example Si) [32]; corrections by more complex mathematical functions empirically determined from a number of experimental density determinations are necessary for glasses containing two or more network formers and/or intermediate ions, where the linearity assumption fails [31].

In this context, the successful example of the QSPR approach for the prediction and interpretation of density of multicomponent silica-based bioglasses offered by the studies of Linati et al. [33] and Lusvardi et al. [34] is of great significance. In fact, a unique QSPR model derived is able to rationalize the variation of density in four series of glasses made up of three to five different components.

The four series of glasses studied have the following general formula:

Series 1 (KZ): $2SiO_2 \cdot 1\,Na_2O \cdot 1CaO \cdot xZnO$ ($x = 0, 0.17, 0.34, 0.68$ mol%);
Series 2 (HZ): $(2 - y)SiO_2 \cdot 1\,Na_2O \cdot 1.1CaO \cdot yP_2O_5 \cdot xZnO$ ($x = 0, 0.16, 0.32, 0.78$ mol%; $y = 0.10$ mol%);
Series 3 (HP5Z): $(2 - y)SiO_2 \cdot 1\,Na_2O \cdot 1.1CaO \cdot yP_2O_5 \cdot xZnO$ ($x = 0, 0.16, 0.36, 0.96$ mol%; $y = 0.20$ mol%);
Series 4 (HP6.5Z): $(2 - y)SiO_2 \cdot 1\,Na_2O \cdot 1.1CaO \cdot yP_2O_5 \cdot xZnO$ ($x = 0, 0.17, 0.36, 0.58$ mol%; $y = 0.26$ mol%);

Among the several structural descriptors derived by MD simulations of the glasses [33, 34] the one which better correlates with the experimental density values is $N_{X\text{–}O\text{–}X}/O_{tot}$, i.e., the total number of Si–O–Si, Si–O–Zn, Si–O–P, P–O–P, P–O–Zn and Zn–O–Zn bridges found in the simulated glasses normalized for the total number of oxygen atoms (O_{tot}). This quantity represents an overall descriptor of the degree of polymerization of the glass network. The QSPR model obtained is reported in Fig. 5.2a and shows that the density increases with the overall packing degree of the ions in the glasses which is promoted by addition of Zn to the parent glass or substitution of P for Si. This is not an obvious result, since the increase in the density values is the effect of the balance between the variation of the weight of the components and of the molar volume of the different glass compositions.

The statistical soundness of this correlation is confirmed by its ability to (a) predict the density values of the training set with an average error of 0.012 g/cm^3); (b) predict the density values of two ternary glasses of significant different compositions (TG1:50.6 SiO_2 · 42.5 CaO · 6.9 ZnO; TG2: 48.6 SiO_2 · 31.7 CaO · 19.7 ZnO) chosen as test set, with a % error comparable to the one obtained for the training set (Fig. 5.2a). Moreover, the QSPR model obtained performs better with respect to the ones obtained by the methods of Priven [35] and Demkina [36] (Fig. 5.2b), especially in the range of high densities (high content of ZnO, more that 0.17 mol%).

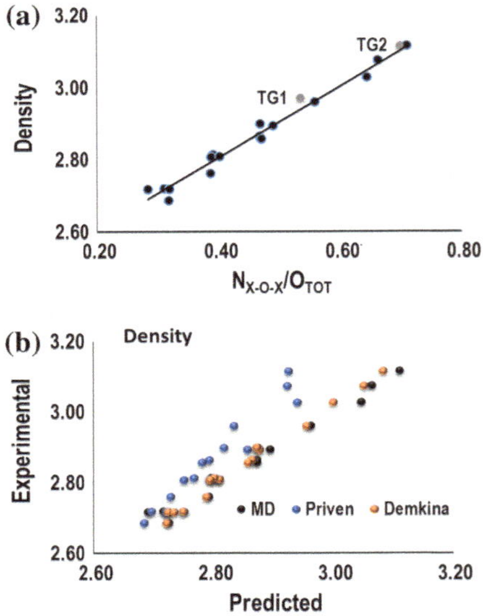

Fig. 5.2 **a** Correlations between the experimental density data values (g/cm^3) and the structural descriptor $N_{X\text{-}O\text{-}X}/O_{tot}$ of multicomponent glasses [34]. The linear regression obtained is: Density = 0.9873 $N_{X\text{-}O\text{-}X}/O_{tot}$ + 2.411 n = 16, R^2 = 0.978, S^2 = 0.012. TG1 and TG2 are used as a test set. **b** Correlations between the experimental density data values and those predicted by means of the $N_{X\text{-}O\text{-}X}/O_{tot}$ descriptor derived by MD, Priven [35] and Demkina [36] methods. The plots are reproduced by the data values reported in [34]

5.3.2 QSPR Models for Glass Transition Temperature and Crystallization Temperature

The invaluable help that computational techniques furnish in the determination of QSPR models for amorphous materials and the importance of utilizing these models as interpretative tools to gain insight difficult to perceive only by the experimental data, is well depicted by the results obtained for complex glasses where two anions are contemporaneously present [37, 38]. In these studies the structural features of Bioactive Fluoro Phospho-Silicate Glasses obtained by classical MD simulations have been used for interpreting the experimental property glass transition temperature (T_g) through a QSPR analysis. The parent compound is the 45S5 Bioglass [39]:

$46.2SiO_2 \cdot 24.3Na_2O \cdot 26.9CaO \cdot 2.6P_2O_5$, hereafter named H. The series of glasses studied are:

Series 1 ($HNaCaF_2$): $46.2SiO_2 \cdot (24.3 - x)Na_2O \cdot 26.9CaO \cdot 2.6P_2O_5 \cdot xCaF_2$ (with x = 0, 5, 10, 15, 20, 24.3 mol%);

Series 2 ($HCaCaF_2$): $46.2SiO_2 \cdot 24.3\,Na_2O \cdot (26.9 - x)CaO \cdot 2.6P_2O_5 \cdot xCaF_2$ (with x = 0, 5, 10, 15, 20, 26.9 mol%);

Series 3 and 4 (HZnO and HP5ZnO): $(2-y)SiO_2 \cdot 1\,Na_2O \cdot 1.1CaO \cdot yP_2O_5 \cdot xZnO$ (x = 0, 0.16, 0.32, 0.78 mol%; y = 0.10, 0.20 mol%); Series 5 (KZnO): $Na_2O \cdot CaO \cdot 2SiO_2 \cdot xZnO$ (x = 0.00, 0.17, 0.34, 0.68 mol%).

The variation of the T_g of silicate glasses upon composition is usually expected to depend on glass polymerization that can be quantified by the Q^n and BO (or NBO) distributions [40–42]; in particular, higher values of glass polymerization are expected to correspond to higher values of T_g. For the series of F-containing glasses analyzed in [37, 38], neither of these two descriptors is able to explain the overall decrease of the T_g data values with respect the F-free H glass for the $HCaCaF_2$ series and the decrease up to 15 % CaF_2 content for the $HNaCaF_2$ series.

The authors overcome this apparent disagreement by invoking simultaneous structural and energetic modifications of glass network upon F addition and they codified this behavior in the F_{net} descriptor. From a structural point of view, the fluoride ions progressively substitute oxygens in metal coordination with a consequent formation of MF_n ionic moieties, that cause the subtraction of Na and Ca ions from the phospho-silicate matrix. This leads to an increment of the polymerization degree of the phospho-silicate portion of the network (increment of %BO and mean $\langle n \rangle$ in the Q^n speciation) [43, 44]. From an energetic point of view, the interaction of the MF_n ionic zones with the phospho-silicate network at low CaF_2 (CaF_2 <15 %) is very weak, being mainly constituted by Na–F neutral pairs, whereas at CaF_2 >15 % the MF_n zones are principally made of Ca–F+ pairs. These link electrostatically the glass matrix, causing an increment in the strength of the glass network with a consequent increment of T_g values.

The F_{net} descriptor is computed as follow:

$$F_{net} = \frac{1}{N}\left[\sum_{i}^{cations}\sum_{j}^{anions} n_i \cdot CN_{ij} \cdot BE_{ij} \cdot m_{ij}\right] \tag{5.3}$$

where N is the total number of atoms, n_i is the number of atoms of the ith species; CN_{ij} is the mean coordination number of ij pairs atoms (i = Si, P, Zn, Na, Ca; j = O2–, F–). BE_{ij} are the bond enthalpies, measured in the gas phase, for each type of bond in the corresponding molecules, as described in [45]. The multiplicative factor m_{ij} represents the maximum number of SiO_4 and PO_4 units linked to the i–O or i–F bonds and is used as fine-tune modulation of the contribution of each bond to the overall network strength.

The linear correlation obtained between the experimental T_g and the F_{net} descriptor is reported in Fig. 5.3; the positive correlation (slope = 0.2851) accounts for the nature of the T_g measurement that represents the temperature necessary to overcome the flow activation energy. The robustness of the QSPR model is corroborated by the variety of glass compositions covered, which envisages ions with different structural role in the different environment of soda-lime-silicate and phospho-silicate glasses.

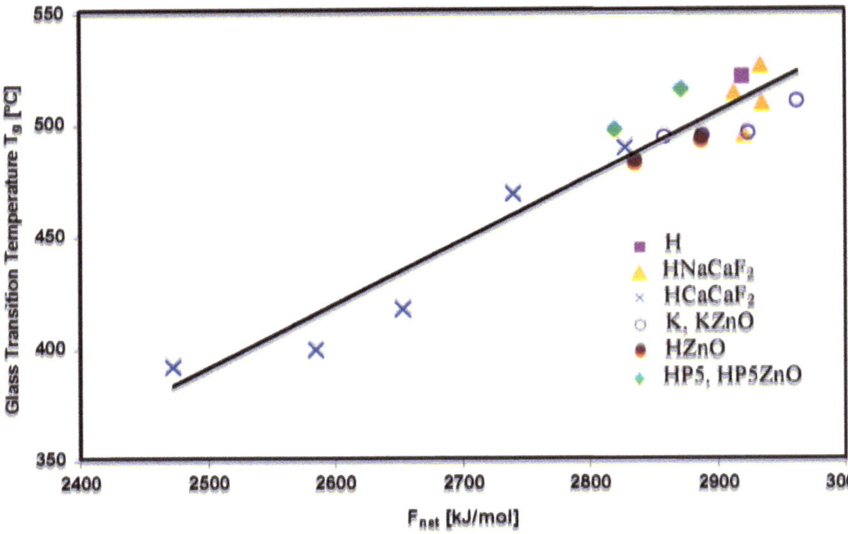

Fig. 5.3 Correlation between experimental glass transition temperature (T_g) and the F_{net} descriptor. The linear regression obtained is: $T_g = 0.2851\ F_{net} - 322.4$, n = 18, $R^2 = 0.912$; $S^2 = 8$. Reprinted with permission from [37] (to which refer for details). Copyright 2009 American chemical Society

The same descriptor F_{net} is able to explains the 68 % of the variation in the crystallization temperature (T_c, first peak) of a series of phospho-silicate glasses doped with ZnO, giving a performance comparable with the descriptor N_{X-O-X}/O_{tot}, which represents the total number of bridges detected in the three-dimensional structure derived by MD simulations, and thus accounts for the polymerization of the glass network [33].

5.3.3 QSAR Models for Leaching and Chemical Durability

The chemical durability of a glass refers to its ability to resist to liquid or atmospheric attacks. The modulation of this physical property of glasses is of fundamental importance in a number of technological area.

Improve durability, i.e. mechanical strength, of glasses would not only enable exciting new applications, but also leads to a significant reduction of material investment for existing applications [46]. However, increasing the durability of a glass by changing its compositions can lead to prohibitively high working temperatures and, therefore, when formulating a commercial glass composition, a compromise is made between durability and workability. On the opposite, the dissolution in body fluid is a major part of the functionality of bioactive glasses [47]. These glasses are designed to create chemical gradients which promote, early in the implantation period, the formation of a layer of biologically active bone-like apatite at the interface. Bone-producing cells, i.e. osteoblasts, can preferentially proliferate on the apatite, and differentiate to form new bones that bond strongly to the implant surface [48]. Glass solubility increases as network connectivity is reduced, consequently, bioactivity occurs only within certain compositional limits and very specific ratios of oxides in the Na_2O–K_2O–CaO–MgO–B_2O_3–P_2O_5–SiO_2 systems [49].

The physico-chemical requirements for biocompatibility and bioactivity in terms of compositional limits and the role of additional ions in tailoring new important mechanical and biological properties for specific clinical applications [39] are poorly known at present. In the following we show, by summarizing the results of two case studies, how the relationships can be established and exploited among the structural role of some key elements that appear to control bioactivity.

5.3.3.1 Zinc-Containing Bioglasses

Zinc added to bioglasses improve their chemical durability, mechanical properties and endows antimicrobial activity; moreover, the release of small concentration of zinc incorporated into an implant material promotes bone formation around the implant and accelerates recovery of the patients, improve adhesion of denture adhesives, etc... Still, it is important to control the Zn releasing rate in order to prevent adverse reactions and to optimize the glass composition to reduce glass degradation without affecting the hydroxyapatite deposition.

The first example of a complete cycle in rational glass design has been reported for these glasses, and is summarized in Fig. 5.4. The authors [33, 50] derived the ratio of Zn/P concentration which produces an optimal dissolution in the body fluid in order to maintain the bioactivity. The QSPR model used accounts for the role of network polymerization on water chemical durability: $\Sigma\%X_i = -1.92\ N_{X\text{–}O\text{–}X}/O_{tot} + 1.33$, n = 6, $R^2 = 0.865$, $S^2 = 0.12$, where $\Sigma\%X_i$, is the total leaching of the glass constituent and the $N_{X\text{–}O\text{–}X}/O_{tot}$ descriptor has been described in the previous paragraph. The number of data points in the data set is small, nevertheless the content of information of the descriptor chosen suggests that solubility is hindered by the zinc tendency to copolymerize with the Si tetrahedral, manifested by a significant increasing of the total number of X–O–X bridges detected in the glass. This model explains the slow rate of zinc dissolution into the media and provides insights into the overall reaction rate reduction of the zinc-containing glasses, regulated by the progressive reduction of the number of NBO species, which ensure the presence of large channels for alkali migration in the network and rapid exchange of Na^+ with H_3O^+ at the glass surface, as summarized by the following linear regressions: $\%P_{(released)} = 0.009\ \%P\text{–}NBO - 0.46$, n = 6; $R^2 \equiv 0.93$, $S^2 = 0.03$; $\%Na_{(released)} = 0.007\ \%Na\text{–}NBO - 0.32$, n = 6; $R^2 = 0.74$, $S^2 \equiv 0.12$; $\%Ca_{(released)} = 0.006\ \%Ca\text{–}NBO - 0.35$, n = 6; $R^2 = 0.84$, $S^2 = 0.03$; where %P–NBO, %Na–NBO, and %Ca–NBO are the percentages of NBOs bonded to P, Na, and Ca ions.

The results of the QSPR study (in silico study) indicated the HZ5 and HP5Z5 as candidates for further studies. Chemical durability tests in water and in-vitro observations in a cellular medium [51, 52] confirmed that the HZ5, HP5Z5, but also

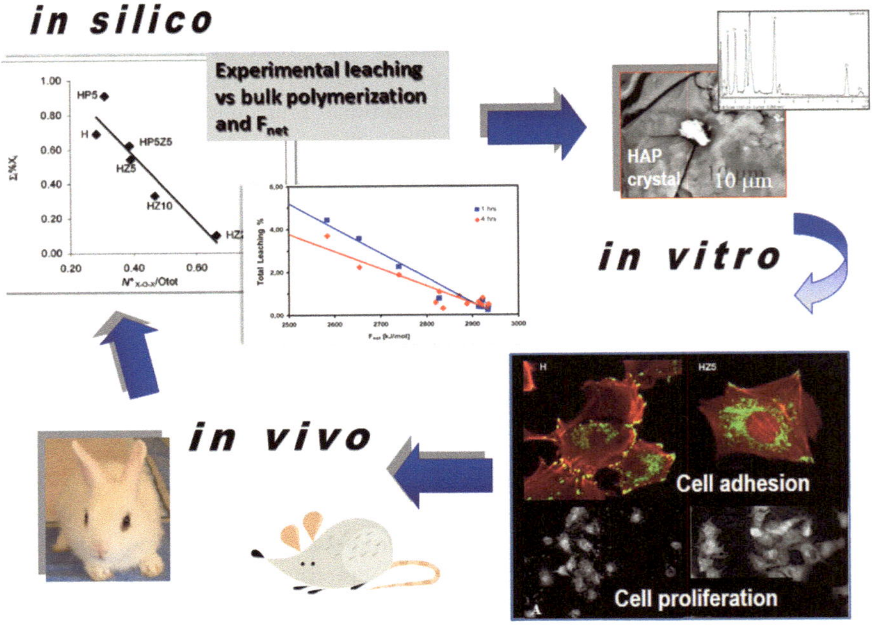

Fig. 5.4 The rational glass design cycle illustrated for Zn-containing bioglasses [33, 50–53]

the HP5Z10 glasses manifest the pre-requisite for bioactivity, since they are able to form a HA layer on their surface after soaking in SBF solution. Moreover, the results of cell culture tests with MC-3T3 osteoblast cells and related cytotoxicity tests allow the selection of the HZ5 and HP5Z5 glasses (not HP5Z10) as the ones with optimal ratio of Zn/P to maintain cell adhesion and cell growth comparable to the parent bioglass (H) used as a control. Finally, in vivo behavior performed on the HZ5 glass [53] matches that in vitro perfectly; they show comparable glass degradation processes and rates, ruled by the amount of zinc in the glass.

These findings triggered further investigations on the chemical durability (express as total leaching % detected after different immersion time in bi-distilled water) of Phospho-modified bioglasses which has been rationalized by means of the F_{net} descriptor defined in the previous paragraph (5.3) [37]. The linear correlations obtained after 1 and 4 h of soaking are: Tot. Leach. % = $-0.01156 \times F_{net} + 34.11$; $n = 9$; $R^2 = 0.965$; $S^2 = 0.020$, and Tot. Leach. % = $-0.00808 \times F_{net} + 23.99$; $n = 13$; $R^2 = 0.851$; $S^2 = 0.105$, respectively (Fig. 5.4). It is worth noting that the correlation coefficients decreases as a function of immersion time ($R^2 = 0.965$, 0.851, 0.682 and 0.640 after 1, 4, 24, 96 h) due to the occurrence of precipitation processes that cannot be taken into account by the F_{net} descriptor. The negative slope of the correlations indicates that the higher F_{net} (i.e. the overall strength of the glass network), the greater the chemical durability.

It is worth noting that the wide range of variation of the correlations is essentially due to the glasses of the $HCaCaF_2$ series (Total Leaching %: 0.77–4.44 mol%,

1 h) which show much higher solubility with respect to the HNaCaF2 series (Total Leaching %: 0.37–0.24 mol%, 1 h). This behavior has been ascribed by the authors [37] to the conversion of Ca^{2+} and Na^+ species to $Ca–F^+$ and NaF ones upon addiction of the fluorine ions with an overall reduction of network complexation.

5.3.3.2 Yttrium-Containing Aluminosilicate Glasses

A key requirement for successful application of glass delivery systems for radiation is a high durability of the glass used to minimize the release and the fatal results of circulation of the radioactive agent in the body. Therefore, also in this case, a deep understanding of the way in which the glass composition controls the glass dissolution is needed.

In a recent work by Christie et al. [54] the specific structural features of the glasses that control the solubility of a series of yttrium aluminosilicate glasses (parent glass composition: $17Y_2O_3–19Al_2O_3–64SiO_2$) have been extracted from MD simulations and used to predict the solubility of these materials. In particular, a linear combination of the following descriptors showed a high correlation with the experimental solubility: (1) CN_{SiOSi}, which is the average O–Si coordination number of oxygen atoms already coordinated to at least another silicon atom. This counts for the connectivity of the silicon atoms in the network; (2) the yttrium clustering ratio R_{YY} using the ratio of the measured Y–Y coordination number (at a cutoff of 5 Å) to the number expected if the yttrium atoms were distributed uniformly (randomly) throughout the available space [55]. Values of $R_{YY} > 1$ denote spatial clustering, while $R_{YY} = 1$ describes a uniform distribution of Y atoms throughout the available space; (3) the number of intratetrahedral O–Si–O bonds per yttrium atom (N_{intra}). In general, any increase in the amount of intratetrahedral Y–O coordination will decrease the number of fragments of the glass network coordinated to yttrium. Because the strong Y–O interaction can be expected to reduce the mobility and increase the resistance to dissolution of these fragments, a positive correlation between the extent of intratetrahedral Y–O coordination and glass solubility can also be expected [54].

The linear combination of these parameters s is given as:

$$s = 0.310\,CN_{SiOSi} + 0.076\,R_{Y\text{-}Y} - 0.136\,N_{intra} \tag{5.4}$$

A good correlation between the solubility of the glasses in water (measured as weight loss) and the descriptor s (Fig. 5.5) is obtained, the correlation coefficients being $R^2 = 0.955$ ($\sigma = 0.97$ mg/cm^2). The negative slope of the regression indicates that as s increases, the solubility of the glass decreases. The coefficients of the first two terms of s are positive, implying that increasing the (Si–)O–Si coordination number and/or increasing the Y–Y clustering ratio will lead to decreasing solubility. Conversely, the sign of the third term of s is negative, implying that an increase in the number of intratetrahedral O–Si–O bonds around the yttrium atoms will increase the solubility.

It is worth noting that the R^2 statistical parameter for the correlation between the solubility s and the CN_{SiOSi} is 0.909 ($\sigma = 1.38$ mg/cm^2), denoting that CN_{SiOSi}

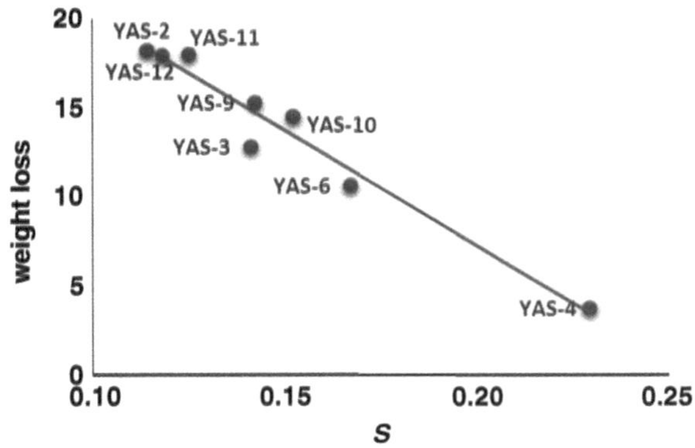

Fig. 5.5 Correlation between weight loss (mg/cm^2) and the structural descriptor *s*. The plot is reproduced by the data values reported in [54] (to which refer for details)

captures most of the experimental trends; the observed small improvement in the linear fit for its combinations with other parameters in part arises from the larger number of parameters in the fit (overfitting).

5.3.4 QSPR Models for Young's Modulus

Elastic properties, specifically Young's modulus E, have attained paramount interest for a variety of glass applications such as accelerated devices, including hard discs and surgery equipment, lightweight construction, and composite materials [46].

From a practical point of view, the mechanical properties of a glass often dictate whether a specific need or application can be met. Therefore, the prediction of these properties according to glass composition is becoming compulsory.

An interesting computational investigation on the composition dependence of mechanical properties of multicomponent glasses has been performed by Pedone et al. [56, 57]. This work represents the first detailed systematic computational study of the mechanical properties of three wide series of alkali silicate glasses, of general formula $xM_2O–(100–x)SiO_2$ (M = Li, Na, K; x = 0, 10, 15, 20, 25, 30 mol%), obtained by means of the MD and energy minimization methods. Besides the correct quantitative calculations of the observable values of the mechanical properties (Young's modulus, shear modulus, bulk modulus and Poisson's ratio), the authors reported an important QSPR model between Young's modulus (E) and the correlation length (L) of the first sharp diffraction peak (FSDP).

Changes in the FSDP as a function of composition have been attributed to variation in the medium range order of the glass [58]. The quantitative rationalization of the Young's modulus modulation by dopant addition reported in Fig. 5.6 promotes the correlation length as an eligible descriptor both for quantitative predictions and interpretation of the structure dependence of the Young's modulus for the alkali silicate glasses. Since the experimental Young's modulus values are reproduced by computational simulations with maximal differences of 4, 4 and 2 % for lithium, sodium and potassium silicate glasses, the statistical significance of the correlations obtained is

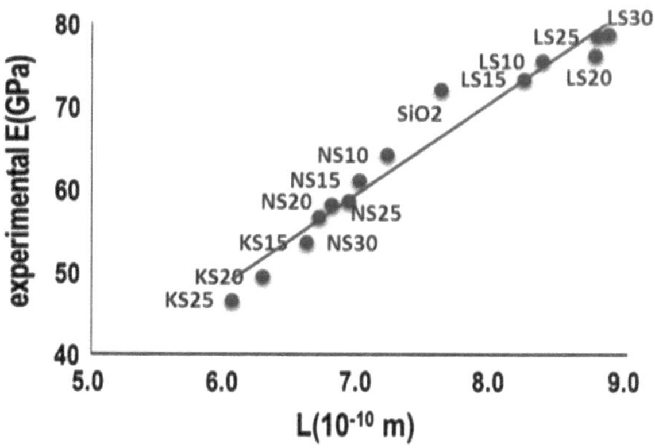

Fig. 5.6 Correlation between the calculated Young's modulus (E) and the correlation length (L) of the first sharp diffraction peak of alkali silicate glasses. The plots is reproduced by the data values reported in [56]

comparable when experimental or computed Young's modulus data values are used: E(GPa)(Exp) = 10.993 L + 16.409 n = 14 R^2 = 0.932 S = 3.025; E(GPa)(Comput.) = 11.177 L + 18.654 n = 16 R^2 = 0.968 S = 2.191. It is worth noting the positive slope and the distribution of the glasses in the E-L space according to the nature of the dopant: the characteristic correlation length decreases as a function of Na and K content, and increases as a function of Li content. Therefore, the intermediate range order decreases with Na and K concentration, whereas the high field strength of Li determine the ordering of the surrounding network and modifier regions.

5.3.5 QSPR Models for NMR Spectra

Solid State nuclear magnetic resonance NMR spectroscopy has been firmly established as a powerful technique for glass structure investigation [60, 61], being very sensitive to the local environment (i.e., bond distances and angles, coordination numbers) and to the nature of the second coordination sphere. Unfortunately, the interpretation of the experimental spectra is hindered by the inhomogeneous broadening of the isotropic line due to the different structural units present in the glass.

In the past, the interpretation of the NMR spectra was based on empirical correlations derived from the study of crystalline materials with known structure, [62] and, successively, on correlations with structural descriptors computed by ab-initio calculations on crystals or model clusters [63–68]. However, crystalline systems generally exhibit a limited diversity of local structures in contrast to the disorder and variety of structural units (different coordination numbers and Q^n distributions) present in multicomponent glasses. Moreover, the cluster approach does not account for the correlations between structural factors that exist in solids and disorder in glasses [69–71].

To overcome these limitations, several studies focused on semi-quantitative comparisons between information derived from NMR spectra and structural features obtained from molecular dynamics simulations on large glass samples. They make use of connectivity between different types of Q^n species and the related descriptors [72–81], but attempts to investigate cation distribution and clustering have also been made [82]. Moreover, the interpretative and predictive relevance of statistical correlations between NMR-derived and MD-derived descriptors (quantitative structure-properties relationships) has also been discussed [25, 33].

A major breakthrough occurred in the early 2000, when the calculation of NMR parameters from first principles [83] and, successively, the simulation of the line widths and shapes of the NMR spectra have become possible [84–86], through the MD-DFT/GIPAW (Gauge Including Projector Augmented Wave) approach. This approach has opened a new route for interpreting NMR parameter distributions and for refining the relationships between NMR parameters and local structural features. In fact, calculations of NMR parameters (chemical shielding and quadrupolar parameters) of each nucleus is performed on the three-dimensional model of the glass obtained by MD simulations and refined by Density Functional Theory (DFT) calculations. Then, comparison between experimental and theoretical spectra features is performed, and, being the results satisfactory, the establishment of quantitative structure-NMR parameter relationships becomes feasible. Accurate relationships between NMR parameters and structural descriptors are extremely useful for the interpretation of experimental data, as they make a reverse approach possible [84, 87–89]. In this way, structural descriptors (i.e. bond and angle distributions) of a glass sample could, in principle, be directly obtained from the experimental NMR parameters distribution (Fig. 5.7).

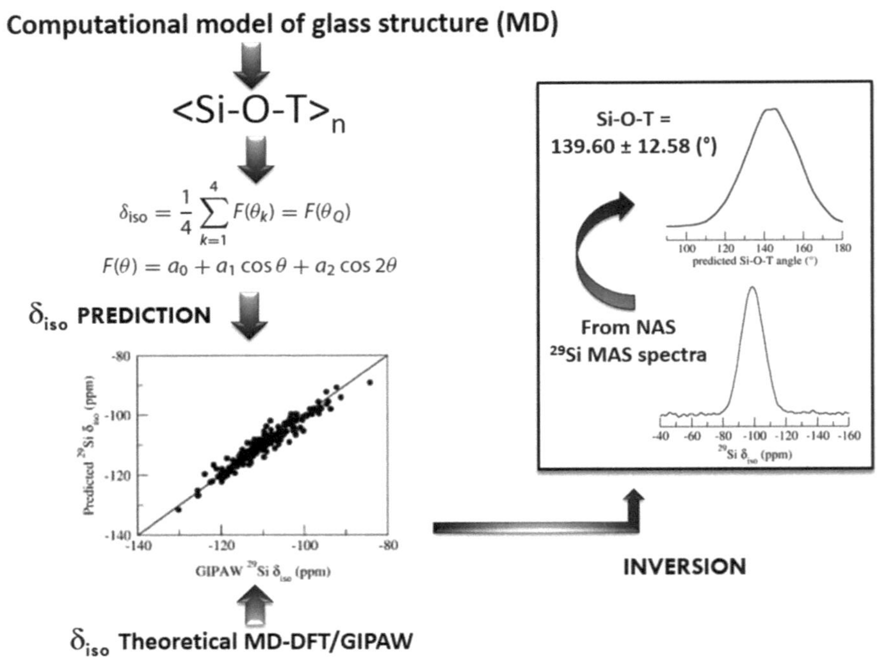

Fig. 5.7 The structural inversion QSPR approach to extract structural distributions from NMR data. The example of the Si–O–T distribution for sodium silicate glasses is given [87]

Table 5.1 QSPR models of multicomponent silicate glasses involving NMR computed parameters and structural descriptors obtained by MD

Glasses	NMR parameters	MD structural descriptors
Alkaline/alkaline earth silicate [85, 87, 89–92]	^{29}Si δ_{iso} of each Q^n species	Mean Si–O–T angle (T denotes the Q^n connected tetrahedron)
	^{29}Si Δ_{cs} of Q^3 species	⟨Si–O⟩BO and ⟨Si–O⟩NBO bond lengths
	^{17}O δ_{iso} of BO and NBO	Average Si–BO, Si–NBO and M–BO, N–NBO distances (M = Na,Ca)
	^{23}Na δ_{iso}	Number of coordinating NBO atoms to a given Na, mean Na–O bond length
Alumino silicate [80, 89, 92, 93]	^{27}Al and ^{29}Si δ_{iso}	⟨Al–O–T⟩ and ⟨Si–O–T⟩ bond angles
	^{29}Si δ_{iso}	Amount of modifier cations in the silicon second coordination sphere
	^{27}Al δ_{iso}	The fractional population of Al polyhedra
Phospho silicate (Bioglasses) [33, 89, 95]	^{29}Si NMR δ_{iso} of each Q^n species	Mean Si–O–T angle
Boro silicate [88, 89]	^{29}Si and ^{11}B δ_{iso}	Mean Si–O–T angle

Some examples of QSPR results involving NMR computed parameters and structural descriptors obtained by MD obtained for a number of multicomponent silicate glasses are summarized in Table 5.1.

Direct information on structural regions dominated by different Q^n species in Alkaline/alkaline earth silicate glasses have been obtained from linear and multilinear regressions. The statistical models achieved an accuracy in prediction of about 2 ppm for the ^{29}Si δ_{iso} [87], 10 ppm for the ^{23}Na δ_{iso} [91], of 2°–4° for the mean value of the Si–O–Si bond angle distribution, 2°–4° [87], and of less than 10 ppm for the chemical shift anisotropy ^{29}Si Δ_{cs} of the Q^3 species [91].

Two of the most investigated relationships in aluminosilicate glasses are those between ^{27}Al and ^{29}Si δ_{iso} and inter-tetrahedral angles [84, 96–98]. In general, poor correlations are obtained unless the connectivity between Si and Al and the different oxygen species (BOs, NBOs, TBOs) is taken into account [94].

In the cases of phosphosilicate glasses (Bioglasses), the analysis of the correlation coefficients obtained for the linear correlations between the theoretical ^{29}Si δ_{iso} and the mean Si–O–T angle (R^2 0.55, 0.62, and 0.89 for Q^1, Q^2, and Q^3 Si species, respectively) clearly indicates that the Q^n distribution of the Si species is controlled by the nonrandom distribution of Na and Ca atoms in the glass [89, 95]. The worst correlations coefficients have to be ascribed to the irregular distributions of Ca ions around the different Si Q^n species (its concentration is maximal around the Q^2 and minimal around the Q^4 species). This is a general trend which has been observed in alkaline and alkaline-earth glasses and aluminosilicate glasses [85, 99].

Finally, an elegant example of structural inverse correlations is reported by Soleilhavoup et al. [88] for borosilicate glasses. The methodology, derived for the first time for vitreous silica [84], consists in extracting the distribution of NMR parameters (i.e. the distribution of isotropic chemical shifts for each boron resonance) from ^{11}B 3QMAS spectra; establishing a quantitative relationship between the ^{11}B isotropic chemical shift and each B–O–B angle; and finally mapping the NMR parameter distribution into a distribution of the B–O–B angle (structural inversion of the ^{11}B NMR spectrum).

5.4 Outlook

The main goal of computational material design is to gain "rational" control of the structure of complex real-life systems at all relevant length scales, thus the optimization and prediction of specific properties which fulfil end-user application requirements become possible. Notwithstanding the great advances achieved in computation, glass design is still in its infancy and constitutes an important avenue for future research.

To this respect, QSPR is a precious tool since it can be used at different steps of the problem-solving strategy for glass design: (a) in a preliminary step, to assist end-users in the choice of the hierarchic level of theory and simulations to provide the most comprehensive description of the glass system at hand; (b) in an intermediate step, to map the amount of information derived from the computations to the space of the glass relevant properties. This might be devised to obtain the correct numerical value of the property or to discover connections, trends and relationships that would otherwise be very difficult to detect by simple observation, i.e. to create a chemical model that is easy to understand; and (c) at the final step to predict properties of new glass formulations in a cheap, efficient and environmentally friendly manner.

Such ambitious tasks require the development of improved atomistic simulation methods and/or new mathematical approaches that enable the quick derivation of specific descriptors for non-covalent amorphous systems at low computational costs.

For the time being, combination of MD and QS PR analysis helps to gain valuable information for the understanding of materials and chemical processes and furnishes a useful tool for predictive purposes.

References

1. A. Eugen, Glasses as engineering materials: a review. Mater. Des. **32**, 1717–1732 (2011)
2. I. Izquierdo-Barba, A.J. Salinas, M. Vallet-Regi, M bioactive glasses: from macro to nano. Int. J. Appl. Glass Sci. **4**, 149–161 (2013)
3. J. Hum, A.R. Boccaccini, Bioactive glasses as carriers for bioactive molecules and therapeutic drugs: a review. J. Mater. Sci. Mater. M. **23**, 2317–2333 (2012)

4. G.N. Greaves, S. Sen, Inorganic glasses, glass-forming liquids and amorphizing solids. Adv. Phys. **56**, 1–166 (2007)
5. SciGlass 6.5 Database and Information System, 2005. http://www.sciglass.info/
6. International Glass Database System Interglad Ver. 7. http://www.newglass.jp/interglad_n/gaiyo/info_e.html
7. A.I. Priven, O.V. Mazurin, Glass property databases: their history, present state, and prospects for further development. Glass: the challange for the 21st century. Adv. Mat. Res. **39–40**, 147–152 (2008)
8. A.I. Priven, General method for calculating the properties of oxide glasses and glass forming melts from their composition and temperature. Glass Technol. **45**, 244–254 (2004)
9. A. Pedone, G. Malavasi, M.C. Menziani, A.N. Cormack, U. Segre, A new self-consistent empirical interatomic potential model for oxides, silicates, and silica based glasses. J. Phys. Chem. B **110**, 11780–11795 (2006)
10. A. Tilocca, N. de Leeuw, A.N. Cormack, Shell-model molecular dynamics calculations of modified silicate glasses. Phys. Rev. B **73**, 104209–104223 (2006)
11. A. Pedone, M. Corno, B. Civalleri, G. Malavasi, M.C. Menziani, U. Segre, P. Ugliengo, An ab initio parameterized interatomic force field for hydroxyapatite. J. Mat. Chem. **17**, 2061–2068 (2007)
12. A. Pedone, G. Malavasi, M.C. Menziani, U. Segre, F. Musso, M. Corno, B. Civalleri, P. Ugliengo, FFSiOH: a new force field for silica polymorphs and their hydroxylated surfaces based on periodic B3LYP calculations. Chem. Mater. **20**, 2522–2531 (2008)
13. A. Tilocca, Short- and medium-range structure of multicomponent bioactive glasses and melts: an assessment of the performances of shell-model and rigid-ion potentials. J. Chem. Phys. **129**, 084504 (2008)
14. M. Salanne, B. Rotenberg, S. Jahn, R. Vuilleumier, C. Simon, P.A. Madden, Including many-body effects in models for ionic liquids. Theor. Chem. Acc. **131**, 1143–1149 (2012)
15. A. Pedone, Properties calculations of silica-based glasses by atomistic simulations techniques: a review. J. Phys. Chem. C **113**, 20773–20784 (2009)
16. M.E. Eberhart, D.P. Clougherty, Looking for design in material design. Nat. Mater. **3**, 659–861 (2004)
17. K. Wu, B. Natarajan, L. Morkowchuk, M. Krein, C.M. Breneman, From drug discovery QSAR to predictive materials QSPR: the evolution of descriptors, methods, and models. pp. 385–410. in *Informatics dor materials science and engineering*, ed. by K. Rajan (Elsevier, Amsterdam, 2013)
18. A.R. Katritzky, U. Maran, V.S. Lobanov, M. Karelson, Structurally diverse quantitative structure-property relationship correlations of technologically relevant physical properties. J. Chem. Inf. Comput. Sci. **40**, 1–18 (2000)
19. W.M. Berhanu, G.G. Pillai, A.A. Oliferenko, A.R. Katritzky, Quantitative structure-activity/property relationships: the ubiquitous links between cause and effect. Chem. Plus Chem. **77**, 507–517 (2012)
20. A.R. Katritzky, M. Kuanar, S. Slavov, C.D. Hall, M. Karelson, I. Kahn, D.A. Dobchev, Quantitative correlation of physical and chemical properties with chemical structure: utility for prediction. Chem. Rev. **110**, 5714–5789 (2010)
21. T. Le, V.C. Epa, F.R. Burden, D.A. Winkler, Quantitative structure-property relationship modeling of diverse materials properties. Chem. Rev. **112**, 2889–2919 (2012)
22. M. Karelson, V.S. Lobanov, A.R. Katritzky, Quantum-chemical descriptors in QSAR/QSPR studies. Chem. Rev. **96**, 1027–1043 (1996)
23. G.S. Henderson, The structure of silicate melts: a glass perspective. Can. Mineral. **43**, 1921–1958 (2005)
24. G. Mountjoy, B.M. Al-Hasni, C. Storey, Structural organisation in oxide glasses from molecular dynamics modeling. J. Non-Cryst. Solids **357**, 2522–2529 (2011)
25. G. Malavasi, A. Pedone, M.C. Menziani, Towards a quantitative rationalization of multicomponent glass properties by means of molecular dynamics simulations. Mol. Simul. **32**, 1045–1055 (2006)

26. L. Adkins, A.N. Cormack, Large-scale simulations of sodium silicate glasses. J. Non-Cryst. Solids **357**, 2538–2541 (2011)
27. G. Malavasi, M.C. Menziani, A. Pedone, U. Segre, Void size distribution in MD-modelled silica glass structures. J. Non-Cryst. Solids **352**, 285–296 (2006)
28. P.X. Liu, W. Long, Current mathematical methods used in QSAR/QSPR studies. Int. J. Mol. Sci. **10**, 1978–1998 (2009)
29. K. Baumann, Cross-validation as the objective function for variable-selection techniques. TrAC Trends Anal. Chem. **22**, 395–406 (2003)
30. S. Inabaw, S. Fujino, Empirical equation for calculating the density of oxide glasses. J. Am. Ceram. Soc. **93**, 217–220 (2010)
31. M.B. Volf, Mathematical Approach to Glass; Glass Science and Technology, vol. 9. (Elsevier, Prague, 1988)
32. I. Priven, O.V. Mazurin, Comparison of methods used for the calculation of density refractive index and thermal expansion of oxide glasses. Glass Technol. **44**, 156–166 (2003)
33. L. Linati, G. Lusvardi, G. Malavasi, L. Menabue, M.C. Menziani, P. Mustarelli, U. Segre, Qualitative and quantitative structure-property relationships (QSPR) analysis of multicomponent potential bioglasses. J. Phys. Chem. B **109**, 4989–4998 (2005)
34. G. Lusvardi, G. Malavasi, L. Menabue, M.C. Menziani, A. Pedone, U. Segre, Density of multicomponent silica-based potential bioglasses: quantitative structure-property relationships (QSPR) analysis. J. Eur. Ceram. Soc. **27**, 499–504 (2007)
35. A.I. Priven, Evaluation of the fraction of fourfold-coordinated boron in oxide glasses from their composition. Glass Phys. Chem. **26**, 441–454 (2000)
36. L.I. Demkina, Investigation of the dependence of glass properties on their composition. Izd. Chimia **23**, 123–127 (1976)
37. G. Lusvardi, G. Malavasi, F. Tarsitano, L. Menabue, M.C. Menziani, A. Pedone, Quantitative structure-properties relationships of potentially bioactive fluoro phospho-silicate glasses. J. Phys. Chem. B **113**, 10331–10338 (2009)
38. G. Lusvardi, G. Malavasi, M. Contrada, L. Menabue, M.C. Menziani, A. Pedone, U. Segre, Elucidation of the structural role of fluorine in potentially bioactive glasses by experimental and computational investigation. J. Phys. Chem. B **112**, 12730–12739 (2008)
39. L.L. Hench, Bioactive materials: the potential for tissue generation. Biomaterials **19**, 1419–1423 (1998)
40. J.E. Shelby, *Introduction to Glass Science and Technology* (RCS Paperbacks, Cambridge, 1997)
41. G. Engelhardt, D. Michel, *High-Resolution Solids State NMR of Silicate and Zeolites* (Wiley, Chichester, 1987)
42. G.N. Greaves, S. Sen, Inorganic glasses, glass-forming liquids and amorphising solids. Adv. Phys. **56**, 1–166 (2007)
43. K. Christie, A. Pedone, M.C. Menziani, A. Tilocca, Fluorine environment in bioactive glasses: ab initio molecular dynamics simulations. J. Phys. Chem. B **115**, 2038–2045 (2011)
44. A. Pedone, T. Charpentier, M.C. Menziani, The structure of fluoride-containing bioactive glasses: new insights from first-principles calculations and solid state NMR spectroscopy. J. Mater. Chem. **22**, 12599–12608 (2012)
45. J.A. Kerr, CRC Handbook of Chemistry and Physics 1999–2000: A Ready-Reference Book of Chemical and Physical Data, 81st edn. ed. by D.R. Lide (CRC Press, Boca Raton, 2000)
46. L. Wondraczek1, J.C. Mauro, J. Eckert, U. Kühn, J. Horbach, J. Deubener, T. Rouxel, Towards ultrastrong glasses. Adv. Mat. **23**, 4578–4586 (2011)
47. L.L. Hench, D.E. Day, W. Höland, V.M. Rheinberger, Glass and medicine. Int. J. Appl. Glass Sci. **1**, 104–117 (2010)
48. T. Kokubo, An introduction to Bioceramics. In Advanced Series in Ceramics, vol. 1. ed. by L.L. Hench, J. Wilson (World Scientific Publishing Co., Singapore, 1993)
49. M. Brink, T. Turunen, R.P. Happonen, A. Yli-Urpo, Compositional dependence of bioactivity of glasses in the system Na_2O-K_2O-MgO-CaO-B_2O_3-P_2O_5-SiO_2. J. Biomed. Mater. Res. **37**, 114–121 (1997)

50. G. Lusvardi, G. Malavasi, L. Menabue, M.C. Menziani, Synthesis, characterization and molecular dynamics simulation of Na_2O-CaO-SiO_2-ZnO glasses. J. Phys. Chem. B **106**, 9753–9760 (2002)
51. G. Lusvardi, G. Malavasi, A. Pedone, L. Menabue, M.C. Menziani, V. Aina, A. Perardi, C. Morterra, F. Boccafoschi, S. Gatti, M. Bosetti, M.F. Cannas, Properties of zinc releasing surfaces for clinical applications. J. Biomater. Appl. **22**, 505–526 (2008)
52. G. Lusvardi, G. Malavasi, L. Menabue, M.C. Menziani, A combined experimental-computational strategy for the design, synthesis and characterization of bioactive zinc-silicate glasses. Key Eng. Mater. **377**, 211–224 (2008)
53. G. Lusvardi, D. Zaffe, L. Menabue, C. Bertoli, G. Malavasi, U. Consolo, In vitro and in vivo behaviour of zinc-doped phosphosilicate glasses. Acta Biomater. **5**, 419–428 (2009)
54. J.K. Christie, A. Tilocca, Molecular dynamics simulations and structural descriptors of radioisotope glass vectors for in situ radiotherapy. J. Phys. Chem. B **116**, 12614–12620 (2012)
55. A. Tilocca, A.N. Cormack, N.H. de Leeuw, The structure of bioactive silicate glasses: new insight from molecular dynamics simulations. Chem. Mater. **19**, 95–103 (2007)
56. A. Pedone, G. Malavasi, A.N. Cormack, U. Segre, M.C. Menziani, Insight into elastic properties of binary alkali-silicate glasses; prediction and interpretation through atomistic simulation techniques. Chem. Mater. **19**, 3144–3154 (2007)
57. A. Pedone, G. Malavasi, A.N. Cormack, U. Segre, M.C. Menziani, Elastic and dynamical properties of silicate glasses from computer simulations techniques. Theor. Chem. Acc. **120**, 557–564 (2008)
58. G. Hauret, Y. Vaills, T. Parot-Rajaona, F. Gervais, D. Mas, Y. Luspin, Dynamic behaviour of (1-x) SiO2-0.5xM2O glasses (M = Na, Li) investigated by infrared and Brillouin spectroscopies. J. Non-Cryst. Solids **191**, 85–93 (1995)
59. M. Eden, NMR studies of oxide-based glasses. Annu. Rep. Prog. Chem. Sect. C Phys. Chem. **108**, 177–221 (2012)
60. J.V. Hanna, M.E. Smith, Recent technique developments and applications of solid state NMR in characterising inorganic materials. Solid State NMR **38**, 1–18 (2010)
61. G. Engelhardt, D. Michel, High-Resolution Solid-State NMR of Silicates and Zeolites (Wiley, New York, 1988)
62. J.A. Tossell, P. Lazzeretti, Ab initio calculations of oxygen nuclear-quadrupole coupling constants and oxygen and silicon NMR shielding constants in molecules containing Si-O bonds. Chem. Phys. **112**, 205–212 (1987)
63. J.A. Tossell, Calculation of NMR shieldings and other properties for 3 and 5 coordinate Si, 3 coordinate-O and some siloxane and boroxol ring compounds. J. Non-Cryst. Solids **120**, 13–19 (1990)
64. X. Xue, M. Kanzaki, An ab initio calculation of ^{17}O and ^{29}Si NMR parameters for SiO_2 polymorphs. Solid State Nucl. Magn. Reson. **16**, 245–259 (2000)
65. T. Clark, P.J. Grandinetti, P. Florian, J.F. Stebbins, An ^{17}O NMR investigation of crystalline sodium metasilicate: implications for the determination of local structure in alkali silicates. J. Phys. Chem. **105**, 12257–12265 (2001)
66. H. Maekawa, P. Florian, D. Massiot, H. Kiyono, M. Nakaruma, Effect of alkali metal oxide on ^{17}O NMR parameters and Si-O-Si angles of alkali metal disilicate glasses. J. Phys. Chem. **100**, 5525–5532 (1996)
67. J.A. Tossell, Quantum mechanical calculation of ^{23}Na NMR shieldings in silicates and aluminosilicates. Phys. Chem. Miner. **27**, 70–80 (1999)
68. A. Pedone, M. Biczysko, V. Barone, Environmental effects in computational spectroscopy. Chem. Phys. Chem. **11**, 1812–1832 (2010)
69. A. Pedone, M. Pavone, M.C. Menziani, V. Barone, Accurate first-principle prediction of ^{29}Si and ^{17}O NMR parameters in SiO_2 polymorphs: the cases of zeolites sigma-2 and ferrierite. J. Chem. Theory Comput. **4**, 2130–2140 (2008)
70. J. Casanovas, F. Illas, G. Pacchioni, Ab initio calculations of ^{29}Si solid state NMR chemical shifts of silcane and silanol groups in silica. Chem. Phys. Lett. **326**, 523–529 (2000)

71. L. Linati, G. Lusvardi, G. Malavasi, L. Menabue, M.C. Menziani, P. Mustarelli, A. Pedone, U. Segre, Medium range order in phospho-silicate bioactive glasses: insights from MAS-NMR spectra, chemical durability experiments and molecular dynamics simulations. J. Non-Cryst. Solids **354**, 84–89 (2008)
72. J.A. Tossell, J. Horbach, O triclusters revisited: classical MD and quantum cluster results for glasses of composition $(Al_2O_3)2(SiO_2)$. J. Phys. Chem. B **109**, 1794–1797 (2005)
73. L. Olivier, X. Yuan, A.N. Cormack, C. Jäger, Combined ^{29}Si double quantum NMR and MD simulation studies of network connectivities of binary Na_2OSiO_2 glasses: new prospects and problems. J. Non-Cryst. Solids **293–295**, 53–66 (2001)
74. C. Leonelli, G. Lusvardi, M. Montorsi, M.C. Menziani, L. Menabue, P. Mustarelli, L. Linati, Influence of small additions of Al_2O_3 on the properties of the Na_2O3SiO_2 glass. J. Phys. Chem. B **105**, 919–927 (2001)
75. B. Park, H. Li, L.R. Corrales, Molecular dynamics simulation of La_2O_3-Na_2O-SiO_2 glasses. II. The clustering of La^{3+} cations. J. Non-Cryst. Solids **297**, 220–238 (2002)
76. L. Barbieri, V. Cannillo, C. Leonelli, M. Montorsi, C. Siligardi, P. Mustarelli, Characterisation of CaO-ZrO_2-SiO_2 glasses by MAS-NMR and molecular dynamics. Phys. Chem. Glass. B **45**, 138–140 (2004)
77. E. Kashchieva, B. Shivachev, Y. Dimitriev, Molecular dynamics studies of vitreous boron oxide. J. Non-Cryst. Solids **351**, 1158–1161 (2005)
78. J. Machacek, O. Gedeon, M. Liska, Group connectivity in binary silicate glasses. J. Non-Cryst. Solids **352**, 2173–2179 (2006)
79. G. Mountjoy, B.M. Al-Hasni, C. Storey, Structural organisation in oxide glasses from molecular dynamics modeling. J. Non-Cryst. Solids **357**, 2522–2529 (2011)
80. A. Jaworski, B. Stevensson, B. Pahari, K. Okhotnikov, M. Eden, Local structures and Al/Si ordering in lanthanum aluminosilicate glasses explored by advanced ^{27}Al NMR experiments and molecular dynamics simulations. Phys. Chem. Chem. Phys. **14**, 15866–15878 (2012)
81. H. Inoue, A. Masuno, T. Watanabe, Modeling of the structure of sodium borosilicate glasses using pair potentials. J. Phys. Chem. B **116**, 12325–12331 (2012)
82. U. Voigt, H. Lammert, H. Eckert, A. Heuer, Cation clustering in lithium silicate glasses: quantitative description by solid-state NMR and molecular dynamics simulations. Phys. Rev. B **72**, 64207 (2005)
83. C.J. Pickard, F. Mauri, All-electron magnetic response with pseudopotentials: NMR chemical shifts. Phys. Rev. B **63**, 245101 (2001)
84. T. Charpentier, P. Kroll, F. Mauri, First-principles nuclear magnetic resonance structural analysis of vitreous silica. J. Phys. Chem. C **113**, 7917–7929 (2009)
85. A. Pedone, T. Charpentier, M.C. Menziani, Multinuclear NMR of $CaSiO_3$ glass: simulation from first-principles. Phys. Chem. Chem. Phys. **12**, 6054–6066 (2010)
86. T. Charpentier, The PAW/GIPAW approach for computing NMR parameters: a new dimension added to NMR study of solids. Solid State Nucl. Magn. Reson. **40**, 1–20 (2011)
87. F. Angeli, O. Villain, S. Schuller, S. Ispas, T. Charpentier, Insight into sodium silicate glass structural organization by multinuclear NMR combined with first-principles calculations. Geochim. Cosmochim. Acta **75**, 2553–2469 (2011)
88. A. Soleilhavoup, J.-M. Delaye, F. Angeli, D. Caurant, T. Charpentier, Contribution of first-principles calculations to multinuclear NMR analysis of borosilicate glasses. Magn. Res. Chem. **48**, S159–S170 (2010)
89. T. Charpentier, M.C. Menziani, A. Pedone, Computational simulations of solid state NMR spectra: a new era in structure determination of oxide glasses. RSC Adv. **3**, 10550–10578 (2013)
90. S. Ispas, T. Charpentier, F. Mauri, D.R. Neuville, Structural properties of lithium and sodium tetrasilicate glasses: molecular dynamics simulations versus NMR experimental and first-principles data. Solid State Sci. **12**, 183–192 (2010)
91. T. Charpentier, S. Ispas, M. Profeta, F. Mauri, C.J. Pickard, First-principles calculation of ^{17}O, ^{29}Si, and ^{23}Na NMR spectra of sodium silicate crystals. J. Phys. Chem. B **108**, 4147–4161 (2004)

92. A. Pedone, E. Gambuzzi, M.C. Menziani, Unambiguous description of the oxygen environment in multicomponent aluminosilicate glasses from ^{17}O solid state NMR computational spectroscopy. J. Phys. Chem. C **115**, 14599–14609 (2012)
93. A. Pedone, E. Gambuzzi, G. Malavasi, M.C. Menziani, First-principles simulations of the ^{27}Al and ^{17}O solid state NMR spectra of a calcium alumino-silicate glass. Theor. Chem. Acc. **131**, 1147–1157 (2012)
94. E. Gambuzzi, A. Pedone, M.C. Menziani, F. Angeli, D. Caurant, T. Charpentier, Probing silicon and aluminium chemical environments in silicate and aluminosilicate glasses by solid state NMR spectroscopy and accurate first-principles calculations. Geochim. Cosmochim. Acta **125**, 170–185 (2014)
95. A. Pedone, T. Charpentier, G. Malavasi, M.C. Menziani, New insights into the atomic structure of 45S5 bioglass by means of solid-state NMR spectroscopy and accurate first-principles simulations. Chem. Mater. **22**, 5644–5652 (2010)
96. F. Angeli, J.M. Delaye, T. Charpentier, J.C. Petit, D. Ghaleb, P. Faucon, Investigation of Al-O-Si bond angle in glass by ^{27}Al 3Q-MAS NMR and molecular dynamics. Chem. Phys. Lett. **320**, 681–687 (2000)
97. E. Lippmaa, A. Samoson, M. Magi, High-resolution ^{27}Al NMR of aluminosilicates. J. Am. Chem. Soc. **108**, 1730–1735 (1986)
98. F. Mauri, A. Pasquarello, B.G. Pfrommer, Y.G. Yoon, S.G. Louie, Si-O-Si bond-angle distribution in vitreous silica from first-principles ^{29}Si NMR analysis. Phys. Rev. B **62**, R4786–R4789 (2000)
99. P. Florian, F. Fayon, D. Massiot, 2J Si-O-Si scalar spin-spin coupling in the solid state: crystalline and glassy wollastonite $CaSiO_3$. J. Phys. Chem. C **113**, 2562–2572 (2009)

Chapter 6
Recrystallization of Silicon by Classical Molecular Dynamics

Evelyne Lampin

Abstract Recrystallization of amorphous silicon is studied by classical molecular dynamics. First, a simulation scheme is developed to systematically determine the amorphous on crystal (a/c) silicon motion and compare it to established measurements by Olson and Roth [1]. As a result, it is shown that MD simulations using Tersoff [2] potential are adapted to simulate solid phase epitaxy, although a temperature shift to high values should be accounted for, while simulations using Stillinger-Weber [3] allows to study liquid phase epitaxy. In a second part, the simulation approach is applied to the case of a nanostructure [4] where classical recipes fail to achieve complete recrystallization. MD simulations are shown to be in agreement with experimental observations. The analysis of the structural evolution with time provide a support to understand the origin of the defects.

6.1 Introduction

There is a long history of using molecular dynamics simulations to prepare and study amorphous silicon (aSi) [5–9]. Amorphous silicon clusters are used to determine the structure of disordered materials, their electronic [8], vibrational [10], mechanical [11] or thermal [12] properties. This interest for aSi is shared by technologists for microelectronic applications. A typical example is the case where crystalline silicon (cSi) is intentionally amorphised in surface by Ge^+ or Si^+ implantation prior to the implantation of light dopants such as B^+. The motivation is to enhance the control of the dopants penetration and hence decrease the p/n junction depth in order to fulfill the requirements for device miniaturisation. The amorphisation step is always followed by an annealing intended to recover crystallinity. During annealing, the amorphous/crystalline (a/c) interface progresses at a velocity that has been measured in the late 80s by Olson and Roth [1] on a wide range of temperatures, from ambient

E. Lampin (✉)
Institute of Electronics, Microelectronics and Nanotechnology,
Avenue Poincaré, CS 60069, 59652 Villeneuve d'Ascq Cedex, France
e-mail: evelyne.lampin@univ-lille1.fr

C. Massobrio et al. (eds.), *Molecular Dynamics Simulations of Disordered Materials*, Springer Series in Materials Science 215,
DOI 10.1007/978-3-319-15675-0_6

to melting. They found that this annealing is a thermally activated process resulting in the following a/c velocity v_{SPE}:

$$v_{\mathrm{SPE}}(\mathrm{nm/s}) = 3.1 \times 10^6 \exp(-\frac{2.68\,\mathrm{eV}}{k_B T}) \quad (6.1)$$

Annealing at 600 °C, a rather low temperature from the Si technologists point of view, results in the total crystallinity regrowth in a few minutes of a layer of typically 150 nm and does not damage the other materials and structures of the transistors. The technique has therefore been widely used. Amorphisation is still used in current technologies although the race to pursue miniaturisation and increase the operating frequency imposes to complexify the fabrication processes and use nanostructuration. In these cases, classical recipes to regrow aSi into defect-free cSi may not be valid, as in the example of a FinFET [4]. Our motivation is to help interpret the discrepancies from the Olson and Roth law (6.1) thanks to atomic scale simulations of aSi recrystallization. We have chosen classical molecular dynamics because it is well suited to follow the evolution of non-crystalline systems of large size although some Monte Carlo lattice techniques have also been adapted to this a priori non-lattice problem [13]. First, we have identified the best-suited interatomic potential, using as a criterium that is should be able to reproduce Olson and Roth's law for the simple case of an amorphous layer on a crystalline bulk. Although previous studies have reported molecular dynamics simulation of amorphous silicon recrystallization [14, 15], the ability to reproduce the experimental law was not systematically tested as in the present work. The procedure we developed to extract the a/c velocity is presented in Sect. 6.2, together with the results for several interatomic potentials. The conclusion is that, although none of the tested potentials is ideal, Tersoff [2] has to be used if the solid phase epitaxy (SPE) is to be studied, while the Stillinger-Weber [3] potential is suited to simulate liquid phase epitaxy (LPE). In Sect. 6.3 we apply this method to study a 1D nanostructure used in FinFET devices. The structure after annealing matchs the different zones observed by TEM, and thanks to the monitoring during annealing by MD, the assumptions on the origin of the incomplete recrystallization are studied.

6.2 Recrystallization of an Amorphous Si Layer

6.2.1 Method

The main question when we started to simulate SPE by molecular dynamics was the choice of the interatomic potential. Crystalline silicon is one of the most studied materials, and numerous formulations of the interaction forces between Si atoms exist [2, 3, 16–24]. These different developments and parametrisation were made because, due to the assumptions used as a basis of their formulation, empirical potential can

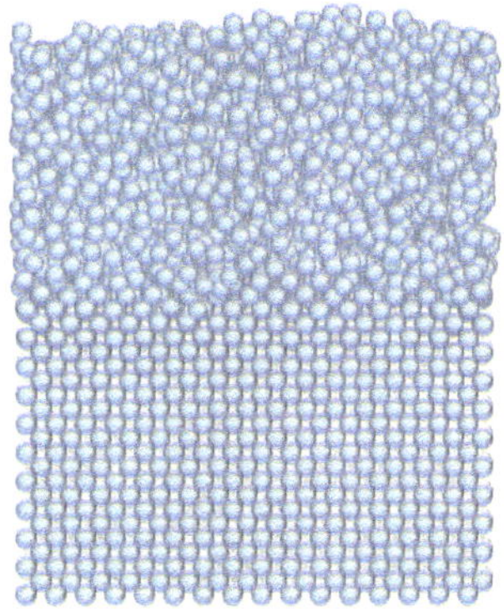

Fig. 6.1 Amorphous on crystalline atomic system. [100] system, 5120 atoms

not contain all the physics of the material. In order to find the best suited interatomic potential for SPE studies, we have chosen to test their ability to give velocities of the a/c interface versus temperature in the simple and largely experimentally validated case of an aSi layer formed at the surface of cSi bulk. The target is the velocity measured by Olson and Roth (6.1). For this purpose, we first have to build the atomic structure of an amorphous on a crystalline stack as shown in Fig. 6.1. The structure is made periodic in the two directions of the interface plane to mimic an infinite layer of amorphous on a crystal.

For our preliminary simulations, we have employed an amorphous cluster previously obtained [25] using the bond switching algorithm of Wooten, Winer and Weare (WWW) [26] with a Keating valence force field [27]. The amorphous cluster is put at a short distance (i.e. inferior to the cutoff radius of the interatomic potential used for the equilibration of the structure) of a cubic crystalline box of the same size. For these first tests, the simulation box contained a relatively small number of atoms ($\approx$1900) but we later show that this is high enough to catch the magnitude of the a/c velocity, although larger systems should be used in case the interface roughness has to be observed. The structural characterization of the amorphous cluster is given in Fig. 6.2, where its radial and angular distributions are plotted in red. The curves present typical features of an amorphous material: a distinct first neighbour peak in the radial distribution, with a clear gap between first neighbours and high order ones, and a radial distribution centered at 109°, the equilibrium bond angle in the diamond lattice of cSi, with a limited distortion of $\pm$15°. The atoms have therefore first neighbours distributed at the crystalline equilibrium distance but with an angular distorsion, while order progressively disappears for second and higher-order neighbors.

The amorphous/crystalline interface is afterwards equilibrated at the target temperature during 4 ps using a 2 fs time step and the velocity scaling algorithm, prior to the use of a Nosé-Hoover thermostat to maintain the desired temperature during

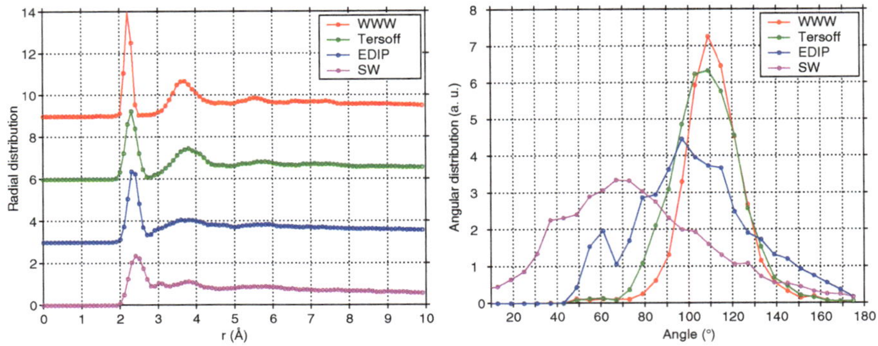

Fig. 6.2 Radial (*left*) and angular (*right*) distributions of the amorphous cluster obtained by WWW (*red*), and equilibrated using Tersoff (*green*), EDIP [21] (*blue*) and Stillinger-Weber (*magenta*) potentials

recrystallization. The position of the a/c interface is monitored during annealing. The position is determined via the computation of a 1D structure factor $S(z)$ defined as:

$$S(z) = \left| \frac{1}{N_z} \sum_{z<z_i<z+dz} e^{j\mathbf{k}.\mathbf{r_i}} \right| \tag{6.2}$$

where N_z is the number of atoms in the interval $[z; z + dz]$, with $dz = 2.25$ Å, $\mathbf{r_i} = (x_i, y_i, z_i)$ denotes the position of atom i, $\mathbf{k}$ is a vector of the reciprocal lattice parallel to the interface and equal to

$$\frac{2\pi}{a_0}(\mathbf{x} + \mathbf{y}) \tag{6.3}$$

in the [100] case, (a_0 is the lattice parameter), to

$$\frac{2\pi}{a_0}(\sqrt{2}\mathbf{x} + \mathbf{y}) \tag{6.4}$$

in the [110] case, and to

$$\frac{2\pi}{a_0}(\sqrt{2}\mathbf{x} + \sqrt{6}\mathbf{y}) \tag{6.5}$$

in the [111] case. Figure 6.3 presents an example of $S(z)$. The structure factor is close to 1 (>0.8 due to finite temperature) in the crystalline part and drop to a value close to 0 (<0.2, due to finite temperature) in the amorphous part. The interface position $z_{a/c}$ is determined as $S(z_{a/c}) = 0.5$. The $z_{a/c}$ extraction is repeated at fix interval during annealing. An example of evolution is shown in Fig. 6.4. First, until 2.8 ns, the a/c interface moves linearly with time, before a plateau is reached which

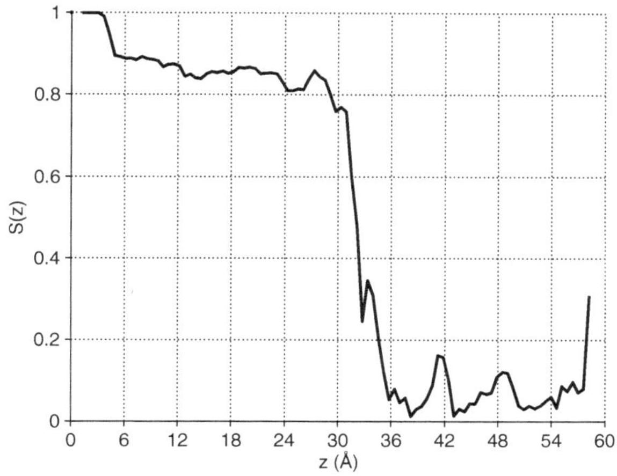

Fig. 6.3 1D-structure factor. [100] system, 5120 atoms, Tersoff

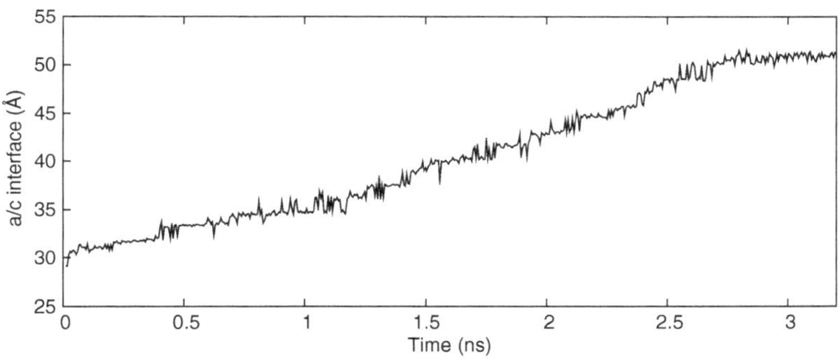

Fig. 6.4 Amorphous/crystalline (a/c) interface position versus time. [100] system, 5120 atoms, 1950 K, Tersoff

corresponds to the complete crystallization of the system. The a/c regrowth velocity, i.e. the slope of the interface position in the linear regime, is extracted by fitting the curve. The procedure is repeated for a range of annealing temperatures depending on the interatomic potential.

6.2.2 Preliminary Results: Two Interatomic Potentials Stand Out

The method presented above has been applied using five interatomic potentials: Stillinger-Weber [3] (SW), Tersoff [2], the environment-dependent interatomic potential [21] (EDIP), a Stillinger-Weber potential formulation with an angular

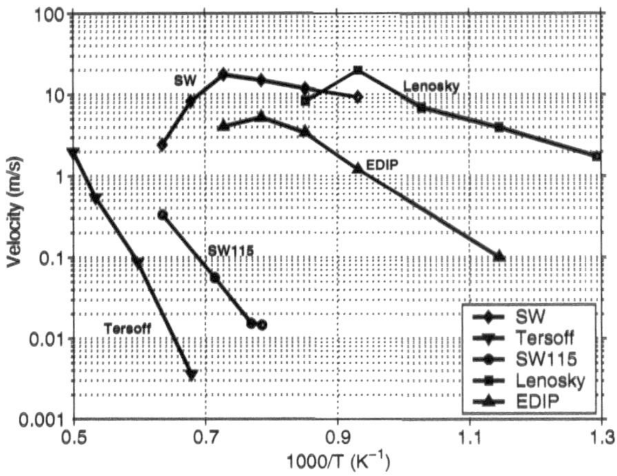

Fig. 6.5 Recrystallization velocity versus inverse temperature for 5 interatomic potentials

parameter λ slightly enhanced (×1.15) to modify the melting temperatures of aSi and cSi (SW115) [23], and the Lenosky potential [22]. The recrystallization velocities obtained with these five potentials are shown in Fig. 6.5. The figure evidences a strong dependence of the recrystallization velocities on the interatomic potential. The temperature range had to be adapted to each interatomic potential, in order to keep values of the velocities high enough to ensure reasonable computation times (Tersoff and SW115), or to have a sufficient number of points in the velocity curve (SW, EDIP, Lenosky). SW, EDIP and Lenosky give fast recrystallization of silicon, with velocities of a few m/s, and they feature two regimes: a first regime at low temperature, with a small activation energy (around 0.5 eV), and a second regime where the velocity reduces when the temperature increases. The transition between the two regimes corresponds to the melting temperature of aSi: above this temperature, the recrystallization is slower until we reach the melting temperature of cSi and the a/c interface moves in the other direction. Using Tersoff and SW115 leads to smaller recrystallization velocities, a shift towards higher temperatures and a higher activation energy, 2–3 eV. The interatomic potentials can therefore be classified into two groups:

- Group 1: SW, EDIP and Lenosky
- Group 2: Tersoff and SW115.

The activation energies of Group 2 are similar to the values measured by Olson and Roth (2.68 eV) while the rapid crystallization obtained with Group 1 are typically obtained in liquid phase epitaxy experiments [28]. An assumption is that the first group could be "LPE group" and the second "SPE group".

In order to investigate this point, we have studied the structure of the disordered part during annealing. Figure 6.2 shows the angular and radial distributions after the step of equilibration. This figure shows that, using Tersoff, the radial and angular distributions are similar to the results obtained for the WWW clusters, and typical of an amorphous material: a distinct first neighbour peak in the radial distribution, with a clear gap between first neighbours and high order ones, and a radial distribution centered at 109°, the equilibrium bond angle in the diamond structure of crystalline silicon, with a limited distortion of ±15°. With the Stillinger-Weber potential on the other hand, the gap in the radial distribution between the first and second neighbour peaks disappears, while the angular distribution is shifted to smaller values and broadened to ±40°. These features are characteristics of a liquid. Therefore, it appears that the initial amorphous silicon cluster is transformed in a sub-melted liquid silicon. Indeed, the Stillinger-Weber potential was originaly parametrised to predict both crystalline and liquid phase of silicon, and previous works have emphasized the difficulty to correctly describe the amorphous phase with the potential [6, 9] and proposed recipes to circumvent its limitations [7]. EDIP modifies the initial amorphous cluster, and creates an intermediate material with characteristics between amorphous and liquid Si. The recrystallization velocities are lower than the values obtained with SW and Lenosky, but the activation energy is too low compared to experimental values for SPE regrowth. We were quite confident in the use of Lenosky and SW115, a priori developed to avoid the weaknesses of more traditional formulations/parametrisations, but it appears that SW115 gives a compromise between SPE and LPE, as EDIP does, while the shift towards low values of the melting temperatures with Lenosky enhances the disagreement with experiments.

The conclusion of these preliminary calculations is that we can retain two interatomic potentials: Tersoff, to simulate SPE, and Stillinger-Weber for LPE.

6.2.3 Consolidated Simulations of SPE and LPE

In a second time, we refined the models of the SPE or LPE recrystallization. For the case of SPE, we made several improvements. First, instead of gluing a WWW amorphous cluster on a crystalline block, we have constructed an amorphous on a crystal stack by freezing the bottom part of a rectangular crystalline box, while the upper part was raised to 3500 K, i.e. above the melting temperature equal to 2500 K with Tersoff potential, and then it was carefully quenched to room temperature with a rate of 10^{11} K/ps. Moreover, the system size was increased from 2000 atoms for our preliminary results to around 5000 atoms. As already stated, 2000 atoms is enough to capture the interface velocity during recrystallization, but these larger systems were used to study the occurrence and type of defects as a function of the orientation. Three kinds of amorphous on crystalline stacks were constructed using the same procedure, along [100], [110] and [111]. The results on the recrystallization defects are detailed elsewhere [29] and will not be presented here. Nevertheless, we have to mention that for some cases that occurred more frequently for the [110] stacks, and even more for the [111] stacks, the annealing did not result in a perfect recrystallization of the amorphous part. In these cases, the determination of the recrystallization velocity was no longer possible because the defects cause the 1D structure factor to be pinned at non-zero values. The results presented in Fig. 6.6 are therefore limited to the cases were the recrystallization is defect-free.

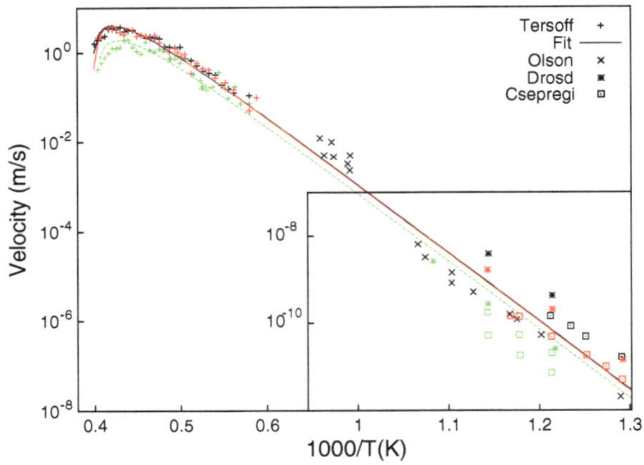

Fig. 6.6 SPE velocity for [100] (*black*), [110] (*red*) and [111] (*green*) oriented systems. Points issued from MD simulations with Tersoff are plotted together with experimental values, but breaks in the axis are introduced to account for the differences in temperature/velocity ranges

Figure 6.6 shows both MD results with the Tersoff potential with a symbol + and measurements by Olson [1] (x), Drosd [30] (⋆) and Csepregi [31] (□). The symbols are drawn in black for [100], red for [110] and green for [111]. The figure evidences that the temperature range is different for experiments, with an upper value of 1420 K that corresponds to the melting temperature of aSi, and for MD calculations which were performed for temperatures ranging from 1700 to 2500 K. Indeed, the Tersoff potential is known to strongly overestimate the melting temperature of silicon, and at the difference of the Stillinger-Weber potential, describes more properly the amorphous phase of Si but not its liquid phase. It seems like the Tersoff potential maintained the amorphous structure even at temperatures where experimentally the melting is observed. This significant shift of the melting is indeed a real advantage from the simulation point of view because the higher the temperature, the faster the recrystallization, the lower the computation time. A recrystallization of 10^{-8} m/s i.e. 10^{-13} nm/ps measured at 950 K would have been completely out-of-reach of the simulation. Nevertheless, the solid lines in Fig. 6.6 clearly demonstrate that the MD results are just extrapolation to high temperatures of the experimental values. Therefore we propose to use Tersoff MD at a temperature around 1700–1800 K in order to investigate particular points of the solid phase epitaxy, as it will be the case in the second part of this chapter for a nanostructure called FinFET, keeping in mind that these effects are expected to occur in reality at much lower temperature.

The molecular dynamics simulations with Stillinger-Weber of the liquid phase epitaxy were also refined. As for the SPE case, bigger systems containing around 5000 atoms were formed by freezing the bottom part of a crystalline box and annealing the upper part above the SW melting temperature (identical to the experimental one, as it was included in the fitting data base) in order to form a liquid on top of a crystal. This was done by using the same interatomic potential that further serves to study the recrystallization. The results for three orientations are presented in Fig. 6.7. The maximum regrowth velocity is higher for [100] due to the smaller distance between {100} planes.

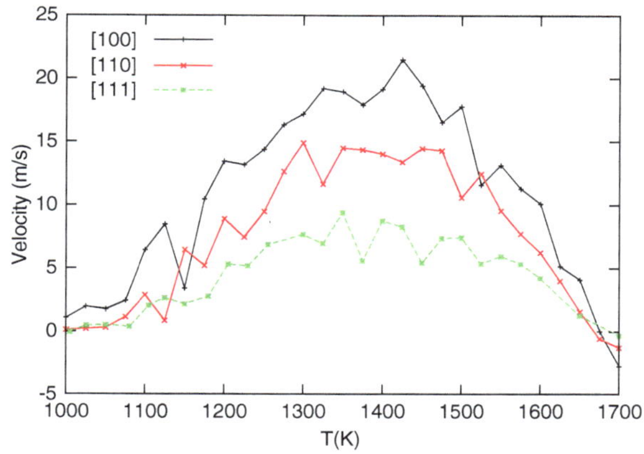

Fig. 6.7 LPE velocity for [100] (*black*), [110] (*red*) and [111] (*green*) oriented systems

6.3 Recrystallization of Amorphous Si in a Nanostructure

6.3.1 Method

Once the conditions of use of molecular dynamics to simulate SPE have been clarified, we have applied the calculations to the case of a nanostructure. Such a kind of nanostructure is fabricated and characterized by Duffy et al. [4]. The nanostructure is schematized in Fig. 6.8. A silicon fin or bar having typical dimensions of 50 nm (height) by 10–15 nm (thickness) and by 150 nm (length) is processed in a SOI (silicon on insulator) wafer to form the active area of the field effect transistors called FinFETs in this case. Moreover, the fin surface is oxidized in order to form the grid oxide. As a result, the silicon core is totally oxide-embedded. The oxide is silica, and is coloured in red in Fig. 6.8. The source and drain junctions are realised using the classical recipe used for former FET generation and already mentioned in Sect. 6.2: preamorphisation by heavy ions such as Si or Ge in order to amorphise silicon prior to the boron implantation (left view in Fig. 6.9). A subsequent rapid thermal annealing (RTA) at 1000 °C is performed that should recrystallize silicon and activate the doping. Nevertheless, this classical recipe fails in the present case. Instead of defect-free crystalline silicon with dopants in substitutional positions, the presence of numerous defects has been observed [4] by TEM. More precisely, as schematized in the right view of Fig. 6.9, three regions can be identified in the final structure: a bottom part has been correctly recrystallized, an upper part is made of grains (polysilicon), and an intermediate region in between is defective, in particular due to the presence of twins. Theses damages are problematic for the device and its electronic performances. It is therefore crucial to understand their origin in the perspective of an optimization of the fabrication process.

The uncomplete recrystallization can have several origins: the effect of the oxide embedement, the small thickness of the crystalline seed after preamorphisation, the

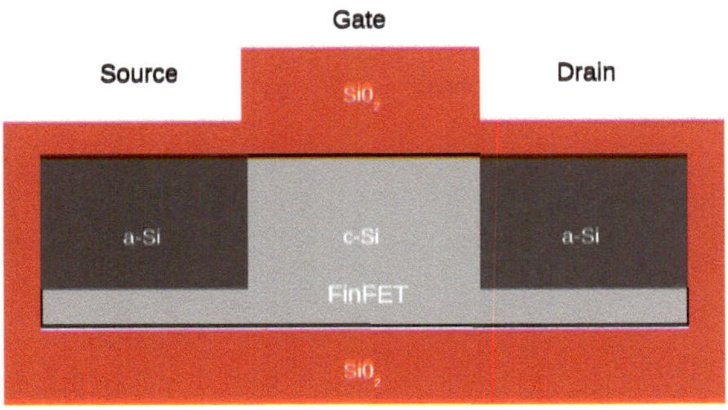

Fig. 6.8 Schematics of a silicon nanostructure (*grey*) embedded in silica (*red*) used in FinFETs

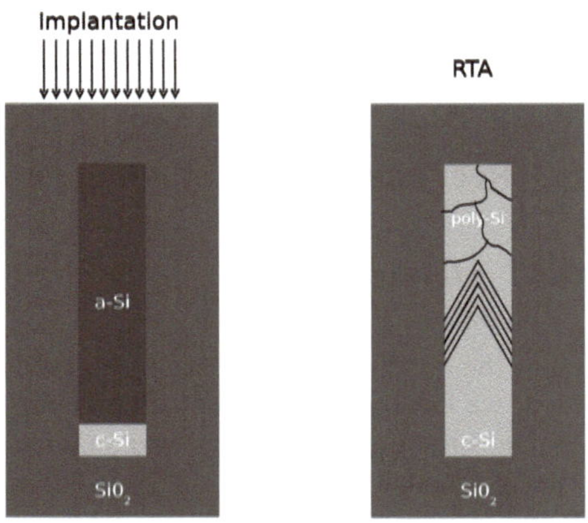

Fig. 6.9 Schematics of the source/drain cross section after amorphising implant (*left*) and successive annealing by RTA (*right*)

low dimensions of the fin. Molecular dynamics simulations are expected to help explain the structure of the fin at the end.

First, an atomic model is constructed. A cross section is given in Fig. 6.10. The box is rectangular, and made periodic in the [001] direction in order to mimic a 1D system. The bottom part is crystalline, while the remaining is amorphous. Amorphous silicon is created using the melting/quenching procedure presented in part 6.2. Tersoff potential is used for all the simulations presented in this part, (see the end of part 6.2). The dimensions of the box are 4.6 nm (width) by 13.8 nm (height). The width on height ratio is therefore the same as for the real fin in [4], but the structure is 4 times smaller (12,800 atoms). Therefore, if size effects are responsible for the uncomplete recrystallization, they should be emphasized in this smallest structure. In the real device, the silicon core is surrounded by silica. Silica could have two possible impacts: the effect of its chemistry, different from silicon, and the effect of the disorder in the material. Although we recently shown [33] that the parametrisation by Munetoh [34] of Tersoff potential allows to build stable Si/SiO_2 interfaces, we decided to limit our study to the impact of the disordered environment on the fin. The assumption will be verified in the following. For this purpose, we have frozen the Si atoms in a 7 Å-thick shell to prevent their recrystallization. The concerned atoms are drawn in red in Fig. 6.10.

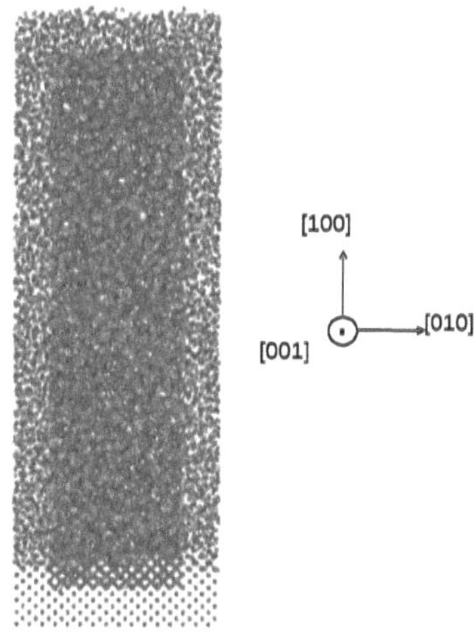

Fig. 6.10 Silicon atomic box constructed to mimic the silica-embedded silicon. Atoms in *red* are frozen

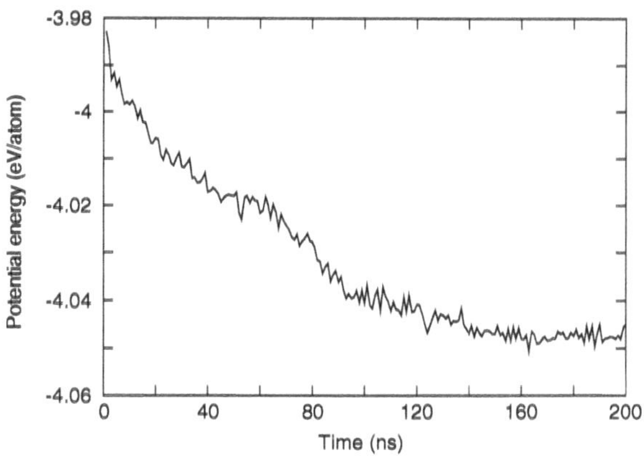

Fig. 6.11 Potential energy per atom during annealing

6.3.2 Results

Once the atomic model is constructed, we performed a molecular dynamics simulation at finite temperature. We have chosen a temperature of 1700 K in order to have a SPE velocity similar to the experimental value at 1000 °C. The potential energy per atom versus time is presented in Fig. 6.11. The major decrease occurs in the interval [0:120 ns]. Therefore we will study the structure at four moments of this interval: 30 ns, 60 ns, 90 ns and 120 ns.

For a purpose of clarity, we have chosen to restrict the representation to crystal-like atoms. We define for each atom i the parameter A_i:

$$A_i = \sum \left(\cos\theta + \frac{1}{3} \right)^2 \tag{6.6}$$

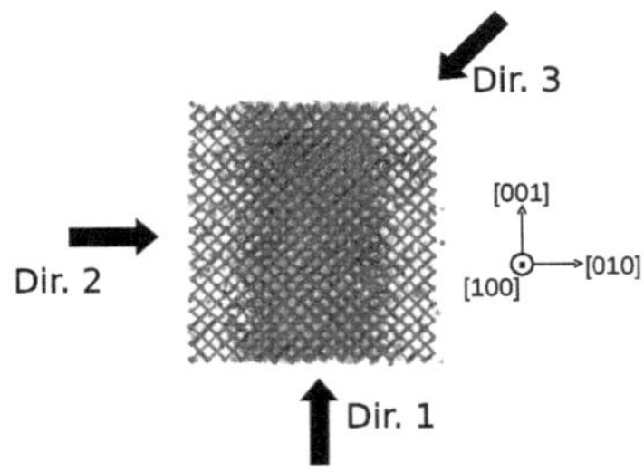

Fig. 6.12 *Top view* of the atomic box. In the following, views of the box during recrystallization will be given in directions *1*, *2* and *3*

where the sum goes on the angles θ formed between atom i and all its pairs of neighbours. If the atom were on the diamond lattice at 0 K, A_i should be equal to 0. Accounting for a deviation due to finite temperature, we adjusted our criterium to values of A_i less than 0.2. In addition, we also studied three different views of the boxes to allow the defects identification. These views are defined in Fig. 6.12. Similarly, we also studied three different views of the boxes to allow the defects identification. These views are defined in Fig. 6.11. View 1 is along [100], View 2 along [010] and View 3 along [111]. The use of the crystal-like atoms representation associated with the combination of three representations allows us to identify grains and twins as shown in the following.

Figure 6.13 shows the structure after 30 ns. The lower crystalline part has grown upside compared to the initial structure (Fig. 6.10) due to SPE initialised by the crystalline seed at the bottom. No modification is noticeable in the upper amorphous part. After 60 ns (Fig. 6.14), the recrystallization by SPE has progressed in the bottom part of the structure and a few additional crystalline planes are visible, especially in View 3. On the other hand, View 2 evidences the apparition of two grains (respectively surrounded by a blue and a green line). The grains have nucleated in the upper part of the amorphous. At 60 ns, their diameter is around a few Å. View 3 also shows a [111] twin in the upper cSi region, just beneath the a/c interface. The twin is highlighted by a magenta line. The presence of this twin explains that the corresponding few Å-thick layer is blurred in View 1. From 60 to 90 ns (Fig. 6.15), the twin goes on growing while SPE proceeds from the bottom. A second twin has appeared, as highlighted by the upper magenta line. As for the two grains in the upper part, their diameter has doubled: after the nucleation they have now entered the regime of growth. The grains go on growing from 90 to 120 ns (Fig. 6.16) until they fill the volume delimited by the embedding shell and the crystal block. In particular, grain 2 has an elongated shape (View 2) because it can only expand along [010]. The progression of the SPE front is stopped in the interval [90:120 ns]. There are two reasons for this first, because the twins have contributed to the formation of a grain misoriented compared to the initial seed. Second, because of the presence of grain 2 that prevents the progression of SPE. From 120 to 200 ns, the energy does not vary significantly (Fig. 6.11), and the structural changes are not observable: only minor rearrangements occur in the grain boundaries. The SPE front does not succeed in progressing upper nor in recovering a crystalline orientation aligned with the initial seed.

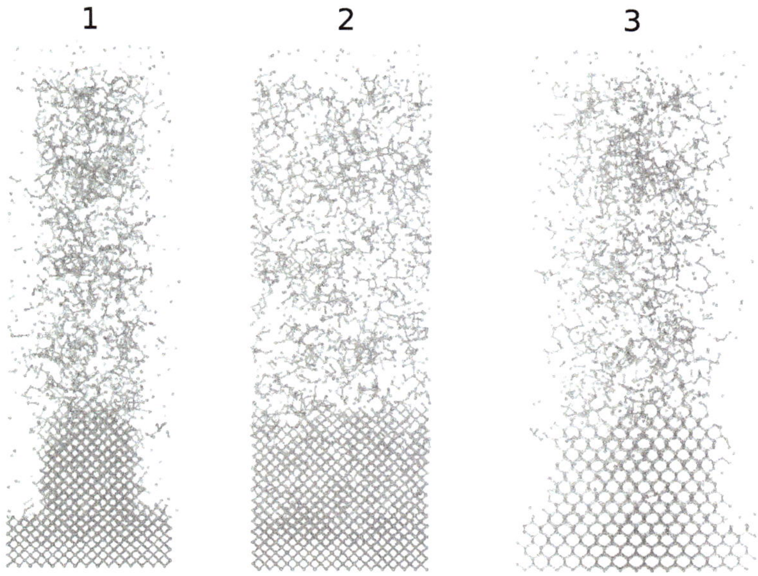

Fig. 6.13 Crystal-like atoms in the fin after 30 ns

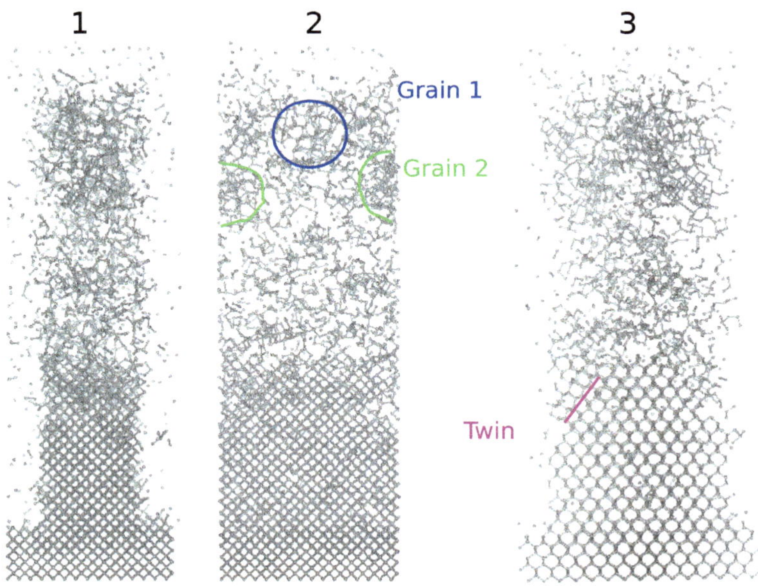

Fig. 6.14 Crystal-like atoms in the fin after 60 ns

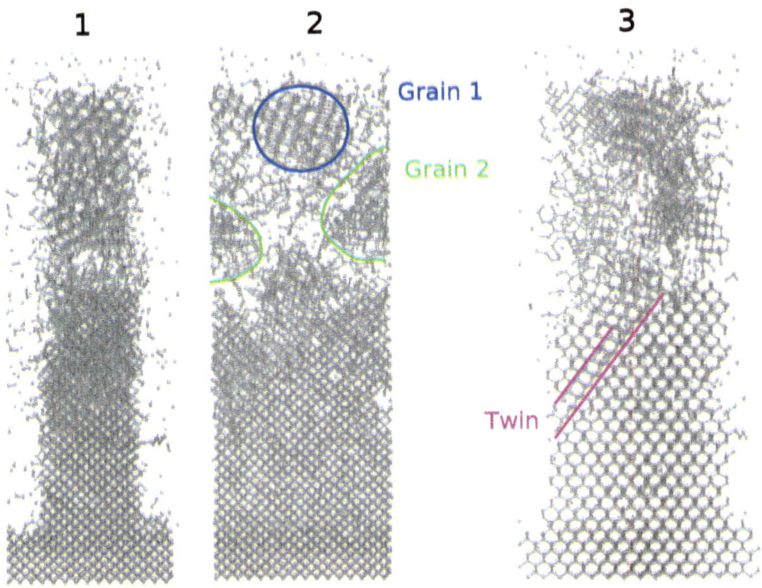

Fig. 6.15 Crystal-like atoms in the fin after 90 ns

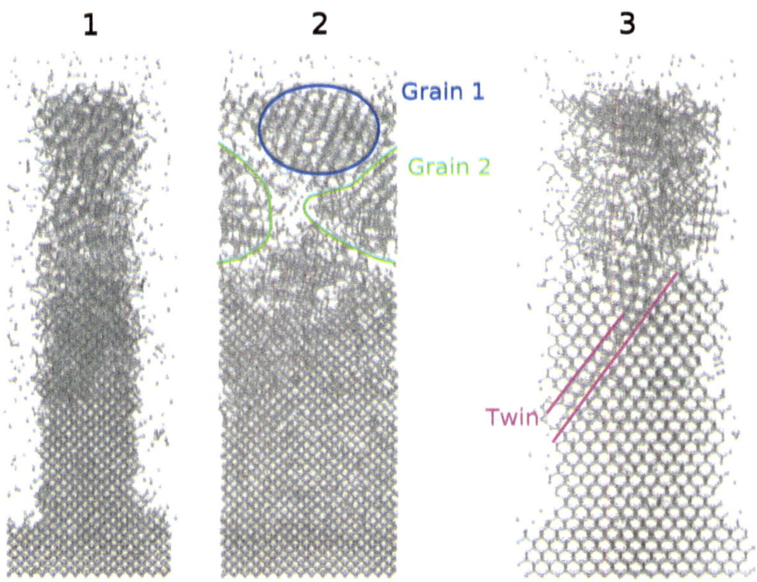

Fig. 6.16 Crystal-like atoms in the fin after 120 ns

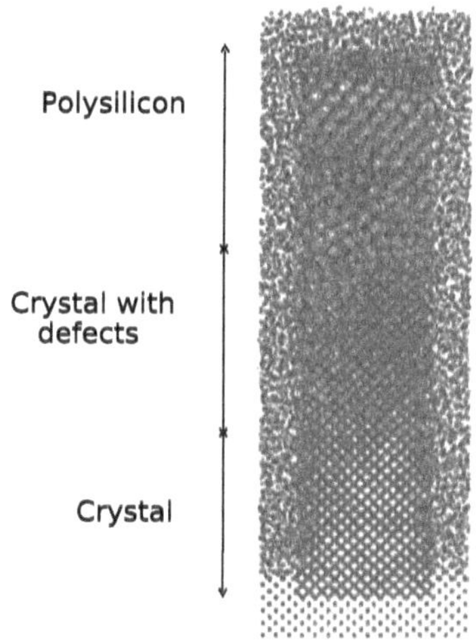

Fig. 6.17 Final structure (200 ns annealing). All atoms are represented. Atoms in *red* are frozen

6.3.3 Discussion

The final structure is displayed in Fig. 6.17. All atoms are plotted in this view, including frozen and non-crystal like ones. Three regions can be delimited:

- a bottom region made of defect free crystalline Si regrown by SPE from the seed
- an upper region made of polysilicon, the volume being filled by the two grains
- an intermediate region made of defective crystalline silicon, with [111] twins that results in a blurring of the crystalline structure.

The final structure obtained by MD is therefore in agreement with the observation by TEM on the FinFET [4]. This agreement validates the approach of simulation as a reliable tool to study recrystallization in complex conditions that occur in nanostructures.

In order to emphasize the role of confinement by the oxide-like shell, we have performed another simulation using the same initial box but without freezing the shell while applying periodic boundary conditions also along [010]. The objective is to compare the results obtained for the fin with the SPE of a 2D layer of the same height. The structures are given in Fig. 6.18 after 60, 120 and 180 ns at the same temperature (1700 K). Unexpectedly, the final structure is not cSi on the total height of the structure, but there is a grain at the surface. The formation of grains was never observed in the preliminary simulations we carried out to build the procedure (Sect. 6.2). However, the amorphous height that recrystallizes was smaller than in the present case. As random nucleation is slower than SPE from the crystal seed, grains had not time to form before the total height had regrown by SPE. A second observation is that the grain is formed at the surface, which suggests the role of the surface in the grain nucleation. Indeed, in the fin, i.e. a configuration presenting more surfaces, we observe the formation of two grains. The surfaces however do not seem

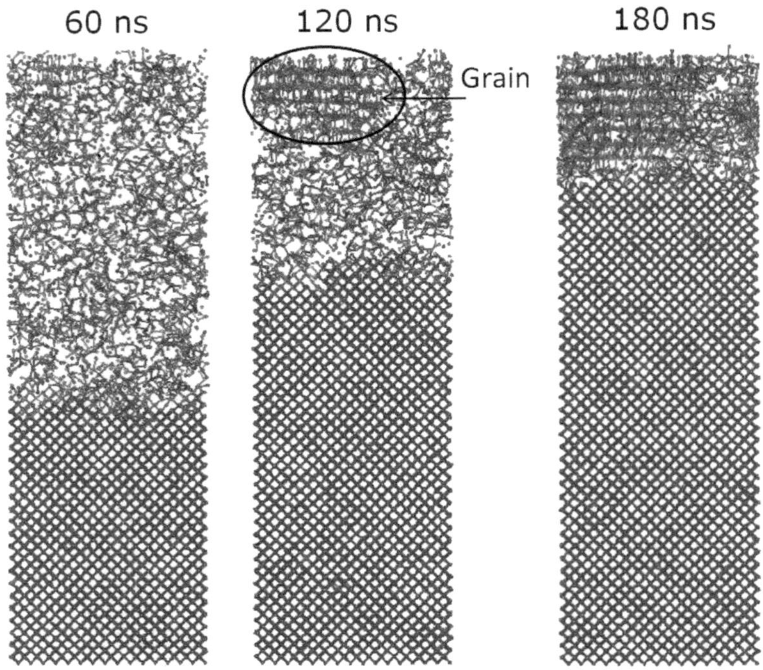

Fig. 6.18 Crystal-like atoms during annealing in the 2D-periodised case

to be responsible for a decrease of the SPE velocity, although this assumption could explain that only a limited portion of the amorphised silicon is recrystallized at the end of the fin annealing. Figure 6.19 presents the two systems, layer and fin, after the same annealing time (60 ns). The a/c interface is roughly at the same position. After 60 ns, the comparison between SPE velocities in the fin and in the layer is not possible anymore since twins have formed in the fin. The twins are formed along the [111] direction. Kuni et al. [35] already reported the formations of such [111] twins in the case of SPE near Si/SiO_2 boundary. We have observed the appearance of the twins during our study of the orientation dependency of the SPE velocity [29]. Figure 6.20 shows a zoom on the structure obtained by recrystallization of a 4.6 nm × 4.6 nm × 6.2 nm amorphous on crystal stack build in the [111] direction and annealed at 1850 K using the Tersoff potential. The dashed lines delimite two twins that occur at a distance of two {111} planes. Above the double twin, the crystalline structure recovers the same orientation as the bottom part. Twins frequently occur when SPE proceeds along [111] (at only one exception over 42 annealings we simulated). The sixfold ring picture of the interface by Drosd and Washburn [30] given in Fig. 6.21 helps to understand why. For the [100] case, the ring in the crystal part only requires one atom to be completed, therefore the probability that a defect occurs is low. This probability increases for the [110] case because two atoms have

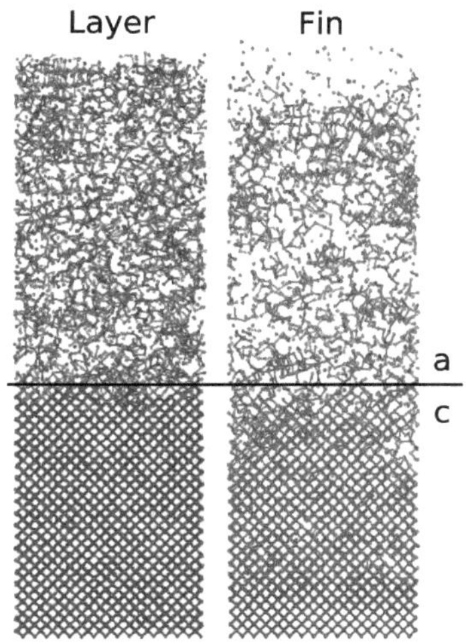

Fig. 6.19 Crystal-like atoms in the layer (*left*) and the fin (*right*) after 60 ns annealing. The amorphous/crystalline interface is visualised by the *line*

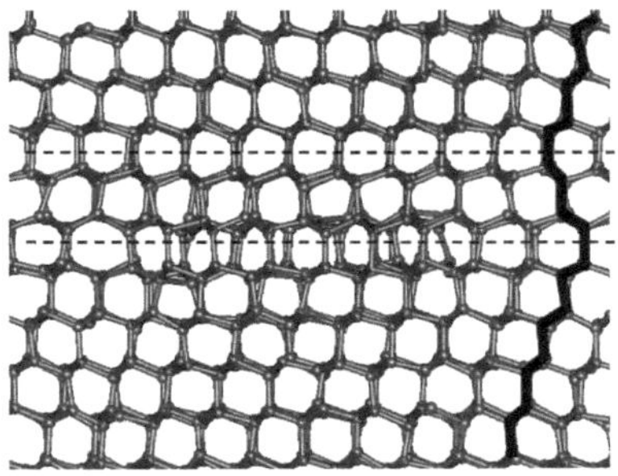

Fig. 6.20 Zoom on a two-planes thick [111] twin (inside the *dashed lines*) obtained after recrystallization of an amorphous on crystal [111]-oriented stack. Tersoff, 1850 K

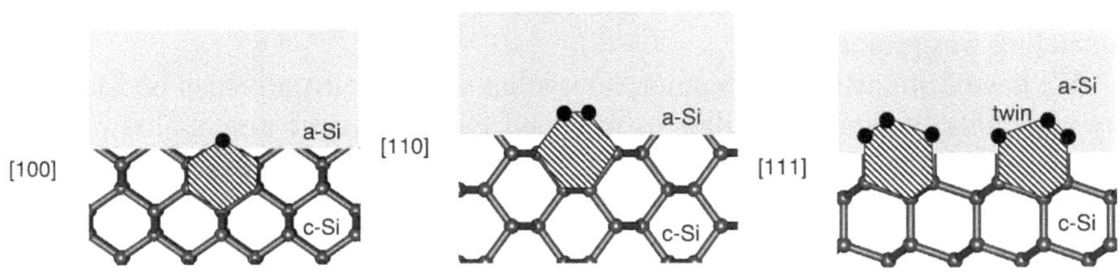

Fig. 6.21 Schematics of the amorphous/crystalline [100], [110] and [111] interfaces

to rearrange to form the ring. For the [111] case, half of the ring has to be created by atoms in the amorphous, and this can be achieved in two ways: the correct way (left side) and the twin way (right side). The twin way does not introduce unfilled bounds and as therefore a high probability to occur. The two configurations have equivalent first-neighbour energies, and therefore the presence of the surface in the case of the fin is probably sufficient to induce the shift from one to the other orientation.

Finally, it has to be noted from the various simulations that for the fin as well as for the layer, it was not possible to cure grains once they have formed, and even in the case of the layer where the SPE front has progressed without defects, it stops as it arrives to the grain. This is due to the fact that the energy inside the grain is low, since it has the diamond crystalline structure, and the grain is too stable to allow the atoms to rearrange to positions oriented with the bottom part.

6.4 Conclusion

The solid phase epitaxy of amorphous on crystalline silicon systems has been studied by molecular dynamics simulations. First a simulation scheme is consolidated in the case of an amorphous layer recrystallization where the a/c velocity is well known from experiments. An atomic model of the a/c interface is constructed and annealed by MD using one suitable interatomic potential for Si–Si interactions. The motion of the amorphous/crystalline interface is extracted and compared to the experimental law by Olson and Roth. Although none of the potentials is in agreement with the experiments, two stand out: Tersoff for SPE but accounting for a shift to higher temperatures compared to the real ones, and Stillinger-Weber, for LPE.

In a second time, the simulation approach is applied to the case of a nanostructure. An atomic model is constructed to mimic the oxide-embedded amorphous on crystalline silicon core. The annealing is simulated using Tersoff. The resulting structure exhibits three distinct regions, as observed experimentally by TEM on a FinFET: a defect-free crystalline region near the seed, grains far from the seed, and crystalline silicon with twins in the intermediate region. The [111] twins are often observed due to the structure of the {111} planes. The presence of a surface enhances the probability to form grains. Once grains are formed, they did not cured even when annealing was prolonged.

We have demonstrated that amorphous silicon recrystallization can be addressed by molecular dynamics simulations provided the interatomic potential is properly chosen. The method can be apply to the study of nanostructures having (nearly) the same size that real nanodevices, and during hundreds of nanoseconds. Therefore it can provide a reliable support to experiments in such configurations where the characterizations are not so trivial.

We have also applied the procedure to germanium [36], and shown that Tersoff also gives SPE for Ge, with a similar shift to higher temperatures as already observed for Si. A possible extension could be the study of recrystallization in Si/Ge systems as used in strained Si/SiGe technologies.

Acknowledgments The results presented in this chapter were obtained during a 5 years collaboration with Dr. Christophe Krzeminski. I acknowledge Pr. Fabrizio Cleri for the numerous discussions on the molecular dynamics technique and on the physics of recrystallization.

References

1. G.L. Olson, J.A. Roth, Kinetics of solid phase crystallization in amorphous silicon. Mater. Sci. Rep. **3**, 1–77 (1988)
2. J. Tersoff, Empirical interatomic potential for silicon with improved elastic properties. Phys. Rev. B **38**, 9902–9905 (1988)
3. F.H. Stillinger, T.A. Weber, Computer simulation of local order in condensed phases of silicon. Phys. Rev. B **31**, 5262–5271 (1985)
4. R. Duffy, M.J.H. Van Dal, B.J. Pawlak, M. Kaiser, R.G.R. Weemaes, B. Degroote, E. Kunnen, E. Altamirano, Solid phase epitaxy versus random nucleation and growth in sub-20 nm wide fin field-effect transistors. Appl. Phys. Lett. **90**, 241912 (2007)
5. R. Biswas, G.S. Grest, C.M. Soukoulis, Generation of amorphous-silicon structures with use of molecular dynamics simulations. Phys. Rev. B **36**, 7437–7441 (1987)
6. J.Q. Broughton, X.P. Li, Phase diagram of silicon by molecular dynamics. Phys. Rev. B **35**, 9120–9127 (1987)
7. W.D. Luedtke, U. Landman, Preparation and melting of amorphous silicon by molecular dynamics simulations. Phys. Rev. B **37**, 4656–4663 (1988)
8. R. Car, M. Parrinello, Structural, dynamical, and electronic properties of amorphous silicon: an ab initio molecular-dynamics study. Phys. Rev. Lett. **60**, 204–207 (1988)
9. M.D. Kluge, J.R. Ray, Velocity versus temperature relation for solidification and melting of silicon: a molecular-dynamics study. Phys. Rev. B **38**, 1738–1746 (1989)
10. E. Kim, Y.H. Lee, Structural, electronic, and vibrational properties of liquid and amorphous silicon: tight-binding molecular dynamics approach. Phys. Rev. B **49**, 1743–1749 (1994)
11. P.L. Palla, S. Giordano, L. Colombo, Interfacial elastic properties between a-Si and c-Si. Phys. Rev. B **78**, 012105-1-4 (2008)
12. Y.H. Lee, R. Biswas, C.M. Soukoulis, C.Z. Wang, C.T. Chan, K.M. Ho, Molecular-dynamics simulation of thermal conductivity in amorphous silicon. Phys. Rev. B **43**, 6573–6580 (1991)
13. L. Pelaz, L.A. Marqus, J. Barbolla, Ion-beam-induced amorphization and recrystallization in silicon. J. Appl. Phys. **96**, 5947–5976 (2004)
14. T. Motooka, K. Nisihira, S. Munetoh, K. Moriguchi, A. Shintani, Molecular-dynamics simulations of solid-phase epitaxy of Si: growth mechanisms. Phys. Rev. B **61**, 8537–8540 (2000)
15. A. Mattoni, L. Colombo, Boron ripening during solid-phase epitaxy of amorphous silicon. Phys. Rev B **69**, 045204-1-8 (2004)
16. E. Pearson, T. Takai, T. Halicioglu, W.A. Tiller, Computer modeling of Si and SiC surfaces and surface processes relevant to crystal growth from the vapor. J. Cryst. Growth **70**, 33–40 (1984)
17. R. Bsiwas, D.R. Hamann, Interatomic potentials for silicon structural energies. Phys. Rev. Lett. **55**, 2001–2004 (1985)
18. B.W. Dodson, Development of a many-body Tersoff-type potential for silicon. Phys. Rev. B **35**, 2795–2798 (1987)
19. D. Vanderbilt, S.H. Taole, S. Narasimhan, Anharmonic elastic and phonon properties of Si. Phys. Rev. B **40**, 5657–5668 (1989)
20. B.C. Bolding, H.C. Andersen, Interatomic potential for silicon clusters, crystals, and surfaces. Phys. Rev. B **41**, 10568–10585 (1990)
21. J.F. Justo, M.Z. Bazant, E. Kaxiras, V.V. Bulatov, S. Yip, Interatomic potential for silicon defects and disordered phases. Phys. Rev. B **58**, 2539–2550 (1998)
22. T.J. Lenosky, B. Sadigh, E. Alonso, V.V. Bulatov, T. Diaz de la Rubia, J. Kim, A.F. Voter, J.D. Kress, Model. Simul. Mater. Sci. Eng. **8**, 825 (2000)
23. E.J. Albenze, P. Clancy, Mol. Simul. **31**, 11 (2005)
24. Y. Lee, G.S. Hwang, Force-matching-based parametrization of the Stillinger-Weber potential for thermal conduction in silicon. Phys. Rev. B **85**, 125204-1-5 (2012)
25. G. Allan, C. Delerue, M. Lannoo, Nature of impurity states in doped amorphous silicon. Phys. Rev. B **61**, 10206–10210 (2000)

26. F. Wooten, K. Winer, D. Weaire, Computer generation of structural models of amorphous Si and Ge. Phys. Rev. Lett. **54**, 1392–1395 (1985)
27. P.N. Keating, Effect of invariance requirements on the elastic strain energy of crystals with application to the diamond structure. Phys. Rev. **145**, 637–645 (1966)
28. G.J. Galvin, J.W. Mayer, P.S. Peercy, Solidification kinetics of pulsed laser melted silicon based on thermodynamic considerations. Appl. Phys. Lett. **46**, 644–646 (1985)
29. E. Lampin, C. Krzeminski, Molecular dynamics simulations of the solid phase epitaxy of Si: growth mechanism and orientation effects. J. Appl. Phys. **106**, 063519 (2009)
30. R. Drosd, J. Washburn, Some observations on the amorphous to crystalline transformation in silicon. J. Appl. Phys. **53**, 397–413 (1982)
31. L. Csepregi, E.F. Kennedy, J.W. Mayer, T.W. Sigmon, Substrate-orientation dependence of the epitaxial regrowth rate from Si-implanted amorphous Si. J. Appl. Phys. **49**, 3906–3911 (1978)
32. E. Lampin, C. Krzeminski, Regrowth of oxide-embedded amorphous silicon studied with molecular dynamics. J. Appl. Phys. **109**, 123509 (2011)
33. E. Lampin, Q.-H. Nguyen, P.A. Francioso, F. Cleri, Thermal boundary resistance at silicon-silica interfaces by molecular dynamics simulations. Appl. Phys. Lett. **100**, 131906 (2012)
34. S. Munetoh, T. Motooka, K. Moriguchi, A. Shintani, Interatomic potential for Si-O systems using Tersoff parametrization. Comput. Mater. Sci. **39**, 334 (2007)
35. Y. Kunii, M. Tabe, K. Kajiyama, Amorphous Si/ crystalline Si facet formation during Si solid phase epitaxy near Si/SiO_2 boundary. J. Appl. Phys. **56**, 279–285 (1984)
36. C. Krzeminski, E. Lampin, Modeling of germanium solid and liquid phase epitaxy by molecular dynamics simulation. Poster at eMRS meeting, spring 2010

Chapter 7
Challenges in Molecular Dynamics Simulations of Multicomponent Oxide Glasses

Jincheng Du

Abstract Despite tremendous progresses made in the past few decades in molecular dynamics simulations of glass and related materials, there exist a number of challenges in MD simulations of multicomponent glasses. This chapter summarizes the progresses in this field and present the challenges that include the reliable and transferable empirical potentials, cooling rate, system size and concentration effect on the simulated glass structures, and the validating structures of multicomponent oxide systems. Several practical examples on multicomponent and technologically important glass systems using classical MD simulations are also given to highlight the capabilities and challenges.

7.1 Introduction

Since its first application on silica glasses about four decades ago [1–3], molecular dynamics (MD) simulations have become an effective and almost indispensable method in studying the atomic structure and structure-property relationships in glass materials. Glass structure lacks long-range order and defies any single experimental method in structural determination, in contrast to crystalline materials where diffraction method can usually uniquely determine the atomic structures. The structure of glasses is also not random or featureless but, on the contrary, there exist plenty of short- and medium-range structure characteristics that play critical roles on the behaviors and properties of glasses. Determining these glass structure characteristics still poses as a significant challenge in modern characterizations methods and remains a frontier of physical science, although significant progresses in methods such as high energy X-ray and neutron diffraction [4], solid state NMR [5], EXAFS [6], and more recently atom tomography [7]. Atomistic simulations, especially MD based methods, have been playing a more and more important role in investigating

J. Du (✉)
Department of Materials Science and Engineering,
University of North Texas, Denton, TX, USA
e-mail: du@unt.edu

C. Massobrio et al. (eds.), *Molecular Dynamics Simulations of Disordered Materials*, Springer Series in Materials Science 215,
DOI 10.1007/978-3-319-15675-0_7

glass structure and structure-property relationships in glass research. This is due to rapid development of simulation methodologies (classical and ab initio MD methods), diversified potential models (partial charge, polarizable, reactive force fields), and accessible high performance computational facilities. As a result, MD simulations are not only practiced in university or national labs but also gradually used in industrial and companies to solve more practical problems. For example, MD simulations have recently been used to study the ion exchange processes in chemically strengthened glasses for display applications [8]. Despite these progresses, there remain several challenges in MD simulations of multicomponent oxide glasses, on which this chapter will be focused on.

Most industrial and technologically important glasses are multicomponent in nature. It is one of the major advantages of glass materials to maintain glass formability in large composition range [9]. The capability to fine-tune thermal-mechanical and other properties, to optimize processing condition, and address cost and environmental friendliness considerations by varying the glass compositions are also important in practical glass applications. It is thus critical to understand the structure of these multicomponent glasses and how each component impacts the glass structure and thus influences the physical, chemical, thermal, mechanical, optical and other properties. Earlier MD simulations involve simpler unary or binary glass compositions but significant progress has been made to study multicomponent glasses.

MD simulations involve studying an assembly of atoms or molecules [10], usually with applied periodic boundary condition for solid and melt, that interact with each other through empirical potentials. With initial positions randomly generated or from crystal structures and velocity assigned from Boltzmann distribution at given temperature, the atoms will move step by step with positions determined through solving Newton's equation of motion at time step in the order of femtosecond. This process goes iteratively until specified time steps are reached. The simulations can be run under different ensembles and the macroscopic thermodynamic properties such as pressure and heat capacity can then be calculated. The most commonly used procedure to generate glass structure is the simulated melt and quench process. Due to the application of periodic boundary condition and relatively small number of atoms as compared to the Avogadro's number used in the simulations, mechanical heating is inevitable (normal melting starts from surface and propagate to the inner part). As a result, higher than the normal melting temperatures (5000–8000 K) are required to ensure equilibration and removal of memory effects of the initial structure. The melt is then gradually cooled down to room temperature through a linear, non-linear or step cooling procedures with nominal cooling rate ranging from 0.1 to 10 K/ps. The trajectories of atoms can be recorded for structural analysis such as pair distribution functions and bond angle distributions or used for further diffusion or dynamic property calculations.

In this chapter, we will discuss some of the major challenges of the multicomponent oxide glasses using classical molecular dynamics simulations. These challengesinclude the development of reliable and transferable empirical potentials, optimization of the glass forming procedures, and experimental validations of simulated structures. A few examples of MD simulations of several multicomponent glasses ranging from soda lime silicate glasses, sodium aluminosilicate glasses, and phosphosilicate glasses and structural information obtained will then be provided. Properties such as mechanical behaviors (Bulk, Young's and shear moduli), ionic diffusion coefficients and electrical conductivity, viscosity, vibrational spectra and others can be calculated from trajectories of structural models generated from MD simulations but they are not the focus of this chapter.

7.2 Current Challenges on MD Simulations of Multicomponent Oxide Glasses

Despite significant progresses and wide applications in different systems, MD simulations of glasses still face a number of challenges in studying glass materials, especially for multicomponent systems. These challenges include reliable and transferable potentials models for multicomponent systems, the cooling rate effect during simulated glass formation, and the system size effect on certain properties and concentration effect on the simulations of minor components. These challenges are discussed in detail below.

7.2.1 Empirical Potentials

Empirical potentials play a critical role in MD simulations. The quality of the potentials determines in large extent the accuracy and validity of the final simulated structures and properties and hence the quality of the simulations. In practice, the availability of potential models is usually the limiting factor whether a system can be simulated. As most of the early MD simulations focused on simpler unary or binary glass systems, the available potentials in the literature are also limited to these simple components and very few potential sets are applicable to multicomponent systems. Fortunately, there are some recent efforts of developing potentials for common oxides that include most of the glass component. Some of these potentials have been tested in wide compositions ranges and can be valuable in studying some practical multicomponent glass systems. One of such set of potentials was initially developed by D.M. Teter, and modified, widely tested and utilized by Cormack, Du et al. [11–21]. The potential set consists of long range Coulombic interactions with fixed partial ionic charges to account for partial covalency in silicate, aluminate, phosphate oxide systems. The charge for oxygen is -1.2 and the cation charges scale proportionally from their formal charges to ensure charge neutrality of the overall system. For example, 2.4 for Si, 1.8 for Al and 0.6 for Na. Short range interactions are in the Buckingham form, which has an exponential repulsion term and a power attractive

Table 7.1 Atomic charge and Buckingham potential parameters for oxide glasses [11–21]

Pairs	A (eV)	ρ (Å)	C (eV Å^6)
$O^{-1.2}$–$O^{-1.2}$	2029.2204	0.343645	192.58
$Si^{2.4}$–$O^{-1.2}$	13702.905	0.193817	54.681
$P^{3.0}$–$O^{-1.2}$	26655.472	0.181968	86.856
$Al^{1.8}$–$O^{-1.2}$	12201.417	0.195628	31.997
$Li^{0.6}$–$O^{-1.2}$	41051.938	0.151160	0.0
$Na^{0.6}$–$O^{-1.2}$	4383.7555	0.243838	30.70
$K^{0.6}$–$O^{-1.2}$	20526.972	0.233708	51.489
$Ca^{1.2}$–$O^{-1.2}$	7747.1834	0.252623	93.109
$Sr^{1.2}$–$O^{-1.2}$	14566.637	0.245015	81.773
$Y^{1.8}$–$O^{-1.2}$	29526.977	0.211377	50.477
$La^{1.8}$–$O^{-1.2}$	4369.39	0.2786	60.28
$Er^{1.8}$–$O^{-1.2}$	58934.851	0.195478	47.651
$Eu^{1.8}$–$O^{-1.2}$	5950.5287	0.253669	27.818
$Ce^{1.8}$–$O^{-1.2}$	11476.9522	0.242032	46.7604
$Ce^{2.4}$–$O^{-1.2}$	31697.724	0.21836	90.659

term to account for dispersion interactions, which can be expressed as

$$V'_{ij}(r) = A_{ij} \cdot \exp\left(-\frac{r}{\rho_{ij}}\right) - C_{ij}/r^6 \tag{7.1}$$

where A, ρ and C are parameters, r is interatomic distance. This set of potential from the published papers is summarized in Table 7.1 [11–21]. It is worth noting that for the A parameter for O–O interaction the value of 1844.7458 eV was also used in the literature but it has been shown the updated value (2029.2233 eV, about 10 % larger than the original one) gave better agreement in terms of Si–O bond length and coordination number for silicon and other cations [12]. A correction term is usually used to modify Buckingham potential at short distances due to much faster increase of the power term than the exponential at term short interatomic distances in the Buckingham potential form. This creates an unphysical, negative infinite potential well at short interatomic distances. For systems running at high temperatures, such as the melt, in constrained dynamics or with unreasonable initial atom configurations, atoms can overcome the barrier and fall into the well that leads to "fusion" of atoms and unphysically high potential energies. This can be corrected by the addition of a separate potential function such as 12–18 Lenard-Jones potential or a splice function for shorter distances to the original Buckingham potential. The correction that Du et al. used was a splice function with the form of [22]

$$V'_{ij}(r) = A_{ij} \cdot r^n + B_{ij} \cdot r^2 \tag{7.2}$$

where A, B, n are parameters. The function will be applied for distance smaller than r_0, where r_0 is defined as the r value when the second derivative of potential energy equals to 0 and the A, B, n parameters were chosen so the potential, force, and first derivative of force for V(r) and $V'(r)$ to be continuous at r_0. Other functional form such as the ZBL potential [23] can also be used for the short range corrections of the Buckingham potential. In such a case, a polynomial function will be needed to link the two potentials in the short distance range.

Water plays a critical role in many properties of glasses. Hence it is important to understand the structural role and mechanism of how water reacts and interacts with oxide glasses. While most of the earlier water and hydroxyl potentials were based on full charge models, Du and Cormack have developed a set of partial charge potentials that are compatible with the above mentioned oxide glass potentials to enable the simulations of hydroxyl groups in the bulk and on the surface of glasses. This potential takes the similar partial charge of the Buckingham form as listed in Table 7.1 for oxide, and uses the Morse potential for O–H interactions together with a screened harmonic three-body term for Si–O–H bond angles. The oxygen atom in hydroxyl group has a different charge (−0.856 vs. −1.2) as compared to the oxygen atoms in the glass [14, 24]. The potential for hydroxyl groups has the Coulomb-subtracted Morse form:

$$V(r) = D\{1 - \exp[-\beta(r - r_0)]\}^2 - D - \frac{e^2 q_O q_H}{4\pi\varepsilon_0 r_{ij}} \tag{7.3}$$

in which D, β and r_0 are parameters. A three-body term was introduced to correctly reproduce the Si–O–H bond angle on silica surfaces. The three-body term has the screened harmonic form,

$$E_{jik}(\theta) = \frac{1}{2}(\theta_{jik} - \theta_0)\exp(-(r_{ij}/\rho_1 + r_{ik}/\rho_2)) \tag{7.4}$$

in which i, j and k represent the center atom and the other two atoms, and r_{ij} and r_{ik} represent the distance between the center atom and its two neighbors, θ_{jik} is the angle between j, i and k, with i at the center. The parameters θ_0, ρ_1 and ρ_2 were obtained by fitting the structure and defect format energies of silicic acid and a number of metal hydroxides. The atomic charges and potential parameters are listed in Table 7.2 [14, 24].

Another comprehensive potential set for oxide glasses was developed by Pedone et al. [25]. This set potential used the same partial charge as the above mentioned potentials but the short range interaction has the form of Morse potential with an additional −12 power term to enhance short range repulsion [25],

$$V_{ij}(r) = D_{ij}\left[\left\{1 - e^{-a_{ij}(r - r_0)}\right\}^2 - 1\right] + \frac{C_{ij}}{r^{12}} \tag{7.5}$$

Table 7.2 Potential parameters of oxide glasses with hydroxyl groups [14, 24]

	Buckingham potential parameters			
	A (eV)	ρ (Å)		C (eV Å^6)
$Si^{2.4}$–$O^{-1.2}$	13702.905	0.193817		54.681
$Si^{2.4}$–$O^{-0.856}$	12443.824	0.193817		54.681
$Na^{0.6}$–$O^{-1.2}$	4383.7555	0.243838		30.7000
$Na^{0.6}$–$O^{-0.856}$	4096.2726	0.243838		30.7000
$Ca^{1.2}$–$O^{-1.2}$	7747.1834	0.252623		93.1090
$Ca^{1.2}$–$O^{-0.856}$	7036.7000	0.252623		93.1090
$Mg^{1.2}$–$O^{-1.2}$	7063.4907	0.210901		19.2100
$Mg^{1.2}$–$O^{-0.856}$	5754.1167	0.210901		19.2100
$H^{0.256}$–$O^{-0.856}$	100.0	0.25		0.0
$O^{-1.2}$–$O^{-1.2}$	1844.7458	0.343645		192.58
$O^{-0.856}$–$O^{-0.856}$	1844.7458	0.343645		192.58
	Morse potential parameters (Coulomb-subtracted)			
	D (eV)	B (Å^{-1})		r_0 (Å)
$H^{0.256}$–$O^{-0.856}$	7.0525	0.190		0.9485
	Screened Harmonic potential parameter			
	E_0 (eV)	θ_0 (°)	ρ_1 (Å)	ρ_2 (Å)
$H^{0.256}$–$O^{-0.856}$–$Si^{2.4}$	12.0	118	2.0	3.2

where D, a, r_0, and c are parameters. This set of potentials has been applied to a several multicomponent glass systems [26–28].

Despite active developments of potentials for oxides, the potentials for simulating multicomponent glasses can be improved in several ways. One of the major issues is the potential transferability. There exist potentials of various oxides in the literature but in most of the cases they are not compatible with each other. For example, the oxygen-oxygen interaction can have different functional forms or different parameters for two sets of potentials. In such a case, parameters of other cation-oxygen interactions are then not compatible and usually cannot be mixed to use together. Simultaneous fitting to a large number of crystal or minerals database can improve the transferability of a set of potential, as in the two previously mentioned potential sets [13, 25].

Another challenge on potentials is due to the fact that most of the current potential fitting for oxides was done in crystalline systems based on the structure and mechanical properties under ambient temperature and pressure. Temperature dependent properties such as melting temperature, thermal expansion coefficient and heat capacity are not commonly used in the fitting. Due to notable differences of the crystalline and glass systems, fine tuning the parameters for glass and melt to correctly describe at wide temperature range is highly desirable for the simulation of glasses, yet only very few sets of potentials in the literature address this issue [29, 30].

Additional challenges exist when modeling glasses with components that exhibit compositional dependent coordination numbers. Boron oxide and aluminum oxide are those examples. NMR studies have shown that a boron anomaly is associated with boron coordination change as a function of composition [31–33]. For example, addition of sodium oxide to boron oxide leads to conversion of three coordinated boron to four coordinated boron. This conversion continues until sodium oxide is about 40 % and then further addition of sodium oxide creates non-bridging oxygen instead of 3–4 coordination conversion for boron ions. This becomes more complicated in borosilicate glasses, where the maximum depends on the boron oxide to silicon dioxide ratio, and in boroaluminosilicate glasses, where there is a competition among the three processes: conversion of 3 to 4-coordinated boron, creation of NBOS, and conversion of 5 and 6 to 4-coordinated aluminum. Several potentials have been applied to the aluminosilicate system which addresses well of the coordination conversion issue [26]. For borate and borosilicate glasses, Cormack and Park developed coordination dependent potentials [34], Huang and Kieffer developed charge equilibration and coordination dependent potentials [35], and more recently, Kieu et al. developed a set of fix charge potentials but the charges are adjusted depending on the composition [36]. Inoue et al. also proposed the usage of fixed charge, pair potential only to solve this issue [37]. The development of better potential for these mixed glass former glasses remain a major challenge in simulating mixed glass former glasses.

Most empirical potentials for oxide glasses adopt the Born model of solids and consist a long range Coulombic term and a short range interactions. Early simulations used formal (or full) charges but more recent potentials mainly use partial charge models. Full charge models are also usually accompanied by three body terms to correctly describe the structural unit geometry and the bond angles between the structural units. Later development based on first principles calculations and empirical fitting found that appropriate partial charges and fine tuning of short range parameters can successfully describe the coordination and bond angles of glass former cations. Another advantage of these partial charge pairwise potentials are their superior computational efficiency that enables simulations of large systems (hundreds thousands to millions of atoms) of oxide glasses. In addition to fixed charge partial charge models, recent developments include the variable charge models with the charges calculated step by step using the charge equilibrium method. This method is especially important in describing the heterogeneous structures or interfaces that include metal/oxide or oxide/water interactions. Examples of these potentials include the ReaxFF [38] and the COMB potentials [39].

It is known that the polarization effect of oxygen ions and some other large cations is important to describe the dielectric behaviors and defects in oxide ceramics. Polarizable potentials have also been used in glass simulations, where ion polarization is commonly treated by using the core-shell model. In the core-shell model, polarizability is treated by the massless shell that is linked to the core through a spring with the spring constant and shell charge determined to reflect the polarizability of the ions. The shells are usually treated in two ways in MD simulations: dynamic shell and adiabatic shell. In a dynamic shell, a small mass is applied to the shell and it moves as other particles in the system. The time step is chosen to be small enough to separate the energy transfer between the shell vibration and the normal ionic motion. For adiabatic shell, the shell position is adjusted or equilibrated after each of MD step. Relatively larger time step can be applied to adiabatic shell simulations but extra steps of shell equilibration are needed. As a result, the polarizable potential simulations have additional computational cost. Tilocca et al. developed a set of shell model polarizable potential for silicate glasses and applied it to the simulation of soda lime silicate and bioactive glasses [40]. It was found that glass structures generated from the polarizable potential gave the silicon and phosphorus Q_n distributions closer to experimental NMR values than those from rigid ion models, although both rigid ion and polarizable potentials generated glass melt structure quite different from those from Car-Parrinello based ab initio MD (AIMD). Polarizable potentials were considered to better describe the melt to glass transition during the cooling process by keeping certain Q_n species and silicon coordination defects for longer time and better resembles those from AIMD simulations [41].

Ab initio based molecular dynamics (AIMD) [42, 43], which obtain the forces from accurate first principles calculations, should be more general and can be applied to any glass system, as long as the first principles theory can adequately describe the chemical bonding in the system. However, AIMD is still limited by the simulation system size and time (up to a few hundred atoms and tens of picoseconds) due to high computational cost. This becomes especially true for multicomponent systems. AIMD has been used in simulating silica, lithium and sodium silicate glasses, and a few other glass systems [42, 43]. The results from AIMD provide accurate trajectories and structures that can serve as a model system for comparison with and those from classical MD simulations and validate the simulation results and potential models. AIMD has also already been used to simulate to several chalcogenide glasses, for which very few empirical potentials are available due to the bonding complexity in the systems. It is expected that AIMD will be applied to more and more systems due to availability of high performance computing facilities but classical MD will not likely to be fully replaced by it in the near foreseeable future. Details regarding ab initio MD simulations can be found in other parts in this book.

7.2.2 Cooling Rate Effect

To ensure integration accuracy of equation of motions, MD simulations use time steps in the order of the femtosecond. This limits the total accessible simulation time in practical simulations ranging from hundreds pico-seconds to a few nano-seconds, which leads to cooling rate of the K/ps level during the simulated cooling process. This cooling rate is still many (6–9) orders of magnitude higher than fastest possible experimental ones. This is also one of the common criticisms of MD simulations of glasses from our experimental colleagues on glass structures generated using computer simulations by the melt-and-quench process.

However, it was observed in the study of sodium silicate glasses that simulated glass structures based on reasonable potentials and procedures are not that much different from the experimental structural data [13], suggesting that direct comparison of cooling rate values from experiment and simulations might be misleading. This was mainly explained by two reasons: firstly the number of atoms used in simulations. Although with applied periodic boundary condition, the number of atoms in glass simulations range from initially a few hundred to now a few to tens of thousands, which is many orders of magnitude smaller than real glass samples. In this situation, the atoms have higher chance to exchange energies with the boundaries so the system can achieve equilibrium faster. As a result, higher cooling rate from simulated process might not lead to drastically different glass structures, which explains the similarities of the simulated as compared to experimentally observed glass structures [44]. With increasing computing power, the trend of cooling rate can be systematically studied using relatively large system sizes.

It is worth mentioning that some structural properties are more sensitive to cooling rate than others. For example, Du and Xiang found in their MD simulations of bioactive glasses (Na_2O–CaO–P_2O_5–SiO_2) that Si Q_n distributions are fairly insensitive to cooling rate effect while minor glass former P Q_n distribution are more sensitive to cooling rate effect [21]. With decreasing cooling rate, the fraction of phosphorus Q_0 species increases (experimental NMR studies suggest that majority of phosphorus are in Q_0 species). This result was later confirmed in MD simulations of the same glass system using polarizable potentials [45]. These results suggest that different structure features equilibrate or freeze in during glass formation at different speed so the cooling rate effect is sensitive to interested structure/property features.

Although most glasses from MD simulations were generated through the melt and quench process, it will be very worthwhile to explore other methods in generating glass structure. Experimentally, the amorphous state can not only be created using the melt and quench method but also using physical or vapor deposition methods, the sol-gel method, pressure or radiation induced armorphizations. Similarly, methods other than melt and quench from simulations would be very valuable to explore. For example, Monte Carlo and MD can be combined in the process of glass formation [46].

7.2.3 Simulation Size and Concentration Effect

In the past few decades, we have witnessed the increase of size of MD simulations in terms of total atom numbers in the simulations. Starting from around 100 atoms in early MD simulations to several thousand of atoms in common practice during the past decade, today's simulations of oxide glasses with hundreds of thousands or even millions of atoms of ionic systems can be achieved. It is worth noting that millions atom simulations for metals has been achieved much earlier. This was due tothe fact that metal potentials such as commonly used Embedded Atom Method (EAM) [47] have short cutoffs and there is no long range Coulombic force. However, simulations of a few thousands atoms are still commonly used in today's MD simulations in the literature as further increasing the system size in simulation will not influence of the structure or properties of interest in many studied systems.

Some properties, however, are more sensitive to system size than the others. One of those that is very sensitive to system size effect is the clustering behavior of minor component rare earth ions in glasses. It is known that rare earth ions have very low solubility in silica [15, 49]. A fraction of a percent will lead to clustering of rare earth ions and deterioration of properties such as optical emission due to non-radiative decay caused by energy transfer between clustered rare earth ions. In order to study the clustering behavior of low concentration components, such as rare earth (RE) oxide in silica glasses, it is necessary to use sufficiently large simulation cells to obtain statistically meaningful results. Early MD simulations of RE contained glasses use several hundred of atoms and a few RE ions. Of course, the clustering tendency cannot be investigated in these systems. In a study of clustering of europium ions in silica and silicate glasses, Kokou and Du have studied the clustering tendency as a function of system size for 1 mol% Eu_2O_3 [19, 48]. It was found that only after certain system size, namely over 6000 atoms, the trend of clustering can be reproducibly obtained in simulations. Figure 7.1 shows comparison of the probability to find a neighboring rare earth ion from random distribution and that from MD simulations. It was shown that for small system sizes (e.g. 1500 and 3000 atoms), the comparison with the random distribution is influenced by statistics and the clustering behavior is not conclusive. However, for larger system sizes where there are sufficient number of europium ions to provide better statistics, the probabilitybased on MD simulated

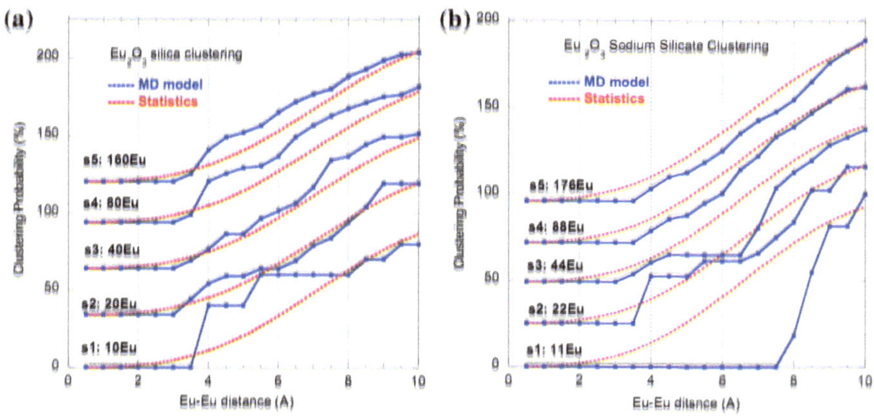

Fig. 7.1 Probabilities of finding neighboring Eu ions as a function of Eu–Eu distance in random distribution (*red curves*) and Eu ions distribution from MD simulation (*blue curves*) in Eu_2O_3 doped **a** silica and **b** sodium silicate glasses. Total number of atoms for glass s1–s5 are 1500, 3000, 6000, 12000, 24000, respectively [48]

structure is consistently higher than the random distribution in silica glass due to europium ion clustering but the probability is consistently lower than the random distribution in sodium silicate glasses. This is a classic example showing the effect of system size on simulated glass properties.

7.2.4 Validating Structure Models from Simulations

Comparing structural information from X-ray and neutron diffraction provides a very valuable way to validate MD simulation results of glasses. In some simple systems, the partial pair distribution function or partial structure factors of all atom pairs can be determined experimentally and they provide excellent validations for simulated structures. However, as the composition becomes more complicated and more elements included, larger number of pair contributions will complicate the comparison and the validation becomes more and more difficult in multicomponent glass systems. For example, for binary oxides, e.g. sodium silicate, there are six partial pair distribution functions, but for a four component systems, for example the bioactive glass composition, there are a total of fifteen partials contributions. The overlap between partial contributions makes it very challenging to assign the peaks and to determine the quality of comparison and hence the validation of the simulated structure models.

The R_X factor proposed by Wright [50] is commonly used to quantify the difference of the simulated and experimental total correlation function T(r), which is a form of pair distribution function and used in comparison with experiments due to symmetric broadening in experiments [51, 52]. The R_X is defined as [50],

$$R_X = \left[\frac{\Sigma_{i=1}^{N}(T^X(r_i) - T^S(r_i))^2}{\Sigma_{i=1}^{N}(T^X(r_i))^2} \right] \tag{7.6}$$

where $T^X(r_i)$ is the total correlation function from experiment and $T^S(r_i)$ is the total correlation function from simulations. In order to compare with experimental diffraction data, the simulated total distribution functions need to be broadened to take into consideration of the limit of momentum transfer (maximum Q value) in the diffraction experiments [51, 52]. The partial correlation function $t_{ij}(r)$ is first broadened by convoluting with the component peak function $p_{ij}^N(r)$, which defines the experimental resolution (for neutron diffraction in this case).

$$t'_{ij}(r) = \int_0^{\infty} t_{ij}(r')[P_{ij}^N(r - r') - P_{ij}^N(r + r')]dr' \tag{7.7}$$

The component peak function is defined as,

$$P_{ij}^N(r) = \frac{\overline{b_i}\,\overline{b_j}}{\pi} \int_0^{Q_{\max}} M(Q)\cos(rQ)dQ \tag{7.8}$$

in which $\overline{b_i}$ is the average neutron scattering length of atom i, $Q_{\max}$ is the maximum value of the experimental scattering vector, $M(Q)$ is the modification function. A common form of modification function is the Lorch function, which gives a gradual cutoff and is defined as,

$$M(Q) = \begin{cases} \frac{\sin \Delta r}{\Delta r Q} & Q \le Q_{\max} \\ 0 & Q > Q_{\max} \end{cases} \tag{7.9}$$

The total correlation function after broadening for the neutron diffraction case is then expressed as,

$$T^N(r) = T^o(r) + D^N(r) = \sum_{i,j} c_i t'_{ij} \tag{7.10}$$

in which $T^o(r)$ equals $4\pi r \rho^o \sum_i c_i \overline{b_i}$, where c_i and b_i are atom fraction and neutron scatter length of element i and ρ^o is the atom number density of the glass.

Figure 7.2 shows the comparison of experimental and simulated total correlation function (T(r)) for 45S5 bioactive glass. The R_x value for this system is 8.0 %. The 45S5 glass has four elements and 15 partial-pair distribution functions (8 of the 15 partial-pair distribution functions are shown in the figure) [21]. The overlap of these partial distributions makes it very difficult to deconvolute the contributions from different pairs. For example, the Si–O and P–O first peak, as well as the Na–O and Ca–O first peak, are largely overlapped; so it is almost impossible to obtain information on the Si–O/P–O or Na–O/ Ca–O bond length directly from neutron diffraction studies of the bioactive glasses.

It is also possible to compare the reciprocal space structure factors, either from neutron or X-ray diffraction measurements. The partial structure factors are first calculated through Fourier transformation the partial pair distribution functions using

$$S_{ij}(Q) = 1 + \rho_o \int_0^R 4\pi r^2 [g_{ij}(r) - 1] \frac{\sin(Qr)}{Qr} \frac{\sin(\pi r/R)}{\pi r/R} dr \tag{7.11}$$

in which $g_{ij}(r)$ is the pair distribution function of atom pair i and j, Q is the scattering vector, ρ_o is the average atom number density, R is the maximum value of the integration in real space which is set to half of the size of one side of the simulation cell. The $\frac{\sin(\pi r/R)}{\pi r/R}$ part of the integrand is a Lorch-type window function, as defined earlier, to reduce the effect of finite simulation cell size. Lorch function reduces the ripples at low Q (the cutoff effect) but it also leads to some broadening of the structure factor peaks. It is valuable to study the cutoff effect by simulating several glasses with different simulation cell sizes. The partial structure factors are then added together with weighting factors to obtain the total neutron structure factor,

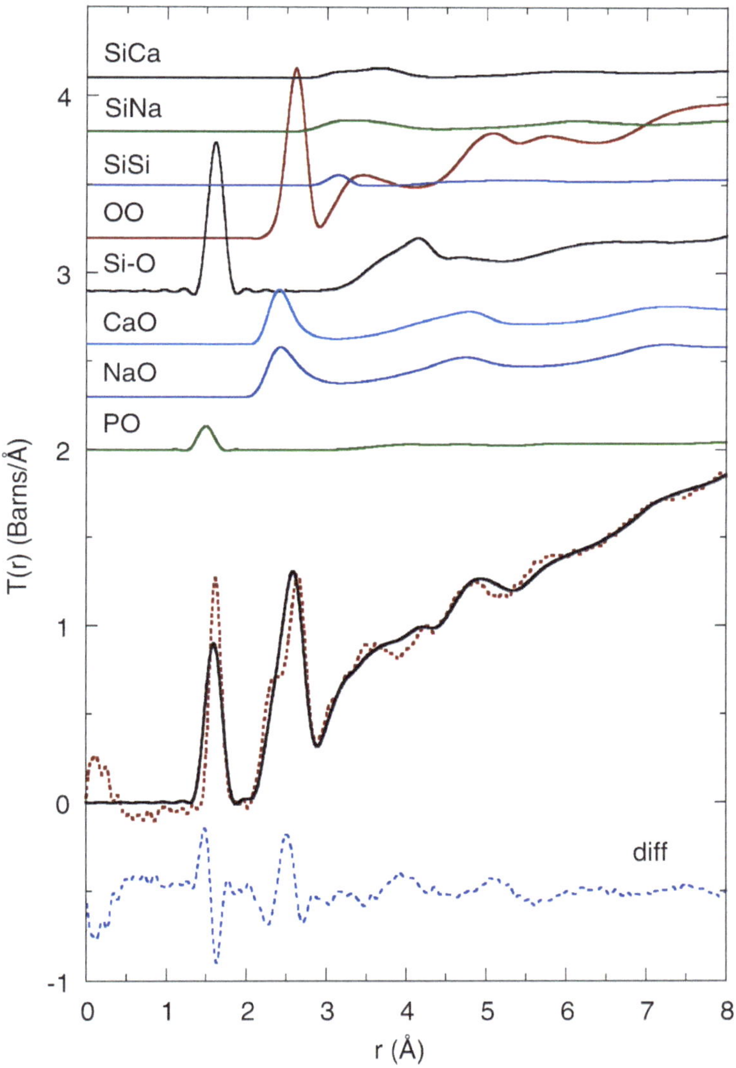

Fig. 7.2 Comparison of experimental (*dotted line*) and simulated (*solid line*) neutron total correlation function T(r) of 45S5. The difference of the two (*dash line*) and eight important (out of total fifteen) partial pair correlation functions from simulations are also shown. Q_{max} of 59 $Å^{-1}$ (the maximum Q value from experiment) and Lorch window function were used in neutron broadening of simulated structures [21]

$$S_N(Q) = (\sum_{i=1}^{n} c_i b_i)^{-2} \sum_{i,j=1}^{n} c_i c_j b_i b_j S_{ij}(Q) \tag{7.12}$$

in which c_l is the fraction of atoms in each of the species, and b_i is the neutron scattering length of the species. The contribution partial structure factors are defined as

$$S^N_{ij(i\leq j)}(Q) = (\sum_{l=1}^{n} c_l b_l)^{-2}(2 - \delta_{ij})c_i c_j b_i b_j S_{ij}(Q) \quad (7.13)$$

in which δ_{ij} is a delta function. By examining the $S^N_{ij}(Q)$, one can determine how each atom pair contributes to the features of the structure factors. b_i and b_j are the neutron scattering lengths of atom type i and j. Figure 7.3 shows the comparison of the experimental neutron structure factor and those calculated from MD simulations of a lanthanum aluminate glass (37.5 % La_2O_3–62.5 % Al_2O_3) [53]. In the figure,

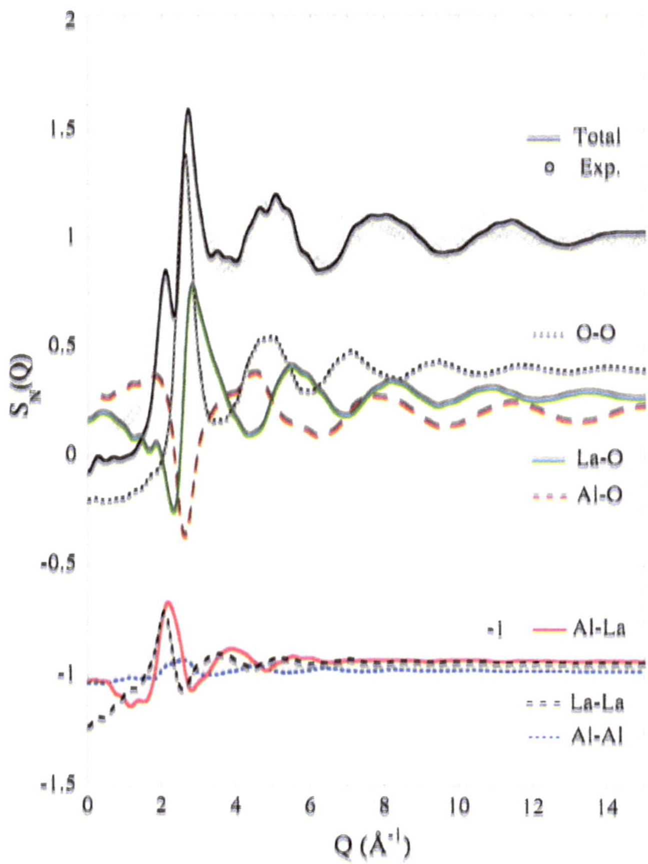

Fig. 7.3 Comparison of experimental and calculated neutron structure factor for lanthanum aluminate glasses [53]. Partial structure factors from MD simulations are also shown

it is also shown the contributions of each of the six partial structure factors. This comparison will not only provide information on how good the agreement is but also what partial contribution leads to the major differences.

There are several site specific experimental techniques such as solids state NMR, EXAFS and Raman spectroscopy that can give additional structural information to be compared directly with simulation results and thus are able to provide further validations. For example, NMR results not only provide Q_n distributions but also how the Q_n species are linked together through double or multi-quantum experiments. This kind of site specific experimental methods is an additional opportunity for detailed structure comparison and validation.

Recently, calculations of the NMR spectra using the GIPAW method based on structure models from classical MD simulations were performed and used to interpret the experimental NMR spectra for various nuclei. Insight on the network structures of complicated bioactive glasses was obtained [54]. Combination of classical MD simulations, DFT relaxation of periodic glass structure models, and GIPAW NMR spectra calculations has also been used to investigate modifier local environment in glasses [55]. For example, strontium local environment in strontium containing bioactive glasses was studied by this combination method with multiple samples. It was found that statistics from multiple glass samples are critical to generate reliable comparison of the NMR spectra [55]. This can become a powerful method in interpreting NMR results to provide rich structural information of glasses [56].

7.3 MD Simulations of Multicomponent Glasses: Practical Examples

7.3.1 Soda Lime Silicate Glasses

Soda lime silicate (Na_2O–CaO–SiO_2) glasses are the basis of a wide range of industrial glass compositions, often with addition of other minor components. Structural understanding of these glasses is thus of both scientific and technological importance. However, there are very few studies of soda lime silicate glasses from MD simulations, possibly due to the additional complexity of CaO introduced to the binary Na_2O–SiO_2 glasses which, together with other alkali silicate glasses, have been very well studied using MD simulations. Cormack and Du [57] have studied the soda lime silicate glasses in the base composition $15Na_2O$–$10CaO$–$75SiO_2$ and investigated the effect of replacing Na_2O with CaO on the short range and medium range structures using MD simulations. Partial charge pairwise potential with BKS [58] parameters were used for the Si–O and O–O interactions and Na–O and Ca–O parameters developed by fitting to the structures and properties of related crystal systems [57].

Although CaO and Na_2O both play the glass modifiers role in silicate glasses (it is confirmed from simulations that each sodium ion creates one NBO and each calcium ion creates two NBOs, by breaking the Si–O–Si linkages), their local coordination environments are quite different. It was found that calcium ions have total coordination number of around six with around 80 % are non-bridging oxygen (NBO), while the coordination number for sodium is around 5 with a little over 50 % of them are NBO. This can be seen from pair correlation functions of Ca–O and Na–O pairs and the deconvoluted BO/NBO contributions. Similar trend was also found in bond angle distributions. This difference is mainly originated from the difference of the field strength of sodium and calcium ions. Stronger bonding to NBOs that create stronger linkage between Si–O network fragments of Ca as compared to Na leads to much improved mechanical property and chemical durability when CaO is introduced to soda silicate glasses. The primitive ring size distribution statistics, a way to measure the medium range structures, was found to be relatively insensitive to the CaO/Na_2O substitution.

Pedone et al. studied soda silicate and soda lime silicate glasses using the partial charge Morse potential with MD simulations [59]. It was found that calcium ions increases the compactness (lower free volume) and increases the cohesion energy due to stronger Ca–O bond as compared to Na–O bond. As a result, soda lime silicate glasses have higher Young's moduli and higher sodium diffusion energy barrier as compared to soda silicate glasses. The substitution of MgO with CaO in soda lime silicate glasses were also studied using MD simulations. It was found that magnesium mainly has fourfold coordination as compared to calcium with mainly sixfold coordination. MgO/CaO substation leads to decrease of diffusion energy barrier and elastic moduli.

7.3.2 Aluminosilicate Multicomponent Glasses

Alumina as a glass-forming component has been commonly used to improve the properties of soda silicate glasses. As a result, aluminosilicate glasses is a common industrial glass system that finds a number of technological applications. For example, the Corning Gorilla© glass that are widely used for display in electronic devices is aluminosilicate glass and the base glass composition of E-glass is also aluminosilicate with other components such as boron oxide. Alumina is generally considered to be a glass former when combined with other glass forming oxides as it cannot form glass by itself and, in some classifications, it is also considered to be an intermediate.

Xiang, Du, Smedskjaer and Mauro recently studied sodium aluminosilicate glasses using classical MD simulations with two sets of potentials [60]. Both sets of potentials use partial charges with -1.2 being assigned to oxygen and the charges of other ions proportionally scaled down from their formal charges. One set uses Buckingham potential format and other one uses Morse potential. Three compositions were studied with similar silica content (around 60 mol%) but different Al/Na ratios covering peralkaline (Al/Na $<$ 1) to peralumina (Al/Na $>$ 1) compositions. Aluminum ions are found to be mainly (over 95 %) fourfold coordinated and are part of the silicon-oxygen glass network. The $[AlO_{4/2}]^-$ units are mainly charge balanced by sodium ions but oxygen triclusters [OAl_3] or [OAl_2Si] are also found to exist in the glasses. The percentage of oxygen triclusters increases from around 1 % to around 10 % with increasing Al/Na ratio. The percentage of five-coordinated Al also increases with Al/Na ratio. Depending on the potentials used, an increase from 0.4 to 1.3 % or from 1.9 to 5.5 % from peralkali (Al/Na = 0.6) to peralumina (Al/Na = 1.5) for the Buckingham and Morse potentials, respectively. Six coordinated aluminum was also observed in the structure generated using the Morse potential.

Connectivity between [AlO_x] polyhedra was analyzed to study the distribution of the glass forming units. It was found that majority linkages are [AlO_x]–[SiO_4] through corner sharing of bridging oxygen, this is followed by [SiO_4]–[SiO_4] linkages through corner sharing. [AlO_x]–[AlO_x] has the lowest possibility. However, no strict aluminum avoidance rule was observed as there are still considerable amount [AlO_x]–[AlO_x] linkage and average of such linage per [AlO_x] increase with increasing Al/Na ratio.

Mechanical properties of these glasses were also calculated based on the structures generated from MD simulations. This was done by calculating the compliance matrix, which is second derivative of potential energy versus strain. Based on the compliance matrix, bulk, Young's and shear moduli, as well as Poisson's ratio, were calculated. The increase trend of bulk and shear moduli with increasing Al/Na ratio was correctly reproduced for both potentials. While the Buckingham potential slightly overestimates both moduli, the Morse potential slightly underestimates the moduli.

Corrales and Du studied the surface of soda lime aluminosilicate glass and melt using MD simulations [61]. The base glass composition was chosen to represent one form of E-glass with an aim to understand sodium ion distribution at the glass fiber surfaces. Constant pressure simulations were used to generate the glass and glass melt. Z-density distribution was used to describe element density in directions perpendicular to the surface. It was found clear segregation and enrichment of sodium ions on the surface. This is associated with non-bridging oxygen ion segregation at the surface. The results help to explain the experimentally observed sodium emission in glass fiber surfaces. Potential of mean force was used to measure the correlation between sodium ions in this paper. Two methods were used to calculate the potential of mean forces: integration of pair distribution functions and constrained dynamic simulations, with both give similar results.

7.3.3 Aluminophosphate and Phosphosilicate Multicomponent Glasses

Phosphate glasses find wide applications as laser media and optical windows, and more recently as bioactive materials. Phosphate glasses are characterized by the chain or network structures formed by [PO_4] tetrahedrons linked through bridging oxygen ions [62]. Compared to [SiO_4] tetrahedron, there exist a terminal double bond P=O in each [PO_4] unit. The other three bridging oxygen can be converted to non-bridging oxygen during the depolymerization of the network, depending on the O/P ratio.

There were concerns on the description of the double bonds in [PO_4] unit using classical MD simulations but it seemed that the mean field approach in pair potentials work well in phosphate glasses as in silicate glasses. Liang et al. [62] studied lithium phosphate binary glasses using MD simulations. It was found that the observed glass transition temperature minimum is related to the abundance of small-membered rings in the structures. An increase in glass stability was observed as the average ring size increased from two- to four-membered rings. Mountjoy studied the structures of rare earth (RE) phosphate glasses, $(R_2O_3)_x(P_2O_5)_{1-x}$ using MD simulations and compared with EXAFS and diffraction experimental methods [63]. It was found that RE coordination number ranges from 6 to 8 and generally decreases with decreasing RE ionic radius [63]. Du et al. studied cerium aluminophosphte glasses ($19Al_2O_3$–$76P_2O_5$–$5CeO_2$) using a set partial charge pairwise potential [20]. It was found that phosphorus ions are mainly four coordinated (99 %) and aluminum ions are four (79 %), five (20 %) and six (1 %) coordinated with an average coordination number of 4.2. Aluminum ions were found to be mostly in the glass forming network for both four and five coordinated states. For phosphorus species, there is a wide distribution of Q_n species.

Phosphosilicate and aluminophosphosilicate glasses were also modeled using MD simulations by Du et al. [20]. Figure 7.4a shows the glass forming cation-oxygen pair distributions. It can be seen that the bond lengths increase from P–O, Si–O and Al–O with peak position in 1.51, 1.62, 1.79 Å respectively. It can also be seen that the first peak of the P–O, Si–O and Al–O decrease in intensity and becomes less symmetric in the sequence, which can be related to the decreasing field strength of the three glass forming cations. Bond angle distribution analyses of the three glass forming cations show that O–P–O BAD is most symmetric at angle of 109° while O–Al–O bond angle distribution is rather broad. It is interesting that five coordinated silicon, which does not exist in usual silicate glasses, was found in cerium phosphosilcate and cerium

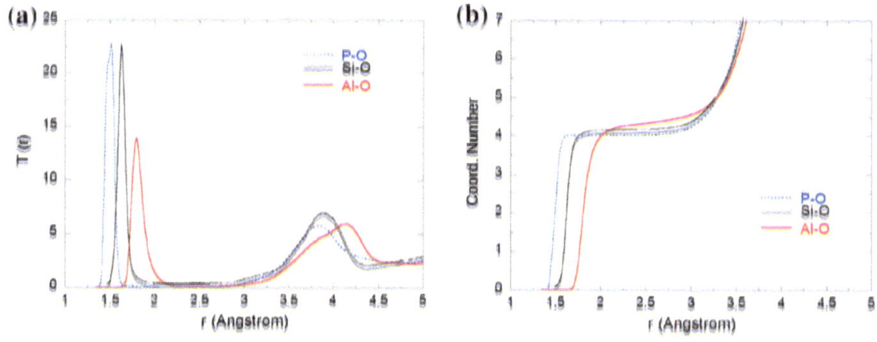

Fig. 7.4 Pair distribution function (**a**) and accumulated coordination numbers (**b**) for P, Si and Al in cerium aluminophosphosilicate glass [20]

aluminophosphosilicate glasses in the range 4–5 %. This was also observed in NMR studies of phosphosilicate glasses. This is shown in the accumulated coordination number of the three network forming cation (Fig. 7.4b).

One of the most interesting findings of the cerium phosphate glasses from MD simulations is the preference of glass forming network unit around cerium ions. Cerium ions have a coordination number of around 7 and the preference in their second coordination shell decrease in the sequence phosphorous, aluminum and silicon. The coordination numbers of Si, Al and P in the second coordination shell are 0.2, 0.5 and 6.6, respectively, for Ce^{3+} and 0, 0, 6.5 for Ce^{4+} [20]. This suggests that cerium and other rare earth ions will be preferentially surrounded by phosphorous and aluminum and forms a solvation shell to separate these high field strength ions from the rigid silicon oxygen network. These simulation results explain well the declustering effect of alumina and phosphorus oxide codopants in rare earth containing optical fibers [20].

Another group of phosphosilicate glasses that have attracted much attention in simulations is the bioactive glasses [64]. These glasses have relatively low phosphorous oxide concentration (2–3 mol%) with large amount (over 20 mol%) of sada and calcia. The compositions can be considered to be invert glass as the silica content is less than 50 %. These glasses can be partially dissolved in body fluid solutions and develop a layer of hydroxyl carbonate apatite (HCP) at the interface and are classified to be bioactive. These complicated quarterly glass system (Na_2O–CaO–P_2O_5–SiO_2) can be further modified by addition or substituting MgO, SrO or ZnO to further fine tune or add additional functionalities. The bioactivity of the glasses originates from their atomic structure and has shown strong composition, and hence structure, dependent. MD simulations have been successfully used to simulate the 45S5 bioglass [64], one of the most bioactive compositions. First reported by Zeitler and Cormack [65] and later followed by Tilocca et al. [41, 66], Pedone et al. [27], and Du and Xiang [8, 21], a number of studies on MD simulations of these and related bioactive glasses have been published. Considerable insights on the atomic structures and their bioactivity correlation have been revealed by these simulations. As a result, this special glass system has become one of the most studied multicomponent glasses.

Zeitler and Cormack [65], and later Du and Xiang [8, 21], used the partial charge pair wise potential with parameters from Table 7.1 to simulate the 45S5 bioactive glass bulk and surface structures. MD simulations using this set of potential correctly have reproduced the basic glass structure features and have shown reasonable agreement with neutron diffraction results. It was found that in 45S5, the glass network structure is highly fragmented with branched chain or small groups, with an average network connectivity of around 1.9. This fragmented network structure is one of the main reason that the glass can be easily dissolved in aqueous solutions that leads to the formation of silanol groups that further polymerize to form amorphous Si–O network, on which calcium and phosphate groups deposit and further crystallize to form HCP crystals. Pedone et al. studied the effect of MgO/CaO substitution on the structure and properties of 45S5 bioactive glasses using MD simulations [27]. Tilloca et al. have reported several papers on MD simulations of 45S5 and related

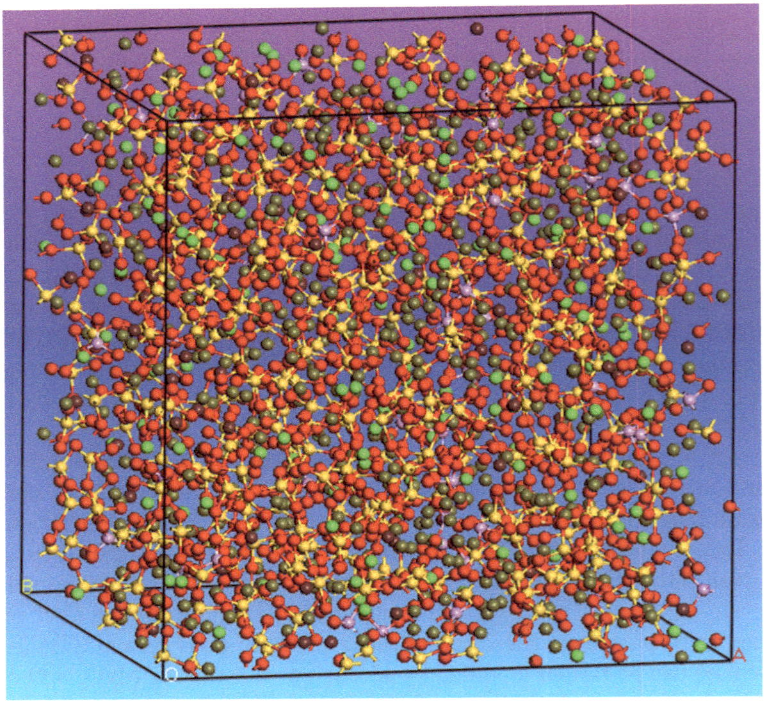

Fig. 7.5 Snapshot of bioactive glass $46.1SiO_2 \cdot 24.4Na_2O \cdot 16.9CaO \cdot 2.6P_2O_5 \cdot 10SrO$ (mol%) from MD simulations. Total number of atoms 2836. Simulation cell size $34 \times 34 \times 34$ Å^3. *Pink ball* Sr; *blue ball* Ca; *green ball* Na; *small yellow ball* Si; *small purple ball* P; *small red ball* O

bioactive glasses using a set of polarizable potentials based on the shell model [41, 66].

Du and Xiang studied the SrO/CaO substitution effect on the glass structure and diffusion of 45S5 bioactive glasses using MD simulations [21]. SrO is introduced to bioactive glasses due to the simultaneous effect of Sr ions to enhance bone growth and inhibit bone absorption. It was found that Sr ions reside in an environment similar to calcium. The substitution leads to linear increase of glass density and decrease of molar volume. However, the substitution does not considerably change the medium range structure such as network connectivity and Qn distribution. The ionic diffusion behaviors including the diffusion coefficients and diffusion energy barriers also remain constant with the substitution. Figure 7.5 shows a snapshot of 45S5 bioactive glass with 10 % SrO/CaO substitution from MD simulations. These simulation results provide further evidence that SrO/CaO substation can be a mechanism to improve efficiency of hard tissue growth and bioactivity while maintain the general dissolution behaviors and other basic physical and chemical behaviors of 45S5 glass [21].

7.4 Concluding Remarks

Classical MD simulations have been successfully used to study the structure and properties of oxide glasses including some multicomponent glass compositions. Structural, diffusion, dynamic and mechanical properties of these glasses can be systematically studied using simulations. With careful experimental validations, these simulations can be used to investigate glass the structure—property relationship and eventually design of new glass compositions. Applications of these simulations methods have now moved beyond academic laboratories and began to be used in industrial environments. There remain, however, several challenges in MD simulations of multicomponent glasses namely the reliability and transferability of empirical potentials, cooling rate and system size effect on the structure features, experimental validation of the simulated structures. Despite these challenges, it is concluded that MD simulations will play a more and more important role in fundamental and practical research of glass materials.

Acknowledgments The author gratefully acknowledges funding support of the NSF GOALI project through the Ceramic Program (project # 1105219) and the DOE NEUP project (project # DE-NE0000748).

References

1. L.V. Woodcock, C.A. Angell, P. Cheeseman, Molecular dynamics studies of the vitreous state: simple ionic systems and silica. J. Chem. Phys. **65**, 1565 (1976)
2. C. Huang, A.N. Cormack, *Computer simulation studies of the structure and transport properties of alkali silicate glasses*. Physics of Non-Crystalline Solids (Taylor & Francis, Cambridge, 1992), pp. 31–35
3. T.F. Soules, A.K. Varshneya, Molecular dynamic calculations of a sodium borosilicate glass structure. J. Am. Ceram. Soc. **64**, 145–150 (1981)
4. A.C. Wright, Neutron scattering from vitreous silica. V. The structure of vitreous silica: what have we learned from 60 years of diffraction studies? J. Non-Cryst. Solids **179**, 84–115 (1994)
5. J.F. Stebbins, Effects of temperature and composition on silicate glass structure and dynamics: SI-29 NMR results. J. Non-Cryst. Solids **106**, 359–369 (1988)
6. G.N. Greaves, EXAFS and the structure of glass. J. Non-Cryst. Solids **71**, 203–217 (1985)
7. S. Gin, J.V. Ryan, D.K. Schreiber, J. Neeway, M. Cabie, Contribution of atom-probe tomography to a better understanding of glass alteration mechanisms: application to a nuclear glass specimen altered 25 years in a granitic environment (2013)
8. Y. Xiang, J. Du, Effect of strontium substitution on the structure of 45S5 bioglasses. Chem. Mater. **23**, 2703–2717 (2011)
9. J.C. Mauro, C.S. Philip, D.J. Vaughn, M.S. Pambianchi, Glass science in the United States: current status and future directions. Int. J. Appl. glass Sci. **5**, 2–15 (2014)
10. M.P. Alan, D.J. Tildesley, *Computer Simulation of Liquids* (Oxford University Press, Oxford, 1989)
11. A.N. Cormack, J. Du, T.R. Zeitler, Alkali ion migration mechanisms in silicate glasses probed by molecular dynamics simulations, in *79th International Bunsen Discussion Meeting*, The Royal Society of Chemistry, UK, vol. 4, pp. 3193–3197 (2002)
12. A.N. Cormack, J. Du, T.R. Zeitler, Sodium ion migration mechanisms in silicate glasses probed by molecular dynamics simulations. J. Non-Cryst. Solids **323**, 147–154 (2003)

13. J. Du, A.N. Cormack, The medium range structure of sodium silicate glasses: a molecular dynamics simulation. J. Non-Cryst. Solids **349**, 66–79 (2004)
14. J. Du, A.N. Cormack, Molecular dynamics simulation of the structure and hydroxylation of silica glass surfaces. J. Am. Ceram. Soc. **88**, 2532–2539 (2005)
15. J. Du, A.N. Cormack, The structure of erbium doped sodium silicate glasses. J. Non-Cryst. Solids **351**, 2263–2276 (2005)
16. J. Du, L.R. Corrales, Compositional dependence of the first sharp diffraction peaks in alkali silicate glasses: a molecular dynamics study. J. Non-Cryst. Solids **352**, 3255–3269 (2006)
17. J. Du, L. Rene Corrales, Understanding lanthanum aluminate glass structure by correlating molecular dynamics simulation results with neutron and X-ray scattering data. J. Non-Cryst. Solids **353**, 210–214 (2007)
18. J. Du, Molecular dynamics simulations of the structure and properties of low silica yttrium aluminosilicate glasses. J. Am. Ceram. Soc. **92**, 87–95 (2009)
19. J. Du, L. Kokou, Europium environment and clustering in europium doped silica and sodium silicate glasses. J. Non-Cryst. Solids **357**, 2235–2240 (2011)
20. J. Du, L. Kokou, J.L. Rygel, Y. Chen, C.G. Pantano, R. Woodman, J. Belcher, Structure of cerium phosphate glasses: molecular dynamics simulation. J. Am. Ceram. Soc. **94**, 2393–2401 (2011)
21. J. Du, Y. Xiang, Effect of strontium substitution on the structure, ionic diffusion and dynamic properties of 45S5 bioactive glasses. J. Non-Cryst. Solids **358**, 1059–1071 (2012)
22. L.R. Corrales, J. Du, Compositional dependence of the first sharp diffraction peaks in alkali silicate glasses: a molecular dynamics study. J. Non-Cryst. Solids **352**, 3255–3269 (2006)
23. J. Du, R. Devanathan, L.R. Corrales, W.J. Weber, A.N. Cormack, Short- and medium-range structure of amorphous zircon from molecular dynamics simulations. Phys. Rev. B (Condens. Matter Mater. Phys.) **74**, 214204-1 (2006)
24. J. Du, Molecular dynamics simulations of the structure of silicate glasses containing hydroxyl groups and rare earth ions, Ph.D. Dissertation, Alfred University, 2004
25. A. Pedone, G. Malavasi, M.C. Menziani, A.N. Cormack, U. Segre, A new self-consistent empirical interatomic potential model for oxides, silicates, and silica-based glasses. J. Phys. Chem. B **110**, 11780–11795 (2006)
26. Y. Xiang, J. Du, M.M. Smedskjaer, J.C. Mauro, Structure and properties of sodium aluminosilicate glasses from molecular dynamics simulations. J. Chem. Phys. **139**, 079904 (2013)
27. A. Pedone, G. Malavasi, M.C. Menziani, Computational insight into the effect of CaO/MgO substitution on the structural properties of phospho-silicate bioactive glasses. J. Phys. Chem. C **113**, 15723–15730 (2009)
28. A. Pedone, G. Malavasi, A.N. Cormack, U. Segre, M.A. Menziani, Elastic and dynamical properties of alkali-silicate glasses from computer simulations techniques. Theor. Chem. Acc. **120**, 557–564 (2008)
29. B. Guillot, N. Sator, A computer simulation study of natural silicate melts. Part I: Low pressure properties. Geochim. Cosmochim. Acta **71**, 1249–1265 (2007)
30. B. Guillot, N. Sator, A computer simulation study of natural silicate melts. Part II: High pressure properties. Geochim. Cosmochim. Acta **71**, 4538–4556 (2007)
31. P.J. Bray, J.G. O'Keefe, NMR investigation of the structure of alkali borate glasses. Phys. Chem. Glasses **4**, 37–46 (1963)
32. J. Zhong, P.J. Bray, Change in boron coordination in alkali borate glasses, and mixed alkali effects, as elucidated by NMR. J. Non-Cryst. Solids **111**, 67–76 (1989)
33. R.E. Youngman, J.W. Zwanziger, Multiple boron sites in borate glass detected with dynamic angle spinning nuclear magnetic resonance. J. Non-Cryst. Solids **168**, 293–297 (1994)
34. A.N. Cormack, B. Park, Molecular dynamics simulations of borate glasses. Phys. Chem. Glasses **41**, 272–277 (2000)
35. L. Huang, J. Nicholas, J. Kieffer, J. Bass, Polyamorphic transitions in vitreous B2O3 under pressure. J. Phys. Condens. Matter **20**, 075107 (2008)
36. Le-Hai Kieu, J. Delaye, L. Cormier, C. Stolz, Development of empirical potentials for sodium borosilicate glass systems. J. Non-Cryst. Solids **357**, 3313–3321 (2011)

37. H. Inoue, A. Masuno, Y. Watanabe, Modeling of the structure of sodium borosilicate glasses using pair potentials. J. Phys. Chem. B **116**, 12325–12331 (2012)
38. A.C.T. van Duin, B.V. Merinov, S.J. Seung, W.A. Goddard III, ReaxFF reactive force field for solid oxide fuel cell systems with application to oxygen ion transport in yttria-stabilized zirconia. J. Phys. Chem. A **112**, 3133–3140 (2008)
39. T. Liang, T. Shan, Y. Cheng, B.D. Devine, M. Noordhoek, Y. Li, Z. Lu, S.R. Phillpot, S.B. Sinnott, Classical atomistic simulations of surfaces and heterogeneous interfaces with the charge-optimized many body (COMB) potentials. Mater. Sci. Eng. R: Rep. **74**, 255–279 (2013)
40. A. Tilocca, N.H. de Leeuw, A.N. Cormack, Shell-model molecular dynamics calculations of modified silicate glasses. Phys. Rev. B (Condens. Matter Mater. Phys.) **73**, 104209-1 (2006)
41. A. Tilloca, Short- and medium-range structure of multicomponent bioactive glasses and melts: an assessment of the performances of shell-model and rigid-ion potentials. J. Chem. Phys. **129**, 084504 (2008)
42. J. Du, L.R. Corrales, Structure, dynamics, and electronic properties of lithium disilicate melt and glass. J. Chem. Phys. **125**, 114702 (2006)
43. A. Tilocca, N.H. De Leeuw, Ab initio molecular dynamics study of 45S5 bioactive silicate glass. J. Phys. Chem. B **110**, 25810–25816 (2006)
44. L.R. Corrales, J. Du, Thermal kinetics of glass simulations. Phys. Chem. Glasses **46**, 420–424 (2005)
45. A. Tilocca, Cooling rate and size effects on the medium-range structure of multicomponent oxide glasses simulated by molecular dynamics. J. Chem. Phys. **139**, 114501 (12 pp) (2013)
46. A. Tandia, K.D. Vargheese, J.C. Mauro, Elasticity of ion stuffing in chemically strengthened glass. J. Non-Cryst. Solids **358**, 1569–1574 (2012)
47. M.I. Baskes, Modified embedded atom method, in *Proceedings of the 1994 International Mechanical Engineering Congress and Exposition*, 6–11 Nov 1994, vol. 42 (ASME, Chicago, 1994), pp. 23–35
48. L. Kokou, J. Du, Rare earth ion clustering behavior in europium doped silicate glasses: simulation size and glass structure effect. J. Non-Cryst. Solids **358**, 3408–3417 (2012)
49. J. Du, A.N. Cormack, Structure study of rare earth doped vitreous silica by molecular dynamics simulation. Radiat. Eff. Defects Solids **157**, 789–794 (2002)
50. A. Wright, The comparison of molecular dynamics simulations with diffraction experiments. J. Non-Cryst. Solids, **159**, 264–268 (1993)
51. A.C. Wright, Neutron and X-ray diffraction. J. Non-Cryst. Solids **106**(1–3), 1–16 (1988)
52. A.C. Wright, Neutron diffraction and X-ray amorphography, in *Experimental Techniques of Glass Science*, ed. by C.J. Simmons, O.H. El-Bayoumi (American Ceramic Society, Westerville, 1993)
53. J. Du, L.R. Corrales, Understanding lanthanum aluminate glass structure by correlating molecular dynamics simulation results with neutron and X-ray scattering data. J. Non-Cryst. Solids **353**, 210–214 (2007)
54. A. Pedone, T. Charpentier, G. Malavasi, M.C. Menziani, New insights into the atomic structure of 45S5 bioglass by means of solid-state NMR spectroscopy and accurate first-principles simulations. Chem. Mater. **22**, 5644–5652 (2010)
55. C. Bonhomme, C. Gervais, N. Folliet, F. Pourpoint, C.C. Diogo, J. Lao, E. Jallot, J. Lacroix, J.-M. Nedelec, D. Iuga, J.V. Hanna, M.E. Smith, Y. Xiang, J. Du, D. Laurencin, 87Sr solid-state NMR as a structurally sensitive tool for the investigation of materials: antiosteoporotic pharmaceuticals and bioactive glasses. J. Am. Chem. Soc. **134**, 12611–12628 (2012)
56. T. Charpentier, M.C. Menziani, A. Pedone, Computational simulations of solid state NMR spectra: a new era in structure determination of oxide glasses. RSC Adv. **3**, 10550–10578 (2013)
57. A.N. Cormack, J. Du, Molecular dynamics simulations of soda-lime-silicate glasses. J. Non-Cryst. Solids **293–295**, 283–289 (2001)
58. B.W.H. van Beest, G.J. Kramer, R.A. van Santen, Force fields for silicas and aluminophosphates based on ab initio calculations. Phys. Rev. Lett. **64**, 1955–1958 (1990)

59. A. Pedone, G. Malavasi, A.N. Cormack, U. Segre, M.C. Menziani, Elastic and dynamical properties of alkali-silicate glasses from computer simulations techniques. Theor. Chem. Acc. **120**, 557–564 (2008)
60. Y. Xiang, J. Du, L.B. Skinner, C.J. Benmore, A.W. Wren, D.J. Boyd, M.R. Towler, Structure and diffusion of ZnO-SrO-CaO-Na_2O-SiO_2 bioactive glasses: a combined high energy X-ray diffraction and molecular dynamics simulations study. RSC Adv. **3**, 5966–5978 (2013)
61. L.R. Corrales, J. Du, *Characterization of Ion Distributions Near the Surface of Sodium-Containing and Sodium-Depleted Calcium Aluminosilicate Melts* (Blackwell Publishing Inc, Beijing, 2006), pp. 36–41
62. J. Liang, R.T. Cygan, T.M. Alam, Molecular dynamics simulation of the structure and properties of lithium phosphate glasses. J. Non-Cryst. Solids **263–264**, 167–179 (2000)
63. G. Mountjoy, Molecular dynamics, diffraction and EXAFS of rare earth phosphate glasses compared with predictions based on bond valence. J. Non-Cryst. Solids **353**, 2029–2034 (2007)
64. L.L. Hench, Bioceramics, a clinical success. Am. Ceram. Soc. Bull. **77**, 67–74 (1998)
65. T.R. Zeitler, A.N. Cormack, Interaction of water with bioactive glass surfaces. J. Cryst. Growth **294**, 96 (2006)
66. A. Tilocca, A.N. Cormack, N.H. de Leeuw, The structure of bioactive silicate glasses: new insight from molecular dynamics simulations. Chem. Mater. **19**, 95–103 (2007)

Chapter 8
Structural Insight into Transition Metal Oxide Containing Glasses by Molecular Dynamic Simulations

Monia Montorsi, Giulia Broglia and Consuelo Mugoni

Abstract In the last years, glass research focused particular attention on transition metal oxide containing systems for semi-conductive applications, for instance glasses for solid-state devices and secondary batteries. In glass matrices, transition metal ions show multiple oxidation states that lead to peculiar structures and to highly complex systems, which produce interesting optical, electrical and magnetic properties. Computational methods have been largely employed as complementary tool to experimental techniques, in order to improve the knowledge on the materials and their performances. In this work, Molecular Dynamic (MD) simulations have been performed on a series of alkali vanado-phosphate glasses in order to gain deep comprehension of the glass structure. The short and medium range order of the V^{4+} and the V^{5+} sites in terms of coordination, pair distribution function, V–O–V linkages, bridging and non-bridging oxygen distributions were calculated and discussed. Finally, the comparison between MD and experimental results shows a very good agreement allowing the validation of the computational model and highlights the correlations between the structure and the conduction mechanism in these glasses. This allows enriching the know-how on these glass systems that result still ambiguous until now.

M. Montorsi (✉) · G. Broglia
Department of Science and Methods for Engineering, University of Modena and Reggio Emilia, via Amendola 2, Pad. Morselli, 42122 Reggio Emilia, Italy
e-mail: monia.montorsi@unimore.it

G. Broglia
e-mail: giulia.broglia@unimore.it

C. Mugoni
Department of Engineering "Enzo Ferrari", University of Modena and Reggio Emilia, Via Vignolese 905/A, 41125 Modena, Italy
e-mail: consuelo.mugoni@unimore.it

C. Massobrio et al. (eds.), *Molecular Dynamics Simulations of Disordered Materials*, Springer Series in Materials Science 215,
DOI 10.1007/978-3-319-15675-0_8

8.1 Introduction

8.1.1 Transition Metal Oxides in Glasses

Transition Metal Oxides (TMOs) containing glasses are known to find large application in solid-state lasers, luminescence solar energy concentrators (LCS) as well as in optical fibres for communication devices [1–7]. Recently large attention has been devoted to the role played by different TMOs when added to a glass formulation for their unique structure and interesting electrical properties [8–13]. Indeed, Transition Metal (TM) ions show a wide range of possible oxidation states in glasses that promote significant changes in the short and intermediate range structure strongly affecting the final properties of glasses. This opens to potential technological applications such as magnetic recording, heterogeneous catalysis, insulators or semiconductors [14]. Several glass matrices such as borosilicate, borate, phosphate or tellurate glasses have been considered until now as well as different TMOs have been added to the base composition to achieve peculiar optical, magnetic and electrical properties [15–18].

The most interesting phenomenon correlated to the semi conductive behaviour in these glasses derives by the presence in the same matrix of partly reduced and oxidized TM species. Therefore filled electron traps (donor level) and empty ones (acceptor level) can coexist in the glasses. Under particular conditions the "trapped" electron can move from the donor site to an acceptor one following a phonon-assisted polaron hopping mechanism [17, 18]. This hopping of unpaired electrons between TM ions in multi valence states leads to an electronic contribution that affects the final conductivity of the material [19–22]. Therefore, the control of the ratio between oxidised-reduced transition metal species in the glass matrix becomes a very critical issue to develop materials for ad-hoc technological applications. Unfortunately, the relative proportion of multiple oxidation states in the glass depends by several factors such as glass composition, melting conditions (temperature and time), oxygen in furnace atmosphere and presence of oxidizing/reducing agents. All these factors leads to a very complex scenario for the material design and optimization in this field of research [1].

8.1.2 Phosphate Glasses

Several glass matrices to which TMOs were added in different concentrations, from low to high content, have been tested. Among them, phosphate glasses were selected for this study due to their wide range of possible compositions, strong glass forming character as well as their easy preparation at low temperature [23]. Moreover, phosphate glasses allow also the obtainment of a stable ionic conductivity of different small ions, if mixed electronic-ionic conductivity would like to be induced [24]. Recently improvements in chemical durability of phosphate glasses have been

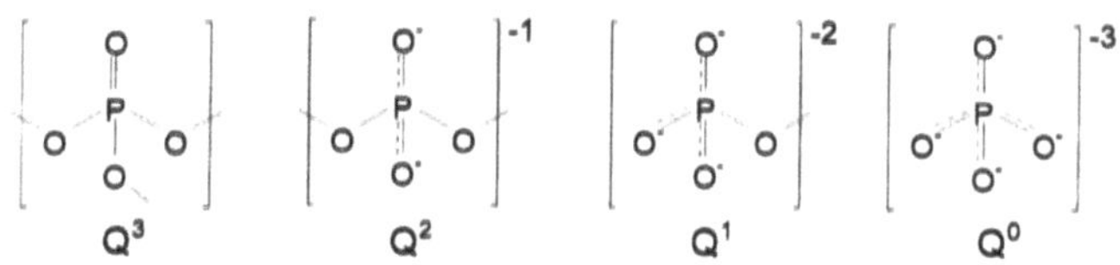

Fig. 8.1 Phosphate tetrahedral units classified in function of the number of bridging oxygen [25]

achieved by adding several extra components in the basic composition [25]. More stable and durable phosphate matrices opened the way to the formulations of more complex systems that can be designed to improve ad-hoc final properties. Functionalized phosphate glasses for technological applications have been investigated. For example, zinc phosphate was used for the obtainment of organic/inorganic composites [26] as well as peculiar improvements in chemical durability and low processing temperature were reported for phosphate glasses derived from nuclear waste [27]. Phosphate based glasses were considered for biomedical applications [28, 29] and lithium containing phosphate compositions were proposed as potential candidate for secondary batteries applications [30, 31].

The peculiar properties that characterize these glasses can be mainly ascribed to the typical phosphate network. Therefore, detailed structural characterization becomes fundamental to investigate the structure evolution associated to compositional changes that directly results in the macroscopic behaviour of the materials [32–34]. The basic structural units of the amorphous phosphate systems are the P-based tetrahedra. Different polymerizations of the phosphate chains can be identified and codified by the Q^n where "n" represents the number of bridging oxygens (BOs) per tetrahedron (shown schematically in Fig. 8.1) [35].

Based on the [O]/[P] ratio [32] in the glass matrices, the number of BO could change and create a different Q_n distribution. A cross-linked network of Q^3 tetrahedra constitutes amorphous P_2O_5 and the introduction of modifier ions depolymerize the phosphate chains, passing from a Q^2 (metaphosphate chains) to a pyro (Q^1) and/or orthophosphate (Q^0) distribution [25] breaking progressively the P–O–P links. Therefore, type and concentration of the modifiers leads to changes in structure that generally results in the final properties [36–38]. For instance, Hudgens et al. working on binary alkali ultra-phosphate glasses $xLi_2O-(1-x)P_2O_5(x <$ 0.5 mol fraction of alkali) observed a minimum in the T_g close to the x = 0.2 while for $x > 0.2$ the T_g rapidly increases [39]. They explain the T_g decrease in terms of loss of the completely polymerized Q^3 phosphate tetrahedral network. At the same time, the increased T_g above $x \approx 0.2$ mol fraction was justified by a sort of restructuring of the medium range order (MRO), which occurs in the glass structure characterized by more alkali-non-bridging oxygen (NBO) interactions. Correlations between the structural features and the glass composition have been extensively investigated by experimental techniques such as Nuclear Magnetic Resonance (NMR) [40, 41], X-ray absorption fine structure (XAFS) [42, 43], magic angle spinning NMR (MAS-NMR) [44, 45] and neutron diffraction [46, 47] as detailed in literature.

8.1.3 Ratio Between Reduced/Oxidised TM Ion in Glasses

Structure characterization becomes very important in TMOs containing glasses due to the different local environments that TM could present depending on its oxidation state in glasses. Multiple oxidation states correspond to different short and medium range environments, which are macroscopically reflected by different electrical properties. Hence, the ratio between reduced and oxidised forms of TM species in the glasses is a key parameter to consider in design and optimization of materials for semi-conductive applications. For example, in phosphate-based glasses with low TMOs concentration, it was generally observed that while chromium, manganese and vanadium are completely reduced in the system, copper, molybdenum and titanium tend to remain at higher valence states [1, 48–50]. The R value defined as $TM^{m+}/TM^{(m+1)+}$ is mainly correlated to the glass basicity and in general it was reported that phosphate glasses show lower basicity, with respect to silicate or borate, which promotes an higher reducing effect on the TM [50, 51]. As an example, a higher content of oxidised species (Cr^{6+} and V^{5+}) was detected in TM doped silicate- with respect to phosphate-based systems [52, 53]. By maintaining fixed the glass matrix and modifying the basicity of the whole system by playing with the type and the amount of network modifier cations, different situations can be observed. Particularly, the R value results significantly affected only in silicate- and borate-based glasses, while in phosphate-based ones it remains practically unaffected. This result was explained, from a structural point of view, considering the presence of the P=O double bond in the 3D network that limits the structural rearrangement associated to a more acid or basic environment. Among the parameters affecting the R value, the melting temperature, the atmosphere (oxidizing or reducing) and the concentration of TM have to be mentioned. For instance, by focusing the attention on phosphate glasses, the increasing of the melting temperature in iron [54] and copper containing systems results into the increase of the R value (i.e. an increase in the amount of the reduces species) that leads to improved electrical properties of the material, associated to the hopping mechanism between reduced and oxidized transition metal species [1, 52, 55–57]. Concerning the correlation between TM concentration and R, it is worth noting that a unique trend cannot be extrapolated due to the peculiar behaviour shown by the different TMs considered in the glass formulations. Generally, it is reported that for copper and iron ions [56, 58] the reduced form is stabilized at higher TMO concentrations, on the contrary for vanadium the oxidised form is prevalent in higher vanadium containing system [59]. For example, in binary copper (II) phosphate glasses, the R ratio initially decreases increasing the copper oxide content, showing a minimum at 0.3 mol%, and then increases confirming the strong dependence of R on the glass composition [36]. It could be generally concluded that structural constraints, coordination and complex ions formation depending on glass composition and experimental conditions strongly modify the oxidized/reduced state of the TM ions leading to possible properties changes in glasses.

8.1.4 TMO Organization in Glasses

The presence of multiple oxidation states of TM leads to different local and medium environments of the species inherent in the 3D network that result also in the final properties of the glass. Evaluation of parameters such as bond distances, angles and coordination distributions of the different species inherent in the glass framework became fundamental to explain how the structure organization affects the performances of the material. In the recent past several experimental characterization such as Electron Paramagnetic Resonance (EPR), X-ray Photoelectron Spectroscopy (XPS) and Extended X-ray Adsorption Fine Structure (EXAFS) and X-ray Absorption near -edge structure (XANES) have been used to investigate and quantify the state of dissimilar TM ions in glasses [12, 60–65]. Several works showed that the coordination of TM is strongly dependent on its concentration because possible distortion of the metal local site as well as the formation of complex ion of TM in phosphate glass could take place. This leads to the shift of the averaged coordination of the TM from 6 to 4 [1, 55, 66–69].

Moreover the TM ions could act as network modifier or former depending by their concentration in the phosphate glasses leading to quite different environment: from a depolymerized 3D framework at low TMO content to a more interconnected glass network for high containing TMO glass formulations. It was detailed also that presence of TM based sharing corner or edge polyhedral leads to the formation respectively of chains or TM ions clusters in the glass network depending on the TMO concentration [1, 69–73].

For example, high percentage of the TMO in phosphate glass can generate an "inverted" glass network in which the PO_4 tetrahedra were substituted by (TM)Ox based polyhedral depending by the TM concentration suggesting the creation of TM based network in high TM containing systems (Fig. 8.2).

Recently, the formation of a TMO based network has been investigated and accurately described for sodium vanado-phosphate glasses [59, 74] by means of Advanced Nuclear Magnetic Resonance (NMR) Spectroscopy and Electron Paramagnetic Resonance (EPR) methods. Changes in the glass structure from infinite phosphate chain at low vanadium content to mixed vanado-phosphate network, and finally to vanadate network (connected to isolated phosphate polyhedra) at high vanadium content are observed. A similar structural model has been proposed by Khattak et al. in vanado-phosphate glasses, in which the addition of V_2O_5 promotes the increase of the P–O–V and V–O–V linkages at the expenses of the P–O–P ones [72]. It is worth noting that peculiar properties in phosphate materials can be found when phosphate network turns into an inverted network (Nasicon glass) [75].

Studies performed on alkali phosphate glasses with low content of transition metals, showed that the TM ions (vanadium, chromium, titanium, iron and others) are predominantly found in octahedral coordination, which can be related to a more modifier character of TM ions, as previously reported [68, 76, 77]. Bianconi et al. [78] observed that at low V_2O_5 concentration (ca. 5 mol%) vanadium exists only as six folded V^{4+}. This is the more common coordination for V^{4+} ions and it is

Fig. 8.2 Formation of vanadate network connected to isolate phosphate polyhedra in glasses containing high V_2O_5 content

the expected configuration for V ions acting as "modifiers" in the glass network. Increasing the vanadium content at 20 mol% the glass structure contains in the first case four and six coordinated V^{4+} and a comparatively lower concentration of V^{5+}. Moving to 50 mol% of vanadium oxide content the glass shows a mixture of V^{4+} and V^{5+} ions, four and five coordinated respectively. Another interesting study, related to the structure of iron in phosphate glasses, has been conducted by Karabulut et al. [54], using Mössbauer and X-ray absorption fine structure (XAFS) spectroscopies. Glasses obtained by melting and quenching procedure and containing 40 mol% of Fe_2O_3 have been prepared by melting the raw materials at different temperatures and atmospheres. The R value has been determined as a function of the varied process factors and it was found that both Fe(II) and Fe(III) are present in all the compositions and the Fe(II) fraction increases with the temperature and using a reducing atmosphere. Regarding the structure of the single cations, the Mössbauer spectroscopy showed that Fe(II) ions occupy a single site whereas Fe(III) ions occupy two distinct sites in all the glasses. XAFS analysis pointed out an average coordination of 4–5 suggesting a mixed tetrahedral–octahedral coordination of this TM ion in these glasses. Similar results have been found in iron-containing sodium [79] and iron containing barium [80] phosphate glasses.

Also copper as TM ion has been largely studied [13, 81, 82] in order to understand how the addition of copper oxide affects the chemical and physical properties of glasses.

Particular attention has been focused on how this TMO could improve mixed electronic-ionic conduction as well as how it could develop fast ion conductivity (FIC) in glasses [36].

Shih et al. [83] analysed the structure of CuO containing sodium poly- and meta-phosphate glasses by Fourier transform infrared (FTIR) and Nuclear Magnetic Resonance (NMR) spectroscopies, observing that the introduction of CuO in the glass composition promotes the formation of P–O–Cu linkages. In detail, in the metaphosphate glass composition, CuO primarily acts as network modifier depolymerizing the phosphate chain, while at higher content as a network former. In poly-phosphate glasses this TMO simultaneously acts both as a network former and modifier. Presence of P–O–Cu linkages in glasses belonging to the $xCuO-(1-x)P_2O_5$ system ($0 < x < 0.5$ mol) [36] was investigated, due to a systematic transformation from a three-dimensional ultraphosphate network, dominated by Q^3 tetrahedra, into an isolated Cu-octahedra. At higher copper content ($x > 0.33$), the Cu-polyhedra begin to share common oxygens to form a sub-network in the phosphate matrix. This clearly increases the density and the glass transition temperature (T_g) experimentally observed. Further investigation on copper containing phosphate glasses points out that Cu^{2+} and Cu^{+} occupy sites with different coordination geometries, 2 + 4 distorted octahedral and 4 + 2 octahedral respectively [84]. Also, UV absorption and emission studies [85] indicate that Cu^{+} ions in calcium phosphate glasses exist in trigonally distorted octahedra.

8.1.5 Vanado-phosphate glasses

Vanado-phosphate glasses (VPGs) are well known for their semi-conductive behaviour [72]. In the last decades they were widely studied in order to understand the mechanism underneath their electronic properties. The electronic conductivity was correlated to small polaron hopping between vanadium ions with different valence state [86], favoured by their electronic configuration [87]. In VPGs, vanadium atoms are mainly characterized by two oxidation states, 4+ and 5+, and the V^{4+} and V^{5+} based interconnected polyhedra become the preferential path for the polaron hopping [88]. Therefore, the relative amount of these two species (V^{4+} and V^{5+}), which is usually expressed as V^{4+}/V_{tot} ratio, influences the final electrical properties of VP systems [86, 89].

The oxidizing power of the V^{4+}/V^{5+} redox couple is enhanced by the vanadium instability in the amorphous phase. Increasing the vanadium content, the systems are more stabilized, and the vanadium reduction decreases. Furthermore, increasing the percentage of vanadium atoms in the glass matrix, the process conditions (for instance the cooling rate) begin to influence the V^{4+}/V_{tot} ratio [90]. Above the melting temperature, the melt tend to lose some oxygen atoms, and V^{5+} ions tend to be reduced in V^{4+} specie.

The analysis of structural rearrangement could be essential to explain the properties that characterize VPG systems. The main features that could affect the electronic behaviour are the number of V^{4+}–O–V^{5+} linkages, the vanadium coordination number and the V–V distance. In our knowledge, a unique description of VPGs structure has not been already reached. For instance, Jordan et al. [36, 91] supposed that a tetrahedral coordination of vanadium atoms allows sharing edges between polyhedra. This configuration should cause a decrease in the V–V length and favour the polaron hopping. However, experimental and simulated evidences show that V polyhedra prefer to share corners, which denote longer V–V distances [47, 59, 90, 92–94].

Several vanadium configurations and geometries could be found in VPG systems, opening to a wide range of final properties for these systems. Coordination is comprised between 4 and 6, and tetrahedral, square pyramidal, trigonal pyramidal, octahedral units are found in VPGs [47, 78, 91]. The structural units are further characterized by distorted structure that could produce intermediate geometries, such as 5 + 1 or 4 + 1 [47].

In the glass network, vanadium plays an intermediate role [59, 78] depending by its content. Generally, for low V_2O_5 content, vanadium ions tend to have higher coordination number and a stronger modifier character. Increasing the content, the coordination decreases and a more former behavior could be associated to vanadium ions.

The introduction of alkaline ions, such as lithium, sodium or potassium [19–22] in amorphous vanado-phosphate matrices includes the ionic contribution to the total conductivity. The mixed electronic-ionic conductivity enables the use of alkali VPGs as cathode material in solid-state batteries [95]. Up to now there is still no general agreement on the predominant conductivity mechanisms, between the ions migration or polarons hopping, in alkali VPGs. The electrical conductivity is strongly influenced by the composition and relative ratio between the various elements involved in the glass matrix [96]. The interactions among the various components of the glasses could modify the predominant contribution (electronic/ionic) or reduce the total conductivity. In literature, several influences were suggested, which could be summarized in the following groups:

- *Alkali content*

Alkaline ions have strong effects on the electrical properties of VPGs. Increasing the network modifier content; the total conductivity could decrease even several orders of magnitude [21]. Various hypotheses are proposed to explain this behaviour. Firstly, high percentages of alkali result in the increase of the number of NBO that could break the paths of the electronic conductivity [21, 96]. Furthermore, modifier ions could interact with polarons, creating neutral entities that inhibited the electronic contribution [21, 96, 97].

Various papers report [19, 97, 98] that increasing the alkali content, electronic contribution tends to decrease, due to the lower amount of V_2O_5 and the effects of modifier ions, until the concentration of alkali is enough to promote the ionic contribution. In some cases, a well defined cross point between the predominance of electronic and ionic contribution could be detected. An example of this behavior is reported by Jozwiak et al. [97] that studied the $xLi_2O-(1-2x)V_2O_5-xP_2O_5$ glasses where x varies from 15 to 45 mol%. They pointed out that the glass with the highest percentage of V_2O_5 (x = 15 mol%) shows predominantly electronic conductivity, due to the highly cross-linked structure, which favors the electron hopping between $V^{4+}-V^{5+}$. On the contrary, high content of alkali ions increases the NBO species in glasses and inhibits the network polymerization, which could cause the break of the useful paths for the polaron hopping that results in the decrease of the electronic contribution. Glass compositions with high percentage of lithium (x = 40–45) are characterized by mainly ionic conductivity, promoted by the depolymerized glass network, where the lithium motion is favored.

- *Alkali type*

In addition to the alkali content, even the alkali type influences the final electrical properties of VPGs. The drop in the conductivity due to the increase of alkali amount is intensified for larger alkali ions [21]. For instance, Assem and Elmehasseb [20] reports further insight in the role of alkali type on electrical properties. Generally, the V–V distances become larger in function of the increase of alkali radius, leading to longer hopping paths that justify the decrease of the electronic contribution to total conductivity experimentally observed. Furthermore, the dimension of alkali ions affects the ionic conductivity. Larger network modifiers show lower mobility in glass matrix, which produces a decrease in the conductivity and an increase of the activation energy [19–21, 97]. For low concentration of modifier ions, the influences on the electrical properties mainly derive from the vanadium content, while increasing the percentages the effects of the alkali amount and type are emphasized [98, 99]. It is worth noting that presence of a network modifier generally promotes the ionic contribution but contemporary presence of more modifiers ions could favor the so called Mixed Alkali Effect (MAE) resulting into an anomalous trend of conductivity as a function of the alkali concentration in VPGs [22, 100].

- $V_2O_5/P_2O_5\,ratio$

For low concentration of alkali ions, the predominant contribution to the total conductivity is electronic. The systems are nearly insensitive to the alkali influence [22] and the main structural changes are due to the V_2O_5/P_2O_5 ratio. The interaction between the two glass formers could interfere in the formation of $V^{4+}{-}O{-}V^{5+}$ paths that allow to suppose a possible inhibition of the electronic conductivity [89]. Structural evidences confirm this hypothesis: as suggested by Tricot et al. for higher V_2O_5/P_2O_5 ratio the glass network is mainly composed by the V ions, but the P ions continue to contribute to the matrix, interposing with the vanadium polyhedra [59].

8.1.6 Why Molecular Dynamics

It is well known that the performance of a material strongly depends on the short and medium environment of the chemical species that characterize the composition. Thus, the structural characterization covers a fundamental role in the understanding of the final material properties. Nowadays, experimental techniques allow to obtain almost full comprehension of the material structure, but the use of computer simulations, as complementary tools, results extremely effective to describe in detail the local and medium environment of the TM ions in phosphate glass matrix.

In particular, techniques such as Molecular Dynamics (MD), allowing the evaluation of structural and dynamical properties, are employed to study the relationships between physical phenomena induced by the TMO addition into phosphate-based glasses. Glasses containing P_2O_5 have been widely studied by MD simulations. Generally, the main area of interest is BioGlass [101–111]. Usually, the percentage of P_2O_5 is lower than 12 mol%, but the effect due to the organization of phosphorus ions in the glass structures strongly influences the bioactivity and the ability to crystallize in hydroxyapatite (HA) [102–104, 112]. Indeed, the P_2O_5 tends to depolymerize the silicate network and accelerate the formation of HA. MD permits to evaluate the degradation of silicate matrix in function of the phosphorous content, leading to a deeper comprehension of this mechanism [103, 104]. Another important topic is phosphate glasses containing Rare Earth (RE) atoms [113–115] for instance Ce, Tb, Dy, etc. REs provide particular optical and magnetic behaviors and their introduction in P_2O_5 based glasses permits to obtain structural rearrangements that improve these properties. In this case, MD allows to study the probability of formation of RE based clusters that strongly influences the final properties and they could be reduced by using the phosphate matrix [113–116]. Furthermore, the coordination number of RE [116] could be analyzed by this computational methods in order to understand the effect of their concentration on this structural property. Last but not least, the structural properties and Li^+ diffusion [117–122] in lithium phosphate glasses have been largely simulated by MD. The alkaline cations, acting as modifier, tend to depolymerize the phosphorus chains, leading to the creation of wider channels that favors the Li^+ movements.

Even the MD could be a suitable technique to study the influences of TMOs on phosphate glass structures; few works on this topic are reported in literature. Zinc oxide could be characterized by changes in local environment in function of its content in glass matrix, which influences the final properties [123, 124]. Tischendorf et al. [124] studied $xZnO-(100-x)P_2O_5$ with $40 < x < 70$ mol% systems respect structural rearrangement and simulated the glass transition temperature (T_g). No variation in the Zn coordination number has been found as expected, and this result could allow to suppose a slight influence of Zn rearrangement on the behavior of T_g. Boiko et al. [123]simulated zinc-sodium phosphate glasses: they discovered that the presence of Na ions strongly affects the Zn local environment. A decrease in Zn coordination has been detected and TM ions tend to show a structural rearrangement more close to the P ions. Al-Hasni et al. [125, 126] investigated the introduction of iron ions in phosphate glasses, where both Fe^{2+} and Fe^{3+} ions were considered, analyzing their effect on the presence of a single or two valence states. The coordination environments of iron are similar, but the two valence states lead to different effects on the phosphorus network.

Several force fields were proposed to simulate glasses containing P_2O_5 that are shortly listed in Table 8.1. Some of the suggested potentials include three body potentials, which permit to better simulate specific angles, but they increase the rigidity of the glass network. The choice of one or another force field is influenced by the features that would be studied.

Table 8.1 Force fields developed for glasses containing P_2O_5

Name	Description	Literature
Tilocca et al.	Shell model based on Buckingham potential and three body potential for O–Si–O	[104, 106–109, 111]
Teter et al.	Buckingham potential and three body potential for O–P–O and P–O–P	[100, 114, 115, 125, 126, 152]
Vessal et al.	Buckingham potential and vessal three body potential for O–P–O, P–O–P, O–Li–O, Li–O–Li	[117, 118, 153]
Pedone et al.	Rigid ionic model based on morse potential	[16, 103, 109]
Du et al.	Buckingham potential	[105, 113]
Karthikeyan et al.	Born mayer potential	[119, 120]
Cygan et al.	Lennard jones potential and three body potential for O–P–O and P–O–P	[121, 122, 124]

In our knowledge, few research groups studied the vanadium-based glasses by simulation techniques. Seshasayee et al. [127] simulated amorphous-V_2O_5 by MD technique that showed a predominance of five coordination for vanadium and a preferential corner-sharing distribution. Hoppe et al. [92] using Reverse Monte Carlo algorithm also evaluated the structure of a-V_2O_5. A good agreement with NMR and a better reproduction of the polyhedron distortion were detected, while the obtained results were inconsistent with the EXAFS. Murawski et al. [93] tried to develop the overlapping polarons (OLP) tunnelling model in VPGs. MD evaluated the V–V distance, which is an important parameter for the OLP. Garofalini et al. [128] studied the interface between a intergranular films of V_2O_5 and lithium silicate glass by MD, in order to simulate the Li diffusion at the electrode/electrolyte. These previous studies take into account only the vanadium ions with a valence state of 5+. Differently, Ori et al. [16] implemented the Pedone potential [129] with parameters codifying both V^{4+}–O and V^{5+}–O interactions. Based on the experimental glasses characterized by Giuli et al. [130], alumino-silicate glasses with different amount of vanadium ions were simulated in order to confirm the goodness of the force field.

8.2 Aim of the Work

Glass-based systems represent one of the most interesting challenges to develop solid-state devices. For this aim, the incorporation into the glass matrix of TMOs, as dopant or as main component, could be used to induce ad-hoc functionalized properties [8, 9, 74]. Recently TM such as V, Fe, Cu have been added to different glass matrices in order to improve functionality and semi-conductive properties [18, 125]. Nowadays, glass and glass-ceramic systems containing V_2O_5 are the object of an extensive research in view of their potential application as cathode materials in secondary batteries [18, 75, 131, 132]. Indeed, the presence of V ions in multiple oxidation states (+3, +4 and +5) induces an electronic conductivity [133–135] in glass via small polaron hopping from low to high valence state of the V ions making them very interesting materials in the solid state field. At the same time, the presence of multiple oxidation state promotes the formation of different geometries [136, 137] and local environments in the glass matrices [16, 60, 130, 138] giving to these glasses a high complexity. For example, in glasses the main oxidation states are +4 and +5, therefore the vanadium-oxygen coordination can vary from tetrahedral to trigonal pyramidal, square pyramidal and distorted octahedral structure units in dependence of the glass composition. For these reasons, vanado-based glasses were extensively investigated by means of experimental methods but the results are not yet conclusive [10, 11, 17]. A work on vanadium phosphate glasses [47] discussed the structure of vanadium in terms of regular and distorted units that shift from a mainly square pyramidal environment to a tetrahedral based network by increasing the amount of vanadium oxide. Depending on the structure environment, the quantification of the non-bridging oxygen (NBO) and interatomic distances of TM ions (V–V) in the glass structure becomes fundamental considering that hopping conduction can occur primarily through vanadium ions along V–O–V linkages [59].

In this work, classical MD simulations are applied in order to provide systematic study of the structure of $Na_2O-V_2O_5-P_2O_5$ glasses (NaVP). To this purpose the selected compositions, $(Na_2O)_{(1-z)/2}-(V_xO_y)_z-(P_2O_5)_{(1-z)/2}(z = 10\text{–}80\text{mol\%})$, experimentally investigated by Tricot et al. [139] by solid state ^{31}P- and ^{51}V-MAS-NMR have been simulated. Moreover two glasses belonging to the $33Li_2O-33V_2O_5-33P_2O_5$ system (33LiVP) with different V^{4+}/V_{tot} ratio are studied with the aim to give detailed information on these glass structures but especially to evaluate the effect of V^{4+}/V_{tot} ratio on the local structure. The interest in this composition derives from its successful use as starting materials for the synthesis of a potential candidate as cathode materials in lithium ions batteries [75, 97, 140].

Table 8.2 summarises all the glass compositions investigated by MD. The compositions reported in italic were also synthesized by melt-quenching method and characterized by X-Ray photoelectron spectroscopy (XPS) and density measures (Mugoni et al. unpublished data) in our lab. XPS analysis was performed in order to quantify the V^{4+}/V_{tot} ratio used for the definition of MD input structures. Density values derive from literature [139] and for the glasses experimentally obtained they were confirmed by using the Priven methods. Clearly the characterization of glass structure as well as the knowledge of vanadium specification (oxidation state and symmetry) become fundamental for further elucidation and knowledge of the physical and chemical properties of vanadium containing materials. Bond distances and bond angle distribution, coordination number distribution as well as the non-bridging oxygen (NBO) have been quantified by using MD. Moreover the medium range order of the glasses was characterized in terms of Q^n species distribution, V–V interatomic distances and V–O–V linkages, because these features are supposed to be related to the small polaron hopping.

Table 8.2 Glass compositions analysed in this work

	Na_2O (Li_2O)	V_2O_5	P_2O_5	V^{4+}/V_{tot}(%)	Density (g/cm^3)
NaVP-10	*44.94*	*10.13*	*44.94*	*29*	*2.54*
NaVP-20	39.99	20.03	39.99	25.8*	2.62
NaVP-35	*32.58*	*34.85*	*32.58*	*15*	*2.73*
NaVP-60	*20.00*	*60.00*	*20.00*	*7.5**	*2.89*
NaVP-80	10.00	80.00	10.00	6.1*	2.98
33LiVP-48	33.29	*33.42*	33.29	48.0*	2.65
33LiVP-26	*33.29*	*33.42*	*33.29*	*26*	*2.65*

The V^{4+}/V_{tot} values with * derives from [139] and the one with # from [75, 125]

8.2.1 Computational Details Place

Molecular Dynamics is a powerful technique; able to yield detailed information on glass structure but its success strongly depends on several factors. Among these, the reliability of the interatomic potentials, the simulation methodology, as well as the simulation box size play fundamental roles in the predictive capability of the model [141, 142]. In this work classical MD simulations were performed using the DLPOLY code in the 2.12 version [143]. A well-validated computational procedure was used [16, 144, 145]. To obtain the glass structures, five different random configurations for each composition, of about 4000 (Table 8.3) atoms, were placed in a cubic box that size length was chosen to reproduce the experimental density As previously reported, several force fields were proposed for phosphate glasses [104, 105, 119, 121, 125, 129, 146], showing limits and goodness depending on the properties that would be analyzed. For the aim of this work, a rigid ionic model with partial charges developed by Pedone et al. has been selected [129]. This potential consists of three main contributions: the first is the long-range Coulombic term, the second one takes into account short-range forces, codified by a Morse function, and the last is an additional repulsive contribution for simulations at high temperature and pressure.

$$U(r) = \frac{(z_i z_j)e^2}{r} + D_{ij}[\{1 - e^{-a_{ij}(r-r_0)}\} - 1] + \frac{C_{ij}}{r^{12}} \tag{8.1}$$

The V–O pair potentials parameters were implemented in this force field using a well-validated refinement procedure, which consists in a combination of conventional and relaxed fitting steps [16, 147], carried out with the general utility lattice program GULP code [148]. Crystal structures containing vanadium in different valence states and coordination were selected and used to refine the vanadium parameters: details on the procedure are discussed by Ori et al. [16]. Table 8.4 reports the parameters used in the MD simulations. Integration of the equation of motion was performed using Verlet Leapfrog algorithm with a time step of 2 fs. Coulombic interactions were calculated by the Ewald summation method with a cutoff of 12 Å and an accuracy of 10^{-4} eV. A cutoff of 5.5 Å has been used for the short-range interactions.

Table 8.3 Details for the preparation of the input simulation boxes

	Na^+ (Li^+)	V^{5+}	V^{4+}	P	O	Total atoms
NaVP-10	*630*	*76*	*33*	*630*	*2146*	*3515*
NaVP-20	555	164	57	555	2189	3520
NaVP-35	*430*	*340*	*60*	*430*	*2260*	*3520*
NaVP-60	240	620	50	240	2370	3520
NaVP-80	113	800	52	113	2443	3521
33LiVP-48	488	172	159	488	2212	3519
33LiVP-26	*460*	*268*	*94*	*460*	*2238*	*3520*

Table 8.4 Parameters used in MD simulations

	D_{ji}(eV)	a_{ij}($Å^{-2}$)	r_0(Å)	C_{ij}($eVÅ^{+12}$)
$Li^{+0.6}$–$O^{-1.2}$	0.0011	3.4295	2.6814	1
$Na^{+0.6}$–$O^{-1.2}$	0.0234	1.7639	3.0063	5
$P^{+3.0}$–$O^{-1.2}$	0.8313	2.5858	1.8008	1
$O^{-1.2}$–$O^{-1.2}$	0.0424	1.379	3.6187	22
V^{+3} $O^{-1.2}$	0.0219	1.4959	3.3985	1
$V^{+2.4}O^{-1.2}$	0.0033	2.1093	2.6636	1

Similar conditions have been already validated and tested successfully on different inorganic glass systems. Similarly to previous works [141, 144, 149–151] based on a rigid ionic model with partial charges, the system was heated at 5000 K, a temperature suitable to bring the system to liquid state in the framework of the adopted force field. The melt was then equilibrated for 100 ps and subsequently cooled continuously from 5000 to 300 K in 1060 ps with a nominal cooling rate of 4.5 K/ps. The temperature was decreased by 0.01 K every time step using a Berendsen thermostat with the time constant parameter for the frictional coefficient set to 0.4 ps. Another 100 ps of relaxation at constant energy and 50 ps of data production were performed at 300 K. Configurations at every 0.1 ps were recorded for structural analysis.

8.3 Results and Discussion

8.3.1 Bond Distances and Coordination Analysis

The interatomic bond distances are investigated in terms of pair distribution function (PDF). The maximum of the first peak in the PDF curves derived by MD corresponds to the average bond distance and this value was obtained for all the cation-oxygen (M–O) and cation-cation (M–M) species in the analysed systems. The values are reported in Table 8.5 and are clearly in agreement with the theoretical and

Table 8.5 Bond distances derived by PDF curves from MD (Å, error within ±0.05)

	P^{5+}–O	V^{5+}–O	V^{4+}–O	Na (Li)–O	V^{5+}–P	V^{4+}–P	V^{5+}–V^{5+}	V^{4+}–V^{5+}	V^{4+}–V^{4+}
NaVP-10	1.50	1.94	1.84	2.32	3.44	3.30	3.68	3.54	–
NaVP-20	1.52	1.92	1.18	2.30	3.44	3.30	3.63	3.50	3.44
NaVP-35	1.52	1.87	1.74	2.35	3.44	3.26	3.65	3.46	3.34
NaVP-60	1.52	1.78	1.73	2.38	3.46	3.28	3.65	3.46	3.34
NaVP-80	1.50	1.73	1.73	2.50	3.45	–	3.63	3.47	3.47
33LiVP-48	1.52	1.84	1.73	1.88	3.45	3.26	3.60	3.49	3.36
33LiVP-26	1.52	1.87	1.74	1.98	3.44	3.28	3.61	3.48	3.39
References	1.43–1.58 [38, 154]	1.57–2.89 [136]	1.57–2.89 [136]	2.34 [111] 1.93–2.00 [155]	3.42 [59]	3.17–3.35 [156]	3.37–3.60 [47]	3.37–3.60 [47]	3.37–3.60 [47]

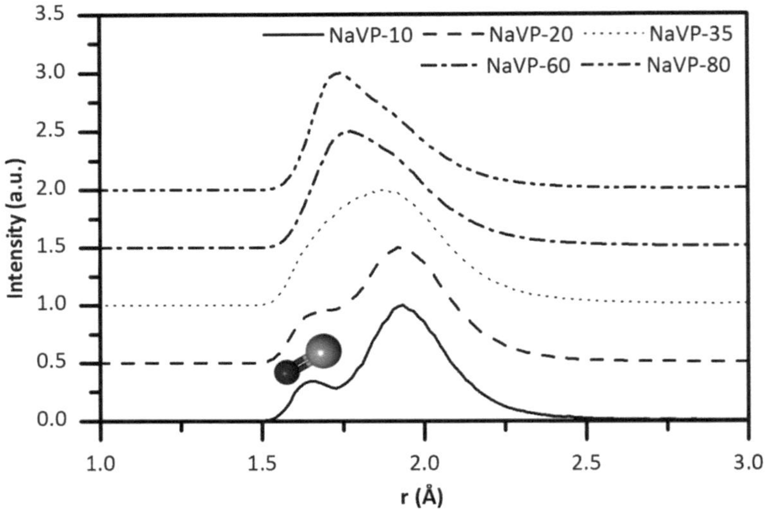

Fig. 8.3 PDF curves of the V^{5+}–O bond in the NaVP series

experimental data found in literature, which allows to validate the computational procedure and the goodness of parameters used in the simulations.

The V^{5+}–O PDF curve, reported in Fig. 8.3, shows the presence of a shoulder at about 1.65 Å that can be associated to the presence of the typical *vanadyl (V=O)* bonds that are found in vanado-phosphate systems.

The increase of the concentration of vanadium ions in the systems leads to the shift of the V^{5+}–O peak from 1.94 to 1.73 Å and the contribution that derives from the vanadyl bonds is overlapped in the peak. Furthermore, the creation of V=O···V bonds tends to elongate the vanadyl bonds, emphasizing the disappearance of the shoulder [47]. The increase of vanadium content in the glass matrix leads to structural changes resulting in the decrease of the V^{4+}–O and V^{5+}–O distances from NaVP-10 to the NaVP-80 system. A value close to 1.73 Å was observed for both the V^{4+} and V^{5+} cations in the higher vanadium containing system.

Regarding the 33LiVP serie, the ratio of V^{4+}/V_{tot} does not play a significant role on both V^{5+}–O and V^{4+}–O distances, showing slightly differences in the range of 1.84–1.87 Å and 1.73–1.76 Å, for 33LiVP-26 and 33LiVP-48, respectively. It is worth noting that the V^{5+}–O and V^{4+}–O distances for the lithium containing glasses are comparable and very close to the NaVP-35 system, which has a composition similar to the 33LiVP glasses, showing that the V^{4+}/V_{tot} ratio as well as the different type of modifier (Na^+ or Li^+) does not significantly affect the local shell of vanadium in terms of averaged V–O interatomic distance. The V–V distances well agree with the data reported in literature and while the V^{4+}–V^{5+} and the V^{5+}–V^{5+} show values close to the higher value of the references data, the V^{4+}–V^{4+} shows lower values. However no significant trend can be observed as a function of the vanadium content in the NaVP series as well as no changes are observed in the 33LiVP series

Table 8.6 Averaged coordination number (error within ±0.03)

	P–O	V^{5+}–O	V^{4+}–O	Na(Li)–O
NaVP-10	4.02	5.56	5.36	5.15
NaVP-20	4.01	5.45	5.18	5.25
NaVP-35	4.00	5.27	4.90	5.51
NaVP-60	4.00	5.12	4.70	5.98
NaVP-80	4.00	4.99	4.58	6.27
33LiVP-48	4.01	5.31	4.98	4.79
33LiVP-26	4.01	5.46	4.78	3.82
References	3.9–4.0 [154]	4.0–6.0 [136]	4.0–6.0 [136]	4–6 [111] (4–5) [119]

depending by the V^{4+}/V_{tot}. The same invariant trend can be observed for the V–P distances notwithstanding the coherence between simulated and literature data can be highlighted.

The averaged coordination number (CN) reported in Table 8.6, which well agrees with the ranges reported in literature, confirms the network former role of the P ion. It is evident that the average CN of V ions slightly decreases from low-to-high vanadium containing systems. This trend suggests an intermediate role played by V^{5+} and V^{4+} ions for increasing concentration of V_2O_5. MD allows to analyze the individual contribution of the various structural units that composed the average CN and, as consequence, further characterize the local structure of vanadium ions in glasses.

Figure 8.4 reports the percentages of the single contributions to the coordination for both V^{5+} and V^{4+} ions along the NaVP series. Figure 8.4a, points out that V^{5+} ions shift from about a 69 % of octahedral ($^{[6]}V^{5+}$) sites in low vanadium containing system (NaVP-10) to a 33 % of six fold coordination in NaVP-80. For NaVP-80, the V^{5+} ions show an homogeneous distribution (about 33 %, each) of four-, five- and six-coordinated structural units. Moreover, in Fig. 8.4b, V^{4+} ions show a relative high percentage of octahedral units($^{[6]}V^{4+}$ ~49 %) and lower percentages of four- and penta-coordination (~13 % and ~37 %, respectively) for the lower vanadium containing system (NaVP-10). For the intermediate compositions (from NaVP-20 to NaVP-60) the main contribution derives from penta-coordinated ($^{[5]}V^{4+}$) ions that decreases to ~26 % in NaVP-80 system. In this glass, V^{4+} ions are mainly four coordinated (~58 %) with respect to the low values of $^{[5]}V^{4+}$ (~25 %) and $^{[6]}V^{4+}$ (~16 %) units. These results confirm the trend previously reported by Bianconi et al., which found a clear change from exa-coordination to mainly four-coordination with the increase of the total vanadium content [78].

The contribution of different coordination environments for the V^{5+} and V^{4+} ions, as a function of the ratio V^{4+}/V_{tot}, is reported in Table 8.7 for 33LiVP systems. The MD results highlight differences in coordination, due to the various percentages of V^{4+} ions in the glass matrix. If the amount of V^{4+} ions is lower, V^{4+} ions tend to be mainly four-folded coordinated with respect to the 33LiVP-48, where the $^{[5]}V^{4+}$

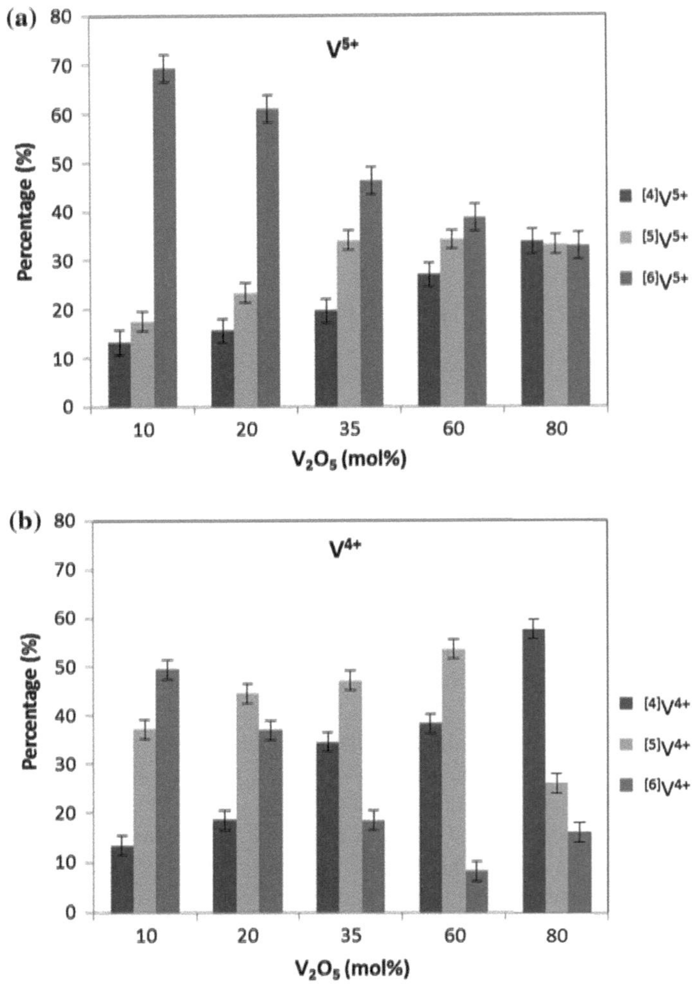

Fig. 8.4 Contributions to the averaged CN of **a** V^{5+} and **b** V^{4+} ions in NaVP series

Table 8.7 Contribution to CN of V^{4+} and V^{5+} ions in 33LiVP systems (error within ±0.3 %)

	V^{4+}		V^{5+}	
	33LiVP-48	33LiVP-26	33LiVP-48	33LiVP-26
[6]	24.7	24.0	49.6	59.3
[5]	48.6	30.0	32.3	27.5
[4]	26.7	46.0	18.1	13.1

contribution is higher. For V^{5+} ions, the differences between the two systems are minor.

The predominant structural unit is the octahedral one, consistent with the results obtained for NaVP systems.

8.3.2 Bond Angle and BO and NBO Distribution

Interconnections between atoms inside the structural units were analyzed in terms of bond angle distributions (BAD) and are detailed in Fig. 8.5. In particular, O–V–O BAD curves point out the presence of a main peak at about $\sim 90°$, related mainly to an exa-coordination that progressively decreases as a function of the increasing content of V_2O_5 in the glasses.

Furthermore, a second peak centered in the range $105° - 110°$ appears with the increase of V_2O_5 content and it can be related to the rise of the contributions that derive from the distorted square pyramidal, distorted trigonal-pyramidal and tetrahedral units. This agrees with the changes of CN of V^{5+} and V^{4+} ions previously discussed in function of the glass composition. The BAD curves clearly show differences between the V^{5+} (Fig. 8.5a) and V^{4+} (Fig. 8.5b) local sites: the peak at 90° decreases more rapidly for O–V^{4+}–O compared to O–V^{5+}–O angle, which confirms the decrease of the average CN. It is worth noting that the second peak in the O–V^{4+}–O BAD for the NaVP series tends to increase with the vanadium content since it becomes the most intense for the higher vanadium containing glass (NaVP-80). Concerning the 33LiVP series, the BAD curves does not show significant differences, which reflect that the V^{4+}/V_{tot} ratio does not affect the interconnections in vanadium local site and confirm the coordination data.

The local configuration of the various ions can be further investigated in terms of BO and NBO distribution (Table 8.8).

The results obtained by MD simulations suggest that in low vanadium containing systems the high presence of P–NBO could be ascribed to the high percentage of network modifiers, which tend to depolymerize the phosphate glass network. Progressive addition of V_2O_5 increases the P–BO contribution, due to the decrease of network modifier content in the glass compositions and the tendency of vanadium polyhedra to create chains interconnected with P ions, according to a more network former role of vanadium atoms in these systems. This hypothesis was already reported in a previous work [59] for sodium vanado-phosphate glasses and confirmed by the higher percentage of V–BO ranging between 88–96 and 96–98% for the V^{5+} and V^{4+} ions respectively, which characterizes the NaVP systems.

8.3.3 Second Shell Coordination Environment

From a structural point of view, the distribution of cations in the second coordination shell of vanadium ions plays an important role to the final properties, for instance

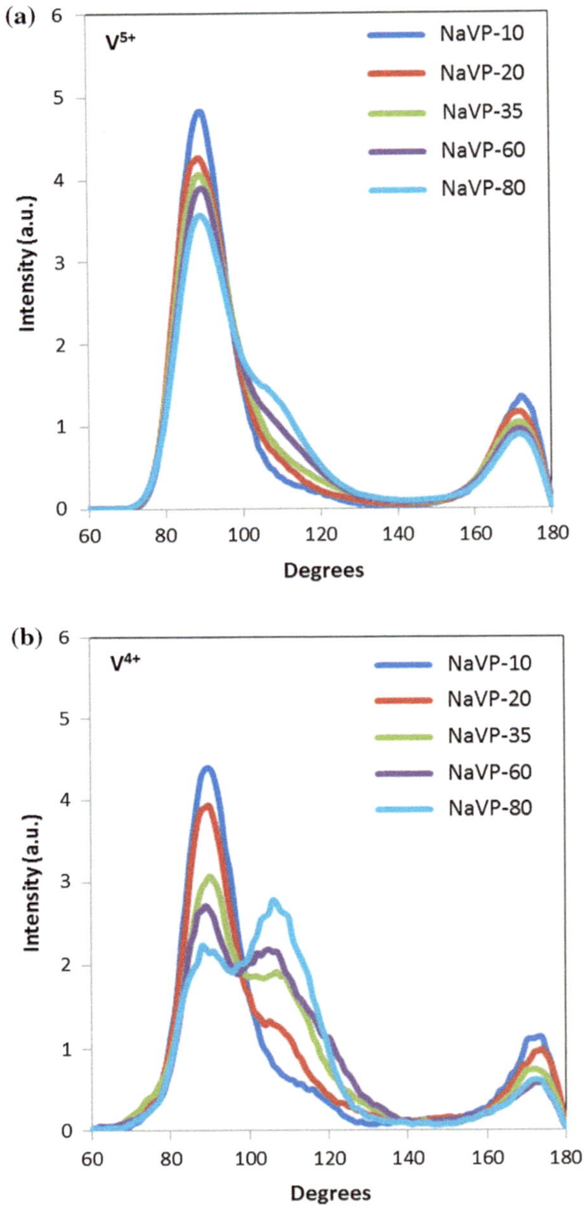

Fig. 8.5 Example of bond angle distribution for **a** V^{5+} and **b** V^{4+} ions in NaVP series

Table 8.8 P–NBO and P–BO distribution along NaVP series (error within ±1.0 %)

	NaVP-10	NaVP-20	NaVP-35	NaVP-60	NaVP-80
P–BO	42.2	59.8	69.2	81.3	84.1
P–NBO	57.8	40.2	37.1	18.7	15.9

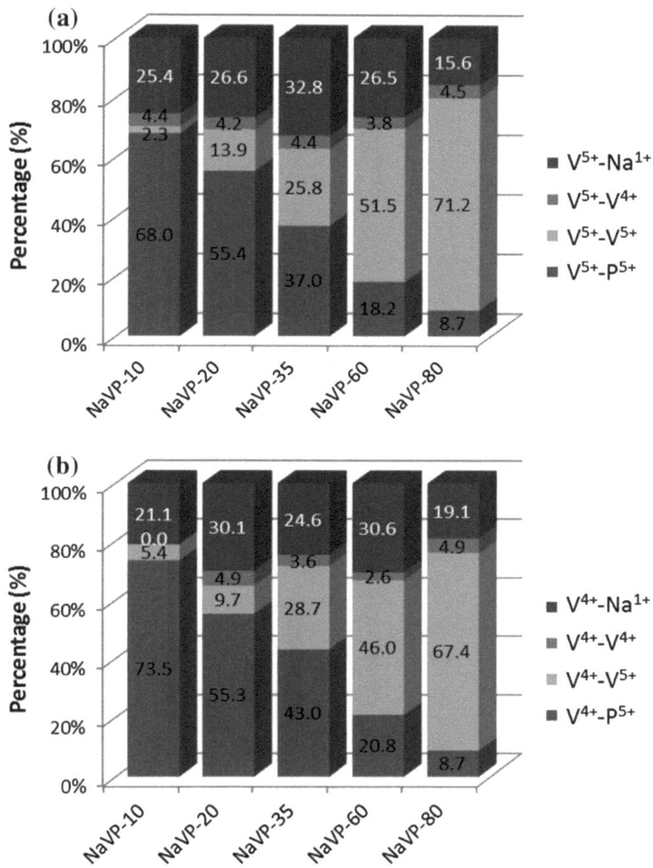

Fig. 8.6 Percentage of cations present in the local site (II shell of coordination) of **a** V^{5+} and **b** V^{4+} ions for the NaVP series (error within ±0.2 %)

the ionic transport or the small polarons hopping. Therefore, a detailed analysis of the medium range order of glasses was performed. Figure 8.6 reports the percentage of the different cations that surround the vanadium species and it is evident the progressive shift towards a vanadium-based network. Indeed increasing the vanadium content, both V^{5+} (Fig. 8.6a) and V^{4+} (Fig. 8.6b) ions tend to be more surrounded by others V ions than P ones. This behavior confirms a typical intermediate character of vanadium ions that could be correlated to the shift from a modifier to a more former character in function of the V_2O_5 content.

The comparison of 33LiVP-26 and 33LiVP-48 systems, reported in Fig. 8.7, shows slight differences between the two structures. The V^{4+}/V_{tot} ratio does not effort great changes to the distribution of ions in the second shell. For 33LiVP-48 (Fig. 8.7a), which is characterized by a V^{4+}/V_{tot} around 48 %, it is possible to notice

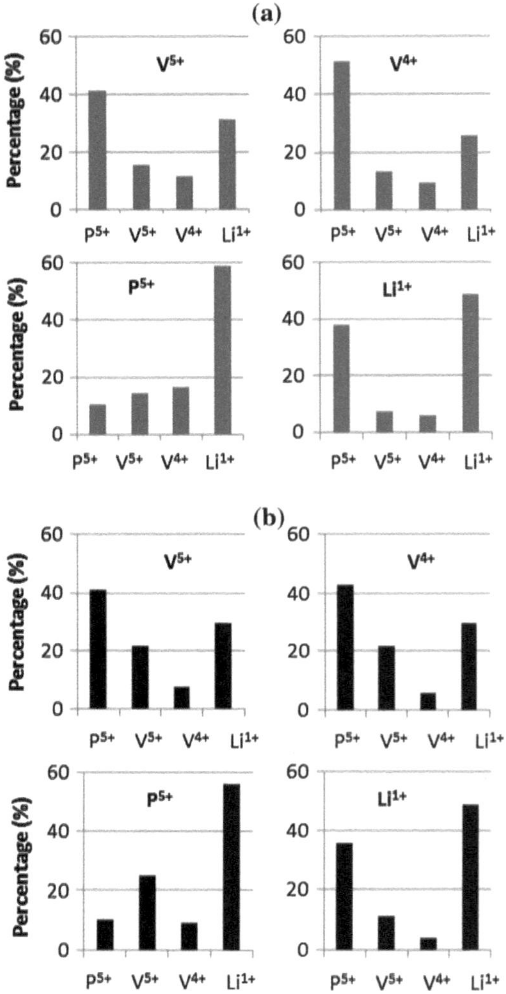

Fig. 8.7 Percentage of cations present in the local site (II shell of coordination) for **a** 33LiVP-48 and **b** 33LiVP-26 (error within ±0.2 %)

that both V^{4+} and V^{5+} ions surround the other ions almost with the same percentage. Differently, for 33LiVP-26 (Fig. 8.7b and $V^{4+}/V_{tot} \approx 26\,\%$) the percentage of V^{5+} ions that surround the other species is higher, as attended due to the higher content in the glass composition.

Q^n analysis can be successfully used to characterize the medium range order of glass structures. Table 8.9 reports the Q^n distribution of P ions in NaVP systems that were calculated only considering the phosphorus ions as network former. The high concentration of Na ions in the NaVP-10 and NaVP-20 leads to a depolymerization of the phosphate network that results in higher values of Q^1 and Q^2 species that is congruent with the BO and NBO distribution previously discussed. At the same time

Table 8.9 Q^n distribution for P for NaVP series (error within ±0.9 %)

	Q^0	Q^1	Q^2	Q^3
NaVP-10	12.5	45.8	34.9	6.8
NaVP-20	31.4	46.5	19.6	2.5
NaVP-35	57.7	31.5	10.2	0.6
NaVP-60	90.8	9.2	0.0	0.0
NaVP-80	96.5	3.5	0.0	0.0

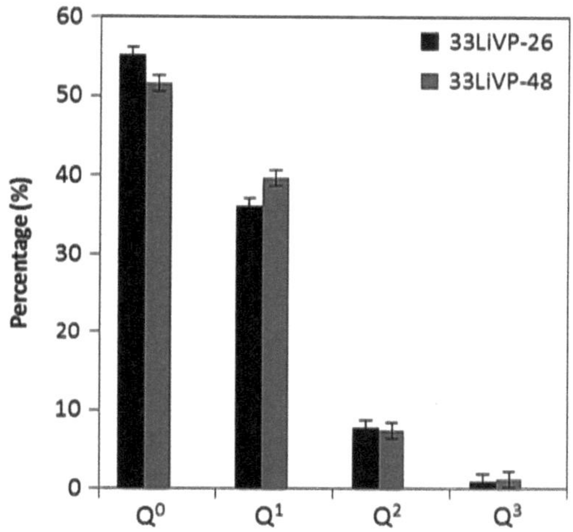

Fig. 8.8 Q^n distribution for P ions for 33LiVP series

the increase of vanadium content along the NaVP series promotes the formation of Q^0 species at the expenses of Q^1 and Q^2. This evidence suggests that P polyhedra tend progressively to be surrounded by the vanadium based ones, up to the P tetrahedra are completely enclosed by vanadium ions (96.5 % of Q^0) in the NaVP-80.

Lithium containing systems have a similar Q^n distribution, as shown in Fig. 8.8, suggesting that the V^{4+}/V_{tot} ratio slightly affects the medium order of phosphorus ions, differently from NaVP systems where the content of V_2O_5 has strong effects on the glass structures.

8.3.4 Cross-Linkages and Electrical Properties

The interconnections among the polyhedra are important to be analyzed in order to understand the peculiarities of the glass network. Experimentally, this information is quite difficult to be extrapolated, while MD simulations allow to quantify the interconnections in terms of P–O–P, P–O–V and V–O–V linkages in the glass structure. Figure 8.9 shows a clear decrease of the P–O–P linkages from 46 (NaVP-10) to 0.1 %

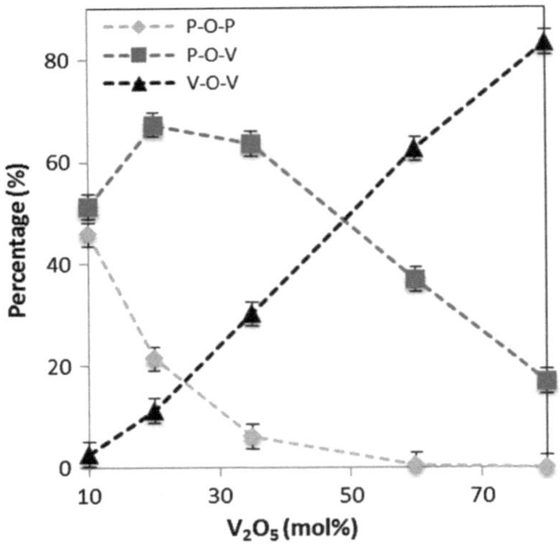

Fig. 8.9 P–O–P, P–O–V and V–O–V total number of linkages along the NaVP-series

(NaVP-80), which corresponds to a progressive increase from 3 to 83 % of the V–O–V linkages. For P–O–V linkages a maximum of 67.3 % is detected for NaVP-20, while for higher amount of V_2O_5 the percentage tends to decrease. The cross-linkages analysis of the NaVP series confirms the transition between a phosphate- to a vanado-based network in function of the increasing content of V_2O_5.

As previously reported, the peculiarity of alkali vanado-phosphate glasses is ascribed to the mixed electronic-ionic conductivity. The electronic conductivity is due to the small polaron hopping between the TM ions with different valence state and it could be related to the presence of V^{5+}–O–V^{4+} species in the glass network. For NaVP systems, Table 8.10 reports the various contributions that are included in the V–O–V linkages. It could be highlighted that the increase of V content results in a major number of V^{5+}–O–V^{4+} linkages, which might be correlated to an increase of the electronic conductivity.

The 33LiVP systems show a slight different behavior in terms of linkages distribution, as reported in Fig. 8.10. As attended, the P–O–V^{4+} and the P–O–V^{5+} linkages show an opposite trend for 33LiVP-26 and 33LiVP-48, because these species are proportional to the relative amount of V^{4+} and V^{5+} ions in the structure. The percentage of V^{5+}–O–V^{4+} linkages is slightly higher than the one calculated for NaVP systems. NaVP-35 has a composition close to the one of 33LiVP, but the amount of V^{5+}–O–V^{4+} linkages for the sodium containing systems is lower. This behavior could be related to presence of different modifier ions that could cause slight changes in the atomic rearrangement.

Table 8.10 Contribution of the various V^{4+}–O–V^{5+}, V^{5+}–O–V^{5+}, V^{4+}–O–V^{4+} in the total percentages of V–O–V linkages in NaVP serie (error within ± 0.8)

	V^{5+}–O–V^{5+}	V^{5+}–O–V^{4+}	V^{4+}–O–V^{4+}
NaVP-10	1.4	1.3	0.0
NaVP-20	6.9	3.5	0.8
NaVP-35	22.3	7.5	0.5
NaVP-60	54.4	7.9	0.2
NaVP-80	73.7	9.1	0.3

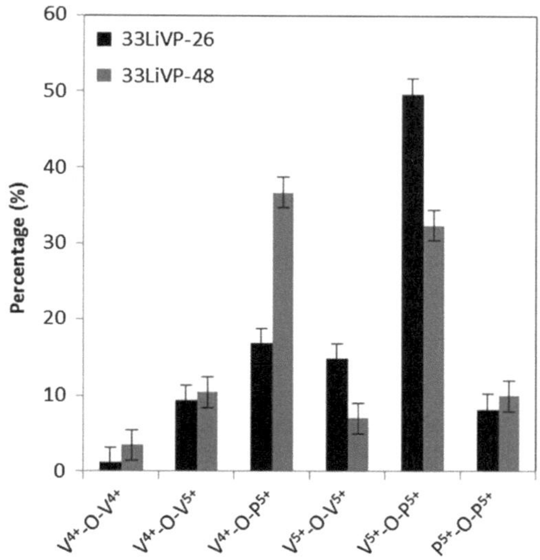

Fig. 8.10 P–O–P, P–O–V and V–O–V total number of linkages along the 33LiVP series

8.4 Conclusions

In the present work, MD simulations were used to study the structure of alkali vanadophosphate glasses providing insight into the short and medium range order of the glass network and to the role of the vanadium ions and their environment in the glass matrices.

The MD results (PDF, BAD and CN) obtained for the NaVP and 33LiVP systems well compare with the experimental data available validating the computational procedure. Differences between V^{4+} and V^{5+} local sites have been highlighted in terms of coordination, interatomic distances and linkages. At the same time the increasing content of the vanadium oxide results into the transition from a phosphate based network to a vanado- based one confirming the network former behavior of V into the investigated compositions. A corner shared distribution between phosphate and vanadate polyhedral is observed and in the higher vanadium content samples higher concentration of V^{4+}–O–V^{5+} linkages are quantified. These linkages are directly involved into the polaron hopping mechanism correlated to the electronic contribution to total conductivity of these glasses and these structural information could be quantified just by using MD simulations pointing out the key role of these methods in design of new functionalized materials. The 33LiVP glasses show higher percentage of V^{4+}–O–V^{5+} with respect to the NaVP series underlining a slight effect of the different network modifier on the structural rearrangement. A complete understanding of the conductivity processes of these glasses is still lacking but the results obtained by MD are promising. This allows considering this technique as new useful instrument to characterize in details the atomistic based features involved into the conduction mechanism in glasses that represent a new challenge in materials for semi-conductive applications.

References

1. C. Mercier, G. Palavit, L. Montagne, C. Follet-Houttemane, A survey of transition-metal-containing phosphate glasses. C. R. Chim. **5**(11), 693–703 (2002)
2. L.E. Bausa, F. Jaque, J. Garcia Sole, A. Duran, Photoluminescence of Ti^{3+} in P_2O_5-Na_2O-Al_2O_3 glass. J. Mater. Sci. **23**(6), 1921–1922 (1988)
3. M. Jamnick, P. Znfigik, D. Tunega, M.D. Ingram, Glass formation and structure in the system Cu_2O-P_2O_5-MoO_3. J. Non-Cryst. Solids **185**, 151–158 (1995)
4. N. Satyanarayana, G. Govindaraj, A. Karthikeyan, Effects of differing ratios of network modifier (Ag_2O) to network formers (MoO_3 + V2O_5) and dopant salt (AgI) concentrations in silver-based superionic glassy compounds. J. Non-Cryst. Solids **136**(3), 219–226 (1991)
5. U. Selvaraj, K.J. Rao, Characterization studies of molybdophosphate glasses and a model of structural defects. J. Non-Cryst. Solids **72**(2–3), 315–334 (1985)
6. T. Minami, K. Imazawa, M. Tanaka, Formation region and characterization of superionic conducting glasses in the systems AgI-Ag_2O-M_xO_y. J. Non-Cryst. Solids **42**(1–3), 469–476 (1980)
7. S.-H. Kim, T. Yoko, Nonlinear optical properties of TeO_2-based glasses: MO_x-TeO_2 (M = Sc, Ti, V, Nb, Mo, Ta, and W) binary glasses. J. Am. Ceram. Soc. **78**(4), 1061–1065 (1995)
8. P. Boolchand, in *Insulating and Semiconducting Glasses*, vol. 17 (World Scientific Publishing Co. Pte. Ltd., Singapore, 2000)
9. S. Bhattacharya, A. Ghosh, Polaron transport in semiconducting silver vanadate glasses. Phys. Rev. B **66**(13), 132203 (2002)
10. G.D. Khattak, N. Tabet, L.E. Wenger, Structural properties of glasses in the series $(SrO)_x(V_2O_5)_{1-x}$, $(SrO)_{0.5-y}(B_2O_3)_y(V_2O_5)_{0.5}$, and $(SrO)_{0.2}(B2O3)_z(V2O5)_{0.8-z}$. Phys. Rev. B. Condens. Matter Mater. Phys. **72**(10), 104203.1–104203.12 (2005)
11. M. Faiz, A. Mekki, B.S. Mun, Z. Hussain, Investigation of vanadium-sodium silicate glasses using XANES spectroscopy. J. Electron Spectros. Relat. Phenom. **154**(3), 60–62 (2007)
12. A. Mekki, G.D. Khattak, D. Holland, M. Chinkhota, L.E. Wenger, Structure and magnetic properties of vanadium-sodium silicate glasses. J. Non-Cryst. Solids **318**(1–2), 193–201 (2003)
13. C. Mugoni, M. Montorsi, C. Siligardi, H. Jain, Electrical conductivity of copper lithium phosphate glasses. J. Non-Cryst. Solids **383**, 137–140 (2014)
14. N.F. Mott, in *Metal Insulator Transistor* (Taylor & Francis, New York, 1990)
15. N.F. Mott, Introductory talk; Conduction in non-crystalline materials. J. Non-Cryst. Solids **8–10**, 1–18 (1972)
16. G. Ori, M. Montorsi, A. Pedone, C. Siligardi, Insight into the structure of vanadium containing glasses: a molecular dynamics study. J. Non-Cryst. Solids **357**(14), 2571–2579 (2011)
17. H.M.M. Moawad, H. Jain, R. El-Mallawany, T. Ramadan, M. El-Sharbiny, Electrical conductivity of silver vanadium tellurite glasses. J. Am. Ceram. Soc. **85**(11), 2655–2659 (2004)
18. H.M.M. Moawad, H. Jain, R. El-Mallawany, DC conductivity of silver vanadium tellurite glasses. J. Phys. Chem. Solids **70**(1), 224–233 (2009)

19. M.C. Ungureanu, M. Lévy, J.L. Souquet, Mixed conductivity of glasses in the P_2O_5–V_2O_5–Na_2O system. Ionics (Kiel) **4**(3–4), 200–206 (1998)
20. E.E. Assem, I. Elmehasseb, Structure, magnetic, and electrical studies on vanadium phosphate glasses containing different oxides. J. Mater. Sci. **46**(7), 2071–2076 (2011)
21. R.J. Barczyński, P. Król, L. Murawski, Ac and dc conductivities in V_2O_5–P_2O_5 glasses containing alkaline ions. J. Non-Cryst. Solids **356**(37–40), 1965–1967 (2010)
22. G.B. Devidas, T. Sankarappa, B.K. Chougule, G. Prasad, DC conductivity in single and mixed alkali vanadophosphate glasses. J. Non-Cryst. Solids **353**(4), 426–434 (2007)
23. R.V.S.S.N. Ravikumar, A.V. Chandrasekhar, L. Ramamoorthy, B.J. Reddy, Y.P. Reddy, J. Yamauchi, P.S. Rao, Spectroscopic studies of transition metal doped sodium phosphate glasses. J. Alloys Compd. **364**(1–2), 176–179 (2004)
24. B. Kang, G. Ceder, Battery materials for ultrafast charging and discharging. Nature **458**(7235), 190–3 (2009)
25. R.K. Brow, Review: the structure of simple phosphate glasses. J. Non-Cryst. Solids **263–264**, 1–28 (2000)
26. C.J. Quinn, G.H. Bell, J.E. Dickinson, in *Proceedings of the XVIth International Congress on Glass*, vol. 4 (1992), p. 79
27. D.E. Day, Z. Wu, C.S. Ray, P. Hrma, Chemically durable iron phosphate glass wasteforms. J. Non-Cryst. Solids **241**(1), 1–12 (1998)
28. J. Du, Y. Xiang, Effect of strontium substitution on the structure, ionic diffusion and dynamic properties of 45S5 bioactive glasses. J. Non-Cryst. Solids **358**(8), 1059–1071 (2012)
29. C. Mercier, C. Follet-Houttemane, A. Pardini, B. Revel, Influence of P_2O_5 content on the structure of SiO_2–Na_2O–CaO–P_2O_5 bioglasses by ^{29}Si and ^{31}P MAS-NMR. J. Non-Cryst. Solids **357**(24), 3901–3909 (2011)
30. J. Fu, Fast Li^+ ion conduction in Li_2O–(Al_2O_3 $Ga_{22}O_3$)–TiO_2–P_2O_5 glass-ceramics. J. Mater. Sci. **33**(6), 1549–1553 (1998)
31. J.B. Bates, N.J. Dudney, G.R. Gruzalski, R.A. Zuhr, A. Choudhury, C.F. Luck, J.D. Robertson, Electrical properties of amorphous lithium electrolyte thin films. Solid State Ionics **53–56**, 647–654 (1992)
32. J.R. Van Wazer, Phosphorus and its compounds Vol. 1. Interscience **73**(15) (1958)
33. Y. Abe, in *Topics in Phosphorus Chemistry, vol. 11*, vol. 14, no. 3 (Wiley, New York, 1983)
34. S.W. Martin, Review of the structures of phosphate glasses. Eur. J. Solid State Inorg. Chem. **28**, 163–205 (1991)
35. F. Liebau, in *Structure and Bonding in Crystals II* (Academic Probation, New York, 1981), p. 197
36. E. Metwalli, M. Karabulut, D.L. Sidebottom, M.M. Morsi, R.K. Brow, Properties and structure of copper ultraphosphate glasses. J. Non-Cryst. Solids **344**(3), 128–134 (2004)
37. J.J. Hudgens, R.K. Brow, D.R. Tallant, S.W. Martin, Raman spectroscopy study of the structure of lithium and sodium ultraphosphate glasses. J. Non-Cryst. Solids **223**(1–2), 21–31 (1998)
38. U. Hoppe, G. Walter, R. Kranold, D. Stachel, Structural specifics of phosphate glasses probed by diffraction methods: a review. J. Non-Cryst. Solids **263–264**, 29–47 (2000)
39. J.J. Hudgens, S.W. Martin, Glass transition and infrared spectra of low-alkali, anhydrous lithium phosphate glasses. J. Am. Ceram. Soc. **76**(7), 1691–1696 (1993)
40. P. Losso, B. Schnabel, C. Jäger, U. Sternberg, D. Stachel, D.O. Smith, ^{31}P NMR investigations of binary alkaline earth phosphate glasses of ultra phosphate composition. J. Non-Cryst. Solids **143**, 265–273 (1992)
41. T.M. Alam, R.K. Brow, Local structure and connectivity in lithium phosphate glasses: a solid-state ^{31}P MAS NMR and 2D exchange investigation. J. Non-Cryst. Solids **223**(1–2), 1–20 (1998)
42. G.K. Marasinghe, M. Karabulut, C.S. Ray, D.E. Day, M.G. Shumsky, W.B. Yelon, C.H. Booth, P.G. Allen, D.K. Shuh, Structural features of iron phosphate glasses. J. Non-Cryst. Solids **222**, 144–152 (1997)
43. D. Bowron, G. Saunders, R. Newport, EXAFS studies of rare-earth metaphosphate glasses. Phys. Rev. B **53**, 5268–5275 (1996)

44. R.K. Brow, R.J. Kirkpatrick, G.L. Turner, The short range structure of sodium phosphate glasses I. MAS NMR studies. J. Non-Cryst. Solids **116**(1), 39–45 (1990)
45. H.S. Liu, T.S. Chin, Low melting PbO-ZnO-P_2O_5 glasses. Part 2. A structural study by Raman spectroscopy and MAS-NMR. Phys. Chem. Glas. **38**(3), 123–131 (1997)
46. J. Swenson, A. Matic, A. Brodin, Structure of mixed alkali phosphate glasses by neutron diffraction and Raman spectroscopy. Phys. Rev. B **58**(17), 331–337 (1998)
47. U. Hoppe, N.P. Wyckoff, M.L. Schmitt, R.K. Brow, A. Schöps, A.C. Hannon, Structure of V_2O_5-P_2O_5 glasses by X-ray and neutron diffraction. J. Non-Cryst. Solids **358**(2), 328–336 (2012)
48. R. Majumdar, D. Lahiri, Redox equilibrium and spectral absorption of some colouring oxides in sodium metaphosphate glass. Trans. Ind. Ceram. Soc. **XXXI**(5) (1972)
49. G. Calas, J. Petiau, *Structure of Non-crystalline Materials II*, pp. 18–23 (Taylor & Francis, London, 1983)
50. G. Calas, J. Petiau, Structure of oxide glasses: spectroscopic studies of local order and crystallochemistry. Geochemical implications. Bull Miner. (1983)
51. K. Moringa, H. Yoshida, H. Takebe, Compositional dependence of absorption spectra of Ti^{3+} in silicate, borate, and phosphate glasses. J. Am. Ceram. Soc. **77**(12), 3113–3118 (1994)
52. D. Ehrt, M. Leister, C. Matthai, C.C. Russel, F. Breitbarth, Determination of the redox states of vanadium in glasses and melts by different methods, in fundamentals of glass science and technology conference, (1997), pp. 204–211
53. T. Murata, M. Torisaka, H. Takebe, K. Morinaga, Compositional dependence of the valency state of Cr ions in oxide glasses. J. Non-Cryst. Solids **220**(2–3), 139–146 (1997)
54. M. Karabulut, G.K. Marasinghe, C.S. Ray, D.E. Day, G.D. Waddill, C.H. Booth, P.G. Allen, J.J. Bucher, D.L. Caulder, D.K. Shuh, An investigation of the local iron environment in iron phosphate glasses having different Fe(II) concentrations. J. Non-Cryst. Solids **306**(2), 182–192 (2002). Aug
55. O. Cozar, I. Ardelean, The local symmetry of Cu^{2+} ions in phosphate glasses. J. Non-Cryst. Solids **92**(2–3), 278–281 (1987)
56. B.-S. Bae, M.C. Weinberg, Ultraviolet optical absorptions of semiconducting copper phosphate glasses. J. Appl. Phys. **73**(11), 7760 (1993)
57. X. Fang, C.S. Ray, A. Moguš-Milanković, D.E. Day, Iron redox equilibrium, structure and properties of iron phosphate glasses. J. Non-Cryst. Solids **283**(1–3), 162–172 (2001)
58. L. Murawski, R.J. Barczyński, D. Samatowicz, Electronic conductivity in Na_2O–FeO–P_2O_5 glasses. Solid State Ionics **157**(1–4), 293–298 (2003)
59. G. Tricot, H. Vezin, Description of the intermediate length scale structural motifs in sodium vanado-phosphate glasses by magnetic resonance spectroscopies. J. Phys. Chem. C **117**(3), 1421–1427 (2013)
60. P. Chaurand, J. Rose, V. Briois, M. Salome, O. Proux, V. Nassif, L. Olivi, J. Susini, J.-L. Hazemann, J.-Y. Bottero, New methodological approach for the vanadium K-edge X-ray absorption near-edge structure interpretation: application to the speciation of vanadium in oxide phases from steel slag. J. Phys. Chem. B **111**(19), 5101–10 (2007)
61. L.D. Bogomolova, M.P. Glassova, O.E. Dubatovko, S.I. Reiman, S.N. Spasibkina, The study of interactions between iron and vanadium ions in semiconducting barium-vanadate glasses doped with Fe_2O_3. J. Non-Cryst. Solids **58**(1), 71–89 (1983)
62. B.-S. Bae, M.C. Weinberg, Oxidation-reduction equilibrium in copper phosphate glass melted in air. J. Am. Ceram. Soc. **74**(12), 3039–3045 (1991)
63. G.E. Brown, G. Calas, G.A. Waychunas, J. Petiau, X-ray absorption spectroscopy; applications in mineralogy and geochemistry. Rev. Mineral. Geochem. **18**(1), 431–512 (1988)
64. J. Wong, F. Lytle, R. Messmer, D. Maylotte, K-edge absorption spectra of selected vanadium compounds. Phys. Rev. B **30**, 5596–5610 (1984)
65. F. Farges, Y. Lefrère, S. Rossano, A. Berthereau, G. Calas, G.E. Brown, The effect of redox state on the local structural environment of iron in silicate glasses: a combined XAFS spectroscopy, molecular dynamics, and bond valence study. J. Non-Cryst. Solids **344**(3), 176–188 (2004)

66. A. Musinu, G. Piccaluga, An X-ray diffraction study of the short-range order around Ni(II) Zn(II) and Cu(II) in pyrophosphate glasses. J. Non-Cryst. Solids **193**, 32–35 (1995)
67. L. Cormier, L. Galoisy, J.-M. Delaye, D. Ghaleb, G. Calas, Short- and medium-range structural order around cations in glasses: a multidisciplinary approach. Comptes Rendus l'Académie des Sci. Ser. IV Phys. **2**(2), 249–262 (2001)
68. R.V.S.S.N. Ravikumar, V. Rajagopal Reddy, A.V. Chandrasekhar, B.J. Reddy, Y.P. Reddy, P.S. Rao, Tetragonal site of transition metal ions doped sodium phosphate glasses. J. Alloys Compd. **337**(1–2), 272–276 (2002)
69. L.A. Farrow, E.M. Vogel, Raman spectra of phosphate and silicate glasses doped with the cations Ti, Nb and Bi. J. Non-Cryst. Solids **143**, 59–64 (1992)
70. O. Cozar, I. Ardelean, V. Simon, L. David, The local structure and interactions between V^{4+} ions in soda-phosphate glasses. Appl. Magn. Reson. **537**, 529–537 (1999)
71. S. Bruni, F. Cariati, A. Corrias, P.H. Gaskell, A. Lai, A. Musinu, G. Piccaluga, Short range order of sodium-zinc, sodium-copper, and sodium-nickel pyrophosphate glasses by diffractometric and spectroscopic techniques. J. Phys. Chem. **99**(41), 15229–15235 (1995)
72. G.D. Khattak, A. Mekki, L.E. Wenger, X-ray photoelectron spectroscopy (XPS) and magnetic susceptibility studies of vanadium phosphate glasses. J. Non-Cryst. Solids **355**(43–44), 2148–2155 (2009)
73. A. Majjane, A. Chahine, M. Et-tabirou, B. Echchahed, T.O. Do, P.M. Breen, X-ray photoelectron spectroscopy (XPS) and FTIR studies of vanadium barium phosphate glasses. Mater. Chem. Phys. **143**(2), 779–787 (2014)
74. G. Ori, C. Mugoni, G. Broglia, C. Siligardi, M. Montorsi, Short- and medium-range order structure of alkali vanado-phosphate glasses: a Molecular Dynamics study (Submitted) (2014)
75. K. Nagamine, T. Honma, T. Komatsu, Selective synthesis of lithium Ion-conductive β-$LiVOPO_4$ crystals via glass-ceramic processing. J. Am. Ceram. Soc. **91**(12), 3920–3925 (2008)
76. C. Albon, D. Muresan, R.E. Vandenberghe, S. Simon, Iron environment in calcium-soda-phosphate glasses and vitroceramics. J. Non-Cryst. Solids **354**(40–41), 4603–4608 (2008)
77. A. Kiani, L.S. Cahill, E.A. Abou Neel, J.V. Hanna, M.E. Smith, J.C. Knowles, Physical properties and MAS-NMR studies of titanium phosphate-based glasses. Mater. Chem. Phys. **120**(1), 68–74 (2010)
78. A. Bianconi, A. Giovannelli, I. Dovoli, S. Stizza, L. Palladino, O. Gzowski, L. Murawski, Xanes (X-ray absorption near edge structure) of V in vanadium-iron phosphate glasses. Solid State Commun. **42**(8), 547–551 (1982)
79. G. Concas, F. Congiu, E. Manca, C. Muntoni, G. Pinna, Mössbauer spectroscopic investigation of some iron-containing sodium phosphate glasses. J. Non-Cryst. Solids **192–193**, 175–178 (1995)
80. P.A. Bingham, E.R. Barney, Structure of iron phosphate glasses modified by alkali and alkaline earth additions: neutron and X-ray diffraction studies. J. Phys. Condens. Matter **24**(17), 175403 (2012)
81. S. Chandra, in *Superionic Solids: Principles and Applications* (North-Holland, Amsterdam, 1981)
82. K. Singh, Ion conducting glasses for solid state electrochemical applications. Indian J. Pure Appl. Phys. **37**, 266–271 (1999)
83. P.Y. Shih, J.Y. Ding, S.Y. Lee, ^{31}P MAS-NMR and FTIR analyses on the structure of CuO-containing sodium poly- and meta-phosphate glasses. Mater. Chem. Phys. **80**(2), 391–396 (2003)
84. A. Musinu, G. Piccaluga, G. Pinna, G. Vlaic, D. Narducci, S. Pizzini, Coordination of zinc and copper in phosphate glasses by EXAFS. J. Non-Cryst. Solids **136**(3), 198–204 (1991)
85. R. Debnath, J. Chaudhury, S.C. Bera, Optical properties and nature of coordination of Cu^{+} ions in calcium metaphosphate glass. Phys. Status Solidi **157**(2), 723–733 (1990)
86. B. Roling, K. Funke, Polaronic transport in vanadium phosphate glasses. J. Non-Cryst. Solids **212**(1), 1–10 (1997)

87. N.F. Mott, Conduction in glasses containing transition metal ions. J. Non-Cryst. Solids **1**(1), 1–17 (1968)
88. N.F. Mott, Electrons in disordered structures. Adv. Phys. **16**(61), 49–144 (1967)
89. S. Gupta, N. Khanijo, A. Mansingh, The influence of V^{4+} ion concentration on the EPR spectra of vanadate glasses. J. Non-Cryst. Solids **181**(1–2), 58–63 (1995)
90. M. Nabavi, C. Sanchez, J. Livage, Structure and properties of amorphous V_2O_5. Philos. Mag. B **63**(4), 941–953 (1991)
91. B. Jordan, C. Calvo, Transport properties and the structure of vanadium phosphate glasses. Can. J. Phys. **55**(5), 436–441 (1977)
92. U. Hoppe, R. Kranold, A reverse monte carlo study of the structure of vitreous V_2O_5. Solid State Commun. **109**(10), 625–630 (1999)
93. L. Murawski, R.J. Barczynski, A. Rybicka, V_2O_5–P_5O_5 glass and its polaron transport properties derived from molecular dynamic simulations of structure. In dielectric and related phenomena: materials physico-chemistry, spectrometric investigations, and applications (1997), pp. 136–141
94. A. Mosset, P. Lecante, J. Galy, J. Livage, Structural analysis of amorphous V_2O_5 by large-angle X-ray scattering. Philos. Mag. B **46**(2), 137–149 (1982)
95. M. Duclot, J.-L. Souquet, Glassy materials for lithium batteries? electrochemical properties and devices performances. J. Power Sources **98**(5631), 610–615 (2001)
96. H. Behzad, M.H. Hekmatshoar, M. Mirzayi, M. Azmoonfar, Activation energy and conductivity of glasses in the P_2O_5–V_2O_5–Li_2O system. Ionics (Kiel) **15**(5), 647–650 (2009)
97. P. Jozwiak, J. Garbarczyk, Mixed electronic-ionic conductivity in the glasses of the LiO–VO–PO system. Solid State Ionics **176**(25–28), 2163–2166 (2005). Aug
98. H. Takahashi, T. Karasawa, T. Sakuma, J.E. Garbarczyk, Electrical conduction in the vitreous and crystallized Li2O-V2O5-P2O5 system. Solid State Ionics **181**(1–2), 27–32 (2010)
99. J. Garbarczyk, M. Wasiucionek, P. Jóźwiak, L. Tykarski, J. Nowiński, Studies of Li_2O-V_2O_5-P_2O_5 glasses by DSC, EPR and impedance spectroscopy. Solid State Ionics **154–155**, 367–373 (2002)
100. T. Sankarappa, G.B. Devidas, M. Prashant Kumar, S. Kumar, B. Vijaya Kumar, Ac conductivity studies in single and mixed alkali vanadophosphate glasses. J. Alloys Compd. **469**(1–2), 576–579 (2009)
101. G. Lusvardi, G. Malavasi, M. Cortada, L. Menabue, M.C. Menziani, A. Pedone, U. Segre, Elucidation of the structural role of fluorine in potentially bioactive glasses by experimental and computational investigation. J. Phys. Chem. B **112**(40), 12730–9 (2008)
102. M.A. Karakassides, A. Saranti, I. Koutselas, Preparation and structural study of binary phosphate glasses with high calcium and/or magnesium content. J. Non-Cryst. Solids **347**(1–3), 69–79 (2004)
103. L. Linati, G. Lusvardi, G. Malavasi, L. Menabue, M.C. Menziani, P. Mustarelli, A. Pedone, U. Segre, Medium-range order in phospho-silicate bioactive glasses: insights from MAS-NMR spectra, chemical durability experiments and molecular dynamics simulations. J. Non-Cryst. Solids **354**(2–9), 84–89 (2008)
104. A. Tilocca, A.N. Cormack, Structural effects of phosphorus inclusion in bioactive silicate glasses. J. Phys. Chem. B **111**(51), 14256–64 (2007)
105. Y. Xiang, J. Du, Effect of strontium substitution on the structure of 45S5 bioglasses. Chem. Mater. **23**(11), 2703–2717 (2011)
106. A. Tilocca, A. Cormack, N. de Leeuw, The formation of nanoscale structures in soluble phosphosilicate glasses for biomedical applications: MD simulations. Faraday Discuss. **136**, 44–55 (2007)
107. D. Di Tommaso, R.I. Ainsworth, E. Tang, N.H. de Leeuw, Modelling the structural evolution of ternary phosphate glasses from melts to solid amorphous materials. J. Mater. Chem. B **1**(38), 5054 (2013)
108. A. Pedone, G. Malavasi, M.C. Menziani, Computational insight into the effect of CaO/MgO substitution on the structural properties of phospho-silicate bioactive glasses. J. Phys. Chem. C **113**(35), 15723–15730 (2009)

109. G. Malavasi, A. Pedone, M.C. Menziani, Study of the structural role of gallium and aluminum in 45S5 bioactive glasses by molecular dynamics simulations. J. Phys. Chem. B **117**(15), 4142–50 (2013)
110. G. Lusvardi, G. Malavasi, L. Menabue, M.C. Menziani, A. Pedone, U. Segre, A computational tool for the prediction of crystalline phases obtained from controlled crystallization of glasses. J. Phys. Chem. B **109**(46), 21586–92 (2005)
111. A. Tilocca, A.N. Cormack, N.H. de Leeuw, The structure of bioactive silicate glasses: new insight from molecular dynamics simulations. Chem. Mater. **19**(1), 95–103 (2007)
112. C.-C. Lin, S.-F. Chen, K.S. Leung, P. Shen, Effects of CaO/P_2O_5 ratio on the structure and elastic properties of SiO_2-CaO-Na_2O-P_2O_5 bioglasses. J. Mater. Sci. Mater. Med. **23**(2), 245–258 (2012)
113. J. Du, L. Kokou, J.L. Rygel, Y. Chen, C.G. Pantano, R. Woodman, J. Belcher, Structure of cerium phosphate glasses: molecular dynamics simulation. J. Am. Ceram. Soc. **94**(8), 2393–2401 (2011). Aug
114. R.A. Martin, G. Mountjoy, R.J. Newport, A molecular dynamics model of the atomic structure of dysprosium alumino-phosphate glass. J. Phys. Condens. Matter **21**(7), 075102 (2009)
115. E.B. Clark, R.N. Mead, G. Mountjoy, A molecular dynamics model of the atomic structure of Tb metaphosphate glass $(Tb_2O_3)_{0.25}(P_2O_5)_{0.75}$. J. Phys. Condens. Matter **18**(29), 6815–6826 (2006)
116. G. Mountjoy, Molecular dynamics, diffraction and EXAFS of rare earth phosphate glasses compared with predictions based on bond valence. J. Non-Cryst. Solids **353**(18–21), 2029–2034 (2007)
117. R.P. Rao, M. Seshasayee, Molecular dynamics simulation of ternary glasses Li_2O–P_2O_5–LiCl. Solid State Commun. **131**(8), 537–542 (2004)
118. R.K. Sistla, M. Seshasayee, Structural study of lithium phosphate glasses by X-ray RDF and computer simulations. J. Non-Cryst. Solids **349**, 22–29 (2004)
119. A. Karthikeyan, P. Vinatier, A. Levasseur, K.J. Rao, The molecular dynamics study of lithium ion conduction in phosphate glasses and the role of non-bridging oxygen. J. Phys. Chem. B **103**(30), 6185–6192 (1999)
120. M. Vogel, Complex lithium ion dynamics in simulated $LiPO_3$ glass studied by means of multitime correlation functions. Phys. Rev. B **68**(18), 184301–11 (2003)
121. J.J. Liang, R.T. Cygan, T. Alam, Molecular dynamics simulation of the structure and properties of lithium phosphate glasses. J. Non-Cryst. Solids **263–264**, 167–179 (2000)
122. T.M. Alam, J.J. Liang, R.T. Cygan, Molecular dynamics simulations of the lithium coordination environment in phosphate glasses. Phys. Chem. Chem. Phys. **2**(19), 4427–4432 (2000)
123. G.G. Boiko, N.S. Andreev, A.V. Parkachev, Structure of pyrophosphate 2ZnO·P_2O_5–$2Na_2O$·P_2O_5 glasses according to molecular dynamics simulation. J. Non-Cryst. Solids **238**(3), 175–185 (1998)
124. B.C. Tischendorf, T.M. Alam, R.T. Cygan, J.U. Otaigbe, The structure and properties of binary zinc phosphate glasses studied by molecular dynamics simulations. J. Non-Cryst. Solids **316**(2–3), 261–272 (2003)
125. B. Al-Hasni, G. Mountjoy, Structural investigation of iron phosphate glasses using molecular dynamics simulation. J. Non-Cryst. Solids **357**(15), 2775–2779 (2011)
126. B.M. Al-Hasni, G. Mountjoy, E. Barney, A complete study of amorphous iron phosphate structure. J. Non-Cryst. Solids **380**, 141–152 (2013)
127. M. Seshasayee, K. Muruganandam, Molecular dynamics study of V_2O_5 glass. Solid State Commun. **105**(4), 243–246 (1998)
128. S. Garofalini, Molecular dynamics simulations of Li transport between cathode crystals. J. Power Sources **110**(2), 412–415 (2002). Aug
129. A. Pedone, G. Malavasi, M.C. Menziani, A.N. Cormack, U. Segre, A new self-consistent empirical interatomic potential model for oxides, silicates, and silica-based glasses. J. Phys. Chem. B **110**(24), 11780–95 (2006)
130. G. Giuli, E. Paris, J. Mungall, C. Romano, D. Dingwell, V oxidation state and coordination number in silicate glasses by XAS. Am. Mineral. **89**(11–12), 1640–1646 (2004)

131. S. Bhattacharya, A. Ghosh, ac relaxation in silver vanadate glasses. Phys. Rev. B **68**(22), 224202 (2003)
132. A. Ghosh, Temperature-dependent thermoelectric power of semiconducting bismuth-vanadate glass. J. Appl. Phys. **65**(1), 227 (1989)
133. E. Denton, H. Rawson, J. Stanworth, Vanadate glasses. Nature **173**(4413), 1030–1032 (1954)
134. A. Feltz, B. Unger, Redox reactions in condensed oxide systems II. Variation of the structure of vanadium phosphate glasses in dependence on the oxidation state of vanadium. J. Non-Cryst. Solids **72**(2–3), 335–343 (1985)
135. L.L. Frazier, P.W. France, Compositional dependence of the electrical conductivity of vanadium phosphate glass. J. Phys. Chem. Solids **38**(7), 801–808 (1977)
136. M. Schindler, F. Hawthorne, W. Baur, Crystal chemical aspects of vanadium: polyhedral geometries, characteristic bond valences, and polymerization of (VO_n) polyhedra. Chem. Mater. **12**(5), 1248–1259 (2000)
137. P.Y. Zavalij, M.S. Whittingham, Structural chemistry of vanadium oxides with open frameworks. Acta Crystallogr. Sect. B Struct. Sci. **55**(5), 627–663 (1999)
138. M. Leister, D. Ehrt, G. von der Gönna, C. Rüssel, F.W. Breitbarth, Redox states and coordination of vanadium in sodium silicates melted at high temperatures. Soc. Glass Technol
139. G. Tricot, L. Montagne, L. Delevoye, G. Palavit, V. Kostoj, Redox and structure of sodium-vanadophosphate glasses. J. Non-Cryst. Solids **345–346**, 56–60 (2004)
140. K. Nagamine, T. Honma, T. Komatsu, A fast synthesis of $Li_3V_2(PO_4)_3$ crystals via glass-ceramic processing and their battery performance. J. Power Sources **196**(22), 9618–9624 (2011)
141. A. Pedone, Properties calculations of silica-based glasses by atomistic simulations techniques: a review. J. Phys. Chem. C **113**(49), 20773–20784 (2009)
142. M. Pota, A. Pedone, G. Malavasi, C. Durante, M. Cocchi, M.C. Menziani, Molecular dynamics simulations of sodium silicate glasses: optimization and limits of the computational procedure. Comput. Mater. Sci. **47**(3), 739–751 (2010)
143. W. Smith, T. Forester, DL_POLY_2. 0: a general-purpose parallel molecular dynamics simulation package. J. Mol. Graph. **14**(3), 136–141 (1996)
144. A. Bonamartini Corradi, V. Cannillo, M. Montorsi, C. Siligardi, A.N. Cormack, Structural characterization of neodymium containing glasses by molecular dynamics simulation. J. Non-Cryst. Solids **351**(14–15), 1185–1191 (2005)
145. J. Du, A.N. Cormack, The structure of erbium doped sodium silicate glasses. J. Non-Cryst. Solids **351**(27–29), 2263–2276 (2005). Aug
146. B. Vessal, Simulation studies of silicates and phosphates. J. Non-Cryst. Solids **177**, 103–124 (1994)
147. N. Afify, G. Mountjoy, Molecular-dynamics modeling of Eu^{3+}-ion clustering in SiO_2 glass. Phys. Rev. B **79**(2), 024202–12 (2009)
148. J. Gale, GULP: capabilities and prospects. Zeitschrift für Krist. **220**(5), 552–554 (2005)
149. J. Du, A.N. Cormack, The medium range structure of sodium silicate glasses: a molecular dynamics simulation. J. Non-Cryst. Solids **349**, 66–79 (2004)
150. M. Montorsi, M.C. Menziani, C. Leonelli, G.C. Pellacani, A.N. Cormack, Molecular dynamics simulations of alumina addition in sodium silicate glasses. Mol. Simul. **24**(1–3), 157–165 (2000)
151. A.B. Corradi, V. Cannillo, M. Montorsi, C. Siligardi, Influence of Al_2O_3 addition on thermal and structural properties of erbium doped glasses. J. Mater. Sci. **41**(10), 2811–2819 (2006)
152. G. Mountjoy, B.M. Al-Hasni, C. Storey, Structural organisation in oxide glasses from molecular dynamics modelling. J. Non-Cryst. Solids **357**(14), 2522–2529 (2011)
153. L. Linati, G. Lusvardi, G. Malavasi, L. Menabue, M.C. Menziani, P. Mustarelli, U. Segre, Qualitative and quantitative structure-property relationships analysis of multicomponent potential bioglasses. J. Phys. Chem. B **109**(11), 4989–4998 (2005)
154. K. Muruganandam, An X-ray RDF study of Li_2O-P_2O_5-LiCl glasses. Solid State Ionics **89**(3–4), 313–319 (1996)

155. T.M. Alam, S. Conzone, R.K. Brow, T.J. Boyle, ^{6}Li, ^{7}Li nuclear magnetic resonance investigation of lithium coordination in binary phosphate glasses. J. Non-Cryst. Solids **258**(1–3), 140–154 (1999)
156. Z. Hiroi, M. Azuma, Y. Fujishiro, T. Saito, M. Takano, F. Izumi, T. Kamiyama, T. Ikeda, Structural study of the quantum-spin chain compound $(VO)_2P_2O_7$. J. Solid State Chem. **146**(2), 369–379 (1999)

Chapter 9
Modelling Networks in Varying Dimensions

Mark Wilson

Abstract Simulation methods and results for two key (related) network-forming systems (SiO_2 and C) are described and reviewed. The application of relatively simple potential models, in which the interaction energies are expressed as functions of atom positions and momenta, are described. The properties of these two key target systems are studied over a range of dimensionalities. Pressure-driven structural changes in glassy SiO_2 are described and a simple ring-closure model developed to map the changes. The phase diagrams (liquid/crystal melting curves) are mapped in both 3- and 2-dimensions for carbon and key structural changes on phase change are studied. A liquid to amorphous phase transformation is identified for carbon in three dimensions and investigated. The two dimensional carbon phase diagram is used to develop methods for generating amorphous structures of two dimensional carbon (amorphous graphene, a-G) and the structures of the materials produced are investigated as a function of the generation conditions. The a-G structures are used as a basis for generating bilayers of SiO_2 and are also folded to form amorphous carbon nanotube (a-CNT) structures.

9.1 Introduction

Network materials may be usefully defined as systems in which significant connectivity may be identified and preserved even in relatively disordered states (such as liquids and glasses). Considering material structure in terms of an underlying network can be traced to Zachariasen [1] who sketched a two-dimensional representation of a network formed from a mixture of two- and three-coordinate sites. The network connectivity may result from the percolation of relatively simple, well-defined units (such as the triangles in Zachariasen's image or the tetrahedral SiO_4 coordination

M. Wilson (✉)
Department of Chemistry, Physical and Theoretical Chemistry Laboratory,
University of Oxford, South Parks Road, Oxford OX1 3QZ, UK
e-mail: mark.wilson@chem.ox.ac.uk

C. Massobrio et al. (eds.), *Molecular Dynamics Simulations of Disordered Materials*, Springer Series in Materials Science 215,
DOI 10.1007/978-3-319-15675-0_9

polyhedra (CP) in SiO_2) or by directionality imposed by covalent bonding (as in, for example, carbon or silicon). These networks may map onto one another, for example, the structure of glassy SiO_2 at ambient pressure can be mapped onto those for amorphous carbon or silicon [2–4]. The structure of materials such as vitreous silica and amorphous carbon may be effectively considered in terms of a continuous random network [5]. The network topologies are controlled by a number of factors, in particular temperature and pressure as well as preparation conditions. Factors controlling topology differences between chemically-related systems may also be identified. For example, systems such as $BeCl_2$, $GeSe_2$, GeO_2 and $ZnCl_2$ can be considered (at ambient pressure) as related to SiO_2 in that they are all essentially constructed from MX_4 tetrahedral CP. The key difference in these systems lies in the angles between neighbouring tetrahedra (which are controlled by the bridging anion polarizabilities [4, 6–8]). A common theme in these networks is the presence of ordering on more than one length-scale (often evidenced by multiple peaks in the experimental scattering functions) and results directly from the topology of the linked tetrahedra. To this end the network ring structure (counting the distributions of shortest paths around the network) represents something of a natural language for discussing ordering beyond the short-range order inherent in the tetrahedra [9]. The application of temperature, and in particular pressure, may alter the network topology (and hence the ring structure) both by modifying the links between the CP and the CP geometries themselves (for example by an increase in nearest-neighbour coordination number). However, whilst the concept of ring structure provides a useful framework for discussing network topology, the rings themselves cannot be easily observed directly, merely inferred from (structural) diffraction data (for example, pair distribution functions) or from (dynamic) Raman data (see [10] and references therein).

Interest in network materials is continuing to increase for a number of reasons. Firstly, improvements in X-ray and neutron diffraction experimental techniques are allowing a far greater range of (pressure-temperature) phase space to be explored than previously accessible (see, for example, Chap. 1 in this volume). Secondly, developments in synthesis techniques have seen a significant growth in metal-organic framework (MOF) materials [11]. In these materials the networks are formed by metal atoms interconnected by organic molecules. The inter-metallic linkages are controlled by the properties of the organic molecules (rather than the properties of single atoms as in systems such as SiO_2) although their structures may follow analogous "rules" to those associated with the atomistic systems. As a result these materials offer potential limitless flexibility of design in not being "constrained" by atomistic properties. Thirdly, developments in experimental synthetic and analysis techniques is making the production of two- (or near-two-) dimensional materials a reality. For example, electron microscopy studies of graphene (in some senses the archetypal two-dimensional system) show amorphous regions characterised by five-, six- and seven-membered rings [12–18]. In addition, thin films of vitreous SiO_2, whose structures have been interpreted as constructed from bilayers of SiO_4 tetrahedra, have been grown on Mo(112) [19], Ru(0001) [20–22] and graphene [23]. Both the amorphous graphene and SiO_2 bilayers are examples of systems in which the ring structure may be observed directly from the electron microscopy images and, as a result, are commensurate with the original sketches of Zachariasen. However, the growth conditions which control the structures of these systems are, at present, not well understood.

In this chapter we will review how amorphous materials may be effectively modelled. Our focus will be on two basic systems; carbon and SiO_2, as archetypal examples. Both systems will be studied in three dimensions. Carbon will also be studied purely in two dimensions (by constructing a phase diagram and generating amorphous graphene structures). SiO_2 will also be studied at reduced dimensionality as bilayers and will be shown to be closely related to the amorphous graphene structures. Finally the amorphous graphene structures will be folded to produced even lower dimensional amorphous nanotube structures.

9.2 Modelling Methodologies

Choosing a modelling methodology to tackle a class of problem is far from trivial. Ideal models will be accurate (in terms of reproducing theoretical and/or experimental properties) and transferable, meaning that a given model may reproduce properties over a wide range of state points and that any model parameters may be altered in a transparent and physically-motivated manner in order to generate models for chemically-related systems. The most accurate and transferable ab initio techniques invoke repeated explicit determination of the electronic structure (effectively solving the Schrödinger equation). However, even for the most efficient density-functional-based methods accessible length- and time-scales are limited. The use of potential models, in which the energy of a given configuration is expressed as a (relatively simple) function of the atomic coordinates, represents a highly useful and computationally efficient simplification.

Despite their relative simplicity potential models may retain significant transferability if the underlying model parameters retain a clear physically-motivated meaning (for example, related to a known property such as the atomic radius). Furthermore, changes in electronic structure may be incorporated in a highly coarse-grained fashion through the explicit inclusion of induced moments on the atom sites, mimicking the motion of the electron density.

The usefulness of potential models goes beyond computational efficiency. The ability to alter parameters in a systematic fashion allows direct contact with limiting theoretical models and allows the roles of different interactions to be uncovered and rationalised. Such control is more difficult to establish in electronic structure calculations. Furthermore, potential model parameters may be obtained by direct reference to higher level ab initio calculations, either applying a "force-fitting" method (attempting to obtain multiple parameter sets in parallel) or by utilising well-directed calculations with the aim of isolating specific parameters (see, for example, [6, 24–26] for examples for ionic models).

9.2.1 Background

In molecular dynamics (MD) simulation atoms are moved in space along their lines of force (which are determined from the first derivative of the potential energy function) using finite difference methods [27, 28]. At each time step the evolution of the energy and forces allow the accelerations on each atom to be determined, in turn allowing the atom changes in velocities and positions to be evaluated and hence allows the system "clock" to move forward, typically in time steps of the order of a few fs. Bulk system properties such as temperature and pressure are easily determined from the atom positions and velocities. As a result simulations can be readily performed at constant temperature and volume (NVT ensemble) or constant temperature and pressure (NpT ensemble). The constant temperature and pressure constraints can be imposed using thermostats and barostat [29–31] in which additional variables are coupled to the system which act to modify the equations of motion.

A recurring problem in amorphous or glassy[1] physics is in generating configurations which reflect these states (rather than a supercooled liquid). Amorphous (glassy) configurations may be generated most commonly from the liquid state by quenching (reducing the system temperature). At the simplest level energy minimisation calculations can be performed using the basic MD methodologies in a steepest descent algorithm in which the atom velocities are systematically quenched (for example when the kinetic energy reaches a maximum). More efficient methods, such as conjugate gradient minimisations, may also be employed [32]. Alternatively, thermostats may be employed to cool the system more gradually by selecting the final glass target temperature and allowing the coupled heat bath to systematically remove heat from the system. As a result the thermostat relaxation time, which controls the rate of heat transfer to and from the coupled heat bath, effectively controls the rate of cooling. The achievable cooling rates are, however, still massive compared to typical experimental values. The rapidity of the cooling rates imposed by the use of atomistic MD warrants investigation of alternative formation pathways. Pathways may be chosen which have direct experimental analogue. For example, a significant number of crystal structures undergo a pressure-induced amorphisation [33, 34] at relatively low temperature. Alternatively, pathways may be devised which have no direct experimental analogue. For example, bonds may be "switched" using a Monte Carlo algorithm (see below) which allows an initially crystalline configuration to become disordered [35].

Monte Carlo (MC) algorithms use random particle (or group of particles) moves and/or changes in cell geometry to explore phase space. Although dynamic information, readily available from MD, is effectively lost, advance sampling methods allow for a much more effective and efficient covering of phase space than is possible in pure MD along [27, 28, 36].

[1] A glass is a form of amorphous material which has been formed by traversing a glass transition.

9.2.2 Potential Models

In constructing potential models to study a useful range of network systems, two basic strategies are utilised. In the first potential models are parameterised for specific systems which allow for the most direct comparison with experimental observation. In the second generic models are employed which may be systematically modified in order to observe the evolution of key properties. A significant advantage of a potential model-based approach (compared with the more computationally demanding, though potentially more accurate, electronic structure methods) is that the individual parameters which control specific aspects of the underlying interactions may be altered in order to develop an understanding of their respective roles in specific processes.

The interaction between atoms in the condensed phase may be expressed as a cluster expansion

$$U(\{\mathbf{R}_i\}) = \sum_i \sum_j U_2(\mathbf{R}_i, \mathbf{R}_j) + \sum_i \sum_j \sum_k U_3(\mathbf{R}_i, \mathbf{R}_j, \mathbf{R}_k) + \cdots, \qquad (9.1)$$

where successive terms represents the two-, three-, four-... body interactions. In the ideal case U_2 represents the *true* interaction potential of an isolated atom dimer, whilst U_3 is the correction required to represent the true three-body energy surface for the isolated trimer etc. In this ideal scenario potential models are completely transferable in that they would, by construction, be able to reproduce the system energetics over all state points. Computationally tractability falls rapidly with cluster size as the time to calculate the energy of a single $n-$body term scales approximately as N^n. It is common, therefore, to truncate the series at the three- or four-body level. Alternatively, higher order terms may be "folded" into the two-body term alone to give an *effective* pair potential (EPP). Pair potentials are relatively computationally tractable and scale well with system size (which is important if long length- and time-scales are to be accessed). However, the transferability, both between state points and different (chemically related) systems, may be compromised. In addition, the potential parameters may lose their physical meaning making systematic model development difficult. For monatomic systems pair potentials will favour close-packed local environments. The inclusion of any directional bonding will require the inclusion of explicit higher order terms in the potential model.

In this chapter we explore a range of networks by considering two (on the face of it relatively simple) systems; C and SiO_2. These two systems present different modelling problems and hence require different approaches.

9.2.2.1 SiO_2

In a chemistry context silica can most usefully be considered as ionic (in the sense that the Si and O atoms have different electronegativities). Experimental diffraction studies on condensed phases support this view, displaying charge-ordered structures (anions surround cations...). As a result, models in which the atoms carry charge, either as formal $\{Si^{4+}, O^{2-}\}$ or as $\{Si^{2\delta+}, O^{\delta-}\}$, where δ is a potential parameter are likely to be most effective.

The most computationally-tractable models are based on an effective pair potential as exemplified by the Fumi-Tosi potential (see [37] and references therein) for which the interaction between a pair of ions labelled i and j is given by

$$U(r_{ij}) = B_{ij}e^{-a_{ij}r_{ij}} + \frac{Q_i Q_j}{r_{ij}} - \sum_{n=6,8,10...} \frac{C_n^{ij}}{r_{ij}^n} f_n(r_{ij}), \tag{9.2}$$

where B_{ij} and a_{ij} are parameters representing the contribution of the ion radii to the repulsive wall and the rate of decay of the repulsion respectively, C_n^{ij} are the dispersion parameters, $f_n(r_{ij})$ are the dispersion damping functions which mimic the loss of asymptotic behaviour in the dispersion energy at short ion-ion separations [38], and $Q_{i(j)}$ is the charge on ion $i(j)$. The electrostatic interactions are long-ranged (in the sense that the energy term decays slowly compared with the system dimensionality) but may be accurately accounted for using methods such as an Ewald summation (which utilises the periodicity enforced by the use of periodic boundary conditions). The exponential dependence of the short-range energy with separation follows that suggested by Born and Mayer [39]. In the Fumi-Tosi potential the pre-exponent is further expanded in terms of the respective ion radii, $\sigma_{i(j)}$, such that $B_{ij} = A_{ij}\exp[a(\sigma_i + \sigma_j)]$. As a result, the unlike and like ion pre-exponents may be varied to change the coordination environment of the ground-state crystal structure (effectively radius-ratio rules). For SiO_2 the relative ion radii results in ambient conditions structures dominated by four coordinate (tetrahedral) SiO_4 local coordination polyhedra.

These EPPs may reproduce the static crystal structures as the high site symmetries may preclude the formation of (low order) induced moments. Once the site symmetry is broken (i.e. at finite temperature) a description of ion polarization effects is required to reproduce key dynamic properties such as phonon mode frequencies. The inclusion of ion polarization is also required to reproduce key liquid static structural and dynamic properties [37].

The use of potential models to study silica has a long history and there are a number of key SiO_2 potentials. For example, the potential of Woodcock et al. (WAC) [40] uses formal valence (Si^{4+}, O^{2-}) ion charges. A result of this formalism is that key vibrational (bending) frequencies are not reproduced, although these modes may be considerably softened by the inclusion of (many-body) anion polarization effects [41]. Sanders et al. [42] developed both rigid-ion and shell models to model both SiO_2 and silicates in general (see [43] for a review). For SiO_2, for example, models have been developed following both strategies [40, 44, 45] by reference to both experimental information (lattice parameters, bulk moduli, phonon model frequencies....) and electronic structure calculations. An alternative strategy for softening the bending modes is to use the ion charges themselves as free parameters as employed in both the TTAM [46, 47] and BKS [44] potentials (the latter being essentially a reparameterization of the former). Both the BKS and TTAM potentials are parameterised with respect to ab initio calculations. More recent models developed are the DCG [48] and TS potentials [45]. The former is a fluctuating charge model in which the ion charges themselves are allowed to adjust to the local environment. The latter is a polarizable-ion model in which the ion charges are also used as additional variables. The suitability of the BKS, TS and DCG potentials to model the phase behaviour of SiO_2 has been considered in depth in [49].

When a pair potential is augmented with a description of ion polarization a many-body character is introduced into the potential. Historically, such effects were incorporated using a shell model [50] in which the dipolar charge displacement is represented by a charged shell connected by a spring to a charged core. The polarizable-ion model (PIM) is a more modern model in which the induced moments are incorporated more directly. The PIM requires additional parameters; the ion dipole polarizabilities, α_i, and the short-range damping parameters (SRDPs) which control the effect of nearest-neighbour overlap interactions on the induced dipole moments [24–26]. PIM parameters can be obtained by reference to ab initio electronic structure calculations [51]. In the MX_2 stoichiometry PIMs have been developed for $ZnCl_2$ [52, 53], $GeSe_2$ [52], GeO_2 [54] and BeF_2 [55]. Generic models have also been developed in which the anion polarizability is used to control the M–X–M bond angles (and hence the overall network topology) [7, 56–58]. Dipoles induced on the bridging anions introduce negative charge to in between pairs of neighbouring cations and so effectively screen the coulombic cation-cation repulsive interaction [6]. As a result, increasing the anion polarizability tends to reduce the mean M–X–M bond angle and may stablize edge-sharing tetrahedra. The control of the M–X–M bond angles has significant implications for the liquid and amorphous structures. For SiO_2 the bond angles are relatively obtuse ($\bar{\theta}_{SiOSi} \sim 145°$) corresponding to a *charge ordered* structure in which the Si–Si nearest-neighbour length-scale is longer than the corresponding O–O scale (reflecting their respective formal charges). As a result, the first peak is the Si–Si partial structure factor lies at $k \sim 1.6\,\text{Å}^{-1}$ compared with $k \sim 2.0\,\text{Å}^{-1}$ for the corresponding O–O function. For systems with more acute bridging angles (for example, $ZnCl_2$ in which $\bar{\theta}_{ZnClZn} \sim 110°$ [59]) the reduction in bond angle corresponds to a reduction in the cation-cation nearest-neighbour length-scale. This results in a depletion of cation density on an intermediate length-scale (so-called intermediate-range order—IRO) and corresponds to a feature in the total scattering function at $k \sim 1\,\text{Å}^{-1}$ (a pre-peak, or first-sharp diffraction peak) [60].

In these models the application of pressure alters the balance between increasing the nearest-neighbouring coordination number (increasing the packing fraction) and a reduction in the polarization energy (as a result of the anions changing coordination number from two to three).

9.2.2.2 Carbon

In the SiO_2 models highlighted above the network structure arises from a combination of the relative ion radii (which effectively controls the local coordination geometry of the coordination polyhedra) and the electrostatic ion-ion interactions (which control how these polyhedra interlink). For a monatomic system such as carbon an EPP will always favour close-packed local environments and so can not effectively generate a network structure based on lower coordinate (four-coordinate in three dimensions and three-coordinate in two) nearest-neighbour environments. In order to stabilize such local topologies higher (>2) order terms must be incorporated into the potential model, for example as a balance between two- and three-body interactions (see [61] for a review). The generic form of such models is

$$U = \sum_{i,j} U_2(r_{ij}) + \lambda \sum_{i,j,k} U_3(r_{ij}, r_{ik}, \theta_{ijk}), \tag{9.3}$$

where U_2 is a pair potential acting over all atoms pairs ij and U_3 is an explicit three-body term which depends on the distance between two pairs of atoms (with atom i at the centre) and angle θ_{ijk} between the two pairs. The parameter λ controls the relative strengths of the two- and three-body terms. For a single component bulk system, in the absence of higher order terms, an attractive two-body term will invariably favour close-packed local environments and hence will maximise the two-body energy. The higher order (many-body) terms often act against this drive to close-packing. In a Stillinger-Weber potential, for example, applied to study Si or Ge [62], the three-body term acts to bias *against* the formation of any set of three bonded atoms with angles differing from the ideal tetrahedron. As a result, any crystal structure which contains non-tetrahedral angles will become thermodynamically disfavoured with respect to the ideal diamond crystal. Again, the λ parameter may be used to effectively control the tetrahedricity of the system (analogous to the manner in which the anion polarizability controls the network topology above) which is of use for obtaining amorphous silicon-like configurations on the (relatively short) simulation time-scales [63, 64].

For carbon, for example, a common atomistic potential model is the three-body (Tersoff II) potential model [65]. This model accounts for the relative stability of the bulk crystalline diamond and graphite structures and account well for the basic mechanical and dynamic properties of single-walled C-NTs [66]. In the Tersoff II potential model [65] the energy of each individual carbon atom is taken to be half that of the bonding pair,

$$U = \sum_{i} U_i = \frac{1}{2} \sum_{i,j \neq i} f_c(r_{ij})\left[f_R(r_{ij}) + b_{ij} f_A(r_{ij})\right], \tag{9.4}$$

where $f_R(r_{ij})$ and $f_A(r_{ij})$ are the repulsive and attraction terms respectively and $f_c(r_{ij})$ is a cut-off function which limits the range of the atom-atom interactions. $f_R(r_{ij})$ and $f_A(r_{ij})$ are given by

$$f_R(r_{ij}) = \alpha_1 e^{-\lambda_1 r_{ij}}$$
$$f_A(r_{ij}) = -\alpha_2 e^{-\lambda_2 r_{ij}}. \tag{9.5}$$

b_{ij} is the bond order of the interaction between atoms i and j which is given by

$$b_{ij} = (1 + \beta^n \xi_{ij}^n)^{\frac{-1}{2n}}. \tag{9.6}$$

and hence increases as the atom coordination number decreases. The effective coordination number, ξ_{ij}, depends on the distance of the coordinating carbons from the central carbon atom as well as having an angular dependence.

$$\xi_{ij} = \sum_{k \neq i,j} f_c(r_{ik}) g(\theta), \tag{9.7}$$

$$g(\theta) = 1 + \frac{c^2}{d^2} - \frac{c^2}{[d^2 + (h - \cos\theta)^2]}, \tag{9.8}$$

where θ is the angle between vectors joining the atom i with atoms j and k. The cut-off function is given by

$$f_c(r_{ik}) = \begin{Bmatrix} 1, & r < R - D \\ \frac{1}{2} - \frac{1}{2}\sin[\frac{\pi}{2}(r - R)/D], & R - D < r < R + D \\ 0, & r > R + D \end{Bmatrix} \tag{9.9}$$

9.3 The Networks

In this chapter we shall consider different network arrangements of SiO_2 and C. We will begin by considering the more traditional problems of modelling the three-dimensional amorphous forms. We will then consider modelling carbon in two-dimensions with the aim of understanding the structure of amorphous graphene followed by a (related) study of pseudo-two-dimensional SiO_2 bilayers. Finally, we consider how the amorphous graphene structures may form even lower dimensional nanotubular structures.

9.3.1 Three-Dimensional Glass: Silica

Molecular dynamics simulation methods are applied to generate amorphous configurations. A number of potential models are employed in order to highlight any structural differences in the configurations obtained from each. The potential is that derived by Tangney and Scandolo (TS) [45] which is an ionic model employing pair potentials to account for the short-range (overlap) and dispersive interactions. Reduced ion charges (of magnitude ~ -1.38e and $+2.76$e for the O and Si ions respectively) are utilised. In addition anion (dipole) polarisation effects are incorporated (as described in [41]) which requires the specification of two parameters, the anion polarizability and a short-range damping parameter (which controls the effect of nearest-neighbour overlap on the induced dipole moments). The inclusion of the polarization terms introduces a many-body character into the potential model. The model parameters are determined by reference to electronic structure calculations.

To assess the role of the method of generation on the final structure glassy configurations are generated by two different procedures. In the first the systems are quenched from the liquid state using Nosé-Hoover thermostats [29, 30] as described above. We use a value of the thermostat relaxation time of $\tau \simeq 1.2$ ps which corresponds to an approximate cooling rate of 10^{15} K s^{-1} simulations. The time between configuration sampling is $\sim$100 ps, significantly longer than the characteristic relaxation time as probed by the intermediate scattering function and so successive configurations can be considered as effectively independent. In the second method the ambient pressure configurations, obtained as described above, are systematically compressed to the required densities At each statepoint 100 configurations are extracted from the liquid state. Simulations are performed on systems containing 999 ions in a constant volume ensemble up to pressures of the order of $p \sim 50$ GPa.

Figure 9.1 shows the simulated total neutron scattering functions obtained from both the compressed and quenched configurations and compared to those obtained from high pressure neutron scattering experiments [67]. The (Ashcroft-Langreth) structure factors are calculated directly from the atom positions via $S^{AL}_{\alpha\beta}(k) = \langle A^*_\alpha(k).A_\beta(k)\rangle$, where $A_\alpha(k) = \frac{1}{\sqrt{N_\alpha}}\sum_{i=1}^{N_\alpha} e^{i\mathbf{k}.\mathbf{r}_i}$ are the fourier components. These Ashcroft-Langreth functions may be converted easily into the Faber-Ziman form [68]. The simulated total scattering functions are obtained directly from the weighted combination of the Faber-Ziman partial structure factors,

$$\begin{aligned} F^{(n)}(k) &= \sum_{\alpha=1}^{n}\sum_{\beta=1}^{n} c_\alpha c_\beta b_\alpha b_\beta \left[S_{\alpha\beta}(k) - 1\right], \\ &= c_O^2 b_O^2 \left(S_{OO}(k) - 1\right) + c_{Si}^2 b_{Si}^2 \left(S_{SiSi}(k) - 1\right) + 2c_O c_{Si} b_O b_{Si} S_{SiO}(k), \end{aligned} \tag{9.10}$$

where $c_{\alpha(\beta)}$ and $b_{\alpha(\beta)}$ are the mole fractions and coherent neutron scattering lengths of chemical species $\alpha(\beta)$ respectively. For a natural isotopic composition, $b_{\mathrm{Si}} = 4.1491(10)$ fm and $b_{\mathrm{O}} = 5.803(4)$ fm [69].

Figure 9.1a shows the TS model to give an excellent representation of the experimental neutron scattering data to high pressure. The functions generated from the configurations produced by compression appear to give significantly better agreement

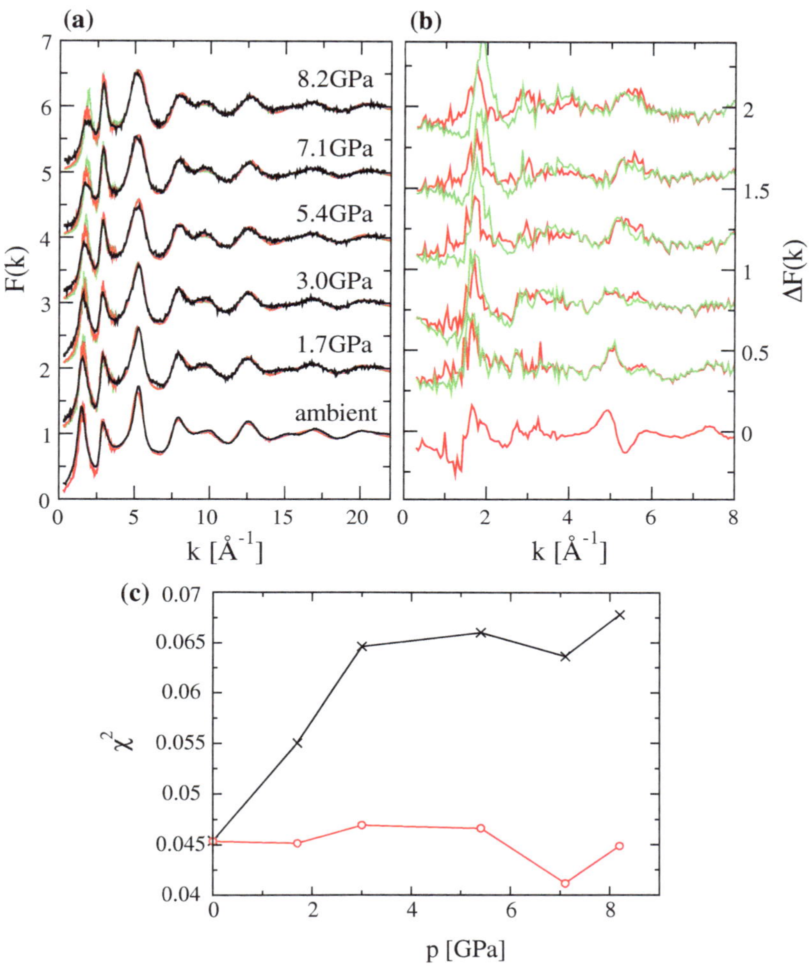

Fig. 9.1 **a** Total neutron scattering function, $F(k)$, for SiO_2 obtained at the pressures indicated from experiment (*black lines*, [67]) and simulation (*green* and *red lines* using configurations generated by liquid quenching and compression respectively). **b** Difference, $\Delta F(k)$, determined between the experimental and simulated total scattering functions in panel (**a**) and determined at the same pressures. The *green* and *red lines* are again for configurations generated by liquid quenching and compression respectively. In both (**a**) and (**b**) successive functions are offset along the ordinate axis for clarity. **c** Quality of fit parameter for the simulated total neutron scattering function with respect to the experimental analogue as a function of pressure for the liquid quench ($\times$) and compressed ($\circ$) configurations respectively

with the experimental data compared to those produced by direct high pressure quenches from the liquid. To quantify this difference we introduce a quality of fit parameter,

$$\langle \Delta F^2 \rangle^{1/2} = \frac{1}{N_Q} \left\{ \sum_{i=1}^{N_Q} [F^{ex}(Q) - F^{sim}(Q)]^2 \right\}^{1/2}, \quad (9.11)$$

where $F^{ex}(Q)$ and $F^{sim}(Q)$ are the total scattering functions obtained from experiment and simulation respectively. Figure 9.1b shows $\langle \Delta F^2 \rangle^{1/2}$ as a function of pressure and highlights the significantly better fit obtained from the compressed configurations compared with the quenched. Furthermore, the quality of fit for the compressed configurations appears to be largely independent of pressure. Figure 9.1c shows the difference between the simulated and experimental total scattering functions,

$$\Delta F(k) = F^{ex}(Q) - F^{sim}(Q), \quad (9.12)$$

determined over the same pressure range as Fig. 9.1a. The greatest differences between the functions generated from the compressed and directly quenched configurations appear in the region of the first peak (at $k \sim 1.6\,\text{Å}^{-1}$) which is associated with structure on intermediate length-scales, that is, structure associated with the inter-connectivity of the SiO_4 tetrahedra. As emphasised in the introduction the ring structure represents a natural language in which to discuss such inter-connectivity.

Figure 9.2a shows the evolution of the mean ring size as a function of pressure for both the compressed and quenched configurations. To relatively moderate pressures ($p \sim 18$ GPa) the mean ring size increases for the quenched configurations but falls for the compressed. For $p > 18$ GPa both methods generate configurations which show a fall in $\langle n \rangle$. Figure 9.2b shows the dependence of the mean ring size on the mean coordination number. Figure 9.2c shows the breakdown of the cation nearest-neighbour coordination numbers as a function of pressure for, for example, the compressed configurations. As pressure increases there is a systematic rise in five-, the at higher pressures six-, coordinate Si atoms. The reduction in the mean ring size is, therefore, correlated with the increase in five- and six-coordinate Si sites.

Previous work indicates that models which preserve the SiO_2 tetrahedra give a rise in mean ring size with pressure [9, 70]. However, models in which both bond angles and nearest-neighbour coordination environments may evolve suggest a more complex ring distribution evolution [71–74] with the emergence of 5-coordinate sites central in promoting the formation of smaller rings [75]. To attempt to account for the behaviour of the compressed configurations we suggest a simple models which links the changes in coordination number with the changes in the ring structure.

9.3.1.1 Ring Closure Model

The initial mean ring size is $\langle n \rangle$, where

$$\langle n \rangle = \frac{\sum n p_n}{\sum p_n}, \quad (9.13)$$

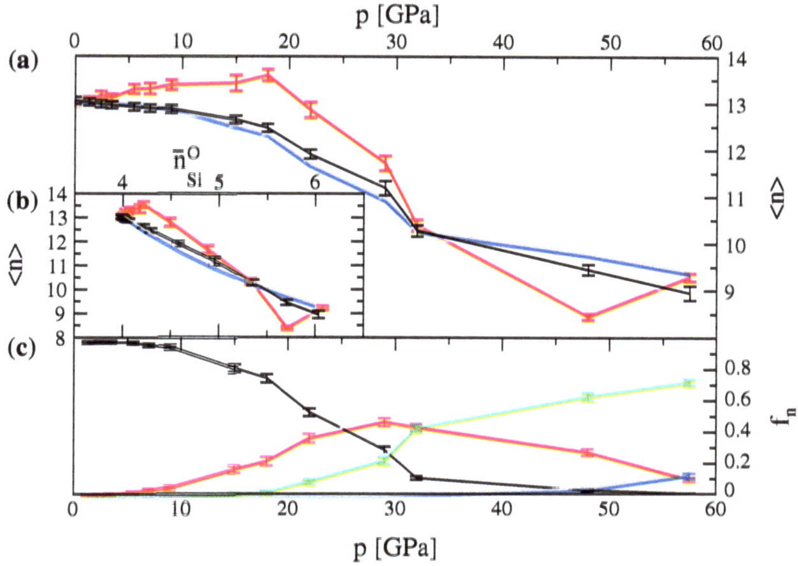

Fig. 9.2 **a** Mean ring size, $\langle n \rangle$, versus pressure for glassy SiO_2 configurations obtained by liquid quenching (*red line*) and compression (*black line*) respectively. The *blue line* shows the prediction of the ring closure model. **b** The same mean rings size data as in panel (**a**) plotted against the mean Si–O nearest-neighbour coordination number, $\bar{n}^{O}_{Si}$. **c** The fraction of 4-, 5-, 6- and 7-coordinate Si (*black*, *red*, *green* and *blue lines* respectively) for compressed configurations as a function of pressure

where p_n is the number of rings of size n and $\sum p_n = N_{tot}$ is the total number of rings. A single ring closure converts one $\langle n \rangle$ ring into two of mean size $(\langle n \rangle/2) + 1$ and increases the total number of rings from N_{tot} to $N_{tot} + 1$.
As a result,

$$(N_{tot} + 1)\langle n' \rangle = (N_{tot} - 1)\langle n \rangle + 2\left\{\frac{\langle n \rangle}{2} + 1\right\}, \tag{9.14}$$

where $\langle n' \rangle$ is the mean ring size after the ring closure. For m such closure events,

$$\begin{aligned}(N_{tot} + m)\langle n' \rangle &= (N_{tot} - m)\langle n \rangle + 2m\left\{\frac{\langle n \rangle}{2} + 1\right\}, \\ &= N_{tot}\langle n \rangle + 2m,\end{aligned} \tag{9.15}$$

and so,

$$\langle n' \rangle = \frac{N_{tot}\langle n \rangle + 2m}{N_{tot} + m}. \tag{9.16}$$

Recall that m represents a single closure event. At a given pressure, for a mix of 4-, 5- and 6-coordinate Si sites only, then

$$m = N_{cat}(f_5 + 2f_6), \tag{9.17}$$

where N_{cat} is the number of cations in the system and f_5 and f_6 are the fractions of 5- and 6-coordinate sites. In the initial stages of closure $f_6 \simeq 0$ and so $N_{cat} f_5$ will simply be the absolute number of 5-coordinate sites. The factor of two in the f_6 term reflects the fact that the formation of a six-coordinate site from a four-coordinate site requires two closure events.

We can obtained a general expression for m as

$$m = N_{cat} \sum_{n=5}^{\infty} (n-4) f_n. \tag{9.18}$$

Substituting 9.18 into 9.16 gives

$$\langle n' \rangle = \frac{N_{tot}\langle n \rangle + 2N_{cat} \sum_{n=5}^{\infty}(n-4) f_n}{N_{tot} + N_{cat} \sum_{n=5}^{\infty}(n-4) f_n}, \tag{9.19}$$

$$= \frac{4\langle n \rangle + 2\sum_{n=5}^{\infty}(n-4) f_n}{4 + \sum_{n=5}^{\infty}(n-4) f_n}, \tag{9.20}$$

where we have taken $N_{tot} = 4N_{cat}$.
The mean Si–O coordination number is given by

$$\bar{n}_{\mathrm{Si}}^{\mathrm{O}} = \sum_{n=1}^{\infty} n f_n,$$

$$= 4\sum_{n=1}^{\infty} f_n + \sum_{n=5}^{\infty} (n-4) f_n. \tag{9.21}$$

Noting that

$$\sum_{n=1}^{\infty} f_n = 1, \tag{9.22}$$

leads to

$$\bar{n}_{\mathrm{Si}}^{\mathrm{O}} = 4 + \sum_{n=5}^{\infty} (n-4) f_n. \tag{9.23}$$

Substituting into 9.20 gives

$$\langle n' \rangle = \frac{4\langle n \rangle + 2(\bar{n}_{\mathrm{Si}}^{\mathrm{O}} - 4)}{\bar{n}_{\mathrm{Si}}^{\mathrm{O}}}. \tag{9.24}$$

Figure 9.2a, b show the dependence of the mean ring size on both pressure and $\bar{n}_{\mathrm{Si}}^{\mathrm{O}}$. The correlation between the simple model results and the results obtained from the compressed configurations (which give the best account of experimental diffraction data) is strong.

9.3.2 Three-Dimensional Monatomic: Carbon

The experimental investigation of the phase diagram of carbon requires access to enormous temperatures and pressures [76, 77]. Shock and ramp wave experiments can reach $p \sim 2000$ GPa [78–82] but are technologically challenging. An understanding of the high pressure behaviour is important. For example, the evolutionary and asteroseismic properties of planets with carbon-rich interiors will depend critically on the high pressure behaviour [78, 79, 83, 84]. Below the melting point carbon displays a "diamond-like" amorphous structure (referred to as "tetrahedral amorphous" carbon, ta-C) [85–91] potentially connecting with the observation of low and high density amorphous states for (chemically-related) systems such as Si and Ge (see, for example [33, 92], and references therein). Here we shall map both the diamond/liquid coexistence curve and the locus of the liquid/amorphous transitions.

9.3.2.1 Liquid/Crystal Coexistence

Figure 9.3a shows the three dimensional diamond/liquid coexistence curve. At high pressures the curve becomes reentrant (negative Clapeyron slope, dp/dT) corresponding to a negative volume change on melting (the liquid is more dense than the diamond crystal). The (four-coordinate) diamond structure has a high pressure stability field and within the same pressure thresholds the liquid evolves from being dominated by three-coordinate sites (at relatively low pressures) to four- and five-coordinate sites (at intermediate pressures) up to eight-coordinate sites (at high pressures). The $\{p, T\}$ at which the Clapeyron slope in infinite (here $\{p, T\} = \{890\,\mathrm{GPa}, 15{,}550\,\mathrm{K}\}$) represents the isochoric transformation. Also indicated on the figure is the location of the isocoordinate change (no change in mean coordination number, $\bar{n}_{\mathrm{C}}^{\mathrm{C}}$, which remains at four). The volume change for the isocoordinate transition is $\Delta V = 0.22\,\mathrm{cm}^3\,\mathrm{mol}^{-1}$ or $\sim 8\,\%$ on melting. Figure 9.4a shows the radial distribution functions at temperatures 1000 K above the respective melting temperatures at pressures corresponding to the isochoric and isocoordinate phase transitions. For the isochoric transition the phase change corresponds to an

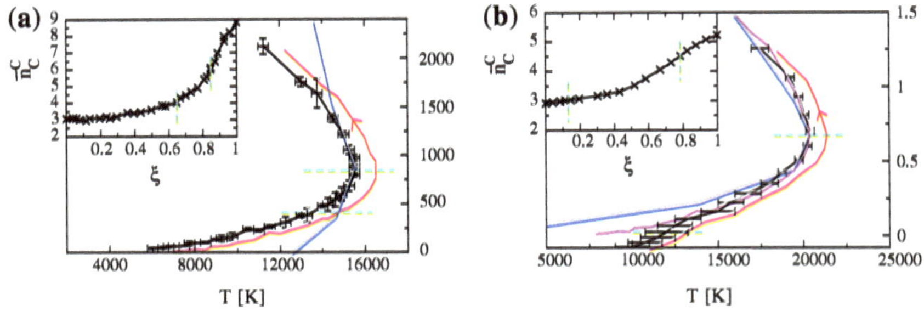

Fig. 9.3 **a** Three- and **b** two-dimensional liquid/crystal coexistence curves (*black lines*). The *dashed lines* indicate the isochoric and isocoordinate transition pressure (stress). The *red lines* indicate pathways around the coexistence curve around which key properties are determined. In both figures the *inset* shows the evolution of the mean coordination numbers about the respective pathways. In both (**a**) and (**b**) the *blue curves* show the scaled gaussian core coexistence curves [93, 94]

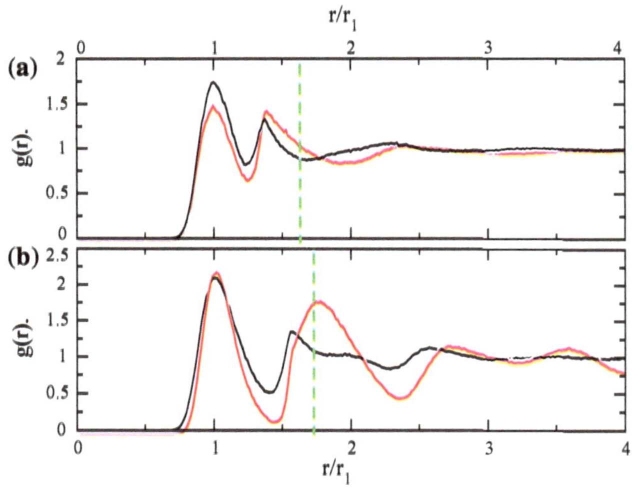

Fig. 9.4 Radial distribution functions obtained from the liquid state simulations in **a** three- and **b** two-dimensions at 1000 K above the respective isocoordinate (*red lines*) and isochoric (*black lines*) crystal to liquid phase transitions. The abscissa length-scale is normalized by the position of the first peak, r_1. The length-scales corresponding to ideal tetrahedral and trigonal C–C–C bond angles are indicated as *dashed vertical lines*

increase in the mean coordination number from 4 to ~5.8. The isocoordinate transition highlights the role of intermediate-range order (order beyond the short-range order imposed by the atom excluded volume). On melting the four-coordinate units are maintained but their packing (and associated ring distribution) changes, resulting in a shorter next-nearest-neighbour length-scale (and hence to a more dense liquid than the corresponding crystal).

To quantify the changes in liquid structure at temperatures near to the coexistence curve system properties are studied along a $\{p, T\}$ path, ζ, defined at each pressure as the melting point, T_m plus 1000 K. The progression along the path ζ is determined by scaling the p and T axes by the highest pressure and temperature studied along the coexistence curve ($p_0 = 2200$ GPa and $T_0 = 16{,}550$ K respectively), $\zeta^2 = (T/T_0)^2 + (p/p_0)^2$, and then normalising to give a total path length of one. The inset to Fig. 9.3a shows, for example, the evolution of $\bar{n}_C^C$ about this pathway. At relatively low p and T $\bar{n}_C^C$ increases slowly reflecting the competing roles of the pressure and temperature which tend to increase and decrease $\bar{n}_C^C$ respectively. As the coexistence curve becomes reentrant $\langle n \rangle$ increases more sharply as the increase in p and decrease in T both tend to increase $\bar{n}_C^C$.

9.3.2.2 Liquid/Amorphous Transition

Figure 9.5a–c shows the evolution of two key structural parameters (the mean coordination number $\bar{n}_C^C$ and mean tetrahedral order parameter, $\bar{q}$) and system volume along three isobaric paths (as indicated in Fig. 9.5d). $\bar{q}$ is given by the mean of

$$q = 1 - \frac{3}{8}\sum_{j=1}^{3}\sum_{k=j+1}^{4}(\cos\psi_{jk} + \frac{1}{3})^2, \tag{9.25}$$

where the atoms j and k are within a given cut off of the central atom i and ψ_{jk} is the angle between vectors $\mathbf{r}_{ij}$ and $\mathbf{r}_{ik}$. Ideal tetrahedral angles result in $\bar{q} = 1$. As the liquid is cooled below the melting point significant changes are observed in $\bar{q}$. At high pressure $\bar{q}$ shows a significant increase from $\bar{q} \sim 0.55$ to $\bar{q} \sim 0.85$ concomitant with a reduction in $\bar{n}_C^C$ from ~6 to ~4 and a rise in the system volume. The changes in $\bar{n}_C^C$ and $\bar{q}$ are indicative of a significant rise in total tetrahedral geometries. At low pressure (corresponding to an isocoordinate transition) the analogous change in $\bar{q}$ occurs more gradually and is concomitant with a decrease in the system volume. At intermediate pressures (corresponding to an isochoric transition) the increase in $\bar{q}$ is associated with a reduction in $\bar{n}_C^C$ from ~5 to ~4. Heating the system back from the high $\bar{q}$ states along the same isobars results in the reverse transformations, occurring at higher temperatures (there is a significant hysteresis). Above the transition in $\bar{q}$ the system shows typically liquid-like atom diffusion coefficients (of the order of $D \sim 10^{-4}\,\mathrm{cm^2\,s^{-1}}$). However, below the transition temperature the diffusion coefficients become vanishingly small. The changes in $\bar{q}$ and D are indicative of transitions between a liquid and an amorphous phase (the latter with a ta-C-like structure). Figure 9.6 shows the evolution of both the static structure factor, $S(k)$, and radial distribution function, $g(r)$, as a function of temperature during the heating cycles for both the high and low pressure simulations. Both the high pressure $S(k)$ and $g(r)$ functions (Fig. 9.6a, c respectively) show a clear change on going from the amorphous to liquid states. In $S(k)$ the sharp low temperature first peak at $k \sim 3.4\,\text{Å}^{-1}$ is replaced by a significantly broader feature at $k \sim 3.5\,\text{Å}^{-1}$ at

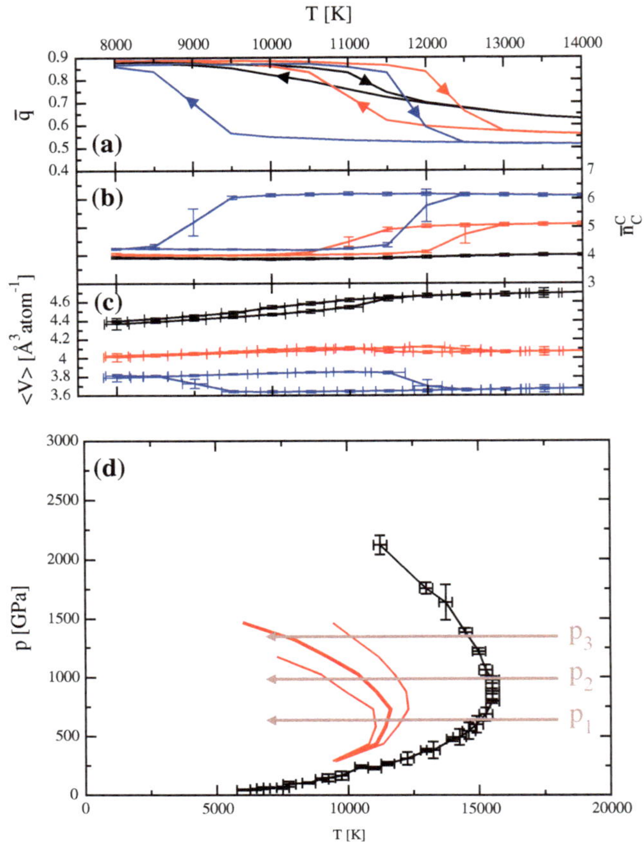

Fig. 9.5 **a** Structural properties calculated along the $\{p, T\}$ paths *highlighted* in panel (**d**) (*black* p_1, *red* p_2, *blue* p_3). The panels show the **a** mean tetrahedral order parameter, $\bar{q}$, **b** mean coordination number, $\bar{n}_{\mathrm{C}}^{\mathrm{C}}$ and **c** mean volume, $\bar{V}$. The *arrows* in the *top panel* highlight the respective heating and cooling cycles. **d** Phase diagram for three-dimensional carbon showing the liquid/diamond coexistence curve (*black line*). The *thick red line* shows the coexistence curve for the liquid↔amorphous transition whilst the *thin red lines* show the spinodal limits obtained from the cooling and heating cycles. The *brown arrows* indicate the three isobaric pathways along which key properties are determined in parts (**a**–**c**)

higher temperature. Similarly, $g(r)$ shows a change from a relatively narrow (tightly defined) first peak (corresponding to $\bar{n}_{\mathrm{C}}^{\mathrm{C}} \sim 4$) to a significantly broader peak centred at higher r (reflecting the increase in nearest-neighbour coordination number). The corresponding low pressure functions show analogous, less well-defined (although still clear) transitions in structure. The low temperature amorphous $g(r)$ functions for both pressures go to zero beyond the first peak indicative of the formation of amorphous states in which there is no diffusion on the simulation time-scale.

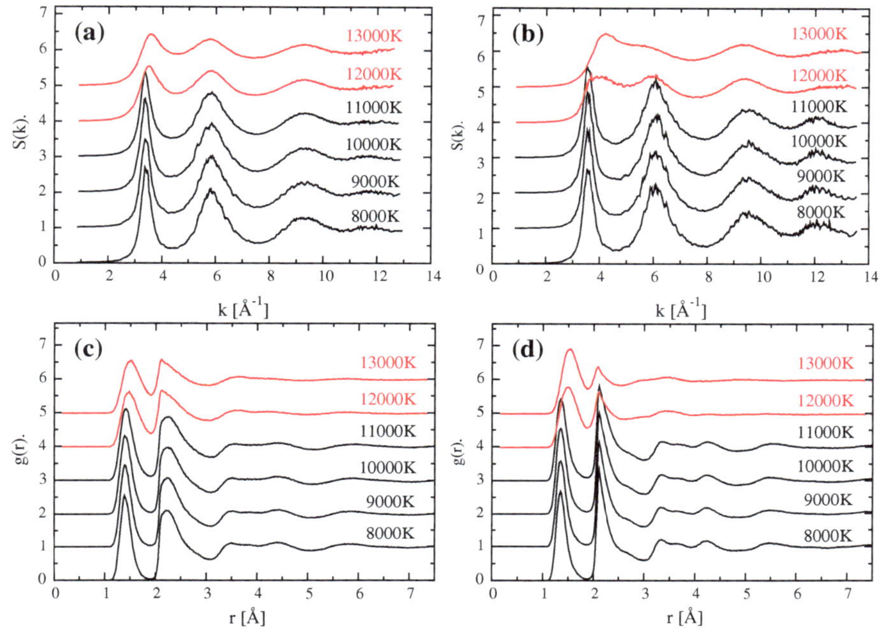

Fig. 9.6 Structure factors and radial distribution functions as a function of temperature (as indicated) at pressures of ∼440 GPa (panels (**a**) and (**c**)) and ∼1030 GPa (panels (**b**) and (**d**)) respectively. In all cases the temperatures are included adjacent to the calculated function. The *black* and *red lines* correspond to functions obtained in the amorphous and liquid regions of the phase diagram respectively

Figure 9.5d shows a reduced phase diagram to highlight the locations in the pT plane of the liquid↔amorphous transitions. The figure shows the locus of transition temperatures (defined as halfway between the temperatures at which $\bar{q}$ changes for the respective cooling and heating cycles). The latter two curves are effectively the spinodal stability limits for the two phases. The temperature range for the liquid↔amorphous transition increases with pressure. Figure 9.7 shows the comparison of the diamond/liquid phase diagram for Si (obtained using a Stillinger-Weber potential [62, 95]) mapped onto that for C. The C and Si melting curves map onto each other consistent with a corresponding states analysis. The figure also shows the liquid/amorphous coexistence curve and associated spinodals (scaled as for the liquid/solid coexistence curve) determined from the isobaric simulations along with the Si LDA/HDA coexistence curve, and associated spinodals, determined from a two-state model [96, 97]. The significant curvature of the carbon liquid/amorphous coexistence curve reflects the enormous pressure range over which coexistence is observed (of the order of ∼1500 GPa for C compared with ∼20 GPa for Si). The temperature axis is scaled by dividing by the temperature at which the Clapeyron slope is infinite (T_{max}(C) = 15,550 K, T_{max}(Si) = 1800 K). The pressures for Si are

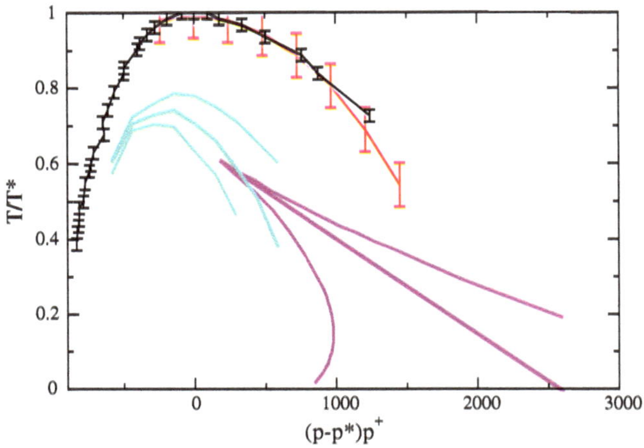

Fig. 9.7 Reduced phase diagram for the diamond crystal/liquid coexistence curves determined for carbon (using a Tersoff-II potential [65]—*black line*) [98] and silicon (using a Stillinger-Weber potential [62]—*red line*) [95]. The pressure and temperature axes are scaled by the respective values at which $dT/dp = 0$ for the coexistence curve as described in the text. The *magenta lines* shows the LDA/HDA coexistence curve for Si (*thick line*) determined from a two state model and the associated spinodals (*thin lines*). The *light blue lines* show the liquid/amorphous coexistence curve determined for carbon (*thick line*) and the associated spinodals (*thin lines*). Both sets of curves are scaled as for the liquid/diamond coexistence curves

mapped onto those for C by subtracting the pressure at which the Clapeyron slope in infinite ($p_{\max}(\text{C}) = 890\,\text{GPa}$, $p_{\max}(\text{Si}) = -3\,\text{GPa}$) and multiplying the Si values by $p^+ = 121$.

9.3.3 Two-Dimensional Glass: Amorphous Graphene

Crystalline graphene, which can be considered as a percolating array of hexagonal (six-membered) rings represents an archetypal two dimensional system and has been the focus of considerable experimental and theoretical studies [99]. Experimental microscopy studies show images of regions of amorphous graphene (a-G) characterised by the presence of five-, six- and seven-membered rings [12–15] and work is on-going to attempt to control the formation of these structures (using, for example, chemical vapour deposition [16] and the electron beam [17, 18]). Our modelling strategy mirrors that used for the three dimensional system. We will establish the liquid/crystal phase diagram and then investigate methods for generating amorphous structures over a range of stresses [100, 101].

9.3.3.1 Liquid/Crystal Coexistence

A useful starting point is to consider the energetics of a set of potential two dimensional crystal structures. Figure 9.8 shows the energy versus area curves for four potential high symmetry crystal structures; the hexagonal lattice (three coordinate, graphene sheet), a close-packed (or trigonal) lattice (six coordinate), a square

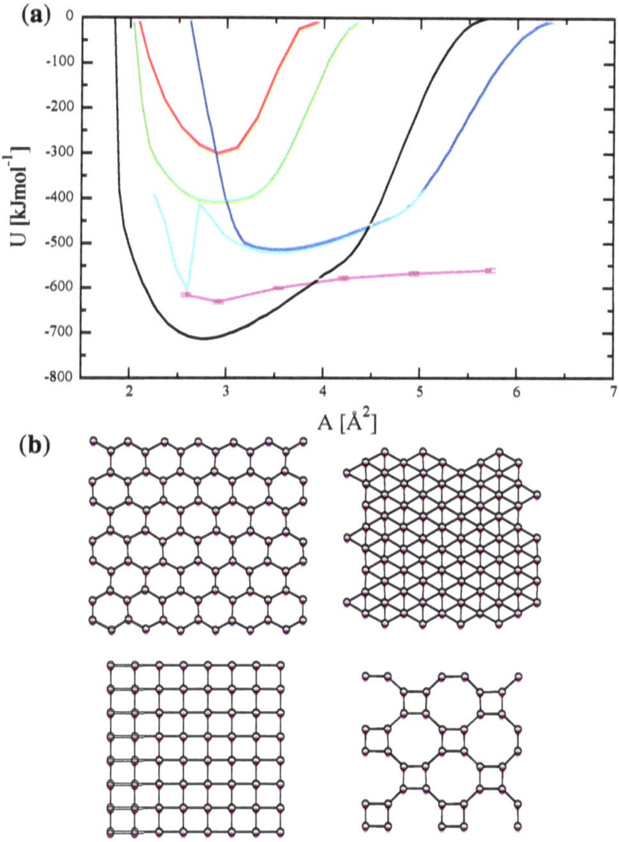

Fig. 9.8 **a** Energy as a function of area for four key two-dimensional crystal structures. Key: *black line* hexagonal lattice, *red line* close-packed lattice, *green line* square net, *blue line* octahedral/square (octsq) net. The energies of the amorphous graphene configurations are also shown as a function of area (*magenta line*). The *error bars* highlight the variance over the configurations generated. The energy of the relaxed octsq structure is shown as the *light blue line*. For $A \gtrsim 2.6\,\text{Å}^2$ the difference in energy reflects the presence of two C–C nearest-neighbour length-scales in the relaxed structure with respect to just one in the unrelaxed. For $A \lesssim 2.6\,\text{Å}^2$ the crystal amorphises as discussed in the text. **b** Structures of four key two-dimensional crystals. *Top left* hexagonal lattice, *top right* close-packed lattice, *bottom left* square net, *bottom right* "octsq", a mix of four- and eight-membered rings

lattice (four coordinate) and a mixed octahedral/square lattice (three coordinate termed "octsq" here and, in some sense, a "two-dimensional clathrate"). The percolating hexagonal lattice structure is energetically favoured over the other polymorphs. Significantly, the relative areas of the hexagonal and close-packed lattices indicates that any thermodynamic phase transition between the two should occur at very high stress corresponding to the hexagonal lattice displaying a large stress stability field. At relatively high area the percolating pattern of octahedra and squares becomes energetically favourable over the graphene sheet indicating the potential to isolate such structures under tension.

Figure 9.3b shows the two dimensional graphene/liquid phase diagram. As with the three dimensional diamond/liquid phase diagram the coexistence curve becomes reentrant at high stress. A negative slope for the coexistence curve can be rationalised by analogue to the three-dimensional case. The two-dimensional Clapeyron slope, $d\sigma/dT$, is given by $\Delta S/\Delta A$ where ΔS and ΔA are the respective entropy and area changes on melting. As a result, the negative slope corresponds to a transformation from the crystal to a more dense liquid. As would be expected, typical two-dimensional system melting points are higher than those obtained in three-dimensions, for example, the maxima in the melting curves at $\{\sigma_{\max}, T_{\max}\} = \{0.70\,\mathrm{J\,m^{-2}}, 20{,}400\,\mathrm{K}\}$ in two dimensions compares with $\{p_{\max}, T_{\max}\} = \{890\,\mathrm{GPa}, 15{,}550\,\mathrm{K}\}$ in three dimensions. Figure 9.3b also shows the effect of scaling the three-dimensional liquid/crystal coexistence curve (Fig. 9.3a) using $\{p_{\max}, T_{\max}\}$. The temperature and pressure axes are rescaled by factors of $T_{\max}(\mathrm{C}-2\mathrm{d})/T_{\max}(\mathrm{C}-3\mathrm{d}) \sim 1.32$ and $\sigma_{\max}(\mathrm{C}-2\mathrm{d})/p_{\max}(\mathrm{C}-3\mathrm{d}) \sim 0.8 \times 10^{-12}\,\mathrm{m}$ respectively.

To highlight the origin of the reentrant behaviour we again follow structural changes in the liquid around a path defined by the determined coexistence curve. The inset to Fig. 9.3b shows the evolution of the mean coordination number around the path defined by the coexistence curve (as described above for the three-dimensional analogue). At low stresses, prior to the maximum in the coexistence curve, the configurations are dominated by local three-coordinate environments. As the stress increases the mean coordination number increases heading towards the close-packed "limit" of six. The origin of the reentrant behaviour mirrors, therefore, that in the three-dimensional diamond/liquid system. The stress stability range for the hexagonal net (graphene) structure is sufficiently large that, over the same pressure range, a large change in the liquid-state coordination environments is stabilised. Again, highlighted are the stresses and which the isochoric[2] and isocoordinate transitions occur. Figure 9.4b shows the radial distribution functions calculated for both the isochoric and isocoordinate systems at 1000 K above their respective melting points. For the isochoric transition the phase change corresponds to an increase in the mean coordination number from 3 to ~4.5. The isocoordinate transition again highlights the change in intermediate-range order with respect to the ideal graphene sheet.

[2]Here *choros* refers to area.

9.3.3.2 Amorphous Structures

Amorphous structures are generated by rapid cooling from the liquid state at each density. The understanding of the location of the liquid/crystal coexistence curve allows quenching to be performed at each density from a set temperature above the respective melting point rather than from a chosen isotherm. Two quenching methods are employed as discussed above using both steepest descent energy minimisation and Nosé-Hoover thermostats [29, 30], changing the rate of cooling by controlling the thermostat relaxation time (which controls the rate of heat transfer between the system and the heat bath). The energies of the final amorphous structures are shown as a function of area in Fig. 9.8. For $A \lesssim 4\,\text{Å}^2$ the amorphous structures are less energetically stable than the ideal graphene crystal although, significantly, the amorphous structures remain energetically favoured with respect to the oct/sq crystal structure. The graphene considered here is ideal (and constrained to a plane) and hence Fig. 9.8 represents the stretch and compression of the C–C bonds. The sheet is not, for example, allowed to fracture at low density.

Figure 9.9a shows the evolution of the mean coordination number and ring size as a function of the system area for both the amorphous and parent liquid structures. The mean coordination number falls from around six (corresponding to pseudo-close-packed local atom environments) to ~3 as the system area increases, concomitant with a rise in the mean ring size from ~3 to ~6. These changes appear more gradual in the liquid configurations when compared with the amorphous structures. For example, for the amorphous configurations a mean ring size of six dominates

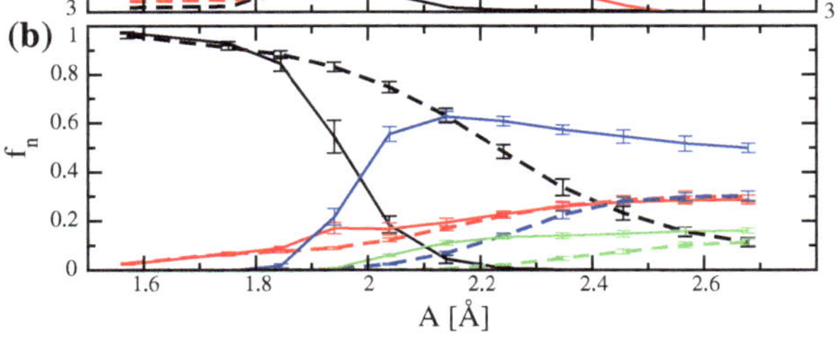

Fig. 9.9 **a** Mean coordination number $\bar{n}_{\mathrm{C}}^{\mathrm{C}}$ (*solid lines*) and mean ring size, $\langle n \rangle$ (*dashed lines*) as a function of area. The *black lines* are determined from the amorphous configurations and the *red lines* from the original (pre-quench) liquid configurations respectively. **b** Fractions of 3-, 5-, 6- and 7-membered rings (*black*, *red*, *blue* and *green lines* respectively) versus area for the amorphous (*solid lines*) and original liquid (*dashed lines*) configurations

for $A \gtrsim 2.2\,\text{Å}^2\,\text{atom}^{-1}$ whilst three-membered rings (close-packed configurations) dominate for $A \lesssim 1.8\,\text{Å}^2\,\text{atom}^{-1}$. Figure 9.9b shows the evolution of the fractions of 3-, 5-, 6- and 7-membered rings. At low areas the three-membered rings dominate the liquid structure. For $A > 2.2\,\text{Å}^2\,\text{atom}^{-1}$ the three-membered rings, present in significant quantities in the liquid state, are almost completely quenched out.

A common problem in obtaining amorphous or glassy configurations is effective suppression of the competing crystallisation process. It is important, therefore, to characterise configurations generated in terms of any inherent crystallinity. Unequivocal specification of a system as amorphous or nanocrystalline is effectively impossible. The problem is exacerbated in two dimensions as the dimensional constraint produces more ordered liquid/amorphous structures with respect to those generated in three dimensions Fig. 9.10a, b shows the structure factors and second moments determined from the centres of the six-membered rings,

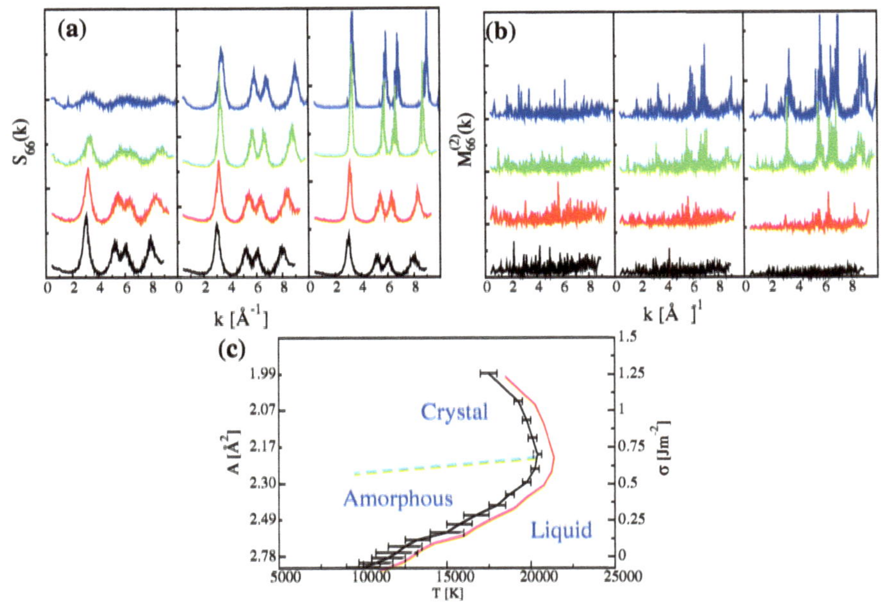

Fig. 9.10 **a** Structure factors and **b** second moments for the two-dimensional carbon calculated using the centres of mass of the six-membered rings. Configurations are generated at four different densities (*black lines* $A \sim 2.57\,\text{Å}^2$, *red lines* $A \sim 2.35\,\text{Å}^2$, *green lines* $A \sim 2.16\,\text{Å}^2$, *blue lines* $A \sim 1.95\,\text{Å}^2$) and at three relaxation rates (*left panel* fastest relaxation [steepest descent], *central panel* intermediate relaxation rate, *right panel* slowest relaxation rate). **c** Phase diagram for two-dimensional carbon highlighting the regions of pT space in which the amorphous and crystalline structures are generated by quenching from the liquid

$$S_{66}(k) = \left\langle \frac{1}{N_6} \sum_{i,j=1}^{N_6} \exp(i\mathbf{k}.\mathbf{R}_{ij}) \right\rangle,$$

$$M_{66}^{(2)}(k) = \frac{1}{N_k} \sum_{p}^{N_k} \frac{|A(\mathbf{k}_p, t)|^2 - S_{66}(k)^2}{S_{66}(k)^2}, \tag{9.26}$$

where $\mathbf{R}_i$ is the position of the centre of mass of a given six-membered ring and where the latter sum runs over the N_k vectos for $|\mathbf{k}_p| = k$ and

$$A(\mathbf{k}_i, t) = \frac{1}{N} \sum_{i,j=1}^{N_6} \exp(i\mathbf{k}.\mathbf{R}_{ij}). \tag{9.27}$$

The rise in $S_{66}(k)$ as $k \longrightarrow 0$ simply reflects the fact that this is a partial structure factor and hence corresponds to an effective "phase separation" in the six-membered ring sub-density only (not in the system as a whole—the sum of the ring-ring partial structure factors would still tend to zero as $k \longrightarrow 0$. The figure shows $S_{66}(k)$ and $M_{66}^{(2)}(k)$ as a function of both stress (area) and quench rate. At the fastest quench rates (corresponding to the application of steepest descent energy minimisation) wholly disorder structures are generated. As the quench rate is reduced more crystalline configurations are generated, in particular at high stresses (low area). The second moment appears a more sensitive measure as to the presence of anisotropic structure than $S_{66}(k)$ alone. Analysis of the crystallinity of the quenched configurations indicates that it is effectively impossible to generate amorphous configurations at all bu the most rapid quench rates (steepest descent energy minimisation) for the high stress environments. Figure 9.10c summarises the regions of the phase diagram in which crystalline and amorphous structures can be generated.

Figure 9.11 shows typical molecular graphics "snapshots" taken under different conditions. Figure 9.11a, b are from the crystallization region indicated in Fig. 9.10c taken at a relatively low areas of $A \sim 1.58\,\text{Å}^2$ and $A \sim 1.95\,\text{Å}^2$. In panel (a) the enormous stress results in structures comprised exclusively from close-packed (six-coordinate, three-membered rings) units. Although distinct nanocrystalline domains are clearly evident the relatively rapid quench rate precludes the formation of a single crystal resulting in a metastable state with a number of grain boundaries. In panel (b) regions of crystalline hexagonal and trigonal net are clear. The formation of a mixture of three- and six-coordinate sites represents the most favourable energetic balance even allowing for the required formation of grain boundaries separating the two environments. A simple calculation using the mean energy of the mixture and those of the close-packed and hexagonal lattices (Fig. 9.8) indicates grain boundary energies of the order of $10^{-8}\,\text{J}\,\text{m}^{-1}$ (corresponding to energies of the order of $10^{-18}\,\text{J}\,\text{atom}^{-1}$ in the boundary). Panel (c) shows the effect of slow cooling at intermediate stress (corresponding to $A \sim 2.15\,\text{Å}^2$ and corresponding to the presence of significant peaks in $M_{66}(k)$ as shown in Fig. 9.10b). Regions of nanocrystalline graphene are

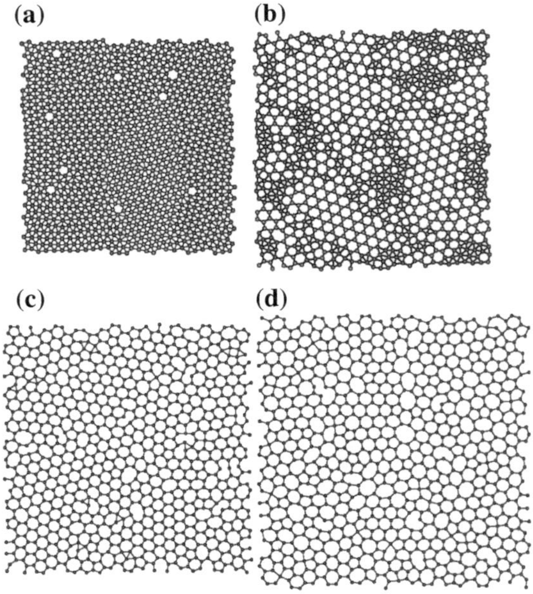

Fig. 9.11 Molecular graphics "snapshots" of configurations generated from the liquid state at four different areas (densities). **a** $A \sim 1.58\,\text{Å}^2$ **b** $A \sim 1.95\,\text{Å}^2$ **c** $A \sim 2.16\,\text{Å}^2$ **d** $A \sim 2.57\,\text{Å}^2$. In panel (**a**) a close-packed structure is formed which consists of well-defined nanocrystallites (of which one is *highlighted*). In panel (**b**) a mixture of nanocrystallites form from close-packed (six-coordinate) and hexagonal (three-coordinate) environments. The two crystallite types are shown in different *colours* to highlight the respective domains. Panels (**c**) and (**d**) are dominated by three-coordinate sites with the former dominated by nanocrystallites and the latter much more amorphous

evident. Panel (d) shows the effect of slow cooling at low stress (corresponding to $A \sim 2.57\,\text{Å}^2$) where a clear disordered, amorphous structure is generated.

Figure 9.8 suggests an alternative pathway to the formation of the a-G structures other than by the traditional quenching route [101]. The figure shows the effect of performing a steepest decent energy minimisation on the octahedron/square structure (which allows for relaxation of the two distinct C–C bond environments). About the energy minimum and at higher areas there is a small relaxation compared to the ideal structures corresponding to two "long" bonds associated with the four-membered rings and one "short" bond which joins these rings (and simultaneously forms the eight-membered rings). At the energy minimum in Fig. 9.8 the difference in bond length is of the order of 7.5 %. At low areas these differences become more dramatic (of the order of 14 % at the lowest stable area) which results in a larger relaxation energy at low area. However, we note that the energies associated with the amorphous structures (which are dominated by 5-, 6- and 7-membered rings) indicate that the oct/sq structures are metastable with respect to these amorphous structures. Indeed, at low areas the oct/sq structure spontaneously amorphises (effectively a stress-induced amorphisation).

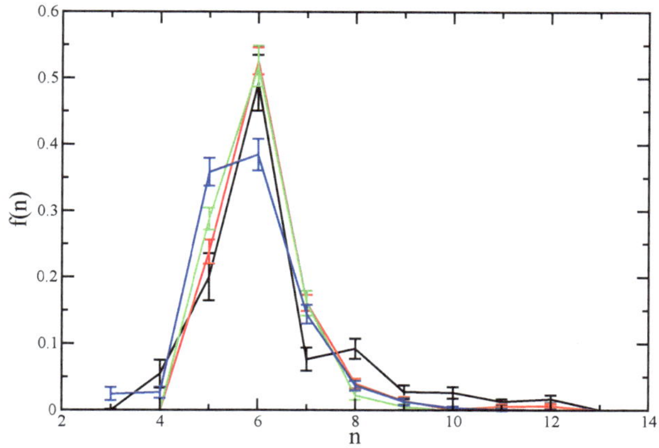

Fig. 9.12 Ring size distributions for $A \sim 2.57\,\text{Å}^2$ obtained from both liquid quenches (*black* and *red lines*) and amorphisation of the octsq crystal (*green* and *blue lines*). The *black* and *blue lines* are obtained using rapid (steepest descent) quenches whilst the *red* and *green lines* are obtained by slower cooling

Figure 9.12 shows the ring distributions obtained from the amorphisation of the oct/sq crystals at $A \sim 2.57\,\text{Å}^2$ compared with those obtained from direct liquid quenches. In each case the temperature is controlled in two ways; from a steepest descent energy minimisation and using thermostats with a relaxation parameter of ~ 0.25 ps (corresponding to an approximate cooling rate of $\sim 10^{14}\,\text{K}\,\text{s}^{-1}$). The ring distributions of the configurations generated from the liquid appear heavily dependent upon the quench rate. A steepest descent cooling, for example, fails to quench out all of the three-membered rings (close-packed units). The configurations generated from the octsq crystal appear less dependent upon quench rate, although the most rapidly cooled configurations show memory effects in the sense that the cooling fails to quench out the four-membered rings and shows an excess of eight-membered rings (both present in the original crystal structures). Critically, relatively slow cooling from either starting point generates broadly similar ring distributions.

To further highlight the differences in the crystalline and liquid states Table 9.1 lists the entropies of melting calculated in both two- and three-dimensions from the present work (taken at the respective coexistence curve maxima) and from previous work focussed on relatively simple models; hard and soft spheres, a Lennard-Jones potential, and the one-component plasma [102]. The absolute values reported here are larger (and consistent with known values for the conformal systems Si and Ge [103]). However, the ratio of the three- and two-dimensional system values appears consistent throughout, reflecting the more ordered nature of the liquid state when confined to two dimensions.

Table 9.1 Entropies of melting for common models in both three- and two-dimensions compared with the current work for carbon and results for Si and Ge [98, 103]

	ΔS (3d)	ΔS(2d)	ΔS(3d)/ΔS(2d)	References
	($\mathrm{J\,K^{-1}\,mol^{-1}}$)			
Hard sphere	9.6	3.4	2.8	[104]
Soft sphere	6.7	2.1	3.2	[102, 105]
Lennard-Jones	14.5	4.2	3.5	[106]
One component plasma	6.8	1.7	4.0	[102, 106, 107]
C	29.4	7.5	4.0	Current work and [98]
Si	27.1	–	–	[103]
Ge	24.8	–	–	[103]

Column 4 shows the ratios of the respective three- and two-dimensional transition entropy changes

9.3.3.3 Mapping onto Simpler Models

One of the simplest models which displays a reentrant crystal/fluid coexistence curve (in both two- and three-dimensions [93, 94]) is a gaussian core model (GCM) [108–111] in which the interaction energy between a pair of particles i and j separated by a distance r_{ij} is expressed as

$$U_{ij}(r_{ij}) = \varepsilon \exp\left(\frac{-r_{ij}^2}{d^2}\right), \tag{9.28}$$

where ε controls the energy scale and d the atom diameter. Solid/fluid coexistence curves have been obtained in both two- [94] and three- [93, 108, 112] dimensions, with respect to a two-dimensional close-packed lattice and bcc crystals respectively. To make contact with the present work the coexistence curves presented in Fig. 9.3 show the superimposed GC results. In each case the parameter set $\{\varepsilon, d\}$ from 9.28 may be related to the coexistence curves presented here by selecting a given point $\{\sigma, T_m\}$ or $\{p, T_m\}$ from the coexistence curve. Here we choose the pressure and temperature at which $dT_m/dp(\sigma) = 0$, $\{p_{\max}, T_{\max}\} = \{890\,\mathrm{GPa}, 15{,}550\,\mathrm{K}\}$, $\{\sigma_{\max}, T_{\max}\} = \{0.70\,\mathrm{J\,m^{-2}}, 20{,}400\,\mathrm{K}\}$, respectively. For the 3d case the required reduced pressure is, $p^* = pd^3/\varepsilon$ and for the 2d case, $\sigma^* = \sigma d^2/\varepsilon$. In both cases the reduced temperature is $T^* = k_B T/\varepsilon$. As a result for the 3d case $d^3 = \frac{k_B T}{T^*}\frac{p^*}{p}$ whilst for the 2d case $d^2 = \frac{k_B T}{T^*}\frac{\sigma^*}{\sigma}$. For the three-dimensional GCM $\{\rho_{\max}, T_{\max}\} = \{0.25, 0.0087\}$ [93] whilst for the two-dimensional model $\{\sigma_{\max}, T_{\max}\} = \{0.15, 0.011\}$ [94] leading to scaling parameters of $\{\varepsilon, d\} = \{1.79 \times 10^6\,\mathrm{K}, 0.99\,\text{Å}\}$ and $\{\varepsilon, d\} = \{1.85 \times 10^6\,\mathrm{K}, 23.0\,\text{Å}\}$ respectively. The two dimensional GCM coexistence curve maps well onto that determined for carbon, whilst the mapping the the three dimensional curve appears less impressive. The successful mapping in two dimensions can be rationalised by considering the dual relationship between the hexagonal and close-packed lattices [113]. To highlight this

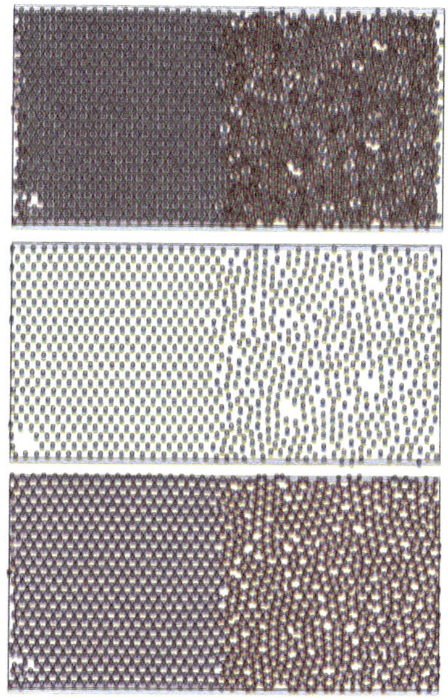

Fig. 9.13 Molecular graphics "snapshots" of a typical stable liquid/solid interface at $A \sim 2.57\,\text{Å}^2$ and the relationship between the atom positions and the ring structure. The *magenta* and *red circles* show the atoms in crystalline and liquid local environments respectively. The *yellow*, *green* and *blue circles* show the centres of mass of the 5-, 6- and 7-membered rings respectively. The *upper panel* shows both atoms and ring positions whilst the *central* and *lower panels* show the locations of the rings and atoms respectively

Fig. 9.13 shows an example stable liquid/crystal interface at $A \sim 2.57\,\text{Å}^2\,\text{atom}^{-1}$ showing both the atom positions and the locations of the centres of mass of the five-, six- and seven-membered rings. The coordinates of the ring centres of mass for the crystalline half of the cell form a close-packed lattice whilst the liquid half forms a disordered pseudo-close-packed array. As a result the system can be considered in terms of effective ring-ring interactions and hence maps effectively onto a GCM (or vice versa). In three dimensions any analogous relationship is less clear and, furthermore, the body-centred (rather than face-centred) crystal structure is thermodynamically stable over a wide density range (and is part of the reentrant liquid/crystal coexistence relation).

9.3.3.4 The Aboav-Weaire Law

The above analysis indicates that the a-G structures are dominated by a mixture of five-, six- and seven-membered rings (Fig. 9.12). Further insight into the topological arrangement of these geometric units can be gained by considering the Aboav-Weaire Law [114, 115] (see also [116] for a review). The Aboav-Weaire Law arises from empirical observations of Lewis [117, 118] and Aboav [114] on the diverse systems of epithelia and MgO crystallite grains. Aboav [114] originally noted that, empirically, that the mean neighbouring ring size, $\langle m_n \rangle$, adjacent to a ring of size n was given by

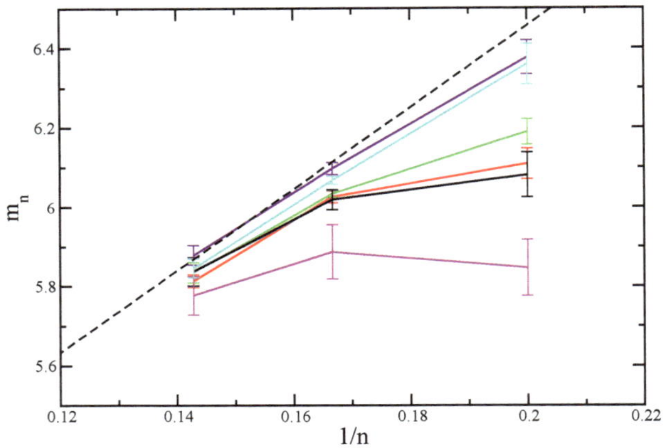

Fig. 9.14 Mean neighbouring ring size, $\langle m_n \rangle$, versus n^{-1} (where n is the central ring size) for configurations generated under a number of conditions (all for $A \sim 2.57\,\text{Å}^2$). The *magenta line* corresponds to the most rapid quench (a steepest descent energy minimisation) whilst the *black*, *red* and *green lines* correspond to successively slower cooling rates. The *cyan* and *violet lines* show the effect of removing the "defect" four- and two-coordinate sites as described in the text. The *dashed line* shows the prediction of the Aboav-Weaire Law

$$\langle m_n \rangle = 5 + \frac{6}{n}. \tag{9.29}$$

Several expanded forms, and subsequent modifications, are present in the literature [119] (also, see [116, 120, 121] for reviews).

Figure 9.14 shows the mean neighbouring ring size $\langle m_n \rangle$ against n^{-1} for $A = 2.57\,\text{Å}^2$ (well into the regime dominated by three-coordinate carbon atoms). The most statistically-significant region is between $n^{-1} = 0.20 - 0.13$ ($n = 5$–7) although the presence of (small numbers of) other ring sizes extends some of the functions beyond this range. The figure highlights the effect of cooling rate on the linearity of dependence of $\langle m_n \rangle$ on n^{-1}. As the cooling rate increases the deviation from linearity also increases. A possible explanation of the deviation from linearity lies in the potential for the networks formed here to contain non-three-coordinate "defect" atom coordinations. It is clear from the above analysis, for example, that simple topological (bond angle) arguments support the concept that large rings will tend to surround smaller rings (and vice versa). However, the presence of, for example, four-coordinate sites will significantly affect this analysis. For example, the presence of a four-coordinate site effectively divides the neighbouring space into four, rather than three, sectors, effectively lowering the bond angles of neighbouring polyhedra (and hence favouring the formation of smaller rings with respect to a purely three-coordinate network).

All basic MD annealing strategies systematically remove kinetic energy. For a-G they generate configurations dominated by five-, six- and seven-membered rings and in which all three- and four-membered rings are quenched out. A small number of four- and two-coordinate coordination environments persist as, unlike the three- and four-membered rings, there is lack of suitable low energy pathways to aid their removal during annealing. A coordination number of four is possible for foam structures whereas a coordination number of two is clearly unstable. In order to establish the topological significance of these "frozen-in" coordination environments they can be removed using the established T1 and T2 mechanisms respectively [121]. In the T1 mechanism the four-coordinate sites are replaced by a pair of three coordinate sites (adding an atom to the simulation cell) while in the T2 process the two-coordinate site is removed. Both procedures will result in the formation of relatively high energy long and short C–C bonds and so the resulting configurations are re-annealed. However, the relaxation procedure itself is still unconstrained with respect to the local coordination environments and so either four- or two-coordinate may re-form. As a result, these procedures do not totally eliminate these local environments but greatly reduce the fraction of such sites to an effective "equilibrium" with percentages of two- and four-coordinate sites of 0.6 and 0.3 % respectively. Figure 9.14 demonstrates the significance of the four-coordinate sites. As the cooling rate decreases fewer such sites are quenching in and the small ring mean neighbouring ring size tends towards the Aboav-Weaire (defect free) value. When the defect sites are (near) removed the linear dependence is recovered.

9.3.4 SiO_2 Bilayers

Recent experimental electron microscopy studies have highlighted the growth of two dimensional bilayers of vitreous silica [20, 23]. The ability to obtain detailed atomistic information allows the ring structure to be resolved in real space for the first time (certainly for this class of material). The thin vitreous SiO_2 films were grown on Mo(112) [19], Ru(0001) [20–22] and graphene [23].

9.3.4.1 Construction Method

The initial SiO_2 bilayer configurations are generated from a-G configurations generated using the "bond-switching" Monte Carlo algorithm [35, 122]. The a-G configurations were constructed with different network sizes and ring statistics. Here we highlight results on a configuration initially containing 200 carbon atoms. Each Si-centred layer is generated from an a-G configuration and joined with oxide anion bridges. Each carbon atom is transformed into a silicon atom which will become the centre of each SiO_4 tetrahedron. Oxygen atoms are placed at the centre of each C–C bond generating a single layer of stoichiometry Si_2O_3 confined to a plane. The single layer is equivalent to a two dimensional network of corner sharing equilateral triangles with oxygen atoms at the vertices and a silicon atom at the centre and is also equivalent to Zachariasen's original network sketch [1]. The Si atoms are then raised out of the confining plane to form a partial tetrahedral unit. The second layer of the

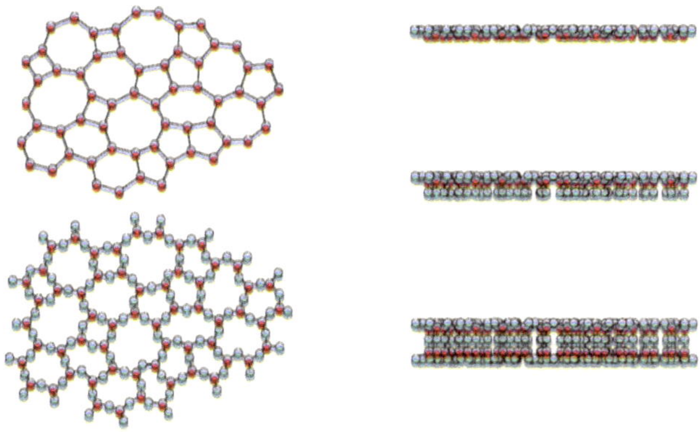

Fig. 9.15 Schematic to show how the SiO_2 bilayer structures are constructed from the a-G configurations. The figure shows the process using the original network structure proposed by Zachariasen as an example (*bottom left*) [1]. The *top left panel* shows the (reverse engineered) a-G configuration. The *top*, *centre* and *bottom right panels* show the monolayer with the Si atoms pulled out of the original plane, the addition of the central O atom layer and the addition of the second, mirror image, monolayer respectively

bilayer is created by producing a mirror image of the first offsetting the layer along the direction perpendicular to the initial plane of confinement to lie above the first bilayer. O atoms are then inserted along the plane central to the two layers to complete the two sets of tetrahedra giving the required SiO_2 bilayer stoichiometry. For the configuration initially containing 200 C atoms the final bilayer contains 400 Si and 800 O atoms respectively. The system super-cell lengths are then re-scaled so as to generate the required Si–O bond lengths. Figure 9.15 shows this procedure graphically using the original Zachariasen configuration as an exemplar [1]. In this case the "original a-G configuration" is obtained by reverse engineering from Zachariasen's original figure.

9.3.4.2 Models

Two forms of potential model are considered. A harmonic potential produces a corner sharing network of identical regular tetrahedra with individual tetrahedra allowed to move and tilt while maintaining the imposed topology but does not impose the reflection symmetry. Harmonic springs join the four nearest-neighbour Si–O and six nearest-neighbour O–O atoms in each individual tetrahedron with the ratios of the O–O and Si–O equilibrium bond lengths chosen so as to produce ideal tetrahedra in isolation. The spring force constants are taken to be equal for both the Si–O and O–O pairs within each tetrahedron. The detail of this interaction is only significant in the sense of allowing for a relatively rapid energy minimization. The harmonic potentials do not preclude different tetrahedra from overlapping as is the case in reality and which limits the motions, for example, in zeolites [123]. In order to prevent this the harmonic potential can be augmented with a repulsive potential which acts between pairs of silicon atoms (effectively acting as an inter-tetrahedron repulsive term). The chosen form is a shifted 24–12 potential,

$$U(r) = 4\varepsilon \left\{ \left(\frac{\sigma}{r}\right)^{24} - \left(\frac{\sigma}{r}\right)^{12} \right\} + \varepsilon, \qquad (9.30)$$

where σ is the atom diameter and ε is the well-depth of the (unshifted) potential. The potential is cut off at the minimum [$r_{\min} = (2)^{1/12}\sigma$] ensuring continuity in both energy and force. The parameter ε is fixed while σ can be varied.

The second model used is the TS potential [45] (Sect. 9.3.1), already shown to give an excellent account of the pressure-driven topology changes in SiO_2 glass.

9.3.4.3 Results

Figure 9.16 shows the energies of the 400 SiO_2 molecule configuration as a function of the number density, n_0 (the number of SiO_2 molecules per unit area). For the purely harmonic potential the energy can be driven to zero above a "critical"

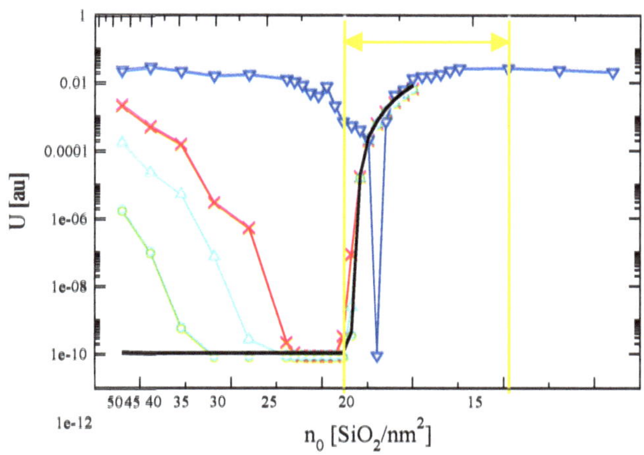

Fig. 9.16 Energy as a function of number density (number of SiO_2 molecules per unit area) for a single (400 molecule) SiO_2 bilayer generated from a 200 atom a-G configuration. The *black line* shows the energy on relaxing with a purely harmonic potential and can be seen to go to zero for $n_0 \gtrsim 20 SiO_2\,nm^{-2}$. Inclusion of a repulsive potential acting between tetrahedron centres introduces an upper density limit above which the energy can no longer be driven to zero and hence introduces as effective density "window". The width of the window is controlled by the range of the inter-tetrahedron interaction (the range increasing in the order *green*, *light blue* and *red lines*). The *blue line* shows the energies obtained using a TS potential with the energy at the minimum subtracted to define the energy zero. The shape of this curve reflects the use of the log energy scale. The *yellow arrows* indicate the range of experimentally-determined densities

density ($n_0 \sim 20 SiO_2\,nm^{-2}$). The numerical value of this critical density varies with the configuration and is hence a function of the detailed morphology. For an ideal crystalline bilayer (constructed as for the amorphous configurations but starting from an ideal graphene sheet) the critical density is $n_0 \sim 17 SiO_2\,nm^{-2}$. The higher critical density for the disordered network reflects the presence of both four- and five-membered rings.

The inclusion on a repulsive potential between Si atoms (effectively acting as a inter-tetrahedral repulsion) introduces an upper "critical" density, above which the energy can no longer be drive to zero. In the absence of any repulsive interactions, once the density is greater than the critical density, then the tetrahedra can always be arranged in such a manner as to drive the total energy to zero (i.e. preserve the ideal local tetrahedral geometry). However, there is nothing to prevent neighbouring tetrahedra from overlapping to yield unphysical configurations, in particular at high density. The inclusion of the repulsive interaction precludes the unphysical overlap of nearest-neighbour tetrahedra. The use of a purely sort-range potential (9.30) defines an effective "density window" over which the system energy may still be driven to zero. The window is the analogue of the "flexibility window" observed in zeolites [123]. The "width" (density range) of the window naturally depends on the range of the short-range repulsion (as governed by the parameter σ). The smaller this term then the larger the distribution of nearest-neighbour tetrahedra configurations which are permitted whilst still driving the energy to zero. The use of a more realistic model (here the TS potential) effectively selects a single density from the flexibility window.

9.3.5 Amorphous Carbon Nanotubes

The ability to generate two-dimensional amorphous carbon structures leads naturally to considering the potential for rolling these sheets to form amorphous carbon nanotubes (a-CNTs). The generation of such structures has a limited literature footprint and generally refers to tubular structures with relatively thick walls (or the order of 10 nm) themselves consisting of three-dimensional amorphous carbon [124–127]. Here we take a-CNT to mean a tubular structure constructed from a single a-G sheet (and hence one atom thick prior to any structural relaxation). Previous work shows how nanotubular structures may be ubiquitous if clear formation pathways can be identified [128]. In the present case it is possible to envisage inducing disorder in pristine crystalline carbon nanotubes in a manner analogous to that employed to induce defects in the ideal graphene sheets.

Initial a-CNT structures are generated by projecting a relaxed planar a-G structure onto a cylinder, the resulting diameter being a simple function of the original sheet dimensions. The cylindrical structures are then allowed to relax using a steepest descent energy minimisation. The relaxation procedure may destroy the original cylindrical symmetry as the carbon atoms are free to move inwards or outwards compared to the original cylinder. As a result the final a-CNT energetics can be considered either in terms of the original cylinder radius, R, or an effective radius,

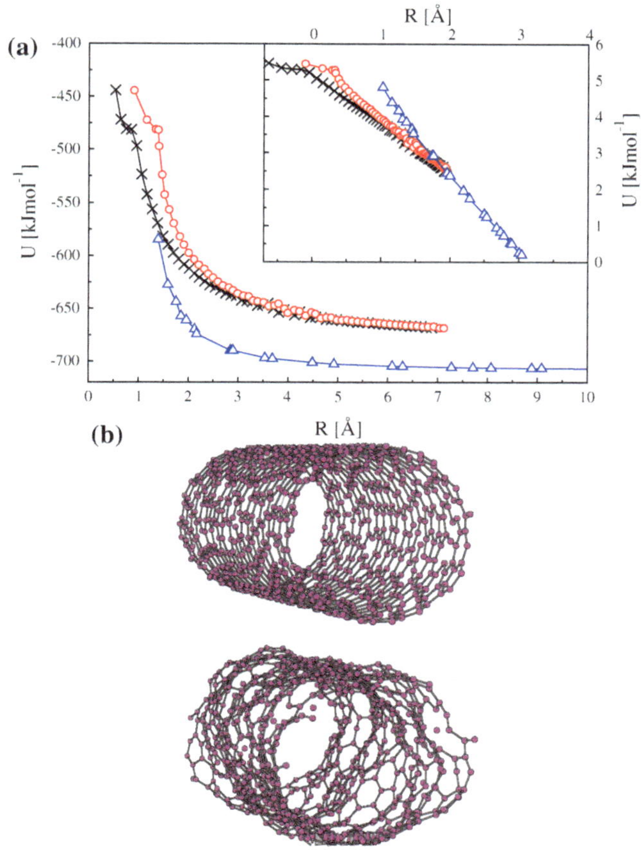

Fig. 9.17 **a** Energies as a function of radius for amorphous carbon nanotube (a-CNT) structures (*black crosses* and *red circles*) compared with those for ideal crystalline nanotubes (*blue triangles* from [66]). The *black crosses* show the a-CNT data as a function of the initial radius of the unrelaxed a-CNT, whilst the *red circles* show the same data as a function of the effective radius, calculated as a mean average radius. The *inset* shows the same data plotted on a log-log scale in order to highlight the different dependencies of the respective energetics on R. **b** Example morphologies in the unrelaxed (*upper panel*) and relaxed (*lower panel*) states for an a-CNT of $R \sim 6.7$ Å ($R_{eff} \sim 6.9$ Å)

R_{eff}, which is a simple mean average of the relaxed carbon atom positions with respect to the centre of mass vector running along the major axis.

Figure 9.17a shows the dependence of the a-CNT energy on both the initial and effective radii and shows the energy of the crystalline CNTs for comparison [66]. The inset to the figure shows the log-log plots in order to highlight the differing dependencies of the CNT energies on radius. The crystalline CNTs show an energetic dependence on radius consistent with a continuum elastic model in which $U \propto R^{-2}$ (they behave in an equivalent fashion to folding a sheet of paper into a cylinder) and show no significant dependence on morphology [66]. In contrast the a-CNTs show a different dependence on the radius irrespective of the choice of that radius ($n \simeq 1.36$ and $\simeq 1.52$ for R and R_{eff} respectively).

The different relaxation behaviour of the a-CNTs compared to the crystalline CNTs of similar radius can be attributed to the larger number of degrees of freedom available for relaxation in the former. For the a-CNTs the presence of non-six-membered rings promotes relaxations out of the original cylindrical geometry (as they promote different local curvatures as, for example, in buckyballs [129]). The ideal crystalline CNTs are constructed soley from six-membered rings and so retain the original cylindrical symmetry. To highlight the extend of this relaxation Fig. 9.17b shows the initial and final structures for an a-CNT of $R \sim 6.7$ Å ($R_{eff} \sim 6.9$ Å). The change in energetic dependence on R at low R reflects in which the relaxation has closed the internal pore along the major cylinder axis (effectively forming amorphous carbon nanowires).

9.4 Summary and Conclusions

The use of potential models, in which the atom-atom interactions are represented in a computationally-tractable fashion, remain significant in the investigation of network-forming materials. Advances in experimental diffraction techniques allow a greater range of phase space to be explored than previously whilst advances in imaging and synthesis allows two- (or near-two-) dimensional systems to be grown and imaged in a controlled manner. Whilst these advances give potentially fantastic insight (for example in terms of observed the mechanism of network collapse or the ordering resulting from specific ring structures) additional modelling is required in order to fully exploit the observations. The simulations described in this chapter highlight a range of different environments for two key network systems; carbon and silica.

Acknowledgments It is a pleasure to acknowledge all those who have helped and supported the development of the simulation strategies in both Oxford and at UCL; Dominik Daisenberger, Avishek Kumar, Jessica Long, Paul McMillan, David Robinson, Philip Salmon, Franziska Schäffel, David Sherrington, Mike Thorpe, Jamie Warner, Anita Zeidler.

References

1. W.H. Zachariasen, J. Am. Chem. Soc. **54**, 3841 (1932)
2. A.C. Wright, G.A.N. Conneli, I.W. Allen, J. Non-Cryst. Solids **42**, 69 (1980)
3. A. Uhlherr, S.R. Elliott, J. Phys. Condens. Matter **6**, L99 (1994)
4. M. Wilson, PCCP **14**, 12701 (2012)
5. M.F. Thorpe, A. Wright, Eighty years of random networks. Phys. Status Solidi B 1–6 (2013)
6. P.A. Madden, M. Wilson, Chem. Soc. Rev. **25**, 339 (1996)
7. M. Wilson, P.S. Salmon, Phys. Rev. Lett. **103**, 157801 (2009)
8. M. Wilson, J. Phys. Condens. Matter **24**, 284114 (2012)

9. C.S. Marians, L.W. Hobbs, J. Non-Cryst. Solids **124**, 242 (1990)
10. A. Pasquarello, R. Car, Phys. Rev. Lett. **80**, 5145 (1998)
11. A.B. Cairns, A.L. Goodwin, Chem. Soc. Rev. **42**, 4881 (2013)
12. F. Banhart, J. Kotakoski, A.V. Krasheninnikov, Structural defects in graphene. ACS Nano **5**, 26–41 (2011)
13. J. Kotakoski, J.C. Meyer, S. Kurasch, D. Santos-Cottin, U. Kaiser, A.V. Krasheninnikov, Stone-wales-type transformations in carbon nanostructures driven by electron irradiation. Phys. Rev. B **83**, 245420 (2011)
14. J. Kotakoski, A.V. Krasheninnikov, U. Kaiser, J.C. Meyer, From point defects in graphene to two-dimensional amorphous carbon. Phys. Rev. Lett. **106**, 105505 (2011)
15. J.C. Meyer, C. Kisielowski, R. Erni, M.D. Rossell, M.F. Crommie, A. Zettl, Nano Lett. **8**, 3582 (2008)
16. J. Zhao, G. Zhu, W. Huang, Z. He, X. Feng, Y. Ma, X. Dong, Q. Fan, L. Wang, Z. Hu, Y. Lü, W. Huang, J. Mater. Chem. **22**, 19679 (2012)
17. A.W. Robertson, C.S. Allen, Y. Wu, K. He, J. Oliviera, J. Neethling, A.I. Kirkland, J.H. Warner, Nat. Commun. **3**, 1144 (2012)
18. J.H. Warner, E.R. Margine, M. Mukai, A.W. Robertson, F. Giustino, A.I. Kirkland, Science **337**, 209 (1985)
19. J. Weissenrieder, S. Kaya, J.-L. Lu, H.-J. Gao, S. Shaikhutdinov, H.-J. Freund, M. Sierka, T.K. Todorova, J. Sauer, Phys. Rev. Lett. **95**, 076103 (2005)
20. L. Lichtenstein, C. Buechner, B. Yang, S. Shaikhutdinov, M. Heyde, M. Sierka, R. Wlodarczyk, J. Sauer, H.-J. Freund, Angew. Chem. Int. Ed. **51**, 404 (2012)
21. M. Heyde, S. Shaikhutdinov, J.-J. Freund, Two-dimensional silica: crystalline and vitreous. Chem. Phys. Lett. **550**, 1 (2012)
22. D. Loeffler, J.J. Uhliruch, M. Baron, B. Yang, X. Yu, L. Lichtenstein, L. Heinke, C. Buechner, M. Heyde, S. Shaikhutdinov, H.-J. Freund, R. Wlodarczyk, M. Sierka, J. Sauer, Phys. Rev. Lett. **105**, 146104 (2010)
23. P.Y. Huang, S. Kurasch, A. Srivastava, V. Skakalova, J. Kotakoski, A.V. Krasheninnikov, R. Hovden, Q. Mao, J.C. Meyer, J. Smet, D.A. Muller, U. Kaiser, Nano Lett. **12**, 1081 (2012)
24. P. Jemmer, P.W. Fowler, M. Wilson, P.A. Madden, J. Chem. Phys. **111**, 2038 (1999)
25. C. Domene, P.W. Fowler, P.A. Madden, M. Wilson, R.J. Wheatley, Chem. Phys. Lett. **333**, 403 (2001)
26. C. Domene, P.W. Fowler, P.A. Madden, J. Xu, R.J. Wheatley, M. Wilson, J. Phys. Chem. A **105**, 4136 (2001)
27. D. Frenkel, B. Smit, *Understanding Molecular Simulation: From Algorithms to Applications*, 2nd edn. (Academic Press, Boston, 2002)
28. M. Allen, D. Tildesley, *Computer Simulation of Liquids* (Oxford Press, Oxford, 1987)
29. S. Nosé, J. Chem. Phys. **81**, 511 (1984)
30. W.G. Hoover, Phys. Rev. A **31**, 1695 (1985)
31. G.J. Martyna, D.J. Tobias, M.L. Klein, J. Chem. Phys. **101**, 4177 (1994)
32. W.H. Press, B.P. Flannery, S.A. Teukolski, W.T. Vetterling, *Numerical Recipies* (Cambridge University Press, Cambridge, 1992)
33. D. Machon, F. Meersman, M.C. Wilding, M. Wilson, P.F. McMillan, Prog. Mater. Sci. **61**, 216 (2014)
34. S.M. Sharma, S.K. Sikka, Prog. Mater. Sci. **40**, 1 (1996)
35. A. Kumar, M. Wilson, M.F. Thorpe. Amorphous graphene: a realization of Zachariasen's glass. J. Phys. Condens. Matter **24**, 485003 (2012)
36. D.P. Landau, K. Binder, *A Guide to Monte Carlo Simulations in Statistical Physics* (Cambridge University Press, Cambridge, 2000)
37. M.J.L. Sangster, M. Dixon, Adv. Phys. **25**, 247 (1976)
38. A.J. Stone, *Theory of Intermolecular Forces* (Oxford University Press, Oxford, 1996)
39. M. Born, J.E. Mayer, Z. Phys. **75**, 1 (1932)
40. L.V. Woodcock, C.A. Angell, P. Cheeseman, J. Chem. Phys. **65**, 1565 (1976)
41. M. Wilson, P.A. Madden, M. Hemmati, C.A. Angell, Phys. Rev. Lett. **77**, 4023 (1996)

42. M.J. Sanders, C.R.A. Catlow, J.V. Smith, J. Phys. Chem. **88**, 2796 (1984)
43. C.R.A. Catlow, A.N. Cormack, Int. Rev. Phys. Chem. **6**, 227 (1987)
44. B.W.H. van Beest, G.J. Kramer, R.A. van Santen, Phys. Rev. Lett. **64**, 1955 (1990)
45. P. Tangney, S. Scandolo, J. Chem. Phys. **117**, 8898 (2002)
46. S. Tsuneyuki, M. Tsukada, H. Aoki, Phys. Rev. Lett. **61**, 869 (1988)
47. S. Tsuneyuki, Y. Matsui, H. Aoki, M. Tsukada, Nature **339**, 209 (1989)
48. E. Demiralp, T. Cagin, W.A. Goddard III, Phys. Rev. Lett. **82**, 1708 (1999)
49. D. Herzbach, K. Binder, M.H. Mueser, J. Chem. Phys. **123**, 124711 (2005)
50. B.G. Dick, A.W. Overhauser, Phys. Rev. **112**, 90 (1958)
51. A. Aguado, L. Bernasconi, S. Jahn, P.A. Madden, Faraday Discuss. **124**, 171 (2003)
52. B.K. Sharma, M. Wilson, Phys. Rev. B **73**, 060201 (2006)
53. B.K. Sharma, M. Wilson, J. Phys. Condens. Matter **20**, 244123 (2008)
54. D. Marrocchelli, M. Salanne, P.A. Madden, C. Simon, P. Turq, Mol. Phys. **107**, 443 (2009)
55. R.J. Heaton, R. Brookes, P.A. Madden, M. Salanne, C. Simon, P. Turq, J. Phys. Chem. B **110**, 11454 (2006)
56. M. Wilson, B.K. Sharma, J. Chem. Phys. **128**, 214507 (2008)
57. M. Wilson, J. Chem. Phys. **124**, 124706 (2006)
58. M. Wilson, J. Chem. Phys. **131**, 214507 (2009)
59. J.A.E. Desa, J. Non-Cryst. Solids **51**, 57 (1982)
60. M. Wilson, P.A. Madden, Phys. Rev. Lett. **72**, 3033 (1994)
61. S. Erkoc, Phys. Rep. **278**, 79 (1997)
62. F.H. Stillinger, T.A. Weber, Phys. Rev. B **31**, 5262 (1985)
63. W.D. Luedtke, U. Landman, Phys. Rev. B **37**, 4656 (1988)
64. W.D. Luedtke, U. Landman, Phys. Rev. B **40**, 1164 (1989)
65. J. Tersoff, Phys. Rev. B **37**, 6991 (1988)
66. C.L. Bishop, M. Wilson, J. Mater. Chem. **19**, 2929–2939 (2009)
67. A. Zeidler, K. Wezka, D.A.J. Whittaker, P.S. Salmon, S. Klotz, H.E. Fischer, M.C. Wilding, C.L. Bull, M.G. Tucker, M. Wilson, Phys. Rev. Lett. XXX (2014)
68. N.E. Cusack, *The Physics of Structurally Disordered Matter* (Adam Hilger, Bristol, 1987)
69. V.F. Sears, Neutron News **3**, 26 (1992)
70. L. Stixrude, M.S.T. Bukowinski, Am. Mineral. **75**, 1159 (1990)
71. L.P. Davila, M.-J. Caturla, A. Kubota, B. Sadigh, T.D. de la Rubia, J.F. Shackelford, S.H. Risbud, S.H. Garofalini, Phys. Rev. Lett. **91**, 205501 (2003)
72. L. Huang, J. Kieffer, Phys. Rev. B **69**, 224203 (2004)
73. L. Huang, J. Kieffer, Phys. Rev. B **69**, 224204 (2004)
74. L. Huang, L. Duffrene, J. Kieffer, J. Non-Cryst. Solids **349**, 1 (2004)
75. Y. Liang, C.R. Miranda, S. Scandolo, Phys. Rev. B **75**, 024205 (2007)
76. F.P. Bundy, The p, t phase and reaction diagram for elemental carbon, 1979. J. Geophys. Res. **85**, 6930 (1980)
77. B.J. Alder, R.H. Christian, Behaviour of strongly shocked carbon. Phys. Rev. Lett. **7**, 367 (1961)
78. M.D. Knudson, M.P. Desjarlais, D.H. Dolan, Shock-wave exploration of the high-pressure phases of carbon. Science **322**, 1822 (2008)
79. J.H. Eggert, D.G. Hicks, P.M. Celliers, D.K. Bradley, R.S. McWilliams, R. Jeanloz, J.E. Miller, T.R. Boehly, G.W. Collins, Melting temperature of diamond at ultrahigh pressure. Nat. Phys. **6**, 40 (2010)
80. D.K. Bradley, J.H. Eggert, R.F. Smith, S.T. Prisbrey, D.G. Hicks, D.G. Braun, J. Biener, A.V. Hamza, R.E. Rudd, G.W. Collins, Diamond at 800 GPa. Phys. Rev. Lett. **102**, 075503 (2009)
81. J. Edwards, K.T. Lorenz, B.A. Remington, S. Pollaine, J. Colvin, D. Braun, B.F. Lasinski, D. Reisman, J.M. McNaney, J.A. Greenough, R. Wallace, H. Louis, D. Kalantar, Laser-driven plasma loading for shockless compression and acceleration of samples in the solid state. Phys. Rev. Lett. **92**, 075002 (2004)
82. J.R. Rygg, J.H. Eggert, A.E. Lazicki, F. Coppari, J.A. Hawreliak, D.G. Hicks, R.F. Smith, C.M. Sorce, T.M. Uphaus, B. Yaakobi, G.W. Collins, Powder diffraction from solids in terapascal regime. Rev. Sci. Instrum. **83**, 113904 (2012)

83. T. Guillot, The interiors of giant planets: models and outstanding questions. Annu. Rev. Earth Planet. Sci. **33**, 493 (2005)
84. N. Madhusudhan, J. Harrington, K.B. Stevenson, S. Nymeyer, C.J. Campo, P.J. Wheatley, D. Deming, J. Blecic, R.A. Hardy, N.B. Lust, D.R. Anderson, A. Collier-Cameron, C.B.T. Britt, W.C. Bowman, L. Hebb, C. Hellier, P.F.L. Maxted, D. Pollacco, R.G. West, A high c/o ratio and weak thermal inversion in the atmosphere of exoplanet wasp-12b. Nature **469**, 64 (2011)
85. D.R. McKenzie, D. Muller, B.A. Pailthorpe, Compressive-stress-induced formation of thin-film tetrahedral amorphous-carbon. Phys. Rev. Lett. **67**, 773 (1991)
86. I.I. Aksenov, V.A. Belous, V.G. Padalka, V.M. Khoroshikh, Sov. J. Plasma Phys. **4**, 425 (1978)
87. P.H. Gaskell, A. Saeed, P. Chieux, D.R. McKenzie, Neutron-sscattering studies of the structure of highly tetrahedral amorphous diamond-like carbon. Phys. Rev. Lett. **67**, 1286 (1991)
88. R. Lossy, D.L. Pappas, R.A. Roy, J.J. Cuomo, V.M. Sura, Filtered arc deposition of amorphous diamond. Appl. Phys. Lett. **61**, 171 (1992)
89. J. Schwan, S. Ulrich, T. Theel, H. Roth, H. Ehrhardt, P. Becker, S.R.P. Silva, Stress-induced formation of high-density amorphous carbon thin films. J. Appl. Phys. **82**, 6024 (1997)
90. K. Ogata, Y. Andoh, E. Kamijo, Crystallization of carbon-films by ion-beam assist technology. Nucl. Instrum. Methods Phys. Res. **B33**, 685 (1988)
91. P.J. Fallon, V.S. Veerasamy, C.A. Davis, J. Robertson, G.A.J. Amaratunga, W.I. Milne, J. Koskinen, Properties of filtered-ion-beam-deposited diamond-like carbon as a function of ion energy. Phys. Rev. B **48**, 4777 (1993)
92. P.F. McMillan, G.N. Greaves, M. Wilson, M.C. Wilding, D. Daisenberger, Adv. Chem. Phys. **152**, 309 (2013)
93. S. Prestipino, F. Saija, P.V. Giaquinta, Phys. Rev. E **71**, 050102(R) (2005)
94. S. Prestipino, F. Saija, P.V. Giaquinta, Phys. Rev. Lett. **106**, 235701 (2011)
95. M. Wilson, P.F. McMillan, Phys. Rev. Lett. **90**, 135703 (2003)
96. E. Rapoport, J. Chem. Phys. **46**, 2891 (1967)
97. E.A. Guggenheim, *Modern Thermodynamics by the Methods of Willard Gibbs* (Methuen & Co. Ltd., London, 1933)
98. D.R. Robinson, M. Wilson, J. Phys. Condens. Matter **25**, 155101 (2013)
99. A.K. Geim, Graphene: status and prospects. Science **324**, 1530–1534 (2009)
100. D.R. Robinson, Part II Thesis, University of Oxford, 2012
101. J. Long, Part II Thesis, University of Oxford, 2013
102. F. Swol, L.V. Woodcock, J.N. Cape, J. Chem. Phys. **73**, 913 (1980)
103. A.R. Ubbelohde, *The Molten State of Matter* (Wiley, Chichester, 1978)
104. W.G. Hoover, F.H. Ree, J. Chem. Phys. **49**, 3610 (1968)
105. W.G. Hoover, M. Ross, K.W. Johnson, D. Henderson, A. Barker, E.C. Brown, J. Chem. Phys. **52**, 4931 (1970)
106. J.-P. Hansen, L. Verlet, Phys. Rev. **184**, 151 (1969)
107. R.W. Hockney, T.R. Brown, J. Phys. C **8**, 1813 (1975)
108. F.H. Stillinger, J. Chem. Phys. **65**, 3968 (1976)
109. F.H. Stillinger, T.A. Weber, J. Chem. Phys. **68**, 3837 (1978)
110. F.H. Stillinger, T.A. Weber, J. Chem. Phys. **70**, 4879 (1979)
111. F.H. Stillinger, Phys. Rev. B **20**, 299 (1979)
112. A. Lang, C.N. Likos, M. Watzlawek, H. Löwen, J. Phys. Condens. Matter **12**, 5087 (2000)
113. H.S.M. Coexter, *Regular Polytopes* (Dover, New York, 1973)
114. D.A. Aboav, Metallography **3**, 383 (1970)
115. D. Weaire, Metallography **7**, 157 (1974)
116. S.N. Chui, Mater. Charact. **34**, 149 (1995)
117. F.T. Lewis, Anat. Rec. **38**, 341 (1928)
118. F.T. Lewis, Anat. Rec. **50**, 235 (1931)
119. J.K. Mason, R. Ehrenborg, E.A. Lazar, J. Phys. A Math. Theor. **45**, 065001 (2012)
120. D. Weaire, S. Hutzler, *The Physics of Foams* (Oxford University Press, Oxford, 2001)
121. D.L. Weaire, N. Rivier, Soap, cells and statistics—random patterns in 2 dimensions. Contemp. Phys. **25**, 59 (1984)

122. T. Aste, D. Sherrington, Glass transition in self-organizing cellular patterns. J. Phys. A Math. Gen. **32**, 7049 (1999)
123. A. Sartbaeva, S.A. Wells, M.M.J. Treacy, M.F. Thorpe, The flexibility window in zeolites. Nat. Mater. **5**, 962 (2006)
124. H. Nishino, C. Yamaguchi, H. Nakaoka, R. Nishida, Carbon **41**, 2159 (2003)
125. H. Nishino, R. Nishida, T. Matsui, N. Kawase, I. Mochida, Carbon **41**, 2819 (2003)
126. L. Ci, B. Wei, C. Xu, J. Liang, D. Wu, S. Xie, W. Zhou, Y. Li, Z. Liu, D. Tang, J. Cryst. Growth **233**, 823 (2001)
127. A. Rakitin, C. Papadopoulos, J.M. Xu, Phys. Rev. B **61**, 5793 (2000)
128. M. Wilson, *Computational Nanoscience* (Royal Society of Chemistry, London, 2011)
129. P.W. Fowler, D.E. Manolopoulos, *An Atlas of Fullerenes* (Oxford University Press, Oxford, 1995)

Chapter 10
Rationalizing the Biodegradation of Glasses for Biomedical Applications Through Classical and Ab-initio Simulations

Antonio Tilocca

Abstract The gradual dissolution of a glass in a living host determines the rate at which processes leading to tissue regeneration can occur, which is of crucial importance for the success of biomedical implants and scaffolds for tissue engineering based on the glass. In-situ radiotherapy applications are also affected—in an opposite way—by the rate at which the glass vector used to deliver radioisotopes will degrade in the bloodstream. This chapter illustrates how a combination of classical and ab-initio simulations techniques, mainly centred on Molecular Dynamics, can shed new light into structural and dynamical features that control the biodegradation of these materials.

10.1 Introduction

A biomaterial is a material able to elicit a favourable reaction from the human body, leading for instance to the repair of damaged or diseased tissues, including but not limited to bones and soft tissues [1, 2]. The high costs and risks associated to auto- and allografts for bone replacement have led to large advances in the development and clinical application of synthetic substitutes. For instance, first-generation bioinert materials are metals, alloys and ceramics such as zirconia and alumina, whose application as bone replacement relies on a tight mechanical fit in the implant site. A superior performance can be achieved by second-generation bioactive materials, able to form chemical bonds with the existing issues, which results in better biocompatibility, integration and stability of the implant [1]. The first biomaterials with these desirable bone-bonding properties were the soda-lime phosphosilicate compositions (such as the 45S5 Bioglass®) introduced by Hench in the 70s [3]. Even though several other bioactive materials (such as crystalline calcium phosphates and silicates [4]) able to bond to tissues have been identified thereafter, their performances have

A.Tilocca (✉)
Department of Chemistry, University College London,
20 Gordon Street, London WC1H 0AJ, UK
e-mail: a.tilocca@ucl.ac.uk

C. Massobrio et al. (eds.), *Molecular Dynamics Simulations of Disordered Materials*, Springer Series in Materials Science 215,
DOI 10.1007/978-3-319-15675-0_10

not reached the level of bioactive glasses (BGs) under many key aspects, such as the rate of growth of new bone [5], or the remarkable ability—unique to specific BG compositions only—to form bonds with soft tissues as well [6]. This enhanced biological activity is likely related to the stimulation of osteogenic processes at the cellular level, triggered by critical concentration of soluble ionic species (such as silicates, phosphates and calcium) released around the implant site by the ongoing dissolution of the BG [7, 8]. This property enables growth of new tissues not only along the implant surface but also away from it: some BGs thus have the potential to trigger regeneration of new tissues. This recognition has greatly renewed the interest in using BGs, as third-generation biomaterials for in vitro tissue engineering [9]. In these applications, suitable cells are seeded and start proliferating within a 3D scaffold, which slowly degrades to non-toxic products, while simultaneously being replaced by natural connective tissues that serve to integrate the new tissue into the target site. The biological and physicochemical properties of the scaffold material are of central importance: besides being biocompatible and able to bond to living tissues, the material must also degrade at a suitable rate, matching the growth rate of new tissue, while at the same time releasing soluble ions and creating the favourable conditions that trigger tissue regeneration [9, 10].

Within the complex picture that describes how these materials work, the dissolution of a bioactive glass is then probably the most critical process, for three different reasons:

(i) A fast initial degradation immediately upon contact with the physiological environment is essential to enable the formation of a stronger bonding interface, as shown by the frequently observed correlation between glass bioactivity and short-term ion release [11];
(ii) The effect, discussed above, that ionic species dissolved from the glass have on cellular processes involved in tissue regeneration [12];
(iii) The long-term degradation of the glass affects its potential for third-generation applications, wherein it is expected that the glass remains initially relatively stable (which is why highly biodegradable materials such as phosphate glasses [13] are not suitable in this context) but gradually disappears after (or while) accomplishing its function.

Optimizing the properties of a biomaterial for tissue-repair and tissue-regeneration applications requires us to move beyond traditional expensive and ineffective trial-and-error approaches, towards more rational schemes that ideally should predict and take into account the expected behaviour of the glass under different conditions. Given the central role of the glass dissolution on its overall activity, a convenient and reliable strategy involves focusing on understanding that process at a fundamental level [14].

This is where computer simulations come into play. The time and space resolution of atomic-scale simulations can hardly be achieved through any other methods. The models are able to reveal elusive structural and dynamical features of a biomaterial that help to explain and predict its behaviour and to rationalize its development. The degradation behaviour of a bioactive glass in a physiological medium depends in first instance on bulk "static" structural features that describe the stable local environment of key soluble ions, or the degree of cross-linking within the silicate glass matrix [15]. At the same time, dissolution is a dynamical process that involves ions (both modifiers and network formers) migrating within the glass matrix, accessing the surface region and interacting with the surrounding aqueous environment; the corrosion of glasses in general, and of bioglasses in particular, results from surface reactions such as ion exchanges and Si–O bond breaking [12, 16–18].

If a more complete understanding of the bioactive behaviour is sought, the simulations must then target both directions: structural (local and medium-range) features and dynamical properties (ion migration and surface reactivity). In the following, we will discuss how classical and ab-initio simulations can be used and in some cases combined to pursue these goals.

10.2 AIMD Versus Classical MD

Molecular Dynamics (MD) simulations are a well-established tool to obtain atomic-scale models of melt-derived amorphous materials, through a straightforward melt-and-quench computational procedure that mimics the experimental one [19–23]. The main limitations that may limit the accuracy of any MD-derived model are: (i) the representation of interatomic forces and (ii) the small size and short time scales (compared to macroscopic processes and samples) of the models. Ab-initio MD simulations [24] are virtually unaffected by (i) but they are much more penalised by (ii) compared to classical MD, which reflects the opposite situation: the accuracy of classical MD simulations strongly relies on the quality of the interatomic potential, but their cheaper computational demands allow them to reach larger samples and longer time scales, which can be crucial to target some specific property covering those scales [25]. This suggests that a convenient strategy to target the behaviour of a biomaterial could in some cases involve combining classical and ab-initio approaches, exploiting the advantages of each one to overcome the corresponding limitations. Some specific examples of this strategy within the realm of bioactive glasses will be also illustrated in the following.

Ab-initio MD (AIMD) can usually target systems of few hundreds atoms and time scales of few tens of picoseconds [26–31]. In the case of glasses, this means that it is possible to perform an AIMD melt-and-quench simulation of a system around 100–200 atoms, using quenching rates around 20–100 K/ps [31–34]. The small system size in AIMD prevents to extract a full description of medium-range order features such as the network connectivity (*NC*) of a glass. However, the AIMD models provide the most accurate description of local structure, such as the coordination of key ions, unbiased from any potential bias introduced by an empirical force field. This could be particularly important in order to target the local structure of species whose description through a classical potential is difficult, such as multi-coordination species like aluminium and boron [35, 36], or to clarify controversial issues that standard classical MD and experimental probes were unable to resolve, such as the role of fluorine and phosphorus in the structure of bioactive silicate glasses [33, 37–42].

While the good quality of available interatomic potentials in the case of bioactive glasses [43–45] in most cases facilitates modelling their bulk structure without having to resort to AIMD, there are two areas where the latter are vital: surface reactivity and ion migration.

Most classical force fields are fitted to structural data of crystalline solids and as a consequence they perform well in reproducing the structure of amorphous materials characterized by short-range atomic environments similar to those found in the crystal. However, the performance of a classical potential are somewhat unpredictable when properties and environments that were not included in the fitting dataset are targeted, such as dynamical processes or distorted/unusual configurations. In other words, even if a certain force field provides an excellent description of the average structure of a glass, diffusive processes are not necessarily reproduced equally well. The same applies to processes occurring at a surface that often exposes different atomic configurations than those found in the bulk, such as defects and undercoordinated ions. AIMD represent a safer way to target ion migration and surface reactivity, because it provides a more accurate description of both the unusual atomic configurations involved, and also of bond-breaking events and electronic rearrangements brought about by ongoing chemical reactions. The latter issue is obviously crucial for the direct modelling of the glass corrosion and biodegradation, that involves hydrolysis of covalent Si–O–Si bonds, as well as formation of new bonds by Si–OH condensation [1, 46, 47].

The migration of a modifier ion such as an alkali or alkaline-earth in silicate glasses does not normally require or produce dramatic changes to the network of covalent bonds forming the silica glass matrix. However, significant changes to the coordination environment of the ion itself do occur during its migration from a site to another, when the ion often visits intermediate configurations which are significantly distorted with respect to the stable initial and final states [26, 48]. For instance, Fig. 10.1 shows the AIMD trajectory of an Na cation migrating along two stable sites in the 45S5 glass: the number and identity of oxygen atoms in the Na coordination shell change during the process, and significant structural rearrangements occur to the coordination shell itself. Na–O links are broken and formed during the hop, often in a concerted fashion, and a correct reproduction of the energies involved in these processes is needed in order to reproduce the diffusive event.

An AIMD approach, where the forces acting on the nuclei are obtained "on the fly" from electronic structure calculations, is intuitively more suited to describe this kind of dynamical transformations to the coordination shell, because any marked changes to the forces experienced by the diffusing ion will be naturally reproduced, at variance with an empirical force field, which may struggle to produce accurate forces in regions of the configuration space displaced away from the ordinary local minima.

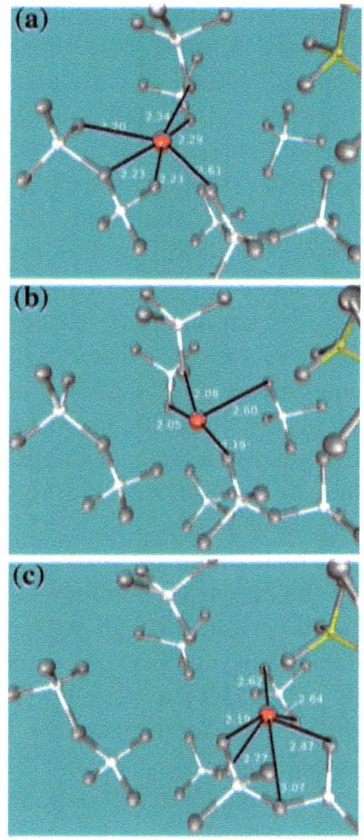

Fig. 10.1 The dynamical changes in the oxygen coordination shell of a Na ion (*red*) diffusing in 45S5 Bioglass(R). *Dark lines* highlight the Na–O interactions within the Na coordination shell, whose corresponding distances are also given. Reprinted with permission from [26], Copyright 2010 American Institute of Physics

10.3 Structural Properties

Atomistic models of melt-derived bioactive glasses have been obtained by both classical MD [43, 49–51] and AIMD [32, 33]. An hybrid approach involves using classical MD to produce an initial structure for a further AIMD refinement or other quantum-mechanical calculations such as electronic, vibrational and NMR spectra [52–55]. It is important to note that while the AIMD refinement can introduce limited local changes to the bonding environments (e.g., interatomic distances and angles) created through classical MD [55–57], substantial medium-range rearrangements to the glass structure (involving for instance different silicate network topology, connectivity and Q_n and ring size distribution) would require overcoming significant energy barriers and are not normally observed over the relatively short AIMD time scale. In other words, the AIMD simulation will conserve the main medium-range structural features determined by the classical potential. This does not represent a serious problem when the purpose of the AIMD calculation is to investigate local structures or properties that depend on them, such as the vibrational and electronic density of states [53, 55, 57]. For this reason, several studies have successfully employed classical

MD models of the glass to target vibrational and electronic properties by ab-initio methods. A remarkable output of these hybrid studies is the identification of vibrational signatures of particular groups, such as those of Si–O–P bonds [53, 55]: this information is essential to assist band assignments and the interpretation of experimental IR and Raman spectra, for instance to identify the sequence of transformations occurring at the glass surface during its degradation [58, 59].

The calculation of the vibrational spectrum from an (AI)MD trajectory involves Fourier-transforming the time-dependent velocity autocorrelation function [60]; an alternative approach involves calculating the phonon frequencies by diagonalizing the Hessian matrix of a model obtained by structural optimization of the classical MD structure [53]. The AIMD-VACF approach naturally include finite-temperature anharmonic effects missing in the Hessian-harmonic approximation, but it does not produce accurate IR intensities (for which an autocorrelation function based on the exact dipole moments would be needed [61–63]). Despite these issues, it turns out that, in the case of 45S5 Bioglass®, the two methods give similar frequencies of the individual modes [53].

The inability of AIMD to modify the initial medium-range structure established by the classical potential can be overcome through a full-AI simulation in which AIMD is used throughout the melt-and-quench process [32]. This is obviously more demanding in terms of computer time but is now becoming increasingly feasible. A full-AIMD approach has been applied to model amorphous materials for which no accurate force fields are available, such as bulk metallic glasses (BMGs) and amorphous $GeSe_2$ [34, 64], but also to bioactive glasses [33].

An accurate model of 45S5 obtained by a full-AIMD simulation highlighted that Si–O–P bonds are indeed present in this key biomaterial [32], confirming that these bonds are not an artefact of the models, possibly produced by a flawed empirical force field, but are instead relatively stable. Whereas previous NMR results had been interpreted in terms of absence of such bonds in the bioglass [65], more recent experiments show that a small fraction of phosphorus is indeed linked to a silicon [40], in agreement with the MD simulations. This now resolved controversy most likely arose from the difficulty of detecting low amounts of Q_1(P) by standard solid-state NMR [66], a problem that obviously does not affect the simulations.

The network connectivity of a glass (average number of bridging oxygen atoms per network former) represents the most common structural descriptor used to link the experimental bioactivity to the glass composition [67]. In general, highly bioactive compositions are invert glasses characterized by a low network connectivity (around 2), denoting a fragmented silicate matrix mainly composed of silicate chains (Fig. 10.2) [15]. This matrix will degrade rapidly in a physiological environment, with a corresponding higher bioactivity, as measured in terms of rate of HA formation in vitro or in terms of effective bonding to tissues in vivo [68]. The link between chain structure and biodegradation [69] reflects the fact that release of $–O–[Si–O–]_n$ chains requires breaking a lower number of Si–O bonds compared to the Q_3 and Q_4 silicate units present in a highly ramified structure, where the motion of these fragments is also more hindered (Fig. 10.3) [70].

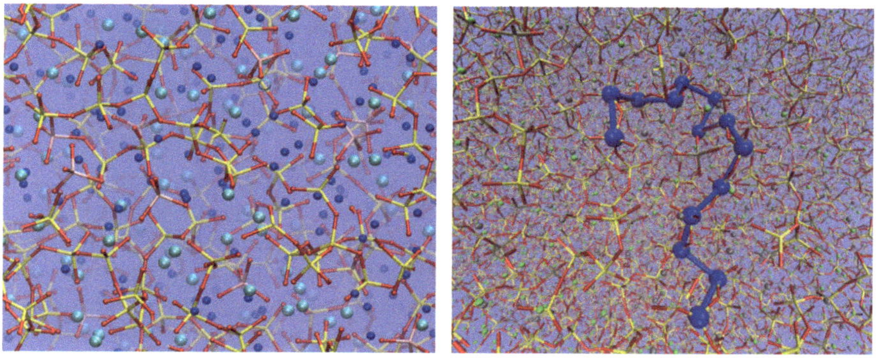

Fig. 10.2 Structure of 45S5 glass. Silicate and phosphate tetrahedra are represented as ball-and-stick, and modifier Na and Ca cations as spheres. The longest silicate chain identified in a model of ~45,000 atoms is highlighted in the *right panel*. Note that most chain fragments in the 45S5 structure are typically much shorter

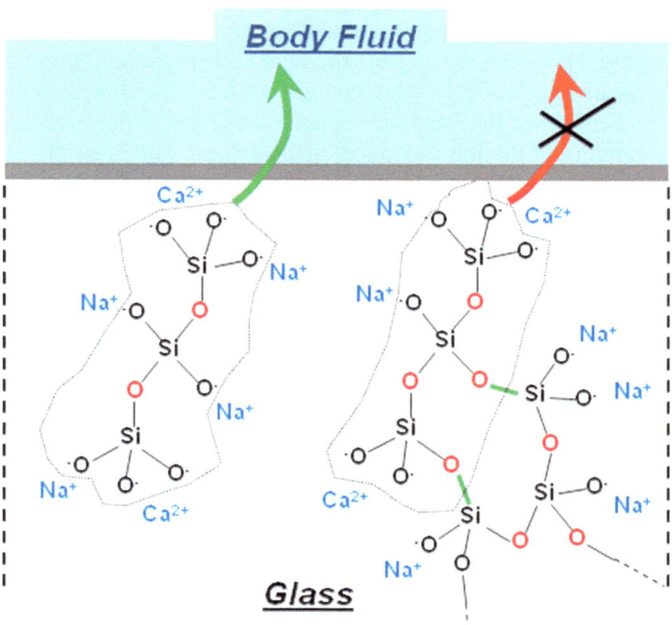

Fig. 10.3 Scheme illustrating the different mobility and release into the surrounding medium of a chain fragment (*left*) versus a more cross-linked fragment (*right*)

Besides the bone-bonding properties, in many biomedical applications one is also interested in the rate of leaching of specific cations from the glass. For instance, active ions such as cerium, gallium, zinc, strontium and cobalt are incorporated in bioactive glasses with the purpose of delivering and gradually releasing them to a target tissue

[42, 71–78]; also, yttrium and other ions are employed as radiation sources in in-situ radiotherapy, and silicate glasses are amongst the most common carriers for these radioisotopes [79, 80]. In the latter applications, it is of vital importance that the radioactive species are confined within the glass carrier during the treatment, because their release in the bloodstream would have adverse effects [81].

Therefore, being able to understand and control the rate of release of specific ions from the glass becomes a key requirement. Again, the network connectivity is obviously a key reference here, as a fast-degrading glass with a low *NC* will rapidly release any incorporated cation. However, the *NC* alone, while providing very useful indications, is not always sufficient to predict the biological and biodegradation behaviour of a glass [82]. This means that other structural parameters linked to the bioactivity must be identified. The simulations play a key role in this search, which involves obtaining accurate models of compositions of known dissolution rate/bioactivity, and identifying specific structural descriptors with a clear correlation to the experimental data [83]. One of these is the degree of clustering and aggregation of ionic species: the MD models highlighted that lower bioactivity (as measured experimentally) in Hench-type bioglasses is often accompanied by the appearance (highlighted in the models) of nanosegregated regions populated with calcium and phosphates [70]. This behavior reflects the marked resistance to dissolution found for phase-separated glasses [84]. which in turn reflects the reduced mobility of ions trapped in clusters, when the latter are spatially separated from each other and thus break the continuity of the ion migration channels needed by fast-diffusing ions in glasses [85].

The simple visual inspection of the models can provide some clues regarding these effects. For instance, Fig. 10.4 highlights a more uniform distribution of all species in the most bioactive composition (45S) compared to a higher-silica (bioinactive 65S) and a higher-phosphate (45S-P12) one [51, 86]. In particular, modifier cations are homogeneously spread across all the available space in 45S, whose silicate network does not contain large gaps, although some small gaps are visible that appear to be populated by small calcium phosphate aggregates. Larger gaps appear in the silicate network of 65S, that the figure suggests as associated with calcium phosphates: modifier cations also appear less uniformly spread in 65 S than in 45S. A higher-phosphate content (right panels in the figure) determines the appearance of large voids in the silicate network, mainly filled with calcium and phosphates, whereas again the sodium distribution appears more uniform compared to calcium [86].

These visual clues, albeit useful, clearly need to be assessed more quantitatively. The ratio R_{M-M} between the M(odifier)–M(odifier) coordination number extracted from the MD model and that expected from a uniform distribution of M cations through the available space represents a reliable measure of the extent of clustering [51, 77, 87]: $R_{M-M} > 1$ denotes M–M clustering, more marked for larger deviations from the uniform $R_{M-M} = 1$ distribution. Moreover, the ratio R_{A-B} calculated in the same way from the A–B coordination number is a useful measure of preferential aggregation between A and B species, for instance with A = Na/Ca and B = Si/P one can assess whether the Ca-P aggregation suggested by the figures is indeed stronger than Na-P. This analysis applied to the MD models shows that, in general, modifier

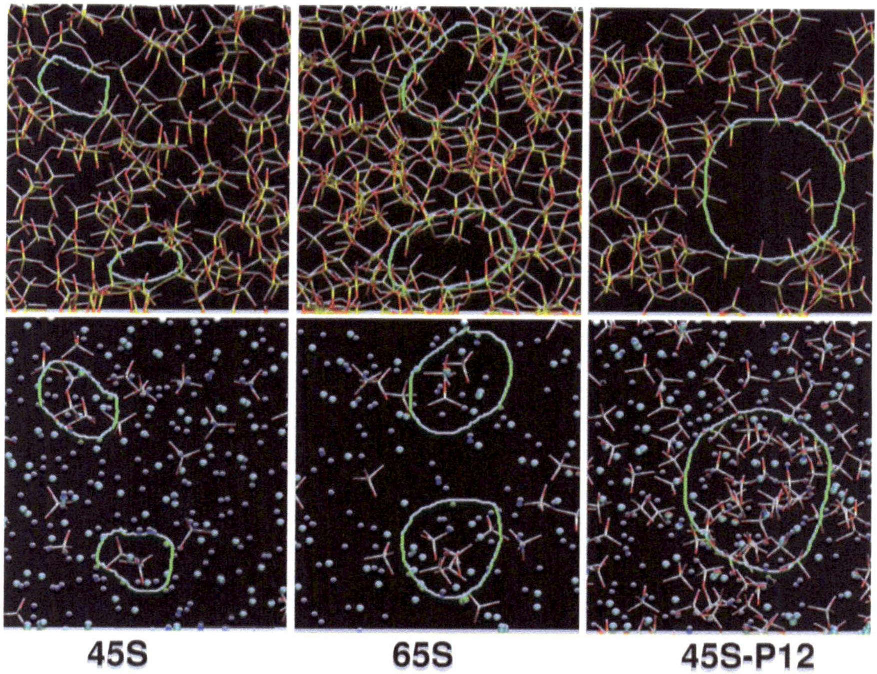

Fig. 10.4 Snapshots of Hench-type bioglass models illustrating formation of calcium phosphate clusters separated from the silicate matrix. Only Si and O atoms are shown in the *top panels*, whereas the *bottom panels* show Na (*blue*) Ca (*cyan*) and phosphate groups. The *green circles* highlight gaps in the silicate network

cations (especially Ca) have a marked preference to associate with phosphate groups [51], which reflects the well-known repolymerization of the silicate network when modifier cations are stripped from it and shifted to the phosphates [88]. The R analysis of the MD models shows that the preference of Ca^{2+} to coordinate phosphate tetrahedra is further enhanced moving to higher-silica and less bioactive compositions; the latter glasses are also characterised by a significantly less uniform distribution of Ca ions [86].

When the ions are confined into clusters spatially separated from each other, ion migration along channels connecting these clusters is inhibited; however, when the size of the clusters increases, their mutual separation will be reduced and fast-conducting migration channels may eventually be re-established, which may explain why in some cases phase separation of modifier-rich phases can lead to enhanced conductivity [89]. The size of the clusters, determined by the glass composition, is thus the key factor to take into account in order to establish whether nanosegregation may inhibit or enhance leaching of the ions in the surrounding environment, and, essentially, biodegradation [90].

Whereas the network connectivity reflects the average strength of the network of Si–O bonds building up the glass matrix, different structural descriptors can be devised to describe the strength of noncovalent interactions between modifier ions and the silicate matrix. Several studies have highlighted that higher field strength cations attract a higher number of non-bridging oxygens in their coordination shell, and form a correspondingly higher number of O–M–O intertetrahedral links, that increase the cohesion between spatially separated silicate fragments "held together" by the same M cation [91–93]. The idea is sketched in Fig. 10.5: the higher the ion strength, the larger the number of O–M–O links where the M cation bridges two different SiO_4 tetrahedra.

The overall strength of modifier-mediated cross-link interactions in the glass can be estimated from the MD structure, by combining the density of inter-tetrahedral links *T* formed by each modifier cation M with the corresponding M–O ionic bond strength [94], and used to complement "standard" descriptors like network connectivity and clustering in order to predict the dissolution behaviour of a glass. Whereas it is possible to use each parameter to understand the behaviour of known compositions and extrapolate this insight to new potentially interesting compositions, a closer link to the experiments and a more consistent description can be built by capitalizing on existing experimental datasets, such as composition/solubility curves. A linear combination $s = \alpha^*NC + \beta^*R_{M-M} + \gamma^*T_M$ of the different structural descriptors *NC*, *R* and *T* discussed above can be fitted to the available experimental data: the resulting best-fit function *s* can then be used to predict the dissolution rate of an hypothetical new composition based on its own *NC*, R_{M-M} and T_M descriptors, extracted from the MD model [95]. Moreover, the final best-fit coefficients α, β and γ will

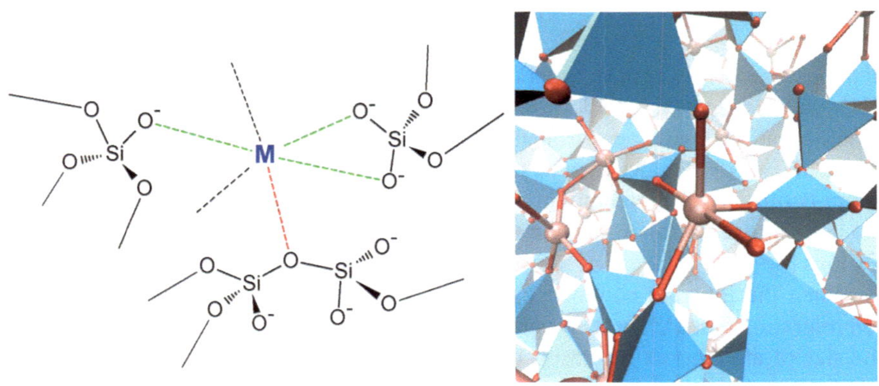

Fig. 10.5 *Left* Scheme of a central modifier cation M coordinating different fragments of the silicate network through M–O interactions. The stronger M–NBO interactions are coloured *green*; the one on the *right* is an example of intra-tetrahedral O–M–O interaction that involves one individual SiO_4 tetrahedron only, whereas all other O–M–O interactions involve two different SiO_4. *Right* Structure of an yttrium-doped bioglass, highlighting the role of yttrium ions (*pink spheres*) joining together the tetrahedra of the silicate backbone. Adapted from [94] with permission from the Royal Society of Chemistry

contain information on *how* each descriptor influences the overall behaviour, and this detailed insight is also precious to unveil the role that each structural feature has in the overall dissolution behaviour. For instance, coefficient of opposite signs denote that the corresponding descriptors have opposite effect on the solubility, and their relative magnitude will determine which structural feature has a larger impact on the glass behaviour. This insight can be used to tailor compositions of higher or lower solubility, depending on the application, by focusing on optimising a higher-impact structural descriptor rather than a low-impact one.

10.4 Surface Reactivity

All the concepts developed to rationalize the degradation behaviour of a bioactive glass discussed so far involve the bulk of the material: the corresponding properties are obtained from periodic 3D models with no surfaces. Nevertheless, because the biodegradation begins at the glass surface, it is essential to consider how the bulk structural features are reflected on the surface, and in general whether conclusions based exclusively on the bulk structure can still be used to explain phenomena that in practice take place mainly at the surface. MD simulations of surface models enable us to translate to this critical region the information obtained for the bulk, and also to highlight particular structural features only formed in the peculiar surface environment and absent from the bulk, that may play an additional role and in some cases modify the expected behaviour of a glass. As also explained above, AIMD simulations are in general more suitable to model the chemistry that follows exposure of a glass surface: this exposure creates unsaturated bonds whose healing will dictate the structure and properties of the stable surface, and an ab-initio treatment is by far the most appropriate to describe formation of new bonds, defects and surface reconstruction. Whereas AIMD simulations of amorphous silica surfaces are rather common [96], not so many studies have concerned the surface of a bioactive glass. An interesting feature that has been identified by experimental (IR) probes on the dry SiO_2 surface is represented by small (e.g., two-membered, 2M) siloxane rings [97]. The high internal strain of these rings makes them some of the strongest chemisorption sites on SiO_2 surfaces, that is, 2M rings represent the sites where the first interaction with water or other molecules in the surrounding medium will likely occur [98]. Several studies have then investigated the reaction of 2M rings with water on a-SiO_2, confirming their high reactivity but also illustrating how the size and features of the surface model itself influence the estimated reactivity of these sites, in terms of reaction barrier and energetics [96, 99]. It turns out that there is a strong dependence of the reaction energies on the local structure, and in particular on the extent to which long-range relaxations are represented in the model [99]. This is particularly important when considering the reactivity of these same rings on the surface of bioglasses, because it suggests that the thermodynamical and kinetic stability of small rings can vary considerably in different surface environments, characterised by different local constraints and rigidity. The insight obtained for the a-SiO_2 surface tells us that a limited local relaxation introduces a higher tension in the ring, that is therefore less stable under such conditions. On the surface of a 45S5 bioactive composition, the local arrangement around 2M and also 3M rings turns out to be more flexible, due to the highly fragmented character of the silicate network [100]. Based on the considerations above, this effect can release some of the internal strain of the small rings, that may therefore be more stable on the 45S5 surface. This is a particularly significant possibility, not last because closed small rings have indeed been proposed as stable surface sites that guide the nucleation of calcium phosphate on the bioactive glass surface [101, 102]. AIMD simulations of the 45S5 surface showed that the kinetic barrier for breaking an exposed 2M-ring is twice that found on the a-SiO_2, [96] confirming the higher intrinsic stability of these rings on the bioactive surface [100]. The stability of small rings in an extended liquid environment was further confirmed by large-scale AIMD runs of the 45S5 aqueous interface at room temperature [103]. The simulations also suggested an additional reason for the higher stability of small rings on the bioactive than the pure silica surface: the hydrolytic opening of a small ring is delayed in the presence of alternative adsorption sites for an incoming water molecule, which are absent on pure silica. Hydrophilic patches populated by modifier cations (Na and Ca) and non-bridging oxygens are more favourable adsorption sites than silica rings, so that water molecules will be displaced from the ring sites towards them. Even though the internal strain of the rings is such that eventually they will be opened, the models show that there is a concrete possibility that some of these strained surface sites may resist hydrolysis long enough to nucleate calcium phosphate precursors, which will then protect them from further reactions [19].

2M rings are absent from the bulk bioglass, whose ring distribution mostly features 4- and 5-membered sites [19], but they can easily form on a freshly exposed surface. The latter will initially contain undercoordinated and very unstable Si atoms with multiple dangling bonds; the formation of a small ring by linking to a neighbouring Si–O unit represents one of the mechanisms by which these unstable Si are healed. Obviously, no 2M rings would result if the dangling bond created by exposing the surface were fully saturated with water [54]. However, surface relaxation is known to occur on a very rapid time scale (1–2 ps), that is, it can be considered complete before contact with water takes place [104, 105]. Therefore, passivating dangling bonds with water before surface relaxation does not seem like a recommended procedure, as in this case the reference dry surface model will miss some key features that would naturally be formed there.

The AIMD models of the surface region represent a necessary starting point for any large-scale MD study of the interface. The limited information available on the atomistic structure of the glass substrate complicates the assessment of the accuracy of surface models obtained by classical force fields. The AIMD models, albeit of limited size, are an essential reference to define the stability of typical surface sites, that should be at least qualitatively reflected in any larger model produced by classical MD simulations. Whereas the latter are generally not suitable to investigate surface reactivity directly, they can certainly be employed to compare the surface structure of large samples of different compositions, *provided* that the structures are qualitatively consistent with the expectations based on AIMD results.

From this perspective, the kind of sites found on bioactive glass surfaces built using shell-model classical MD are the same as those predicted by AIMD [106]. Now, because AIMD data enable us to accurately estimate the bioreactivity of each individual site, one can use the surface density of each site found in large-size classical MD models for a quantitative comparison between compositions of different bioactivity, highlighting how specific surface features can be correlated to the bioactivity. This represents an effective *multiscale* strategy, wherein the power of each methods is exploited to fill the appropriate piece of the overall puzzle: the site reactivity through AIMD [100], and the corresponding density by classical MD [106]. One can also estimate the surface density of silanols that would be produced by full hydration of the dry surface upon contact with a physiological environment: an interesting and somewhat unexpected feature revealed by classical MD models is that there is only a small difference in the expected Si–OH population of a highly bioactive and a bioinactive surface [106]. This essentially means that a high bioactivity is not just the result of a higher silanol concentration, as it has been often suggested [107, 108].

Another somewhat puzzling finding of the simulations is the high density of small rings found on bioinactive surfaces, which would seem to dispute their central role as nucleating templates for the apatite layer [106]. A possible explanation is that, because the local environment of a higher-silica bioinactive composition is closer to that of pure *a*-SiO_2, the stability of small rings is lower than in the bioactive case, following the ideas exposed above on the effect of local relaxation on the ring tension. Therefore, on the bioinactive surface, only a lower fraction of these small-ring templates would effectively resist hydrolysis and remain available for the time needed to nucleate the apatite nuclei. Further calculations are needed to assess this hypothesis.

A similar approach can be applied to investigate the causes of the enhanced activity of nanosized samples of bioglasses [109–111]. Classical MD simulations are required to produce realistic nanoparticle models of several nanometres [112, 113]. So far, it has been possible to look at how the reduced size affects the surface sites involved in the bioreactivity of a glass, by comparing the surface structure of a dry 45S5 nanoparticle to that of a compact sample exposing a flat surface [114]: the reduced size and the surface curvature appear to enhance key features such as the fragmentation of the silicate matrix and the high density of small rings, as well as the mobility of modifier cations, contributing to the large (though still unexploited) potential of nanosized bioactive glasses in biomedicine [115, 116]. Figure 10.6 shows examples of free orthosilicate (Q_0) groups and strained ring sites exposed at the surface of a 45S5 nanoparticle of realistic size: these sites are found more frequently on the nanoparticle than in the bulk or on the flat surface of the material.

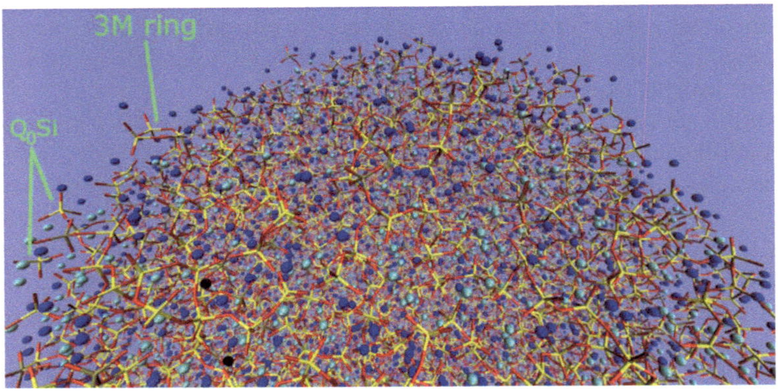

Fig. 10.6 Enlarged *top view* of a 10 nm diameter 45S5 bioactive glass nanoparticle, highlighting relevant surface sites

10.5 Final Remarks

The performances of a glass in a biomedical context depend on its biodegradation. This applies not only to conventional, melt-derived invert compositions of the Hench type, but also to glasses of very different compositions (such as phosphate- and borate-based systems [13, 117] and metallic glasses [118]) and obtained through different routes (such as sol-gel [119]). Therefore, a fundamental understanding of the composition-biodegradation relationships of a well-known class of biomaterials such as the Hench bioglasses is likely to be relevant also for other classes of glass for biomedicine. As an example, general structural descriptors that can be used to understand leaching of yttrium radioisotopes from non-bioactive aluminosilicate glass vectors can also be applied to predict the performances of yttrium-doped bioactive glasses [95]. Atomic-scale modelling has become one of the most powerful tool available at present to obtain a high-resolution picture of bulk and surface structure: Molecular Dynamics simulations are suitable not only for producing reliable structural models but also to examine directly the dynamics of key processes for the biodegradation, such as surface reactivity and ion migration.

References

1. J.R. Jones, Review of bioactive glass: from Hench to hybrids. Acta Biomater. **9**, 4457–4486 (2013)
2. L.L. Hench, Bioceramics. J. Am. Ceram. Soc. **81**(7), 1705–1728 (1998)
3. L.L. Hench et al., Bonding mechanisms at the interface of ceramic prosthetic materials. J. Biomed. Mater. Res. **2**(6), 117–141 (1971)
4. M. Vallet-Regí, Ceramics for medical applications. J. Chem. Soc. Dalton Trans. **2**, 97–108 (2001)

5. H. Oonishi et al., Quantitative comparison of bone growth behavior in granules of bioglass ®R, A-W glass-ceramic, and hydroxyapatite. J. Biomed. Mater. Res. **51**(1), 37–46 (2000)
6. L.L. Hench, Bioceramics-from concept to clinic. J. Am. Ceram. Soc. **74**(7), 1487–1510 (1991)
7. J.R. Jones, P. Sepulveda, L.L. Hench, Dose-dependent behavior of bioactive glass dissolution. J. Biomed. Mater. Res. **58**(6), 720–726 (2001)
8. I.D. Xynos et al., Gene-expression profiling of human osteoblasts following treatment with the ionic products of bioglass ®R 45S5 dissolution. J. Biomed. Mater. Res. **55**(2), 151–157 (2001)
9. L.L. Hench, J.M. Polak, Third-generation biomedical materials. Science **295**(5557), 1014–1017 (2002)
10. R.A. Martin et al., Characterizing the hierarchical structures of bioactive sol-gel silicate glass and hybrid scaffolds for bone regeneration. Phil. Trans. R. Soc. A **2012**(370), 1422–1443 (1963)
11. M. Ogino, F. Ohuchi, L.L. Hench, Compositional dependence of the formation of calcium-phosphate films on bioglass. J. Biomed. Mater. Res. **14**(1), 55–64 (1980)
12. P. Sepulveda, J.R. Jones, L.L. Hench, In vitro dissolution of melt-derived 45S5 and sol-gel derived 58S bioactive glasses. J. Biomed. Mater. Res. A **61**(2), 301–311 (2002)
13. J.C. Knowles, Phosphate based glasses for biomedical applications. J. Mater. Chem. **13**(10), 2395–2401 (2003)
14. A. Hoppe, N.S. Güldal, A.R. Boccaccini, A review of the biological response to ionic dissolution products from bioactive glasses and glass-ceramics. Biomaterials **32**(11), 2757–2774 (2011)
15. A. Tilocca, Structural models of bioactive glasses from molecular dynamics simulations. Proc. R. Soc. A **465**(2104), 1003–1027 (2009)
16. A.E. Clark, C.G. Pantano, L.L. Hench, Auger spectroscopic analysis of bioglass corrosion films. J. Am. Ceram. Soc. **59**(1–2), 37–39 (1976)
17. B.C. Bunker, Molecular mechanisms for corrosion of silica and silicate glasses. J. Non-Cryst. Solids **179**, 300–308 (1994)
18. C. Cailleteau et al., Insight into silicate-glass corrosion mechanisms. Nat. Mater **7**(12), 978–983 (2008)
19. A. Tilocca, Models of structure, dynamics and reactivity of bioglasses: a review. J. Mater. Chem. **20**(33), 6848 (2010)
20. Y.B. Wang et al., Atomistic insight into viscosity and density of silicate melts under pressure. Nat. Commun. **5** (2014)
21. K. Vollmayr, W. Kob, K. Binder, Cooling-rate effects in amorphous silica: a computer-simulation study. Phys. Rev. B **54**(22), 15808–15827 (1996)
22. C.D. Huang, A.N. Cormack, Structural differences and phase-separation in alkali silicate-glasses. J. Chem. Phys. **95**(5), 3634–3642 (1991)
23. C. Massobrio, A. Pasquarello, Short and intermediate range order in amorphous $GeSe_2$. Phys. Rev. B **77**(14), 144207 (2008)
24. R. Car, M. Parrinello, Unified approach for molecular-dynamics and density-functional theory. Phys. Rev. Lett. **55**(22), 2471–2474 (1985)
25. A. Tilocca, Cooling rate and size effects on the medium-range structure of multicomponent oxide glasses simulated by molecular dynamics. J. Chem. Phys. **139**(11), 114501 (2013)
26. A. Tilocca, Sodium migration pathways in multicomponent silicate glasses: Car-Parrinello molecular dynamics simulations. J. Chem. Phys. **133**(1), 014701–014710 (2010)
27. J.H. Los et al., First-principles study of the amorphous In3SbTe2 phase change compound. Phys. Rev. B **88**(17), 174203 (2013)
28. K. Nishio, T. Miyazaki, H. Nakamura, Universal medium-range order of amorphous metal oxides. Phys. Rev. Lett. **111**(15), 155502 (2013)
29. G. Spiekermann et al., Vibrational properties of silica species in MgO–SiO_2 glasses obtained from ab initio molecular dynamics. Chem. Geol. **346**, 22–33 (2013)
30. A. Aliano, A. Catellani, G. Cicero, Characterization of amorphous In_2O_3: an ab initio molecular dynamics study. Appl. Phys. Lett. **99**(21), 211913 (2011)

31. A.J. Parsons et al., Neutron scattering and ab initio molecular dynamics study of cross-linking in biomedical phosphate glasses. J. Phys. Condens. Matter **22**(48), 485403 (2010)
32. A. Tilocca, Structure and dynamics of bioactive phosphosilicate glasses and melts from ab initio molecular dynamics simulations. Phys. Rev. B **76**(22), 224202 (2007)
33. J.K. Christie et al., Fluorine environment in bioactive glasses: ab initio molecular dynamics simulations. J. Phys. Chem. B **115**(9), 2038–2045 (2011)
34. L. Giacomazzi, C. Massobrio, A. Pasquarello, Vibrational properties of vitreous $GeSe_2$ with the Becke-Lee-Yang-Parr density functional. J. Phys. Condens. Matter **23**(29), 295407 (2011)
35. L.P. Huang, J. Kieffer, Thermomechanical anomalies and polyamorphism in B_2O_3 glass: a molecular dynamics simulation study. Phys. Rev. B **74**(22), 224107 (2006)
36. Q.J. Zheng et al., Structure of boroaluminosilicate glasses: impact of Al_2O_3/SiO_2 ratio on the structural role of sodium. Phys. Rev. B **86**(5), 054203 (2012)
37. T. Schaller et al., Fluorine in silicate glasses: a multinuclear nuclear magnetic resonance study. Geochimica et Cosmochimica Acta **56**(2), 701–707 (1992)
38. D.S. Brauer et al., Structure of fluoride-containing bioactive glasses. J. Mater. Chem. **19**(31), 5629–5636 (2009)
39. G. Lusvardi et al., Elucidation of the structural role of fluorine in potentially bioactive glasses by experimental and computational investigation. J. Phys. Chem. B **112**(40), 12730–12739 (2008)
40. F. Fayon et al., Evidence of nanometric-sized phosphate clusters in bioactive glasses as revealed by solid-state ^{31}P NMR. J. Phys. Chem. C **117**, 2283–2288 (2013)
41. M.D. O'Donnell et al., The effect of phosphate content on the bioactivity of soda-lime-phosphosilicate glasses. J. Mater. Sci. Mater. Med. **20**(8), 1611–1618 (2009)
42. K. Fujikura et al., Influence of strontium substitution on structure and crystallisation of bio-glass ®*R* 45S5. J. Mater. Chem. 22(15) 7395–7402 (2012)
43. A. Pedone et al., Role of magnesium in soda-lime glasses: insight into structural, transport, and mechanical properties through computer simulations. J. Phys. Chem. C **112**(29), 11034–11041 (2008)
44. Y. Xiang, J. Du, Effect of strontium substitution on the structure of 45S5 bioglasses. Chem. Mater. **23**(11), 2703–2717 (2011)
45. A. Tilocca, Short- and medium-range structure of multicomponent bioactive glasses and melts: an assessment of the performances of shell-model and rigid-ion potentials. J. Chem. Phys. **129**(8), 084504 (2008)
46. Ö.H. Andersson, K.H. Karlsson, On the bioactivity of silicate glass. J. Non-Cryst. Solids **129**(1–3), 145–151 (1991)
47. A.E. Clark, C.G. Pantano, L.L. Hench, Auger spectroscopic analysis of bioglass corrosion films. J. Am. Ceram. Soc. **59**(1–2), 37–39 (1976)
48. A.N. Cormack, J. Du, T.R. Zeitler, Alkali ion migration mechanisms in silicate glasses probed by molecular dynamics simulations. Phys. Chem. Chem. Phys. **4**(14), 3193–3197 (2002)
49. C. Bonhomme et al., Sr-87 solid-state NMR as a structurally sensitive tool for the investigation of materials: antiosteoporotic pharmaceuticals and bioactive glasses. J. Am. Chem. Soc. **134**(30), 12611–12628 (2012)
50. R.N. Mead, G. Mountjoy, Modeling the local atomic structure of bioactive sol-gel-derived calcium silicates. Chem. Mater. **18**(17), 3956–3964 (2006)
51. A. Tilocca, A.N. Cormack, Structural effects of phosphorus inclusion in bioactive silicate glasses. J. Phys. Chem. B **111**(51), 14256–14264 (2007)
52. A. Pedone, E. Gambuzzi, M.C. Menziani, Unambiguous description of the oxygen environment in multicomponent aluminosilicate glasses from O-17 solid state NMR computational spectroscopy. J. Phys. Chem. C **116**(27), 14599–14609 (2012)
53. M. Corno et al., B3LYP simulation of the full vibrational spectrum of 45S5 bioactive silicate glass compared to v-silica. Chem. Mater. **20**(17), 5610–5621 (2008)
54. E. Berardo et al., DFT modeling of 45S5 and 77S soda-lime phospho-silicate glass surfaces: clues on different bioactivity mechanism. Langmuir **29**(19), 5749–5759 (2013)

55. A. Tilocca, N.H. de Leeuw, Ab initio molecular dynamics study of 45S5 bioactive silicate glass. J. Phys. Chem. B **110**(51), 25810–25816 (2006)
56. L. Giacomazzi, C. Massobrio, A. Pasquarello, First-principles investigation of the structural and vibrational properties of vitreous $GeSe_2$. Phys. Rev. B **75**(17), 174207 (2007)
57. S. Ispas et al., Vibrational properties of a sodium tetrasilicate glass: ab initio versus classical force fields. J. Non-Cryst. Solids **351**(12–13), 1144–1150 (2005)
58. M. Cerruti et al., Surface modifications of bioglass immersed in TRIS-buffered solution. a multitechnical spectroscopic study. J. Phys. Chem. B **109**(30), 14496–14505 (2005)
59. M. Cerruti, D. Greenspan, K. Powers, Effect of pH and ionic strength on the reactivity of Bioglass ®*R* 45S5. Biomaterials **26**(14), 1665–1674 (2005)
60. M.P. Allen, D.J. Tildesley, *Computer Simulation of Liquids* (Clarendon, Oxford, 1989)
61. E. Fois et al., Water molecules in single file: first-principles studies of one-dimensional water chains in zeolites. J. Phys. Chem. B **105**(15), 3012–3016 (2001)
62. W. Chen et al., Role of dipolar correlations in the infrared spectra of water and ice. Phys. Rev. B **77**(24), 245114 (2008)
63. P.L. Silvestrelli, M. Bernasconi, M. Parrinello, Ab initio infrared spectrum of liquid water. Chem. Phys. Lett. **277**(5–6), 478–482 (1997)
64. H. Tian et al., First-principles study of the binary Ni60Ta40 metallic glass: the atomic structure and elastic properties. J. Non-Cryst. Solids **358**(15), 1730–1734 (2012)
65. I. Elgayar et al., Structural analysis of bioactive glasses. J. Non-Cryst. Solids **351**(2), 173–183 (2005)
66. R. Mathew et al., Toward a rational design of Bioactive glasses with optimal structural features: composition-structure correlations unveiled by solid-state NMR and MD simulations. J. Phys. Chem. B **118**(3), 833–844 (2014)
67. Z. Strnad, Role of the glass phase in bioactive glass-ceramics. Biomaterials **13**(5), 317–321 (1992)
68. O. Peitl, E.D. Zanotto, L.L. Hench, Highly bioactive P_2O_5-Na_2O-CaO-SiO_2 glass-ceramics. J. Non-Cryst. Solids **292**(1–3), 115–126 (2001)
69. D. Arcos, D.C. Greenspan, M. Vallet-Regí, Influence of the stabilization temperature on textural and structural features and ion release in SiO_2-CaO-P_2O_5 sol-gel glasses. Chem. Mater. **14**(4), 1515–1522 (2002)
70. A. Tilocca, Models of structure, dynamics and reactivity of bioglasses: a review. J. Mater. Chem. **20**(33), 6848–6858 (2010)
71. J. Du, Y. Xiang, Effect of strontium substitution on the structure, ionic diffusion and dynamic properties of 45S5 Bioactive glasses. J. Non-Cryst. Solids **358**(8), 1059–1071 (2012)
72. A. Goel, R.R. Rajagopal, J.M.F. Ferreira, Influence of strontium on structure, sintering and biodegradation behaviour of CaO-MgO-SrO-SiO_2-P_2O_5-CaF_2 glasses. Acta Biomater. **7**(11), 4071–4080 (2011)
73. A. Goel et al., Structural role of zinc in biodegradation of alkali-free bioactive glasses. J. Mater. Chem. B **1**(24), 3073–3082 (2013)
74. J.M. Smith et al., Structural characterisation of hypoxia-mimicking bioactive glasses. J. Mater. Chem. B **1**(9), 1296–1303 (2013)
75. A.J. Salinas et al., Substitutions of cerium, gallium and zinc in ordered mesoporous bioactive glasses. Acta Biomater. **7**(9), 3452–3458 (2011)
76. G. Lusvardi et al., In vitro and in vivo behaviour of zinc-doped phosphosilicate glasses. Acta Biomater. **5**(1), 419–428 (2009)
77. G. Malavasi, A. Pedone, M.C. Menziani, Study of the structural role of gallium and aluminum in 45S5 bioactive glasses by molecular dynamics simulations. J. Phys. Chem. B **117**(15), 4142–4150 (2013)
78. J.C. Du et al., Structure of cerium phosphate glasses: molecular dynamics simulation. J. Am. Ceram. Soc. **94**(8), 2393–2401 (2011)
79. M. Kawashita, Ceramic microspheres for in situ radiotherapy of cancer. Mater. Sci. Eng. C **22**(1), 3–8 (2002)

80. E.M. Erbe, D.E. Day, Chemical durability of Y_2O_3-Al_2O_3-SiO_2 glasses for the in vivo delivery of beta radiation. J. Biomed. Mater. Res. **27**(10), 1301–1308 (1993)
81. J.K. Christie, A. Tilocca, Aluminosilicate glasses As Yttrium vectors for in situ radiotherapy: understanding composition-durability effects through molecular dynamics simulations. Chem. Mater. **22**(12), 3725–3734 (2010)
82. R. Hill, An alternative view of the degradation of bioglass. J. Mater. Sci. Lett. **15**(13), 1122–1125 (1996)
83. T. Le et al., Quantitative structure-property relationship modeling of diverse materials properties. Chem. Rev. **112**(5), 2889–2919 (2012)
84. H.-F. Wu, C.-C. Lin, P. Shen, Structure and dissolution of CaO-ZrO_2-TiO_2-Al_2O_3-B_2O_3-SiO_2 glass (II). J. Non-Cryst. Solids **209**(1–2), 76–86 (1997)
85. P. Jund, W. Kob, R. Jullien, Channel diffusion of sodium in a silicate glass. Phys. Rev. B **64**(13), 134303 (2001)
86. A. Tilocca, A.N. Cormack, N.H. de Leeuw, The structure of bioactive silicate glasses: new insight from molecular dynamics simulations. Chem. Mater. **19**(1), 95–103 (2007)
87. R.N. Mead, G. Mountjoy, A molecular dynamics study of the atomic structure of (CaO)x(SiO_2)1-x glasses. J. Phys. Chem. B **110**(29), 14273–14278 (2006)
88. M.W.G. Lockyer, D. Holland, R. Dupree, NMR investigation of the structure of some bioactive and related glasses. J. Non-Cryst. Solids **188**(3), 207–219 (1995)
89. A. Pradel et al., Mixed glass former effect in the system 0.3Li_2S-0.7[(1-x)SiS_2-xGeS_2]: a structural explanation. Chem. Mater. **10**(8), 2162–2166 (1998)
90. T.R. Stechert, M.J.D. Rushton, R.W. Grimes, Predicted mechanism for enhanced durability of zinc containing silicate glasses. J. Am. Ceram. Soc. **96**(5), 1450–1455 (2013)
91. J.K. Christie, J. Malik, A. Tilocca, Bioactive glasses as potential radioisotope vectors for in situ cancer therapy: investigating the structural effects of yttrium. Phys. Chem. Chem. Phys. **13**(39), 17749–17755 (2011)
92. Y. Vaills, Y. Luspin, G. Hauret, Two opposite effects of sodium on elastic constants of silicate binary glasses. Mater. Sci. Eng. B **40**(2–3), 199–202 (1996)
93. C.C. Lin et al., Composition dependent structure and elasticity of lithium silicate glasses: effect of ZrO2 additive and the combination of alkali silicate glasses. J. Eur. Ceram. Soc. **26**(16), 3613–3620 (2006)
94. J.K. Christie, A. Tilocca, Integrating biological activity into radioisotope vectors: molecular dynamics models of yttrium-doped bioactive glasses. J. Mater. Chem. **22**(24), 12023–12031 (2012)
95. J.K. Christie, A. Tilocca, Molecular dynamics simulations and structural descriptors of radioisotope glass vectors for in situ radiotherapy. J. Phys. Chem. B **116**(41), 12614–12620 (2012)
96. P. Masini, M. Bernasconi, Ab initio simulations of hydroxylation and dehydroxylation reactions at surfaces: amorphous silica and brucite. J. Phys.: Condens. Matter **14**(16), 4133 (2002)
97. T.A. Michalske, B.C. Bunker, Slow fracture model based on strained silicate structures. J. Appl. Phys. **56**(10), 2686–2693 (1984)
98. B.A. Morrow, I.A. Cody, L.S.M. Lee, Infrared studies of reactions on oxide surfaces. 7. mechanism of adsorption of water and ammonia on dehydroxylated silica. J. Phys. Chem. **80**(25), 2761–2767 (1976)
99. A. Rimola, P. Ugliengo, A quantum mechanical study of the reactivity of (SiO_2)-defective silica surfaces. J. Chem. Phys. **128**(20), 204702 (2008)
100. A. Tilocca, A.N. Cormack, Exploring the surface of bioactive glasses: water adsorption and reactivity. J. Phys. Chem. C **112**(31), 11936–11945 (2008)
101. V. Bolis et al., Surface properties of silica-based biomaterials: Ca species at the surface of amorphous silica as model sites. J. Phys. Chem. C **112**(43), 16879–16892 (2008)
102. N. Sahai, M. Anseau, Cyclic silicate active site and stereochemical match for apatite nucleation on pseudowollastonite bioceramic-bone interfaces. Biomaterials **26**(29), 5763–5770 (2005)
103. A. Tilocca, A.N. Cormack, Modeling the water-bioglass interface by ab initio molecular dynamics simulations. ACS Appl. Mater. Interfaces **1**(6), 1324–1333 (2009)

104. A.S. D'Souza, C.G. Pantano, Mechanisms for silanol formation of amorphous silica fracture surfaces. J. Am. Ceram. Soc. **82**(5), 1289–1293 (1999)
105. E.A. Leed, C.G. Pantano, Computer modeling of water adsorption on silica and silicate glass fracture surfaces. J. Non-Cryst. Solids **325**(1–3), 48–60 (2003)
106. A. Tilocca, A.N. Cormack, Surface signatures of bioactivity: MD simulations of 45S and 65S silicate glasses. Langmuir **26**(1), 545–551 (2010)
107. T. Kokubo, Surface chemistry of bioactive glass-ceramics. J. Non-Cryst. Solids **120**(1–3), 138–151 (1990)
108. H. Takadama et al., Mechanism of apatite formation induced by silanol groups—TEM observation. J. Ceram. Soc. Jpn **108**(2), 118–121 (2000)
109. S. Labbaf et al., Spherical bioactive glass particles and their interaction with human mesenchymal stem cells in vitro. Biomaterials **32**(4), 1010–1018 (2011)
110. S.K. Misra et al., Comparison of nanoscale and microscale bioactive glass on the properties of P(3HB)/Bioglass (R) composites. Biomaterials **29**(12), 1750–1761 (2008)
111. F. Quintero et al., Laser spinning of bioactive glass nanofibers. Adv. Funct. Mater. **19**(19), 3084–3090 (2009)
112. A. Shekhar et al., Collective oxidation behavior of aluminum nanoparticle aggregate. Appl. Phys. Lett. **102**(22), 221904 (2013)
113. D. Spagnoli, J.P. Allen, S.C. Parker, The structure and dynamics of hydrated and hydroxylated magnesium oxide nanoparticles. Langmuir **27**(5), 1821–1829 (2011)
114. A. Tilocca, Molecular dynamics simulations of a bioactive glass nanoparticle. J. Mater. Chem. **21**(34), 12660–12667 (2011)
115. M. Mackovic et al., Bioactive glass (type 45S5) nanoparticles: in vitro reactivity on nanoscale and biocompatibility. J. Nanopart. Res. 14(7), 966–987 (2012)
116. M. Erol, A.R. Boccaccini, Nanoscaled bioactive glass particles and nanofibres. Bioact. Glasses: Mater. Prop. Appl. 129–161 (2011)
117. W.H. Huang et al., Kinetics and mechanisms of the conversion of silicate (45S5), borate, and borosilicate glasses to hydroxyapatite in dilute phosphate solutions. J. Mater. Sci. Mater. Med. **17**(7), 583–596 (2006)
118. B. Zberg, P.J. Uggowitzer, J.F. Loffler, MgZnCa glasses without clinically observable hydrogen evolution for biodegradable implants. Nat. Mater. **8**(11), 887–891 (2009)
119. B.B. Yu et al., Effect of calcium source on structure and properties of sol-gel derived bioactive glasses. Langmuir **28**(50), 17465–17476 (2012)

Chapter 11
Topological Constraints, Rigidity Transitions, and Anomalies in Molecular Networks

M. Micoulaut, M. Bauchy and H. Flores-Ruiz

Abstract In this chapter, we present the first connection between realistic atomic scale simulations and topological constraint theory which has been introduced in the context of rigidity transitions of network glasses. Such rigid constraints can be computed rather simply by changing composition at low temperature but their estimates as a function of temperature or pressure remains challenging. We introduce and describe a method based on the calculation of standard deviations of relevant neighbor or partial bond-angle distributions which allows to estimate with confidence atomic stretching and bending topological constraints. The counting is illustrated from several archetypal liquids and glasses, including oxides and chalcogenides (SiO_2, Ge_xSe_{1-x},...). These results permit connecting the role of mechanical constraints in disordered systems to elucidating some of its most intruiging features (adaptation), with calculated anomalies in structural and dynamic properties.

11.1 Introduction

Detailed compositional studies of materials are cumbersome and challenging given that the number of possible compositions is infinite, this statement being true for both experimental and theoretical investigations. For instance, in glass science only a microscopic fraction of the compositional phase space has been explored, starting from reference compositions of proven fundamental or technological interest.

M. Micoulaut (✉) · H. Flores-Ruiz
Laboratoire de Physique Thorique de la Matire Condense, 4, Place Jussieu, 75252 Paris Cedex 05, France
e-mail: mmi@lptl.jussieu.fr

H. Flores-Ruiz
e-mail: hugomarcelofr@gmail.com

M. Bauchy
Department of Civil and Environmental Engineering, University of California, 5731B Boelter Hall, Los Angeles, CA 90095-1593, USA
e-mail: bauchy@ucla.edu

C. Massobrio et al. (eds.), *Molecular Dynamics Simulations of Disordered Materials*, Springer Series in Materials Science 215,
DOI 10.1007/978-3-319-15675-0_11

Currently many theorists have turned their attention to atomistic simulations to determine the structure of such disordered systems. However, while the available computing power has increased exponentially over the past several decades, a large enough computing power for direct molecular dynamics (MD) simulations of glass on a realistic laboratory time scale will not be available in a foreseeable future. One has therefore to rely on approaches that permit to anticipate compositional trends in physical and chemical properties using a minimal set of insightful input parameters. Rigidity theory [1] provides such a possibility because it focuses only on the key microscopic physics governing the thermal, mechanical, and rheological properties of glasses, while filtering out unnecessary details which ultimately do not affect the overall properties. Rigidity theory has enabled accurate predictions of glass compositions where elastic phase transitions can be expected [2, 3]. A certain number of limitations exist, however.

In the present chapter, we describe a very general method using the statistical mechanical behavior of relevant atomic-scale quantities (radial and angular distributions), which allows to compute such topological constraints without any prerequisites on coordination numbers, bonds or chemical interactions. We rely on a precise atomic-scale approach (as molecular dynamics) which is able to substantiate and enrich the general trends of rigidity theory via the explicit account of the local structure. This allows to bridge the gap between constraint counting algorithms based mostly on coordination numbers and phenomenological approaches which attempt to extend the counting by taking into account additional dependencies. Using these new tools, one is now able to investigate effects of other thermodynamic variables such as pressure and temperature, and to establish new correlations between the constraint behavior, and structural and dynamical quantities that are usually calculated from atomic simulations.

We review a certain number of results that have emerged in this field during the past years with a special emphasis on network glasses and network glass-forming liquids. In terms of systems, we focus here on a set of different chalcogenides and oxide glasses that are investigated as a function of composition, temperature or pressure: Ge_xSe_{1-x}, As_xSe_{1-x}, SiO_2, GeO_2, and 2 SiO_2–Na_2O. For oxide stoichiometric compositions, we show that such systems contain tetrahedra which act as rigid units having a much smaller angular excursion for the Group IV (Ge,Si) element than their chalcogenide counterparts. We also find that such excursions for oxygen in oxides are much larger than those for selenium in $GeSe_2$, suggesting that oxygen bending constraints are, in fact, *broken* (i.e. ineffective) in SiO_2 and GeO_2. This provides a microscopic rationale for the large excursions of bond-angles around oxygen in oxide networks, consistent with experimental evidence [4].

When investigated as a function of composition, we show that structural changes are mostly noticeable in the angular environment of the As/Ge atoms when moving from a flexible to a stressed-rigid phase. It is found that the amplitude of the angular excursions around Ge atoms is increased in the stressed-rigid phase with an increased distortion of the tetrahedra. Finally, we establish a correlation between the softening of certain constraints and computed anomalies in structural and dynamic properties in rigidity tuned by pressure. This leads ultimately to the definition of the pressure analogue of the Boolchand phase [2] which is observed experimentally in rigidity transitions driven by composition.

11.2 Topological Constraint Counting

11.2.1 Rigidity Transitions: Successes and Limitations

Rigidity theory provides an elegant atomic scale approach to understanding the physico-chemical behavior of network glasses. It is mostly based on the pioneering contributions of Lagrange and Maxwell [5, 6] who introduced the concept of constraints in mechanics and its application to the stability of structures such as trusses and bridges in terms of their mean number of constraints (Fig. 11.1). These ideas were extended to disordered atomic networks by Phillips [7–9] who recognized that glass forming tendency of covalent alloys was optimized for particular compositions. Specifically, it was recognized that stable glasses have an optimal connectivity, or mean coordination number $\bar{r} = \bar{r}_c$, which satisfies exactly the Maxwell stability criterion of mechanically isostatic structures, or the condition, $n_c = n_d$, where n_c is the count of atomic constraints per atom and n_d the network dimensionality, usually 3.

In covalent glasses the dominant interactions are usually near-neighbor bond-stretching (BS) and next-near-neighbor bond-bending (BB) forces, the number of constraints per atom can be exactly computed in a mean-field way, and is given by:

$$n_c = \frac{\sum\limits_{r\geq 2} n_r[\frac{r}{2} + 2r - 3]}{\sum\limits_{r\geq 2} n_r} \tag{11.1}$$

Fig. 11.1 An atomic structure of $SnSe_2$ obtained from First-Principles Molecular Dynamics [10]. The resulting structure can be viewed in terms of atomic positions but can be also be viewed as a bar (bonds) and node (atoms) network that can be analyzed using Maxwell-Lagrange constraints

where n_r is the concentration of species being r-fold coordinated. The contribution of the two terms in the numerator is obvious. Each bond is shared by two neighbors, and one has $r/2$ BS bond-stretching constraints for a r-fold atom. For BB (angular) constraints, one notices that a twofold atoms involves only one angle, and each additional bond needs the definition of two more angles, leading to the estimate of $(2r-3)$. For one-fold terminal atoms, a special count [10] is achieved as no BB constraints are involved. By defining the network mean coordination number $\bar{r}$ of the network by:

$$\bar{r} = \frac{\sum_{r\geq 2} r n_r}{\sum_{r\geq 2} n_r} \tag{11.2}$$

one can reduce (11.1) to the simple equation:

$$n_c = \frac{\bar{r}}{2} + 2\bar{r} - 3 \tag{11.3}$$

Applying the Maxwell stability criterion, isostatic glasses ($n_c = 3$) are expected to be found at the magic number [7] of $\bar{r} = 2.40$ in 3D, corresponding usually to a non-stoichiometric composition where glass-forming tendency has been found to be optimized experimentally [12, 13].

In addition, the deeper origin of this stability criterion has been revealed from the vibrational analysis of bond-depleted random networks constrained by bond-bending and bond-stretching interactions [14]. It has been found, indeed, that the number of zero frequency (floppy) modes f (i.e. eigenmodes of the dynamical matrix) is vanishing for $\bar{r} = 2.38$ when rigidity percolates in the network. The condition $n_c = n_d$ therefore defines a mechanical stiffness transition above which redundant constraints produce a stiff and internally stressed networks, identified with a stressed-rigid phase [15, 16]. For $n_c < n_d$ however, floppy modes can proliferate and lead to a flexible phase where local deformations with a low cost in energy (typically 5 meV [17]) are possible. With the prediction of such thresholds and their observation in various chemical and physical properties in chalcogenide glasses [18, 23], the connectivity related flexible to stressed-rigid elastic phase transition has become the focus of modern theory of network glasses as it allows to analyze in depth compositional trends of their properties.

Experimental signatures of this peculiar transition has been found essentially in chalcogenide glasses from Raman scattering [18], stress relaxation and viscosity measurements [19], vibrational density of states [17], Brillouin scattering [20], Lamb-Mossbauer factors [21], resistivity [22], and Kohlrausch fractional exponents [23].

11.2.2 Intermediate Phases

New features in the field of rigidity theory have emerged from experiments by Boolchand that have opened new questions and areas of investigations and subsequent perspectives. First, the underlying nature of the onset of rigidity in glasses has been challenged because two thresholds have been found experimentally [24, 25] instead of the previously reported solitary flexible to rigid transition [15]. These two thresholds define an intervening region, an intermediate phase (IP), between the flexible and the stressed rigid phase. Theory from simple lattice [26] or cluster models [27] has shown that the thresholds were associated with percolation of rigidity ($n_c = 3$ on average) at the low connectivity end, and with percolation of stressed rigidity ($n_c > 3$) at the high connectivity end of the IP. This automatically provides a finite width for nearly isostatic compositions.

Glasses having such IP compositions display some unusual properties, the most prominent one being revealed in calorimetric measurements, which show a nearly vanishing non-reversing enthalpy of relaxation of the glass transition endotherm (reversibility windows, [28–30]). It captures most of the relaxational events during the glass transition (Fig. 11.2, left panel). For many other glasses, spectacular properties have been observed such as weak ageing phenomena [33] that are essentially driven by the nearly vanishing of the enthalpic relaxation and, correspondingly, fragility minima of the melts [19, 34, 35], absence of internal stress [36], anomalous mechanical properties [37], enhanced thermal stability [38, 39]. In certain glasses

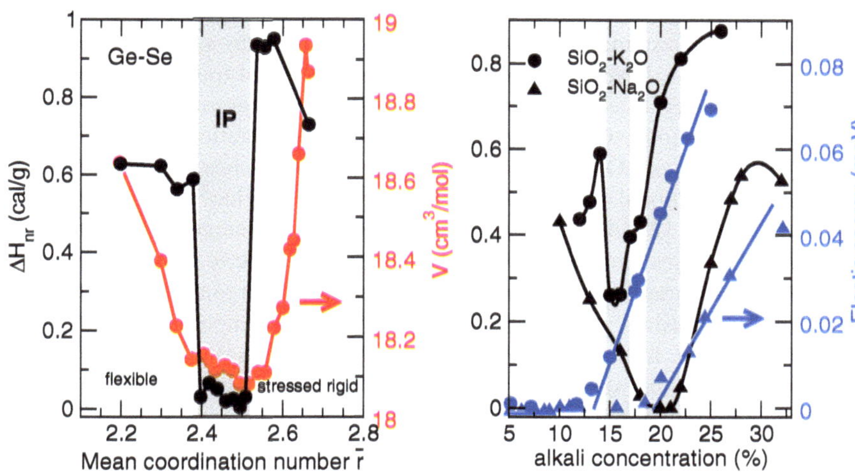

Fig. 11.2 Non-reversing heat flow ΔH_{nr} in Ge–Se (*left panel*, *black symbols* [28–30]) and alkali silicates (*right panel*, *black symbols* [31, 32]), compared to molar volume (*left panel*, *right axis*, *red symbols*) and released elastic energy under annealing (*right panel*, *right axis*, *blue symbols*). The *gray zone* defines approximatively the IP. Note that the global minimum ΔH_{nr} found in Ge–Se (*left*) results from samples of proven homogeneity, in contrast to e.g. potassium silicates (*right*)

such as Group IV selenides or sulphides (Ge–S, Si–Se), the location of these compositional zones or windows is very close to the value $\bar{r} = 2.4$, but in many other glasses one can not infer simply what aspects of structure lead to a shift of the IP with respect to the mean-field rigidity transition [40–42]. These results indicate that glasses can be now classified both from their mechanical and from their thermal properties; the IP was discovered in a mechanical context, and subsequently has been identified from temperature-dependent properties of supercooled melts, and especially the reversibility of the glass transition itself. However, these findings are not restricted to glasses and the glass transition as links between the IP and protein folding [43], high-temperature superconductors [44] or computational phase transitions [45] have been stressed that go much beyond simple analogies. This simply underscore the ability of a complex network to rearrange by adapting internal thermodynamic variables [46] to applied constraints, stress, or conditions. The understanding of the IP is therefore of broad interest as it appears to be a generic feature of disordered networks, and finally encourages the reinvestigation of archetypal glasses within this framework.

Another striking feature is that these observations, and the concepts from rigidity theory, are not restricted to chalcogenide alloys as initially believed. The same reported transitions, phenomena, anomalies are now observed and characterized in oxides ([31, 32, 47], Fig. 11.2, right panel), heavy-metal-oxide glasses [48] or hydrogenated silicon [49]. Given the dominant presence of silicate glasses in glass industry, it also opens the perspective to design applications by alloying elements in order to obtain desired IP properties.

The discovery of the IP has not been without controversy due, in part, to the absence of any structural signature from diffraction [50] (see however [51, 52]) or Nuclear Magnetic Resonance [53], whereas calorimetric, thermal, and optical signatures have been systematically found. Most of the criticisms against the IP and the reversibility window have not relied on any solid theoretical framework that can be carefully examined (i.e. technically) and reproduced. However, such qualitative models have continued to be waved in the literature as a demonstration for the absence of the intermediate phase. On the other hand, rigorous approaches on model networks using either a vibrational eigenmode analysis [26], the spin glass cavity method [46], cluster expansions [27, 45], Pebble game algorithms on triangular lattices [55] or Molecular Dynamics simulations of glasses [56–59], all have arrived to the conclusion that structural features including the possible self-organization, will lead to two transitions/thresholds defining a window over some compositional interval. At this stage however, it is not clear what is the underlying nature of the associated transitions, rigidity and stress, and if these transitions manifest in discontinuities of thermodynamic functions.

11.2.3 Limitations

Even in its mean-field version, rigidity theory contains some obvious limitations that can be detected by the simple inspection of (11.1).

In its most general mean-field version, rigidity theory does not make a distinction between isovalent systems [60], nor does it take into account the role of atomic sizes. These are limitations that largely come from the topological nature of the theory which relies essentially on the coordination number r of the atoms. As an immediate consequence, rigidity effects in Ge–Se should be similar to those in the isovalent Si–Se, or those in sodium silicates similar to rubidium silicates, in obvious contrast with what is observed experimentally, at least in silicates [61]. This has opened new challenges in the application of constraint counting in modified oxide glasses. In fact, in the search of applications of oxide glasses, physical and chemical properties are often tuned by simply changing the nature of the modifier (usually an alkali or alkaline earth oxide). At a first glance, and without considering the nature of the modifier atom, it is obvious that the depolymerization of a silica-rich network must reduce network connectivity through the creation of non-bridging oxygen (NBO) which must lead at some point to a flexible network [32]. However, if one wants to predict elastic thresholds with accuracy, additional details are necessary since constraint counting is not as straightforward as in the covalent chalcogendides. In the latter, the 8-N rule (where N is the number of outer shell electrons) is usually found to uphold [62], and the coordination number can be deduced with confidence. Moreover all interactions constraining the network (BS and BB) can be considered as intact [18, 23]. When moving to other non-chalcogenide glass networks, a central issue is therefore to properly define coordination numbers, and mechanically effective coordination numbers that contribute to constraints.

One also realizes that the enumeration of bonding constraints is performed over a fully connected network, i.e. at $T = 0\,K$ when neither bonds nor constraints are broken by thermal activation. This situation may hold as long as one is considering strong covalent bonds or when the viscosity η is very large at $T < T_g$, given that η is proportional to the bonding fraction, but equation is obviously not valid in a high temperature liquid. Similarly, in (11.1) it is implicitly assumed that the average constraint count is performed over all the atoms of the network which implies homogeneity of the system at the macroscopic scale, while neglecting the possibility of atomic-scale phase separation. Attempts have been reported which propose a modification of (11.1) that incorporate the effect of a decoupling of the network backbone in stressed rigid glasses [63] to account for specific observed features.

However, NMR-related relaxational phenomena in Ge–Se indicate that the low temperature rigidity concept can be extended from the glass to the liquid in binary chalcogenide melts with confidence [64]. Building on this idea, Gupta and Mauro have extended topological constraint counting to account quantitatively for thermal effects [9] *via* a two state thermodynamic function q(T), which quantifies the number of rigid constraints as a function of temperature [37] and subsequently modifies (11.1) to become:

$$n_c = \frac{\sum\limits_{r\geq 2} n_r [q_r^{\alpha}(T)\frac{r}{2} + q_r^{\beta}(T)(2r-3)]}{\sum\limits_{r\geq 2} n_r}, \qquad (11.4)$$

where q_r^{α}(T) and q_r^{β}(T) are step functions associated with BS and BB interactions of a r-coordinated atom. Equation (11.4) and n_c now depends explicitly on temperature. This function has two obvious limits because all relevant constraints can be either intact at low temperature ($q(T)=1$) or entirely broken ($q(\infty)=0$) at high temperature. At a finite temperature however, a fraction of these constraints are broken by thermal activation. Different forms can be proposed for q(T) based either on an energy landscape approach [65] or involving a simple activation energy for broken constraints [66]. A simple step-like function allows obtaining analytical expressions for fragility and glass transition temperature [67], heat capacity [68] and hardness [69] as a function of composition for binary and multicomponent glasses, and has set the basis of "topological engineering" of multicomponent glassy materials.

At this point, one may wonder if such topological constraints can be defined as a function of another obvious thermodynamic variable, for example pressure. However, in contrast with the previous derivations of constraints from composition or temperature (11.4), one can not compute straightforwardly the number of rigid constraints n_c(P) with pressure, given that the coordination change with pressure is highly system dependent. However, the issue of establishing a topological constraint count for densified liquids and glasses appears to be quite attractive as anomalous behaviors are found in structural, dynamical or thermodynamic properties [70, 71, 127] which bear similarities with those observed in the intermediate phase at ambient pressure [29, 72, 73], underscoring a possible common physical origin. But before a unified picture can be proposed, one needs to derive a precise computation of n_c(P). Rigidity induced by pressure has been obtained in simple network formers from Molecular simulations [56], although an explicit estimate of n_c(P) at the same simple and elegant level as the one used in (11.1) and (11.4) for x and T was not reported.

11.3 Motion Instead of Forces

A general and alternative approach can be proposed in order to establish the number of topological constraints $n_c(x, T, \mathrm{P})$ for any thermodynamic condition using Molecular Dynamics (MD). In both MD's versions, classical or First Principles (FPMD) using the Car-Parrinello scheme, Newton's equation of motion is solved for a system of N atoms or ions, representing a given material. Forces are either evaluated from a model interaction potential which has been fitted to recover some materials properties, or directly calculated from the electronic density in case of a quantum mechanical treatment using density functional theory (DFT).

Can topological constraints be extracted from the interaction potentials involved in such brute force methods? Obviously not since most of them will not lead to a reduction of solely BS and BB constraining interactions which are the relevant ones for the problem of our interest. As in classical mechanics however, instead of treating mathematically the forces and querying about motion which is of course the standard procedure of MD simulations, and the one of any textbook mechanical problem, one may follow the opposite direction and try to relate the atomic motion to the absence of a restoring force, whatever its origin. Indeed, in following this path that moves us away from the "Culture of force" analyzed by Wilcek [74], one does not necessarily need to formulate the physical origin of the forces as only the resulting motion is considered. However, at an early stage of the analysis it is checked that the upstreamed integration of motion from model forces has been correctly achieved, and that the resulting structure accurately reproduces a certain number of experimental observables.

11.3.1 Radial and Angular Standard Deviations

In the case of atomic scale systems, since one attempts to enumerate BS and BB constraints, one is actually not seeking for motion arising from large radial and angular excursions, but for the opposite behavior and also for atoms displaying a small motion readily maintaining corresponding bonds and angles fixed around a mean value. These can ultimately be identified with a BS or BB interaction constraining the network structure at the molecular level.

This is the starting consideration of the present approach that is combined with Molecular Dynamics simulations. Having generated the atomic scale configurations at different thermodynamic conditions from MD, we then apply a structural analysis in relation to rigidity theory as discussed in the following. Before, one needs to set up the simulation scheme.

11.3.1.1 Setting Up Simulations

The forthcoming analysis and discussions are based on atomic scale trajectories obtained at various thermodynamic conditions. We focus on two families of amorphous and liquid chalcogenides with different compositions (Ge–Se and As–Se) at ambient pressure by using FPMD simulations, and on network forming (GeO_2 and SiO_2) or modified oxides (2 SiO_2–Na_2O, NS2 hereafter), the latter being investigated from classical MD as a function of pressure and temperature.

For both germanium-selenium (120 atoms) and arsenic-selenium (200 atoms) systems, FPMD simulations have been performed for different compositions, including IP compositions, at constant volume. A periodically repeated cubic cell has been used, and the sizes used allowed to recover the experimental density of the glasses [28, 40] with pressures that do not exceed 0.5 GPa. The electronic structure has been described within DFT and evolved self-consistently during the motion [75]. We have adopted here a generalized gradient approximation using the exchange energy obtained by Becke [76], and the correlation energy according to Lee, Yang and Parr (BLYP) [77]. Valence electrons have been treated explicitly, in conjunction with norm conserving pseudopotentials of the Trouiller-Martins type to account for core-valence interactions. The wave functions have been expanded at the Γ point of the supercell on a plane wave basis set with an energy cutoff $E_c = 20$ Ry. A fictitious electron mass of 200–400 a.u. has been used for the first-principles molecular dynamics (FPMD) approach, and the time step has been set to $\Delta t = 0.12$ fs to integrate the equations of motion. Temperature control has been implemented for both the ionic and electronic degrees of freedom by using Nosé-Hoover thermostats. Additional details on the simulation can be found in [78, 79].

The initial coordinates of the systems have been obtained from previous simulations [80, 81] on either the stoichiometric compound ($GeSe_2$) or from a random setup in a high (As_2Se_3) liquid prior to a long initial stage (100 ps) devoted to the memory loss of the initial configuration. For off-stoichiometric compositions, atoms have been randomly replaced by Se ones, prior to an initial simulation of 25 ps at 2000 K in order to loose, again, memory of the initial configuration. Additional runs of each 25 ps have been performed at temperatures of $\simeq$1400 and 1200 K. At the latter temperature, several independent configurations separated by a time interval of 5 ps have been chosen and have served as starting configurations of independent quenches to inherent structures, performed at an average cooling rate of $\simeq$10 K/ps. Finally, for each composition statistical averages have been performed at 300 K over the quenched samples accumulated over a time interval of 60–84 ps.

Oxides (SiO_2, GeO_2 and NS2) have been simulated on much larger systems (800–3000 atoms) from MD simulations using a classical Born-Mayer force field of the form:

$$V_{ij}(r_{ij}) = \frac{q_i q_j}{r_{ij}} + A_{ij} e^{-B_{ij} r_{ij}} - \frac{C_{ij}}{r_{ij}^6} \tag{11.5}$$

with parameters reported in the literature [82–84], and which have been fitted to reproduce physical properties of SiO_2, GeO_2 and NS2 in the crystalline and/or amorphous phase. The equations of motions for atoms have been solved with a time step of 1–2 fs by the leap-frog algorithm. For instance, liquid NS2 has been simulated by placing 667 silicon, 667 sodium, and 1666 oxygen atoms in a cubic box with periodic boundary conditions and changing densities. As the atoms are bearing charges (ions), the long range coulombic forces are evaluated by an Ewald sum.

11.3.1.2 Comparison with Experiments

Structure We first compare the simulated structures with available experimental data on structure. Figure 11.3 shows the total structure factor S_T(k) for a selection of systems such as GeO_2, $GeSe_3$ and As_2Se_3. Results for NS2 are provided below. An excellent agreement of the calculated S_T(k) with the experimental counterparts [85–87] is found, with all typical features being reproduced over the entire range of wavevectors k: first principal peak (PP1) at $\simeq$1–2 Å^{-1} and second principal peak (PP2) at $\simeq$4–5 Å^{-1}, depending on the system. The high wavector region ($k >$10–12 Å^{-1}) is also very well reproduced and leads to a very good agreement when the

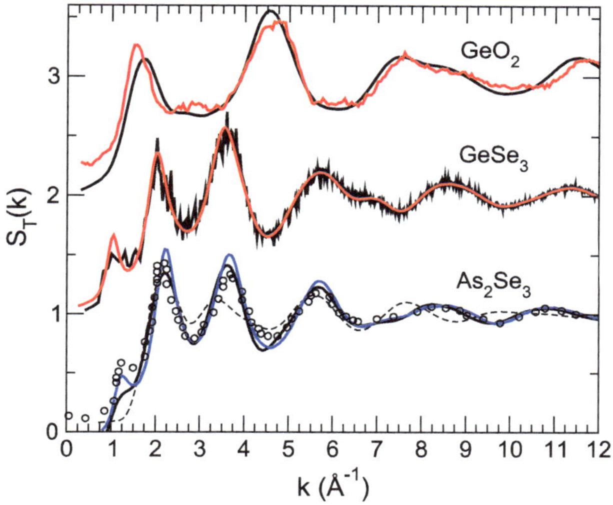

Fig. 11.3 Total simulated neutron-weighted structure factor S_T(k) for three selected glassy compounds, compared to experimental data: GeO_2 [85], $GeSe_3$ [86], and As_2Se_3 [87, 88]. The *broken curve* (for As_2Se_3) is a result from a classical force field [89]

interference function I(k) = k[S_T(k)–1] (which blows up the oscillations found at higher k value) is considered [5]. Similar agreement is found for the partial structure factors when such quantities are available [81, 85, 90]. Note that the first sharp diffraction peak, while rather small in certain experiments [81, 90], reduces in some simulations to a simple shoulder (As_2Se_3, [51, 91]) on the low wavevector side of the main peak, or can be fully acknowledged as in $GeSe_3$. On the other hand, one notices that force fields used in classical MD [89] fails to reproduce the main features of the measured structure factor in As_2Se_3. This simply underscores the fact that the structure of chalcogenides with covalent bonding can be hardly reproduced from a classical treatment of the inter-atomic interaction which implicitly neglects the possibility of charge transfer during the simulation, and such charge transfers obviously govern the presence of homopolar defects [92, 93] which affect the structure factor. However, even for FPMD simulations, it has been shown that the reproduction of all the structural features, including the function S_T(k) and the coordination and chemical defects, was depending substantially on the chosen electronic model for the exchange-correlation [80, 94]. An analysis of the partial structure factors in connection with rigidity transitions is detailed below.

In real space, we represent structural correlations for, again, a selected number of compounds (Fig. 11.4) such as As_2Se_3 and $GeSe_2$, and the positive conclusions holding for the reciprocal space ar still valid for the total pair correlation functions g(r) (Fig. 11.4c). In panel a), the simulation result for $GeSe_2$ can be directly compared to measurements from isotopic substituted neutron diffraction [95], and these show a very good agreement. For all pairs (Ge–Ge, Ge–Se and Se–Se), most features

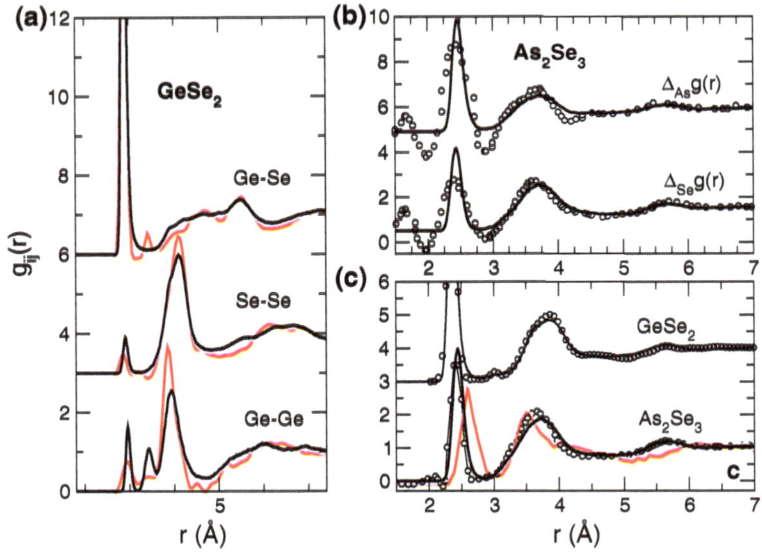

Fig. 11.4 **a** Calculated partial pair correlation functions $g_{ij}(r)$ (*black*) of amorphous $GeSe_2$, compared to experiments (*red* [95]). **b** Calculated partial differential pair distribution functions $\Delta_{As}g(r)$ and $\Delta_{Se}g(r)$ (*solid lines*) in amorphous As_2Se_3, compared to experimental data (*circles*, [97]) obtained from anomalous X-ray scattering. **c** Calculated total pair correlation function g(r) for amorphous $GeSe_2$ and As_2Se_3, and compared to experimental data [86, 88]. The *red curve* is the result from a classical force field [89]

(peak position and peak widths) are very well reproduced as are also the pre-peaks reflecting homopolar bondings. Distances found for such bondings (2.44 and 2.37 Å for Ge–Ge and Se–Se, respectively) compare very favourably with the experimental estimate (2.42 and 2.32Å, respectively). The partial g_{GeGe} (Fig. 11.4a) also exhibits two other peaks at 3.03 and 3.68 Å which can be unambiguously assigned [81] to Ge–Ge correlations appearing respectively in edge-sharing (ES) and corner-sharing (CS) tetrahedra (Fig. 11.5), as also identified by Salmon and co-workers [86], the former being highly sensitive to the residual pressure as recently demonstrated [96]. The partial g_{SeSe} has essentially two main contributions : the Se–Se homopolar prepeak

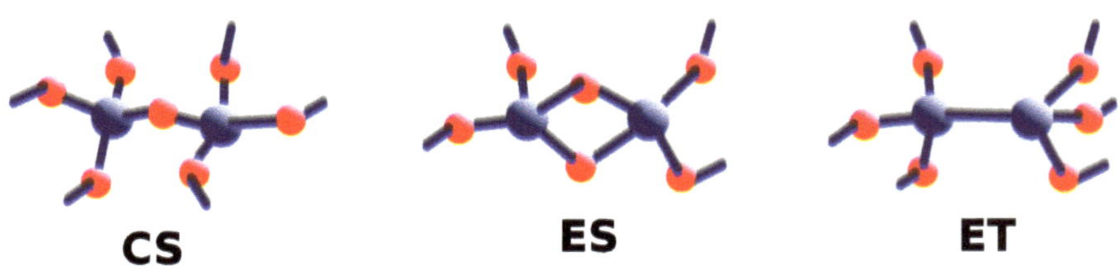

Fig. 11.5 Typical bonding types encountered in Ge–Se glasses: corner-sharing (CS) tetrahedra, edge-sharing (ES) tetrahedra, and ethane-like (ET) units involving a homopolar Ge–Ge bond

found at 2.37 Å and the main peak at 3.88 Å which is associated with the distance defining the edge of a $GeSe_{4/2}$ tetrahedron. Finally, the main feature of the function g_{GeSe} consists in a very intense peak at 2.36 Å, the Ge–Se distance defining the tetrahedra, which is very well separated from the secondary contributions.

Figure 11.4b shows the computed differential pair distribution functions [91] for amorphous As_2Se_3 which agree successfully with results from anomalous X-ray scattering [97, 98]. The peaks found in the differential function $\Delta_{As}g(r)$ actually result from a homopolar distance at 2.59 Å that is evidenced in the corresponding As–As partial, and from the distance As–Se (2.46 Å) defining the pyramidal $AsSe_{3/2}$ unit. Similarly, the low distance behavior of $\Delta_{Se}g(r)$ is due to the homopolar distance Se–Se found at 2.35 Å, whereas the main peak at 3.75 Å (experimentally, 3.64 Å [97]) is related to the Se–Se distance being also part of the pyramidal unit. The secondary peak at 3.70 Å in both the total pair correlation function and in $\Delta_{As}g(r)$ is associated with As–As correlations between two pyramids, as also proposed by Hosokawa and co-workers [97, 98]. These obtained bond distances are found to be in excellent agreement with those determined experimentally: 2.42 Å for As–Se, and 3.7 Å for the secondary peak from neutron [87], and 2.42 Å for As–Se from X-ray diffraction [97]. One should finally note that as the As–As homopolar bond distance (2.59 Å) is quite close to the As–Se distance (2.46 Å), one can not detect such As–As motifs from the total pair distribution function (Fig. 11.4c) which is dominated by the high intensity of the first As–Se peak.

Dynamics and transport: In the liquid state, we first compute the mean-square displacement of an atom of type α in the melt, given by

$$\langle r^2(t)\rangle = \frac{1}{N_\alpha}\sum_{i=1}^{N_\alpha}\langle|\mathbf{r}_i(t) - \mathbf{r}_i(0)|^2\rangle, \tag{11.6}$$

and extract from the dependence of $\langle r^2(t)\rangle$ the long time behavior where the dynamics becomes diffusive. Using the Einstein relation at long times $\lim_{t\to\infty}\langle r^2(t)\rangle/6t = \mathrm{D}$, one can have access to the diffusion constants D for various species from the mean square displacement $\langle r^2(t)\rangle$. When experimental data is available (e.g. NS2 [99, 102]), self-diffusion constants are plotted for the system of interest as a function of the inverse temperature, revealing an Arrhenius behavior. Because of the slowing down of the dynamics at low temperature, the computation of D is here restricted to $T > 1500$ K. The diffusion of sodium atoms is found to be remarkably close (Fig. 11.6) to the experimental data obtained by Negodaev et al. [99], and Gupta and King [100], with a very good agreement around $10^4/T \simeq 6$. The same level of agreement holds for the network forming ions (Si, O) as the computed diffusion constants compare very favourably to the experimental data of Truhlarova et al. [102]. The chosen Teter potential [84] appears thus to be highly accurate in reproducing not only the structure but also the self-diffusion of elements in the NS2 melt, and appears to be substantially better in this respect when compared to alternative potentials [103] which lead to self-diffusion coefficients that are found to be at least one order of magnitude lower than the experimental values [104].

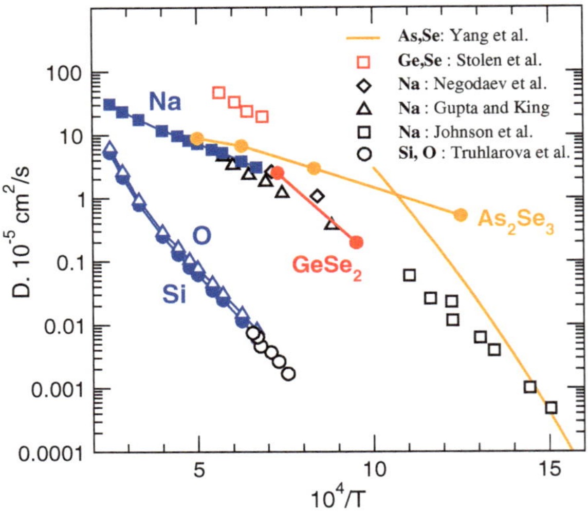

Fig. 11.6 Computed diffusion constants D_{Na}, D_{Si} and D_O in NS2 (*blue curves* and *symbols*), D_{Ge}, and D_{Se} in $GeSe_2$ (*red curves* and *symbols*), and D_{Se} in As_2Se_3 (*orange curve* and *symbols*) as a function of inverse temperature. The NS2 is compared to experimental data for D_{Na} [99–101] and D_{Si}, D_O [102] (*open symbols*). The chalcogenides are compared to estimates using viscosity data from Stolen et al. ($GeSe_2$, *open red squares* [105]) and Yang et al. (As_2Se_3, *orange curve* [106]). See text for details

We are not aware of any experimental data for diffusion in chalcogenides but can estimate rather accurately a mean diffusion constant from viscosity data using the Eyring relation $D = k_B T/\lambda\eta$, where λ a typical hopping length for the diffusing atom [61]. In silicate melts, this equation holds very well for deeply supercooled melts [107] with $\lambda = 2.8$ Å, a distance typical of Si–Si and O–O separation. From reported viscosity data on Ge–Se and As–Se systems [105, 106], we proceed similarly and derive an average diffusion coefficient by postulating a reasonable value for the hopping length corresponding to the Se–Se bond distance (3.7 Å [90]). We then obtain diffusion coefficients from viscosity data [105, 106] that can be compared to the values computed from the simulated trajectories using the Einstein equation (Fig. 11.6).

For the largest system (NS2), the computation of viscosity can be performed by using the Green-Kubo (GK) formalism [108] which is based on the calculation of the stress tensor auto-correlation function, given by:

$$\eta = \frac{1}{k_B T V} \int_0^\infty \langle P_{\alpha\beta}(t) P_{\alpha\beta}(0) \rangle \tag{11.7}$$

using off-diagonal components $\alpha\beta$ $(\alpha, \beta) = $(x,y,z) of the molecular stress tensor $P_{\alpha\beta}(t)$ defined by:

$$P_{\alpha\beta} = \sum_{i=1}^{N} m_i v_i^{\alpha} v_i^{\beta} + \sum_{i=1}^{N} \sum_{j>i}^{N} F_{ij}^{\alpha} r_{ij}^{\beta} \quad \alpha \neq \beta, \tag{11.8}$$

where the brackets in (11.7) refer to an average over the whole system. In (11.8), m_i is the mass of atom i, and F_{ij}^{α} is the component α of the force between the ions i and j, r_{ij}^{β} and v_i^{β} being the β component of the distance between two atoms i and j, and the velocity of atom i, respectively.

Figure 11.7 shows the viscosity of the NS2 liquid at ambient pressure as a function of inverse temperature, and it can be seen that the computed value [107] is in very good agreement with experimental measurements from Bockris et al. [108] and from Neuville [109]. To gain additional insight into the relaxational behavior, we use a functional form for the viscosity-temperature behavior given by the Vogel-Fulcher-Tamman law [111]:

$$\log_{10} \eta(T) = \log_{10} \eta_{\infty} + \frac{(12 - \log_{10} \eta_{\infty})^2}{\mathscr{M}(T/T_g - 1) + (12 - \log_{10} \eta_{\infty})}, \tag{11.9}$$

where T_g is the glass transition temperature defined by $\eta(T_g)$=10^{12} Pas.s, $\mathscr{M}$ is the fragility index, and $\log_{10} \eta_{\infty}$ is the extrapolated viscosity at infinite temperature. It is found that the parameters obtained from a fit of the numerical data using equation (11.9) agree quite well with those determined experimentally. We find indeed $\mathscr{M} = 26.1$, to be compared with $\mathscr{M} = 25.5$ from Bockris' data [109], and $\mathscr{M} = 44.3$ from Neuville's data [110], whereas the MD related glass transition temperature (calculated at $\eta = 10^{12}$ Pa.s) is found to be 575 K, to be compared with 632 and 726 K, respectively. One should stress that the reproduction is excellent although it is well-known that the reproduction of viscosity using the GK formalism is difficult

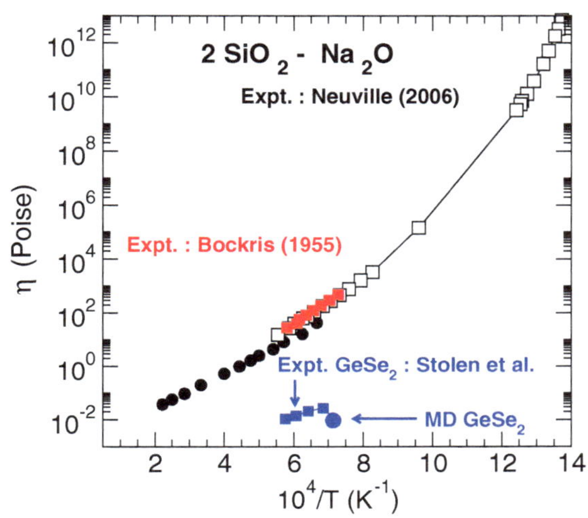

Fig. 11.7 Simulated viscosity of the NS2 liquid at zero pressure (*filled black circles* [107]), compared to experimental data from Bockris et al. (*red squares* [109]), and Neuville (*open squares* [110]). Simulated viscosity of a 1400 K $GeSe_2$ liquid (*blue circle*), compared to experimental data by Stølen et al. (*blue squares* [105])

[107]. Given the small size of the chalcogenide systems, we have not considered so far calculated viscosities except for a large system sized (480 atoms [112]) $GeSe_2$ for which one finds $\eta(1400\,K) = 0.010 \pm 0.005$ Pa.s at an equilibrium pressure of 0.6 GPa, which is very close to be measured value of Stolen et al. [105] who found $\eta \simeq 0.02$ Pa.s for 1450 K (Fig. 11.7).

In conclusion, these comparisons with experiments show that excellent structural models can be obtained either from FPMD simulations, or from classical MD using an accurate force field. This is a prerequisite before any investigation in the context of rigidity transitions. The reproduction of transport coefficients appears to be more challenging, and this is highlighted for the case of simulated liquid silica which is known to have diffusivities which are highly force field dependent, and lead to a spread in simulated data spanning over several orders of magnitude [113]. The excellent agreement found in the present systems allows one to be reasonably confident in the ability of exploring details of the dynamics in connection with rigidity transitions.

11.3.2 Bond-Stretching

Having obtained structural models of high accuracy, we can now turn to our main purpose, the enumeration of topological constraints from the atomic scale trajectories obtained by MD simulations (Fig. 11.8).

To obtain the number of bond-stretching interactions we focus on neighbour distribution functions (NDFs) around a given atom i [114]. A set of NDFs can be defined by fixing the neighbor number n (first, second etc.), the sum of all NDFs yielding the usual i-centred pair correlation function $g_i(r)$ whose integration up to the first minimum gives the coordination numbers r_i, and hence the corresponding number of bond-stretching constraints $r_i/2$ [7, 14]. Figure 11.9 shows such application to the As_2Se_3 glass and the 600 K $GeTe_4$ liquid [115]. In As_2Se_3, three NDFs (colored

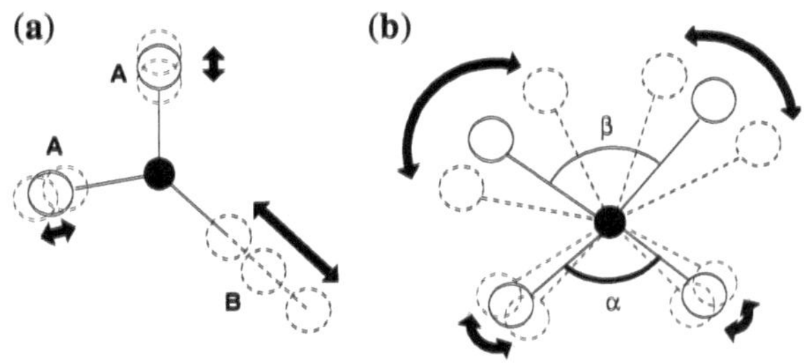

Fig. 11.8 Method of constraint counting from MD-generated configurations. Large (small) radial (**a**) or angular (**b**) excursions around a mean value are characterized by large (small) standard deviations on bond B or angle (small on A or a), representing broken (intact) constraints

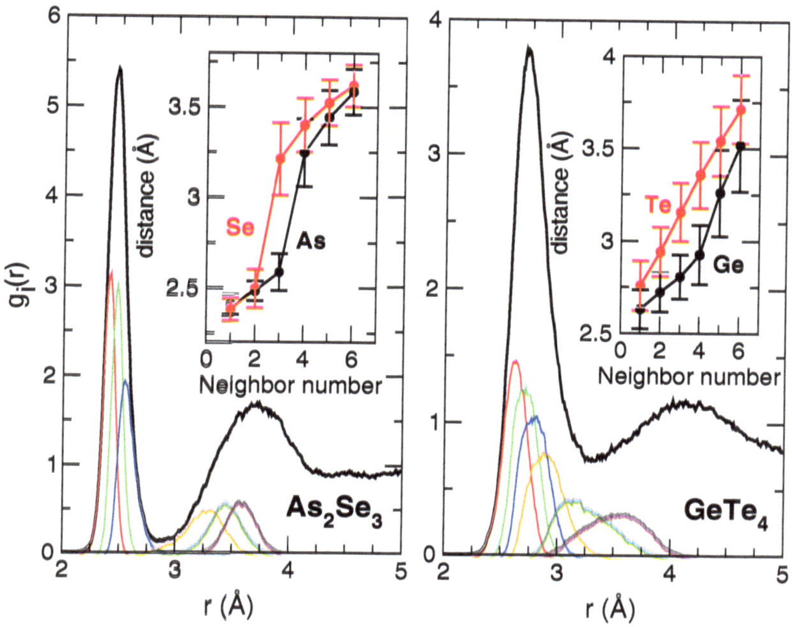

Fig. 11.9 Decomposition of partial pair correlation functions $g_i(r)$ into neighbor distributions in amorphous As_2Se_3 and liquid (450 K) $GeTe_4$. The *insets* show the positions (first moments) of the neighbor distributions and their standard deviations (second moments represented as *error bars*)

curves) contribute to the first peak of $g_{As}(r)$, very well separated from the second shell of neighbours, and indicative of the presence of three neighbours around an As atom. It is to be noticed that at the minimum of $g_{As}(r)$, one has already the fourth NDF, indicating that fourfold As atoms should be present in the glass [40], but in quite small amounts, typically less than 10 %. For the $GeTe_4$ liquid, one finds much more neighbors at the corresponding minimum ($r_i > 4$).

The separation between first and second shell of neighbours can be also characterized by plotting the NDF peak positions as a function of the neighbour number (insets of Fig. 11.9). For e.g. As_2Se_3, one remarks that there is a clear gap in distance between the third and the fourth neighbor, the first three NDFs displaying furthermore a much lower radial excursion (error bars, see Fig. 11.9) as compared to the NDFs of the next neighbors. Similar observations can be made for the Se atom. Thus, we find $r_{As} = 3$ and $r_{Se} = 2$ leading to 1.5 and 1 BS constraints in As_2Se_3. For the case of liquid $GeTe_4$, it is seen that the shell structure tends to disappear as the NDFs positions grow continuously with the neighbor number.

11.3.3 Bond-Bending

11.3.3.1 Average Behavior

Bond-bending (BB) constraint counting is based on partial bond angle distributions (PBADs) $P(\theta_{ij})$ and defined as follows: for each type of a central atom 0, the N first neighbours i are selected, leading to N(N − 1)/2 possible angles $i0j$ ($i=1\ldots N-1$, $j=2\ldots N$), i.e. 102, 103, 203, etc. The standard deviation $\sigma_{\theta_{ij}}$ (written as σ_θ or σ_i hereafter) of each distribution P(θ_{ij}) gives a quantitative estimate of the angular excursion around a mean angular value, and provides a measure of the bond-bending strength [116–118]. Small values for σ_θ correspond to an intact bond-bending constraint which maintains a rigid angle at a fixed value, whereas large σ_θ correspond to a bond-bending weakness giving rise to an ineffective or *broken* constraint.

Figure 11.10 shows the PBADs for glassy $GeSe_2$ and GeO_2. Broad distributions are found in most of the situations, together with a certain number of sharper distributions (colored) which are identified as intact angular constraints given that these arise from weak motions around an average bond angle. For instance, the relevant

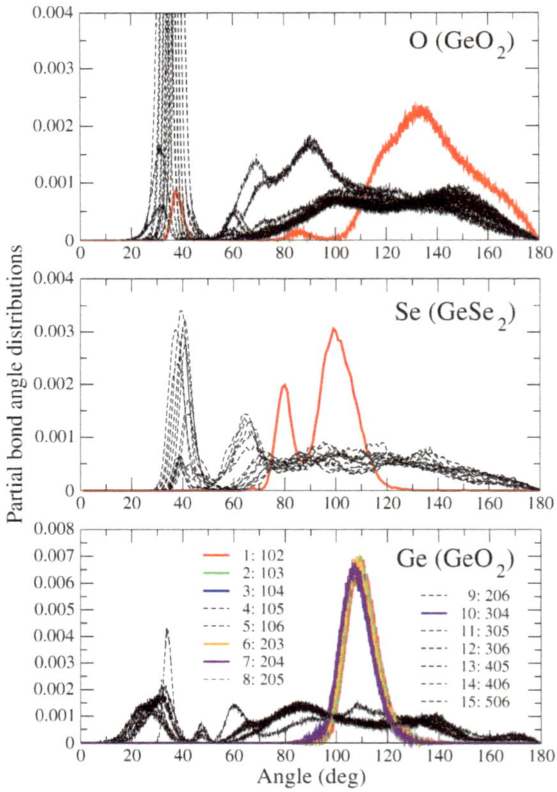

Fig. 11.10 From *top* to *bottom* oxygen, selenium and germanium partial bond angle distributions (PBAD) in GeO_2 and $GeSe_2$ for an arbitrary N = 6 [118]. The *colored curves* correspond to PBADs having the lowest standard deviation(s) σ_θ. The sharp peaks at $\theta \simeq 40°$ correspond to the hard-core repulsion. Labels defined in the *bottom panel* are used throughout the text

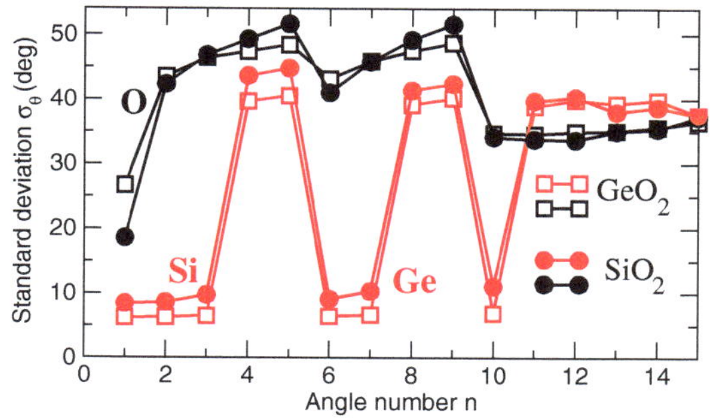

Fig. 11.11 Oxygen and Si/Ge standard deviations computed from 15 PBADs in vitreous germania and silica (see labelling in Fig. 11.10)

angle 102 around oxygen in germania (see labels in panel c) is found to be centred at 135°, close to the value obtained from experiments [119]. All other angles display broad variations and correspond to angles defined by next-nearest neighbor shells. One may thus anticipate from this trivial example that only one BB interaction constrains the oxygen atom. Similarly, the corresponding selenide distribution shows a bimodal distributions with peaks at 80° and 100°, indicative of edge- and corner-sharing tetrahedra [90], respectively (Fig. 11.5), a feature that is absent in the oxides. Figure 11.10c shows the results for the Ge-centred PBADs in germania. Among the 15 possible angles which are considered for the chosen set of N = 6 possible neighbor, only six angles have nearly identical and sharp distributions. These are the six angles defining the tetrahedra, and they are centred close to the "tetrahedral" angle of 109°.

From the N(N − 1)/2 different PBADs, one can now compute a second moment (or standard deviation) for an arbitrary set of triplets (i0j) with (i, j = 1…N). Figure 11.11 shows corresponding results for the standard deviations σ_θ for the stoichiometric oxide glasses. For all systems, the PBADs relative to the Group IV (Si, Ge) atom have a low standard deviation σ_θ, of the order of 10–20°. One finds e.g. $\sigma_{Ge} \simeq 7°$ for the PBAD 102 (angle number 1) of GeO_2, which is substantially smaller as compared those computed from other distributions (105, 106, etc.) which have $\sigma_\theta \simeq 40°$.

One also finds that oxide glasses have all the σ_θ nearly equal for the six relevant (Ge, Si) distributions which are associated with bending motions around the tetrahedral angle of 109°. A different situation occurs in the stoichiometric chalcogenides [118] which exhibit increased bending for the angles defining the tetrahedra, as discussed below. These results exemplifies the difference in the bending nature of the tetrahedra in these two families of networks, pointing to a higher rigidity of the tetrahedra in oxides, as all angular excursions are maintained at the same low value (typically 7°).

11.3.3.2 Individual Constraints

To gain deeper insights into which constraints are relevant from those which are irrelevant, we follow a given angle individually during the course of the simulation

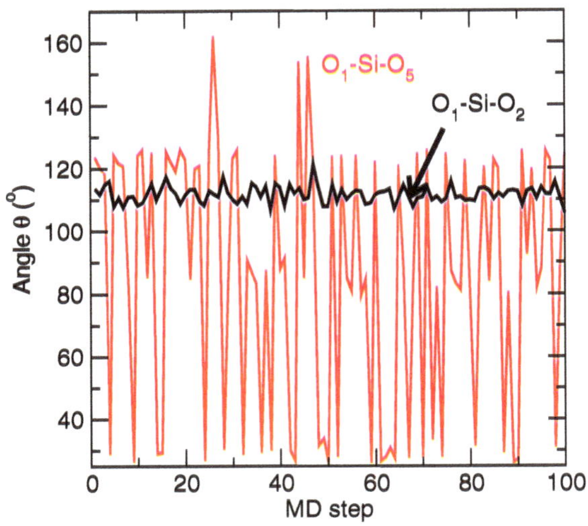

Fig. 11.12 Time evolution (in MD steps) of two typical angles in glassy NS2 defined by either the first two oxygen neighbors around a silicon atom (O_1–Si–O_2) or by neighbor 1 and neighbor 5 (O_1–Si–O_5)

(Fig. 11.12) at a given thermodynamic condition. For each individual atom k, the angular motion over the time trajectory leads indeed to a single bond angle distribution $P_k(\theta)$ characterized by a mean $\bar{\theta}_k$ (the first moment of the distribution), and a second moment (or standard deviation σ_{θ_k}). The latter represents, once again, a measure of the strength of the underlying BB interaction. In fact, if σ_{θ_k} is large (one has usually $\sigma_{\theta_k} > 15$–$20°$ [117]), it suggests that the BB restoring force which maintains the angle fixed around its mean value $\bar{\theta}_k$ is ineffective. As a result, the corresponding BB topological constraint will be broken, and will not contribute to network rigidity. The opposite reasoning can be applied for low values of σ_{θ_k} which will give rise to an intact BB constraint and contribute to n_c. The average over the whole system then leads to a distribution $f(\sigma)$ of standard deviations which can be analyzed and followed under different thermodynamic conditions.

This improved scheme (individual constraints) permits to separate effects which may arise from disorder, from those which are originated by the radial or angular motion and which enter in the constraint counting analysis. In fact, if averaged over time and system, the former can indeed lead to increased values of σ simply because there is an increased tendency to have different angles and bond lengths which will in turn broaden corresponding distributions. By following angles and distances with time (Fig. 11.12), one avoids this problem, and one can now safely enumerate constraints.

Figure 11.13 shows the distribution $f(\sigma)$ of angular standard deviations for a bridging oxygen in the NS2 liquid for increasing temperatures. The assignment of the peaks is rather obvious. In fact, at ambient pressure and at elevated temperatures (4000 K), all constraints must be broken because of thermal activation so that $f(\sigma)$ displays a broad distribution centred at 25°. On the opposite, at low temperature (300 K) the standard deviations are found inside a sharp distribution centred at low σ ($<10°$),

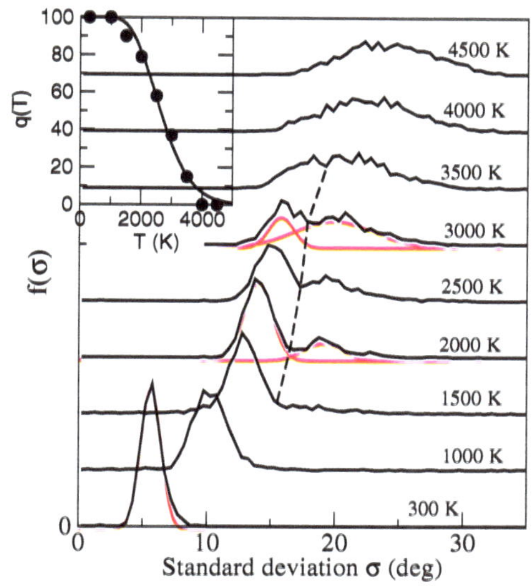

Fig. 11.13 Behaviour of the bridging oxygen centred standard deviation distributions f(σ) in a NS2 liquid. Note the bimodal distribution occuring at T$\simeq$2000 K. The *broken line* defines a boundary between broken and intact constraints, estimated to be about $\sigma_\theta = 15°$ at low temperature. Gaussian fits (*red curves*) are shown for selected temperatures.The *inset* shows the fraction q(T) of intact BO constraints as a function of temperature. The *solid curve* is a fit using the Mauro-Gupta function [37]

indicating that corresponding BB constraints are active. Interestingly, there is a temperature interval at T$\simeq$2000 K at which both contributions vary, and lead to a bimodal distribution. The high and low contributions of this bimodal distribution can thus be assigned to broken and intact constraints, respectively. The fraction of intact BB constraints can be computed (inset of Fig. 11.13), and exhibits a broad step-like behavior having all the features of the Mauro-Gupta q(T) function [37] introduced in (11.4). An approximate limit can be established (minimum of the bimodal distribution) that serve to separate intact constraint contributions from intact ones.

Having set the basis of Molecular Dynamics based topological constraint counting, we now review certain results obtained within this new framework.

11.4 Rigidity with Composition

We first apply such methods to Ge–Se network glasses which are probably the most well documented systems in the field of rigidity transitions. We then use the same tools to investigate the constraint behavior in the liquid.

11.4.1 Topological Constraints

Fig. 11.14 shows the constraint counting analysis for Ge–Se with different Ge content, similarly to what has been obtained for oxides (Fig. 11.11). When the standard

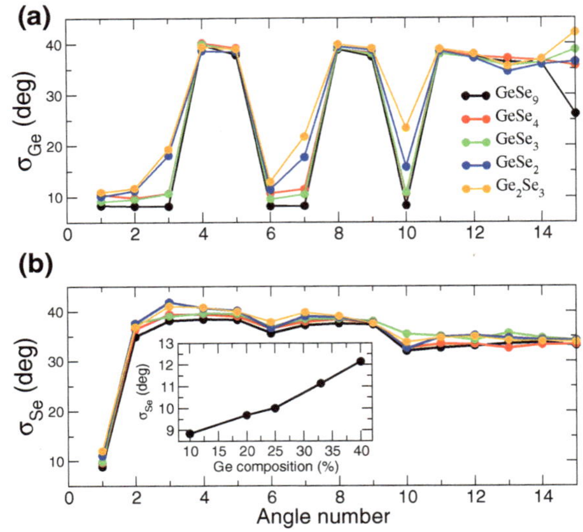

Fig. 11.14 Standard deviation σ_{Ge} and σ_{Se} extracted from the partial bond angle distributions (PBAD) for five selected compositions in glassy Ge_xSe_{1-x}

deviations σ_{Se} (a) and σ_{Ge} (b) are represented as a function of the angle number (see labelling in the bottom panel of Fig. 11.10), one observes that six of the 15 possible standard deviations (having set N = 6 possible neighbors) display a low value for σ_{Ge} of about 10°, and four times smaller than all the others. One thus recovers the result found for the stoichiometric oxides for which these six standard deviations were those associated with angular bending inside a tetrahedron. However, a more detailed inspection shows that there is a clear difference between compositions having the six standard deviations σ_{Ge} (1, 2, 3, 6, 7, 10) nearly equal (x = 10, 20, and 25 %) and compositions belonging to the stressed rigid phase (33, 40 %) which have have increased σ_{Ge} for selected angles. This behavior contrasts with the one found for the single Se standard deviation (inset of Fig. 11.14).

As the Ge content is increased, the angular excursion inside the tetrahedra is growing for selected angles, as highlighted for $GeSe_2$ by the growth of σ_{Ge} from less than 10° in the n = 3 (104) PBAD while leaving the stiffest angle (102) constant. When the six standard deviations σ_{Ge} defining the tetrahedra are plotted as a function of the Ge content (Fig. 11.15, left), it is found that the angular motion involving the fourth neighbour (PBADs 104, 204, 304) exhibits a substantial increase once the system is in the stressed rigid phase, while the others (102, 103, 203) are left with a similar angular excursion close to the one found for oxides where we have concluded on the presence of rigid tetrahedra.

The present findings underscore the fact the quantity σ_{Ge} is an indicator of stressed rigidity [118]. Moreover, the presence of stress will lead to asymmetric intratetrahedral bending motions involving an increased bending motion for selected triplets of atoms. This indicates that some BB constraints have softened to accommodate stress. The origin of this softening can be sketched from a simple bar network (Fig. 11.15, left) by considering stretching motion instead, and connects to the well-known

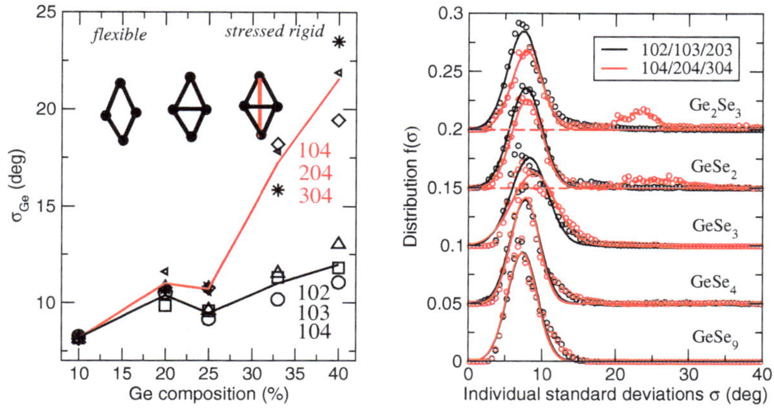

Fig. 11.15 *Left* Standard deviations σ_{Ge} as a function of Ge composition in the Ge–Se system [118], splitted into a contribution involving the fourth neighbour (*red line*, average of 104, 204 and 304) and the other contributions (*black line*). The shaded area corresponds to the Boolchand intermediate phase [28]. A simple bar structure represents the nature of the different elastic phases (see text for details). *Right* Distribution of Ge angular standard deviations using individual constraints [90]. The total distributions have been splitted, depending on the neighbor number: angles involving the first three neighbors (102, 103, 203, *black symbols*), and the fourth neighbor (104, 204, 304, *red symbols*). The *solid curves* are Gaussian fits which serve to estimate the population of broken constraints at high x content

relationship between stressed rigidity and bond mismatch in highly connected covalent networks [120]. In such systems, atoms having a given coordination number can indeed not fulfill all their bonds at the same length because of a too high bond density. In the toy structures of Fig. 11.15, all bars can have the same length in flexible and isostatic (intermediate) networks but once the structure becomes stressed rigid, at least one bar (the red bar in Fig. 11.15) must have a different length. A similar argument holds for angles. In the stressed rigid Ge–Se, because of the high network connectivity $GeSe_{4/2}$ tetrahedra must accommodate for the redundant cross-links which force softer interactions (i.e. angles) to adapt and to break a corresponding constraint. This leads to increased angular excursions for atomic Se–Ge–Se triplets (104, 204, 304, see Fig. 11.15) involving the farthest (fourth) neighbour of a central Ge atom.

When such systems are analyzed from individual constraints (Fig. 11.15, right), one recoveres the present enunciated conclusions, and one can compute the fraction of angles that have softened under stress. For the flexible $GeSe_9$ and intermediate $GeSe_4$ compositions, a single distribution for all six involved angles (102,...304) is found, located at low $\sigma \simeq 8$–$9°$ which indicates of a weak angular intra-tetrahedral motion. However, at higher Ge content corresponding angles involving the fourth neighbor (104, 204, 304) partly soften and produce a bimodal distribution (red curve), indicative that some angular constraints ($\sigma \simeq 22°$) are now broken. An enumeration shows that the fraction ξ of broken constraints is 17.2 and 21.4 % for $GeSe_2$

and Ge_2Se_3, respectively [90]. This implies a reduction of the number of Ge BB constraints by a quantity $\Delta n_c = 3x\xi$, i.e. 0.17 and 0.26 for the aforementioned compositions so that n_c reduces from 3.67 to 3.5, and from 4.00 to 3.74 for $GeSe_2$ and Ge_2Se_3, respectively.

11.4.2 Behavior in the Liquid Phase

How do constraints in chalcogenide melts behave at high temperature? For the $GeSe_2$ system, we investigate the behavior of topological constraints as a function of temperature. Figure 11.16 shows the analysis for 300 K (same as Fig. 11.15) and 1050 K, using either an average count (left) or individual constraints giving the distribution f(σ) (right). Obviously, changes are weak indicating that a counting at low temperature still holds in the high temperature liquid and thus close to the glass transition temperature. From Fig. 11.16, we find that the fraction of broken constraints (about 17.2 % in the 300 K $GeSe_2$) remains of the same order (18 % computed at the minimum of the bimodal distribution 104/204/304) in the liquid thus indicating that in chalcogenides thermal effect on topological constraints are weak. This situation actually contrasts with the findings obtained for oxides (Fig. 11.13). Our conclusion is consistent with recent results from neutron spin-echo spectroscopy showing

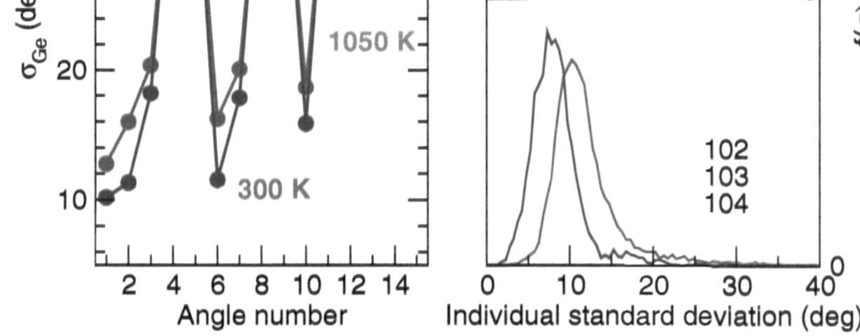

Fig. 11.16 Effect of temperature on standard deviations in liquid (1050 K, *red*) and amorphous (300 K, *blue*) $GeSe_2$. *Left* Standard deviation σ_{Ge} extracted from the partial bond angle distributions (PBAD). *Right* Distribution of Ge angular standard deviations using individual constraints. The total distributions are splitt, depending on the neighbor number: angles involving the first three neighbors (102, 103, 203), and the fourth neighbor (104, 204, 304, shifted)

that the rigidity concept can be extended from the glass to the liquid [121]. In this work, parameters giving the temperature dependence of the relaxation patterns of binary chalcogen melts have indeed shown to be linearly dependent with the low temperature mean coordination number $\bar{r}$, and thus to follow the count achieved at low temperature. Similarly, relaxational phenomena in Ge–Se using the constraint approach have also been reported from liquid-state NMR [64].

11.4.3 Dynamic Anomalies

Experimentally observed thresholds with composition in glassy chalcogenides are actually correlated with diffusivity anomalies in the liquid state. We calculate the mean square displacement following (11.6), and then apply the Einstein relationship to obtain diffusion constants in the long-time limit (See Fig. 11.6). Here, we follow the behaviour of the diffusion constant D_i with composition along an isotherm for different systems.

In Ge–Se, we find that the diffusion actually maximizes for the composition having 20 % Ge (Fig. 11.17), in a manner similar to that obtained in As–Se for which a maximum of $D_{Se} = 0.72 \times 10^{-5}$ cm^2/s occurs at 35 % As [51]. Furthermore, when represented as a function of inverse temperature, both D_{As} and D_{Se} exhibit an

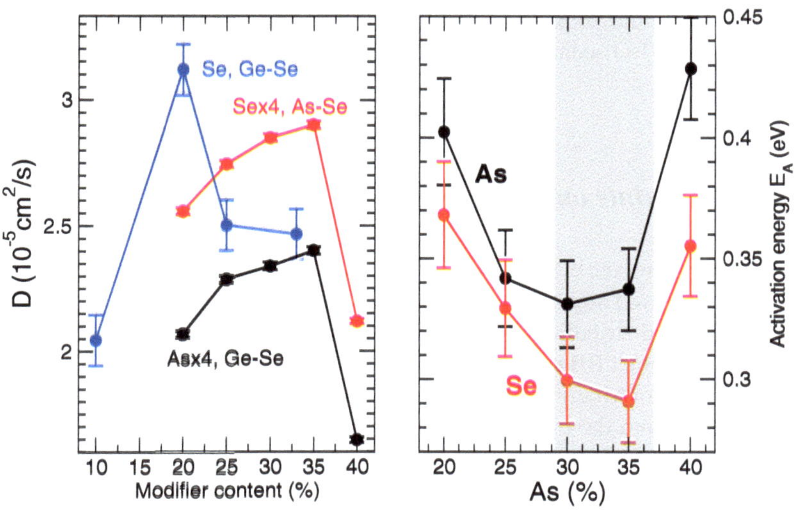

Fig. 11.17 *Left* Calculated Selenium diffusion constant in Ge–Se liquids (*blue*, 1073 K) and As–Se (*red*, 800 K), and As diffusion constant in As–Se (*black*) as a function of Ge/As content. *Right* Computed activation energy E_A for diffusion of As (*black symbols*) and Se (*red symbols*) as a function of As content in liquid As–Se [51]. The *gray zone* indicates the approximate location of the reversibility window [40]

Arrhenius behavior, with an activation energy E_A that is represented in Fig. 11.17 as a function of composition. Both quantities exhibit a minimum at 30–35 % As (0.29 and 0.34 eV for Se and As, respectively), that is found in the region where the diffusivity anomaly is obtained.

Using a simplified Kirkwood-Keating model of the glass transition [122], it has been shown that chalcogenides satisfying the condition $n_c \simeq 3$ are strong glass-forming liquids having an activation energy for viscosity/relaxation time or diffusion that is minimum at this same $n_c \simeq 3$ value. This suggests that the As–Se system must be stress-free at 30–35 % As. As mentioned earlier, this conclusion is consistent with calorimetric and optical methods [40] determining the boundaries of the stress-free IP [123], located between 29 and 37 % As. IP compositions have been found to be stress-free, and when the corresponding interval is represented (gray zone in Fig. 11.17), one can clearly notice that the diffusivity anomaly is obviously located in this compositional window.

11.5 Rigidity with Pressure

In the following, we investigate the effect of pressure on rigidity transitions and on the topological constraint count. There has been some previous investigations in this field from Trachenko and Dove [56, 124] showing that both GeO_2 and SiO_2 display a pressure window where the isobaric dilatation coefficient becomes negative. In a similar system, the sodium disilicate NS2, one follows the BB constraint of a bridging oxygen (BO)-centred angle (similarly to Fig. 11.13) but now as a function of pressure.

11.5.1 Constraints on (T, P) Maps

Figure 11.18a shows the effect of increasing pressures ($0 < P < 20$ GPa) on the distribution of BO-centred angular standard deviations in a 2000 K liquid. One first starts from the bimodal distribution found at ambient pressure (Fig. 11.13) out of which the fraction q(T) of intact BB constraints can be estimated (q(2000 K) = 0.77) given the presence of two contributions at σ_i and σ_b. With increasing P, two phenomena take place as shown from the evolution of the bimodal distribution. The intensity of the peak centred at $\sigma = \sigma_b$ is first increasing (at e.g. 4 GPa) before reducing and finally vanishing at higher pressures. At low pressure, it is a clear indication that thermally activated broken constraints are restored.

Secondly, the low σ contribution associated with the intact constraints has a nearly constant intensity but shows a continuous shift of the peak position σ_i with pressure, indicating, as for the case of Ge–Se (Fig. 11.15), the presence of increased stress.

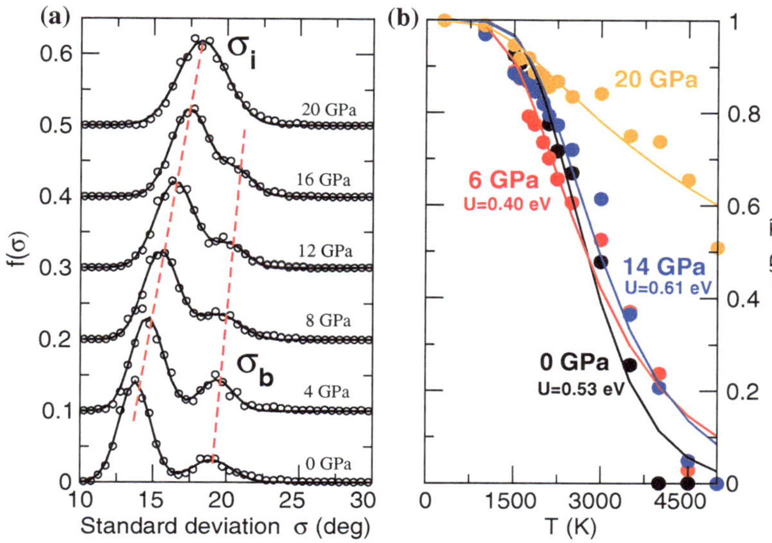

Fig. 11.18 **a** Behaviour of the bridging oxygen centred angular standard deviation distributions f(σ) in a NS2 liquid at 2000 K as a function or pressure. **b** Isobaric fraction q(P_0, T) of intact angular (BB) constraints for different isobars in simulated NS2 liquids (*symbols*). The lines represent least-squares fit using the Mauro-Gupta function [37]

Along an isobar at P = P_0, it is now possible to evaluate the fraction of intact constraints q(P_0, T), and Fig. 11.18b shows such computed fractions at different fixed pressures P_0, established from the respective populations contributing to the bimodal distribution (Fig. 11.18a). At moderate pressures ($P \leq 14$ GPa), all display a broad step like behavior, with q(P_0, T) decaying to zero as the temperature is increased. For very high pressures (e.g. $P_0 \simeq 20$ GPa), q(P_0, T) is found to decrease only weakly with temperature, and does not come to zero even at the most elevated temperatures (5000 K). A least-squares fit (solid lines in Fig. 11.18b) to the numerical data using the Mauro-Gupta function [37, 57] shows that the activation energy U necessary to break a constraint is equal to 0.61, 0.40 and 0.53 eV for 0, 4 and 14 GPa, respectively. These data, thus, indicate that thermally activated constraints can break more easily in a certain pressure interval because *U* passes through a minimum.

Such trends can be generalized when represented on a (P,T) map (Fig. 11.19) or with pressure along an isotherm (black curves, Fig. 11.20a). In Fig. 11.19, a clear minimum is found at low temperatures in a pressure window approximatively located over the interval 2.5 < P < 12 GPa for 2000 K. Note that the width of the window decreases with decreasing temperature, and is found to be centred at around 5 GPa at 1500 K, i.e. at the same pressure where a thermodynamic anomaly has been found in silica (negative dilatation coefficient, [124]).

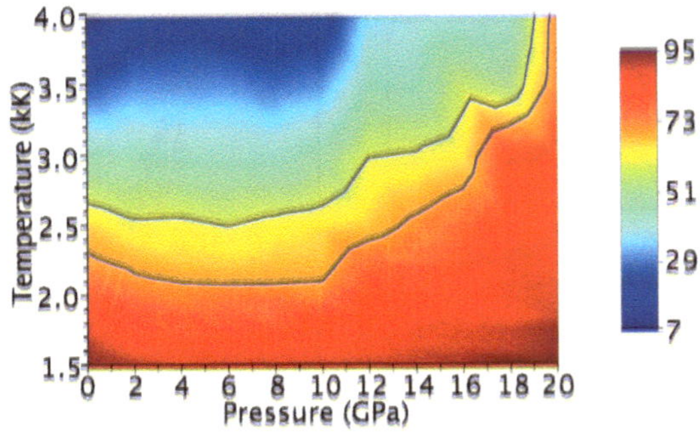

Fig. 11.19 Contour plot of the fraction (in %) of intact constraints q(T, P) of liquid NS2 in the (T, P) map

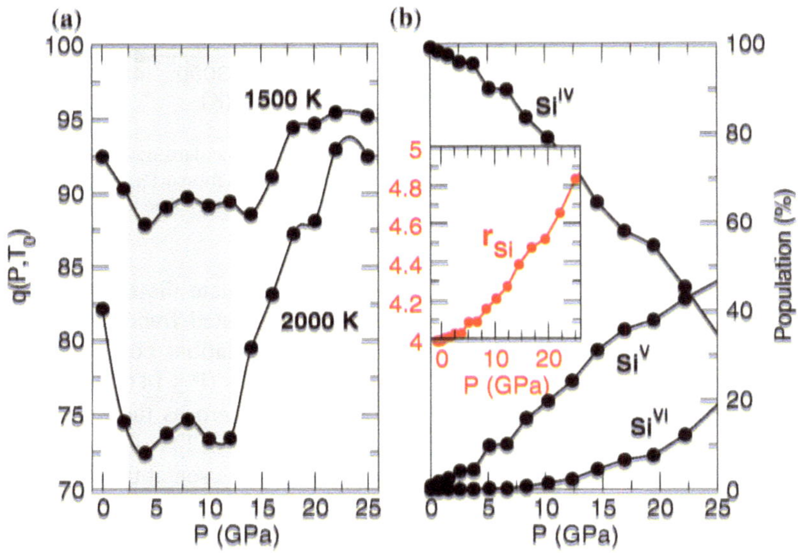

Fig. 11.20 **a** Computed behavior of the fraction of intact constraints q(P, T) as a function of pressure along two isotherms (1500 and 2000 K). **b** Population of four, five and sixfold coordinated Si species (Si^{IV}, Si^{V}, Si^{VI}) as a function of pressure in a 2000 K liquid. The *inset* shows the mean Si coordination number r_{Si} (*red curve*)

11.5.2 Adaptive Constraints

The obtained anomaly (Figs. 11.19 and 11.20a) results from the increase of connectivity arising from the tetrahedral to octahedral silicon coordination change [125] under pressure (Fig. 11.20b). This increase in connectivity leads in fact to additional stress induced by an increase of the number bond-stretching constraints equal to $n_c^{BS}(Si)s = r_{Si}/2$ which is found to increase from 2 at ambient pressure to 2.05 at

6 GPa (inset of Fig. 11.20). To counterbalance such effects and accommodate stress, BO bond-angles increase their angular excursion which leads to partial softening of the BB interaction and reduces the number of corresponding constraints. At ambient pressure, the NS2 glass and corresponding liquids are flexible ($n_c = 2.56$ at low temperature [31, 32]). As seen from Fig. 11.20, the average silicon (and incidentally oxygen [126]) coordination number r_{Si} increases with pressure, resulting from the growth of Si^V, Si^{VI} species.

At network level, the global increase of $n_c^{BS}(Si)$ arising from the growth of r_{Si} can be reduced in a certain pressure interval by breaking the energetically softer BB constraints of the BO atoms so that q(P, T) decreases at low pressure (Figs. 11.19 and 11.20a). This is network adaptation in a fashion that is very similar to what has been found in rigidity tuned by composition [26, 27, 46, 55] at ambient pressure. In fact, the observed features display striking similarities with the self-organized intermediate phase found in network glasses [24, 25, 28–30] where the growing stress induced by cross-linking atoms is partially released over a finite compositional interval having nearly $n_c \simeq 3$. Similarly, in the present NS2 liquid as both r_{Si} and r_O) continue to increase with pressure, the adaptive behavior can only hold up to a certain point for e.g. $\simeq$12 GPa at 2000 K found in Fig. 11.19. Beyond, the softening of BB constraints can no longer accommodate the steadily increasing pressure-induced stress resulting in onset of stiffness under pressure, which is manifested by an increase of q(P, T) for pressures greater than $\simeq$12 GPa. Ultimately, one is left with a pressure window made possible by constraints becoming adaptive.

11.5.3 Link with Water-Like Anomalies

We, furthermore, find that the adaptive behavior of q(P, T) is strongly tied with anomalies in transport coefficients (Fig. 11.21a). In the pressure window where BB constraints partially soften and reduce from e.g. q(P, 2000 K) = 0.82 at $P = 0$ to 0.72 at P = 6 GPa (Fig. 11.20a) in order to adapt for the increasing pressure-induced stress, and we find that viscosity and diffusion display a minimum and maximum respectively. It has been furthermore found that the corresponding activation energies for viscous flow (E_A^{η}) or oxygen diffusion (E_A^D) show a minimum in approximately the same pressure window [52, 57]. These data underscore the fact that BB constraint adaptation is directly related to the lowering of energy barriers involved in viscous flow and transport.

The obtained diffusivity anomaly for the present NS2 liquid actually also relates to the well-known transport anomalies reported for densified tetrahedral liquids [70, 127] as exemplified by the well-known example of densified water [71]. The correlation becomes obvious when we represent the oxygen diffusivity D_O of NS2 (Fig. 11.21b) as a function of the system density [126], and compare the trend with corresponding results for densified silica (2500 K [127]) and water (220 K [71]).

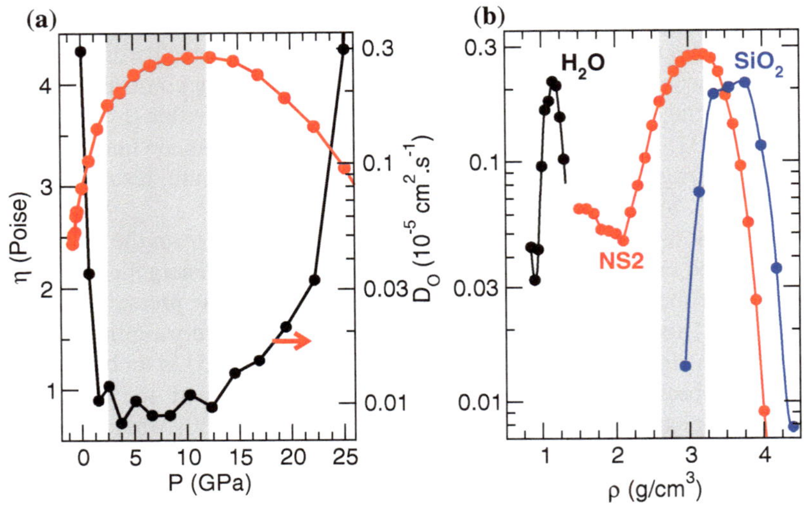

Fig. 11.21 **a** Computed viscosity $\eta(P)$ as a function of pressure for T = 2000 K in the NS2 liquid. *Right axis* Oxygen diffusion constant (*red symbols*) $D_O(P)$ in NS2 at 2000 K. The *grey area* represents an approximate delimitation of the adaptative phase, based on the window defined by the trend in q(P, 2000 K) (see Fig. 11.20a). **b** Oxygen diffusion constant (same as panel **a**) in NS2 at 2000 K but now plotted as a function of the system density, compared to diffusivity anomalies in silica (*blue symbols*, 2500 K [127]) and water (*black symbols*, 220 K [71]

It has been stressed [71, 127] that defining local structural order parameters [128] in such dense liquids could help in understanding the relationships between such diffusivity anomalies, structural and thermodynamic anomalies under temperature and density change. One order parameter (translational) measures the tendency of pairs of molecules to be separated by a preferential distance, while a second order parameter (orientational) measures the tendency of a molecule and its nearest neighbours to adopt preferential orientations. As liquids are followed under density and temperature change, we find that diffusivity anomalies under compression and temperature-induced space-filling tendency are deeply related to a change in the local structural order, and also to the excess entropy [70]. From the present investigated NS2, one now recognize that anomalies arise from the constraint behavior (Fig. 11.20), and that stress adaptation is the dominant feature which drives the evolution of transport coefficient under density and temperature change. The location of the adaptive pressure window is found in pressure/density ranges where D_O maximizes, so that one can conjecture that a similar behavior may well be obtained in water around 1.15 g/cm^3 for the considered isotherm (220 K), and in silica around 3.5 g/cm^3 at 2500 K. We view these anomalies as consequences of structural rearrangements [127] driven by stress adaptation in select pressure or density windows when the underlying networks are isostatic. We address the issue below, and begin by focusing on structure factors which are directly accessed from scattering experiments.

11.5.4 First Sharp Diffraction Peak Anomalies

We examine now structural anomalies that correlate to those detected previously, and follow the total neutron weighted structure factor $S_N(k)$ of glassy NS2 with pressure. In Fig. 11.22 is represented the total computed structure factor $S_N(k)$, obtained from a linear combination of the partial structure factors $S_{ij}(k)$, defined by:

$$S_N(k) = \frac{\sum_{i,j} c_i c_j b_i b_j S_{ij}(k)}{\left(\sum_i c_i b_i\right)^2} \tag{11.10}$$

where $S_{ij}(k)$ have been calculated from the Fourier transform of the pair distribution functions obtained from the simulated trajectories [52]. At ambient pressure, and similarly to what has been reported before (Fig. 11.3), the computed $S_N(k)$ reproduces rather well the neutron diffraction results [129], in particular the measured position of the FSDP ($k_{FSDP} = 1.85\,Å^{-1}$) and its width Δk_{FSDP}. As the pressure is increased, the FSDP or k_{FSDP} are found to steadily upshift from 1.85 to $2.3\,Å^{-1}$ for 25 GPa (Fig. 11.23). The principal peak found at $5.3\,Å^{-1}$ remains nearly constant in intensity and wavevector position whereas the second PP at $3\,Å^{-1}$ increases and becomes dominant at elevated pressures.

The most revealing partial structure factor (Si_{Si-O}) is represented for the three selected pressures in the inset of Fig. 11.22. For all ($S_N(k)$, $S_{ij}(k)$), a Lorentzian fit is used in order to extract the position k_{FSDP} and width Δk_{FSDP}. An anomalous behavior in the FSDP for some of the partial structure factors is found (Fig. 11.23), as the position k_{FSDP} maximizes for the partials involving the network-forming species

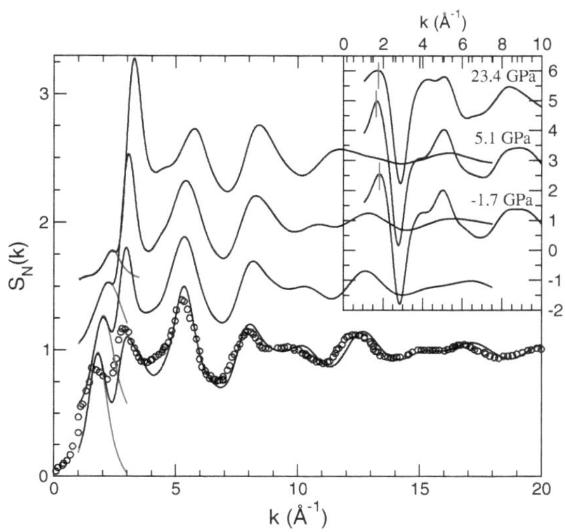

Fig. 11.22 Total simulated structure factor $S_N(k)$ of amorphous NS2 at increasing pressures. From *bottom* to *top*: 0, 3, 12, 50 GPa. The *circles* are experimental data by Misawa et al. [129]. The *red curves* are the Lorentzian fits (see text for details). The *inset* shows the Si–O partial structure factor at three selected pressures, together with the position k_{FSDP}

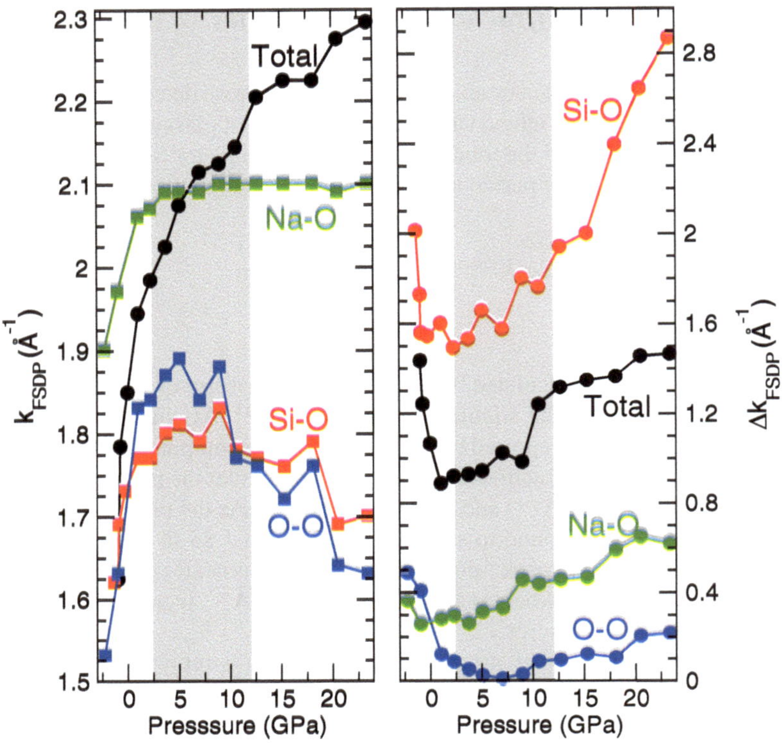

Fig. 11.23 *Left* Position k_{FSDP} of the FSDP (in $Å^{-1}$) of different partial structure factors of NS2. *Right* Width Δk_{FSDP} of the FSDP (in $Å^{-1}$). The *gray zones* correspond to pressure regions of dynamic and rigidity anomalies (see text for details)

(Si–O, O–O, Fig. 11.23, left panel) with a FSDP increasing from 1.53 to 1.9 $Å^{-1}$ at 5 GPa for $S_{O-O}(k)$, and then decreasing at higher pressures. The width of the FSDP Δk_{FSDP} (Fig. 11.23 right) also exhibits anomalies in the same pressure interval, and a pronounced minimum is obtained for Δk_{FSDP} in the total $S_N(k)$ and the partial $S_{Si-O}(k)$.

The position of the FSDP usually reflects some repetitive characteristic distance between structural units [130]. The present findings of Fig. 11.23, thus, indicates that in the same pressure interval where BB constraints soften to accommodate the increase of stress due to the connectivity increase, a typical length scale of distance $L = 7.7/k_{FSDP}$ (4.05 Å for the O–O correlations[1]) emerges, and then decreases. Similarly, the broadening of the FSDP is indicative of a characteristic correlation length emerging in the same pressure window. This follows from the well-known

[1] We relate the position r of a peak in real space to the position k of a corresponding peak in Fourier space by using the relation $k \cdot r \approx 7.7$, which identifies the location of the first maximum of the spherical Bessel function $j_0(kr)$

Scherrer equation for microcrystals, which connects the width of a Bragg peak with the average size of the microcrystals. This leads to the definition of a correlation length defined by $\xi = 7.7\Delta k_{FSDP}$. From Fig. 11.23, one realizes that in the pressure window all atomic pairs give rise to a typical correlation length ξ. For instance, this correlation length is equal to 6.61 and 4.05 Å for the $S_N(k)$ and $S_{Si-O}(k)$, respectively. Other partials display a Δk_{FSDP} with a broader minimum.

Such correlations between anomalies in dynamic properties and anomalies in the FSDP parameters have a one to one correspondence with results [51] obtained from a systematic study on amorphous As–Se where rigidity is tuned by composition. Here all FSDP anomalies are also located in or close to an intermediate phase reported experimentally [40]. This underscores, therefore, not only the generic behavior of such trends but also the fact that there is clearly a structural signature for the intermediate phase, in contrast with previous statements based solely [50] on the analysis of the total structure factor. As seen from Fig. 11.23 and from [51], FSDP parameters of the total neutron or X-ray $S_N(k)$ may not necessarily exhibit such a structural signature.

11.6 Conclusion

In this chapter, we have shown that structural information obtained from molecular dynamics is able to provide a rigorous atomic-scale basis to phenomenological constraint counting concepts applied to network glasses. An initial step is the production of realistic models of glasses whose structure has to agree with what is found from experiments. In addition, the dynamic properties of melts and liquids can be compared, whenever possible, to viscosity or diffusivity measurements.

Building on the notion of standard deviation which quantify the radial and angular excursions of bonds and angles, one can then enumerate quite precisely the number of bond-stretching and bond-bending interactions which constrain the network at a molecular level, and which provide clues on the way rigidity onsets under various thermodynamic conditions: composition, temperature or pressure. It is found that many of the rigidity-induced changes are found in angular motion, as most of the coordinations and bond-stretching constraints can be obviously determined.

In systems undergoing a rigidity transition with composition such as in Ge_xSe_{1-x}, it is found that the angles defining the $GeSe_{4/2}$ tetrahedron soften with increasing Ge content to accommodate stress, while the Se angular constraint is slightly increasing. The occurence of stress at high Ge content results in an asymetric bending motion inside the Ge tetrahedra, and this also appears in densified systems. In fact, for a typical silicate system (NS2) we have shown that a subtle interplay can take place between pressure-induced stress and the softening of topological constraints by thermal activation. It now becomes clear that, ultimately, a rigidity map containing the flexible, intermediate and stressed-rigid phases can be established for any iono-covalent compound from the estimate of the number of constraints as a function of composition, temperature, and pressure. At present however, only some limited zones of such a map have been characterized. In sodium silicates, these have been limited to a fixed composition (NS2, [57]), at zero pressure and zero temperature [31], and at zero pressure and fixed composition [117]. For the Ge_xSe_{1-x}, all investigations in the context of rigidity transitions have been performed at zero pressure (see however [86]), but preliminary information does exist as a function of temperature [64, 121]. For the As–Se system, simulations have been performed only at ambient pressure [51, 91]. For this particular system, a strong correlation has been established between the structure in real and reciprocal space and the intermediate phase, which has been also found in densified silicates [52], underscoring the generality of the findings.

Finally, such structural anomalies are obviously located in pressure/composition windows where the network structure adapts to avoid the occurence of stress. This in turn leads to transport anomalies, and the well-known water-like diffusivity anomaly [71] of densified tetrahedral liquids which have been previously reported. Network adaptation or self-organization is at the core of the identified isostatic intermediate phase [28], and the present framework relating Molecular Dynamics and topological constraint counting, allows studying such phenomena in detail.

Acknowledgments Support from Agence Nationale de la Recherche (ANR) (Grant No. 09-BLAN-0109-01 and Grant No. 11-BS08-0012) is gratefully acknowledged. MM acknowledges support from the French-American Fulbright Commission, and from International Materials Institute (H. Jain). GENCI (Grand Equipement National de Calcul Intensif) is acknowledged for supercomputing access. The authors thank C. Massobrio, S. Le Roux, M. Boero, C. Bichara, P. Boolchand, S. Boshle, K. Gunasekera, M. Malki, G.G. Naumis, A. Pasquarello, J.-Y. Raty, G. Ferlat, J.C. Phillips, P. Simon, M. Salanne, D. De Sousa-Meneses, R. Vuilleumier, S. Ravindren, N. Mousseau, S. Chakraborty for stimulating discussions.

References

1. M.F. Thorpe, P.M Duxbury (eds.), *Rigidity Theory and Applications* (Kluwer Academic, Plenum Publishers, New York, 1999)
2. M. Micoulaut, M. Popescu (eds.), *Rigidity and Boolchand Intermediate Phases in Nanomaterials* (INOE Publishing House, Bucarest, 2009)
3. M.F. Thorpe, J.C. Phillips (eds.), *Phase Transitions and Self-organization in Electronic and Molecular Networks* (Kluwer Academic, Plenum Publishers, New York, 2001)
4. M. Zhang, P. Boolchand, Science **266**, 1355 (1994)
5. J.L. Lagrange, *Mécanique Analytique* (Courcier, Paris, 1788)
6. J.C. Maxwell, Philos. Mag. **27**, 294 (1864)
7. J.C. Phillips, J. Non-Cryst. Solids **34**, 153 (1979)
8. J.C. Phillips, Phys. Rev. B **28**, 7038 (1983)
9. J.C. Phillips, Phys. Today **35**, 27 (1982)
10. M. Micoulaut, W. Welnic, M. Wuttig, Phy. Rev. B **78**, 224209 (2008)
11. P. Boolchand, M.F. Thorpe, Phys. Rev. B **50**, 10366 (1994)
12. W.J. Bresser, P. Suranyi, P. Boolchand, Phys. Rev. Lett. **56**, 2493 (1986)
13. R. Azoulay, H. Thibierge, A. Brenac, J. Non-Cryst. Solids **18**, 33 (1975)
14. M.F. Thorpe, J. Non-Cryst. Solids **57**, 355 (1983)
15. H. He, M.F. Thorpe, Phys. Rev. Lett. **54**, 2107 (1985)
16. M.F. Thorpe, J. Non-Cryst. Solids **76**, 109 (1985)

17. W.A. Kamitakahara, R.L. Cappelletti, P. Boolchand, B. Halfpap, F. Gompf, D.A. Neumann, H. Mutka, Phys. Rev. B **44**, 94 (1991)
18. X. Feng, W. Bresser, P. Boolchand, Phys. Rev. Lett. **78**, 4422 (1997)
19. M. Tatsumisago, B.L. Halfpap, J.L. Green, S.M. Lindsay, C.A. Angell, Phys. Rev. Lett. **64**, 1549 (1990)
20. A.N. Sreeram, A.K. Varshneya, D.R. Swiler, J. Non-Cryst. Solids **128**, 294 (1991)
21. P. Boolchand, W.J. Bresser, M. Zhang, Y. Wu, J. Wells, R.N. Enzweiler, J. Non-Cryst. Solids **82**, 143 (1995)
22. S. Asokan, M.Y.N. Prasad, G. Parthasarathy, Phys. Rev. Lett. **62**, 808 (1989)
23. R.A. Bohmer, C.A. Angell, Phys. Rev. B **45**, 1091 (1992)
24. D. Selvenathan, W. Bresser, P. Boolchand, Solid State Commun. **111**, 619 (1999)
25. D. Selvenathan, W. Bresser, P. Boolchand, Phys. Rev. B **61**, 15061 (2000)
26. M.F. Thorpe, D.J. Jacobs, M.V. Chubynsky, J.C. Phillips, J. Non-Cryst. Solids **266–269**, 859 (2000)
27. M. Micoulaut, J.C. Phillips, Phys. Rev. B **67**, 104204 (2003)
28. S. Bhosle, P. Boolchand, M. Micoulaut, C. Massobrio, Solid State Commun. **151**, 1851 (2011)
29. S. Bhosle, K. Gunasekera, P. Boolchand, M. Micoulaut, Int. J. Appl. Glass Sci. **3**, 205 (2012)
30. S. Bhosle, K. Gunasekera, P. Boolchand, M. Micoulaut, Int. J. Appl. Glass Sci. **3**, 189 (2012)
31. Y. Vaills, T. Qu, M. Micoulaut, F. Chaimbault, P. Boolchand, J. Phys. Condens. Matter **17**, 4889 (2005)
32. M. Micoulaut, Am. Mineral. **93**, 1732 (2008)
33. S. Chakravarty, D.G. Georgiev, P. Boolchand, M. Micoulaut, J. Phys. Condens. Matter **17**, L7 (2005)
34. K. Gunasekara, S. Bhosle, P. Boolchand, M. Micoulaut, J. Chem. Phys. **139**, 164511 (2013)
35. G. Yang, O. Gulbiten, Y. Gueguen, B. Bureau, J-Ch. Sangleboeuf, C. Roiland, E.A. King, P. Lucas, Phys. Rev. B **85**, 144107 (2012)
36. F. Wang, S. Mamedov, P. Boolchand, B. Goodman, M. Chandrasekhar, Phys. Rev. B **71**, 174201 (2005)
37. P.K. Gupta, J.C. Mauro, J. Chem. Phys. **130**, 094503 (2009)
38. R.G. Srinivasa, A. Arunbabu, S. Asokan, J. Solid State Chem. **184**, 3345 (2011)
39. B.J. Madhu, H.S. Jayanna, S. Asokan, Eur. Phys. J. B **71**, 21 (2009)
40. D.G. Georgiev, D.G. Georgiev, P. Boolchand, M. Micoulaut, Phys. Rev. B **62**, R9228 (2000)
41. P. Chen, C. Holbrook, P. Boolchand, D.G. Georgiev, M. Micoulaut, Phys. Rev. B **78**, 224208 (2008)
42. D. Novita, P. Boolchand, M. Malki, M. Micoulaut, Phys. Rev. Lett. **98**, 195501 (2007)
43. D.J. Jacobs, A.J. Rader, L.A. Kuhn, M.F. Thorpe, Proteins **44**, 150 (2001)
44. J.C. Phillips, Phys. Rev. Lett. **88**, 216401 (2002)
45. R. Monasson, R. Zecchina, S. Kirkpatrick, B. Selman, L. Troyansky, Nature **400**, 133 (1999)
46. J. Barré, A.R. Bishop, T. Lookman, A. Saxena, Phys. Rev. Lett. **94**, 208701 (2005)
47. R. Rompicharla, D.I. Novita, P. Chen, P. Boolchand, M. Micoulaut, W. Huff, J. Phys. Condens. Matter **20**, 202101 (2008)
48. S. Chakraborty, P. Boolchand, M. Malki, M. Micoulaut, unpublished
49. S.W. King et al., J. Non-Cryst. Solids **379**, 67 (2013)
50. M.T.M. Shatnawi, C.L. Farrow, P. Chen, P. Boolchand, A. Sartbaeva, M.F. Thorpe, S.J.L. Billinge, Phys. Rev. B **77**, 094134 (2008)
51. M. Bauchy, M. Micoulaut, M. Boero, C. Massobrio, Phys. Rev. Lett. **110**, 165501 (2013)
52. M. Micoulaut, M. Bauchy, Phys. Status Solidi **250**, 976 (2013)
53. B. Bureau, J. Troles, M. Le Floch, F. Smektala, J. Lucas, J. Non-Cryst. Solids **326–327**, 58 (2003)
54. M. Micoulaut, Phys. Rev. B **74**, 184208 (2006)
55. M.V. Chubynsky, M.A. Brière, N. Mousseau, Phys. Rev. E **74**, 016116 (2006)
56. K. Trachenko, M.T. Dove, Phys. Rev. B **67**, 212203 (2003)
57. M. Bauchy, M. Micoulaut, Phys. Rev. Lett. **110**, 095501 (2013)
58. F. Inam, G. Chen, D.N. Tafen, D.A. Drabold, Phys. Status Solidi B **246**, 1849 (2009)

59. G. Chen, F. Inam, D.A. Drabold, Appl. Phys. Lett. **97**, 131901 (2010)
60. B. Effey, R.L. Cappelletti, Phys. Rev. B **59**, 4119 (1999)
61. B.O. Mysen, P. Richet, *Silicate Glasses and Melts: Structure and Properties* (Springer, Berlin, 2005)
62. P. Boolchand (ed.), *Insulating and Semi-Conducting Glasses* (World Scientific, Singapore, 2000)
63. L. Cai, P. Boolchand, Philos. Mag. B **82**, 1649 (2002)
64. E.L. Gjersing, S. Sen, R.E. Youngman, Phys. Rev. B **82**, 014203 (2010)
65. J.C. Mauro, R.J. Loucks, P.K. Gupta, J. Phys. Chem. A **111**, 7957 (2007)
66. C.A. Angell, B.E. Richards, V. Velikov, J. Phys. Condens. Matter **11**, A75 (1999)
67. J.C. Mauro, R.J. Loucks, J. Chem. Phys. **130**, 234503 (2009)
68. M. Smedskjaer, J.C. Mauro, R.E. Youngman, C.L. Hogue, M. Potuzak, Y. Yue, J. Phys. Chem. B **115**, 12930 (2011)
69. M.M. Smedskjaer, J.C. Mauro, Y. Yue, Phys. Rev. Lett. **105**, 115503 (2010)
70. B. Shadrack Jabes, M. Agarwal, C. Chaktavarty, J. Chem. Phys. **132**, 234507 (2010)
71. J.R. Errington, P.G. Debenedetti, Nature **409**, 318 (2001)
72. U. Senapati, A.K. Varshneya, J. Non-Cryst. Solids **197**, 210 (1996)
73. U. Senapati, A.K. Varshnaya, J. Non-Cryst. Solids **185**, 289 (1995)
74. F. Wilcek, Phys. Today **57**, 10 (2004)
75. R. Car, M. Parrinello, Phys. Rev. Lett. **55**, 2471 (1985)
76. A.D. Becke, Phys. Rev. A **38**, 3098 (1988)
77. C. Lee, W. Yang, R.G. Parr, Phys. Rev. B **37**, 785 (1988)
78. S. Nosé, Mol. Phys. 52, 255 (1984)
79. W.G. Hoover. Phys. Rev. A **31**, 1695 (1985)
80. M. Micoulaut, R. Vuilleumier, C. Massobrio, Phys. Rev. B **79**, 214204 (2009)
81. C. Massobrio, M. Micoulaut, P.S. Salmon, Solid State Sci. **12**, 199 (2010)
82. S. Tsuneyucki, M. Tsukada, H. Aoki, Y. Matsui, Phys. Rev. Lett. **61**, 869 (1988)
83. R.D. Oeffner, S.R. Elliott, Phys. Rev. B **58**, 14791 (1998)
84. A.N. Cormack, J. Du, T.R. Zeitler, Phys. Chem. Chem. Phys. **4**, 3193 (2002)
85. P.S. Salmon, A.C. Barnes, R.A. Martin, G.J. Cuello, J. Phys. Condens. Matter **19**, 415110 (2007)
86. P.S. Salmon, J. Non-Cryst. Solids **353**, 2959 (2007)
87. S. Xin, J. Liu, P.S. Salmon, Phys. Rev. B **78**, 064207 (2008)
88. M. Fabian, E. Sváb, V. Pamukchieva, A. Szekeres, S. Vogel, U. Ruett, J. Phys. Condens. Matter **253**, 012053 (2010)
89. J.C. Mauro, A.K. Varshneya, J. Non-Cryst. Solids **353**, 1226 (2007)
90. M. Micoulaut, A. Kachmar, M. Bauchy, S. Le Roux, C. Massobrio, M. Boero, Phys. Rev. B **88**, 054203 (2013)
91. M. Bauchy, M. Micoulaut, J. Non-Cryst. Solids **377**, 34 (2013)
92. C. Massobrio, A. Pasquarello, R. Car, Phys. Rev. B **64**, 144205 (2001)
93. C. Massobrio, A. Pasquarello, Phys. Rev. B **77**, 144207 (2008)
94. C. Massobrio, A. Pasquarello, R. Car, J. Am. Chem. Soc. **121**, 2943 (1999)
95. I. Petri, P.S. Salmon, H.E. Fischer, Phys. Rev. Lett. **84**, 2413 (2000)
96. A. Bouzid, C. Massobrio, J. Chem. Phys. **137**, 224201 (2012)
97. S. Hosokawa, Y. Wang, W.-C. Pilgrim, J.-F. Bérar, S. Mamedov, P. Boolchand, J. Non-Cryst, Solids **352**, 1517 (2006)
98. S. Hosokawa, Y. Kawakita, W.-C. Pilgrim, F. Hensel, J. Non-Cryst. Solids **293–295**, 153 (2001)
99. G.D. Negodaev, I.A. Ivanov, K.K. Evstop'ev, Elektrokhim. **8**, 234 (1972)
100. Y.P. Gupta, T.B. King, Trans. Metall. Soc. AIME **239**, 1701 (1967)
101. J.R. Johnson, R.H. Bristow, H.H. Blau, J. Am. Ceram. Soc. **34**, 165 (1951)
102. M. Truhlarova, O. Veprek, Silikaty **14**, 1 (1970)
103. G.J. Kramer, A.J.M. de Man, R.A. van Santen, J. Am. Chem. Soc. **113**, 6435 (1991)
104. J. Horbach, W. Kob, K. Binder, Chem. Geol. **174**, 87 (2001)

105. S. Stolen, T. Grande, H.-B. Johnsen, Phys. Chem. Chem. Phys. **4**, 3396 (2002)
106. G. Yang, T. Rouxel, J. Troles, B. Bureau, C. Boussard-Plédel, P. Houizot, J.-C. Sangleboeuf, J. Am. Ceram. Soc. **94**, 2408 (2011)
107. M. Bauchy, B. Guillot, M. Micoulaut, N. Sator, Chem. Geol. **346**, 47 (2013)
108. M.P. Allen, D.J. Tildesley, *Computer Simulations of Liquids* (Oxford University Press, Oxford, 1987)
109. J.O.M. Bockris, J.D. Mackenzie, J.A. Kitchener, Trans. Faraday Soc. **51**, 1734 (1955)
110. D.R. Neuville, Chem. Geol. **229**, 28 (2006)
111. J.C. Mauro, Y. Yue, A.J. Ellison, P.K. Gupta, D.C. Allan, Proc. Natl. Acad. USA **106**, 19780 (2009)
112. M. Micoulaut, S. Le Roux, C. Massobrio, J. Chem. Phys. **136**, 224504 (2012)
113. M. Hemmati, C.A. Angell, J. Non-Cryst. Solids **217**, 236 (1997)
114. J.Y. Raty, C. Otjacques, J.P. Gaspard, C. Bichara, Solid State Sci. **12**, 193 (2010)
115. M. Micoulaut, A. Kachmar, Th Charpentier, Phys. Status Solidi B **249**, 1890 (2012)
116. M. Micoulaut, C. Otjacques, J.-Y. Raty, C. Bichara, Phys. Rev. B **81**, 174206 (2010)
117. M. Bauchy, M. Micoulaut, J. Non-Cryst. Solids **357**, 2530 (2011)
118. M. Bauchy, M. Micoulaut, M. Celino, M. Boero, S. Le Roux, C. Massobrio, Phys. Rev. B **83**, 054201 (2011)
119. R. Hussein, R. Dupree, D. Holland, J. Non-Cryst. Solids **246**, 159 (1999)
120. N. Mousseau, M.F. Thorpe, Phys. Rev. B **52**, 2660 (1995)
121. F.J. Bermejo, C. Cabrillo, E. Bychkov, P. Fouquet, G. Ehlers, W. Hussler, D.L. Price, M.L. Saboungi, Phys. Rev. Lett. **100**, 245902 (2008)
122. M. Micoulaut, J. Phys. Condens. Matter **22**, 285101 (2010)
123. P. Chen, P. Boolchand, D.G. Georgeiv, J. Phys. Condens. Matter **22**, 065104 (2010)
124. K. Trachenko, M.T. Dove, V. Brazhkin, F.S. El'kin, Phys. Rev. Lett. **93**, 135502 (2004)
125. M. Bauchy, M. Micoulaut, Phys. Rev. B **83**, 184118 (2011)
126. M. Bauchy, J. Chem. Phys. **137**, 044510 (2012)
127. M. Scott Shell, P.G. Debenedetti, A.Z. Panagiotopoulos, Phys. Rev. E **66**, 011202 (2002)
128. P.L. Chau, A.J. Hardwick, Mol. Phys. **93**, 511 (1998)
129. M. Misawa, D.L. Price, K. Suzuki, J. Non-Cryst. Solids **37**, 85 (1980)
130. P.S. Salmon, R.A. Martin, P.E. Mason, G. Cuello, Nature **435**, 75 (2005)

Chapter 12
First-Principles Modeling of Binary Chalcogenides: Recent Accomplishments and New Achievements

Assil Bouzid, Sébastien Le Roux, Guido Ori, Christine Tugène, Mauro Boero and Carlo Massobrio

Abstract This contribution is focussed on a set of first-principles molecular dynamics results obtained over the past fifteen years for disordered chalcogenides. In the first part, we sketch and review the historical premises underlying research efforts devoted to the understanding of structural properties in liquid and glassy Ge_xSe_{1-x} systems. We stress the importance of selecting well performing exchange-correlation functionals (within density functional theory) to achieve a correct description of short and intermediate range order. In the second part, we provide a specific, comparative example of structural analysis for chalcogenide GeX_4 systems differing by the chemical identity of the X atom. We are able to demonstrate that the correct account of differences between the coordination environments of the two corresponding glasses requires system sizes substantially larger than ~100 atoms.

A. Bouzid (✉) · S.L. Roux · G. Ori · C. Tugène · M. Boero · C. Massobrio
Institut de Physique et Chimie des Matériaux de Strasbourg,
University of Strasbourg—CNRS UMR 7504,
23 rue du Loess BP 43, 67034 Strasbourg, France
e-mail: assil.bouzid@ipcms.unistra.fr

S.L. Roux
e-mail: sebastien.leroux@ipcms.unistra.fr

G. Ori
MultiScale Materials Science for Energy and Environment,
CNRS-MIT UMI 3466, 77 Massachusetts Avenue, Cambridge, MA 02139, USA
e-mail: Guido.ori@ipcms.unistra.fr

G. Ori
Institut Charles Gerhardt Montpellier,
CNRS UMR 5253, University of Montpellier II, ENSCM,
8 Rue de l'Ecole Normale, 34296 Montpellier, France

C. Tugène
e-mail: Christine.Tugene@ipcms.unistra.fr

M. Boero
e-mail: mauro.boero@ipcms.unistra.fr

C. Massobrio
e-mail: carlo.massobrio@ipcms.unistra.fr

C. Massobrio et al. (eds.), *Molecular Dynamics Simulations of Disordered Materials*, Springer Series in Materials Science 215,
DOI 10.1007/978-3-319-15675-0_12

Finally, the role played by the pressure in altering the structural properties of glassy $GeSe_2$ is invoked, in light of recent studies devoted to a density-driven structural transformation occurring in this system.

12.1 Introduction

This contribution aims at providing an example of atomic-scale studies devoted to prototypical disordered networks: those belonging to the Ge_xSe_{1-x} family. For these systems, atomic-scale simulation approaches have to account for the details of electronic structure since deviations from the perfect tetrahedral order (homopolar bonds and miscoordinations), heavily affecting the structural topologies, are quite sensitive to the level of descripion of chemical bonding. Nowadays, there is a striking evidence that models based on interatomic potentials merely fitted on a set of experimental parameters are bound to fail for these systems. However, in the past (late eighties), molecular dynamics models based on interatomic potentials (herafter termed classical), were employed to achieve a first, qualitative description of disordered chalcogenides, by reaching an undoubted succes. When such efforts were deployed, no unambiguous experimental evidence existed on any remarkable difference among the structures of different disordered networks, all of them being thought as consisting of interconnected undefective tetrahedra. This conferred legitimacy to the classical approaches, as those proposed by the Vashishta team, that can be taken as the first milestone of atomic-scale modelling for chalcogenides. Within this context, we refer the interested reader to a number of relevant papers issued from the Vashishta team and devoted to liquid and glassy $GeSe_2$ [1–3].

Our purpose here is to trace the historical development of first-principles molecular dynamics when applied (mostly by the present team) to disordered chalcogenides. This account is complemented by a presentation of recent results. We shall refer almost exclusively to first-principles molecular dynamics (FPMD) simulations fully based on (or at least highly inspired by) Density Functional Theory (DFT). The focus is on the disordered Ge_xSe_{1-x} family of materials, in which the nature of the network arrangement and the bonding strongly depends on the composition x. Within this family, glassy materials form easily for $x < 0.43$. In the following, we present the main results and historical steps toward the understanding of the atomic scale properties of binary Ge_xSe_{1-x} chalcogenides by considering the cases of $x = 0.33$ and $x = 0.2$ at ambient conditions. For $x = 0.33$ ($GeSe_2$), the effect of the choice of the exchange-correlation functional and the sensitivity to the system size will be analyzed in detail. By turning to glassy $GeSe_4$, its structure will be rewieved and described in light of a comparison with the properties of glassy GeS_4. This chapter ends with the study of pressure effects on the topology of glassy $GeSe_2$. Concerning the theoretical framework underlying these calculations, as a general rule for this chapter, methodological details are not systematically provided, since the general framework of first-principles molecular dynamics is described elsewhere in this issue. However, for specific cases (as the comparison between glassy GeS_4 and glassy $GeSe_4$) we found appropriate to mention some specific technical features, essential to appreciate the interest of the comparison.

12.2 Towards an Accurate Description of Binary Chalcogenide Materials using First-Principles Molecular Dynamics

12.2.1 The $GeSe_2$ System: $x = 0.33$

The first "DFT-inspired" molecular dynamics results on binary $GeSe_2$ were obtained in the mid-90s by Cappelletti and Co. [4] who studied the vibrational properties of the glassy state and compared them to neutron scattering experiments. They used the computational scheme developed by Sankey and Co. [5] based on a multicenter tight-binding model for DFT-based molecular dynamics. In this method the electronic band structure is calculated using an approximate set of wave functions that are assumed to be a superposition of wave functions of isolated atoms located at each atomic site. The DFT approximated energy is then based on the non self-consistent Harris [6] and Foulkes [7] approximation, for which a reference electronic density is taken as a sum of neutral-atom spherical atomic density. By using this scheme, an energy functional could be expressed as the one based on the Local Density Approximation (LDA) exchange-correlation recipe.

The vibrational properties calculated via the Harris functional were in good agreement with the measured one. Despite the fact that the model was very small (63 atoms), it reproduced successfully the regions of the A_1 and A_{1c} vibrational modes characteristic of the $GeSe_2$ glass. The computed structure factor was in good agreement with the neutron-scattering experiment and a non negligible fraction of wrong bonds was reported. A second similar study [8, 9] within the same theoretical framework and with a larger model (216 atoms), gave improved structural and vibrational properties.

While the quality and the interest of these approaches are undeniable, the search of a quantitative description of these binary alloys required to go a step further by resorting to simulations based on a fully self consistent theoretical framework.

12.2.1.1 From LDA to GGA Based First-Principles Molecular Dynamics

The first set of fully self consistent FPMD calculations on a disordered $GeSe_2$ system was published by Massobrio, Pasquarello and Car [10, 11] on the structure of the liquid $GeSe_2$. The first issue addressed was the quantitative reach of the model employed for chalcogenides. In particular, the authors expressed some concern on the choice of the exchange and correlation functional to be used. Although at that time these results were not immediately published, the first calculations in this respect have been carried out using the local density approximation (LDA) for the exchange-correlation (XC) part of the DFT total energy. Such LDA results were unable to reproduce essential features of the experimental data, such as the FSDP in the total neutron structure factor. A few years later the LDA results were shown and compared to those obtained by using the Generalized Gradient Approximations (GGA) for the XC functional, in particular the Perdew-Wang (PW) functional [12]. As shown in Fig. 12.1 the LDA $GeSe_2$ liquid does not feature the FSDP in any of the Faber-Ziman (FZ) partial structure factors.

In order to understand why drastically different results for the structural properties can occur when two different XC functionals were used, a detailed rationale was put forth by the same authors [14]. This work highlighted the effects of changes in the modelling of the electronic structure and how this affects the valence charge distribution, as an electronic property, and the intermediate range order (IRO). Since the LDA and the PW approximations account differently for the XC energy, different degrees of ionicity (and, in principle, of covalency) have been introduced by performing distinct calculations in the two separate frameworks. Also, information was gathered by optimizing the electronic structure (separately via both the LDA and PW schemes) for specific configurations obtained within one given framework. To correlate the changes in the bonding character to a given functional, the electronic structure and the valence charge distribution had been analyzed for selected configuration of both models by focusing on a specific Se–Ge–Se trimer (e.g. see Fig. 12.2).

The investigation of the valence charge profiles (e.g. see Fig. 12.2) revealed an accumulation of charge around the selenium atoms and a depletion around the germanium atoms resulting from the higher electronegativity of selenium. In addition, the existence of lobes pointing along the bond directions indicates the covalent contribution to the bonding character in both approximations. When looking carefully

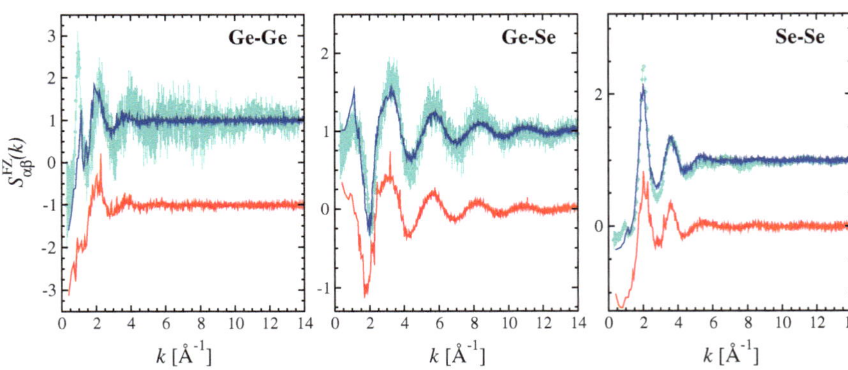

Fig. 12.1 Faber-Ziman partial structure factors for liquid $GeSe_2$. Calculations for PW (*solid blue line*) and LDA (*solid red line* shifted downwards by −2) FPMD models from [12] are compared to the experiment (*cyan dots* with *error bars*) from [13]

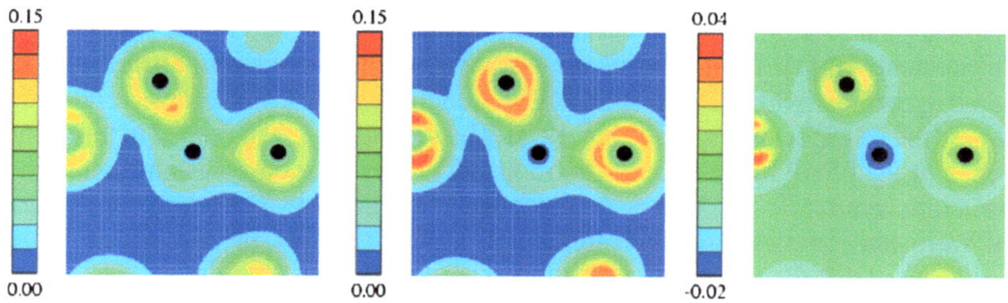

Fig. 12.2 Contour plots of the valence electronic charge density (a.u.) for a selected Se–Ge–Se trimer within liquid $GeSe_2$. From *right* to *left* plots for LDA, PW and the difference (PW–LDA) are shown

at the contour plots, the GGA favors the charge transfer compared to the LDA, as a manifestation of the higher degree of ionicity. This is confirmed by the contour plot of the difference between the charge densities obtained with the two functionals (e.g. see Fig. 12.2). The spherical nature of the difference plot has important implications. The distinction between the two schemes lies in the way the ionic character is described. The PW favors a higher degree of ionicity, however the covalency remains equivalent in the two schemes. As a consequence, the occurrence of the first sharp diffraction peak (FSDP) in the neutron structure factor can be related to an increase in the ionic character of the bonds, since the LDA drastically fails to reproduce its intensity and position (e.g. see Fig. 12.3).

In the quoted paper [14] it was emphasized that the LDA scheme compares only qualitatively with experiment, and is inadequate to describe the Short Range Order (SRO) as well as the IRO. For this reasons the PW was adopted to generate and study the structural properties of binary chalcogenides.

The first applications of this theoretical recipe were devoted to liquid $GeSe_2$ [10, 11]. These works provided an atomic scale description of the $GeSe_2$ liquid of unprecedented quality. The first focus was on the reproducibility of the FSDP that character-

Fig. 12.3 Calculated neutron structure factor for liquid $GeSe_2$, GGA–PW (*green line*) and LDA results from [14] (*blue line*, shifted downwards by 0.4) are compared to the experiment from [15] (*black circles*)

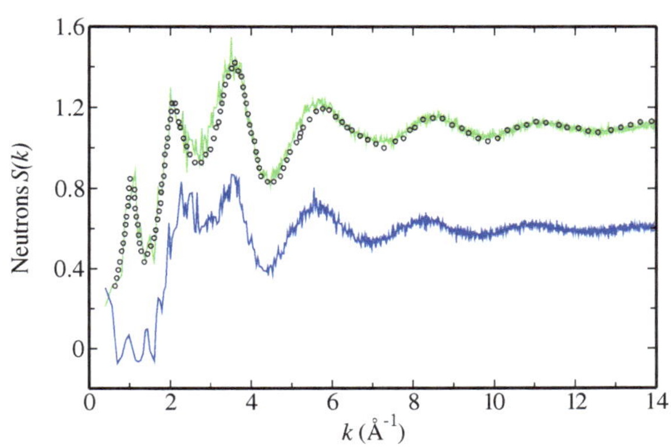

izes the total structure factor measured experimentally [13] and it was obtained also in classical MD [1]. The total structure factor was nicely reproduced, featuring a very good agreement with the experimental data, including the low wave vectors region where the FSDP occurs (e.g. see Fig. 12.3). The second issue was related to the occurrence of the FSDP in the Bhatia-Thornton (BT) concentration-concentration partial structure factor, $S_{cc}^{BT}(k)$. This peak was found experimentally [13], unlike in the case of Vashishta's studies based on empirical potentials [1, 2, 16]. In those papers, it was conjectured that the absence of the FSDP in the $S_{cc}^{BT}(k)$ partial structure factor could be a property generic to all binary glasses. The significance of chemical order was also a matter of concern. Indeed, classical MD studies stated that liquid and glassy $GeSe_2$ were mainly ordered networks of interlinked tetrahedra. This was in contrast with experimental evidence, showing that a departure from chemical order occurs in both the liquid and the amorphous states [13].

The topology of the $GeSe_2$ liquid network was found to be more subtle than the theoretical descriptions provided by the Continuous Random Network (CRN) and the crystalline like models. Ge atoms were found in different environments, 63 % of Ge atoms being fourfold and 30 % being undercoordinated. Se atoms were found in twofold and threefold environments (74 % and 23 %, respectively). In contradiction with the classical MD simulation, a non negligible fraction of Ge (10 %) and Se (39 %) was found to exist in homopolar configurations, confirming the departure from the chemical order. The calculated Faber-Ziman (FZ) partial structure factors, $S_{\alpha\beta}^{FZ}(k)$, reproduced the shape of their experimental counterparts with a high accuracy. However, the intensity of the FSDP remained underestimated for the Ge−Ge correlations whereas it remained overestimated for the Ge−Se correlations. In addition, no trace was found of the appearance of the FSDP in $S_{cc}^{BT}(k)$. Overall, this work provided a remarkable improvement of the microscopic description of the structure of liquid $GeSe_2$ and a good agreement with the diffraction data. However, it did not contribute to account for the physical origin of the FSDP in the $S_{cc}^{BT}(k)$, which still remained (at the moment of that specific publication) an open issue.

As a corollary motivation for this work, emphasis was given to the origin of the IRO through the appearance of FSDP [17]. The FSDP has a peculiar behavior with temperature, especially when the temperature is raised highly beyond the melting point [13, 18]. To check this statement, attention was devoted to the correlation between the number of predominant tetrahedral units, the intensity of the FSDP and the temperature. By using the same FPMD scheme, a high temperature model was generated at 1373 K and was compared to the experimental counterpart [13, 18]. This comparison revealed a good agreement in the reproduction of the FSDP, its intensity being caracterized by a drastic reduction at high temperature (e.g. see Fig. 12.4).

The decrease of the FSDP was correlated to the reduction in the number of $GeSe_4$ tetrahedral units. Thus, while the proportion of fourfold Ge centered motifs decreases, the proportion of three-fold and miscoordinated Ge atoms increases. Also, the proportion of Se−Se homopolar bonds increases along with temperature generating an important change in the Se local environment. This study showed that the breakdown of the IRO is due to the appearance of various structural motifs, leading to a network in which the tetrahedral predominant coordination is replaced by a coexistence of

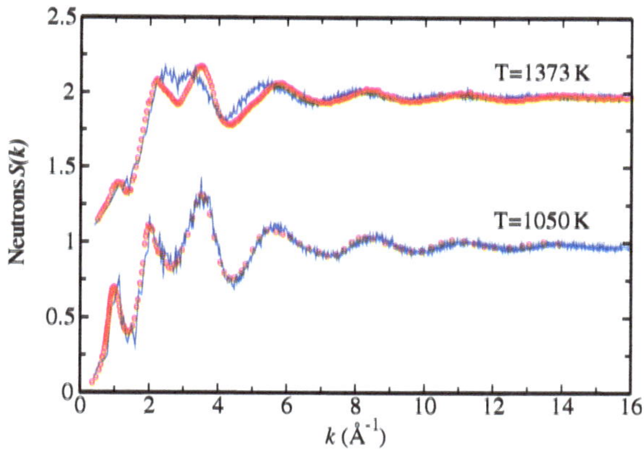

Fig. 12.4 Neutrons structure factors for liquid $GeSe_2$ at T = 1050 K and T = 1373 K. Experimental data sets: T = 1050 K from [13] and T = 1373 K from [18] (*open red circles*). GGA-PW data sets: T = 1050 K from [10] and T = 1373 K from [17] (*solid blue line*)

several coordination units, none of them accounting for more than 40–50 % of the Ge atoms.

As an additional contribution toward the understanding of the origin of the IRO, a particular attention was given to the range of real space correlations responsible for the appearance of the FSDP. For a given partial Pair Correlation Function (PCF), $g_{\alpha\beta}(r)$, this range can be determined by truncating $g_{\alpha\beta}(r)$ at decreasing distances r_c and by monitoring the behavior of the corresponding Fourier-transformed structure factor $S_{\alpha\beta}^{FZ}(k)$. The two quantities are related by:

$$S_{\alpha\beta}^{FZ}(k) \;=\; 1 \;+\; 4\pi\rho \int_0^{r_c} dr\, r^2\, \frac{\sin kr}{kr} \left(g_{\alpha\beta}(r) - 1\right) \tag{12.1}$$

In [12] the $S_{\text{Ge-Ge}}^{FZ}$ partial structure factor was calculated for an extended cell of 3240 atoms, constructed by replicating the 120 atoms cell in the three directions of space. The calculation was performed by Fourier transforming the corresponding PCFs, $g_{\text{GeGe}}(r)$, using different r_c. The results of these calculations were compared to the exact Reciprocal Space (RS) calculation (e.g. see Fig. 12.5). For the maximum possible cut-off distance in real space, r_{max}, the comparison shows a pattern very close to the RS calculation. By reducing the cutoff radius ($r_c < r_{max}$) (e.g. see Fig. 12.5), a well defined peak is visible at the FSDP position for $r_c = 10.5$ Å. However, for $r_c = 6$ Å only a residual bump is found at the FSDP position, while for $r_c = 5$ Å this feature disappears completely. As a consequence, it was concluded that the FSDP in the $S_{\text{GeGe}}^{FZ}(k)$ structure factor is mainly due to correlations beyond 5 Å in real space. A similar analysis was applied to the total structure factor, confirming this result. However, the persisting shoulder between 1 and 1.5 Å^{-1} for $r_c = 5$ Å proved that the FSDP carries as well some contributions from Ge–Se and Se–Se correlations at a shorter range.

A number of hypothesis were proposed to correlate the FSDP to the presence of specific structural motifs. Two proposals are worth of mention. The first links the FSDP to the presence of crystalline-like signatures, its position being related

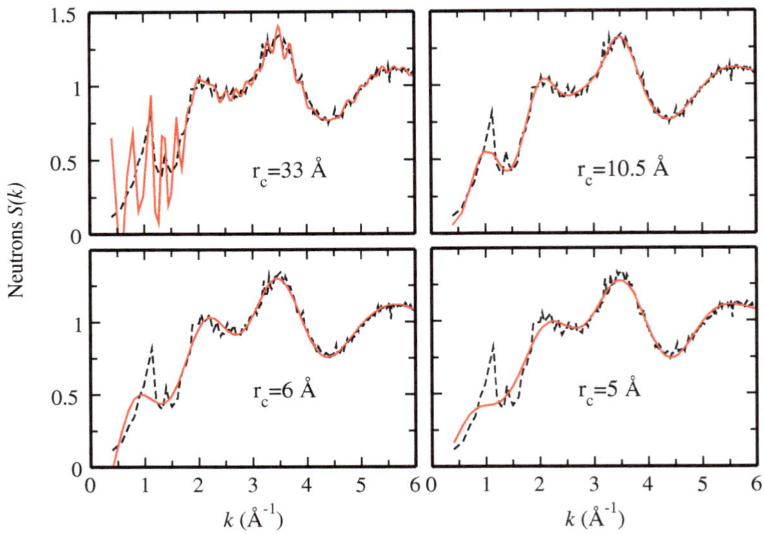

Fig. 12.5 Total neutron structure factor for liquid $GeSe_2$ (*solid red lines*) obtained by Fourier integration of the calculated PDF within given a integration ranges [0–r_c]. The *black dashed lines* correspond to the neutrons $S(k)$ directly calculated in reciprocal space [12]

to the inter-layer separations in the crystalline phase [19, 20]. The second invokes the occurrence of cluster regions separated by interstitial voids giving rise to correlation distances typical of the IRO [21]. The reliability of these hypothesis was tested by studying the liquid structures of $GeSe_2$ and SiO_2 in the same theoretical framework [22], via the calculation of the second order moment of the structure factor. It appeared than none of these schemes is totally convincing in establishing the atomic-scale origins for the appearance of the FSDP.

12.2.1.2 Influence of the GGA Exchange-Correlation Functional

Overall, PW (GGA) calculations on liquid $GeSe_2$ were instrumental in setting the scene for a quantitative description of disordered chalcogenides based on FPMD. However, the quest for a real predictive power required further improvements to be obtained, especially when focusing on Ge–Ge correlations. To address this issue, the choice was to explore the effect of different XC functionals. The GGA family of DFT based XC functionals does contain different proposals. For instance, the generalized gradient approximation after Becke (B) for the exchange energy and Lee, Yang and Parr (LYP) for the correlation energy [23, 24] is an alternative to go beyond the PW description. The reliability of BLYP has been assessed in different DFT calculations. It has been shown that the BLYP functional performs very well for properties encompassing equilibrium geometries, vibrational frequencies, and atomization energies of nanostructured systems. The success of BLYP is in part due to the LYP correlation energy, in particular to the correlation energy formula due to Colle and Salvetti [24] which was recast in terms of the electron density and of a suitable Hartree-Fock density matrix, thus providing a correlation energy and a correlation potential. This scheme is expected to enhance the localized behavior of the electron density at the expense of the electronic delocalization effects that favor the metallic character. In the case of liquid $GeSe_2$, it has been shown that choosing BLYP improves upon PW and LDA results [25]. In particular BLYP minimizes the electronic delocalization effects that are unsuitable in any bonding situation characterized by competing ionic and covalent contributions.

The simulations based on the BLYP functional have shown a substantial improvement of the short-range properties of liquid $GeSe_2$. The shapes of the Ge–Ge and to a lesser extent the Ge–Se pair-correlation functionals were more structured in a way consistent with a higher level of tetrahedral organization (e.g. see Fig. 12.6). The BLYP average coordination number is only 3.5 % higher than the experimental one against 7.7 % in the PW case. The bond angle distributions (BADs) are sharper, the Se–Ge–Se one being more symmetric around the tetrahedral angle 109°, and the Ge–Se–Ge one showing two distinct peaks resulting from Edge-Sharing (ES) and Corner Sharing (CS) connections. Despite these improvements, the impact of the BLYP scheme on the properties of the IRO remains elusive. Smaller changes are found for the intensity and the position of the FSDP in the partial structure factors [26].

The performances of the BLYP XC functional have been also assessed through the study of the amorphous $GeSe_2$ [26, 27]. In this case, a clear improvement has been found on both the real space properties and the reciprocal space properties with respect to the PW scheme. This was substantiated by the good agreement between the proportion of homopolar bonds calculated for the FPMD model and the experiment.

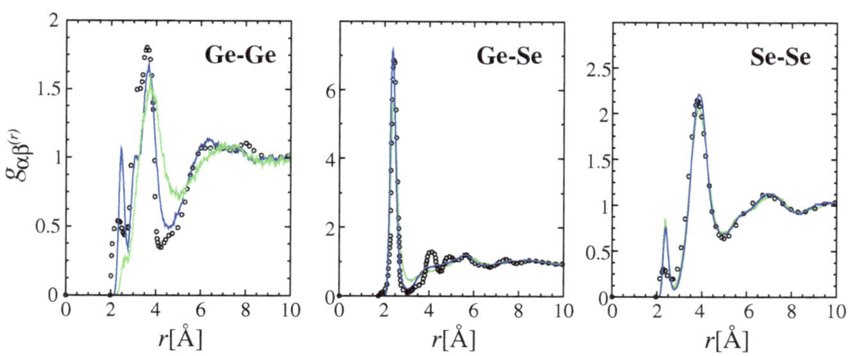

Fig. 12.6 Partial pair-correlation functions for liquid $GeSe_2$: BLYP results from [25] (*solid blue line*), PW results from [12] (*solid green line*), experimental results from [13] (*open black circles*)

12.2.1.3 Size Effects

Common to all calculations mentioned so far is the use of a model consisting of N = 120 atoms (80 Se and 40 Ge). in a periodic cubic box. As to the length of the temporal trajectories, recent achievements [28–30] proved that, for this system size, statistical averages can be routinely taken on temporal trajectory lasting (at least) as much as 100 ps on affordable computers. Given this situation, and the amount of results obtained in this context with N = 120, it is worthwhile to determine whether the use of a larger periodic simulation box has some impact on the predictive power of the available DFT-FPMD schemes. Understanding the occurrence of possible size effects on periodic molecular dynamics simulations is a longstanding issue that has received a great deal of attention ever since the first attempts to achieve sensible modelling of materials [31–46]. Very large simulation boxes have been adopted to monitor the (possible) onset of such effects [38–46].

Since no FPMD studies were available to compare structural data on disordered network-forming systems pertaining to two sizes for the simulation cell (of about 100 and 500 atoms) and compare them at the same level of statistical accuracy, the work by Micoulaut and Co. [47] on liquid $GeSe_2$ is worth of mention. These authors performed calculations on a periodic simulation cell containing N = 480 atoms, four times larger than what it was done before.

In this article the authors payed a particular attention to the possible occurrence of size effects on the IRO, by confirming that N = 120 was an adequate size to capture most of the structural features. The results for N = 480 were indeed very similar to the ones obtained for the smaller size system, although a few moderate changes and/or improvements were pointed out in the larger model. In particular, the short range order was found to be characterized by a higher degree of chemical order, with a larger number of $Ge(Se_4)$ tetrahedra. However, this tendency toward a reduction of chemical disorder had a little effect on either the diffusion coefficients or the electronic density of states. The results of this work suggested that in the case of liquid $GeSe_2$ size effects are not relevant when increasing the number of atoms from $\simeq$100 up to $\simeq$500, therefore calling for simulations on even larger systems to settle the issue in a conclusive manner.

12.2.1.4 Summary

In summary, the success in modeling the prototypical network forming $GeSe_2$ system, results from three well focusses, and yet different research efforts. First, interest has been given to choose the better XC functional and it was emphasized that the GGA family is suitable to improve the predictive power of the models. An optimal account of bonding ionicity through the use of specific GGA functionals was found to be crucial improve the comparison with experiments. Second, interest has been devoted to practical application of the established theoretical scheme to study disordered chalcogenides. In particular, pieces of evidence have been provided for the high temperature effect, the origin and the reproducibility of the FSDP in the partials structure factor and in the total structure factor. Third and last, the focus was turned to the XC functionals within the the GGA family where the BLYP proved to give better descriptions of fundamental structural properties, and reproduces more accurately delicate features of the liquid and glassy states.

12.2.2 The $GeSe_4$ System: $x = 0.2$

In parallel to the work performed to elucidate the atomic scale properties of disordered $GeSe_2$ significant efforts have been made to untangle the microscopic structure of disordered $GeSe_4$. Overall the details of the FPMD calculations to describe the $GeSe_4$ system evolved concurrently to the progress made to describe the $GeSe_2$ system.

In the first paper describing the atomic scale structure of liquid $GeSe_4$ [11], the FPMD simulation were performed using the PW functional. In this work Haye and Co. [11] found a total structure factor in good agreement with the diffraction data for a similar concentration ($x = 0.2$), including the appearance of a FSDP indicating that the IRO in the liquid state was well described. The calculated coordination numbers were also very close to those predicted by the Chemically Ordered Network (CON) model, and revealed a nice agreement with the $8 - N$ phenomenological rule. The Ge atoms were found to be mostly four fold coordinated (86 %) with small proportions of three and five fold coordinations, whereas most of the Se atoms were found in a two fold conformations (89 %). Two distinct topologies could then be considered, the key difference between them lying in the relative percentages of coexisting Ge–Se–Se, Ge–Se–Ge and Se–Se–Se triads, respectively referred to as AB, AA and BB units hereafter (e.g. see Fig. 12.7). In the first case Ge atoms are connected by Se dimers (AB configurations). BB and AA configurations are absent since no Se atoms are left available to form chains longer than Se_2 or to link two Ge atoms as nearest neighbors. In the second case no AB connections exist since a phase separation occurs between two coexisting domains having as compositions $GeSe_2$ and Se_n, leading to an equal amount of AA and BB units.

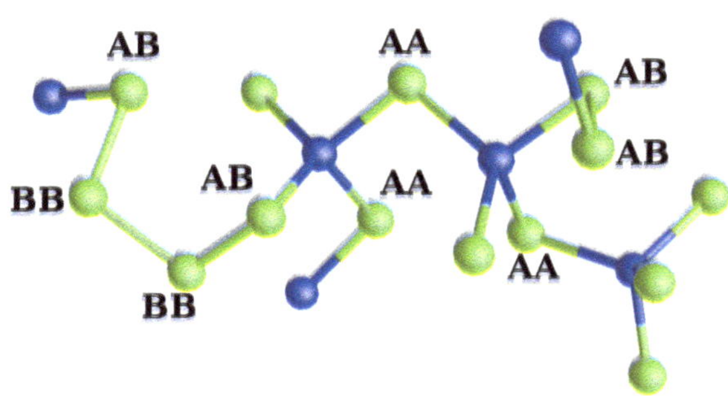

Fig. 12.7 Representative subset of Ge and Se atoms in amorphous $GeSe_4$ where Ge atoms are dark (*blue*) and Se atoms are light (*green*). Se atoms along a connection path between two Ge atoms are labelled as AA, those between one Ge atom and one Se atom are labelled as AB, and those between two Se atoms are labelled as BB

The calculated proportions of AA, AB and BB units were respectively equal to 30, 30 and 40 % refuting the phase separation hypothesis. The microscopic structure of $GeSe_4$ was described as a network of a four fold Ge centred atoms, interconnected with either Se chains or Se atoms. A further analysis showed that 49 % of Ge were connected on a ES fashion and 51 % in a CS fashion. The Se chains were found to be either dimers (20 %), trimers (10 %), 4-mers (14 %) or higher than 4-mers (28 %).

Well after the appearance of these results, the existence of such a phase separation was invoked to put forth a bimodel phase percolation model for Ge_xSe_{1-x} glasses in the $0 \leq x \leq 0.33$ range [48]. However at the same time a second FPMD based study [49], comparing the structure of glassy $GeSe_4$ to that of glassy $SiSe_4$, substantiated the absence of the phase separation in glassy $GeSe_4$. This was due to the significant proportions of AA and BB motifs, found to be respectively 23 and 26 %. This FPMD model was more chemically ordered than the one generated by Tafen and Drabold [50] few years earlier using non-self consistent Harris functional DFT based calculations. Extremely high quench rates were employed in that case. The FPMD model of glassy $GeSe_4$ obtained in paper [49] was later used to compute the density-functional NMR chemical shifts [51]. In this work Kibalchenko and Co. [51] compared the theoretical NMR spectrum to the experimental one and derived a composition of the studied glass in terms of local coordination. This study appeared consistent with a decomposition in terms of structural motifs which coherently accounts for both NMR and neutron diffraction experiments. About 20 % of the Se atoms were found in AB type, Se–Se–Ge linkages. For $GeSe_4$ in particular, these results were not consistent with the phase-separation model, supporting the occurrence of a fully bonded network in which Se–Se–Ge linkages ensure the connection between Se–Se–Se chains and Ge–Se–Ge tetrahedral arrangements.

Following the ideas developed in the case of the $x = 0.33$ system, the atomic structure of glassy $GeSe_4$ has been further studied using different GGA exchange-correlation functionals. In the work of Sykina and Co. [52] the Perdew, Burke, Ernzerhof (PBE) XC functional was used, while shortly after that Micoulaut and Co. [53] presented models generated using the BLYP functional. The partial PCFs of the different GGA based calculations are shown and compared in Fig. 12.8. The Short Range Order (SRO) structure in glassy $GeSe_4$ was only moderately modified when switching from the PW to the PBE or the BLYP functionals. Whereas the choice of the XC functional had almost no effects on the $g_{SeSe}(r)$ and $g_{GeSe}(r)$ PCFs, small changes on both position and intensity were observed for the peaks of the $g_{GeGe}(r)$ PCF (e.g. see Fig. 12.8). The impacts of the PBE and the BLYP recipes were reflected by more intense peaks and deeper minima than in the PW case. This changes were associated with the increase of the fraction of ES motifs from 22 % for the PW model to 30 and 33 % the PBE and the BLYP models respectively (e.g. see Fig. 12.8).

Interestingly the PBE calculations highlight the presence of extended Se chains, up to $n = 12$, which are compatible with a network structure in which $GeSe_4$ and Se_n units of various lengths coexist in a fully interconnected fashion. Neither of the PBE or the BLYP models presented any sign of phase separation, thus closing the debate on the possible occurrence of this phenomenon.

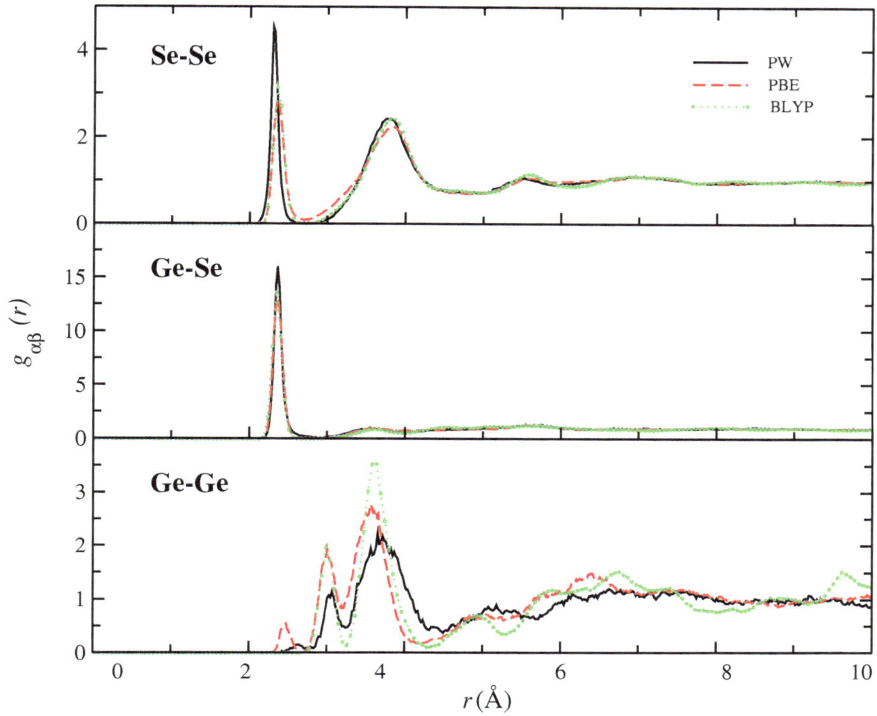

Fig. 12.8 Partial pair correlation functions for glassy $GeSe_4$. Results obtained using the PW functional (*black line*) are compared to results obtained using the PBE [52] (*red dashed line*) and BLYP [53] (*green symbols*) functionals

12.3 Comparison Between Glassy $GeSe_4$ and GeS_4

In what follows, we provide a specific example of a structural study focussed on the analogies and differences between two chalcogenides glasses having the same composition, namely glassy GeS_4 (g-GeS_4) and glassy $GeSe_4$ (g-$GeSe_4$). Our calculations are based on FPMD models of g-GeS_4 and g-$GeSe_4$ made of N = 480 (96 Ge + 384 S/Se) and N = 120 (24 Ge and 96 S/Se) atoms. Recently, a careful analysis of size effects was undergone for liquid $GeSe_2$ (N = 120 and N = 480), by concluding that most structural properties are essentially unchanged when moving from N = 120 to N = 480 [47]. However, the limited extent of the trajectory sampling for configurations quenched from the liquid state calls for a study of this kind on glassy systems, since the statistical accuracies of the results are dissimilar when comparing averaged properties in the liquid and in the glass. For this reason, we selected two different values of N to ascertain the occurrence of system size effects, for which no stringent molecular dynamics results are known in the case of glasses at least within the first-principles framework.

Let us consider first the case of g-GeS_4. We started with the largest system, N = 480, by using a periodically repeated cubic cell of 23.52 Å corresponding to the experimental density, $\rho_0(exp)$, equal to 0.0317 atom Å^{-3} [54, 55].

The initial coordinates were obtained by a configuration extracted from the fully equilibrated trajectories produced for g-GeS_2 [56]. To achieve the correct composition, the number of S atoms was changed to 384 by modifying the chemical identity of 64 randomly chosen Ge atoms. We produced a single trajectory for N = 480 in the NVT ensemble, by following a well documented setup having an extended record of reliability [25, 57–59]. Accordingly, to loose the memory of the initial configuration, the system was randomized at 2000 K during ≃25 ps and then quenched by steps of 500 K, lasting 10 ps each, the last step bringing the system from T = 500 to 300 K. At this temperature, the value of the pressure associated with $\rho_0(exp)$ was non-negligible (1 GPa). By expanding the box size to 24.82 Å (ρ_0 = 0.0313 Å^3) and further relaxing the whole structure, we were able to lower the pressure to less than 0.1 GPa. In the N = 120 case, the initial configuration was created by first selecting one configuration of liquid $GeSe_2$ and then by changing the chemical identity of the chalcogen atoms (from Se to S). As a choice for the density, we took the one corresponding to ∼0 pressure for N = 480 (ρ_0 = 0.0313 Å^3, see above). Again, the system was randomized at 2000 K during ≃25 ps and then quenched by steps of 500 k, lasting 10 ps each, the last step bringing the system from T = 500 to 300 K. In both cases (N = 120 and N = 480) data collection is carried out at T = 300 K (10 ps) after completion of a last thermal cycle involving an heating step at T = 700 K (10 ps).

In the case of the $GeSe_4$ glass, and only for N = 120, we carried out the calculations in the isothermalisobaric (NPT) ensemble, by keeping fixed the shape of the simulation cell that adjusts to the imposed pressure in an isotropic manner. This choice was motivated by a recent study on the structural properties of amorphous $GeSe_2$ where it was shown that even residual small pressures can affect its equilibrium properties [60]. Such investigation was a prerequisite to the study of glassy $GeSe_2$ under pressure, ultimately enabling to unravel a density driven mechanism for a structural transformation of the network [61]. Within the framework of that investigation, only the size N = 120 was selected, due to the computational burden of considering a large set of different reduced densities and pressures. The above choice for the methodological framework was motivated by two additional reasons. First, working in the NPT ensemble instead of the NVT one offers the opportunity to produce a new, statistically independent set of results on g-GeS_4. Also, the present scheme differs from previous ones not only by the constraint of a fixed volume but also by the selection of a different exchange-correlation functional (BLYP at the place of PW and PBE, see references).

In the case of $GeSe_4$ and N = 120, the availability of previous calculations [62] allowed us to start from an amorphous configuration. After releasing the residual pressure at T = 0 K, with a new density found to be equal to ρ_0 = 0.0317 Å^{-3}, box size = 15.58 Å, (initial values equal to 0.0339 atom Å^{-3}, box size 15.27 Å), the model generated undertook to a thermal cycle in the NPT ensemble. Within the fixed shape, variable size version of the Parrinello-Rahman technique [63] we allowed for isotropic variation of the cell dimensions by running at several temperatures as follows: 16 ps at 100 K, 14 ps at 400 K, 10 ps at 600 K, 38 ps at 900 K, 26 ps at 600 K and 81 ps at 300 K. Overall, the results presented hereafter for g-$GeSe_4$ are averaged over $\simeq$70,000 configurations. For N = 480, we took advantage of configurations available for glassy $GeSe_2$, we changed the chemical identities to make our system compatible with the new composition and we proceeded as detailed above (see the case of glassy GeS_4) in order to loose memory of the initial configuration and produce a statistically meaningful trajectory. The density of the system was taken to be equal to the value $\rho_0 = 0.0317$ Å^{-3} found to correspond to zero pressure in the N = 120 case.

For the four models under investigation, the system sizes allow to cover a k-region significantly smaller than the position of the FSDP, $k_{FSDP} \simeq 1$ Å^{-1} [64].

FPMD was employed in conjunction with the Generalized Gradient Approximation (GGA) due to Becke (B) for the exchange energy and Lee, Yang and Parr (LYP) for the correlation energy [23, 24]. The wave functions were expanded at the Γ point of the supercell on a plane wave basis set with an energy cutoff $E_c = 30$ Ry for GeS_4 and $E_c = 20$ Ry for $GeSe_4$. A fictitious electron mass of 1000 a.u. (i.e. in units of $m_e a_0^2$ where m_e is the electron mass and a_0 is the Bohr radius) was used throughout. Two distinct time steps of $\Delta t = 0.17$ and 0.12 fs are adopted to integrate the equations of motion for the GeS_4 and the $GeSe_4$ model respectively. Temperature control was implemented for both the ionic and electronic degrees of freedom by using Nosé-Hoover thermostats [65–67].

12.3.1 Neutron Total Structure Factor and Total Pair Correlation Function

The total neutron structure factor $S_T(k)$ is defined by

$$S_T(k) - 1 \equiv \sum_{\alpha=1}^{2}\sum_{\beta=1}^{2} \frac{c_\alpha c_\beta b_\alpha b_\beta}{\langle b\rangle^2}\left[S_{\alpha\beta}^{FZ}(k) - 1\right] \tag{12.2}$$

where α and β denote either one of the two chemical species, c_α and b_α are the atomic fractions and coherent neutron scattering lengths of Ge, Se or S, $\langle b\rangle = c_{Ge}b_{Ge} + c_X b_X$ is the mean coherent neutron scattering length, and $S_{\alpha\beta}^{FZ}(k)$ is a Faber-Ziman (FZ) partial structure factor [68].

The coherent neutron scattering lengths, b, for Ge, Se and S of natural isotopic abundance are equal to 8.185, 7.97 and 2.847 fm respectively [69].

The corresponding real space information is given by the total pair correlation function:

$$\begin{aligned} g_T(r) - 1 &= \frac{1}{2\pi^2 n_0 r}\int_0^\infty dk\, k\, [S_T(k) - 1]\sin(kr) \\ &= \sum_{\alpha=1}^{n}\sum_{\beta=1}^{n} \frac{c_\alpha c_\beta b_\alpha b_\beta}{\langle b\rangle^2}\left[g_{\alpha\beta}(r) - 1\right] \end{aligned} \tag{12.3}$$

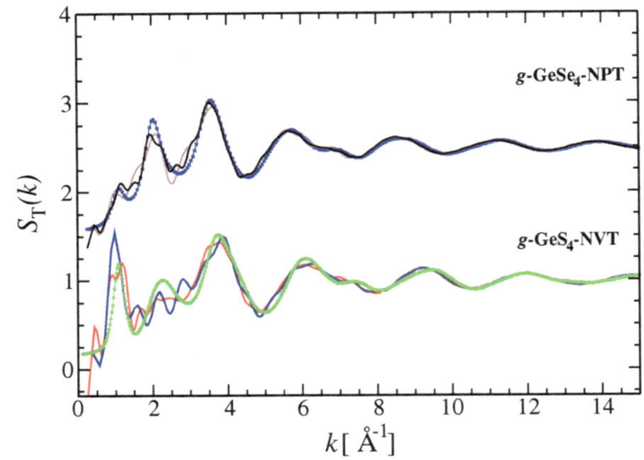

Fig. 12.9 Fourier transformed total neutron structure factors of the FPMD models at 300 K for g-GeS_4: N = 120 atoms (*blue line*), N = 480 atoms (*red line*) and g-$GeSe_4$: N = 120 atoms (*brown line*), N = 480 atoms (*black line*) compared to the experimental measurements for g-GeS_4 from [54] (*solid green line* with *circles*) and g-$GeSe_4$ from [55, 70] (*solid blue line* with *squares*)

where n_0 is the atomic number density and $g_{\alpha\beta}(r)$ is a partial pair distribution function.

In Fig. 12.9 the calculated total neutron structure factors $S_T^{\text{th}}(k)$ are presented for the FPMD models of GeS_4 and $GeSe_4$ along with their experimental counterparts. The data are Fourier transformed from direct space through the calculation of the corresponding total pair correlation functions. The improvement arising from the use of a larger model is particularly noticeable in the case of g-GeS_4, for which the N = 120 and N = 480 sets of results differ dramatically in the region between the peaks located at k $\sim$ 1 and 4 Å^{-1}. Spurious oscillations found for N = 120 are replaced by a largely steady profile, underestimating the intensity of the second peak. For models of glassy systems, the importance of relying on several independent trajectories to improve the statistics has been underscored in previous publications [71] and it turns out to be fully confirmed here. For g-GeS_4, a spurious peak appears clearly at $k \simeq 0.5$ Å^{-1}, that is at the left of the First Sharp Diffraction Peak (FSDP) region, located around $k \simeq 1$ Å^{-1}. The position of this peak bears no physical meaning, as being related to some persistent and yet statistically not significant long range correlation found in the quenched configuration. This interpretation is consistent with the presence of this peak in both total structure factors calculated directly and via Fourier transform.

Turning to the results in direct space, the calculated total pair correlation functions $g_T(r)$ are shown for g-GeS_4 and g-$GeSe_4$ in Fig. 12.10 along with the corresponding diffraction data. Experimental data were obtained by Fourier transforming the total neutron structure factors and using upper limits of integration of 19.95 Å^{-1} for g-$GeSe_4$ and 49.95 Å^{-1} for g-GeS_4. [54, 70] These limits result from the finite measurement window function of the diffractometer and lead to oscillations at r less $\sim$2 Å, more pronounced for g-$GeSe_4$.

The overall agreement is quite remarkable, the differences being limited to the higher intensity of the calculated $g_T(r)$ (g-$GeSe_4$) and to the large bump found for $3 < r < 5$ Å in g-GeS_4, this feature being moderately shifted at higher values, as shown by its maximum located at 3.59 Å instead of 3.39Å (experimental data). We recall that

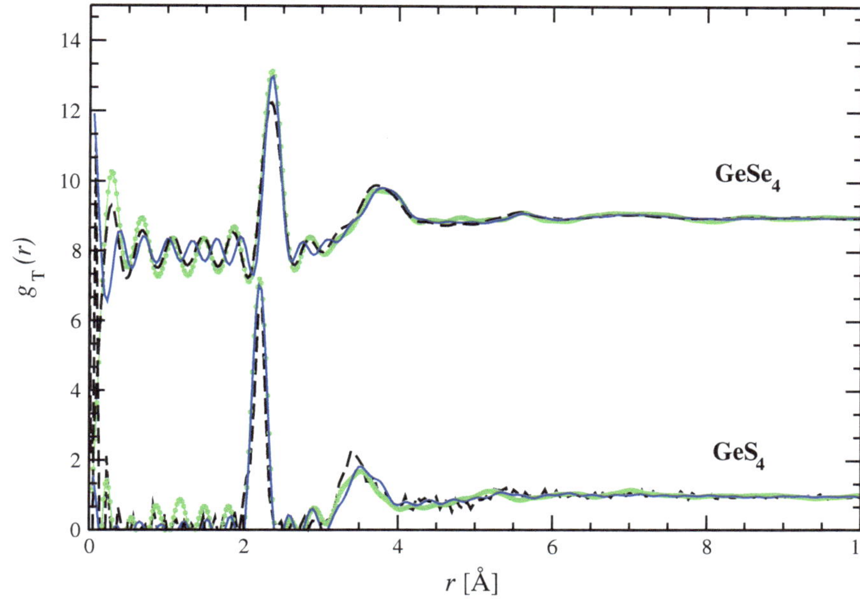

Fig. 12.10 Total pair distribution functions of the FPMD models at 300 K for g-GeS_4 and g-$GeSe_4$ (shifted by +8). For both systems: N = 120 (*solid green line* with *circles*), N = 480 (*solid blue line*) compared to the experimental measurements for g-GeS_4 from [54] (*black brocken line*) and g-$GeSe_4$ from [55, 70] (*black brocken line*)

in these glasses the basic building blocks are the Ge–X_4 (X = S, Se) tetrahedra, the first peak in $g_T(r)$ being a signature of the intra-tetrahedra Ge–X (X = S, Se) bonds, while the second peak corresponds to the inter-tetrahedra interatomic distances.

12.3.2 Faber-Ziman Partial Structure Factors

The calculated Faber-Ziman (FZ) partial structure factors of g-GeS_4 and g-$GeSe_4$ are shown in Fig. 12.11. For g-$GeSe_4$, the changes associated with the larger system size are minimal and limited to the intensities of the first and third peak in the Ge–Ge partial structure factor. In line with the above observations for the total neutron structure factor of g-GeS_4, it appears that for N = 480 the height of the FSDP intensities is reduced in the Ge–Ge partial structure factor, although such peak takes an undesirable split pattern, due to statistical noise. The most evident consequence of a somewhat limited statistical accuracy is the presence of the spurious peak at $k \simeq 0.5$ Å^{-1} for which no sign is found in either $S^{FZ}_{GeSe}(k)$ or $S^{FZ}_{SeSe}(k)$ of g-GeS_4 (this feature is found also in the total netron structure factor, see above). In both systems, the intermediate range order (to be associated to features around $k \simeq 1$ Å^{-1}) is due to the Ge–Ge and Ge–Se (or Ge–S) correlations since no deviations from the basisline profile of $S^{FZ}_{SeSe}(k)$ (or $S^{FZ}_{SS}(k)$) are found in the FSDP region.

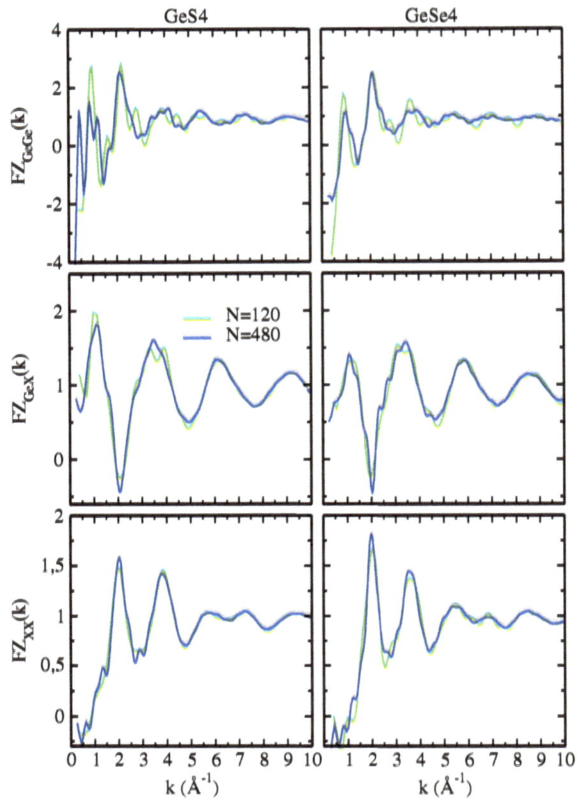

Fig. 12.11 (*Color online*) The Faber-Ziman partial structure factors $FZ_{GeGe}(k)$ (*top panel*), $FZ_{GeX}(k)$ (*middle panel*) and $FZ_{XX}(k)$ (*bottom panel*). From *left* to *right*: for the amorphous GeS_4 (*solid black lines*) and $GeSe_4$ (*solid orange lines*) models

12.3.3 Real Space Properties

12.3.3.1 Pair Correlation Functions

In Fig. 12.12 we display the calculated partial pair correlation functions $g_{\alpha\beta}(r)$ for g-GeS_4 and g-$GeSe_4$. For N = 480, the partial pair correlation functions $g_{GeGe}^{GeS_4}(r)$ and $g_{GeGe}^{GeSe_4}(r)$ exhibit a first small peak at $r \simeq 2.45$ Å, signature of the presence of homopolar Ge–Ge bonds. Interestingly, such feature is absent in both systems for N = 120. At larger distances and for both system sizes the pair correlation functions have in common the peaks representative of edge-sharing (ES) and corner-sharing (CS) tetrahedra, located at $r \simeq 2.9$ Å, $r \simeq 3.6$ Å ($g_{GeGe}^{GeS_4}(r)$) and $r \simeq 3.0$ Å, $r \simeq 3.72$ Å ($g_{GeGe}^{GeSe_4}(r)$) respectively. However, the intensity of the CS-related peak behaves differently when moving from N = 120 to 480, with a sharp increase in the g-GeS_4 case and a sharp decrease in the g-$GeSe_4$ case. The impact of a larger size on the pattern taken by $g_{GeGe}^{GeS_4}(r)$ and $g_{GeGe}^{GeSe_4}(r)$ is striking for r > 5 Å. For the larger systems, a clear maximum at around 6–6.5 Å takes the place of an oscillating profile

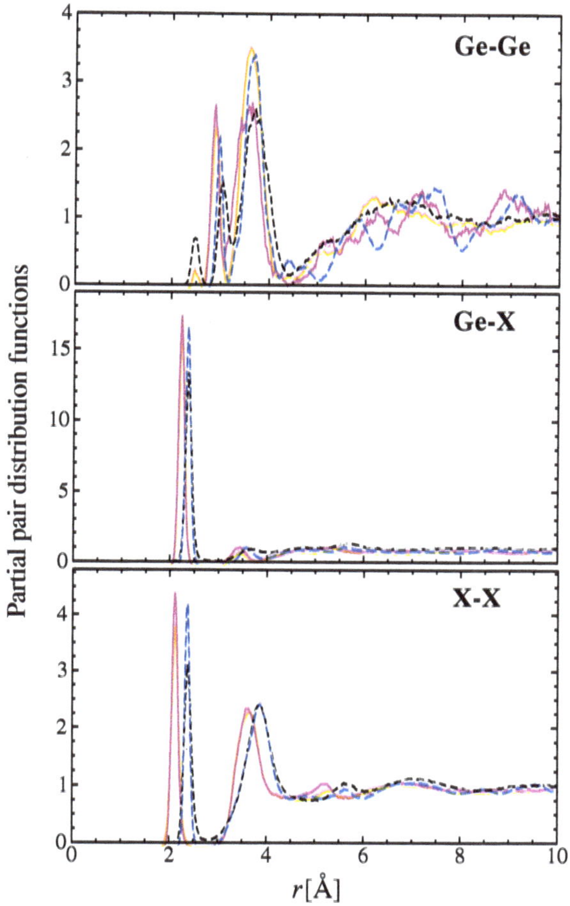

Fig. 12.12 Partial pair distribution functions $g_{GeGe}(r)$ *(top panel)*, $g_{GeX}(r)$ *(middle panel)* and $g_{XX}(r)$ *(bottom panel)*, for the GeX_4 (X = S, Se) FPMD models. The results for *g*-GeS_4: N = 120 *(magenta line)*, N = 480 *(orange line)* and *g*-$GeSe_4$: N = 120 *(broken black lines)*, N = 480 *(broken blue lines)* models are compared

where several maxima and minima are clearly discernible. Once again, we are able to detect an unambiguous effect due to the adoption of a large system size, to which the minority species (Ge) is particularly sensitive.

The partial correlations functions $g_{GeX}^{GeS_4}(r)$ and $g_{GeX}^{GeSe_4}(r)$ are very similar, differing only by the position of the first peak (2.23–2.24 Å as a signature of the Ge–S bonds and 2.36–2.37 Å as a signature of the Ge–Se bonds). These observations are consistent with the overwhelming predominance of Ge-centered tetrahedral motif in these systems. Similarly, the first peaks at $r \simeq 2.11$–2.12 Å in $g_{XX}^{GeS_4}(r)$ and $r \simeq 2.37$–2.38 Å in $g_{XX}^{GeSe_4}(r)$ (bottom panel in Fig. 12.12) are indicative of the presence of homopolar X–X bonds being part of chains. For $r > 3$ Å, the $g_{XX}^{GeS_4}(r)$ and $g_{XX}^{GeSe_4}(r)$ behave identically indicating a very similar Se/S sub-networks. The peaks at $r \simeq 3.63$ Å in *g*-GeS_4, and at $r \cong 3.85$–3.87 Å in the *g*-$GeSe_4$ glasses are the signatures of the intra-tetrahedral X–X (X = S, Se) connections.

12.3.3.2 Coordination Numbers and Structural Units

The coordination numbers $\bar{n}_{\alpha\beta}$ are listed in Table 12.1. These are defined as the mean number of nearest neighbors of type β located around an atom of type α. As an integration range, we have taken the first minimum of the total pair correlation function. The total coordination numbers for Ge and X (Se or S) are given by $\bar{n}_{\mathrm{Ge}} = \bar{n}_{\mathrm{GeGe}} + \bar{n}_{\mathrm{GeX}}$ and $\bar{n}_{\mathrm{X}} = \bar{n}_{\mathrm{XX}} + \bar{n}_{\mathrm{XGe}}$, respectively, where $\bar{n}_{\mathrm{XGe}}/c_{\mathrm{Ge}} = \bar{n}_{\mathrm{GeX}}/c_{\mathrm{X}}$. The average coordination number irrespective of chemical species type is given by the expression $\bar{n} = c_{\mathrm{Ge}}(\bar{n}_{\mathrm{GeGe}} + \bar{n}_{\mathrm{GeX}}) + c_{\mathrm{X}}\,(\bar{n}_{\mathrm{XX}} + \bar{n}_{\mathrm{XGe}})$.

The coordination numbers are consistent with networks made of overwhelming proportions of tetrahedra, the only noticeable deviation from this arrangement being the n_{GeGe} value (0.36) for g-$GeSe_4$, N = 480. The lower value of n_{GeGe} in g-GeS_4, N = 480, together with the absence of such nearest-neighbor Ge–Ge contacts in the N = 120 models suggests two considerations, i.e. first, g-GeS_4 is more chemically ordered than g-$GeSe_4$ and, second, the occurrence of homopolar bonds in the more

Table 12.1 The first peak position (FPP) and second peak position (SPP) in $g_{\alpha\beta}(r)$, and nearest neighbor coordination numbers $\bar{n}_{\alpha\beta}$ obtained for the FPMD models of g-GeS_4 and g-$GeSe_4$

$g_{\alpha\beta}(r)$	Model	FPP (Å)	SPP (Å)	$\bar{n}_{\alpha\beta}$
$g_{\mathrm{GeGe}}(r)$	g-$GeSe_4$-120	–	2.97–3.69	–
	g-$GeSe_4$-480	2.47	3.70	0.36
	g-GeS_4-120	–	2.77–3.64	–
	g-GeS_4-480	2.47	2.89–3.60	0.03
$g_{\mathrm{GeX}}(r)$	g-$GeSe_4$-120	2.36	3.58	3.96
	g-$GeSe_4$-480	2.37	3.67	3.85
	g-GeS_4-120	2.22	3.43	4.00
	g-GeS_4-480	2.23	3.43	3.98
$g_{\mathrm{XGe}}(r)$	g-$GeSe_4$-120	2.36	3.58	0.99
	g-$GeSe_4$-480	2.37	3.67	0.96
	g-GeS_4-120	2.22	3.43	1.00
	g-GeS_4-480	2.23	3.43	0.99
$g_{\mathrm{XX}}(r)$	g-$GeSe_4$-120	2.38	3.87	0.99
	g-$GeSe_4$-480	2.37	3.85	1.04
	g-GeS_4-120	2.11	3.63	1.00
	g-GeS_4-480	2.12	3.63	0.99
		$\bar{n}_{\mathrm{Ge}}$	$\bar{n}_{\mathrm{X}}$	$\bar{n}$
	g-$GeSe_4$-120	3.96	1.99	2.386
	g-$GeSe_4$-480	4.21	2.00	2.44
	g-GeS_4-120	4.00	2.00	2.40
	g-GeS_4-480	3.99	1.98	2.385

The coordination numbers $\bar{n}_{\alpha\beta}$ were obtained by using an integration range of 0–2.7 Å where the upper limit corresponds to the first minimum in the total pair distribution functions for the GeX_4 models chemically disordered system ($GeSe_4$) is somewhat prevented by the limited number of Ge atoms when N = 120 (24 Ge atoms only).

To provide a more complete description of the network we calculated the average percentages of the individual α-l structural units where an atom of species α (Ge, X = S or Se) is l-fold coordinated to other atoms. To clarify this notation, Ge–GeS_3 represents a Ge atom that is connected to 1 other Ge atom and 3 S atoms while Ge–S_4 represents a Ge atom that is connected to 4 S atoms. Bonds are deemed to be formed when the interatomic distance for a given pair of atoms is smaller than 2.7 Å, this value corresponds to the first minimum in the total pair distribution functions for both systems. The proportion of units $\bar{n}_\alpha(l)$ are summarized in Table 12.4.

Tetrahedra Ge(X_4) motifs are largely predominant for both systems, even though their percentage lowers due to the occurrence of homopolar bonds in g-$GeSe_4$, N = 480. Interestingly, the deviation from chemical order, found to be more important in g-$GeSe_4$, is confirmed by the presence of threefold Ge-coordinated units for both N = 120 and N = 480. These units are essentially absent in g-GeS_4, where the only deviation from chemical order manifests itself through the formation of fourfold Ge–GeX_3 units.

The percentages of structural units relative to the Se atoms exemplify the impact of two concomitant effects. Focussing on the results for N = 120, one notice that AA, AB and BB occur with similar weights in g-GeS_4 and in g-$GeSe_4$, AB connections being way more numerous (about 43 %, more than 10 % larger than BB and AA ones). The above percentages changes when considering the N = 480 models and notable differences arise among the two set of values relative to g-GeS_4 and g-$GeSe_4$. In g-$GeSe_4$, AB connections remains largely predominant (38 %). On the contrary, g-GeS_4 features close percentages of AA, AB and BB connections. This conjecture is in line with experimental evidence [72], pointing to an important population of S_n structures for low content of Ge atoms in g-Ge_xS_{1-x} glasses. These pieces of evidence should be confirmed by a further set of simulations for each one of the four cases, in order to produce statistical errors. However, past statistical uncertainties found for the AA, AB and BB populations appear to be smaller than the variations reported above (Table 12.2).

Turning to the relative weights of corner-sharing and edge-sharing connections, both networks are characterized by an unambiguous majority of Ge connected in a corner-sharing fashion. This is consistent with the relative intensities and widths of the second and third peaks in the corresponding partial Ge–Ge pair correlation functions (see Fig. 12.12).

12.3.4 *g*-GeS_4 Versus *g*-$GeSe_4$: Conclusions

First-principles molecular dynamics studies of glassy GeS_4 and glassy $GeSe_4$ have shown than both systems are characterized by tetrahedral connections interlinked with X–X (X = S or Se) homopolar bonds. From the methodological point of view,

Table 12.2 Proportion, $\bar{n}_\alpha(l)$ of the different coordination units in amorphous GeS_4 and $GeSe_4$

		Proportion $\bar{n}_\alpha(l)$ (%)			
		g-$GeSe_4$-120	g-$GeSe_4$-480	g-GeS_4-120	g-GeS_4-480
Ge atom					
$l = 2$					
	X_2	–	4.83	–	0.39
$l = 3$					
	GeX_2	–	–	–	–
	X_3	3.62	3.59	–	0.71
$l = 4$					
	GeX_3	–	6.17	–	2.03
	X_4	96.36	85.00	99.95	96.73
$l = 5$					
	Ge_2X_3	–	–	–	–
	GeX_4	–	–	0.05	0.08
	X_5	–	–	–	–
X atom					
$l = 1$					
	Ge	–	1.45	–	0.51
	X	1.04	1.73	–	1.31
$l = 2$					
	X_2 (BB)	28.12	30.74	28.12	32.02
	XGe (AB)	43.61	38.24	43.75	33.95
	Ge_2 (AA)	27.08	26.48	28.12	31.79
$l = 3$					
	X_2Ge	–	0.54	–	<0.1
	XGe_2	0.13	0.32	–	0.15
	Ge_3	–	0.32	–	0.23
	N_{Ge-Ge}	0.0	6.20	0.20	2.65
	N_{X-X}	71.87	70.14	71.87	67.43
	N_{Ge}(ES)	25	25.44	29.16	32.76
	N_{Ge}(CS)	75	68.34	70.62	64.58

We also provide the number of Ge atom involved in homopolar bond(s) N_{GeGe} and the fraction of X atom (S or Se) involved in a homopolar bond(s) N_{XX}. These quantities have been calculated including neighbors separated by less than 2.7 Å

details related to the nearest neighbor arrangements of these glasses are found to be sensitive to the size of the periodic system adopted, either N = 120 or N = 480. An important reduction of Ge–X–X connections occurs for N = 480, proving that, for specific systems, intermediate range properties can be (at least moderately) sensitive to the number of atoms composing the periodic simulation cell.

12.4 Binary Chalcogenides Under Pressure

12.4.1 Overview of the Experimental Findings

As summarized in the previous sections, the development of a reliable methodology to build and study realistic binary chalcogenides has been a stepwize process. After solving the main issues related to the FSDP and the IRO, it is natural to focus on exploring the phase diagram of these disordered materials. Indeed the modifications of the structural and the electronic properties of these systems under high pressure and/or temperature are of particular interest.

Experimentally, the crystallization and the amorphization of binary amorphous systems under pressure have already been the subjects of many studies [73–81]. In 1965, Prewitt and Young [73] found two new crystalline phases of Germanium and Silicium disulfides at T = 300 K under pressure. Later, Shimada and Dachille [74] studied the crystallization of amorphous $GeSe_2$ under pressure and found that three different crystalline phases exist depending on temperature and pressure. They also quoted that the crystallization was inhibited for temperatures lower than T = 573 K.

Using Raman scattering and energy-dispersive X-ray diffraction Grzechnik, Grande and Slolen [75] studied the inverse process, the pressure induced amorphization of the crystalline β-$GeSe_2$. Their work showed that no phase transition from crystal to known polymorphs of $GeSe_2$ was observed until 11 GPa. However, between 11 and 14 GPa, the sample showed a darkening possibly related to a transition of the coordination during compression, from 2 to 3 for the Se atoms and from 4 to 6 for the Ge atoms. At room temperature, the transition from four fold to six fold coordination is kinetically inhibited in crystalline materials, and in the range of 7–14 GPa a large disordered domain forms, the β-$GeSe_2$ is no longer stable. It was concluded that the pressure induced amorphization of an AB_2 system results from its tendency to reach higher coordinated phases at different pressures and depends on the parent structure as well as the ionicity of the system.

More light was brought to the $GeSe_2$ crystalline and glassy phases by Grande and Co. [76] using X-ray diffraction measurements under pressure up to 7.7 GPa and temperature up to T= 1100 °C. The crystalline phase of $GeSe_2$ at ambient pressure was found to be a 2D structure with 50 % of edge and 50 % of corner sharing Ge atoms.

In contrast with the work by Shimada [74] Grande and Co. [76] observed that amorphous $GeSe_2$ at T= 250 – 300 °C partly crystallized in two phases when compressed, the first at 3.0 GPa and the second at 7.7 GPa. Also, during quenching performed at 7.7 GPa an orthorhombic polymorph of $GeSe_2$ was found at 500 °C.

In further work the same team [82] used angle-dispersive X-ray diffraction in the pressure range 2–6 GPa to study crystalline $GeSe_2$. This confirmed the transition to a three-dimensional crystalline structure above 2 GPa at 698 K. This transition was explained by an anisotropic lattice distortion due to the cooperative tilting of rigid $Ge(Se_4)$ tetrahedra. The authors claimed that a similar transition, including anomalous compressibility and thermal expansion phenomena, could be observed in the high pressure and high temperature phases of disordered germanium dichalcogenides. In analogy to crystalline polymorphism, this behaviour could be called "amorphous polymorphism".

The $GeSe_2$ system has also been the subject of X-ray diffraction study under pressure in the liquid phase [77]. It was found that under compression the connectivity of the liquid changes from a 2D to 3D. The 2D structure is characterized by a mixture of ES and CS tetrahedra, while the 3D one is only made by CS tetrahedra. This conversion from a mixture of ES/CS to a full CS network is accompanied by a breakdown of the IRO, completely lost above 2.5 GPa.

The occurrence of a structural transition in the liquid and the crystalline states, in addition to the two amorphous structures which mimic the 2D–3D transition observed by Shimada, suggest that a first order transition between two liquids $GeSe_2$ with different densities could be possible. Furthermore, measurements based on high pressure Raman spectroscopy on glassy $GeSe_2$ [78] showed that the Raman spectra are reversible. In addition no structural transition was reported up to 9 GPa and only a change in the sample color was noticed.

The topological changes in glassy $GeSe_2$ have been investigated by high-energy X-ray and neutron diffraction at pressures up to 9.3 GPa [79] (e.g. see Fig. 12.13).

Using these measurements the authors interpreted the densification mechanism as a continuous increase of the mean coordination number with a conversion from edge to corner sharing tetrahedra. The diffraction data showed that the conversion is accompanied by an elongation of the mean Ge–Se bond length and a substantial distortion of the $GeSe_4$ tetrahedra. Moreover, the evolution of the PCF as a function of pressure highlighted a decrease of the IRO with increasing the pressure up to 9.3 GPa, accompanied by a decrease in the intensity of the FSDP in the corresponding structure factor. Recently, glassy $GeSe_2$ has been the subject of acoustic measurements using synchrotron radiation under pressure up to 9.6 GPa [80]. The outcome of these measurements showed a minimum in the network rigidity around 4 GPa. The system responds differently depending on the applied pressure. Thus until 3 GPa the ES

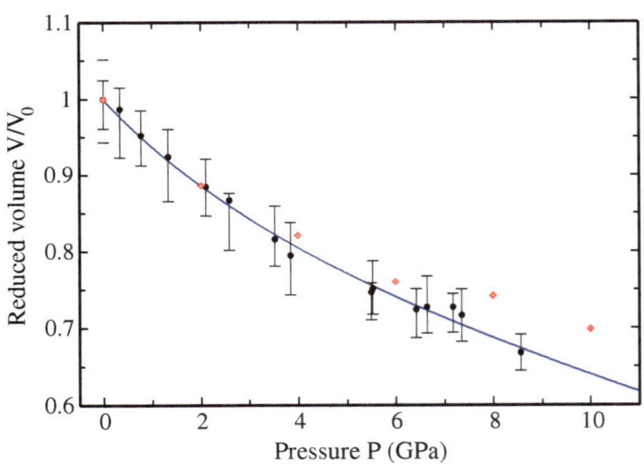

Fig. 12.13 Pressure-volume equation of state for amorphous $GeSe_2$. Experimental compression in a hydrostatic medium, a 4:1 methanol:ethanol mixture [79] (*solid black circles*); simulation data from [83] (*red open diamond*); third-order isothermal Birch-Murnaghan equation of state fit to the experimental compression data (*solid black line*)

convert to CS tetrahedra maintaining the flexible character of the network. However, for higher pressures, a continuous increase on the density and thus the coordination persists, while the ratio ES and CS tetrahedra remain constant.

As summarized above, many experimental efforts were devoted to study and understand the behaviour of disordered phases of $GeSe_2$ under pressure. From a theoretical point of view the efforts deployed so far to understand the behavior of these glasses under pressure are by far not comparable to the availability of experimental data. To date, the only work on this subject is the one by the team of Drabold [83]. In this work, the structural changes of the amorphous $GeSe_2$ under pressure up to 75 GPa were investigated. The simulations were performed using non self-consistent LDA-Harris functional molecular dynamics, the pressure being applied using the Parinello-Rahman method. In agreement with the experimental work, this study revealed a continuous and reversible structural transition upon compression into a metallic amorphous phase. The average coordination of the Ge and the Se atoms changes gradually from 4 to 6 and 2 to 4, respectively. For high pressures, Se clusters appear in a chain like fashion. The FSDP was found to shift to large k values with a decreasing intensity. Regarding the structure, the mean bond length does not change considerably until 12 GPa while in the range of 12–16 GPa it increases substantially. Similarly, the mean coordination number increases until 12 GPa. The pressure was also found to suppress the homopolar Ge–Ge bonds.

The nature of the density-driven network collapse in $GeSe_2$ glass is elusive. Three interpretations of the different set of results at pressures up to 9 GPa could be proposed: (i) a continuous structural change in which the mean coordination number $\bar{n}$ steadily increases [79, 83] as CS motifs are initially replaced by ES motifs [84]; (ii) a two steps mechanism [80], before 3 GPa the network is found to be flexible as the ES are replaced by CS motifs, beyond the threshold of 3 GPa the mean coordination number $\bar{n}$ increases leading to a rigid network; (iii) or more simple mechanism where the network topology remains unchanged [78].

As in the case of glassy $GeSe_2$, the structural changes of amorphous $GeSe_4$ under pressure have been tracked by high-energy X-ray diffraction experiments up 8.6 GPa [81]. The main outcome of this work rely on finding an invariant mean coordination number with increasing the pressure. This result is in striking contrast with the case of $GeSe_2$ where a gradual increase of the coordination was found (e.g. see Fig. 12.14). The densification mechanism in amorphous $GeSe_4$ was interpreted in terms of shifts of the second neighbor and the higher coordination shells. The peaks related to these shells in the PCFs shift to shorter distances and increase in intensity indicating a tendency of the network to fill the voids present at ambient pressure.

Within the Ge_xSe_{1-x} family, the origin of the densification mechanism depends on the bonding and the chemistry of the network. Many possible interpretations were proposed to explain and describe this process. In order to discriminate between all the proposed mechanisms and to provide a clear atomic scale description of the modification of the structural and the electronic properties of the material under the effect of pressure, it is highly desirable to perform first-principles self-consistent simulations.

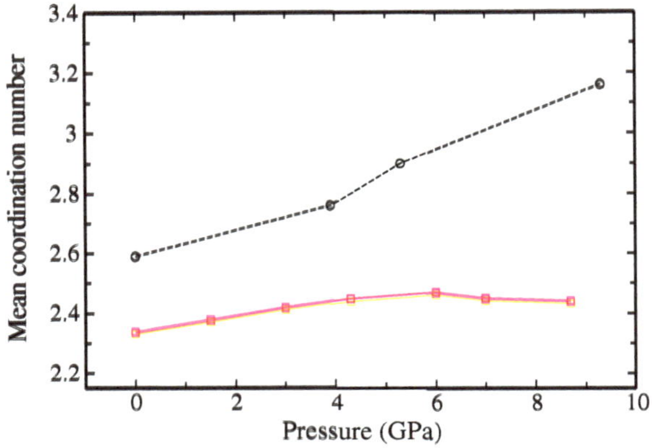

Fig. 12.14 Evolution of the average coordination number $\bar{n}$, integrated from the first peak of $g(r)$, as a function of pressure. The *black open circles* are $GeSe_2$ data from [79]. The *red open squares* are $GeSe_4$ date from [81]

12.4.2 Amorphous $GeSe_2$ Under Pressure

The initial $GeSe_2$ configuration was taken from the system studied in [26]. Simulations were performed within the FPMD framework, namely the Car-Parrinello method as implemented in the CPMD code. We adopted a generalized gradient approximation (GGA) for the exchange and correlation part of the total energy. In particular, the Becke-Lee-Yang-Parr (BLYP) has been used.The Troullier-Martins norm-conserving pseudopotentials are employed for the core-valence interactions description. The largest cutoff used in the pseudopotentials construction was equal to 1.06 Å. A fictitious electron mass of 1000 a.u. and a time step of $\Delta t = 0.24$ fs were adopted to integrate the equations of motion, ensuring good control of the conserved quantity. The wave functions are expanded in plane waves basis set at the Γ point of the super cell, with a cutoff energy of 20 Ry. A temperature control is implemented for both ionic and electronic degrees of freedom by using the Nosé-Hoover thermostats. To construct our amorphous model and achieve optimal statistical sampling we implemented a thermal cycle featuring 40 ps at T = 300 K, 50 ps at T = 600 K, 150 ps at T = 900 K (to allow significant diffusion) and, on cooling, 70 ps at T = 600 K and 150 ps at T = 300 K, with statistical averages taken over a final portion of the trajectory length 50 ps at T = 300 K.

12.4.2.1 Enhancing Ambient Pressure Model

It is of great interest to make sure that available $GeSe_2$ models (studied at fixed volume, corresponding to the experimental density) are indeed available at a vanishing (ambient) pressure. Previous theoretical studies carried out at the experimental density, under the assumption of ambient pressure, were instrumental in providing a clear picture of the network structure. Briefly, the structure consists of a variety of structural units, encompassing $GeSe_4$ tetrahedra, homopolar bonds, and defective Ge and Se coordinations, accounting for a moderate departure from chemical order.

Analysis of the methodology employed within FPMD reveals that most simulations have been performed at a fixed density. Even though these simulations are intended to be stress-free (zero pressure) no information on the actual values of the pressure were available. Account of the pressure is worthwhile since the density vs pressure relationship holding experimentally might not be exactly reproduced by the DFT model [85]. As a consequence, undesirable pressure effects altering the room temperature equilibrium properties can be observed. The starting point of the present study corresponds to the observation that a set of FPMD simulations were carried out at 300 K for a number density $\rho^a = 0.034\,Å^{-3}$, slightly larger than the value quoted in the experimental work $\rho^{exp} = 0.0334\,Å^{-3}$ [86, 87]. In order to evaluate the pressure on the cell we have simulated a new system at the experimental density ρ^{exp}, in addition we have also computed the pressure on the previously simulated system at ρ^a. A residual pressure of ~1 GPa has been found in both model, showing that the model was not entirely stress-free for a volume (density) equal to the experimental value, this latter corresponding to ambient pressure. By expanding the simulation box and further relaxing the whole structure, we were able to lower the pressure to less than ~0.1 GPa. The new corresponding density is $\rho^b = 0.0326\,Å^{-3}$. The final structure at the relaxed density was selected to start a thermal annealing cycle as described above. Releasing this residual pressure has an important effect on the structural rearrangement. Figure 12.15 shows the new set of partial pair correlation

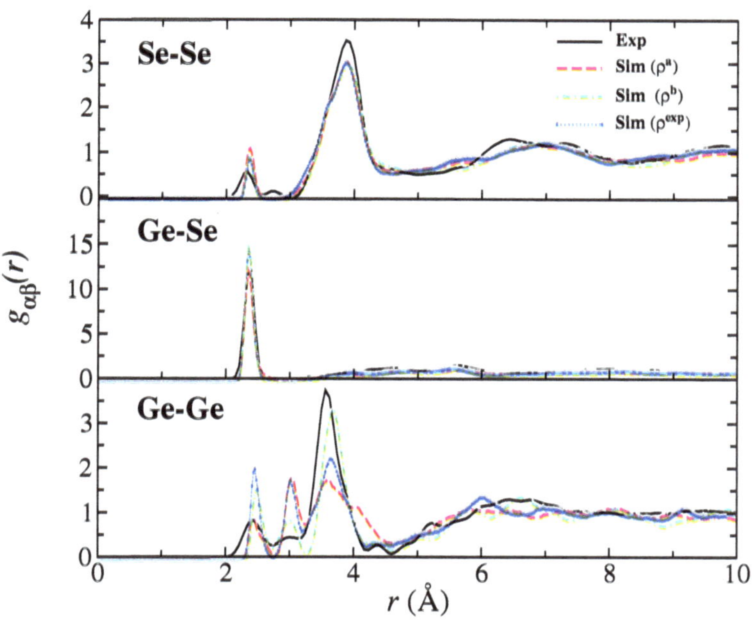

Fig. 12.15 Partial pair correlation functions for glassy $GeSe_2$. The experimental results of [88] (*solid curve*) are compared with the results of [26] with ρ^a (*dashed red curve*) and with the new simulation at the experimental density ρ^{exp} (*dotted blue curve with stars*) and the present relaxed structure results at ρ^b (NVT ensemble, *dashed-dotted green curve*)

Table 12.3 First (FPP), second (SPP) and third (TPP) peak positions are provided for all models

$\alpha\beta$	density	FPP (Å)	$n_{\alpha\beta}$	SPP (Å)	$n'_{\alpha\beta}$	TPP (Å)	$n''_{\alpha\beta}$
Ge–Ge	exp	2.42	0.25	3.02	0.34	3.57	3.2
	ρ^a	2.43	0.20	3.11	0.69	3.61	3.15
	ρ^b	2.47	0.28	3.0	0.37	3.67	3.19
	ρ^{exp}	2.45	0.30	3.02	0.52	3.64	2.68
Ge–Se	exp	2.36	3.71	–	–	–	–
	ρ^a	2.35	3.58	–	–	–	–
	ρ^b	2.34	3.64	–	–	–	–
	ρ^{exp}	2.35	3.65	–	–	–	–
Se–Se	exp	2.32	0.20	–	–	3.89	9.3
	ρ^a	2.37	0.30	–	–	3.89	9.9
	ρ^b	2.34	0.20	–	–	3.89	9.93
	ρ^{exp}	2.37	0.20	–	–	3.86	9.65

$n_{\alpha\beta}$ corresponds to the partial coordination number obtained by integrating the corresponding PCF first peak up to the first minimum. Results for previous simulations at ρ^a from [26] and experiment from [88] compared to new simulation at ρ^{exp} and at the relaxed structure ρ^b

functions compared to previous simulation results, simulation at the experimental density and the experiment results.

Regarding the previous simulations and experiment, the present $g_{\alpha\beta}$ for both ρ^b and ρ^{exp} present a small shifts in the peaks positions within the typical statistical fluctuations of the order 0.03–0.05 Å (see Table 12.3). When looking at the general shape, g_{GeSe} and g_{SeSe} are less affected by the change of the density, while g_{GeGe} exhibits large changes. Adjusting the density to the experimental one enhances the shape of the third peak, and a clear minimum appears. The intensities of the first and the second peaks remain always largely over-estimated. The relaxed density model (ρ^b) provides a better agreement with the experimental g_{GeGe}. The discrepancies between the second and the third peak intensities and the measured one considerably reduced, since the second peak intensity decreases and the third intensity peak increases. A quantitative comparison between these peaks could be provided by the partial coordination numbers $n_{\alpha\beta}$, as obtained by integrating the corresponding $g_{\alpha\beta}$ peaks (see Table 12.3), also a striking improvement of these numbers is found for ρ^b.

Not only the simulation at the experimental density provide a better agreement with the measurements than the old ρ^a one, but also the different coordination numbers as obtained from ρ^b simulation feature provides an unprecedented improvement.

The three main peaks in the g_{GeGe} refers as r increases to Ge atoms involved in homopolar bonds, edge-sharing and corner-sharing connection respectively. The following table present the different Ge environment fractions as obtained from direct analysis:

The current results within the relaxed model (ρ^b) shows a very good reproduction of the experimental fraction of Ge atoms involved in homopolar bonds. The fraction of corner sharing Ge atoms is improved upon previous simulation, this stems from the substantial change of the shape of the third g_{GeGe} peak.

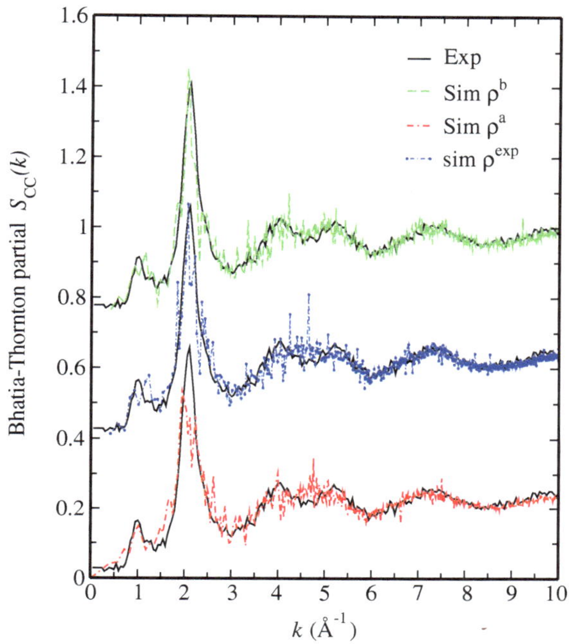

Fig. 12.16 Bhatia-Thornton concentration-concentration partial structure factor $S_{cc}^{BT}(k)$ for glassy $GeSe_2$. Experimental results of [88] (*black solid line*), simulation results with ρ^a [26](*red dashed curve*), simulation results with ρ^{exp} (*blue dotted curve* and simulation results of the calibrated system at ρ^b (*green dashed curve*)

The reciprocal space properties also are improved by releasing the residual pressure. Figure 12.16 shows the Bhatia-Thornton concentration-concentration $S_{cc}^{BT}(k)$ partial structure factors. The correct reproduction of $S_{cc}^{BT}(k)$ proved to be a stringent test of the FPMD approaches. Comparison between all the simulated models and the measurement shows that the calibrated system gives a better account of the $S_{cc}^{BT}(k)$. The intensity of the main peak located at $k \sim 2\,\text{Å}^{-1}$ is very close to the experimental one. This same height was underestimated when working at the ρ^a. In addition a minimum appears in the range $4\ \text{Å}^{-1} < k < 5.5\ \text{Å}^{-1}$, when simulated with ρ^b. The new set of data improves the position and the shape of the first sharp diffraction peak (FSDP) (Table 12.4).

Table 12.4 $N_{\text{Ge-Ge}}$ ($N_{\text{Se-Se}}$) is the percentage of Ge (Se) atoms in Ge–Ge (Se–Se) homopolar bonds, N_{Ge}(ES) is the percentage of Ge atoms forming edge-sharing connections and N_{Ge}(CS) is the percentage of Ge atoms forming corner-sharing connections

	$N_{\text{Ge-Ge}}$ (%)	$N_{\text{Se-Se}}$ (%)	N_{Ge}(ES) (%)	N_{Ge}(CS) (%)
Simulation at ρ^a [26]	20	30	58	22
Simulation at ρ^{exp}	27	19	54	19
Simulation at ρ^b	23	18	35	42
Simulation [89]	17	30	38	45
Experiment [88]	25	20	34	41

Note that in [89], a molecular dynamics approach was used in conjunction with a reverse Monte Carlo method

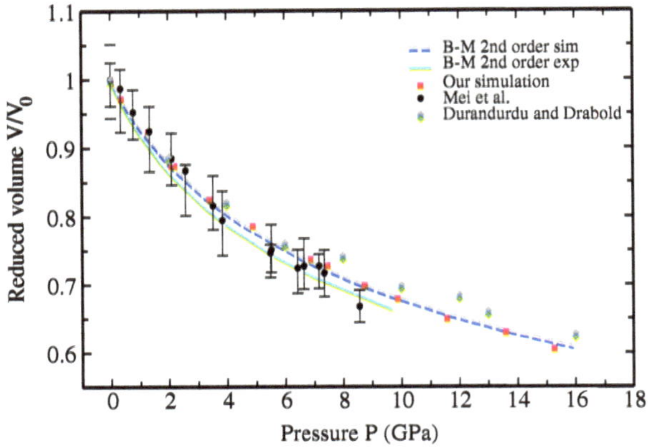

Fig. 12.17 The pressure-volume equation of state for $GeSe_2$ glass under compression where V is the volume at pressure P and V_0 is the volume under ambient conditions. The measured data from Mei et al. [79] (*black filled circles* with *vertical error bars*) are compared to the results obtained from FPMD in the present work (*filled red squares*) [61] and in the work by Durandurdu and Drabold [83] (*blue diamond*). The measured and simulated data are fitted to a second-order Birch-Murnaghan equation of state (*solid green* and *dashed blue curve* respectively)

Overall, the new calibrated system shows an enhanced chemical order with as many as 92 % of Ge atoms being fourfold coordinated, 72 % of them within a $GeSe_4$ tetrahedron. In the previous simulation, these values were equal to 78.1 and 62.5 % respectively.

In conclusion, account of residual pressure effects at room temperature leads to a better agreement between atomic-scale models and experiments. Therefore, the relaxed configuration at ρ^b has been used as initial model to generate, study and predict the structural changes under pressure.

The previous research opened the door to the possiblity of an accurate atomic scale investigation of the structural transformation of $GeSe_2$ under high pressure. In the same sperit of FPMD, simulations were carried out covering a pressure ranging up to ~16 GPa. The resulting equation of state is displayed in Fig. 12.17. We refer to our paper by Wezka, Bouzid et al. [61] for details on the pressure-induced structural transition recorded therein and the peculiar role played by homopolar atoms.

References

1. P. Vashishta, R.K. Kalia, I. Ebbsjo, Phys. Rev. B **39**(9), 6034–6047 (1989)
2. P. Vashishta, R.K. Kalia, A.G. Antonio, Phys. Rev. Lett. **62**(14), 1651–1654 (1989)
3. H. Iyetomi, P. Vashishta, R.K. Kalia, Phys. Rev. B **43**, 1726–1734 (1991)
4. R.L. Cappelletti, M. Cobb, D.A. Drabold, W.A. Kamitakahara, Phys. Rev. B **52**, 9133–9136 (1995)
5. O.F. Sankey, D.J. Niklewski, Phys. Rev. B **40**(6), 3979–3995 (1989)
6. J. Harris, Phys. Rev. B **31**(4), 1770–1779 (1985)

7. W.M.C. Foulkes, R. Haydock, Phys. Rev. B **39**(17), 12520–12536 (1989)
8. M. Cobb, D.A. Drabold, R.L. Cappelletti, Phys. Rev. B **54**(17), 12162–12171 (1996)
9. M. Cobb, R.L. Cappelletti, D.A. Drabold, J. Non-Cryst. Solids. **222**, 348353 (1997)
10. C. Massobrio, A. Pasquarello, R. Car, Phys. Rev. Lett. **80**, 2342–2345 (1998)
11. M.J. Haye, C. Massobrio, A. Pasquarello, A. De Vita, S.W. De Leeuw, R. Car, Phys. Rev. B **58**, R14661–R14664 (1998)
12. C. Massobrio, A. Pasquarello, R. Car, Phys. Rev. B **64**(14), 144205 (2001)
13. I.T. Penfold, P.S. Salmon, Phys. Rev. Lett. **67**(1), 97–100 (1991)
14. C. Massobrio, A. Pasquarello, R. Car, J. Am. Chem. Soc. **121**(12), 2943–2944 (1999)
15. S. Susman, K.J. Volin, D.L. Montague, D.G. Price. J. Non-Cryst. Solids **125**, 168–180 (1990)
16. P. Vashishta, R.K. Kalia, J.P. Rino, Phys. Rev. B **41**(17), 12197–12209 (1990)
17. C. Massobrio, F.H.M. van Roon, A. Pasquarello, S.W. De Leeuw, J. Phys.: Condens. Matter **12**, L697 (2000)
18. I. Petri, P.S. Salmon, W.S. Howells, J. Phys.: Condens. Matter **11**(50), 10219 (1999)
19. P.H. Gaskell, D.J. Wallis, Phys. Rev. Lett. **76**(1), 66 (1996)
20. M. Wilson, P.A. Madden, Phys. Rev. Lett. **80**(3), 532 (1998)
21. S.R. Elliott, J. Non-Cryst. Solids **182**(1), 40–48 (1995)
22. C. Massobrio, A. Pasquarello, J. Chem. Phys. **114**, 7976–7979 (2001)
23. A.D. Becke, Phys. Rev. A **38**(6), 3098–3100 (1988)
24. C. Lee, W. Yang, R.G. Parr, Phys. Rev. B **37**(2), 785–789 (1988)
25. M. Micoulaut, R. Vuilleumier, C. Massobrio, Phys. Rev. B **79**, 214205 (2009)
26. C. Massobrio, M. Micoulaut, P.S. Salmon, Solid State Sci. **12**, 199–203 (2010)
27. L. Giacomazzi, C. Masobrio, A. Pasquarello, J. Phys.: Condens. Matter **23**, 295401 (2011)
28. S. Le Roux, A. Zeidler, P.S. Salmon, M. Boero, M. Micoulaut, C. Massobrio, Phys. Rev. B **84**(13), 134203 (2011)
29. S. Le Roux, A. Bouzid, M. Boero, C. Massobrio, Phys. Rev. B **86**(22), 224201 (2012)
30. S. Le Roux, A. Bouzid, M. Boero, C. Massobrio, J. Chem. Phys. **138**(17), 174505 (2013)
31. M.J. Mandell, J. Stat. Phys. **15**(4), 299–305 (1976)
32. L.R. Pratt, S.W. Haan, J. Chem. Phys. **74**(3), 1864–1872 (1981)
33. A.R. Denton, P.A. Egelstaff, Zeitschrift für Physik B Condensed Matter **103**(3), 343–349 (1997)
34. L. Verlet, Phys. Rev. B **159**(1), 98 (1967)
35. S.M. Foiles, N.W. Ashcroft, L. Reatto, J. Chem. Phys. **81**(12), 6140–6145 (1984)
36. L. Reatto, M. Tau, J. Chem. Phys. **86**(11), 6474–6485 (1987)
37. M. Rovere, D.W. Hermann, K. Binder, EPL (Europhys. Lett.) **6**(7), 585 (1988)
38. D.P. Sellan, E.S. Landry, J.E. Turney, A.J.H. McGaughey, C.H. Amon, Phys. Rev. B **81**(21), 214305 (2010)
39. M.J. Mandell, J.P. McTague, A. Rahman, J. Chem. Phys. **66**(7), 3070–3075 (1977)
40. J.N. Cape, J.L. Finney, L.V. Woodcock, J. Chem. Phys. **75**(5), 2366–2373 (1981)
41. N. Choudhury, S.K. Ghosh, Phys. Rev. E **66**(2), 021206 (2002)
42. F. Delogu, Phys. Rev. B **79**(18), 184109 (2009)
43. S.S. Jang, W.A. Goddard, J. Phys. Chem. B **110**(15), 7992–8001 (2006)
44. M.H. Müser, K. Binder, Phys. Chem. Miner. **28**(10), 746–755 (2001)
45. J.J. Salacuse, A.R. Denton, P.A. Egelstaff, Phys. Rev. E **53**, 2382–2389 (1996)
46. J.J. Salacuse, A.R. Denton, P.A. Egelstaff, M. Tau, L. Reatto, Phys. Rev. E **53**(3), 2390 (1996)
47. M. Micoulaut, S. Le Roux, C. Massobrio, J. Chem. Phys. **136**(22), 224504 (2012)
48. P. Lucas, E.A. King, O. Gulbiten, J.L. Yarger, E. Soignard, B. Bureau, Phys. Rev. B **80**, 214114 (2009)
49. C. Massobrio, M. Celino, P.S. Salmon, R.A. Martin, M. Micoulaut, A. Pasquarello, Phys. Rev. B **79**(17), 174201 (2009)
50. D.N. Tafen, D.A. Drabold, Phys. Rev. B **71**, 054206 (2005)
51. M. Kibalchenko, J.R. Yates, C. Massobrio, A. Pasquarello, J. Phys. Chem. C **115**(15), 7755–7759 (2011)
52. K. Sykina, E. Furet, B. Bureau, S. Le Roux, C. Massobrio, Chem. Phys. Lett. **547**, 30–34 (2012)

53. M. Micoulaut, A. Kachmar, M. Bauchy, S. Le Roux, C. Massobrio, M. Boero, Phys. Rev. B **88**, 054203 (2013)
54. E. Bychkov, M. Miloshova, D.L. Price, C.J. Benmore, A. Lorriaux, J. Non-Cryst. Solids **352**, 63–70 (2006)
55. P.S. Salmon, J. Non-Cryst. Solids. **353**, 2959–2974 (2007)
56. M. Celino, S. Le Roux, G. Ori, B. Coasne, A. Bouzid, M. Boero, C. Massobrio, Phys. Rev. B **88**(17), 174201 (2013)
57. M. Bauchy, M. Micoulaut, M. Celino, S. Le Roux, M. Boero, C. Massobrio, Phys. Rev. B **84**, 054201 (2011)
58. S. Le Roux, A. Zeidler, P.S. Salmon, M. Boero, M. Micoulaut, C. Massobrio, Phys. Rev. B **84**, 134203 (2011)
59. S. Le Roux, A. Bouzid, M. Boero, C. Massobrio, Phys. Rev. B **86**, 224201 (2012)
60. A. Bouzid, C. Massobrio, J. Chem. Phys. **137**, 046101 (2012)
61. K. Wezka, A. Bouzid, K.J. Pizzey, P.S. Salmon, A. Zeidler, K. Klotz, H.E. Fischer, C.L. Bull, M.G. Tucker, M. Boero, S. Le Roux, C. Tugène, C. Massobrio, Phys. Rev. B **90**(5), 054206 (2014)
62. C. Massobrio, M. Celino, P.S. Salmon, R.A. Martin, M. Micoulaut, A. Pasquarello, Phys. Rev. B **79**(17), 174201 (2009)
63. M. Parrinello, A. Rahman, J. Appl. Phys. **52**, 7182 (1981)
64. P.S. Salmon, J. Liu, J. Phys.: Cond. Mat. **6** 1449 (1994)
65. S. Nosé, **52**(2), 255–258 (1984)
66. W.G. Hoover, Phys. Rev. A **31**, 1695–1697 (1985)
67. P.E. Blöchl, M. Parrinello, Phys. Rev. B **45**, 9413–9416 (1992)
68. T.E. Faber, J.M. Ziman, Phil. Mag. **11**(109), 153–173 (1965)
69. http://www.ncnr.nist.gov/resources/n-lenghs/
70. I. Petri, P.S. Salmon, Phys. Chem. Glasses **43C**, 185 (2002)
71. C. Massobrio, A. Pasquarello, Phys. Rev. B **77**, 144207 (2008)
72. S. Chakraborty, P. Boolchand, J. Phys. Chem. B **118**(8), 2249–2263 (2014)
73. C.T. Prewitt, H.S. Young, Science **149**, 535 (1965)
74. M. Shimada, F. Dachille, Inorg. Chem. **16**(8), 2094–2097 (1977)
75. A. Grzechnik, T. Grande, S. Stølen, J. Solid State Chem. 141(1), 248–254 (1998)
76. T. Grande, M. Ishii, M. Akaishi, S. Aasland, H. Fjellvåg, S. Stølen, J. Solid State Chem. **145**(1), 167–173 (1999)
77. W.A. Crichton, M. Mezouar, T. Grande, S. Stølen, A. Grzechnik, Nature **414**(6864), 622–625 (2001)
78. P.V. Teredesai, Phys. Chem. Glasses: Eur. J. Glass Sci. Technol. Part B **47**(2), 240–243 (2006)
79. Q. Mei, C.J. Benmore, R.T. Hart, E. Bychkov, P.S. Salmon, C.D. Martin, F.M. Michel, S.M. Antao, P.J. Chupas, P.L. Lee, S.D. Shastri, J.B. Parise, K. Leinenweber, S. Amin, J.L. Yarger, Phys. Rev. B **74**(1), 014203 (2006)
80. S.M. Antao, C.J. Benmore, B. Li, L. Wang, E. Bychkov, J.B. Parise, Phys. Rev. Lett. **100**, 115501 (2008)
81. L.B. Skinner, C.J. Benmore, S. Antao, E. Soignard, S.A. Amin, E. Bychkov, E. Rissi, J.B. Parise, J.L. Yarger, J. Phys. Chem. C **116**(3), 2212–2217 (2011)
82. A. Grzechnik, S. Stølen, E. Bakken, T. Grande, M. Mezouar, J. Solid State Chem. **150**(1), 121–127 (2000)
83. M. Durandurdu, D.A. Drabold, Phys. Rev. B **65**, 104208 (2002)
84. W. Fei, S. Mamedov, P. Boolchand, B. Goodman, M. Chandrasekhar, Phys. Rev. B **71**, 174201 (2005)
85. D. Alfe, G.D. Price, M.J. Gillan, Phys. Rev. B **65**(16), 165118 (2002)
86. I. Petri, P.S. Salmon, Phys. Chem. Glasses **43**, 185–190 (2002)
87. R. Azoulay, H. Thibierge, A. Brenac, J. Non-Cryst. Solids **18**(1), 33–53 (1975)
88. I. Petri, P.S. Salmon, H.E. Fischer, Phys. Rev. Lett. **84**(11), 2413 (2000)
89. D.N. Tafen, D.A. Drabold, Phys. Rev. B **71**(5), 054206 (2005)

Chapter 13
Molecular Modeling of Glassy Surfaces

Guido Ori, Carlo Massobrio, Assil Bouzid and B. Coasne

Abstract Progress in computational materials science has allowed the development of realistic models for a wide range of materials including both crystalline and glassy solids. In recent years, with the growing interest in nanoparticles and porous materials, more attention has been devoted to the design of realistic models of glassy surfaces and finely divided materials. The structural disorder in glassy surfaces, however, poses a major challenge which consists of describing such surfaces using computer simulations. In this paper, we show how atomic-scale simulations can be used to develop and investigate the properties of glassy surfaces. We illustrate how both first principles calculations and classical molecular mechanics can be used to follow the trajectory at finite temperature of these systems, and obtain statistical thermodynamic averages to compare against available experiments. Both glassy oxide (silica) and non-oxide (chalcogenide) surfaces are considered.

G. Ori (✉) · B. Coasne
Multiscale Materials Science for Energy and Environment, CNRS-MIT (UMI 3466),
77 Massachusetts Avenue, Cambridge, MA 02139, USA
e-mail: Guido.Ori@ipcms.unistra.fr

G. Ori · B. Coasne
Institut Charles Gerhard Montpellier, CNRS (UMR 5253), ENSCM,
Université Montpellier 2, 8 Rue de L'Ecole Normale, 34296 Montpellier Cedex 5, France

G. Ori · C. Massobrio · A. Bouzid
Institut de Physique et Chimie des Matériaux de Strasbourg, CNRS (UMR 7504),
23 Rue du Loess, 67034 Strasbourg Cedex 2, France
e-mail: Carlo.Massobrio@ipcms.unistra.fr

B. Coasne
Department of Civil and Environmental Engineering, Massachusetts Institute of Technology,
77 Massachusetts Avenue, Cambridge, MA 02139, USA
e-mail: coasne@mit.edu

C. Massobrio et al. (eds.), *Molecular Dynamics Simulations of Disordered Materials*, Springer Series in Materials Science 215,
DOI 10.1007/978-3-319-15675-0_13

13.1 Introduction

Surfaces and divided materials with high surface area, such as porous materials and nanoparticles, are useful for many applications in catalysis, adsorption, electronics, etc. [1, 2]. As a subclass of these materials, surfaces with a glassy structure are receiving increasing attention. Owing to their large surface area and tunable pore size, glassy porous solids such as siliceous, carbonaceous, and chalcogenide materials are at the heart of important applications: heterogeneous catalysis, energy (hydrogen storage, lithium batteries), environment (water treatment, nuclear waste storage) and Earth science (the soil porosity ensures exchange between soil and atmosphere) [3, 4]. As a result, searching for "surface" and "glassy surface" in ISI Web of Knowledge reveals a regular increase in the number of entries over the period 1995–2013 (Fig. 13.1).

Progress in computational materials science over the last decades has allowed the development of realistic molecular models for a wide range of materials. The interest in silica lies in its ubiquity nature as well as in man-made materials. Silica is used as filler in composites and tires, for purification of gases and liquids, and catalyst support. More recently, silica nanoparticles and gels have found applications in drug delivery, biosensors, and cosmetics that exploit interactions with organic molecular species [1, 5]. Engineering complex silica-containing materials requires quantitative understanding of the role of the silica precursors, surface chemistry of silica, and competitive interactions with solvents and molecules. Many questions related to the silica surface and its interfacial interaction with other molecular entities remain difficult to answer using available instrumentation. This lack of complete understanding is even more important for non-oxide surfaces such as glassy chalcogenides, which are defined as made up of chalcogen elements (S, Se, and/or Te) possessing a glassy architecture at the molecular level. As a subclass of these materials, chalco-

Fig. 13.1 (*color* online) Record counts versus publication year (1992–2013) for the words "surface" (*red*) and "glassy (or amorphous) surface" (*blue*) in article topic. Raw data obtained from ISI Web of Science, Databases = SCIEXPANDED, SSCI, A&HCI, CPCI-S, and BKCI-S, search date, 05-12-2013

genide glassy films are useful for many applications such as in optoelectronics and nonvolatile memory devices [6, 7]. As an alternative prototype of disordered systems strongly dependent on surface properties, investigations have also focused on the synthesis and characterization of nanochalcogenides such as nanoporous chalcogenide glasses (also referred to as chalcogels) [8, 9]. Recently, chalcogels have been shown to be efficient materials for gas separation and environmental remediation.

Molecular simulation can provide insights and accelerate rational material design for both oxide and non-oxide solids. Available simulation techniques comprise methods for specific length and time scales including Quantum Mechanics, Molecular Dynamics (MD) and Monte Carlo techniques, and coarse-grained methods. The key to the success of any simulation using electronic structure, empirical forcefields, or coarse-grained models is the accuracy of the energy expression. The energy expression delivers the intra- and inter-molecular potential, influences the motion of the atoms, and determines the computed thermodynamic and dynamic properties. It is consequently the major assumption to ensure the correlation of simulation results with laboratory measurements.

Commonly, the structure of a glass-like material is obtained in a computer following strategies which bring a crystalline starting phase (whose structure is known) toward the glassy structure by decreasing the long range order [10]. This type of disorder can be achieved with a variety of computational techniques such as MD carried out at very high temperature in order to melt the material followed by a quenching step to kinetically freeze the structure in a glassy state. This procedure is repeated an appropriate number of times until the physicochemical features of the bulk material (density, angle distribution, radial distribution function, etc.) are in good agreement with experiments. To be representative of the real glassy material, the simulation set up must consist of a unit cell large enough. For this reason, the simulation of glassy materials is usually much more expensive than that of crystalline materials, particularly when quantum techniques are adopted. It is worth mentioning that other procedures can be used to build and study glassy systems with high surface area. This is the case, for example, of Reverse Monte Carlo technique that has been widely used to model porous carbon materials [11–13].

In the present work, both glassy oxide (silica) and non-oxide (chalcogenide) surfaces obtained using first principles and classical calculations are considered. The remainder of this paper is organized as follows. A brief state of the art is presented in Sect. 13.2 for both silica and chalcogenide surfaces. In Sect. 13.3 a detailed account of a silica-based material with high surface area such as mesoporous silica MCM-41 and its adsorption properties are reviewed. In Sect. 13.4 we discuss the structure and electronic properties of a glassy chalcogenide surface. Our conclusions and perspectives are summarized in Sect. 13.5.

13.2 State of the Art

13.2.1 Silica Surfaces

The simulation of bulk and surface properties of amorphous silica has received a lot of interest since the emergence of numerical computer methods [14–16]. However, despite the huge progress in defining accurate forcefields to simulate glassy silica, limited progress has been reported on the simulation of silica surfaces. Several simple two-body potentials are nowadays available for the simulation of bulk silica with good agreement with experiments in terms of structure, dynamics, and mechanical properties [17–20] . Some of these forcefields have been parameterized against first principles calculations or directly against experimental data. However, for the case of silica, despite the fact that several forcefields are already widely used to model glassy silica surfaces [18–21], their ability to reproduce features such as surface H-bonding of glassy surface is somewhat approximate. As a result, in specific cases, such as the study of silica surface in contact with complex organic molecules, [5] first principles calculations are needed to describe the interactions involved. One of the widely used first principles techniques is the Car-Parrinello MD (FPMD) method [22]. The FPMD method has gained enormous popularity in many different fields of molecular physics and chemistry. FPMD calculations have been used to study silica-vacuum [23, 24] and glass-water interfaces [25] for instance.

In what follows a brief summary of the principal structural features of glassy silica are presented. For more details the reader is referred to [5, 10, 26]. Silica (SiO_2) is a solid material with a density between 2 and 3 g/cm^3 and a high melting point (ca. 1700 °C). In most crystalline silica polymorphs, a silicon atom is bound to four oxygen atoms and each oxygen is bound to two silicon atoms. The silicon atoms are at the centers of regular tetrahedra of oxygen atoms [SiO_4] which are connected through an oxygen vertex. Glassy silica consists of a continuous and disordered network of edge- and corner-sharing [SiO_4] tetrahedra [27]. The SiO_2 structure is neutral because the formal oxidation numbers for silicon and oxygen are (+4) and (−2), respectively, and each oxygen atom in the structure bridges two [SiO_4] tetrahedra so that the silicon-oxygen ratio is 1:2. Thus, the [SiO_4] tetrahedron represents the building block of almost all silica polymorphs, which only differ in the connectivity of the tetrahedral framework. The different linking patterns for each polymorph lead to different structural, physical, and chemical properties. The Si-O bond shows a predominantly covalent nature as shown by the electron density maps obtained from X-ray diffraction and by the bond population analysis obtained from quantum mechanical calculations [28, 29]. Both experimental and computational results assign a charge of ca. +1.4 on silicon atoms and −0.7 on oxygen atoms [28, 29]. The [SiO_4] unit is rigid because changing the O-Si-O angle from the equilibrium value of 109° is energetically very unfavorable. In contrast, the Si-O-Si angle that connects two tetrahedra can easily vary, for instance, as a function of temperature or geometrical constraints. Indeed the Si-O-Si angle is very flexible, since it costs no energy to change the angle value between 130° and 180°, as demonstrated in a number of computational studies [30–32].

The flexibility of the Si-O-Si angle leads to the large number of all-silica materials, envisaging all possible silica polymorphs from dense crystalline and amorphous structures (quartz or glasses) to porous systems (gels). Microporous crystalline zeolites and mesoporous materials also belong to the very large family of silica polymorphs. When silica is fractured under ultrahigh vacuum conditions a complex reconstruction of the broken Si-O-Si bonds occurs in order to self-heal the dangling bonds at the surface. In normal laboratory conditions, the most common termination is given by two main functional groups: the siloxane links (Si-O-Si) and the silanol groups (Si-OH). The siloxane bonds, which are also present in the bulk structure (although not accessible), result from the sharing of the [SiO_4] tetrahedra. Silanol groups are the result of an incomplete condensation during the polymerization that forms the massive solid. Protons or hydroxyl groups due to the presence of water in the surrounding environment (such as in the form of moisture) are the most suitable candidates to saturate the unfilled Si-O bonds. The common picture of a silica surface is a system made up of regions of siloxane links interrupted by silanol groups (Fig. 13.2) [5, 26, 33]. Silanols groups are responsible for the hydrophilic properties

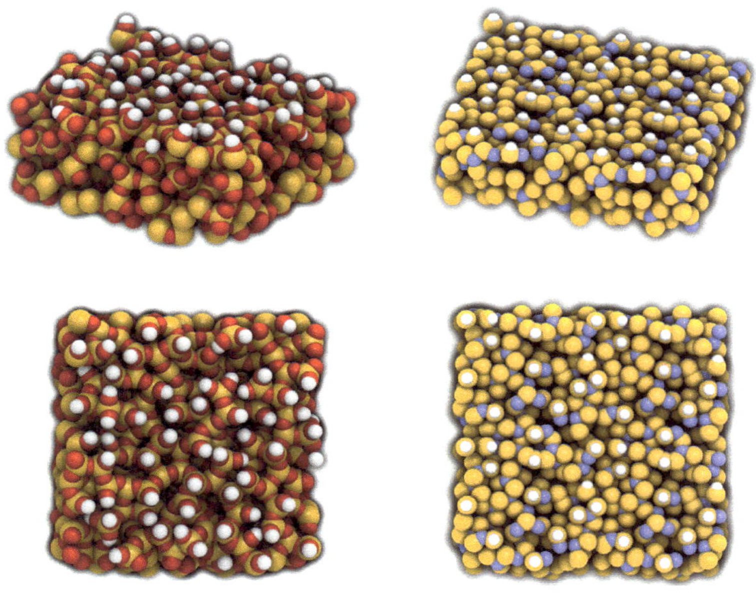

Fig. 13.2 (*color* online) Side (*top*) and top (*bottom*) views of glassy silica (*left*) and chalcogenide (*right*) slab models. The silica surface possesses a 5 OH/nm^2 surface density of hydroxyl groups while the chalcogenide surface possesses a 2.3 SH/nm^3 of thiol groups. *color* code: silica: Si *yellow*, O *red*, and H *white*; chalcogenide: S *yellow*, Ge *blue*, and H *white*

of silica since they can interact with polar groups or molecules via hydrogen bond interactions (H-bond). Silanol groups that are separated by more than ~3.3 Å can be considered isolated as they are unable to establish mutual hydrogen bond. This type of silanol group is then free to establish H-bond interactions with adsorbates as both H-bond donors and acceptors. In contrast, two silanols that do not belong to connected tetrahedra but are closer than ~3.3 Å establish H-bonds.

According to quantum mechanical calculations on cluster models, two silanols belonging to tetrahedra that share a common oxygen are usually not involved in a H-bond. When highly hydrated silica is heated, the main effect is the progressive condensation of pairs of surface silanols involved in H-bond interactions and the formation of a siloxane bond. Thermal treatments constitute a simple way to tune the adsorption properties of silica-based materials. The area density of silanol groups can vary between 0 and 9.4 groups per nm^2. The average surface density of OH groups obtained by common methods is 5 OH/nm^2. Zhuravlev [34] has proposed a "master curve" showing the concentration of the silanols at the surface of silica as a function of the sample thermal treatment (Fig. 13.3).

The different level of accuracy on the reproduction of the surface features (such as surface H-bonding and ionization grade) of glassy silica is what differentiates the available classical forcefields. For instance, forcefields such as BKS [15], CHIK [20], CLAYFF [18], and INTERFACE [19] are receiving increasing attention to model silica surfaces and their interaction with other molecular entities (such as water or organic molecules). Most of these forcefields have been fitted on experimental and/or first principles data obtained on bulk silica. Although these forcefields are based on "simple" pair potentials, they yield good agreement with experimental data. Of particular interest, the forcefield developed by Pedone et al. FFSiOH [21] which has been fitted on first principles calculations and contains a 3-body term for the accurate description of the surface H-bonding.

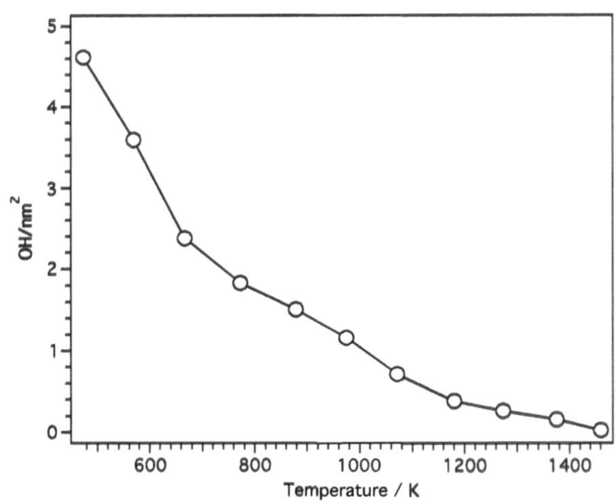

Fig. 13.3 Average concentration of total OH groups per square nanometer as a function of the silica sample treatment temperature. From data found in [5, 34]

Mesoporous silica MCM-41 is a particular case of silica-based material with high surface. Such mesoporous materials are crucial for catalytic support for gas separation, drug delivery, etc. This material is made up of a hexagonal array of cylindrical pores where the pore distribution is narrow with an average value that can be varied from 1.5 up to 10 nm (depending on the experimental conditions). Ugliengo et al. [35] have developed a realistic model of MCM-41 pores by using density functional approach based on the rather accurate hybrid B3LYP functional. Other models of mesoporous silica reported in the literature rely on the use of classical forcefields which provide an approximated description of the chemical bonds. In particular, Bhattacharya and coauthors [36–38] using lattice Monte Carlo simulations developed several types of mesoporous silica. Recently, the MCM-41 model developed by Ugliengo et al. [35] has been used to investigate its structural, morphological, and adsorption properties [39]. The resulting model consists of a numerical model with surface chemistry (defects, surface terminations, etc.) obtained using a method that allows taking into account the quantum nature of chemical bonding. Other approaches have been proposed to model mesoporous materials based on different techniques. A more detailed overview of these approaches is discussed in the introduction of Sect. 13.3.

13.2.2 Chalcogenide Surfaces

Chalcogenide glassy films find applications in optoelectronics, nonvolatile memory devices, etc. For example, glassy $Ge_2Sb_2Te_5$ and GeS_2 films prepared by deposition techniques can be used as phase-change materials [40, 41]. The structure and chemical order of these materials are found to be strongly dependent on the chalcogenide composition and conditions used for the deposition. In some cases, these structures largely differ (in terms of quantity of defects) from those obtained by common melt-quenching methods used for the preparation of bulk glasses. It appears that glassy GeS_x (with $x = 2, 4, 6$) films, obtained by pulsed laser deposition from the pristine bulk glasses, show a significant content of defects and "wrong bonds" (such as homopolar bonds) [42–44]. In this case, the departure from perfect chemical order is due to the fact that as-deposited g-GeS_x films are far away from the equilibrium state [42–44]. For the case of glassy GeS_2 (g-GeS_2, g stands hereafter for glass) the perfect chemical order corresponds to the absence of any undercoordinated or overcoordinated Ge atoms or S atoms ($\neq 4$ and $\neq 2$ for Ge and S, respectively) [45]. As an alternative prototype of disordered systems strongly dependent on surface properties, investigations have also focused on the synthesis and characterization of nanochalcogenides such as nanoporous chalcogenide glasses (also referred to as chalcogels). In this case, common synthesis methods are based on nucleation-to-growth or linker-driven assembly of building units where the final structures can reach a state close to equilibrium. Obtaining in a controlled way such materials, which exhibit a large surface area from $\sim$10 to 500 m^2/g made up of highly polarizable atoms, can lead to breakthroughs for applications relying on the surface properties of the host compound (photocatalysis and gas separation for instance) [9]. Recently, chalcogels have proven to be efficient sorbents for environmental remediation from gaseous and water waste media [46, 47]. The complexity in modeling the atomic structure of glassy chalcogenides arises from the inability of simple classical forcefields to accurately reproduce the mixture of purely covalent and iono-covalent bonding that characterizes most chalcogenide systems. Only few classical forcefield have been tentatively developed and tested for chalcogenides; [48, 49] however, they are difficult to transfer to other chalcogenides. On the other hand, first principles molecular dynamics (FPMD) based on fully self-consistent density functional theory have been well assessed and successfully employed to model liquid and glassy chalcogenides. Recently, we were able to produce a bulk model for g-GeS_2 in remarkable agreement with neutron scattering data, thereby legitimating its further application to surface studies [45]. This bulk model was obtained by simulated melt quenching. Looking at previous available atomic-scale modeling in this domain, amorphous $GeSe_2$ surfaces were considered by the team of Drabold [50, 51]. Chalcogenide surfaces were found to be characterized by a slight atomic expansion and a number of ring structures larger than in the bulk. One of the main features of bulk chalcogenide properties was also recovered, namely the intermediate range order (IRO) through the appearance of the first sharp diffraction peak (FSDP) in the total neutron structure factor. Akola et al. [52] studied a computer-aided deposition (AD) of $Ge_2Sb_2Te_5$ by first principles Molecular Dynamics. In comparison to the MQ structure, the AD model showed a different local environment for the Ge atoms (tetrahedral rather than the typical distorted octahedral found for this system) and the presence of "wrong bonds" (homopolar and Ge-Sb bonds). The structure factor and electronic properties of the two models were found to be very similar.

In addition to bare chalcogenide surfaces, the study of hydrogenated chalcogenide surfaces is also relevant to their final application. Furthermore, although glassy chalcogenides cannot easily adsorb hydrogen due to its low density of dangling bonds, hydrogenation of M-Se_xS_y- and M-Te (with M = Ge and Sb) has been performed by playing on the structure and morphology of deposited films [53, 54]. Both the optical and electronic properties of chalcogenide films were found to be affected by surface hydrogenation [55, 56]. However, little is known about the effect of hydrogenation on the adsorption properties of molecular species on chalcogenide surfaces (Fig. 13.2 shows a glassy GeS_2 surface where undercoordinated S atoms are saturated with H atoms with a thiol surface density of 2.3 SH/nm^2). The availability of a surface model obtained by FPMD is a necessary prerequisite to construct simplified, yet accurate, schemes for the study of chalcogenide glasses at the interface with other systems.

13.3 Modeling of Mesoporous Silica and Its Adsorption Properties

The simulation of porous materials using fully atomistic approaches allows taking into account surface roughness and morphological disorder in a realistic way. As a result, several procedures have been considered in order to build realistic silica surface and porous silica models [57]. For instance, starting from a bulk silica sample obtained through melt quenching by FPMD or classical MD simulation, a single surface [23] or cylindrical [58] and slit [59] pores made up of glassy silica can be obtained by carving out a specific part of the sample (see Fig. 13.4). Part of the surface dangling bonds is then saturated with H atoms in order to obtain the desired silanol density. One possible approach is to use the method proposed by Coasne and coworkers [60, 61] where realistic models of glassy silica surfaces with different silanol densities are built using a simulated annealing procedure.

Another possible approach to model mesoporous materials is to mimic the synthesis process of the real material. This strategy has been used by Gelb and Gubbins [62] to develop realistic models for Vycor and controlled pore glasses. The input of such an approach is the representation of the templating surfactants using simplified potentials to describe the interactions involved in the system. Siperstein et al. [63]

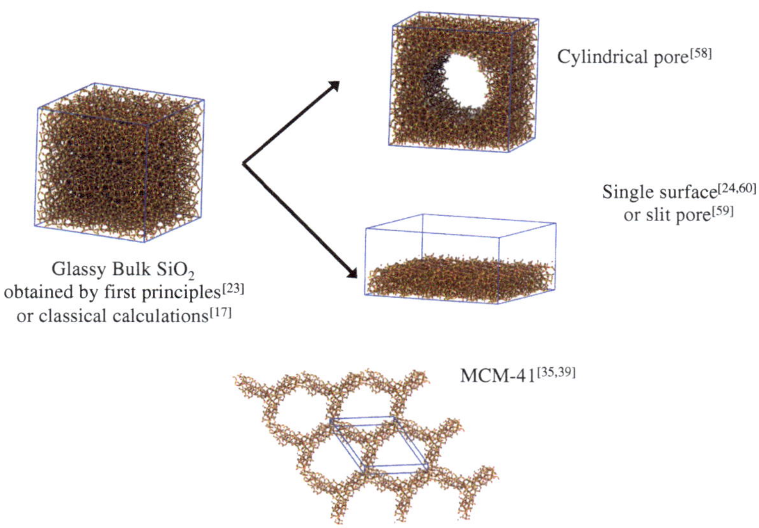

Fig. 13.4 (*color* online) (*top*, *left*) Typical configuration of bulk silica as obtained from melt quenching technique using first principles [23] and classical calculations [17]. (*top*, *right*) Cylindrical pore of glassy silica obtained by classical simulation (silanol surface density equal to 5.3 OH/nm^2) [58]. (*top*, *right*) Single silica surface which can be used to build slit pore of silica. The single surface can be produced using either first principles [24, 60] or classical calculations [59]. *Bottom* Model of mesoporous silica MCM-41 obtained by a first principles-based approach [35, 39]. The *yellow*, *red*, and *white* spheres are the Si, O, and H atoms of the surfaces, respectively

reported on-lattice Monte Carlo simulations of surfactant-solvent-silica systems that were able to reproduce the formation of hexagonal structure, resembling the arrangement of silica MCM-41 pores. Using this approach Coasne et al. [36, 64] reported the preparation of atomistic silica mesopores that keep the morphological features of the MCM-41 or SBA-15 models. It is worth mentioning that other methods are available in the literature, such as those based on reconstructing disordered materials. The pioneering work of Levitz [65] consisted of using off-lattice formalism allowing the 3D reconstruction of correlated disordered porous systems.

Here we review the characterization of the surface properties of the realistic model of MCM-41 developed by Ugliengo et al. [35]. The model of mesoporous silica (MCM-41) was prepared according to the following procedure (full details can be found in [36, 39]). Starting from a hexagonal supercell of α-quartz containing about 600 atoms ($a = b = 4.06$ nm and $c = 1.22$ nm), high temperature classical molecular dynamics was conducted to force amorphization of the structure, carefully avoiding bond rupture and the formation of strained silica rings. An empirical partial charge rigid ionic potential, [17] which has been shown to properly reproduce the structure, transport and mechanical properties of oxides, silicates and silica based glasses was used to obtain the glassy state. In a second step, a hole of approximate hexagonal symmetry was created within the MCM-41 unit cell, and the unfilled valences of the inner hole walls were saturated with OH groups, resulting in $Si_{142}O_{335}H_{102}$ unit cell and a tetrahedral-site (T-site) density of 8.2 T-sites/nm^3. As a final step, the internal coordinates of the starting structure were fully optimized at the B3LYP/6–31G(d,p) level. Figure 13.4 shows a $3x \times 3y$ supercell of the atomistic model of mesoporous silica MCM-41. Further analysis of the structure reveals that the MCM-41 model has a surface density of silanol groups of 7.2 OH/nm^2. This value is close to the typical value of models obtained by carving pores out of atomistic silica blocks [57, 66]. With the aforementioned procedure significant relaxation occurs upon optimization, however the pore remains overall with an hexagonal shape. Nevertheless, since the hexagonal shape of the pore was used as a starting point, it cannot be asserted that it corresponds to the most stable pore morphology in real MCM-41 samples.

The structural and morphological properties of the MCM-41 model reviewed in this work have been reported in [39]. Figure 13.5 shows the small angle neutron scattering spectrum (SANS) of the MCM-41 model. This model shows a simulated X-ray diffraction pattern [39] and a calculated SANS that are in good agreement experiments [67].

While the X-ray diffraction patterns provide information about the spatial arrangement of the pores in the porous network, SANS allows assessing the textural properties of the surface of porous materials. In particular, the diffused intensity $I(Q)$ obtained at small Q values by means of SANS provides crucial information regarding the surface properties of the material in the Porod range, i.e. $QD > 1$ (D is the pore width) [68]. As shown in Fig. 13.5, peaks obtained at large momentum transfer ($Q > 1$ Å^{-1}) are in very good agreement with those obtained for bulk amorphous silica: ~1.5, ~3.0, ~5.25, and ~8.0 Å^{-1} [67]. The SANS spectrum in Fig. 13.5 was fitted using the algebraic decay law $I(Q) \sim Q^{-x}$ over the range 0.1–0.8 Å^{-1} in order to determine the Porod exponent that characterizes the surface roughness of the

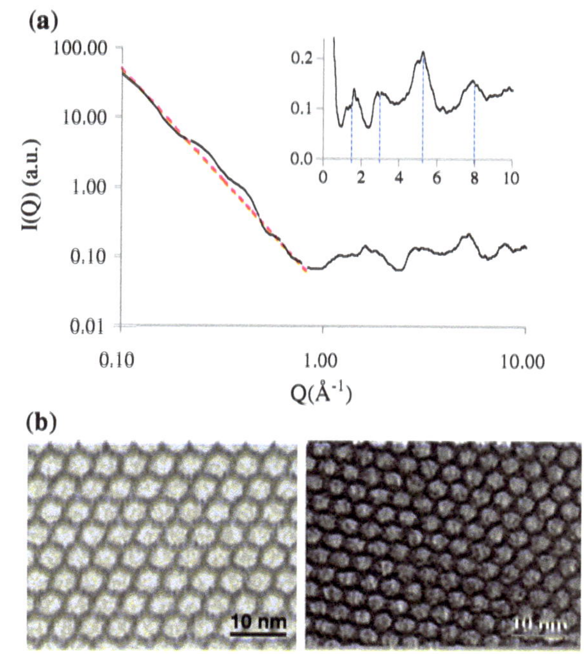

Fig. 13.5 (*color* online) **a** Small angle neutron scattering spectrum of the atomistic model of MCM-41 materials (*black line*). The *dashed red line* corresponds to an algebraic decay over the range 0.1–1 Å^{-1} (*see text*). The *inset* displays the same small angle neutron spectrum in a linear scale. **b** (*left*) Simulated TEM image of the atomistic structural model of MCM-41 materials. (*right*) Experimental electron microscopy of MCM-41 having a pore diameter $D = 2.0$ nm (adapted from Kruk et al. [67]). Note that the structural model has larger pores than in the experiments

porous solid at length scales ~8–62 Å. x was found equal to 3.2 ± 0.2. This Porod exponent, which is lower than the value $x = 4$ for cylindrical pores having a smooth pore/void interface with atomistic surface roughness only, is typical of disordered porous silicas such as Vycor [69] or silica gels [70]. Such a value is also in agreement with SAXS or SANS experiments for MCM-41 materials [71].

Figure 13.5 also shows the experimental [67] and simulated transmission electron microscopy images (TEM) of MCM-41. The relative intensity $\Delta I/I$ of each histogram point is calculated by applying the Beer-Lambert law to each histogram bin:

$$\frac{\Delta I}{I} = 1 - \Sigma_i \exp(-\alpha_i N_i) + \xi \tag{13.1}$$

where N_i is the total number of atoms of type i = Si, O, or H projected to the histogram bin and α_i is a constant proportional to the elastic scattering cross section of species i. In this work, the simulated images, were obtained for α_i equal to the atomic number of the element i (Z_i). ξ in 13.1 consists of a random number in the range [0, 0.333] which is added in order to simulate the electronic noise generated by the

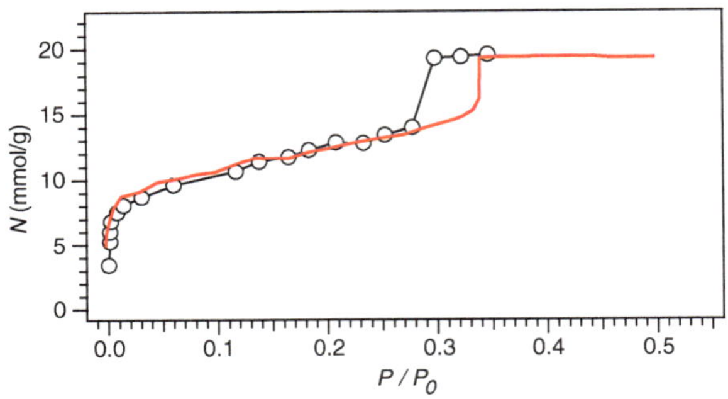

Fig. 13.6 Nitrogen adsorption isotherms at 77 K in the atomistic model of MCM-41 materials. The *red line* corresponds to the theoretical model based on Derjaguin's approach (see text)

TEM apparatus. In Fig. 13.5 the darker areas in both TEM images represent the pore walls while the lighter areas represent the pore voids and areas with low densities. The TEM image for the MCM-41 for the atomistic sample of the mesoporous silica captures all the features of the TEM image for the real sample.

Figure 13.6 shows the N_2 adsorption isotherm at 77 K in the atomistic model of mesoporous silica MCM-41. Such an adsorption isotherm conforms to the classical picture of adsorption and capillary condensation in nanopores. At low pressures, a molecular thick film is adsorbed at the pore surface. The thickness of the adsorbed film increases with increasing pressure in the multilayer adsorption regime. Then, at a pressure $P = 0.25P_0$ much lower than the bulk saturating vapor pressure, a sharp increase in the adsorbed amount is observed as capillary condensation occurs within the nanopores. The isosteric heat of adsorption Q_{st} as a function of the adsorbed amount of N_2 in the atomistic model of MCM-41 pores shows a curve which is characteristic of adsorption of simple gases on heterogeneous surfaces; Q_{st} ~14 kJ/mol at low loading (when strongly adsorbing sites are being filled) and then decreases in a continuous way to a value close to the heat of liquefaction of nitrogen (7 kJ/mol) as further adsorption takes place. As expected, Q_{st} increases as condensation occurs due to the heat released as the gas/liquid interface within the pore disappears. At low pressure, the data obtained is in nice agreement with the data obtained experimentally by Jaroniec and co-workers [67] for N_2 adsorption at 77 K in MCM-41 materials having a pore size (D = 3.6 nm) similar to that of the atomistic model considered in the present work. This result suggests that the surface of the atomistic model accurately describes the specific interaction between nitrogen and hydroxylated silica surfaces.

Nitrogen adsorption at low temperature is a routine characterization technique of nanoporous materials. For instance, the specific surface of porous materials is usually assessed from adsorption experiments (prior to capillary condensation of the fluid) on the basis of the Brunauer, Emmett, and Teller (BET) method. The BET model corresponding to the N_2 adsorption isotherm at 77 K in the atomistic model of MCM-41 materials fits very well the simulated data with a correlation coefficient $R^2 = 0.999$ (see [39] for the comparison). We found S_{BET} ~1000 m^2 /g (the latter value is obtained by considering as the surface area occupied by an adsorbed N_2 molecule, $A_{N2} = 0.162\,nm^2$) and C = 100. The value obtained for C is consistent with the values that are usually reported in the literature for oxide surfaces, C ~80–150. It is interesting to note that the BET surface, S_{BET}, is in reasonable agreement with the one determined from the chord length distribution and/or accessible surface area (1034 m^2/g). This result shows that the BET model provides reasonable estimates of the surface area of porous materials. The fact that the BET surface overestimates the surface area obtained from mathematical procedures (920 m^2/g) supports the suggestion made by [57] to use a smaller value for A_{N2} (0.135 nm^2 instead of 0.162 nm^2).

Coasne and Ugliengo [39] provided also a theoretical picture of adsorption in regular silica pores. In particular, they show that the simulated adsorption/desorption isotherm can be described using available theoretical models. In case of capillary condensation (such as for N_2 in the present work), the thermodynamically approach known as the Derjaguin-Broekhoff-DeBoer model [72, 73], provides a comprehensive picture of adsorption and capillary condensation in nanopores.

13.4 First Principles Simulations of Chalcogenide Surfaces

We now report on the development of realistic atomic models of the *g*-GeS_2 surface by means of first-principles molecular dynamics (FPMD) based on fully self-consistent density functional theory. This approach has been well assessed and successfully employed to model bulk liquid and glassy chalcogenides [74–76]. Very recently, we were able to produce a bulk model for *g*-GeS_2 that featured an unprecedented agreement with neutron scattering data, thereby legitimating its further application to surface studies [45]. This bulk model was obtained by simulated melt quenching. The details of the theoretical model and computational framework employed in this work are described in the following paragraph. A detailed account of the structure of the *g*-GeS_2 surface model is then given in terms of pair correlation functions, structure factors, coordination numbers, and bond angle distributions. Special attention is then devoted to the nature of chemical bonding as well as to the electronic and charge properties of the *g*-GeS_2 surface. In particular, the chemical bonding is analyzed by means of maximally-localized Wannier Functions. With the aim of constructing an interatomic potential based on first-principles data, we also focus on the derivation of atomic charges for S and Ge atoms.

13.4.1 Model Building

We adopted the method by Car and Parrinello [22] to ensure a self-consistent evolution of the electronic structure during molecular dynamics motion. The electronic structure was described in the framework of density functional theory (DFT) with the generalized gradient approximation (GGA) due to Becke (B) for the exchange energy and Lee, Yang and Parr (LYP) for the correlation energy

[77–79]. For the case of chalcogenides, we refer to [80] for a detailed account of the reasons underlying the better performances of the BLYP approach when compared, for instance, to the Perdew and Wang scheme. Here, we just recall that the BLYP exchange-correlation functional provides a better description of valence electron localization effects, which are crucial in the case of iono-covalent systems. Since van der Waals (vdW) interactions are found to be significant in some cases, for the present study the BLYP functional was combined with the dispersion correction proposed by Grimme [81]. Such a correction is a thorough DFT-based empirical correction self-consistently tuned on different functional, from PBE to B3LYP, and benchmarked on a wealth of different systems ranging from simple molecules to complex reactive surfaces. No experimental parameter is included in the construction of this vdW correction and its inclusion does not affect at any stage the Kohn-Sham equations, thus preserving the first-principle character of the electronic structure calculations. In our work, the valence electrons were treated explicitly, in conjunction with norm conserving pseudopotentials of the Trouiller-Martins type to account for core-valence interactions. The wave functions were expanded at the γ point of the supercell on a plane wave basis set with an energy cutoff $E_c = 20$ Ry. This energy cutoff value has already been shown to be fully adequate to attain converged properties for the relevant physical quantities of the Ge-S dimer (cohesive energy, interatomic distance, vibrational frequency). A fictitious electron mass of 1200 a.u. and a time step of $\Delta t = 0.12$ fs are adopted to integrate the equations of motion. Simulations are performed for a fixed volume (NVT ensemble) for each step of sample generation and data collection. We start from a bulk sample of g-GeS_2 which was obtained in our previous work by using the same theoretical framework [45]. Such a structural model can be safely considered as the best available for g-GeS_2, in spite of the fact that some peak intensities and features in both the total pair correlation function and the total neutron structure factor moderately differ from the experimental patterns.

In particular, we use the bulk model produced by the procedure labeled FPMD(1) in [45]. This model, which was equilibrated at 300 K, is labeled as g-GeS_2(b) in what follows (b standing hereafter for bulk). The g-GeS_2(b) sample is made up of 480 atoms, and has a volume of $23.58 \times 23.58 \times 23.58$ Å^3. A g-GeS_2 surface was created by adding at the top and bottom of the g-GeS_2(b) glass (along the z direction) empty volumes of a height 12 Å. In so doing, one obtains a g-GeS_2 slab exhibiting two surfaces embedded in a simulation box of a volume $23.58 \times 23.58 \times 47.58$ Å^3. To produce a surface model at finite temperature we gradually heated to 300 K the system by increasing the temperature in a stepwise manner with temperature intervals of 50 K. Temperature control was implemented for both ionic and electronic degrees of freedom using Nosé-Hoover thermostats. The system was then equilibrated at 300 K for 20 ps and the last 15 ps were used for data collection. This surface is labeled hereafter as g-GeS_2(s) where s stands for surface. Surface bond rearrangements along the equilibration at 300 K promote a further local stabilization with respect to mere optimization without annealing. However, this further local stabilization does not alter the S:Ge ratio as a function of the slab height, which remains close to stoichiometry ($\sim$2). More details about the surface model preparation can be found in [82].

13.4.2 Results and Discussion

13.4.2.1 Structural Characterization of *g*-GeS 2 Glassy Surface

We first determined the structural properties of *g*-GeS_2(s) by calculating the partial pair correlation functions $g_{\alpha\beta}$(r) with α, β = Ge or S. In order to compare the partial pair correlation functions for *g*-GeS_2(b) and *g*-GeS_2(s), $g_{\alpha\beta}$(r) for *g*-GeS_2(s) have been corrected for the finite size of the sample: [83]

$$g_{\alpha\beta}(\mathrm{r})* = g_{\alpha\beta}(\mathrm{r})/f(\mathrm{r}) \quad \mathrm{with} f(\mathrm{r}) = 1 - r/(2h) \tag{13.2}$$

where h is the thickness of the slab. While this correction is not needed to compare the position of the peaks in the $g_{\alpha\beta}$(r) functions, it allows correcting the peak amplitudes for the finite size of the sample. The partial pair correlation functions for Ge-S, Ge-Ge, and S-S pairs are shown in Fig. 13.7. *g*-GeS_2(b) and *g*-GeS_2(s) show very similar Ge-S bond lengths, which is identified by the position of the first peak (2.20 Å). This value is close to the experimental Ge-S bond for *g*-GeS_2(b) (2.20–2.23 Å) [84, 85]. The shape of the pair correlation functions for *g*-GeS_2(b) and *g*-GeS_2(s) are similar. The large amplitude of the first g_{GeS}(r) peak for *g*-GeS_2(b) and *g*-GeS_2(s) indicates that heteropolar Ge-S bonding is the most common type of bonds in these systems. Differences are found in the amplitude of some of the peaks. For instance, the first peak in the g_{GeS}(r) function for *g*-GeS_2(s) has a larger amplitude than its bulk counterpart. This difference is indicative of a larger content of Ge-S bonds with respect to the total numbers of bonds for *g*-GeS_2(s).

The distributions of the coordination number (CN) around Ge and S atoms for the bulk and surface models are shown in Fig. 13.8. For the surface model the CN distribution have been obtained by considering the top external layer within the last 10 Å. For both the bulk and surface models, CN is determined by computing the average number of atoms (Ge or S) within a cutoff distance from the S or Ge atoms.

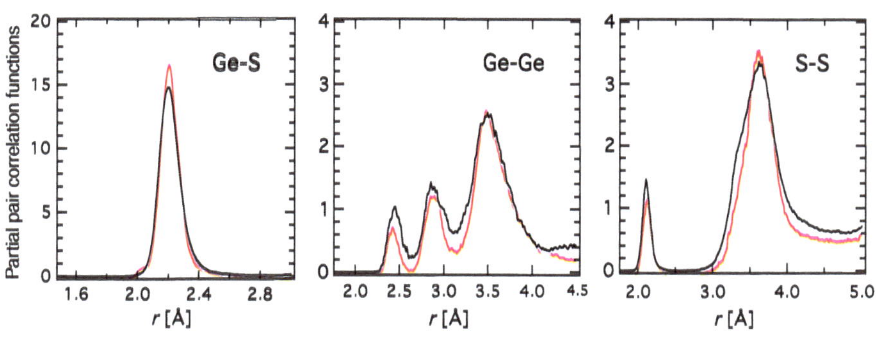

Fig. 13.7 (*color* online) Ge-S (*left*), Ge-Ge (*centrer*), and S-S (*right*) pair correlation functions: $g_{\alpha\beta}$(r) for *g*-GeS_2(b) (*black line*) and $g_{\alpha\beta}$(r) for *g*-GeS_2(s) (*red line*) at 300 K; the data for *g*-GeS_2(s) were corrected for the finite size of the sample (see text). Adapted from [82]

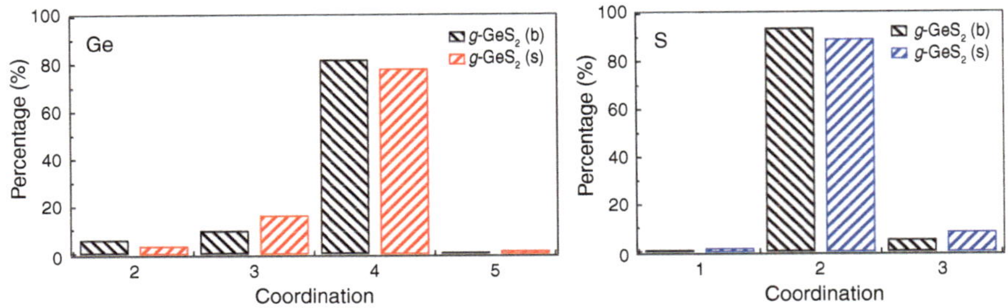

Fig. 13.8 (*color* online) Percentage of *l*-coordinated Ge (*left*) and S (*right*) atoms for *g*-GeS_2(b) and *g*-GeS_2(s). Adapted from [82]

The cutoff distance was extracted from the pair distribution functions in Fig. 13.7 as the position of the minimum between the first and the second peaks. *g*-GeS_2(b) and *g*-GeS_2(s) are mainly made of tetrahedrally coordinated Ge and twofold coordinated S. The average CN of Ge and S are close to those of bulk (Ge: 3.79 and S: 2.05). However, differences can be noted in the distribution of the individual coordinating units (see [82] for details). When compared to the bulk counterpart, *g*-GeS_2(s) shows a decrease in the numbers of fourfold (−3.6 %) and twofold (−2.5 %) coordinated Ge, while there is an increase in the threefold coordination (+6.6 %). Furthermore, twofold coordinated S decreases by 4.4 % which is mainly balanced by an increase of the threefold coordinated S atoms (+3.4 %). These distributions show that *g*-GeS_2(s) posses a slightly lower chemical order than *g*-GeS_2(b). On the other side, *g*-GeS_2(b) and *g*-GeS_2(s) show a similar content of homopolar bond, intended as Ge-Ge and S-S bonds. Figure 13.9 shows the S-Ge-S and Ge-S-Ge bond angle distributions (BAD) for *g*-GeS_2(b) and *g*-GeS_2(s). The S-Ge-S BAD exhibits a broad peak at about 110° with a left shoulder at 98°.

These peaks correspond to angles in corner-sharing tetrahedra (109°) and angles in edge-sharing tetrahedra at 98° and 110°. *g*-GeS_2(s) shows a slightly larger peak at 110° and a slightly smaller peak at 98°. This result is indicative of a small increase in corner-sharing tetrahedra for *g*-GeS_2(s) when compared to *g*-GeS_2(b). The Ge-S-Ge BAD show a peak at 83° and one at 103°. The first one arises form edge-sharing tetrahedra while the one at 103° is due to corner-sharing tetrahedra. This result is consistent with the S-Ge-S BAD analysis above, and underlines a slightly higher corner-sharing/edge-sharing ratio in *g*-GeS_2(s) with respect to *g*-GeS_2(b).

More detail about the contributions of corner-sharing and edge-sharing tetrahedra to the S-Ge-S and Ge-S-Ge BAD can be found in [82]. Figure 13.9 (right) shows the comparison between the structure factor $S_T(k)$ for *g*-GeS_2(b) and *g*-GeS_2(s). As a consequence of the close patterns for the Faber-Ziman partial structure factors (see [82] for details) the total structure factor $S_T(k)$ for *g*-GeS_2(b) and *g*-GeS_2(s) are very similar. The peak at $k \sim 1$ Å^{-1} (known as first sharp diffraction peak, FSDP) arises predominantly from the Ge-Ge correlations. The Ge-S correlations also contribute to the FSDP but to a much lower extent. The S-S correlations does not contribute to

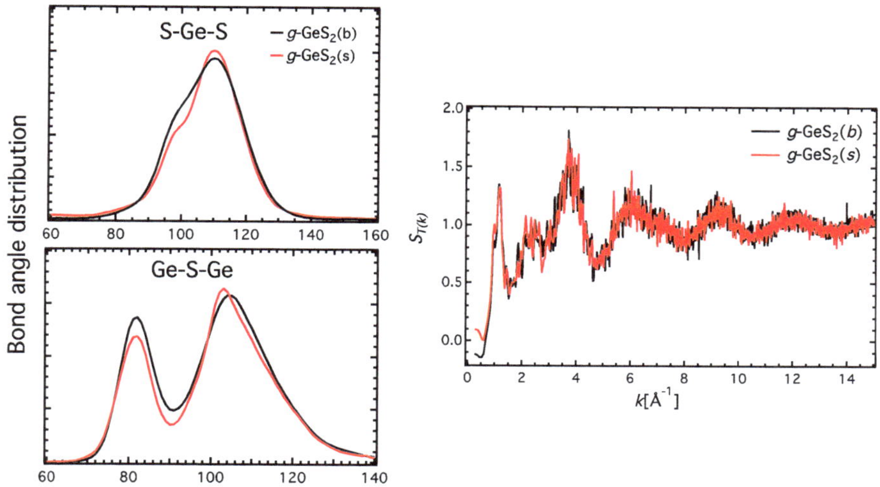

Fig. 13.9 (*color* online) S-Ge-S (*left top*) and Ge-S-Ge (*left bottom*) bond angle distributions (BAD). The *black* and *red lines* correspond to the data for g-GeS_2(b) and g-GeS_2(s), respectively. (*right*) Total structure factors for g-GeS_2(b) (*black line*) and the g-GeS_2(s) model (*red line*). Adapted from [82]

the peak at $k \sim 1$ Å^{-1}, which suggests that for both g-GeS_2(b) and g-GeS_2(s) surface, Ge atoms account for most of the intermediate range structural order in amorphous GeS_2.

13.4.2.2 Electronic Properties

The electronic properties of g-GeS_2(s) were investigated by determining the electronic density of states (EDOS) (Fig. 13.10 left). The EDOS for g-GeS_2(s) bears some resemblance with the experimental valence spectrum obtained for g-GeS_2(b) [86]. The main difference between the EDOS for g-GeS_2(b) and g-GeS_2(s) concerns the pseudogap around the Fermi level, where for g-GeS_2(b) is deeper.

In order to probe the chemical bonding in the g-GeS_2 surface, we determined its electronic structure through the position of the maximally-localized Wannier function centers (WF) [87]. The center of the Wannier orbital indicates the maximum probability for the location of an electron (or electron pair) in a quantum system. The analysis of the WF centers with respect to the nuclear positions allows gaining insight into the chemical bonding involved in systems such as water [88, 89], glassy silicon [90], and oxides [91]. This analysis has been extended to germanium selenides g-Ge_xSe_y where a complex mixture of ionocovalent and purely covalent bonds was found [74–76]. Typically fourfold coordinated Ge atoms are characterized by four WF centers resulting from the bonds established between Ge and its S neighboring atoms. For S, the existence of six valence electrons and the twofold coordination of

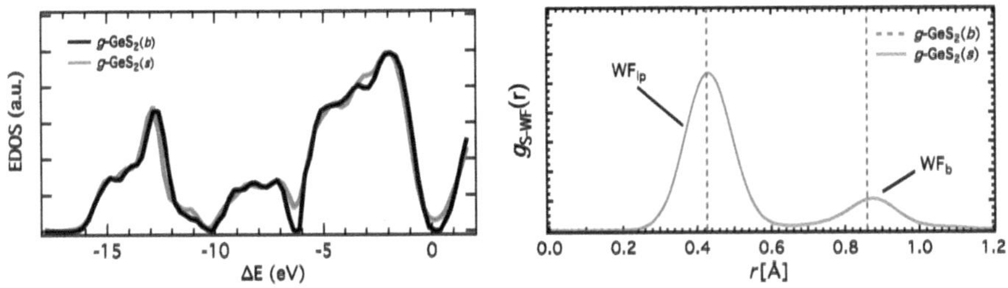

Fig. 13.10 (*color* online) (*left*) Electronic density of states extracted from the Kohn-Sham eigenvalues. The *black* and *grey lines* correspond to the data for g-GeS_2(b) and g-GeS_2(s), respectively. (*right*) Correlation functions of S-WF pair for g-GeS_2(s). The *dashed lines* show the position of the peaks for the g-GeS_2(b). WF_{lp} and WF_b indicate the peaks due to the correlation between S atoms and the WF lone pairs and WF bonds, respectively. Adapted from [82]

S is at the origin of a specific pattern; two WF centers are localized close to the S atom, representing the two lone pairs of electrons not involved inn chemical bonding (WF_{lp}). The other two WF centers which are localized along the S-X (with X = Ge or S) bonds, reflect interatomic bonding (WF_b). Typically, when the location of the WF_b center is taken with respect to half the bond distance, each WF center is found to be closer to the S atom than to Ge atoms (S-WF distance: 0.89 Å). This result illustrates tha fact the Ge-S bonds are ionocovalent, since a sizeable electron transfer occurs towards the more electronegative atom (S).

For the case of homopolar bonding (whether Ge-Ge or S-S) there is a WF center for each homopolar bond located in the middle of the bond, as expected due to the covalent character of Ge-Ge and S-S homopolar bonds. Figure 13.10 (right) shows the pair correlation function $g_{SWF}(r)$ between the S atoms and the WF centers. $g_{SWF}(r)$ shows two peaks: a first peak centered at 0.435 Å and a second peak centered at 0.875 Å. The first peak corresponds to the distance between the S atoms and the WF_b centers. This value is lower than the value obtained for Se-WF_b (1 Å) in g-$GeSe_2$ [45]. This result is consistent with the lower ionic character of the Ge-Se bond with respect to the Ge-S bond.

With the aim of constructing an interatomic potential based on first-principles data, we also focus on the derivation of atomic charges for S and Ge atoms as obtained from various techniques. Details about the techniques used can be found in [82]. The final output of this comparison based on both classical and first-principles methods, show us that the Qeq [92] (EQeq [93]) and Bader [94] methods capture the effect of the local coordination on the partial charges n chalcogenide materials. However, the absolute charges obtained by the Bader method are too large to be used, eventually, in potentials for classical simulations. These high values would confer to the Ge and S atoms a nearly pure ionic character. For instance, these values are greater than most of the commonly used partial charges for silica. The charge-coordination correlation found in the Qeq method seems more appropriate to develop a forcefield describing the interactions with a g-GeS_2 glassy surface. Overcoordinated Ge(S) atoms possess a large positive (negative) charge with respect to the corresponding to stoichiometric coordination. This is directly related to the higher valence state which, in a formalism purely based on formal ionic charges (cations (Ge) and anions (S)), result in an increased charge localization. The Qeq approach seems to have the best suited technique to describe changes in the valence (charge) state for different coordinations. As far as this issue is concerned, the Qeq (EQeq) method produces a useful set of charges able to describe the structural order of g-GeS_2 materials.

13.5 Summary and Perspectives

The nature and properties of glassy surfaces such as silica and chalcogenide surfaces have been reviewed and discussed. For the case of silica, we focused the attention on mesoporous silica MCM-41. The study of the interaction between such glassy silica surfaces or porous silica-based materials and more complex molecules than simple gases (such as organic molecules) represents an open field where the level of accuracy in describing the interactions involved is crucial. For the case of glassy chalcogenide surfaces, we reported in this paper the development of a realistic model of GeS_2 glassy surfaces using first principles molecular dynamics simulations. We addressed the description of the chemical bonding involved, the structure, and electronic properties of the chalcogenide system. We also propose a set of partial atomic charges which are dependent on the local chemical order of the atoms constituting the system. This result represents the first step towards the development of a classical forcefield for chalcogenides in order to scale up the simulation of glassy chalcogenides to larger scale.

References

1. J. Le Bideau, L. Viau, A. Vioux, Chem. Soc. Rev. **40**, 907 (2011)
2. H.F. Hamann, M. O'Boyle, Y.C. Martin, M. Rooks, H.K. Wickramasinghe, Nat. Mater. **5**, 383 (2006)
3. P. Simon, Y. Gogotsi, Nat. Mater. **7**, 845 (2008)
4. S.L. Brock, H. Yu, in *Chalcogenide Aerogels*. Aerogels Handbook, ed. (Springer, New York, 2011), p. 367
5. A. Rimola, D. Costa, M. Sodupe, J.-F. Lambert, P. Ugliengo, Chem. Rev. **113**, 4216 (2013)
6. A. Zakery, S.R. Elliott, in *Chalcogenide Glasses and Their Applications in Optical Non-linearities* (Springer, Berlin, 2007)
7. B.J. Eggleton, B. Luther-Davies, K. Richardson, Nat. Photon. **5**, 141 (2011)
8. S. Bag, A.F. Gaudette, M.E. Bussell, M.G. Kanatzidis, Nat. Chem. **1**, 217 (2009)
9. G.S. Armatas, M.G. Kanatzidis, Nat. Chem. **8**, 217 (2009)
10. A. Pedone, J. Phys. Chem. C **113**, 20773 (2009)
11. R.L. McGreevy, L. Pusztai, Mol. Simul. **1**, 359 (1988)
12. D.A. Keen, R.L. McGreevy, Nature **344**, 423 (1990)
13. J.P. Pikunic, C. Clinard, N. Cohaut, K.E. Gubbins, J.-M. Get, R.J.-M. Pellenq, I. Rannou, J.-N. Rouzaud, Langmuir **19**, 8565 (2003)
14. M.J. Sanders, M. Leslie, C.R.A. Catlow, J. Chem. Soc. Chem. Commun. 1271 (1984)

15. B.W.H. van Beest, G.J. Kramer, R.A. van Santen, J. Phys. Rev. Lett. **64**, 1955 (1990)
16. J.R. Hill, J. Sauer, J. Phys. Chem. **98**, 1238 (1994)
17. A. Pedone, G. Malavasi, M.C. Menziani, A.N. Cormack, U. Segre, J. Phys. Chem. B **110**, 11780 (2006)
18. R.T. Cygan, J.J. Liang, A.G. Kalinichev, J. Phys. Chem. B **108**, 1255 (2004)
19. H. Heinz, T.-J. Lin, R.K. Mishra, F.S. Emami, Langmuir **29**, 1754 (2013)
20. A. Carré, J. Horbach, S. Ispas, W. Kob, EuroPhys. Lett. **82**, 17001 (2008)
21. A. Pedone, G. Malavasi, M.C. Menziani, U. Segre, F. Musso, M. Corno, B. Civalleri, P. Ugliengo, Chem. Mater. **20**, 2522 (2008)
22. R. Car, M. Parrinello, Phys. Rev. Lett. **55**, 2471 (1985)
23. C. Mischler, W. Kob, K. Binder, Comp. Phys Comm. **147**, 225–225 (2002)
24. C. Mischler, J. Horbach, W. Kob, K. Binder, J. Phys.: Condens. Matter **17**, 4005 (2005)
25. A. Tilocca, A.N. Cormack, ACS Appl. Mater. Interfaces **1**, 1324 (2009)
26. R.K. Iler, *The Chemistry of Silica*, 2nd edn. (Wiley, New York, 1979)
27. W.H. Zachariasen, J. Am. Chem. Soc. **54**, 3841 (1932)
28. R. Bruckner, Encycl. Appl. Phys. **18**, 101 (1997)
29. C.R.A. Catlow, A.N. Cormack, Int. Rev. Phys. Chem. **6**, 227 (1987)
30. J.B. Nicholas, R.E. Winans, R.J. Harrison, L.E. Iton, L.A. Curtiss, A.J. Hopfinger, J. Phys. Chem. **96**, 7958 (1992)
31. M.R. Bär, J. Sauer, Chem. Phys. Lett. **226**, 405 (1994)
32. B.T. Luke, J. Phys. Chem. **97**, 7505 (1993)
33. H. Knözinger, in *The Hydrogen Bond. Recent Developments in Theory and Experiments* ed. by P. Schuster, G. Zundel, C. Sandorfy, vol. 3 (North-Holland, Amsterdam), p. 1263
34. L.T. Zhuravlev, Colloids Surf. A **173**, 1 (2000)
35. P. Ugliengo, M. Sodupe, F. Musso, I.J. Bush, R. Orlando, R. Dovesi, Adv. Mater. **20**, 4579 (2008)
36. S. Bhattacharya, B. Coasne, F.R. Hung, K.E. Gubbins, Langmuir **25**, 5802 (2009)
37. S. Bhattacharyya, G. Lelong, M.-L. Saboungi, J. Exp. Nanosci. **1**, 375 (2006)
38. S. Bhattacharyya, K.E. Gubbins, J. Chem. Phys. **123**, 134907 (2005)
39. B. Coasne, P. Ugliengo, Langmuir **28**, 11131 (2012)
40. M. Frumar, B. Frumarova, P. Nemec, T. Wagner, J. Jedel-sky, M. Hrdlicka, J. Non-Cryst. Solids **356**, 544 (2006)
41. F. Jedema, Nat. Mater. **6**, 90 (2007)
42. R.K. Pan, H.Z. Tao, H.C. Zang, C.G. Lin, T.J. Zhang, X.J. Zhao, J. Non-Cryst. Solids **2358**, 357 (2011)
43. W.-S. Lim, S.-W. Kim, H.-Y. Lee, J. Korean Phys. Soc. **1764**, 51 (2011)
44. C. Lin, H. Tao, Z. Wang, B. Wang, H. Zang, X. Zheng, X. Zhao, J. Non-Cryst. Solids **438**, 355 (2009)
45. M. Celino, S. Le Roux, G. Ori, B. Coasne, A. Bouzid, M. Boero, C. Massobrio, Phys. Rev. B **88**, 174201 (2013)
46. B.J. Riley, J. Chun, J.V. Ryan, J. Matyas, X.S. Li, D.W. Matson, S.K. Sundaram, D.M. Strachan, J.D. Vienna, RSC Adv. **1**, 1704 (2011)
47. B.J. Riley, J. Chun, W. Um, W.C. Lepry, J. Matyas, M.J. Olszta, X. Li, K. Polychonopoulou, M.G. Kanatzidis, Environ. Sci. Technol. **47**, 7540 (2013)
48. J.C. Mauro, A.K. Varshneya, J. Am. Ceram. Soc. **90**, 192 (2007)
49. P. Vashishta, R.K. Kalia, I. Ebbsjö, J. Non-Cryst. Solids **106**, 301 (1988)
50. X. Zhang, D.A. Drabold, Phys. Rev. B **62**, 15695 (2000)
51. F. Inam, D.A. Drabold, J. Non-Cryst. Solids **354**, 4 (2008)
52. J. Akola, J. Larrucea, R.O. Jones, Phys. Rev. B **83**, 094113 (2011)
53. S.-W. Kim, W.-S. Lim, T.-W. Kim, H.-Y. Lee, Jpn. J. Appl. Phys. **47**, 5337 (2008)
54. A.K. Wesley, A.G. Clare, W.C. LaCourse, J. Non-Cryst. Solids **181**, 231 (1995)
55. G.C. Chern, I. Lauks, J. Appl. Phys. **54**, 4596 (1983)
56. W.-S. Lim, S.-W. Kim, H.-Y. Lee, J. Korean Phys. Soc. **51**, 1764 (2007)
57. B. Coasne, A. Galarneau, R.J.M. Pellenq, F. Di Renzo, Chem. Soc. Rev. **42**, 4141 (2013)

58. B. Coasne, L. Viau, A. Vioux, J. Phys. Chem. Lett. **2**, 1150 (2011)
59. G. Ori, F. Villemot, L. Viau, A. Vioux, B. Coasne, Mol. Phys. **112**, 1350 (2014)
60. B. Siboulet, B. Coasne, J.F. Dufrêche, P. Turq, J. Phys. Chem. B **115**, 7881 (2011)
61. B. Coasne, F. Di Renzo, A. Galarneau, R.J.M. Pellenq, Langmuir **24**, 7285 (2008)
62. L.D. Gelb, K.E. Gubbins, Langmuir **14**, 2097 (1998)
63. F.R. Siperstein, K.E. Gubbins, Langmuir **129**, 2049 (2003)
64. F.R. Hung, S. Bhattacharya, B. Coasne, M. Thommes, K.E. Gubbins, Adsorption **13**, 425 (2007)
65. P.E. Levitz, Adv. Colloid Interface Sci. **71**, 76 (1998)
66. R.J.M. Pellenq, P.E. Levitz, Mol. Phys. **100**, 2059 (2002)
67. M. Kruk, M. Jaroniec, Y. Sakamoto, O. Terasaki, R. Ryoo, C.H. Ko, J. Phys. Chem. B **104**, 292 (2000)
68. J.D.F. Ramsay, Adv. Colloid Interface Sci. **76–77**, 13 (1998)
69. P. Levitz, G. Ehret, S.K. Sinha, J.M. Drake, J. Chem. Phys. **95**, 6151 (1991)
70. J.M. Drake, P. Levitz, J. Klafter, Isr. J. Chem. **31**, 135 (1991)
71. K.J. Edler, P.A. Reynolds, J.W. White, J. Phys. Chem. B **102**, 3676 (1998)
72. B.V. Derjaguin, N.V. Churaev, J. Colloid Interface Sci. **54**, 157 (1976)
73. E. Charlaix, M. Ciccotti, in *Capillary Condensation in Confined Media*. Handbook of Nanophysics: Principles and Methods, ed. by K. Sattler (CRC Press, Boca Raton, 2010)
74. C. Massobrio, M. Celino, P.S. Salmon, R.A. Martin, M. Micoulaut, A. Pasquarello, Phys. Rev. B **79**, 174201 (2009)
75. S. Le Roux, A. Zeidler, P.S. Salmon, M. Boero, M. Micoulaut, C. Massobrio. Phys. Rev. B **84**, 134203 (2011)
76. M. Bauchy, M. Micoulaut, M. Celino, S. Le Roux, M. Boero, C. Massobrio, Phys. Rev. **84**, 134203 (2011)
77. A.D. Becke, Phys. Rev. A **38**, 3098 (1988)
78. C. Lee, W. Yang, R.G. Parr, Phys. Rev. B **37**, 785 (1988)
79. J.P. Perdew, Y. Wang, Phys. Rev. B **45**, 13244 (1992)
80. M. Micoulaut, R. Vuilleumier, C. Massobrio, Phys. Rev. B **79**, 214205 (2009)
81. S. Grimme, J. Comput. Chem. **27**, 1787 (2006)
82. G. Ori, C. Massobrio, A. Bouzid, M. Boero, B. Coasne, Phys. Rev. B **90**, 045423 (2014)
83. B. Coasne, J.T. Fourkas, J. Phys. Chem. C **115**, 15471 (2011)
84. A. Ziedler, W.E. Drewitt, P.S. Salmon, A.C. Barnes, W.A. Crichton, S. Klotz, H.E. Fischer, C.J. Benmore, S. Ramos, A.C. Hannon, J. Phys.: Condens. Matter **21**, 474217 (2009)
85. A. Bytchkov, G.J. Cuello, S. Kohara, C.J. Benmore, D.L. Price, E Bychkova, Phys. Chem. Chem. Phys. **15**, 8487 (2013)
86. D. Foix, H. Martinez, A. Pradel, M. Ribes, D. Gonbeau, Chem. Phys. **323**, 606 (2006)
87. G.H. Wannier, Phys. Rev. **52**, 191 (1937)
88. R. Scipioni, D.A. Schmidt, M. Boero, J. Chem. Phys. **130**, 024502 (2009)
89. M. Boero, M. Parrinello, S. Huffer, H. Weiss, J. Am. Chem. Soc. **122**, 501 (2000)
90. P.L. Silvestrelli, N. Marzari, D. Vanderbit, M. Parrinello, Solid State Commun. **107**, 7 (1998)
91. S.V. Sukhomlinov, K.S. Smirnov, J. Phys. Condens. Matter **24**, 475501 (2012)
92. A.K. Rappe, W.A. Goddard III, J. Phys. Chem. **95**, 5 (1991)
93. C.E. Wilmer, K.C. Kim, R.Q. Snurr, J. Phys. Chem. Lett. **3**, 5 (2012)
94. R.F.W. Bader, *Atoms in Molecules: A Quantum Theory* (Oxford University Press, New York, 1990)

Chapter 14
Rings in Network Glasses: The B_2O_3 Case

Guillaume Ferlat

Abstract There has been a considerable debate, in particular since the emergence of atomistic simulations, about the structure of glassy B_2O_3, a prototypical network-forming system based on trigonal units. Some intermediate-range order in the form of threefold rings, present in the glass but not in the crystalline phases, has remained so far very difficult to reproduce in atomistic simulations. After a brief summary of the evidences accumulated regarding the boroxol rings, a review of the numerical studies of liquid and glassy B_2O_3 is provided. The reasons for the failure of the *quench-from-the-melt* techniques are stressed and a methodology, based on *first-principles* calculations of experimental observables (diffraction, NMR, Raman, IR, heat capacity) from various glassy models is devised to provide incontrovertible answers to the debate. This allows assessing not only the content of boroxol rings but also the sensitivity of each observable to this quantity. The presence of threefold rings in the glass is then showed to have ramifications for the understanding of the crystalline and liquid phases. This includes the prediction of yet unknown B_2O_3 polymorphs structurally close to the glass, the understanding of the so-called *crystallisation anomaly* and the evidencing of structural transitions in the liquid. Finally, the discussion is extended to parent systems such as B_2S_3.

14.1 Introduction: Rings in Glasses

Although considerable progress has been made, characterising the atomic structure of glasses remains quite challenging for both theory and experiments. Since the short-range order in a glass is essentially the same as in the corresponding crystals, the glass structural specificity is to be found in the medium-range order, i.e. that beyond the first-coordination shell. In principle the full characterisation of the atomic order is revealed by determining the complete set of the n-body distribution functions. In practice standard (X-ray, neutron or electron) scattering techniques [1] only access 2-body distribution functions (or even only a subset of these functions)

G. Ferlat (✉)
IMPMC, Université Pierre et Marie Curie, 4 Place Jussieu, 75005 Paris, France
e-mail: ferlat@impmc.upmc.fr

C. Massobrio et al. (eds.), *Molecular Dynamics Simulations of Disordered Materials*, Springer Series in Materials Science 215,
DOI 10.1007/978-3-319-15675-0_14

and very few techniques are able to probe higher-order correlations. These include X-ray absorption spectroscopy [2], coherent X-ray diffraction [3–5] and fluctuation electron microscopy [6, 7]. However, the latter two techniques have become available only recently and applications have been limited so far to a few cases [8]. Even in the case of the more established X-ray absorption spectroscopy, the extraction of angular correlations in disordered systems has remained difficult and essentially limited to simple systems [9] and/or to first-neighbours angular correlations [10] due to the intrinsic short-range nature of this technique. Other techniques such as nuclear magnetic resonance (NMR) can sometimes bring very useful insight to triplet arrangements but through indirect correlations with the NMR parameters [11].

Atomistic simulations are therefore essential tools since they provide 3D atomic structures from which about any n-body correlation function can be computed. Beyond the traditional two-body correlation functions, $g_2(r)$, it is also quite common to visualise angular correlations.

However, the information contained in the higher-order correlations often needs to be recast in more practical ways (e.g. rings statistics, channels sizes, voids distribution) depending on the type of systems and the problematic under investigation. In the case of network-forming systems, such as silicon (Si) or silica (SiO_2), it is quite natural to describe the topological order by the connectivity of atoms (or polyhedra) and the rings they form. Being able to assess the rings statistics of a given glass should allow to relate its structure to some of its properties such as density, porosity, viscosity, glass forming ability, etc.

In oxide systems, one usually defines the *n*th order of a ring by the number of network-former cations (Si for silica, B for boron oxide) involved in the closed loop. Since each polyhedra (SiO_4 tetrahedra in silica, BO_3 triangles in boron oxide) is corner-shared by an oxygen, a n-fold ring contains 2n cation-oxygen bonds.

As compared to crystals, the rings distribution in a glass is usually broader as a result of the larger free-volume available and of the greater angular variability generally found in the glassy structure. For instance in silica, the structure of cristobalite consists entirely of sixfold rings, that of α-quartz contains solely six and eightfold rings, that of coesite has four, six and eightfold rings whereas the silica glass structure, supposedly peaked at the six and sevenfold rings also includes smaller (threefold) and possibly larger ($n = 9, 10, \ldots$) rings [12–15].

From the modelling point of view, the rings distribution is a feature characterising the quality of the obtained structure. Fulfilling the experimentally-known constraints on the rings distribution is a much more stringent test of the model than the sole reproduction of the radial distribution functions (or equivalently the structure factors) since it involves information contained in n-body correlations with $n > 2$. In order to get reliable rings statistics from the simulations, the potential should be accurate enough to give a good account of the many-body effects, the system size should be large enough to accommodate the largest rings of interest, and the quenching rate should be slow enough to allow for a proper structural reorganisation from the liquid to the glassy phase.

Unfortunately, there is no experimental technique that can probe the entire rings distribution in glasses. Direct imaging of the rings has been recently achieved thanks to scanning tunneling microscopy and transmission electron microscopy but this has been so far restricted to two-dimensional glasses such as bilayer vitreous silica [16]. However, valuable experimental information regarding specific values of the distribution have been obtained in some systems. This is the case for the smaller rings (i.e. two, three and fourfold rings) which are characterised by specific angles and/or vibrational modes. Rings larger than fivefold rings are relatively strain-free and tend to be spectroscopically indistinguishable. In glassy silica, Raman information [17–20] has played a key role in establishing the existence of three and fourfold rings although these represent a minority ($\lesssim 1\,\%$) of the rings. Their most noticeable spectral signatures arise from breathing modes of oxygens towards the ring center, which are well decoupled from other modes [21]. These rings tend to be planar for energetic reasons [22]. It is also fair to say that while the existence of these rings was quite clear from the experiments, in turn the simulations have been very valuable to quantify their proportion [20].

Since we are mostly interested in the smaller rings, we sweep under the carpet that several possible definitions of rings exist which can affect the results for the larger rings distribution [23–26]. A discussion of these definitions and of the associated technical issues can be found in [25].

The rest of this chapter is mostly devoted to the structure of vitreous boron oxide (v-B_2O_3), which together with SiO_2 and P_2O_5 is an archetypal network-forming system [27] and also the end member of the large family of borate glasses [28]. Citing Elliott [29], the possible presence of threefold rings in v-B_2O_3 is perhaps the most well-documented case of intermediate-range order. However, quantitative estimates have been controversial for many decades, especially since the apparition of the first molecular dynamics simulations of this system. Although this issue is now hopefully settled, the obtention of a good numerical model of v-B_2O_3 using conventional (i.e. quench from the melt) techniques is still extremely challenging. We thus use the B_2O_3 example as a (possibly exacerbated) illustration of the challenges to be faced when simulating the medium-range order of a glass. After having introduced the problematic, I will show how a strategy was devised to assess the rings content. Then, I will show that the specific rings found in v-B_2O_3 have direct implications for our understanding of the crystalline and liquid phases, not only of B_2O_3 but also of parent systems such as B_2S_3.

14.2 Boroxol Rings in Vitreous B_2O_3

I hereafter briefly summarise the structural works on v-B_2O_3 taking advantage of several good reviews which appeared at various times [28, 30–32]. A special focus to atomistic simulations is then given in the next section.

In pure boron trioxide, the basic building unit is the planar BO_3 triangle [33] (see Fig. 14.1). The first proposed structural model of v-B_2O_3 is that of a continuous random network of such corner-linked triangles [27]. This view was however questioned, as soon as 1953 [34], by the observation of an extremely sharp peak at

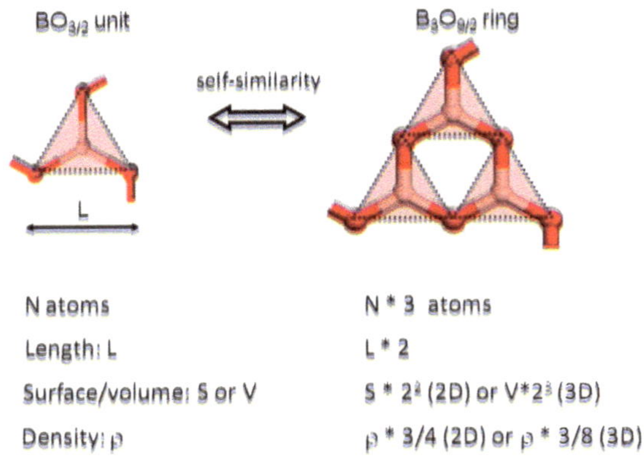

Fig. 14.1 *Left* The BO_3 triangle (indicated here $BO_{3/2}$). *Right* The B_3O_6 boroxol ring (indicated here $B_3O_{9/2}$). Both units are self-similar: the replacement of all *triangles* by boroxol units in a given 2D- or 3D-structure, increases the number of atoms by a factor of 3, the surface by 2^2 and the volume by 2^3; the density is thus lowered from the initial one by a factor $\frac{3}{4}$ in 2D or $\frac{3}{8}$ in 3D

808 cm^{-1} in the Raman spectrum (see Fig. 14.16), which was attributed to the presence of planar threefold rings (B_3O_6) of triangles, called boroxols (see Fig. 14.1). This mode was indeed later unambiguously assigned to the in-phase breathing of oxygens inside the rings, in particular thanks to elegant measurements in isotopically substituted glasses [35]. The sharpness of the peak, which is different from the broad features usually observed for glasses, is a consequence of the strongly localised character of this mode [36]. Unfortunately, the quantitative determination of the proportion of atoms involved in these rings is not trivially derived from the peak area since it would require an accurate knowledge of the Raman cross-section for the mode under consideration; these calculations have become available from *first-principles* only recently [37, 38]. In any case, the existence of these rings in the glass is, at least at this point, very disturbing since there are no such rings in any of the two known crystals of B_2O_3: the closest polymorph, B_2O_3-I, is made of independent BO_3 triangles. I will come back to this point in Sect. 14.6.

However, other experimental evidences for boroxols have been accumulated over the years. NMR and nuclear quadrupolar resonance experiments do evidence the presence of two populations in both ^{17}O and ^{11}B spectra [39–46] which were assigned to atoms *inside* and *outside* the rings. These assignments were strongly supported by the comparison of the spectra with borate crystals of known structure [47] as well as by molecular orbitals calculations on cluster models [48–50]. Quantitative measurements have consistently established that the ratio of the two populations (rings/non-rings) is typically about 1:1, i.e. that there are about as many boroxol units as independent BO_3 groups, confirming the model originally proposed by Krogh-Moe [30]. Note that the convention in the literature is to report the fraction of boron atoms involved in boroxols, f. The 1:1 model corresponds to $f = 75\,\%$ (while the corresponding fraction of oxygens in rings, f_O, is 50 %). More precisely,

NMR derived values range from $f = 65$–$85\,\%$. Hereafter, structures in this range of f will be refered to as the *boroxol-rich model*.

Diffraction data analyses have been used by both sides to support [31, 32, 51] or dispute [52–54] the boroxol-rich model: f values ranging from essentially 0–80 % were reported. This should come as no surprise since the 3-body information is largely washed out in these data. However, both camps had seemingly good arguments. It has been claimed that some peaks in the partial distribution functions are narrower in the presence of large amounts of boroxols and thus that they could serve as signatures of the rings [32]. However, Swenson and Börjesson (SB) argued [54] that previous diffraction analyses lack an essential constraint, namely the density: using reverse Monte Carlo (RMC) simulations, they came to the conclusion that it is not possible to produce structural models containing more than 30 % of boroxols that simultaneously reproduce the experimental density and the neutron and X-ray diffraction data [54]. This claim is however overlooking an earlier RMC study by Bionducci et al. [55] which showed that various hand-made models, containing from 0 to 50 % boroxols, gave equivalent and good fits of the diffraction data. In their simulations both the density and the boron coordination were constrained. It was rightly concluded that in the absence of interatomic potentials, it is very improbable that the random movements of the particles organise to a chemical order starting from a disordered configuration [55]. We will further discuss (and refute [38, 56]) the arguments of SB [54] in the Sect. 14.4.

Indeed, the density is a key quantity. Early hand-made *ball-and-stick* models had difficulties producing reasonable models at the correct glass density. The BO_3-made model of [57] has a too high density while that of the boroxol-made (and lamellar) model of [58] is too low. To this extent, SB [54] are right: having a model at the correct density is mandatory before any claim is made regarding the structure. Because of simple steric considerations, a structure made of boroxols will tend to have a much lower density than the corresponding structure made solely of BO_3 units. This has been used by both pros- and opponents of the boroxol model. On one hand, the density of the glass ($1.84\ \mathrm{g\,cm^{-3}}$) is surprisingly lower (by $\sim$40 %) than that of the closest crystal polymorph, B_2O_3-I ($2.54\ \mathrm{g\,cm^{-3}}$): this is indicative of a very different structure, compatible with a significative amount of rings in the glass. On the other hand, a structure topologically identical to B_2O_3-I but entirely made of boroxols should have a density reduced by $\frac{3}{8}$ (i.e. 62.5 %), that is even lower than the glass density. This value is derived by simple scaling arguments using the self-similarity between the BO_3 triangle and the B_3O_6 supertriangle (Fig. 14.1), first noted by Bell and Carnevale [58]: the replacement of all BO_3 by B_3O_6 units in a given 3D-structure, everything else being unchanged, increases the volume by a factor 2^3 but the number of atoms only by a factor of 3; the resulting density is thus lowered from the initial one by a factor $\frac{3}{8}$. That is to say, to match the glass density, a structure topologically equivalent to B_2O_3-I would seemingly incorporate a maximum amount of $f \sim 44\,\%$. We will further explore this question in Sect. 14.6. To conciliate simultaneously a large amount of boroxols and a high enough density, the possibility of interpenetrating networks, as commonly found in borate crystals, has been suggested [30]. For instance, the structure of caesium enneaborate has two interlocking twin networks from which a very good *hand-made* model of v-B_2O_3 can be made, as first pointed out by Krogh-Moe [30] and further used by Takada et al. [59, 60] and others [38, 61]. I will show however in Sect. 14.6 that the interpenetrating networks hypothesis is not a necessary condition, i.e. that single networks of boroxols can match the glass density.

It has been suggested [29, 62, 63] that Raman and NMR could be particularly sensitive to boroxol groups. Given its anomalous intensity, there has been suspicion that the Raman peak at 808 cm^{-1} is strongly affected by matrix-element enhancement effects since the corresponding peak in inelastic neutron scattering (INS) data, closer to the bare vibrational density of states, is much smaller [29]. However, more recent analyses of high-resolution INS spectra [64, 65] and the comparison of these data with those obtained from borate crystals for which f is known, fully support the view that $f > 67\,\%$ [65]. Teter claimed that the intensity of the the Raman peak at 808 cm^{-1} corresponds to $f = 10.7\,\%$, using a normalisation to the corresponding intensities from sodium metaborate solutions containing known fractions of metaborate rings [66]. However, this claim, which is at variance with the rest of the Raman litterature, is likely erroneous because of some of the assumptions made. In particular, this value is derived assuming that the Raman scattering for the metaborate ring peak at 745 cm^{-1} in aqueous solutions is the same than that of the boroxol rings at 808 cm^{-1} in v-B_2O_3. This is very unlikely given the different nature of the rings (the metaborate involves fourfold coordinated boron atoms) and given the different dielectric constant of the materials. A direct assessment of the Raman peak sensitivity to f will be provided in Sect. 14.4.

The very same peak (808 cm^{-1}) has been used by various authors to monitor the amount of rings in pressure- or temperature-induced transformations [67–75]. This has been done by assuming that the peak area (after appropriate normalisation) is proportional to the number of oxygen atoms in boroxols.[1] Note that these measurements do not provide absolute values of f but relative ones (to that used for the absolute scaling). However, when cross-checking was possible, very good agreement with values derived from NMR [73, 76] was obtained [74]. The application of either high pressure or high temperature conditions leads to a marked decrease of the Raman peak, reflecting a structural reorganisation from a boroxol-rich to a boroxol-poor network. In the pressure-induced transitions (at ambient temperature), the complete disappearance of the boroxols occur at ~11–14 GPa [71, 72].

The decrease of f with pressure, which is supported by other types of experiments [72, 73, 76–78] and simulations [78, 79], can be understood by simple steric arguments (Fig. 14.1): because of their poor packing efficiency, boroxols tend to be replaced with increasing density by more compacted units, first the BO_3 triangle

[1] Being related to a vibration of the oxygen atoms in the rings, the peak area is proportional to f_O, which is directly related to f, the number of boron atoms in the rings, by $f = 1.5 f_O$.

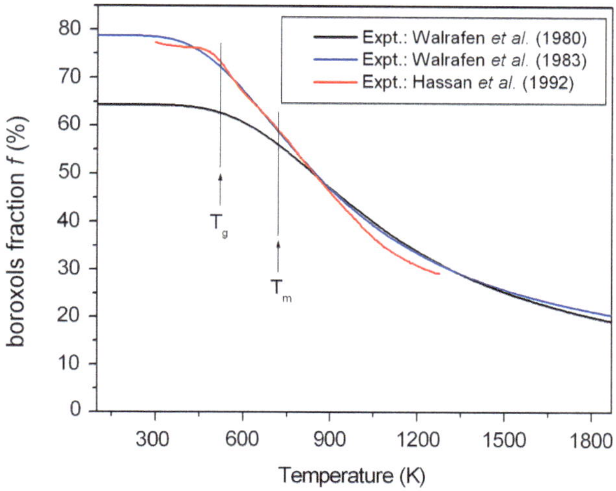

Fig. 14.2 Temperature dependence of the number of boron atoms in boroxol rings, f, as obtained from the Raman peak at 808 cm^{-1}. The *solid lines* represent graphical interpolations of experimental data from various authors [67–69] using different normalisation procedures. The experimental data of Walrafen et al. [67, 68] were modelled by these authors with the function $ln\{\frac{f(T)}{A-f(T)}\} = B/T + C$ where A, B, C are constants (A is the value of f at T_g). The values of $\{A, B, C\}$ are {0.644, 3237.66, −2.58893} in [67] and {0.7882, 2490.5, −2.3734} in [68]. The data from Hassan et al. which were digitised from Fig. 9 of [69] originally provide f_{at}, the number of atoms in boroxol rings and are represented here as f, using the relationship $f = \frac{7.5}{6} f_{at}$. The *vertical arrows* indicate the glass transition (T_g) and melting (T_m) temperatures

then the BO_4 tetrahedra. The BO_3 to BO_4 transformation mirrors that occurring in crystals from the B_2O_3-I to B_2O_3-II polymorphs.

The temperature dependence of the Raman peak (808 cm^{-1}) has been measured [67–69] from 77 up to 1867 K, i.e. from the glass to the liquid state (see Fig. 14.2). The rings concentration is essentially constant from room temperature up to the glass transition. Then at T_g (~470–540 K), boroxols start to open, and above T_g the number of rings steeply decreases [69]. Neutron diffraction measurements in the liquid state are consistent with this view [80].

Using a simple two-state model and some additional assumptions, the Raman measurements have been used to derive an energy for the (temperature-induced) boroxol-BO_3 transformation. Walrafen et al. reported values of 6.4 ± 0.4 kcal/(mol boroxol) [67] and 5.6 ± 1.0 kcal/(mol boroxol) [68], the former value being more robust according to the authors. Using other data from an older Raman investigation, a value of 11.8 kcal/(mol boroxol) was derived [30]. The change with temperature of various properties like density [81–83], heat capacity [84], viscosity [73, 83, 85–88] and surface tension [89] have also been claimed to be consistent with a marked structural change taking place between room temperature and 1300 K [30].

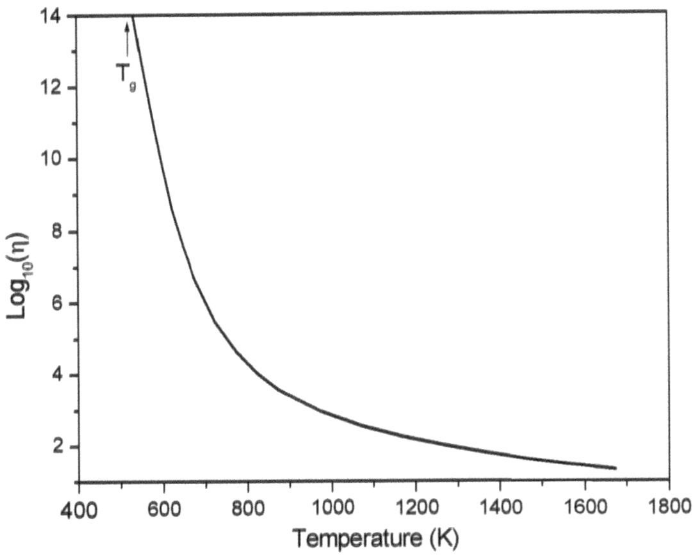

Fig. 14.3 Temperature dependence of the viscosity in B_2O_3. The line, taken from [85] is a graphical interpolation of actual data points [83]. Note that there exist several other sets of experimental data [73, 83, 85–88], for which absolute values can differ by up to an order of magnitude; the trends are however similar to the one illustrated here

In other words, boroxol rings essentially form in the temperature range 500–1300 K as the liquid is cooled. In this range, the viscosity increases by typically 12 orders of magnitude, as shown in Fig. 14.3. This is an important aspect to keep in mind as we are now going to review molecular simulations of B_2O_3.

14.3 Atomistic Simulations of Liquid and Vitreous B_2O_3

From the experimental point of view, an overwhelming body of evidences for boroxols has been built over the years and there is little, if any, room for controversy. However, a resurgence of criticisms appeared in the early eighties with the emergence of atomistic simulations.

Molecular dynamics (MD) simulations of v-B_2O_3 using empirical 2-body potentials were pioneered by Soules [90–92] and Amini et al. [93], followed later by others [62, 94–96]. In all cases, a random network of BO_3 triangles without any boroxols was obtained. The authors thus challenged the boroxol model though the simulations were at the most confronted either to neutron diffraction only [93] or to limited X-ray diffraction results [96]. At least in one case were the limitations in the simulations clearly pointed out [91], namely the short simulation times (10 ps) and the absence of directional or covalency forces in the interatomic potentials.

Incorporation of three-particle interactions in the MD simulations were then considered [59, 97–105] by various groups with however varying degrees of success. In most cases, these interactions were modelled in the form of a B–O–B and/or O–B–O bond bend term. In some cases, four-particle interactions in the form of a torsion angle were also added [101, 106, 107]. The fraction of boroxols obtained in these works vary from 0 to 53 %. Higher amounts were obtained when using biased sampling [108] or tweaked initial configurations [60].

No boroxols at all were found in the simulations of Verhoef and Den Hartog [98, 99]. Their simulated structures do not compare favourably with either the neutron or X-ray diffraction data. Calculated infra-red spectra were claimed to reproduce the most salient features (although the explicit comparison with experiments was not provided). Raman spectra were calculated using a bond polarisability model and showed a peak at $\sim$870 cm^{-1} arising from the BO_3 breathing mode, which the authors identify with the experimental peak at 808 cm^{-1}, although the calculated peak is a factor 10 too large [98]. Subsequently [100], the authors went as far as to reinterpret the Raman data in isotopically substituted samples [35] despite the poor agreement of their calculations with experiments (which were not even shown in [100]). One should further note that these potentials give poor results in the crystalline phases, in particular regarding the B–O–B bond angles [59].

In order to improve the agreement with the experimental static structure factor, Bermejo et al. [102, 103] proposed a refinement of the two- and three-body terms initially constructed by Verhoef and Den Hartog [99]. The addition of a four-body term was also considered [106]. However, the obtained data still show some drastic mismatch with the experiments, in particular for values of the scattering vector q below 5 $Å^{-1}$. The authors reported that a nearly perfect match could be obtained on either the low- or high-q range but never simultaneously on both ranges. With potentials limited to three-body interactions [102, 103], no boroxols were detected while a small amount, $f \equiv 3.6\,\%$, was obtained in the simulations involving the four-body term [106]. In our opinion, not only may these potentials suffer from intrinsic limitations due to e.g. their analytical form, but also may there be an even stronger bias in the strategy used for the equilibration. Indeed, the liquid was equilibrated with only two-body interactions down to a temperature as low as 900 K (i.e. quite close to the experimental glass transition temperature, 540 K, and very likely well below the numerical glass transition temperature corresponding to the quenching rate actually used in the simulations) and only afterwards were the higher-order interactions switched on for a limited time (100 ps). Within such a short time, and given the extremely slow dynamics at these temperatures (see Figs. 14.3 and 14.6), the system is unable to re-equilibrate and thus the rings are unlikely to form, even if their presence is favoured by the higher-order interactions.

Still with a potential of 3-body type, the first MD simulation for which the presence of boroxols was reported is actually that by Inoue et al. [97]. The predicted B–O–B bond angles distribution (peaked at 120°) is in much better agreement with the experimental value ($\sim$130°), derived indirectly from X-ray scattering [51] and from NMR [39], than in the previously mentioned MD [62, 91, 93, 99] ($\sim$150–160°). The obtained number of boroxols in the glass, $f = 23\,\%$, is however too small and the reason was attributed to the too high quenching rate (2.3×10^{14} K s^{-1}).

Later, Takada and co-workers [59, 101, 109] put the emphasis on constructing interatomic potentials from the crystalline phases which are transferable to the vitreous one. The two known polymorphs [110, 111] were used: B_2O_3-I, which is made of corner-shared BO_3 triangle units (i.e. three-coordinated boron atoms, $^{[3]}$B, and two-coordinated oxygens $^{[2]}$O) and B_2O_3-II, which involves BO_4 tetrahedral units (four-coordinated borons, $^{[4]}$B, and a mixing of both two- and three-coordinated oxygens, $^{[2]}$O and $^{[3]}$O). A series of potentials were parametrised using the energy surfaces from Hartree-Fock calculations [59, 101]. In addition to three-body (and in some cases four-body) terms, the introduction of coordination-dependent terms was proposed for the two-body potentials. More specifically, these latter terms depend on the oxygen coordination ($^{[2]}$O or $^{[3]}$O) [101]. Regarding the boroxols, significant amounts were obtained, from $f = 25$–$53\,\%$, depending mostly on the potential but also on the statistical ensemble (*NPT* or *NVT*) [101]. However, two serious problems arose: using *NPT* simulations, the final glass densities were significantly underestimated (from -10 to $-36\,\%$) while using *NVT* simulations at the glass densities led to the formation of unrealistic coordination defects, such as $^{[4]}$B or $^{[3]}$O, in small but sizeable amounts (1–6 %). Thus, the authors pointed out the difficulty to realise simultaneously a high proportion of boroxol rings and the experimental density.

Another significant contribution of Takada et al. [59, 60] is the *hand-made* construction of B_2O_3 crystalline structures with high boroxol contents, obtained from topological modifications of related compounds such as HBO_2-III or $Cs_2O-9B_2O_3$. The obtained crystals were shown to have structural and vibrational characteristics similar to those of the glass [59]. In the initial report [59], the models densities were still lower than that of the glass. However, additional polymorphs were subsequently generated with the same strategy [60] and among them, one with a density of 1.85 g cm^{-3} and $f = 75\,\%$, which are the typical values expected for the glass. The next section will show a detailed testing of this model using *first-principles* calculations. Still a bit later [108], Takada followed a different route to produce a glass model of B_2O_3 by using a hybrid MD/MC simulation scheme: the MC part is used to bias the acceptance of boroxol rings and the obtained structures are then relaxed within MD simulations using the potentials that were constructed in earlier works [59]. A genuinely amorphous structure of 1500 atoms could be obtained with $f = 74\,\%$ at the glass density. However, this model, although providing a very good match of the experimental structure factor, was found by the present author to be unstable when relaxed within *first-principles* calculations (unpublished work).

Cormack used a coordination-dependent potential supplemented by three-body terms [105, 112], slightly modified from Takada's potential. A value of $f \sim 15\,\%$ was obtained. A discrepancy with the experimental total radial distribution function in the region around 3.7 Å was noted and seen as an indication of a too small value of boroxol rings in the simulations. The origin was ascribed to finite-size effects (systems of 1010 atoms at the most were used) and possibly to the glass-forming process [105].

A different route to incorporate the many-body, non-additive interactions comes from polarisable models [66, 78, 113–115]. A first attempt in this direction was made by Teter [66] who used a simple representation of the oxygen polarisation by assigning four auxiliary charges in tetrahedral symmetry to each oxygen ion. The parameters were calibrated from ab-initio calculations using the LDA exchange-correlation functional. Within this crude representation, the fraction of boroxols was found to increase from 1 to 12 % as the magnitude of the auxilliary charges was increased from zero. However, it should be noted that the resulting glass incorporates coordination defects which do not exist in the experimental glass. This is likely a reflection of the flaws of the LDA functional which uncorrectly predicts the high pressure crystal (B_2O_3-II) to be more stable than B_2O_3-I (as visible in Fig. 14.3 of [66]).To the best of our knowledge, this is the only MC study of B_2O_3 (apart from the hybrid MD/MC simulation of Takada [108]). Note that the work of Teter [66] is sometimes referred as being of ab-initio type [61, 63], although this is actually not the case: this is an empirical model whose parameters have been calibrated from ab-initio calculations.

A more realistic representation of polarisability was then provided by the group of Maranas et al. [113–115]: this model incorporates induced dipoles arising from charges and from other induced dipoles on oxygen atoms. The polarisation effects of oxygen atoms were found to be important: in particular, boron atoms within boroxol rings were found to have a slightly lower energy than those outside of the rings and this effect originates from the polarisation part of the potential [113]. This is a crucial findings confirmed very recently by Salanne et al. using other types of polarisable force-fields [78, 116]. Regarding the glass preparation, Maranas et al. used an original yet fancy protocol aimed at simulating the formation of the B_2O_3 network from the dehydration of boric acid, H_3BO_3: forces were imposed so as to act as semipermeable membranes that selectively pass or block individual atomic species, thus creating, according to the authors, a *chemical reactor containment vessel* [113]. On one hand, this is an interesting procedure in the sense that it gives some control over the network formation in particular through the temperature of dehydration and equilibration. On the other hand, the dehydration of boric oxide is the route usually followed experimentally to obtain the crystalline phase B_2O_3-I, rather than the glass. In any case, the formation of boroxols was observed and interestingly, the amount of rings was found to depend upon the temperature during the network formation. More specifically, 2 procedures were used in which after a high temperature run (used for dehydration) and prior to the final cooling, the sample was held at either 2000 or 1800 K for about 100 ps (i.e. until a network of BO_3 units had formed). The fraction of rings varies from about 8–33 % depending on whether the sample is quenched from 2000 K or from 1800 K. The authors pointed out that the latter value is consistent with experimental estimates (40 % at 1000 K)[2] [69]. I however note that the values claimed are subjected to some ambiguity since some of the boron atoms are not three-coordinated as they should because of the method used in the simulations to form the glassy network [113]: the 8 and 33 % values,

[2]The incorrect 32 % value quoted in [113] actually corresponds to f_{at} in [69] and thus to $f = 40\,\%$.

obtained when considering the $^{[3]}B$ atoms only, become respectively 3.75 and 15 % if taking into account all the boron atoms in the simulated samples. This modest increase may not be statistically significant given the small total number of boron atoms (81). However, the authors further show that as the temperature is decreased below 2000 K, ring structures become energetically favoured (by 1.6 % at ~1800 K). In summary, the authors conclude that boroxols form at temperatures less than about 2000 K. Using a higher preparation temperature prevents the formation of rings; as soon as the network is formed, ring formation won't occur even if the system is subsequently cooled into the favourable range, perhaps because of an energy barrier or kinetic limitations for the transformation within the network. They also note that the temperature range over which the glass transition is observed in their simulations, 1500–3000 K, is significantly displaced from the corresponding experimental range (500–1200 K). This was expected in light of the very short simulation times [113]. I strongly adhere to these conclusions which are fully supported by the *first-principles* simulations detailed in the next section. One can regret however that the performances of the polarisable potentials of Marana et al., regarding e.g. the structure [113] or the dynamics [114], were never assessed by any comparison with experimental data.

The potentials developed recently by Kieffer et al. [61, 117, 118] combine coordination-dependent and charge redistribution features in addition to explicit three-body terms. Both two- and three-body interactions depend on the effective number of nearest neighbours of an atom. A charge transfer term controls the extent of charge polarisation in a covalent bond, as well as the amount of charge transferred between atoms upon rupture or formation of such a bond. Conditional three-body terms, i.e. which depend on the degree of covalency in atomic interactions, constrain both B–O–B and O–B–O angles. The potential parameters were parametrised so as to match structural and vibrational data for the liquid and crystal phases. The authors used the methodology of Takada to produce an initial glass structure having $f = 75\,\%$ at the glass density [60]. This structure was equilibrated at 2500 K, under which conditions the boroxols gradually dissolved. Thus, by heat treating the system for various periods of time (up to a nanosecond), liquid structures having $f = 75, 63, 50$ and 10 % were generated and quenched (at a rate of 2.5×10^{12} K s^{-1}). No additional boroxols were generated upon cooling, and here again this was attributed to the fast quenching rates [61]. The resulting densities vary from 1.75 to 1.81 g cm^{-3} in good agreement with the experimental values. The static structure factors for both the $f = 10$ and 63 % glasses were shown to reproduce very well the experimental data [61, 119]. Another interesting findings of this study is the possibility to generate a new polymorph, B_2O_3-0, by applying either temperature or negative pressure to B_2O_3-I. There are no boroxol rings in B_2O_3-0 but the presence of larger rings, which can be either *puffed* or *puckered* were related to the thermomechanical anomalies of B_2O_3 (i.e. the existence of a minimum in the mechanical modulus in the molten state) [61, 119].

Using a potential with three- and four-body terms, Kashchieva et al. [107] intended to explore the effects of cooling rates and of system sizes. They reported results for 10 systems from 100 to 2000 atoms and used, for each of them, two cooling rates differing by 2 orders of magnitude, respectively 3.3×10^{12} and 3.3×10^{10} K ps^{-1}, between 1300 and 300 K. With the *fastest* cooling rate, a low value $f \sim 10$–15 %

was obtained for all glasses, with no apparent size-dependency. With the *slowest* cooling rate, a maximum value of $f \sim 33\,\%$ was obtained for the 400 atoms system and an almost linear decrease down to $f \sim 17\,\%$ for the larger systems sizes. In my opinion, this latter trend is due to an incomplete relaxation of the larger systems: indeed a strictly equivalent simulation time (800 ps at 1300 K + 30,000 ps from 1300 to 300 K) was allowed for each system irrespectively of their sizes, whereas it should be larger for bigger systems. This is supported by the results obtained for the density: with the *fastest* cooling rate, a much too low density ($\sim$1.64 g cm^{-3}) is obtained with no size-dependency beyond statistical errors whereas with the *slowest* cooling rate, one observes an (almost linear) decrease from reasonable values ($\sim 1.77\ g\ cm^{-3}$) for the smallest systems down to much poorer values ($\sim 1.68\ g\ cm^{-3}$) for the larger systems. Thus, this study is rather an additional illustration of the importance of the sampling given the slow rate of appearance of boroxols at low temperatures: a too fast quench (on a small system) has the same detrimental consequences than has an improper equilibration (of a big sample). In my opinion, finite-size effects per se are likely not a crucial issue at least for the quantity of interest here, f. In their structural and thermomechanical study, Kieffer et al. [61] noted no size-dependency in excess of statistical errors for systems of 640 and 2560 atoms [61].

First-principles molecular dynamics (FPMD) simulations, based on the density functional theory (DFT), were recently applied to B_2O_3 by different groups [37, 38, 120–124]. The obtained fraction of rings in the glass, quenched from the melt, varies from about 6 [121] to 22 % [38]. The first report, that of Umari and Pasquarello [37], actually revived the controversy [63]. While the configuration obtained from the liquid quench contained only 9 % of boroxols, strong evidences were provided from Raman and ^{11}B NMR analysis that the *true* value should be $\sim$75 % (the Raman peak at 808 cm^{-1} being underestimated and the population ratio being incorrect in the NMR spectra) [37]. However, this extrapolation has been immediately criticized by SB [63]: these authors argued that (a) it is difficult to estimate f from calculated Raman spectra (b) the Raman peak at 808 cm^{-1} is actually consistent with $f = 11\,\%$ (Teter's argument [66]) (c) other techniques such as NMR are not directly sensitive to f (d) they have shown, using RMC simulations, that it is not possible to produce boroxol-rich models which would simultaneously reproduce the experimental density and the diffraction data. I have already mentioned that the argument (d) is actually denied by different works which came either before [55] or after [38, 61, 125] its formulation. Regarding the points (a), (b) and (c), it is a fact that some underlying assumptions (such as the proportionality of the Raman signal with f) behind the indirect determination [37] were arguable. However, we subsequently provided a boroxol-rich model able to reproduce all the experimental information within a *first-principles* scheme [38]: this work answers all the previous criticisms made by SB [63] and is detailed in the next section.

Finally, Salanne et al. recently devised several types of polarisable models [78, 116] which parameters were calibrated using previous *first-principles* calculations [38]. These models were shown to provide a very good account of experimental data (structure factors, density) from ambient to high-pressure conditions. Starting

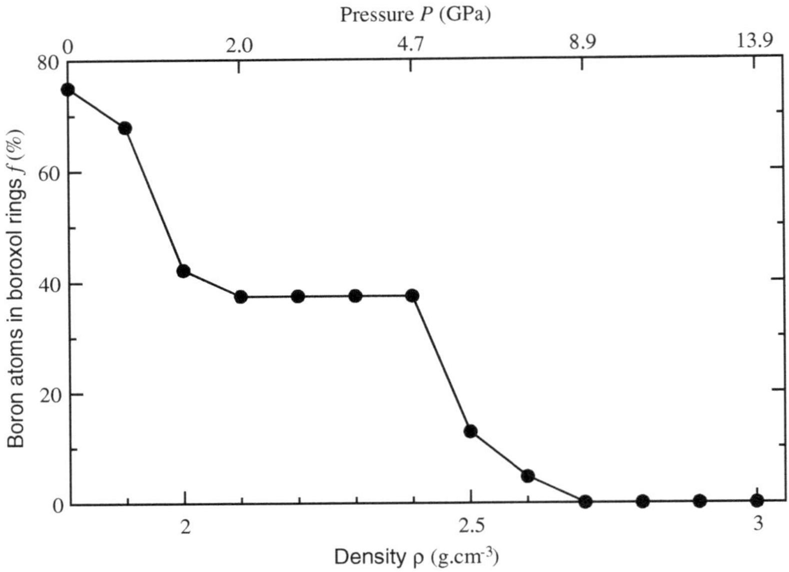

Fig. 14.4 The pressure (or density) dependence of the fraction of boron atoms within boroxol rings f (%) as obtained from MD simulations using the Aspherical Ion Model (AIM) from Salanne et al. [78] and the boroxol-rich model (f = 75 %) at ambient pressure [38]

from a boroxol-rich model [38], the progressive disappearance of the boroxol rings with pressure could be evidenced [78], see Fig. 14.4.

Other modelling studies not relying on interatomic potentials include the statistical model of agglomeration of Micoulaut et al. [126, 127]. With a boroxol formation energy of 5.3 ± 0.7 kcal/mol as a fitted parameter, a boroxols fraction of 83 % was predicted.

Thus, although the MD results in the literature are scattered, several preliminary conclusions can already be drawn. The inclusion of many-body interactions is mandatory in order to generate boroxol rings. The origin of the rings stabilisation is to be found in the oxygen polarisation. However, this does not mean that the potential should explicitly be of a polarisable type since polarisation effects may be accounted implicitly by, e.g. angular three-body terms or by an additional rings stabilisation term.

The incorporation of many-body effects is of course a necessary but not a sufficient condition. As important as the *goodness* of the potential, is the glass formation process. This has been stressed in the majority of the studies reviewed above. The problem comes from the fact that in B_2O_3, the marked structural change accompanying the increase of f as the liquid is cooled is occurring just a few hundreds of Kelvin prior to the glass transition (Fig. 14.2): most of this change occurs in a temperature range where the viscosity is dramatically increasing (Fig. 14.3). Thus, following numerically this change is extremely challenging for atomistic simulations since this requires an increase in the simulation time by several orders of magnitude.

Typically, at 2000 K, the viscosity is of the order of 1 Pa.s, a value comparable to that of oil, and from our own experience (see Fig. 14.6) it requires simulations times of at least 100 ps to escape from the caging regime (and thus to enter in the diffusive regime). Even with a nanosecond timescale, the lowest temperature that can be explored ergodically is ~1800 K. At this temperature, f is typically 20 % or less (Fig. 14.2) whereas to get a chance to observe rings fractions greater than 50 % would require an extensive sampling at about 800 K. However, should the simulation time scales proportionally to the viscosity, an increase of at least 3 orders of magnitude would be needed from 1800 to 800 K. Note that this is a minimal estimate, only accounting for the time needed to reach the diffusive regime. Longer simulation times may be needed for a full structural relaxation. Indeed, the rings stabilisation energy being quite small (of the order of 5–10 kcal/mol), the system may stay trapped for long times in high energy configurations before a new basin is found (corresponding here to a network re-organisation to a more boroxol-rich configuration).

The time scale of all the simulations reviewed above is typically a 1 ns or less. With only one exception [107], the quenching rates are in the range 10^{12}–10^{14} K ps^{-1}. Such time scales give rise to high fictive temperatures, typically above 1500 K which are thus considerably displaced from the corresponding experimental temperatures. It is therefore no surprises that the values reported for f rarely exceed 30 %, the value expected for a structural arrest at 1500 K (Fig. 14.2). In my opinion, the values reported from the various MD studies are better understood when taking account the time spent in the *mildly high* temperature range, i.e. in between the onset of rings formation ($\lesssim$2000 K) and the lowest temperature that could be sampled ergodically within the afforded time ($\gtrsim$1500 K) before the structural arrest occurs. This is further supported by reporting the f values obtained from the various MD studies *not* at the final temperature (usually the room temperature) but at the lowest temperature sampled prior to the final quench (thus assuming that the quench is essentially instantaneous). As seen in Fig. 14.5, there is an overall good match between the simulated and experimental values. I note however that there are cases where such a report is difficult to made and thus subjected to some ambiguities. Further, a proper quantitative comparison with the experiments would require some corrections to be accounted for, in particular for the density mismatch or the residual pressure in the simulated models [128]. I thus intend this comparison to be qualitative. Notwithstanding, Fig. 14.5 shows that some force-fields in the literature may be good enough. Contrary to the prevailing belief, some potentials might even tend to *over-estimate* the fraction of boroxols. Now that classical force-fields of quality comparable to *first-principles* ones are available (e.g. [78]), I think that the efforts should be put on longer sampling and/or better glass processing. It thus would be very desirable to use methods allowing for accelerated dynamics, such as e.g. metadynamics (using the rings fraction as an order parameter) [129, 130]. Alternatives to the standard *quench-from-the-melt* procedure such as *bond switching* [131] methods could also be of high interest.

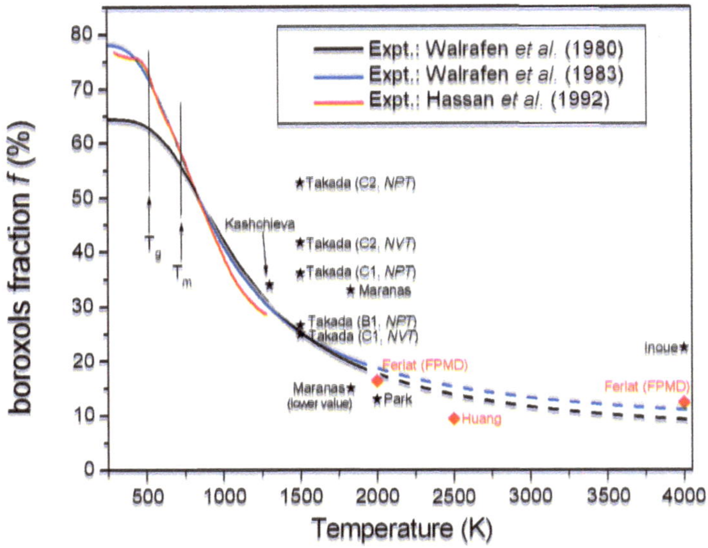

Fig. 14.5 Values for the fraction of boroxol rings, f, reported from various MD simulations studies [38, 61, 97, 101, 112, 113], indicated by the first author's name. The *solid lines* represent graphical interpolations of experimental data [67, 68] while the *dotted lines* are extrapolations of these data to temperatures higher than those actually measured, using the formulas given in [67, 68] (see also the caption of Fig. 14.2). Lozenge symbols indicate values from simulations obtained in the liquid phase which are therefore directly comparable to the experiments. Values shown by *stars* were initially reported in the glass at 300 K but are represented here at the lowest temperature prior to the final quench (see text). Maranas stands for the value quoted in [38] while the lower value is that obtained when all boron atoms are considered (see text)

14.4 Assessing the Fraction of Boroxol Rings from First-Principles

As seen in the previous section doubts about the boroxol-rich model have been expressed from time to time in the literature, the latest being that of SB [63]. In order to provide clear-cut answers we recently assessed the performances of models with varying amount of boroxol rings, within a *first-principles scheme* [38].

Unless stated otherwise, the calculations were carried out using the SIESTA code [132]. The electronic structure is described within a generalized gradient approximation to density functional theory using the PBE functional [133]. We used norm-conserving pseudopotentials [134] and DZP basis sets. The real-space grid was defined by a 280 Ry cutoff. The basis quality has been extensively tested by comparing the results with those from fully converged plane-wave basis sets (mostly using the CPMD code [135]): the systems used for benchmarking included known borate crystals [136] as well as pure B_2O_3 liquid. The time step in the MD simulations is 1 fs.

Before discussing the glass structure, results in the liquid phase are briefly reported. Several conditions of temperatures and densities were explored in the *NVT*

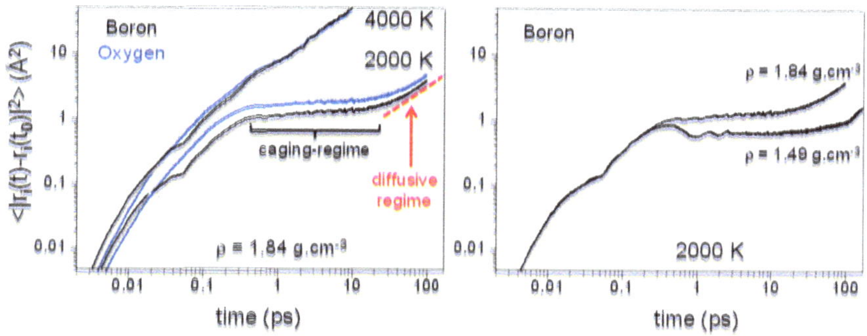

Fig. 14.6 Mean-squared displacement (MSD) obtained from the *first-principles* MD simulations in liquid B_2O_3. *Left* oxygen and boron MSD at temperatures of 4000 and 2000 K for a density of 1.84 $g\,cm^{-3}$ (corresponding to the glass density at 300 K). *Right* boron MSD at 2000 K for densities of 1.84 and 1.49 $g\,cm^{-3}$. The latter density is that of the liquid at ~2000 K

ensemble, using system-sizes of mostly 100 atoms (up to 320 atoms). Let's first consider liquid simulations at the glass density (1.84 $g\,cm^{-3}$), as it is quite common in the literature to use *NVT* simulations at the final target density.

Figure 14.6 shows the mean-squared displacement of the atoms at various conditions. As seen in the left panel of Fig. 14.6, a severe slowing down of the dynamics is occurring as the temperature is lowered from 4000 to 2000 K: whereas it takes about 1 ps to reach the diffusive regime at the higher temperature, a much longer time of the order of 100 ps is needed at 2000 K to escape from the caging regime. The diffusion coefficient, obtained from the slope in the diffusive regime, is reduced by more than 2 orders of magnitude in this temperature range. As a measure of the equilibration time, let's define, following [137], τ as the time at which $\sqrt{\langle r_O(t)^2 \rangle} = 5.6$ Å, i.e. the average time it requires for an O ion to move twice its diameter of 2.8 Å: τ is ~5.8 ps at 4000 K and ~650 ps at 2000 K. This sets up a typical time scale for the simulations: ideally one would like to simulate over several τ for a reliable (i.e. ergodic) sampling of the liquid. It is seen here that the value of τ obtained at 2000 K is already quite demanding for *first-principles* MD simulations within present-day's computational resources. Even for a classical force-field, it seems rather challenging, at least in the near future, to sample temperature conditions lower than, say 1500–1800 K, given the exponential increase of the viscosity with decreasing temperature (Fig. 14.3).

In any case, the instantaneous evolution of the boroxols fraction, f, along a liquid simulation of 300 ps at 2000 K (and at the glass density, 1.84 $g\,cm^{-3}$) is shown Fig. 14.7. Significant fluctuations are observed between 0 and 30 %, the average being $\langle f \rangle \sim 9\,\%$. Several quenches to 300 K were branched from the simulation at 2000 K, using quenching rates in the range 10^{13}–10^{14} $K\,ps^{-1}$. The value of f in the glassy samples shows dependency upon the starting liquid configuration; however no (or marginal) change with temperature was observed (beyond that expected from the liquid trajectory, Fig. 14.7). This reflects that the system is unable during the

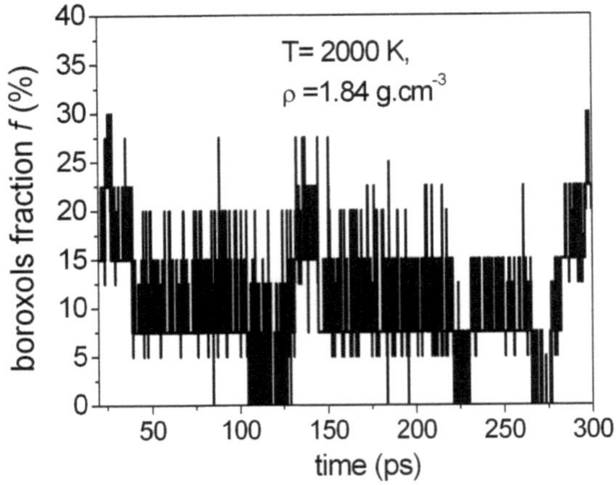

Fig. 14.7 Instantaneous evolution of the boroxols fraction f, in the liquid at 2000 K and $\rho = 1.84\ \mathrm{g\,cm^{-3}}$. The density is that of the ambient glass

quench to escape from the liquid initial *inherent structure*. In the following, one of the obtained glassy samples, containing $f = 22.5\,\%$, is used for a more detailed study: it is hereafter refered as the boroxol-poor (BP) model.

A final note about the liquid is made regarding the statistical ensemble. Experimentally the density of B_2O_3 increases from ~1.5 to 1.84 $\mathrm{g\,cm^{-3}}$ as the temperature is decreased from 2000 to 300 K and this variation should be accounted for during the numerical quench. Thus in principle the *NPT* ensemble in which the system's volume adapts itself to the target temperature and pressure should be favoured. However, given the small systems sizes affordable within FPMD simulations, the pressure is hardly well estimated and in practice this results in poorly driven volumes. Thus, FPMD simulations are commonly realised in the *NVT* ensemble all along the quench using the glass density. However, this induces a bias since the liquid prior to the quench is then simulated at a too high density, resulting in a net residual pressure which, in B_2O_3, is of the order of 1 GPa.[3] Now it should be reminded that high pressure conditions are unfavourable to the presence of boroxols (Fig. 14.4) [71, 72, 74, 75, 77, 78]. To check the magnitude of this bias, simulations at 2000 K at the corresponding liquid density (~1.5 $\mathrm{g\,cm^{-3}}$) were carried out. The average fraction of boroxols $\langle f \rangle$ increases from ~9 % at the glass density to ~22 % at the liquid density (see Fig. 14.27 in Sect. 14.7). This variation is slightly above the error bars (±7 %, as defined by the root mean square of the distributions of f): this shows the importance of being at the *correct* equilibrium density (or at zero-pressure conditions) [128].

[3] Another source of errors in DFT-based calculations comes from the exchange-correlation functional used which is an approximation of the unknown exact one. Gradient-generalised approximation (GGA) based functionals, such as PBE [133] used here, tend to overestimate the equilibrium volume and thus lead to positive residual pressures in simulations at the experimental density. This problem is significantly reduced in some recently proposed functionals by the incorporation of van der Waals contributions. However, no systematic nor significant variation of $\langle f \rangle$ (beyond the statistical error bars, ±7 %) were observed in the liquid phase using the PBE-D2 [138] functional.

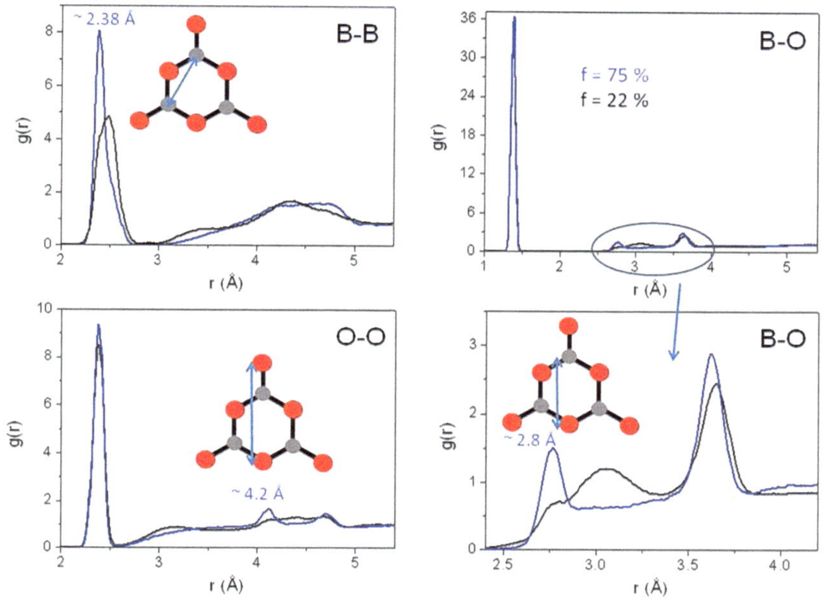

Fig. 14.8 Boron-boron, boron-oxygen and oxygen-oxygen partial radial distribution functions obtained at 300 K for the boroxol-poor (BP, $f = 22\,\%$) and boroxol-rich (BR, $f = 75\,\%$) models. Interatomic distances more pronounced in the BR model are highlighted

Note that the dynamics are even more sluggish at the *actual* liquid density (Fig. 14.6, right panel): the relaxation time τ previously defined is estimated to be $\sim$2300 ps. In other words the viscosity decreases with increasing density: this behaviour, albeit *anomalous* is well known in silica and water and has been recently discovered in B_2O_3 [73, 123, 124, 139].

In summary, the results obtained from the FPMD simulations in the liquid phase essentially confirm the findings from others authors which were summarised in the previous section. At this point, it should be rather clear that glassy structures obtained from conventional numerical quench are doomed to underestimate f given the affordable simulation times. We thus sought for a glassy model produced by an alternative method [60] which has both $f = 75\,\%$ and the correct glass density. This model was originally constructed by Takada et al. [60] from topological modifications of the Cs_2O–$9B_2O_3$ crystal structure so as to remove the ceasium atoms and delete/create some bonds, this low-alkali compound being the closest known analogue to the v-B_2O_3 structure. It has been used in other MD studies [60, 61, 78, 79] using empirical force-fields and subjected in our work [38] to *first-principles* simulations. FPMD simulations were carried out at 300 K on systems containing 80 and 320 atoms for durations of 20 ps. No size-effects were observed and the obtained structures are hereafter referred as the boroxol-rich (BR) model.

Figure 14.8 shows the partial radial distribution functions obtained for the BP and BR glassy models. Note that the B–O and O–O first peaks (at $r_{BO} \sim 1.37$

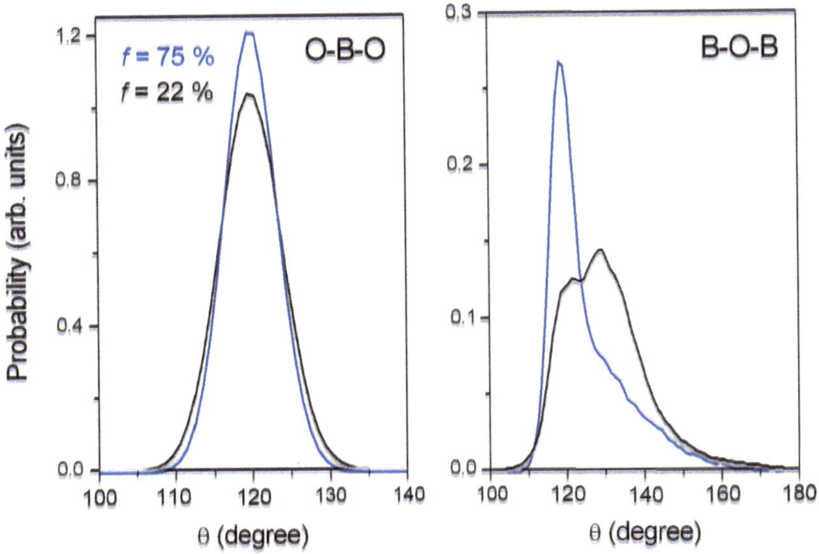

Fig. 14.9 O–B–O and B–O–B bond angular distributions obtained at 300 K for the boroxol-poor ($f = 22\,\%$) and boroxol-rich ($f = 75\,\%$) models

and $r_{OO} \sim 2.37$ Å, respectively) are almost identical in both models. The main differences lie in the shape and the position of the B–B first peak, which is sharper and shifted to shorter distances, by about 0.1 Å, in the case of the boroxol-rich model. Other less marked differences include more prominent peaks in the BR case, at ∼2.75 Å in the B–O partial and at ∼4.1 Å in the O–O partial. These differences reflect the greater coplanarity of the BO_3 triangles in the BR model. The former peak arises from oxygen second neighbours (of boron atoms) within a same ring (at a position $\sim 2 \cdot r_{BO}$) [32]. However, as mentioned by SB [54], the peak at ∼3.6 Å in the B–O partial is only slightly different for BR and BP models and thus is a poor signature of the boroxol rings fraction contrary to previous claims [32].

The O–B–O and B–O–B bond angular distributions are shown in Fig. 14.9. The former one is a single symmetric peak centered at ∼120°and is essentially similar for both models, albeit slightly narrower in the BR case. The B–O–B distribution, which is related to the connection between the BO_3 units, shows a much wider spread from ∼110 to 170° and differs for each model. In the BR model, a sharp peak at ∼120° and a shoulder-like contribution from ∼130° are visible. The former contribution arises from triplets within the boroxol rings while the shoulder corresponds to oxygens bridging different units. In comparison, the BP model shows a redistribution from the 120 to the 130° region as a result of the smaller amount of rings.

To gain deeper insights, a structural analysis was carried out according to the different atomic sites. There are two possible sites for boron atoms: *inside* or *outside* a boroxol ring. For oxygens, up to 4 different sites, labelled *A*, *B*, *C*, *D* in [42] and in Fig. 14.10 can be defined. The results, which are similar in both BP and BR

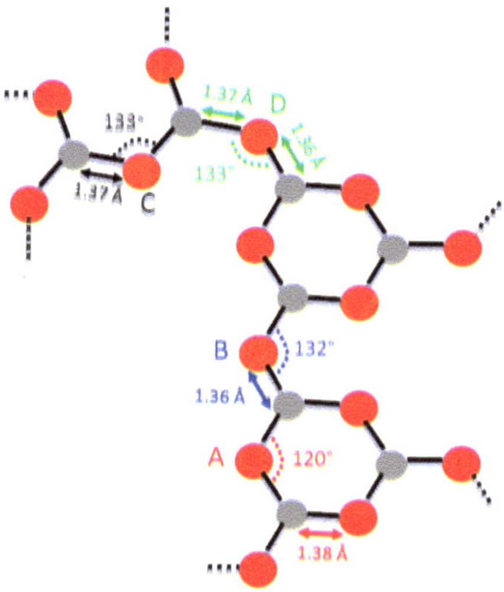

Fig. 14.10 A schematic 2D representation of the different atomic sites encountered in the simulations of v-B_2O_3. The structural values reported are those obtained with the PBE functional. Boron atoms are shown as small (*gray*) balls while oxygens are big (*red*) balls. Up to 4 different types of oxygen environments (A, B, C, D) can be identified

models, are summarised schematically in Fig. 14.10. A non-ring BO_3 unit has three slightly nonequivalent B–O bonds of average length ~1.37 Å and three O–B–O angles of ~120°. A boroxol unit has 6 larger *internal* B–O bonds, ~1.38 Å, and 3 shorter *external* bonds, ~1.36 Å. B–O–B angles *internal* to the rings equal ~120° while these angles for oxygens which are bridging different units (B, C, D) tend to be much larger, ~133°. Although the absolute values of the reported angles and interatomic distances slightly vary with the exchange-correlation functional used in the simulations, the mentioned trends were found to be independent of this choice.

From the obtained simulations, a vast number of experimental observables can be computed. Among them, the static structure factor is an obvious first basic test of the models. It is computed here for both models from the density fluctuations in Fourier space and compared to available neutron diffraction data in Fig. 14.11. As can be seen, both models reproduce very well the experimental data in the whole q range available, including the so-called first-sharp diffraction peak (FSDP) at ~1.6 Å^{-1}. A very slight overstructuration and a frequency misfit are visible at high q for both models which result from the well-known tendencies of GGA functionals such as the one used here (PBE) to overestimate the bond lengths (by typically 0.01 Å). However, given the absence of free-parameters in the simulations, the agreement is quite good. From this sole comparison, no superiority of either model can be evidenced, confirming that the total structure factor is not a good probe of f, as already mentioned in other works [37, 61]. The same conclusions hold when considering the x-ray structure factors, not shown here.

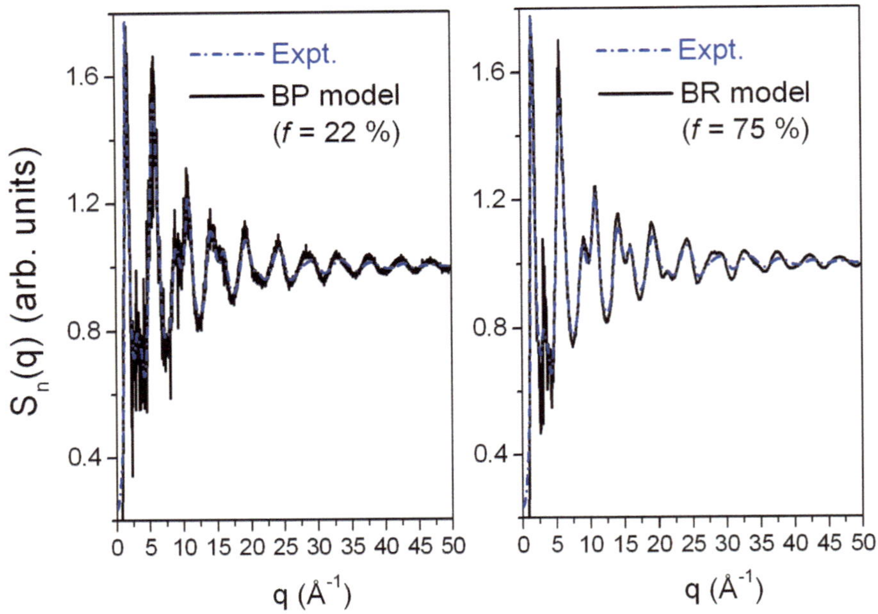

Fig. 14.11 Static neutron structure factors calculated for the boroxol-poor ($f = 22\,\%$) and boroxol-rich ($f = 75\,\%$) models compared to the experimental data [32]

This is in contrast with the results of SB [54]: boroxol-rich models generated from RMC simulations were shown to give a slightly but significantly less good agreement than boroxol poor models as a result of some structural artifacts in the former ones. These artifacts, mainly visible as an unrealistic peak at ~50° in the O–B–O and dihedral angle distributions, were due to significant distortions of the rings from planarity in boroxol-rich models. According to SB [54], these distortions were the result of the imposed constraints in the model generation procedure to simultaneously reproduce the diffraction data and the glass density. We stress however that none of these artifacts were observed in our BR model (although it was produced at the correct glass density). Thus, these artifacts are certainly not specific of BR models but instead were due to the method used in [54].

In any case, our work clearly invalidated the claim that it is not possible to simultaneously reproduce the diffraction data and the density for high values of f [54, 63]. This has been further confirmed in a recent investigation by Soper [56, 125], using the empirical potential structure refinement (EPSR) method, a variant of RMC which makes use of potentials in the fitting procedure. The work of Soper [56, 125] complements ours by providing larger models (2000 atoms), which were not made from any underlying crystalline network. The desired amounts of boroxols were introduced ad-hoc and, again, a very good representation of the diffraction data (including simultaneously X-ray and neutron data) could be obtained from a large range ($f = 5$ to $80\,\%$) of structural models [56, 125]. Although still not perfect, the agreement with

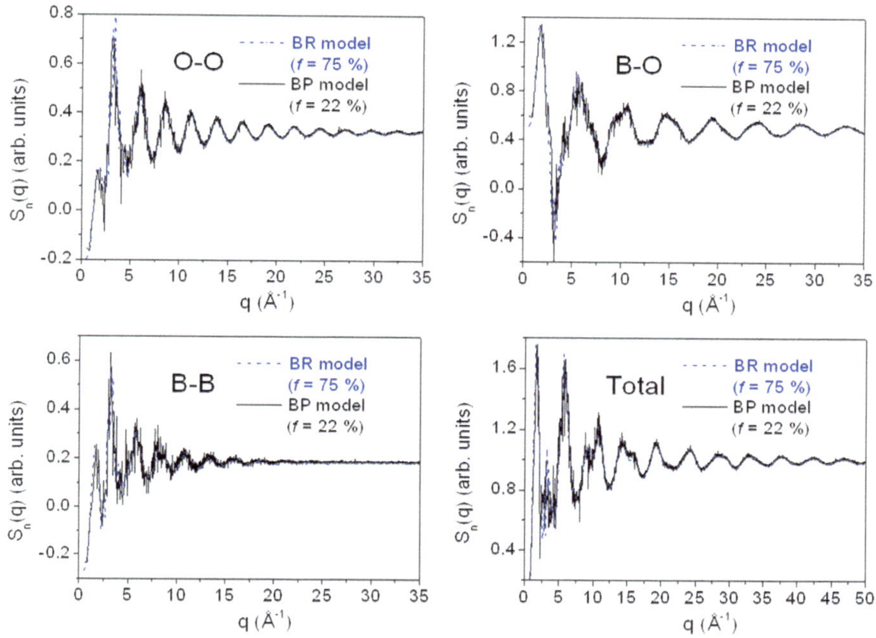

Fig. 14.12 Static neutron partial structure factors calculated for the boroxol-poor ($f = 22\,\%$) and boroxol-rich ($f = 75\,\%$) models: O–O, B–O, B–B. The total signals are also shown

experiments was even better than those shown in Fig. 14.11 since EPSR (as RMC) is by construction designed to provide the best possible fits, thus removing the systematic discrepancies due to incorrect interatomic distances as typically obtained in a fixed force-field simulation. However, the main point here is that a similar level of agreement could be obtained for any type of model, precluding the possibility to distinguish them from diffraction information alone.

Unfortunately, in the case of B_2O_3 only the (neutron and X-ray) total structure factors have been measured and no partials are available. We provide however in Fig. 14.12 a comparison of the neutron weighted Faber-Ziman partials obtained from our models. It is seen that the differences are very tiny and essentially limited to a small frequency shift in the B–B partial. Clearly, a quantitative assessment of f is unlikely to be made from these measurements.

NMR has been extensively used in borates (see [140] for a review) and has played a key role for assessing the existence and proportions of different sites in both crystals and glasses [76, 140–143]. In v-B_2O_3 in particular, the NMR various techniques (among which $^{[11]}$B MAS, $^{[11]}$B DAS and $^{[17]}$O DOR) have been essential tools for the quantitative determination of the fraction of atomic sites in boroxol rings [38, 39, 41–44, 47, 144–146]. Because of their locally different environments, the B and O atoms which are constituents of the rings and those which comprise the non-ring units give rise to different contributions [48, 49, 147] which are, in most cases, readily observable experimentally.

We therefore tested our BP and BR models by computing $^{[11]}$B DAS and $^{[17]}$O MAS spectra, shown in Figs. 14.13 and 14.14 respectively. The NMR chemical shifts and electric field gradients were calculated within the Gauge Included Projector Augmented Wave formalism as in [148] using the PARATEC code. DAS and MAS NMR spectra of the central transition ($-\frac{1}{2} \leftrightarrow \frac{1}{2}$) were simulated taking into account both the quadrupolar and chemical shift interactions as described in [149] and including spinning sidebands.

For both $^{[11]}$B DAS and $^{[17]}$O MAS spectra, it is clear that the BR model provides a significantly better representation of the experimental data [41, 42] than the BP one. This is most evident in the $^{[11]}$B DAS comparison and this better agreement is not fortuitous, as revealed by a closer inspection of the contributions from nuclei inside or outside the boroxol rings. These contributions give rise to significantly different NMR signatures as a result of the NMR sensitivity to the O–B–O and B–O–B angles. While the O–B–O angle distribution is essentially the same for in- and out-of rings sites, the B–O–B one differs: average values of ~120 and ~130° were obtained respectively as shown before, Fig. 14.10. Thus, the NMR observables provide stringent tests of the models' bond angle distributions and an indirect but quantitative way to access the value of f. Although this sensitivity had been exploited before, the present comparison provides the first direct confirmation of the adequacy of boroxol-rich models and they confirm the earlier claims made from the indirect analysis of Umari and Pasquarello [37].

Once could also think of using local spectroscopies such as the X-ray absorption spectroscopy (XAS) or the Inelastic X-ray Spectroscopy (IXS). These techniques are sensitive to the very local environment of the excited atom and, close to the absorption edge, to the electronic structure. We carried out calculations at both the boron and oxygen K-edges using the XSPECTRA module of the QUANTUM ESPRESSO package [150]. As in germanates [151], we found that the spectra showed signatures related to the triplet angles involving the absorbing atom: small differences between the BP and BR models were obtained at the O K-edge, as a result of the differing B–O–B angles distributions while essentially no differences were obtained at the B K-edge, as a result of the similar O–B–O angles distribution. In any case, the differences were found to be much smaller than the experimental resolution of presently available data [152, 153].

We now turn to the vibrational properties. The phonons were computed at the Γ point only from the finite-displacement method using the CPMD program [135]. The obtained vibrational density of states, v-DOS, for both models are shown Fig. 14.15. Despite the relatively poor statistics due to the limited number of atoms, several differences are apparent. In particular, the BR model shows a much marked contribution at ~800 cm^{-1} as could be anticipated. The analysis of the modes from our calculations reveals that the contributions at this energy result solely from the breathing mode of the oxygens inside the boroxol rings, as already well established by previous works [35, 37]. Other noticeable differences are visible in the regions 100–200, 400–500, 650–800, 1100–1300 cm^{-1}.

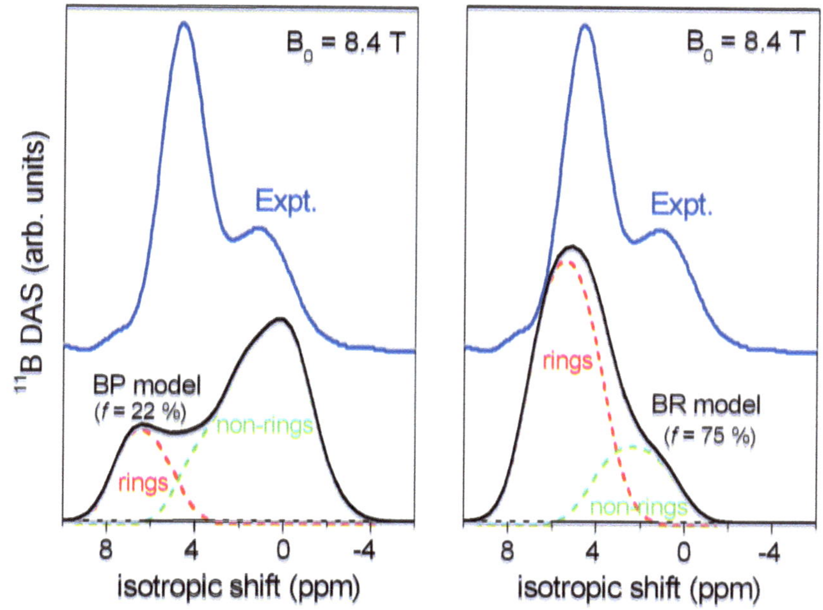

Fig. 14.13 [11]B DAS (isotropic projection) NMR spectra obtained for the boroxol-poor ($f = 22\,\%$) and boroxol-rich ($f = 75\,\%$) models compared to the experimental spectrum [41]. Contributions from nuclei inside or outside the boroxol rings are shown as *dashed lines*. A Gaussian broadening of 100 Hz is applied

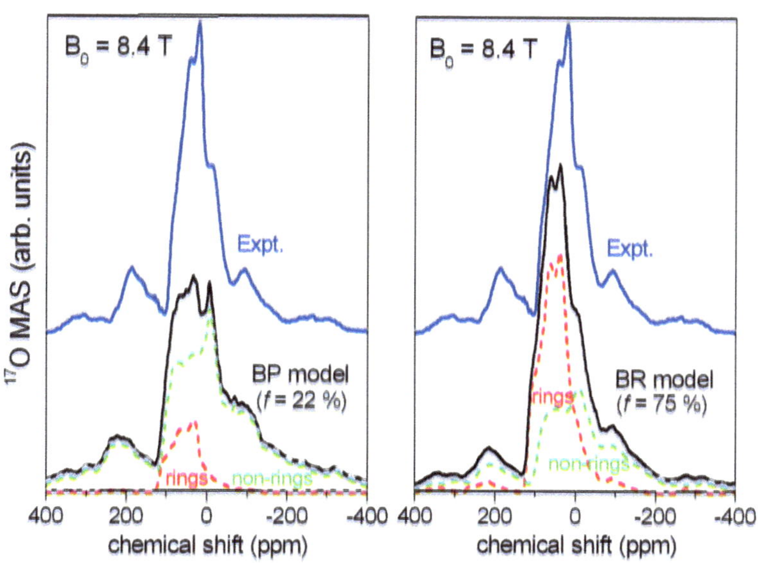

Fig. 14.14 ^{17}O MAS NMR spectra obtained for the boroxol-poor ($f = 22\,\%$) and boroxol-rich ($f = 75\,\%$) models compared to the experimental spectrum [42]. Contributions from nuclei inside or outside the boroxol rings are shown as *dashed lines*. A Gaussian broadening of 100 Hz is applied

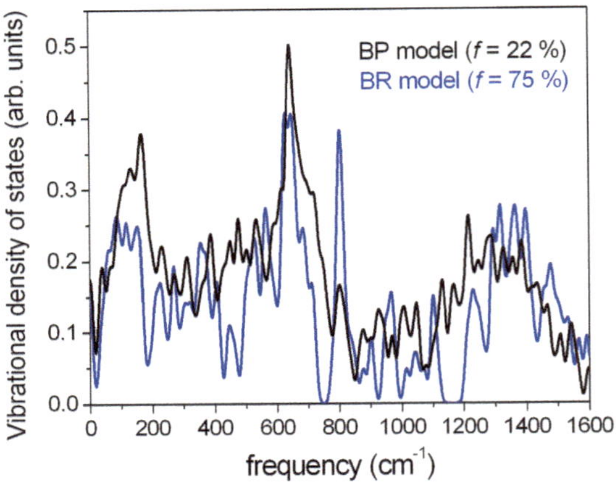

Fig. 14.15 Vibrational density of states obtained for the boroxol-poor ($f = 22\,\%$) and boroxol-rich ($f = 75\,\%$) models

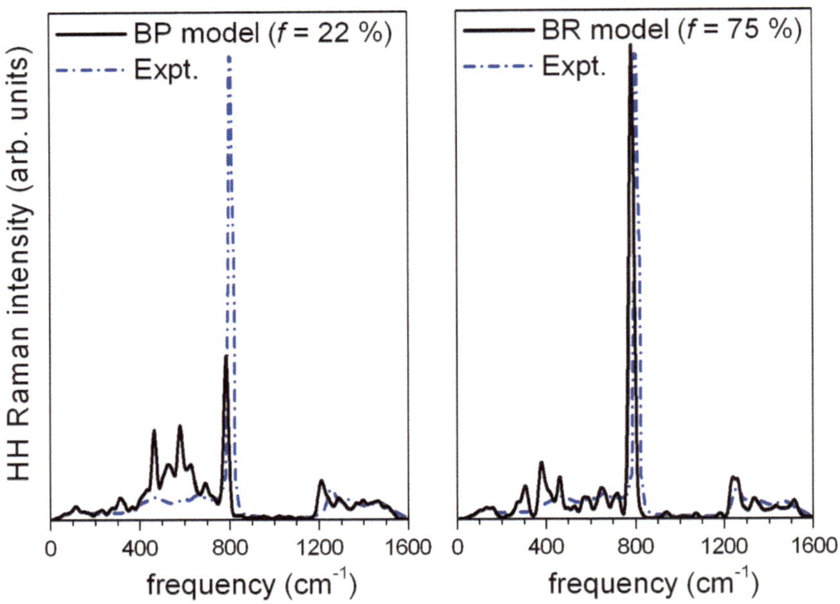

Fig. 14.16 Reduced horizontal-horizontal (HH) Raman spectra calculated for the boroxol-poor ($f = 22\,\%$) and boroxol-rich ($f = 75\,\%$) models (*solid lines*) compared with the experimental spectrum (*dash-dotted lines*) [154]. A Gaussian broadening of $10\,\mathrm{cm}^{-1}$ is applied

The infra-red and Raman intensities were calculated as in [21] using the PWSCF code, part of the QUANTUM ESPRESSO package [150]. The Raman spectra obtained for BP and BR models, Fig. 14.16 are drastically different, in particular regarding

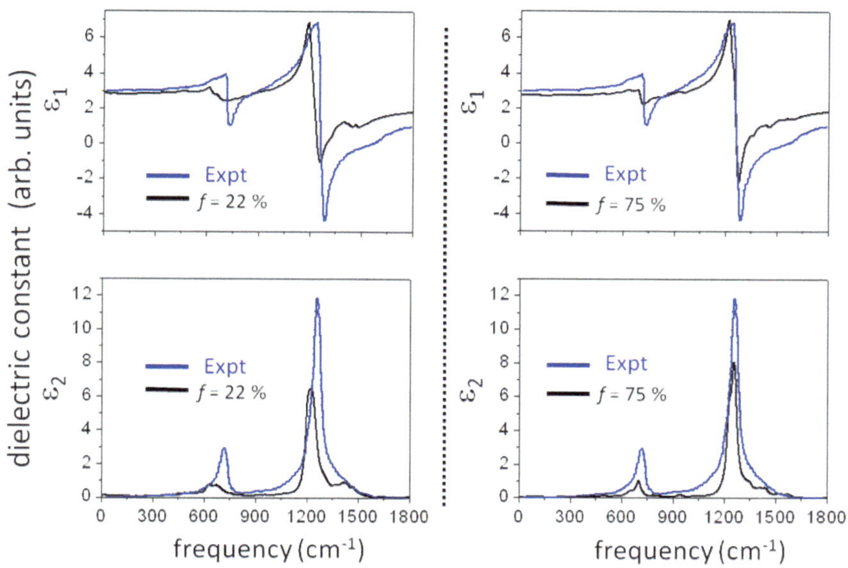

Fig. 14.17 Infra-red spectra calculated for the boroxol-poor ($f = 22\,\%$, *left* panels) and boroxol-rich ($f = 75\,\%$, *right* panels) models compared with the experimental spectrum [154]. The *upper* and *lower* panels show respectively the real and imaginary parts of the dielectric constant

the intensity of the peak at $\sim$800 cm^{-1}. The area under this peak is found to be proportional to f, validating a posteriori the hypothesis made in [37] (note that it is f_O which is actually probed and $f_O = \frac{2}{3}f$). Thus, while it is true that the marked enhancement of this peak as compared to the other modes is due to a scattering-matrix enhancement effect (as seen from comparing Figs. 14.15 and 14.16), it is also true that it can be reliably used as a measure of f: there are no significant non-linear effects in the range $f = 22$–$75\,\%$. The area of this peak is dramatically underestimated in the BP case whereas it matches the experimental one in the BR case within the error bar associated to this comparison (9 %, thus $f_O = 50 \pm 9\,\%$ and $f = 75 \pm 15\,\%$). Note that the superiority of the BR model is not limited to this peak but is also apparent in the frequency regions 400–600, and 1200–1300 cm^{-1}. Contributions in the latter region have been attributed to modes involving the connection of the boroxols with the rest of the network [69]. We point out that, to our knowledge there are no models in the literature which agree at this level of accuracy over such a large frequency range.

As compared to Raman, the IR calculations, Fig. 14.17, show a deceptively small variation with the models: both reproduce reasonably well the experimental data, the BR model being however slightly superior in the region 1200–1300 cm^{-1}. This shows that the vibrational modes probed by IR are mostly not sensitive to f and should not be used to assess the model quality, contrary to the claims made in [98–100]. Note that the present level of agreement with experiments is unprecedented and much superior to that shown in [98–100].

From the $3N$ vibrational modes ω_i, it is easy to compute at any temperature the heat capacity at constant volume within the harmonic approximation:

$$C_v(T) = \sum_i^{3N} k_b \left(\frac{\hbar\omega_i}{k_b T}\right)^2 \frac{exp(\frac{\hbar\omega_i}{k_b T})}{(exp(\frac{\hbar\omega_i}{k_b T}) - 1)}, \tag{14.1}$$

where k_b is the Boltzmann constant. As a benchmark calculation, we show Fig. 14.18 the obtained C_v for the B_2O_3-I crystal with the experimental data [155] for the heat capacity at constant pressure, C_p, in the temperature range from 5 to 250 K. The good agreement between the two curves shows that, in this temperature range, C_p is reasonably approximated by C_v and it validates the scheme used in the calculations. The equivalent comparison between our BP and BR models and the experimental data [84, 156] for the B_2O_3 glass is shown in Fig. 14.19. An overall good agreement is obtained between the calculations and the experiment, albeit a slight deviation is visible at high temperature, most probably due to the fact that the harmonic approximation becomes insufficient. However, the main point here is that the values obtained for the BP and BR models are almost identical. Thus, here again, these data cannot be used to infer anything about the fraction of boroxol rings. This invalidates the claim that the difference between the glass and the crystal heat capacities is due to the existence of boroxol rings in the glass [30, 156].

As a summary of this section, the BR model passes more than reasonably well all the experimental tests while the BP one definitely fails for some of them, most noticeably the Raman and NMR tests. This confrontation has demonstrated straightforwardly which observables are sensitive or not to the presence of boroxol rings in the glassy structure: Raman and NMR are by far the best probes allowing for a quantitative assessment while infra-red, heat-capacity, XAS/IXS and diffraction (NRD or XRD) data are weakly or even not sensitive to f at all. Given the incontrovertible superiority of the BR structure, it seems difficult to maintain that v-B_2O_3 can have anything other than a high content of boroxol rings. In finer details, there are some indications (mostly in the Raman data) that the value used in the BR model ($f = 75\,\%$) is slightly too large. By using the same strategy (i.e. computations of Raman and NMR spectra) for glassy models with varying amounts of boroxols in the range $f = 0$ to $75\,\%$ (those used in Fig. 14.20), we have assessed the *optimal* value to be typically $f = 65 \pm 10\,\%$.

Already at this stage, this work hopefully puts an end to the boroxols controversy by invalidating the common arguments in favour of BP models: previous claims for BP structures were essentially derived from the models' ability to reproduce IR [98–100] and/or diffraction data [54, 63], which are shown here to be necessary but however not sufficient conditions. Any proposition of a new model of glassy boron oxide will have to reproduce not only the structure factors but also the Raman and NMR data at a level at least as good as the one obtained for the present BR model. This said, even diffraction data is not entirely free from subtle rings' signatures: a recent high-pressure study of v-B_2O_3 using high quality neutron diffraction data [78] evidenced the progressive disappearance with pressure of the small peak at 2.75 Å

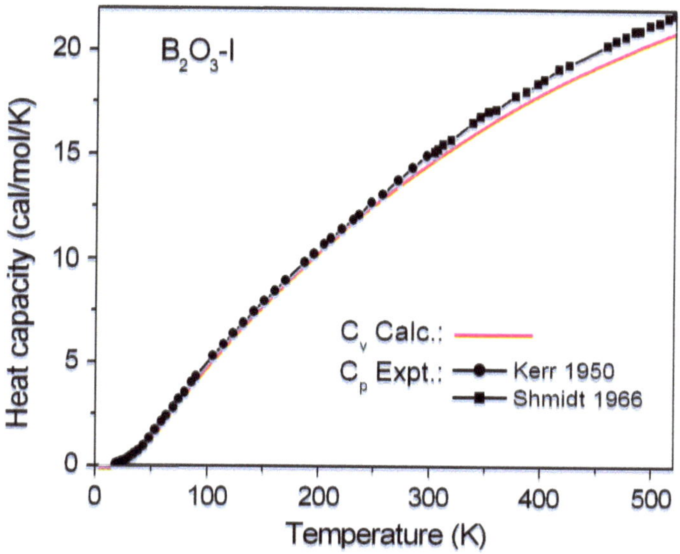

Fig. 14.18 Heat-capacity of B_2O_3-I calculated (at constant volume) and compared to experimental data (at constant pressure) [84, 155]

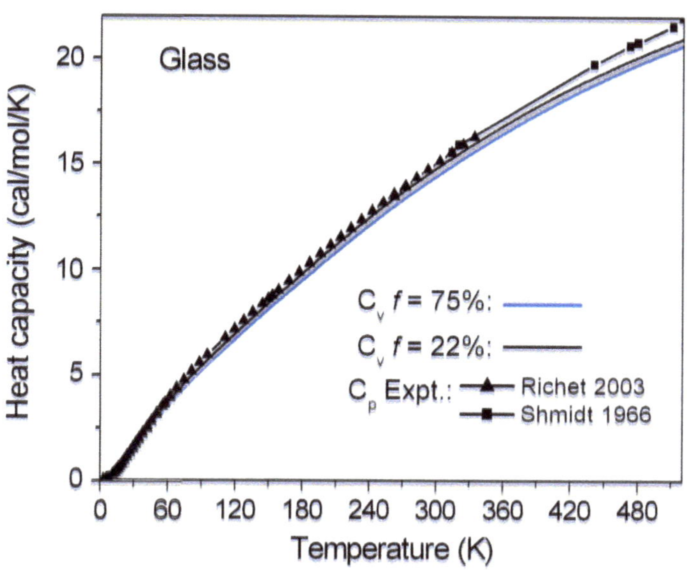

Fig. 14.19 Heat-capacity calculated (at constant volume) for the boroxol-poor ($f = 22\,\%$) and boroxol-rich ($f = 75\,\%$) models compared to experimental data (at constant pressure) [84, 156]

in the total radial distribution function, which as shown in Fig. 14.8 is more marked in BR models, thus allowing for a qualitative (but not quantitative) follow-up of the rings' presence.

So if experiments and simulations are now reconciled about the fact that the boroxol content in v-B_2O_3 is quite high, the reason why it is so remains at this point a genuinely open question that we shall now address in the next sections.

14.5 Rings and Energy

Thanks to the simulations, attempts to correlate the energy of the system with the the amounts of boroxol rings can be pursued [157]. Structural models with varying amounts of boroxol rings, from $f = 0$–$75\,\%$, were obtained by gradually melting the BR model, i.e. by increasing the temperature at constant density, a strategy previously employed in [61, 117, 118]. From these trajectories, instantaneous configurations were picked randomly and their internal positions relaxed at 0 K at the fixed glass density, providing *glassy* models, also known as *inherent structures* [158]. The obtained total energies are plotted versus the amount of boroxol rings in Fig. 14.20. Within statistical scattering, a monotonic decrease of the total energy with increasing f is observed. Using a linear fit of the data, a slope of -6.6 ± 1 kcal/(mol B_2O_3) was initially found (from structures relaxed using a DZP basis set and the PBE functional) [38], revised to -5.6 ± 1 kcal/(mol B_2O_3) using PW basis sets. A possibly more accurate value of -4.7 ± 1 kcal/(mol B_2O_3) is obtained using the dispersion corrected PBE-D2 [138] functional.

We report on the same figure the energy of a hypothetical crystalline polymorph, B_2O_3-0 whose density is close to that of the glass ($1.81\,\mathrm{g\,cm^{-3}}$) at ambient conditions [61]: interestingly it falls well on our data at $f = 0$.

Thus, at the glass density, boroxol-rich structures are more stable than boroxol-poor ones. This may be understood by the simple steric hindrance argument already mentioned (Fig. 14.1): because of their large volume, boroxol rings are favoured in low-density structures (whereas BO_3-made structures are favoured at higher density, as in B_2O_3-I). Now, the energy decrease with increasing boroxol content, at fixed low-density, is likely a reflection of the situation that occurs as the system is quenched (Fig. 14.2), from the high temperature (boroxol-poor) liquid to the low temperature (boroxol-rich) glass.[4] According to this identification, the slope obtained in Fig. 14.20, 4.7 ± 1 kcal/(mol B_2O_3) = 7 ± 1.5 kcal/(mol boroxol) is related to the boroxol stabilisation enthalpy at 0 K for which values of 5.6 ± 1.0 and 6.4 ± 0.4 kcal/(mol boroxol) were derived from experimental Raman investigations [67, 68].

The origin of this stabilisation energy remains unclear [28, 157]: it has been proposed that it includes electronic contributions such as π-bonding from p orbitals

[4] Although the liquid and glass densities are not identical, it may be a good enough approximation at this point.

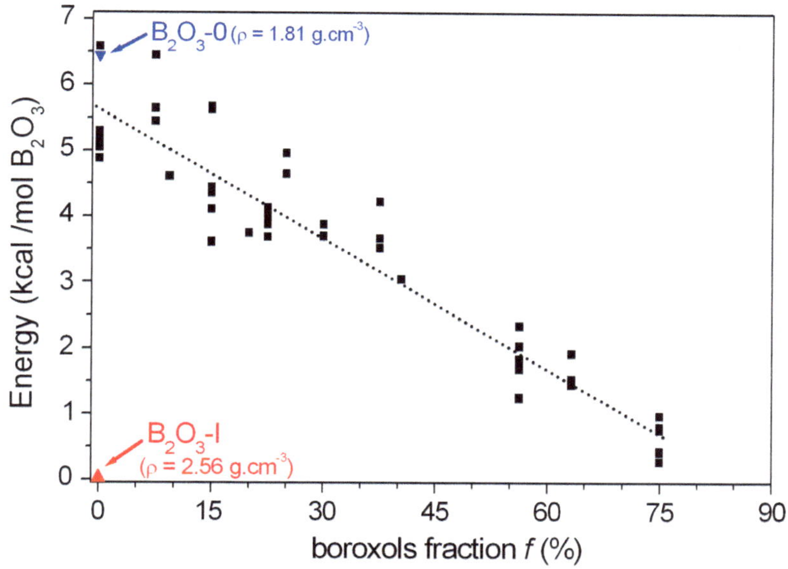

Fig. 14.20 Energy for configurations of varying boroxol amount at 0 K and $\rho = 1.84\ \mathrm{g\,cm^{-3}}$. The energy reference is the energy obtained for the B_2O_3-I crystal

of atoms within the boroxol rings, i.e. some sort of *aromaticity*. We have attempted to characterise the electronic localisation in our configurations using maximally localised Wannier functions [159]; however no clear differences were observed for atoms inside or outside boroxol rings. Nonetheless, using the very same configurations and a classical force-field, we found that the negative slope of Fig. 14.20 could be reproduced only if the oxygen polarisation is turned on [116]. This confirms an electronic character to the boroxol stabilisation energy although it may not necessarily be of aromatic type. Further, we note that this electronic effect may have a topological origin: invoking the steric argument made above, boroxols are progressively incorporated into the liquid to compensate the negative pressure (tensile stress) that would otherwise occur with decreasing temperature at (approximately) constant density.

In any case, the system is eventually quenched in a low-density and low-energy structure. The next section intends to further explore the reason why.

14.6 Boroxol Rings in Crystalline Structures: Predictions of New B_2O_3 Polymorphs from First-Principles

Given the trend observed in Fig. 14.20, it seems natural to ask the following questions: (i) what would be the energy of a structure entirely made of boroxol rings ($f = 100\,\%$)? Could it compete with that of the known crystal, B_2O_3-I? (ii) what is the driving force

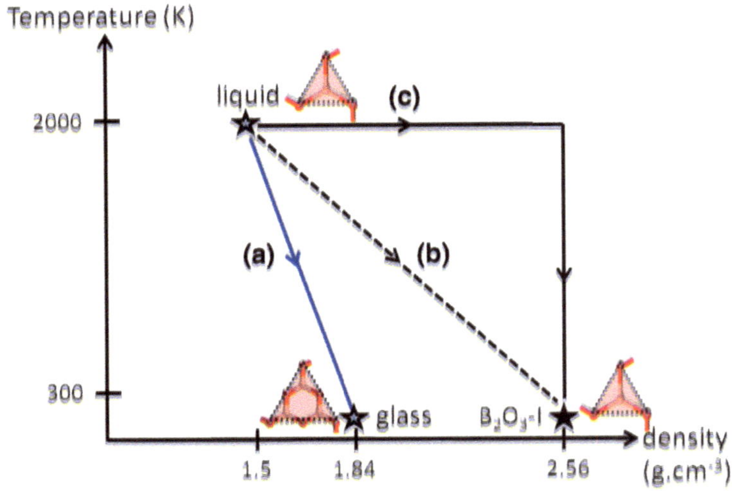

Fig. 14.21 Schematic illustration of the *crystallisation anomaly*. In B_2O_3, cooling the liquid at ambient pressure has so far always resulted in a glassy phase of low density, as shown schematically by the path (A). The crystallisation from the melt, path (B), is never observed under ambient pressure: the obtention of B_2O_3-I requires the melt to be pressurised before being cooled, path (C)

for the energy decrease in Fig. 14.20 and for the fact that the liquid, as it is quenched, follows path (A) in Fig. 14.21 (vitrification in a low-density structure, irrespectively of the quenching rate, even at the lowest experimentally explored) rather than path (B) (crystallisation in the B_2O_3-I polymorph)?

As we shall see, there are strong indications that our knowledge of the B_2O_3 crystalline polymorphism is incomplete and that there are probably yet unknown B_2O_3 crystals to be discovered which would explain the supercooled liquid behaviour and the glass properties.

In contrast to the wide diversity of crystalline structures found in silicates and metal-containing borates, the polymorphism of pure B_2O_3, is seemingly very poor. Indeed, only two different polymorphs, referred as B_2O_3-I and B_2O_3-II, have so far been reported experimentally [110, 111], which are made of BO_3 and BO_4 units respectively (see for instance Figs. 1 and 2 in [109]). Of particular importance for our discussion, none of these polymorphs contain any boroxol rings at all, whereas these rings can be found in large amounts in several other crystalline metal-containing borates, such as $K_3B_3O_6$ or Cs_2O–$9B_2O_3$ [28]. The fact that vitreous B_2O_3 incorporates such regular superstructural units suggests that these might be relevant to form other B_2O_3 crystalline structures.

The lack of *ambient-pressure* polymorphism in B_2O_3 is in stark contrast to the situation observed in most simple oxide systems. By *ambient-pressure* polymorphs, we mean crystals built upon the same structural unit and thus in which the network-forming cations have the same coordination as the ambient glass. In silica for instance, more than 20 polymorphs (quartz, coesite, cristobalite, keatite, moganite, tridymite and pure SiO_2 zeolites such as ferrierite or faujasite) made solely of tetrahedral SiO_4 units have been reported [160]. The fact that B_2O_3-I is the only known BO_3-based crystal is in itself anomalous as compared to silica: one expects the same kind of topological diversity in both systems as a result of the many various ways to connect the building units.

Also very intriguing are the conditions required to obtain these polymorphs from the melt: crystallisation is observed only if a small pressure is applied (typically above 0.4 and 2.0 GPa for the obtention of B_2O_3-I and B_2O_3-II, respectively), a behaviour known as the *crystallisation anomaly* [161–163]. Even if the melt is seeded with crystals and maintained for several months at various temperatures below the melting point, no crystal growth is observed at ambient pressure at any imposed cooling rates ($|dT/dt| > 10^{-5}\,\mathrm{K\,s^{-1}}$) [161–163]. In other words, cooling the B_2O_3 liquid at ambient pressure has so far always resulted in a glassy phase, v-B_2O_3, of density ($\rho = 1.84\,\mathrm{g\,cm^{-3}}$) significantly lower than that of B_2O_3-I ($2.55\,\mathrm{g\,cm^{-3}}$). The fact that pressure is required to crystallise B_2O_3-I from the melt[5] casts some doubts on whether it is the true ambient polymorph.

The structural and density differences between B_2O_3-I and v-B_2O_3 have been early recognised [30] and have been a major motivation for the prediction of boroxol-containing crystals [58–60]. Pioneering the computer synthesis of B_2O_3 polymorphs [59, 60], Takada et al. used *hand-made* modifications of known crystalline structures, HBO_2-III, Cs_2O–$9B_2O_3$ and B_2O_3-I, to produce pure B_2O_3 structures: by inserting or deleting BO_3 units into the parent compounds (after removal of the unwanted atoms, H or Cs), 7 new crystals of varying amount of boroxol rings were generated [60]. These works showed that it is possible to generate boroxol-rich structures at the glass density, at a time when it was controversial. Other numerical predictions of B_2O_3 polymorphs include the work of Kieffer and Huang [61, 117]. By applying either positive or negative pressure on B_2O_3-I, two polymorphs, referred as B_2O_3-III and B_2O_3-0 were generated. The former is a high-coordinated ($^{[4]}$B) phase that we shall not consider further since we are interested here in *ambient-pressure* polymorphism. The latter one, B_2O_3-0, is actually a structure of same topology but different symmetry than B_2O_3-I. It has a low density, close to that of the glass, but does not contain any boroxol rings.

We have recently engaged in a more systematic determination of *ambient-pressure* B_2O_3 polymorphs by using topological design principles and *first-principles* calculations [166]. Given that we were only interested in low-energy polymorphs, i.e. those occurring under ambient-like or low-pressure conditions, the search can be considerably simplified by taking advantage of the following considerations: (i) as a result of its strong and directional bonds, the relevant building block is the BO_3 triangle unit, as in B_2O_3-I (ii) being a supertriangle homothetic to the building block (Fig. 14.1), the boroxol ring B_3O_6 can itself play the role of a building block, as in the glass (iii) since in both units, boron atoms are threefold coordinated only, an efficient and systematic search can be obtained from the decoration of known 3-connected

[5] As an alternative to the high-pressure synthesis, crystalline B_2O_3-I can also be prepared by the stepwise dehydration of orthoboric acid (H_3BO_3)[164] or by seeding a melt with borophosphate [165].

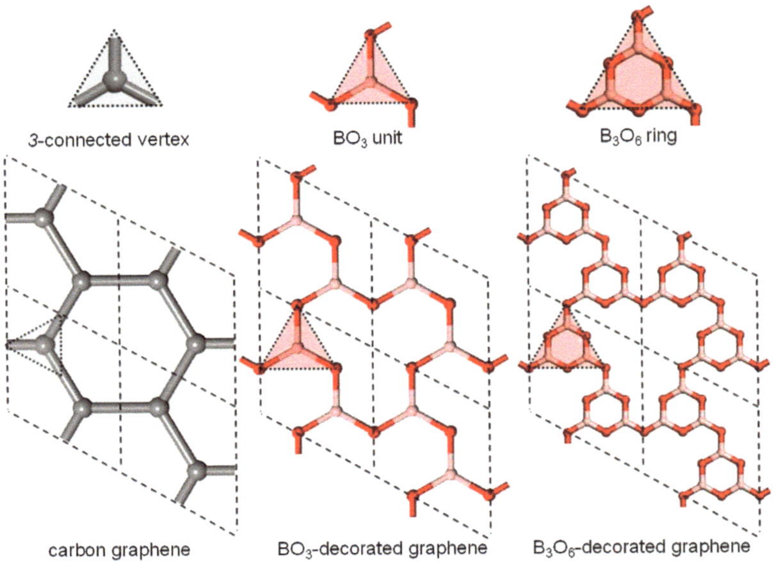

Fig. 14.22 Construction of B_2O_3 polymorphs. Using the self-similarity between a 3-connected vertex (*left*), the BO_3 *triangle* (*middle*) and the B_3O_6 ring (*right*), new polymorphs can be constructed from the decoration of known 3-connected networks, as illustrated here in the case of a graphene layer. The T0 and T0-*b* polymorphs were obtained by stacking the BO_3- and B_3O_6-decorated layers respectively. The decoration is illustrated here for clarity on $2D$ structures but note that all the other polymorphs of Fig. 14.23 are $3D$ (fully connected) structures. Adapted from Fig. 2 of [166]

networks by the relevant units. We thus used all possible 3-connected networks, fully connected in $3D$ space, with up to six vertices in the unit cell, as obtained from an exhaustive mathematical search, based on graph theory and originally applied to sp^2-carbon polymorphs [167] (see Fig. 14.22). The lamellar network of graphite was also added in the database, providing a total of 13 topologically different networks. Vertices of these networks were then decorated by BO_3 triangle units. Among the generated structures, the known B_2O_3-I polymorph was recognised; the 12 remaining novel structures are labelled T0 to T11 in Fig. 14.23. In T8 and T10, 50 % of the boron atoms belong to threefold rings, i.e. boroxol rings. Further to expand the search and to investigate the role of the boroxol ring as a structural motif, 13 additional structures, indicated by the extension -*b*, were generated by replacing the BO_3 units by the B_3O_6 ones. In this way, structures which are made of 100 % boroxol units were obtained, taking advantage of the self-similarity between a BO_3 and a B_3O_6 unit (Figs. 14.1 and 14.22). All structures (atomic positions and lattice cell) were then relaxed at 0 K within the DFT framework using the PBE [133] functional, ultrasoft pseudopotentials [168] and the CASTEP plane-wave code [169].[6]

---

[6]As compared to [166], the calculations were repeated with tighter (more accurate) pseudopotentials and a larger plane-wave basis-set cutoff of 784 eV. This resulted in some very small differences in the results shown Fig. 14.23.

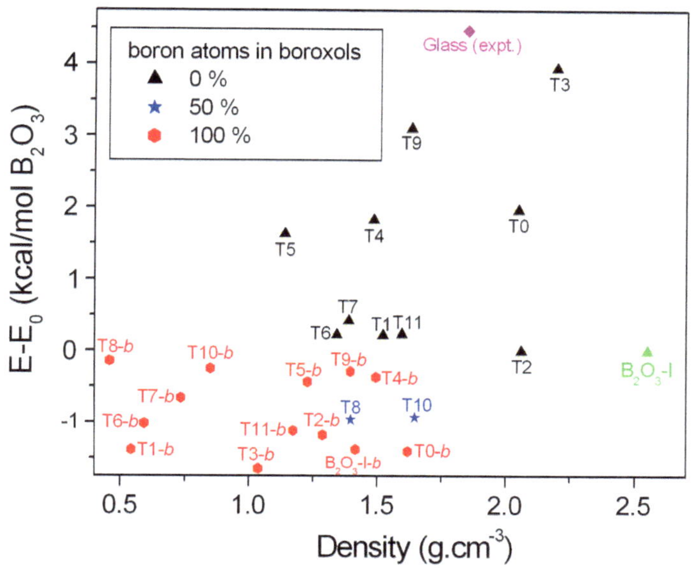

Fig. 14.23 Energy as a function of density for the B_2O_3 polymorphs. The energy reference, E_0, is that obtained for B_2O_3-I. The proportion of boron atoms in boroxol rings in the polymorphs are indicated by different symbols. The experimental energy for the glass is taken from [84]

This resulted in the prediction of 25 new polymorphs containing either 0, 50 or 100 % of boron atoms in boroxol rings. As can be seen in Fig. 14.23, all these crystals have a low-density, that is, lower than B_2O_3-I and, for most of them, comparable to that of the glass. Some have a very low density ($<1.0\ \mathrm{g\,cm^{-3}}$), which in silicates would correspond to framework densities typical of zeolites.

As could be anticipated, most of the crystals based upon boroxol rings (-*b*) tend to have lower densities than their triangle-based equivalents. The density ratio however largely varies from a topology to another.[7] Many obtained structures, in particular those made of boroxol rings, are very flexible, showing large volume changes upon small pressure variation.

Less trivial but also of great interest is the fact that in many cases the energy is lowered in the triangle-boroxol substitution. The net effect, as seen in Fig. 14.23, is that for densities lower than c.a. $2.0\ \mathrm{g\,cm^{-3}}$ the lowest-energy crystals are those

[7] If there are no relaxations at all in the triangle-boroxol substitution, that is for exactly homothetical $3D$-topologies, one expects the density ratio to be $\frac{3}{2^3} = 0.375$. This is close to the value obtained for instance in the T1 to T1-*b* case (0.35). However, the density ratio can be much higher because of structural relaxations, in particular in the directions orthogonal to the boroxol rings plane. For a topology with a strong lamellar character, the triangle-boroxol substitution will double the intra-layers lengths while keeping unchanged the inter-layer distance, resulting in this case in a density ratio of $\frac{3}{2^2} = 0.75$. This is indeed very close to the value obtained in the T0 to T0-*b* case (0.76). There are a few cases for which the density is unchanged (as in the T4 to T4-*b* case); these correspond to initially porous geometries which contain large rings and for which a more efficient packing in the final relaxed structure was achieved by folding the largest rings.

having a high proportion of boroxol rings. Further, a large number of the predicted crystals tend to be *clustered* in a narrow density and energy range, typically in between 1.5 and 2.0 $g\,cm^{-3}$ and -1.5 to 0.5 kcal/mol. This density range typically encompass that in between the high-temperature liquid (ρ_{liquid}(2000K) $\sim$1.5 $g\,cm^{-3}$) and the ambient glass (ρ_{glass}(300 K) $\sim$1.84 $g\,cm^{-3}$).

We thus argue that these new polymorphs are much more relevant than is B_2O_3-I for understanding the glass properties and that their energy quasi-degeneracy is responsible for the ease of vitrification: such a situation with many local minima in the energy landscape is expected to prevent the nucleation of a given phase by favouring a disordered structure instead [170]. We also note that for a given topology, several minima were often found (only the lowest energy minimum is shown Fig. 14.23): this further reflects a rugged energy landscape.

As pressure is applied (typically 1 GPa), the energy degeneracy is progressively lifted (not shown here) and the energy difference between B_2O_3-I and any other polymorph becomes marked [166], rendering the crystallisation of B_2O_3-I increasingly favourable. This thus provides a scenario to understand the B_2O_3 *crystallisation anomaly*, i.e. the fact that pressure is required to crystallise B_2O_3-I from the melt.

Among the lowest energy polytypes and those closest to the glass density, one finds e.g. B_2O_3-I-*b*, the boroxol-made equivalent of B_2O_3-I, T0-*b*, the boroxol-made equivalent of graphite, T8 and T10 for which $f = 50\,\%$. Interestingly, these crystals share greater structural similarities with the glass than does B_2O_3-I. This is illustrated by the structure factors obtained at finite temperature, Fig. 14.24, in particular in the low-q region, sensitive to the medium-range order. In the T10 structure, shown Fig. 14.25, the dihedral angle distribution between adjacent ring and non-ring units is well in line with the one measured in the glass by two-dimensional NMR [44].

These findings, thus, confirm that boroxol rings are essential to stabilise low density structures [38, 171]. As in the previous section, one could ask if this a purely geometrical effect (larger rings are favoured at low density to fill the empty space) or if there is an additional electronic contribution to the formation of the boroxol rings. The triangle-to-boroxol substitution used here can help to gain insight on this question. We used a 2D monolayer (topologically equivalent to graphene) to generate triangle- and boroxol-made single sheets, Fig. 14.22. If both structures are unrelaxed (and therefore strictly equivalent topologically), the boroxol-made sheet is favoured by about 11 kcal/mol B_2O_3: this strongly suggests that there exists an intrinsic electronic origin of that order (c.a. 10 kcal/mol B_2O_3) to the rings stabilisation energy. However, if the internal coordinates are allowed to relax, the gain is reduced to only 1 kcal/mol since there are less degrees of freedom in the boroxol-made sheet than in the triangle-made one. Thus, both electronic and geometric effects compete in the energies shown in Fig. 14.23.

Most of the predicted crystals form cage- or channel-like structures. This could make these interesting materials as guest host, in particular for hydrogen storage [172] or for molecular sieves applications [173]. This nanoporosity may also be reminiscent in the glass structure. Indeed, large specific areas and high hydrogen uptakes were reported both in B_2O_3 fine powders [172] and in amorphous samples [174]. The presence of relatively large voids has also been observed in glassy configurations

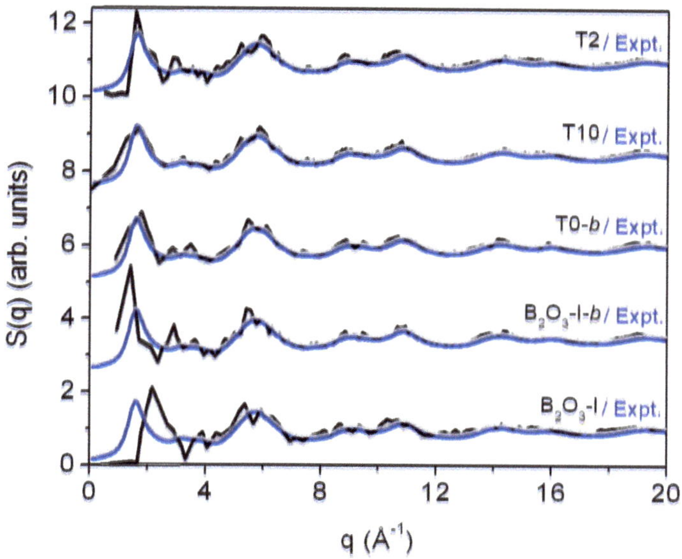

Fig. 14.24 Structure factors obtained from MD simulations at finite temperature (T_g = 540 K) for some selected polymorphs compared with the experimental data for the glass [32]

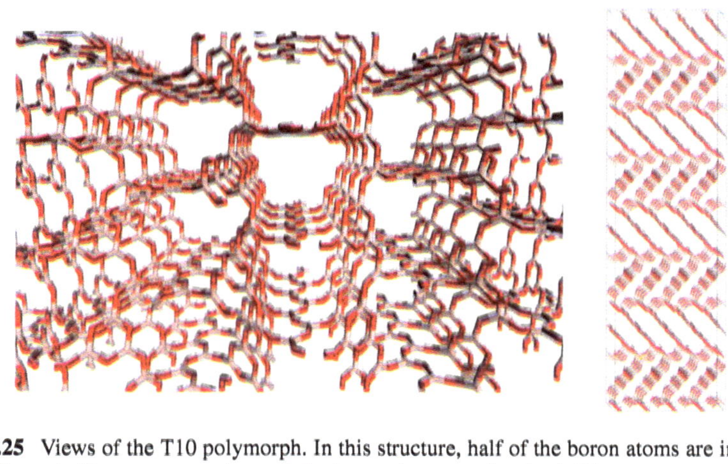

Fig. 14.25 Views of the T10 polymorph. In this structure, half of the boron atoms are in boroxol rings. Channels-like arrangements, made by larger rings, are clearly visible

derived from RMC simulations [175]. Further investigations of these aspects could be all the more promising given the recent discovery that glassy silica is a nanoporous material which can accommodate gases such as He or Ne [176, 177].

The structure search used here was limited in two aspects, namely by the maximum number of nodes (6) per unit cell, and by the units used for the decoration (all the nodes were decorated by a single unit type, either the triangle or the boroxol ring. Larger self-similar units such as threefold rings of threefold rings, could also be used). The search in principle could be pushed further by extending these limitations (larger cells and/or decoration with a mix of triangle/boroxol or any self-similar units). As a matter of fact, a fewer additional polymorphs predictions have since been made [178].

These results reveal a much richer polymorphism in B_2O_3 than expected before. In this sense, the behavior of B_2O_3 appears now much more similar to that from other well-known glassy systems such as silica. The small differences in energy between the polytypes comes from the variability of the B–O–B angle linking the structural units, just as in silica Si–O–Si angles bridging the tetrahedra. In B_2O_3, the tendency to form threefold rings allows for the formation of porous structures. The large angular flexibility of the B–O–B bridges linking the rings gives rise to several energy minima for a given topology. These two aspects, the large number of possible topologies and the many geometrical variations among them, both contribute to a very rugged energy landscape and are therefore essential ingredients to the aptitude to vitrify. This, however, does not exclude that other aspects, related to the nucleation or the crystal growth kinetics, might also be at play [163].

Whether the predicted B_2O_3 crystals can be experimentally synthesised remains an open question. On one hand, most of the structures investigated fall within the energy range of thermodynamically accessible polymorphs (in silica, the highest energy zeolite, ISV, is 3.44 kcal/mol above quartz [160]). On the other hand, kinetics or experimental difficulties may prevent their observation. Part of the difficulty may well lies in the fact that one needs to avoid the glass transition resulting from the polymorphic degeneracy in the energy-density diagram, i.e. one has to find a method which would favour one of the polymorphs over the others. Interestingly, one can find several mentions of low density crystalline structures in the literature [179–181], although in most cases it was barely noticed and not recognised as a new polymorph discovery (since this was not the priory motivation of these works [180, 181]). Their density varies from 1.8 down to 0.69 $\mathrm{g\,cm^{-3}}$ and thus these putatively unidentified new polymorphs could correspond to some of our predictions. Note that these were not obtained from a liquid quench but rather from chemical routes such as the dehydration of H_3BO_3 ([179, 180]) or chemical vapour deposition [181]. Notwithstanding that these could be false results, further experimental studies dedicated to crystal synthesis would be very much indicated. Promising directions may include the various methods used to produce thin films of B_2O_3, such as chemical vapor deposition [181] and its variants [182, 183], magnetron sputtering [184], infrared heating [185] or sol-gel techniques [186] although so far, in most of the cited cases, the obtained structures were either amorphous or nano-objects such as wires. Physical methods able to generate negative pressure conditions (tensile stress) [187] would be highly desirable although technically challenging.

14.7 Back to the Liquid: Structural Transitions Under Tensile Stress (or How to Generate High Proportions of Rings)

In summary of the previous section, the polymorphs prediction allows to unravel some of the seemingly anomalous properties of B_2O_3. As the liquid is quenched, it vitrifies in a low density glass (path (A) of Fig. 14.21) because of the many underlying crystals in this region of the phase diagram. It acquires some of their structural characteristics, such as a high proportion of boroxol rings. The formation of these rings compensates the negative pressure that would otherwise exists in the glass at low temperature. As already stated, the structural transition occurring from the high-temperature liquid to the low-temperature glass (Fig. 14.2) is extremely challenging to follow numerically because of the exponentially growing viscosity. It has actually remained so far impossible and it is likely to remain so for a long time with brute-force methods (standard MD/MC techniques) given the affordable computational resources. However, given that low density conditions favour the formation of large rings, it is tempting to follow a different route in the phase diagram, schematically indicated as (1) in Fig. 14.26, i.e. going to negative pressure at constant temperature in the liquid. To do so, we produced simulations in the *NVT* ensemble starting from the liquid at the glass density ($T = 2000$ K, $\rho = 1.84$ g cm^{-3}) and slowly increase the volume of the simulation box (at a constant rate over a ~20 ps timescale) until a density of $\rho = 0.9$ g cm^{-3} was achieved. Then simulations (of at least 300 ps) were branched at different densities (path (1) in Fig. 14.26), including the liquid at the experimental density ($\rho_{\text{liq}} = 1.49$ g cm^{-3}) and a much smaller one ($\rho_{\text{low}} = 1.13$ g cm^{-3}). The reverse process was also carried out, i.e. compressing the liquid from ρ_{low} to ρ_{glass}, with long simulations sampling intermediate densities ($\rho = 1.29$ g cm^{-3} and $\rho = 1.72$ g cm^{-3}), along path (2) in Fig. 14.26.

The instantaneous evolution of the boroxols fraction f, at fixed density, monitored along the trajectories which were branched from path (1) are shown Fig. 14.27. For the lowest density ($\rho_{\text{low}} = 1.13$ g cm^{-3}), a clear step-like increase is observed as time is running. Interestingly, corresponding jumps in the instantaneous pressure were observed: initially at ~ -1.1 GPa at the beginning of the trajectory, the final pressure is reduced to ~ -0.5 GPa. This reflects a correlation between the internal pressure and the boroxols fraction. Note that the final value obtained, $f \sim 52\,\%$, is quite high given the relatively high temperature, 2000 K.

The liquid obtained in these conditions is extremely viscous: the diffusive regime could not be reached within the time afforded. This means that a simulations' timescale of several hundreds of pico-seconds, already quite long by ab-initio standards, is still insufficient for a proper relaxation of the system towards its equilibrium.

Not surprisingly, the reverse behaviour is observed for trajectories branched along the reverse path (2): as the density is increased and as the simulation time runs, the boroxols fraction tends to decrease from an initially too high value (reminiscent of the starting density, $\rho_{\text{low}} = 1.13$ g cm^{-3}). The final f values obtained for all the sampled densities along both paths are shown in the inset of Fig. 14.26. A sort of

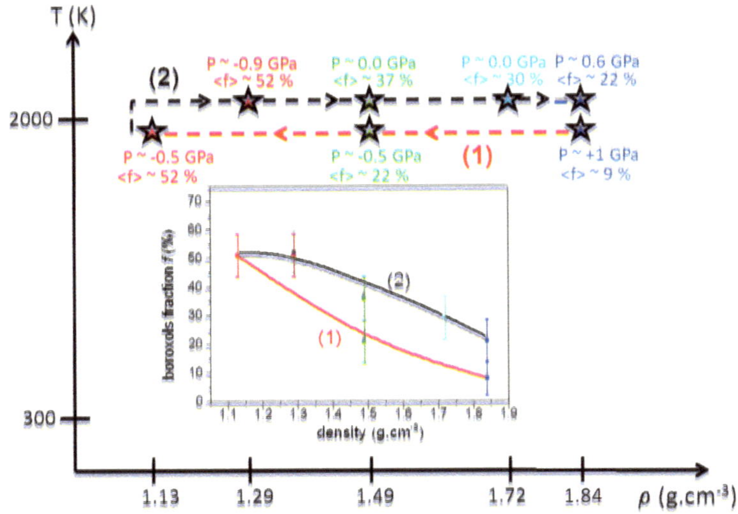

Fig. 14.26 Density conditions explored in the liquid at 2000 K. Starting from $\rho = 1.84\ \mathrm{g\,cm^{-3}}$ (over-pressurised liquid at the glass density), lower densities along path (1) were sampled corresponding to typically *ordinary* ($\rho = 1.49\ \mathrm{g\,cm^{-3}}$) and *stretched* ($\rho = 1.13\ \mathrm{g\,cm^{-3}}$) conditions. The system was then compressed back to the starting density, path (2). Values for the average pressure and fraction of boroxols $\langle f \rangle$ are indicated. The inset shows the evolution of $\langle f \rangle$ along these paths

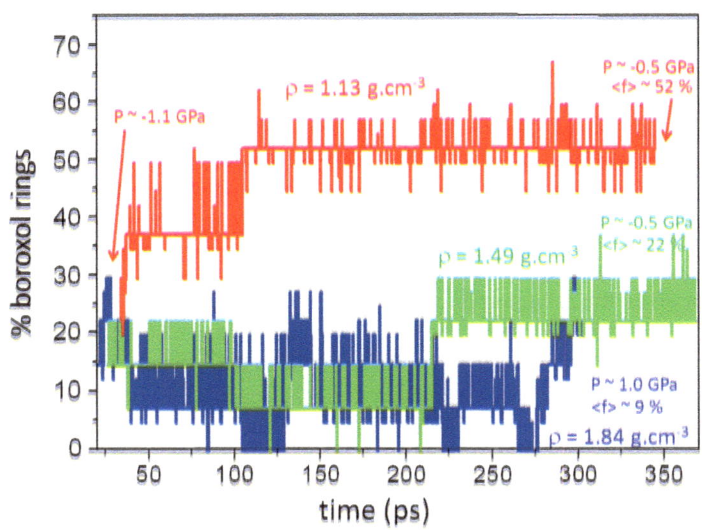

Fig. 14.27 Instantaneous evolution of the boroxols fraction f, along three trajectories in the *NVT* ensemble at 2000 K. The three densities explored correspond typically to a slightly pressurised, an ordinary and a stretched liquid

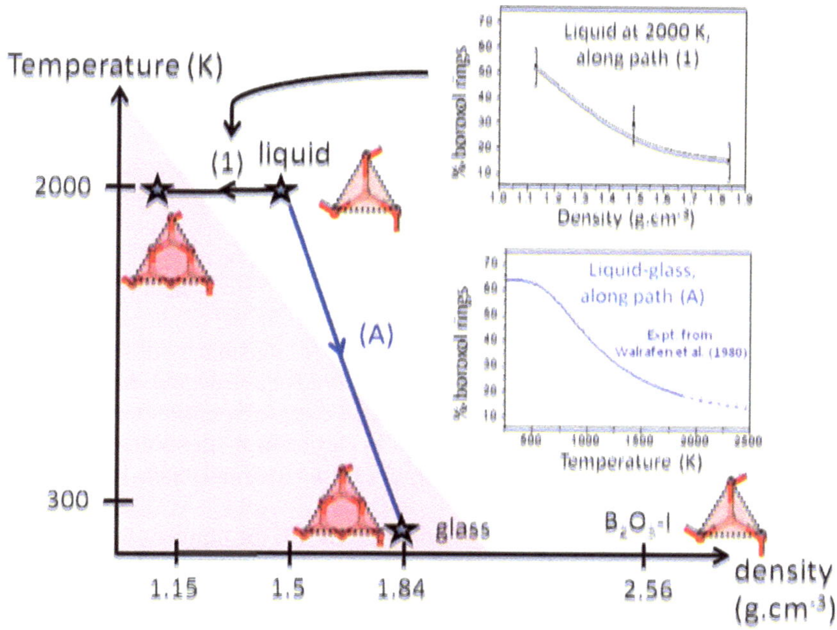

Fig. 14.28 Schematic temperature-density phase diagram of B_2O_3. As the liquid is cooled and approaches the glass transition, path (A), it undergoes a structural transition from a BO_3- to a boroxol-dominated network. A similar structural transition can be induced by stretching the liquid, path (1). Along both paths, boroxols rings form to compensate the internal negative pressure

hysteresis curve is obtained as a result of the limited sampling. Therefore, the set of values obtained along path (1) (expansion from $\rho = 1.84$ g cm^{-3}) likely represents a lower bound (as the starting conditions correspond to a pressurised liquid and are therefore unfavourable to the formation of boroxols) while the values obtained along path (2) correspond to an upper bound.

These findings provide useful insights and open interesting questions. A methodology is provided to enhance the content of rings in the liquid. The structural evolution obtained as the liquid is stretched at constant temperature (Fig. 14.26) resembles that observed experimentally as the liquid-glass transition is approached (Fig. 14.2): along both paths, the boroxols rings develop so as to compensate the internal negative pressure, Fig. 14.28. From a methodological point of view, the first transition (stretching the liquid at fixed temperature) is easier to follow with MD than the second one (glass transition) since the ergodicity is in principle better preserved. In addition, by quenching both the temperature and the volume from the structure obtained at ρ_{low}, it should be possible to recover a new boroxol-rich model of glassy B_2O_3.

More fundamentally, these findings raises the possibility that B_2O_3 is experiencing a *liquid-liquid transition* along path (1), Fig. 14.28. This transition, in the stretched liquid, is in competition with the liquid-vapor one and possibly with the crystallisation in one of the new predicted B_2O_3 polymorphs [166]. We can safely discard that what we observed is the system cavitating: this happened once in a *NPT* simulation at large negative pressure and showed marked signatures of the gas phase, such as a strong increase of the diffusion and of the proportion of under-coordinated atoms. Instead the obtained phase at ρ_{low} is very viscous and fully polymeric (99.9 % of boron are threefold coordinated by oxygens as in the ordinary, i.e. zero-pressure liquid). Its structure is fully disordered, as evidenced by the usual structural markers (radial distribution functions, structure factors, angular distributions). No signs of crystallisation were observed, although more work, unfortunately much beyond what is feasible within an ab-initio framework [188, 189] would be needed to confirm this point. In any case, it would be interesting to investigate further the thermodynamics and the similarities (structure, porosity) of this phase with the predicted low-density crystals [166]. Its high amount of rings is very likely responsible for its high viscosity. It also likely impacts its surface tension. The two liquid phases would share a common short-range order (that defined by the boron and oxygen coordination) but mostly differ in their medium-range one (as defined in particular by the boroxols fraction). Thus, this transition would somehow ressemble the putative and much debated liquid-liquid transition of water [190, 191].

Back to generalities about simulating the structure of the liquid or the glass, a last methodological point should be stressed. In addition to the need for long simulation times, the correlation between the rings structure and the internal pressure raises some warnings. In the *NVT* ensemble (or any fixed volume ensemble), care should be given that the stress is fully relaxed in the final structure: in the present case, a positive residual pressure (above the target one) will likely reflect an overestimation of the boroxols fraction and vice versa. Simulation in a *NPT* ensemble ensures by definition that the target pressure is reached; however the same kind of caveat applies if the density is unknown: the target pressure might be fulfilled by various structures differing in their rings topology, schematically by a boroxol-rich structure at low density or by a boroxol-poor structure at high density. In this case, the knowledge of either the density or the rings fraction is an essential constraint.

14.8 Rings in Other Borates and Thioborates

A close analog to B_2O_3 is the chalcogenide compound B_2S_3 which thus provides comparative insight concerning the structural chemistry. B_2S_3 is iso-electronic to B_2O_3 and it is well established that the molecular building block in both crystalline and glassy B_2S_3 is the planar BS_3 triangle unit. Interestingly, it is also clear that threefold rings, called borosulphols or thioboroxols, which are the equivalent of the boroxol rings in vitreous B_2O_3 are largely present in both the glass [192–200] and the B_2S_3-I crystal [201]. A specificity however of the chalcogenide is the existence of twofold (edge-sharing) rings. In B_2S_3-I, the proportion of borons involved in two and threefold rings are $f_2 = 25$ and $f_3 = 75$ %, respectively. The corresponding values in the glass are less clear.

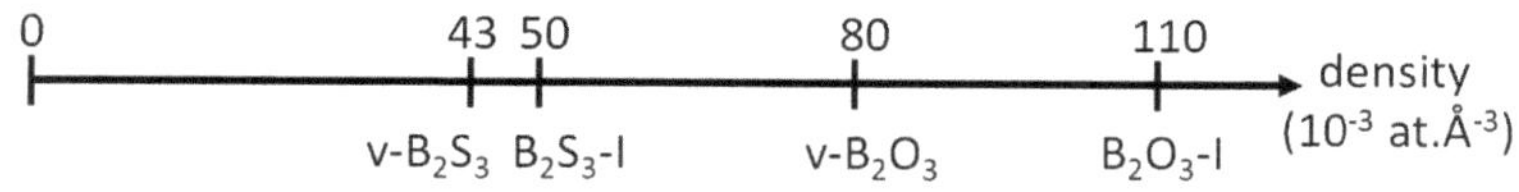

Fig. 14.29 Atomic densities of the vitreous and crystalline phases in the B_2S_3 and B_2O_3 systems

We have recently undertaken an extensive study of the liquid and glassy B_2S_3 phases using the methodology described previously for B_2O_3, namely computation of structure factors, NMR and Raman data from different models, varying by their rings structure and obtained by using different thermal histories. Preliminary results can be found in [200]. As was the case for B_2O_3, the structure factors from all tested models reproduced very well the experimental data available [197]. Note however that the structure factor test is sufficient to rule out a chain-like model of twofold rings previously obtained from classical MD [202]. This illustrates both the higher degree of realism of the *first-principles* models provided in [200] and the fact that the reproduction of the total structure factor is a necessary but not sufficient condition. Calculation of the Raman data appeared to be the most instructive test of the models. It is noteworthy that, unlike the case of B_2O_3, both two and threefold rings appeared spontaneously in large amounts in the MD trajectories, without having to tweak the models. This can be understood by the comparison of the glass and crystal (atomic) densities for both systems, Fig. 14.29. First, in B_2S_3 the glass density is very similar to that of the known crystal polymorph. Second, the B_2S_3 (atomic) densities are much smaller, typically halves those of B_2O_3. This is well in line with the findings of the previous sections: low density favours the formation of these rings.

As seen, although chemically simple, B_2O_3 and B_2S_3 show rich structural behaviours. Of course, the complexity can be dramatically increased in related systems such as borates, thioborates, boro-silicates or boro-phosphates to name just a few. In the case of alkali borates, $(M_2O)_x–(B_2O_3)_{(1-x)}$, the presence of the alkali cation M (Li, Na, K, Rb or Cs) induces the presence of novel structural units such as non-bridging oxygens and/or BO_4 tetrahedral units, depending on the alkali concentration x and, to a much lesser extent to the type of alkali. Typically, at low concentration ($x \lesssim 40\,\%$), all introduced alkali cations will convert BO_3 into BO_4 units. Higher content of alkali leads to the formation of units involving non-bridging oxygens. These are the basic mechanisms used to explain the presence of a maximum in the fraction of fourfold-coordinated boron atoms at typically $x \sim 40\,\%$, clearly evidenced by NMR. However, other quantities such as the density, the glass transition temperature, the viscosity, the thermal expansion coefficient also show an extremum as a function of x but not at the same composition; this is the so-called *borate anomaly*. It has been proposed that the determination of the extremum requires a structural caracterisation beyond the short-range basic units, i.e. the knowledge of the various possible superstructural units: these include many types of rings such as diborate, triborate, pentaborate, metaborate groups and their variants [28]. Since the composition x can be seen as a kind of chemical pressure, the composition-induced structural changes are likely to be mirrored in pressure-induced ones. There has been however only a few numerical investigations [62, 94, 95, 203–214] of borates up to now and much remain to be done.

In boro-silicates or boro-phosphates with a low content of modifiers, the network topology has to accommodate both trigonal and tetrahedral units. This raises interesting questions related to the glass homogeneity: are the two types of units randomly distributed or are there locally enriched regions? Not surprisingly, microscopic simulations are still very scarce [92, 105, 112, 215].

Acknowledgments I would like to thank all the co-workers involved in the [38, 78, 121, 136, 166, 200], in particular Francesco Mauri, Thibault Charpentier, Ari P. Seitsonen, Mathieu Salanne, Axelle Baroni and Matthieu Micoulaut. I also thank Pascal Richet for providing me with the experimental data of [156], François-Xavier Coudert for interesting discussions, Sara and Eva Lacarce for support and patience. Large support from the French supercomputers (GENCI-CINES/IDRIS) has made possible the results presented in this contribution (Grant x2014081875). I also acknowledge support from French state funds (managed by the ANR under reference ANR-11-IDEX-0004-02, cluster of Excellence MATISSE) and support from the HPC resources of The Institute for scientific Computing and Simulation (financed by Region Île de France and the project Equip@Meso under reference ANR-10-EQPX-29-01).

References

1. H.E. Fischer, A.C. Barnes, P.S. Salmon, Rep. Prog. Phys. **69**, 233 (2006)
2. A. Filipponi, J. Phys.: Condens. Matter **13**, R23 (2001)
3. S.O. Hruszkewycz et al., Phys. Rev. Lett. **109**, 185502 (2012)
4. P. Wochner et al., Proc. Natl. Acad. Sci. USA **106**, 11511 (2009)
5. M. Altarelli, R.P. Kurta, I.A. Vartanyants, Phys. Rev. B **82**, 104207 (2010)
6. M.M.J. Treacy, K.B. Borisenko, Science **335**, 950 (2012)
7. J.M. Gibson, M.M.J. Treacy, T. Sun, N.J. Zaluzec, Phys. Rev. Lett. **105**, 125504 (2010)
8. P. Wochner, M. Castro-Colin, S.N. Bogle, V.N. Bugaev, Int. J. Mater. Res. **102**, 874 (2011)
9. A. Filipponi et al., Phys. Rev. B **40**, 9636 (1989)
10. G. Ferlat et al., Phys. Rev. B **73**, 214207 (2006)
11. T. Charpentier, M.C. Menziani, A. Pedone, RSC Adv. **3**, 10550 (2013)
12. J.P. Rino et al., Phys. Rev. B **47**, 3053 (1993)
13. K. Vollmayr, W. Kob, K. Binder, Phys. Rev. B **54**, 15808 (1996)
14. L.W. Hobbs, C.E. Jerusum, V. Pulim, B. Berger, Phil. Mag. A **78**, 679 (1998)
15. R.M. Van Ginhoven, H. Jónsson, L.R. Corrales, Phys. Rev. B **71**, 024208 (2005)
16. P.Y. Huang et al., Science **342**, 224 (2013)
17. S.K. Sharma, J.F. Mammone, M.F. Nicol, Nature **292**, 140 (1981)
18. F.L. Galeener, R.A. Barrio, E. Martinez, R.J. Elliott, Phys. Rev. Lett. **53**, 2429 (1984)
19. A. Pasquarello, R. Car, Phys. Rev. Lett. **80**, 5145 (1998)
20. P. Umari, X. Gonze, A. Pasquarello, Phys. Rev. Lett. **90**, 027401 (2003)
21. M. Lazzeri, F. Mauri, Phys. Rev. Lett. **90**, 036401 (2003)
22. F.L. Galeener, Solid State Commun. **44**, 1037 (1982)
23. D.S. Franzblau, Phys. Rev. B **44**, 4925 (1991)
24. K. Goetzke, H.J. Klein, J. Non-Cryst, Solids **127**, 215 (1991)
25. X. Yuan, A.N. Cormack, Comput. Mater. Sci. **24**, 343 (2002)
26. S. Le Roux, P. Jund, Comput. Mater. Sci. **49**, 70 (2010)
27. W.H. Zachariasen, J. Am. Chem. Soc. **54**, 3841 (1932)
28. A. C. Wright, Phys. Chem. Glasses: Eur. J. Glass Sci. Technol. B **51**, 1 (2010)

29. S.R. Elliott, J. Non-Cryst, Solids **97–98**, 159 (1987)
30. J. Krogh-Moe, J. Non-Cryst, Solids **1**, 269 (1969)
31. P.A.V. Johnson, A.C. Wright, R.N. Sinclair, J. Non-Cryst, Solids **87**, 5071 (1982)
32. A.C. Hannon et al., J. Non-Cryst. Solids **177**, 299 (1994)
33. B.E. Warren, H. Krutter, O. Morningstar, J. Am. Ceram. Soc. **19**, 202 (1936)
34. J. Goubeau, H. Keller, Z. Anorg, Allg. Chem. **272**, 303 (1953)
35. C.F. Jr, Windisch, W. M. Jr Risen. J. Non-Cryst. Solids **48**, 307 (1982)
36. R.A. Barrio, F.L. Castillo-Alvarado, F.L. Galeener, Phys. Rev. B **44**, 7313 (1991)
37. P. Umari, A. Pasquarello, Phys. Rev. Lett. **95**, 137401 (2005)
38. G. Ferlat et al., Phys. Rev. Lett. **101**, 065504 (2008)
39. G.E.J. Jellison, L.W. Panek, P.J. Bray, G.B.J. Rouse, J. Chem. Phys. **66**, 802 (1977)
40. S.J. Gravina, P.J. Bray, J. Magn. Res. **89**, 515 (1990)
41. R.E. Youngman, J.W. Zwanziger, J. Non-Cryst, Solids **168**, 293 (1994)
42. R.E. Youngman et al., Science **269**, 1416 (1995)
43. S.-J. Hwang et al., Solid State Nucl. Magn. Res. **8**, 109 (1997)
44. C. Joo, U. Werner-Zwanziger, J.W. Zwanziger, J. Non-Cryst, Solids **261**, 282 (2000)
45. S. Wang, J.F. Stebbins, J. Am. Ceram. Soc. **82**, 1519 (1999)
46. S. Kroeker, P.S. Neuhoff, J.F. Stebbins, J. Non-Cryst, Solids **293–295**, 440 (2001)
47. S. Kroeker, J.F. Stebbins, Inorg. Chem. **40**, 6239 (2001)
48. J.A. Tossell, J. Non-Cryst, Solids **183**, 307 (1995)
49. J.A. Tossell, J. Non-Cryst, Solids **215**, 236 (1997)
50. J.W. Zwanziger, Solid State Nucl. Magn. Res. **27**, 5 (2000)
51. R.L. Mozzi, B.E. Warren, J. Appl. Cryst. **3**, 251 (1970)
52. F.M. Dunlevey, A.C. Cooper, Bull. Am. Ceram. Soc. **51**, 374 (1972)
53. E. Chason, F. Spaepen, J. Appl. Phys. **64**, 4435 (1988)
54. J. Swenson, L. Börjesson, Phys. Rev. B **55**, 11138 (1997)
55. M. Bionducci et al., J. Non-Cryst. Solids **177**, 137 (1994)
56. A.K. Soper, J. Phys.: Condens. Matter **22**, 404210 (2010)
57. S.R. Elliott, Philos. Mag. B **37**, 435 (1978)
58. R.J. Bell, A. Carnevale, Philos. Mag. B **43**, 389 (1981)
59. A. Takada, C.R.A. Catlow, G.D. Price, J. Phys.: Condens. Matter **7**, 8659 (1995)
60. A. Takada, C.R.A. Catlow, G.D. Price, Phys. Chem. Glasses **44**, 147 (2003)
61. L. Huang, J. Kieffer, Phys. Rev. B **74**, 224107 (2006)
62. W. Soppe, C. Van Der Marel, H.W. Den Hartog, J. Non-Cryst, Solids **101**, 101 (1988)
63. J. Swenson, L. Börjesson, Phys. Rev. Lett. **96**, 199701 (2006)
64. A.C. Hannon, A.C. Wright, J.A. Blackman, R.N. Sinclair, J. Non-Cryst, Solids **182**, 78 (1995)
65. R.N. Sinclair et al., Phys. Chem. Glasses **41**, 286 (2000)
66. M. Teter, in *Borate Glasses, Crystals and Melts*, ed. by A.C. Wright, S.A. Feller, A.C. Hannon (Soc. Glass Technol, Sheffield, 1997), p. 407
67. G.E. Walrafen, S.R. Samanta, P.N. Krishnan, J. Chem. Phys. **72**, 113 (1980)
68. G.E. Walrafen, M.S. Hokmabadi, P.N. Krishnan, S. Guha, J. Chem. Phys. **79**, 3609 (1983)
69. R. Hassan, E.S. Campbell, J. Chem. Phys. **97**, 4326 (1992)
70. S. Guha, G.E. Walrafen, J. Chem. Phys. **80**, 3807 (1984)
71. M. Grimsditch, A. Polian, A.C. Wright, Phys. Rev. B. **54**, 152 (1996)
72. J. Nicholas, S. Sinogeikin, J. Kieffer, J. Bass, Phys. Rev. Lett. **92**, 215701 (2004)
73. V.V. Brazhkin et al., Phys. Rev. Lett. **105**, 115701 (2010)
74. G. Carini Jr, E. Gilioli, G. Tripodo, C. Vasi, Phys. Rev. B **84**, 024207 (2011)
75. G. Carini Jr et al., Phys. Rev. Lett. **111**, 245502 (2013)
76. S.K. Lee et al., Phys. Rev. Lett. **94**, 165507 (2005)
77. A.C. Wright et al., Phys. Chem. Glasses **41**, 296 (2000)
78. A. Zeidler et al., Phys. Rev. B **90**, 024206 (2014)
79. A. Takada, Phys. Chem. Glasses **45**, 156 (2004)
80. M. Misawa, J. Non-Cryst, Solids **122**, 33 (1990)
81. J.D. Mackenzie, J. Phys. Chem. **63**, 1875 (1959)

82. P.B. Macedo, W. Capps, T.A. Litovitz, J. Chem. Phys. **44**, 3357 (1966)
83. A. Napolitano, P.B. Macedo, E.G. Hawkins, J. Am. Ceram. Soc. **48**, 613 (1965)
84. N.E. Shmidt, Russ. J. Inorg. Chem. **11**, 241 (1966)
85. P.B. Macedo, A. Napolitano, J. Chem. Phys. **49**, 1887 (1968)
86. E.F. Riebling, J. Am. Ceram. Soc. **49**, 19 (1966)
87. L.L. Sperry, J.D. Mackenzie, Phys. Chem. Glasses **9**, 91 (1968)
88. N.S. Srinivasan, J.M. Juneja, S. Seetharaman, Metall. Mater. Trans. A **25**, 877 (1994)
89. P.G. Pittoni, Y.-Y. Chang, S.-Y. Lin, J. Taiwan Inst. Chem. Eng. **48**, 613 (1965)
90. T.F. Soules, J. Chem. Phys. **71**, 4570 (1979)
91. T.F. Soules, J. Chem. Phys. **73**, 4032 (1980)
92. T.F. Soules, A.K. Varshneya, J. Am. Ceram. Soc. **64**, 145 (1981)
93. M. Amini, S.K. Mitra, R.W. Hockney, J. Phys. C **14**, 3689 (1981)
94. Q. Xu, K. Kawamura, T. Yokokawa, J. Non-Cryst, Solids **104**, 261 (1988)
95. W. Soppe, H.W. Den Hartog, J. Non-Cryst, Solids **108**, 260 (1989)
96. W. Soppe, C. Van Der Marel, W.F. van Gunsteren, H.W. Den Hartog, J. Non-Cryst, Solids **103**, 201 (1988)
97. H. Inoue, N. Aoki, I. Yasui, J. Am. Ceram. Soc. **70**, 622 (1987)
98. A.H. Verhoef, H.W. Den Hartog, Radiat. Eff. Defects Solids **119**, 493 (1991)
99. A.H. Verhoef, H.W. Den Hartog, J. Non-Cryst, Solids **146**, 267 (1992)
100. A.H. Verhoef, H.W. Den Hartog, J. Non-Cryst, Solids **180**, 102 (1994)
101. A. Takada, C.R.A. Catlow, G.D. Price, J. Phys.: Condens. Matter **7**, 8693 (1995)
102. R. Fernández-Perea, F.J. Bermejo, E. Enciso, Phys. Rev. B **53**, 6215 (1996)
103. F.J. Bermejo, J. Dawidowski, R. Fernández-Perea, J.L. Martínez, Phys. Rev. B **54**, 244 (1996)
104. R.E. Youngman, J. Kieffer, J.D. Bass, L. Duffrène, J. Non-Cryst, Solids **222**, 190 (1997)
105. L. Cormier, D. Ghaleb, J.-M. Delaye, G. Calas, Phys. Rev. B **61**, 14495 (2000)
106. R. Fernández-Perea, F.J. Bermejo, M.L. Senent, Phys. Rev. B **54**, 6039 (1996)
107. E. Kashchieva, B. Shivachev, Y. Dimitriev, J. Non-Cryst, Solids **351**, 1158 (2005)
108. A. Takada, Eur. J. Glass Sci. Technol. B **47**, 493 (2006)
109. A. Takada et al., Phys. Rev. B **51**, 1447 (1995)
110. G.E. Gurr, P.W. Montgomery, C.D. Knutson, B.T. Gorres, Acta Crystallogr. **B26**, 906 (1970)
111. C.T. Prewitt, R.D. Shannon, Acta Crystallogr. **B24**, 869 (1968)
112. B. Park, Ph.D. thesis, Alfred University, New-York, 1998
113. J.K. Maranas, Y. Chen, D.K. Stillinger, F.H. Stillinger, J. Chem. Phys. **115**, 6578 (2001)
114. S.K. Fullerton, J.K. Maranas, J. Chem. Phys. **121**, 8562 (2004)
115. S.K. Fullerton, J.K. Maranas, Nano. Lett. **5**, 363 (2005)
116. A. Baroni et al., in preparation (unpublished)
117. L. Huang, M. Durandurdu, J. Kieffer, J. Phys. Chem. C **111**, 13712 (2007)
118. L. Huang, J. Nicholas, J. Kieffer, J. Bass, J. Phys.: Condens. Mater. **74**, 224107 (2006)
119. J. Kieffer, Phys. Chem. Glasses: Eur. J. Glass Sci. Technol. B **50**, 294 (2009)
120. V.V. Brazhkin et al., Phys. Rev. Lett. **101**, 035702 (2008)
121. K. Trachenko et al., Phys. Rev. B **78**, 172102 (2008)
122. S. Ohmura, F. Shimojo, Phys. Rev. B. **78**, 224206 (2008)
123. S. Ohmura, F. Shimojo, Phys. Rev. B. **80**, 020202 (2009)
124. S. Ohmura, F. Shimojo, Phys. Rev. B. **81**, 014208 (2010)
125. A.K. Soper, J. Phys.: Condens. Mater. **23**, 365402 (2011)
126. M. Micoulaut, R. Kerner, D.M. Dossantosloff, J. Phys.: Condens. Matter **7**, 8035 (1995)
127. R.A. Barrio, R. Kerner, M. Micoulaut, G.G. Naumis, J. Phys.: Condens. Matter. **9**, 9219 (1997)
128. A. Bouzid, C. Massobrio, J. Chem. Phys. **137**, 046101 (2012)
129. D. Donadio, M. Bernasconi, Phys. Rev. B **71**, 073307 (2005)
130. D. Donadio, P. Raiteri, M. Parrinello, J. Phys. Chem. B **109**, 5421 (2005)
131. P. Kroll, J. Non-Cryst, Solids **351**, 1127 (2005)
132. J.M. Soler et al., J. Phys.: Condens. Matter **14**, 2745 (2002)
133. J.P. Perdew, K. Burke, M. Ernzerhof, Phys. Rev. Lett. **77**, 3865 (1996)
134. N. Troullier, J.L. Martins, Phys. Rev. B **43**, 1993 (1991)

135. http://www.cpmd.org
136. G. Ferlat et al., Phys. Chem. Glasses: Eur. J. Glass Sci. Technol. B **47**, 441 (2006)
137. E. Lascaris et al., J. Chem. Phys. **140**, 224502 (2014)
138. S. Grimme, J. Comput. Chem. **27**, 1787 (2006)
139. J. Diefenbacher, P.F. McMillan, J. Phys. Chem. A **105**, 7973 (2001)
140. P.J. Bray, Inorg. Chim. Acta **289**, 158 (1999)
141. S. Wang, J.F. Stebbins, J. Non-Cryst, Solids **231**, 286 (1998)
142. S.K. Lee, C.B. Musgrave, P. Zhao, J.F. Stebbins, J. Phys. Chem. B **105**, 12583 (2001)
143. O.L.G. Alderman et al., Phys. Chem. Chem. Phys. **15**, 8506 (2013)
144. I. Hung et al., J. Magn. Res. **197**, 229 (2009)
145. I. Hung et al., J. Solid State Chem. **182**, 2402 (2009)
146. A. Wong et al., Phys. Chem. Chem. Phys. **11**, 7061 (2009)
147. J.W. Zwanziger, Solid State Nucl. Magn. Reson. **27**, 5 (2005)
148. C.J. Pickard, F. Mauri, Phys. Rev. B **63**, 245101 (2001)
149. T. Charpentier, J. Virlet, Solid State Nucl. Magn. Reson. **12**, 227 (1998)
150. P. Giannozzi et al., J. Phys.: Condens. Matter **21**, 395502, 19 pp (2009)
151. D. Cabaret, F. Mauri, G.S. Henderson, Phys. Rev. B **75**, 184205 (2007)
152. S.K. Lee et al., Nat. Mater. **4**, 851 (2005)
153. X. Liu, M.E. Fleet, Phys. Chem. Min. **28**, 421 (2001)
154. F.L. Galeener, G. Lucovsky, J.C. Mikkelsen Jr, Phys. Rev. B **22**, 3983 (1980)
155. E.C. Kerr, H.N. Hersh, H.L. Johnston, J. Am. Chem. Soc. **72**, 4738 (1950)
156. P. Richet, D. de Ligny, E.F. Westrum Jr, J. Non-Cryst, Solids **20**, 315 (2003)
157. B. Park, E. Bylaska, L.R. Corrales, Phys. Chem. Glasses **44**, 174 (2003)
158. F.H. Stillinger, T.A. Weber, Phys. Rev. A **25**, 978 (1982)
159. N. Marzari et al., Rev. Mod. Phys. **84**, 1419 (2012)
160. P.M. Piccione et al., J. Phys. Chem. B **104**, 10001 (2000)
161. F.C. Kracek, G.W. Morey, H.E. Merwin, Am. J. Sci. **35A**, 143 (1938)
162. D.R. Ulhmann, J.F. Hays, D. Turnbull, Phys. Chem. Glasses **8**, 1 (1967)
163. M.J. Aziz, E. Nygren, J.F. Hays, D. Turnbull, J. Appl. Phys. **57**, 2233 (1985)
164. L. McCulloch, J. Am. Chem. Soc. **59**, 2650 (1937)
165. D. Kline, P.J. Bray, H.M. Kriz, J. Chem. Phys. **48**, 5277 (1968)
166. G. Ferlat, A.P. Seitsonen, M. Lazzeri, F. Mauri, Nat. Mater. **11**, 925 (2012)
167. B. Winkler, C.J. Pickard, V. Milman, G. Thimm, Chem. Phys. Lett. **337**, 36 (2001)
168. D. Vanderbilt, Phys. Rev. B **41**, 7892 (1990)
169. S.J. Clark et al., Z. Kristallogr. **220**, 567 (2005)
170. S.L. Price, Acc. Chem. Res. **42**, 117 (2009)
171. F. Claeyssens, N.L. Allan, N.C. Norman, C.A. Russell, Phys. Rev. B **82**, 094119 (2010)
172. S.-H. Jhi, Y.-K. Kwon, K. Bradley, J.-C.P. Gabriel, Solid State Commun. **129**, 769 (2004)
173. C. Barboiu et al., J. Membr. Sci. **326**, 514 (2009)
174. S.-H. Jhi, Y.-K. Kwon, K. Bradley, J.-C. P. Gabriel, Boron oxide and related compounds for hydrogen storage. US Patent 7479240 B2, 2009
175. J. Swenson, L. Börjesson, J. Non-Cryst, Solids **223**, 223 (1998)
176. T. Sato, N. Funamori, T. Yagi, Nat. Commun. **2**, 345 (2011)
177. C. Weigel et al., Phys. Rev. Lett. **109**, 245504 (2012)
178. F. Claeyssens, J.N. Hart, N.C. Norman, N.L. Allan, Adv. Funct. Mater. **23**, 5887 (2013)
179. S.S. Cole, N.W. Taylor, J. Am. Ceram. Soc. **18**, 55 (1935)
180. S. Kocakusak et al., Chem. Eng. Proc. **35**, 311 (1996)
181. Q. Yang et al., Physica A **27**, 319 (2005)
182. M. Putkonen, L. Niinistö, Thin Solid Films **514**, 145 (2006)
183. O. Moon, B.-C. Kang, S.-B. Lee, J.-H. Boo, Thin Solid Films **464–465**, 164 (2004)
184. D. Buc et al., Thin Solid Films **515**, 8723 (2007)
185. R. Ma, Y. Bando, Chem. Phys. Lett. **374**, 358 (2003)
186. A. Martyła, B. Olejnik, P. Kirszensztejn, R. Przekop, Int. J. Hydrogen Energy **36**, 8358 (2011)
187. F. Caupin et al., J. Phys.: Condens. Matter. **24**, 284110 (2012)

188. D.T. Limmer, D. Chandler, J. Chem. Phys. **135**, 134503 (2011)
189. J.C. Palmer et al., Nature **510**, 385 (2014)
190. P.H. Poole, F. Sciortino, U. Essmann, H.E. Stanley, Nature **360**, 324 (1992)
191. C.A. Angell, Nature **510**, 673 (2014)
192. A.E. Geissberger, F.L. Galeener, in *The Structure of Non-Crystalline Materials 1982*, ed. by P.H. Gaskell, J.M. Parker, E.A. Davis (Taylor and Francis, London, 1983), p. 381
193. M. Menetrier et al., Phys. Chem. Glasses **33**, 222 (1992)
194. S.-J. Hwang et al., J. Am. Chem. Soc. **120**, 7337 (1998)
195. M. Royle, J. Cho, S.W. Martin, J. Non-Cryst, Solids **279**, 97 (2001)
196. J. Cho, S.W. Martin, J. Non-Cryst, Solids **182**, 248 (1995)
197. R.N. Sinclair et al., J. Non-Cryst. Solids **293**, 383 (2001)
198. W. Yao, S.W. Martin, V. Petkov, J. Non-Cryst, Solids **351**, 1995 (2005)
199. M. Micoulaut, in Current Problems in Condensed Matter, Theory and Experiment, edited by J. L. Moran-Lopez (Plenum Press, 1998), p. 339
200. G. Ferlat, M. Micoulaut, Phys. Chem. Glasses: Eur. J. Glass Sci. Technol. B **50**, 284 (2009)
201. H. Diercks, B. Krebs, Angew. Chem. **89**, 327 (1977)
202. S. Balasubramanian, K.J. Rao, J. Phys. Chem. **98**, 9216 (1994)
203. N. Umesaki et al., J. Non-Cryst. Solids **177**, 200 (1994)
204. A.H. Verhoef, H.W. den Hartog, J. Non-Cryst, Solids **182**, 235 (1995)
205. J. Swenson, L. Börjesson, W.S. Howells, Phys. Rev. B **57**, 13514 (1998)
206. A.N. Cormack, B. Park, Phys. Chem. Glasses **41**, 272 (2000)
207. C.-P.E. Varsamis, A. Vegiri, E.I. Kamitsos, Phys. Rev. B **65**, 104203 (2002)
208. A. Vegiri, C.-P.E. Varsamis, E.I. Kamitsos, Phys. Rev. B **80**, 184202 (2009)
209. A. Vegiri, E.I. Kamitsos, Phys. Rev. B **82**, 054114 (2010)
210. M.A. González et al., J. Non-Cryst. Solids **354**, 203 (2008)
211. M.M. Smedskjaer et al., J. Chem. Phys. **133**, 154509 (2010)
212. M.M. Smedskjaer et al., J. Phys. Chem. B **115**, 12930 (2011)
213. T. Ohkubo et al., J. Phys. Chem. B. **117**, 5668 (2013)
214. D.L. Sidebottom, S.E. Schnell, Phys. Rev. B. **87**, 054202 (2013)
215. S. Le Roux et al., J. Phys.: Condens. Matter **23**, 035403 (2011)

Chapter 15
Functional Properties of Phase Change Materials from Atomistic Simulations

Sebastiano Caravati, Gabriele C. Sosso and Marco Bernasconi

Abstract Chalcogenide alloys are materials of interest for optical recording and electronic nonvolatile memories. These applications rest on an ensemble of functional properties: a fast and reversible transformation between the amorphous and the crystalline phase upon heating and a strong optical and electronic contrast between the two phases that allow discriminating the two states of the memory. We discuss the insights gained from atomistic simulations based on Density Functional Theory on the functional properties of the prototypical phase change compounds $Ge_2Sb_2Te_5$ and GeTe. We review the results on the structural and bonding properties of the crystalline and amorphous phases, the origin of the optical and electronic contrast between the two phases and the source of the fast crystallization of the supercooled liquid. The results on the crystallization kinetics obtained from large scale simulations with interatomic potentials based on Neural Network methods are also discussed.

15.1 Introduction

Chalcogenide alloys such as GeSbTe-based compounds are of interest for applications in rewritable optical media (Digital Versatile Disc, Blu-ray disc) and in non-volatile electronic memory devices [1–5]. Both applications rest on a fast and reversible transformation between the crystalline and amorphous phases which represent the two states of the memory. The phase change is induced by heating, either

M. Bernasconi (✉) · S. Caravati
Department of Materials Science, University of Milano-Bicocca,
Via R. Cozzi 55, 20125 Milano, Italy
e-mail: marco.bernasconi@mater.unimib.it

S. Caravati
e-mail: sebastiano.caravati@mater.unimib.it

G. C. Sosso
London Centre for Nanotechnology, Faculty of Maths and Physical Sciences,
University College London, 17-19 Gordon Street, London WC1H 0AH, UK
e-mail: g.sosso@ucl.ac.uk

C. Massobrio et al. (eds.), *Molecular Dynamics Simulations of Disordered Materials*, Springer Series in Materials Science 215,
DOI 10.1007/978-3-319-15675-0_15

due to laser irradiation in optical media or to Joule effect in electronic phase change memories (PCMs).

The materials of choice for both PCM and Blu-ray discs lie on the pseudobinary GeTe-Sb_2Te_3 tie-line with the $Ge_2Sb_2Te_5$ (GST) composition used for PCM and $Ge_8Sb_2Te_{11}$ used for the optical discs. However, the binary compound GeTe with different doping [6], InSbTe [7], InGeTe [8], and GaSbTe [9] alloys are also under scrutiny for their higher crystallization temperature of interest for applications at high temperatures, e.g. in automotive electronics. A PCM is essentially a resistor of a thin film of GST with a low field resistance that changes by several orders of magnitude depending on the state of chalcogenide alloy, metallic in the crystalline form and insulating in the amorphous phase. In memory operations, cell read out is performed at low bias. Programming the memory requires instead a relatively large current to heat up the GST and induce the phase change, either the melting of the crystal and subsequent amorphization or the recrystallization of the amorphous. The PCM was included in the International Technology Roadmap for Semiconductor (ITRS) in 2005, due to its promising scalability beyond the 45 nm node as opposed to limitations of Flash downscaling. In 2012 the company Micron has actually started producing 45 nm feature-size PCM and downscaling is in progress.

A peculiar ensemble of properties makes materials in this class suitable for these applications. First, the amorphous phase displays a very high crystallization speed which can complete on the 10–100 ns scale in the electronic memory once the system is brought above the glass transition temperature. Still the amorphous phase is very stable at the operation condition of the device (<85 °C for PCM) which assures data retention for above 10 years. Second, the two phases show a very large contrast in the optical properties and in electronic conductivity which allows discriminating the two states of the memory from the measurement of the optical reflectivity in optical disc or of the electrical resistance in PCM. Finally, in PCM the crystallization of the amorphous at viable voltages is possible thanks to the phenomenon of the threshold switching first discovered by Ovshinsky [10] in the late sixties in the $Ge_{10}Si_{12}As_{30}Te_{48}$ alloy also called Ovonic switching after his name. This effect appears in the amorphous phase as a sudden, reversible transition from a high-resistivity to a low-resistivity state once the applied critical voltage exceeds a critical (threshold) voltage.

Although optical discs and phase change electronic memories are now commercial products, the microscopic origin of the functional properties listed above and exploited in the devices has remained elusive and controversial. In the last few years, atomistic simulations based on density functional theory (DFT) have provided important insights on the structure and the chemical bonding in materials in this class which allowed also to shed light on the origin of optical and electronic contrast between the amorphous and crystalline phases and on the microscopic origin of the fast crystallization speed. In the following, we will briefly review these outcomes starting from a discussion of the structural and bonding properties of the crystalline and amorphous phases. We will focus mostly on the prototypical $Ge_2Sb_2Te_5$ compound and the binary GeTe system which shares many features with the ternary GeSbTe alloys. More detailed information on the structural properties of amorphous GST can also be found in the chapter by Akola et al. [11], and by Raty et al. [12]. Regarding the study of crystallization, although pioneering DFT simulations of this process have been performed [13–16], the limitations in size and simulation time hinder a complete characterization of the kinetics of the phase change by full DFT simulations. A route to overcome these limitations has been demonstrated recently by Sosso et al. [17] who developed an interatomic potential for GeTe with a close to DFT accuracy by fitting a large DFT database with a flexible neural network (NN) scheme. The results of large scale NN simulations of the crystallization kinetics of GeTe will be reviewed in Sect. 15.5. We also mention that a Tersoff-like [18] classical potential has been recently developed for GeTe with good results on the structural properties of the liquid and amorphous phases [19].

15.2 Structure and Bonding of the Crystalline and Amorphous Phases

15.2.1 Crystalline GeTe

GeTe has two crystalline phases at normal pressure, a trigonal ferroelectric phase (space group $R3m$) [20] and a cubic paraelectric phase (space group $Fm\bar{3}m$) above the Curie temperature of 705 K [21]. The trigonal phase, with two atoms per unit cell, can be seen as a distorted rocksalt geometry with an elongation of the cube diagonal along the <111> direction and an off-center displacement of the inner Te atom along the <111> direction giving rise to a 3 + 3 coordination with three short and stronger bonds (2.84 Å) and three long and weaker bonds (3.17 Å) in a distorted octahedral configuration. In the conventional hexagonal unit cell the structure can be also seen as an arrangement of GeTe bilayers along the c direction with shorter intrabilayer bonds and weaker interbilayers bonds (see Fig. 15.1a). In the cubic phase, the alternation of long and short bonds partially survives in a disordered manner along all equivalent <111> directions as revealed by EXAFS measurements [22].

The strong bonds within a bilayer can be seen as formed by the σ_{pp} bonds from p-type orbitals of Ge and Te. Since Ge has only two p electrons, the closing of the p-shell is possible thanks to a dative bond from the Te p lone pair leading to a threefold coordination with short bonds for both Ge and Te atoms. The backbonds of the short bonds give then rise to the weaker interbilayer bonds. The latter may also be seen as resulting from a partial resonant bonding between the short and long bond in a sort of 3-centers-2-electrons (3c–2e) bond. This idea, first proposed for GeTe by Lucovsky and White [23], was elaborated further in [26–28] to explain the optical contrast between the amorphous and crystalline phase of phase change materials at large. The resonance is only partial, since as opposed to e.g. aromatic bonds, a symmetry breaking is present in GeTe and in related phase change materials leading to the formation of shorter and longer bonds. The long range order is a necessary requisite for the resonant bonding to exist which implies that the structural disorder in the amorphous phase would lead to the disappearing of the partial resonance.

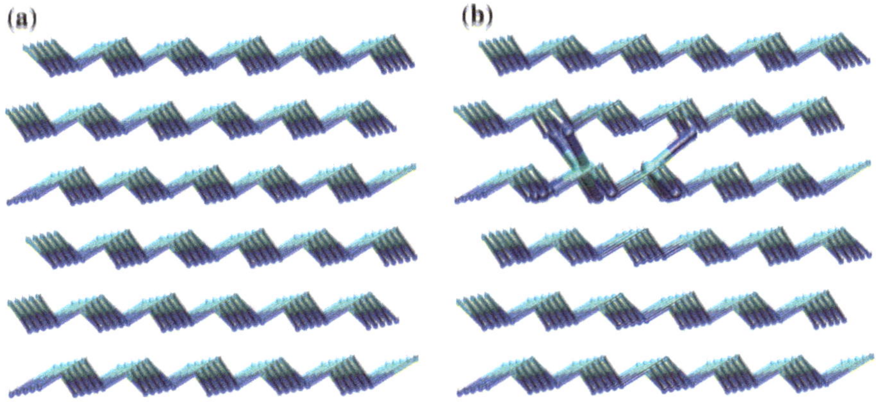

Fig. 15.1 **a** Geometry of the GeTe crystal seen as a stacking of bilayers along the *c* axis of the conventional hexagonal unit cell with three short intrabilayers bonds and three long interbilayers bonds. **b** Reconstruction due to the the presence of a Ge vacancy in trigonal GeTe; three long weaker interlayer bonds turn into three short bonds [25] (see Sect. 15.3)

15.2.2 Crystalline $Ge_2Sb_2Te_5$

The amorphous phase of $Ge_2Sb_2Te_5$ crystallizes in the metastable rock salt cubic phase (c-GST) [29] which only above 250 °C turns into the hexagonal phase (space group $P\bar{3}m1$) stable at normal conditions. In both structures atoms are in an octahedral-like coordination. In c-GST one sublattice is fully occupied by Te atoms and Ge, Sb, and 20 % of vacancies randomly occupy the other sublattice of the face-center-cubic (fcc) crystal [29]. $Ge_2Sb_2Te_5$ and most of the other compositions along the pseudobinary GeTe-Sb_2Te_3 tie-line give rise to metastable cubic phases ideally described as $(GeTe)_x(\square_1 Sb_2Te_3)_y$ with one vacancy on the Sb/Ge sublattice per Sb_2Te_3 formula unit. In these structures we recognize the presence of three *p* electrons per lattice site which allows the closing of the *p*-shell of all atoms [30, 31]. To this end lone pairs (LP) of Te are involved in dative bonds similarly to GeTe [32]. However, the presence of Sb (with three *p* electrons) would in principle make only a fraction of these dative bonds from Te lone-pairs necessary to close the *p* shell of Ge atoms. It has been shown [33] that a Ge atom can actually receive two (or more) dative bonds from neighboring Te atoms giving rise to a symmetric Te-Ge-Te 3-centers-4-electrons bond (3c-4e), probably similar to the nearly symmetric Te-Sb-Te bonds in the hexagonal Sb_2Te_3 crystal [34] which also displays an octahedral-like coordination. The formation of this type of bonding has been identified by DFT calculations on a model of trigonal GeTe in which three Ge atoms have been converted into two Sb atoms plus a vacancy [33]. It remains to be seen which fraction of Te lone pairs is actually involved in 3c-4e bonds and how many lone pairs may instead be considered as free lone pairs pointing towards neighboring vacancies in c-GST (see also Sect. 15.3).

15.2.3 The Amorphous Phase

Models of the amorphous phase of GeTe and GST alloys generated by quenching from the melt within DFT molecular dynamics have been obtained by several groups [13–15, 35–41] with very similar results. In the amorphous phase, Ge and Sb atoms are mostly fourfold coordinated, while Te atoms are mostly threefold coordinated. All Te and Sb and the majority of Ge atoms are in a defective octahedral-like geometry with octahedral bonding angles but a lower than six coordination (see Fig. 15.2d). Actually, the distribution of the bond lengths for Ge and Sb atoms in octahedral-like sites shows a bimodal shape with three shorter bonds and other longer ones. The defective octahedral-like environment of Ge/Sb recalls a $3 + n$ ($n = 0, 1, 2$) geometry similar to the $3+3$ bonding coordination in crystalline trigonal GeTe [42]. The same analysis in c-GST shows a much less pronounced separation in longer and shorter bonds with respect to the amorphous phase [42]. The shorter bonds in a-GST are in fact shorter than the shorter bonds in c-GST in agreement with extended x-ray absorption fine structure (EXAFS) and x-ray absorption near edge structure (XANES) measurements [44]. Some Ge atoms are, however, tetrahedrally coordinated, as inferred from EXAFS data [44] although only in the fraction of about 27–33 % [35, 37, 38]. The distribution of different local environments shows that the tetrahedral coordination of Ge is favored by wrong Ge-Ge and Ge-Sb bonds, absent in the crystal. This feature can be rationalized as follows: the sp3 bonding configuration is stronger than the spp bonding orbitals for the Ge-Ge bond while the reverse is true for the Ge-Te bonds [45]. Finally, there are no chains of Te atoms but a few dimers and trimers. The partial coordination numbers in the model of amorphous GST of [38] are reported in Table 15.1. Direct simulation of the deposition process has actually revealed that models of as-deposited $Ge_2Sb_2Te_5$ show a significantly larger fraction of "wrong bonds" and consequently a larger fraction of tetrahedra [45]. The fraction of Ge atoms in a tetrahedral geometry increases from about 27 % in melt-quenched models [38] to about 54 % in as-deposited (asdep) models [45]. In the DFT models, the Ge-Te bond length in tetrahedra (2.71 Å) are shorter than the Ge-Te bonds (2.81 Å) in defective octahedra [37]. An increase in the fraction of tetrahedra therefore brings the average Ge-Te bond length in closer agreement with the value inferred from EXAFS data which is as short as 2.61 ± 0.01 [44]. Note that the Ge-Ge bond in tetrahedra is only 2.62 Å long in the melt-quenched models of [38]. Due to the higher fraction of tetrahedra and a lager fraction of homopolar bonds, the EXAFS spectrum computed from the asdep model is in much better agreement with the experimental EXAFS measured on asdep samples than those computed from the melt-quenched models [45]. However, a similar bond length of 2.61 Å is also obtained from EXAFS on laser amorphized GST [44]. Therefore, a discrepancy between the average bond length predicted for melt quenched models from DFT and those inferred from EXAFS data still persists. This misfit might be ascribed to deficiencies in the mostly used Perdew-Becke-Ernzerhof (PBE) [46] generalized gradient approximation (GGA) to the exchange-correlation functional. Actually, slightly shorter bonds are found in [47] by using the meta-GGA functional by Tao et al. [48]. It has also been suggested that the discrepancies with experiments might originate from an underestimation of

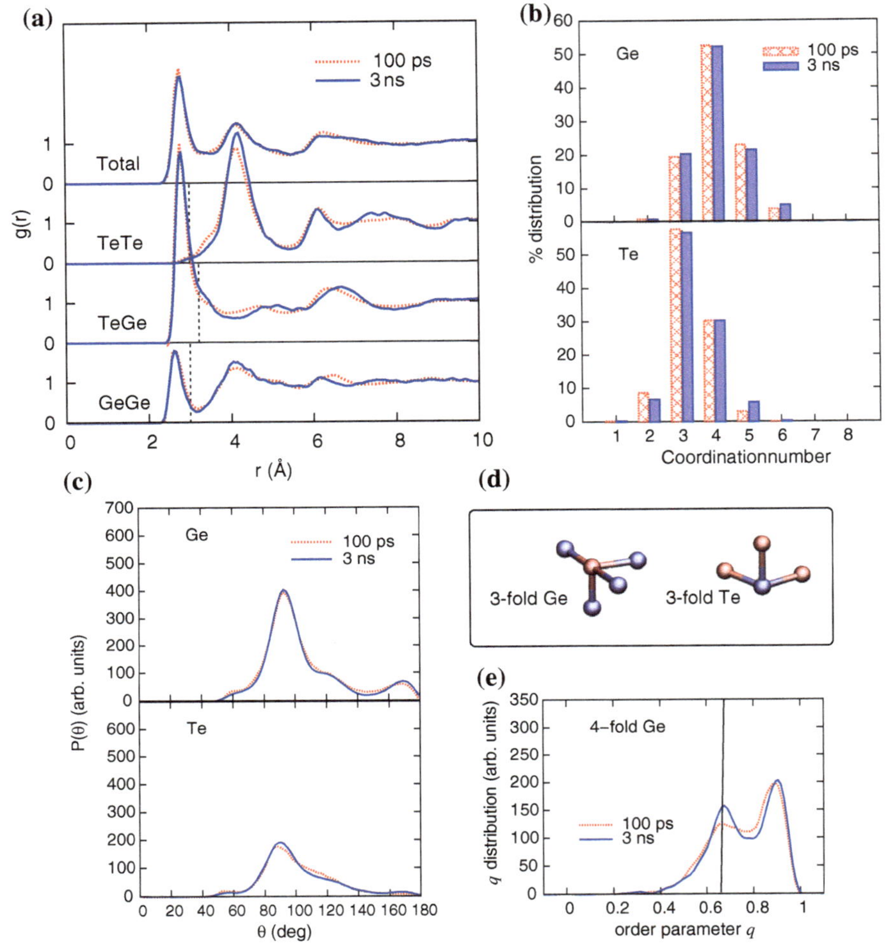

Fig. 15.2 Structural properties at 300 K of two 4096-atom models of amorphous GeTe generated by quenching from the melt in 100 ps (*dashed line*) or 3.1 ns (*continuous line*) in Neural Network molecular dynamics simulations. **a** Total and partial pair correlation functions. The *vertical lines* are the interatomic distance thresholds used to define the coordination numbers. **b** Distribution of coordination numbers of Ge and Te atoms. **c** Partial angle distribution functions referring to X-Ge-Y and X-Te-Y triplets (X,Y = Ge or Te). **d** Snapshot of 3-fold coordinated Te and 4-fold coordinated Ge in defective octahedra. **e** Order parameter q for tetrahedricity for fourfold coordinated Ge atoms [37] defined by $q = 1 - \frac{3}{8}\sum_{i>k}(\frac{1}{3} + \cos\theta_{ijk})^2$ where the sum runs over the couples of atoms bonded to a central atom j and forming a bonding angle θ_{ijk}. q=1 for the ideal tetrahedral geometry, q=0 for the six-coordinated octahedral site, and q=5/8 for a four-coordinated defective octahedral site (*vertical line*)

the fraction of tetrahedra in DFT models of melt-quenched GST [49], possibly due to limitations in the exchange and correlation functional. Recent simulations [49] with hybrid functionals [50] and to a lesser extent with the Becke-Lee-Yang-Parr (BLYP)

Table 15.1 Average coordination numbers of the model of amorphous $Ge_2Sb_2Te_5$ discussed in [38]

	With Ge	With Sb	With Te	Total
Ge	0.29	0.36	3.31	3.96
Sb	0.36	0.43	3.36	4.15
Te	1.33	1.34	0.30	2.97

[52] functional have indeed lead to models of amorphous GST with a larger concentration of tetrahedra and a better agreement with the experimental bond lengths from EXAFS [49]. These latter simulations have, however, made use of relatively small simulations cell (72 atoms) with limited testing with larger (216-atom) models [49]. Since sizeable fluctuations are expected in the fraction of tetrahedra due to finite size effects (see e.g. [17]), an improved statistics with larger simulations cells and several independent models is needed to assess the dependence of the structure of amorphous phase on the choice of the exchange and correlation functional [54]. To this end, we have recently generated four independent 459-atom models of a-GST by quenching from the melt with the BLYP functional. The resulting average fraction of tetrahedra turns out to be 32.8 % which is very close to the outcome of PBE simulations, 27 % [38, 54]. It remains to be seen whether the use of hybrid functionals with large cells might confirm this outcome. Another factor which might potentially affect the fraction of tetrahedra is the protocol used in DFT simulation to quench from the melt. The quenching time from the melting temperature to 300 K is usually limited to lie below 100 ps due to the computational load of DFT simulations. In order to assess possible effects of the quenching time on the structure we have generated a large 4096-atom model of a-GeTe by quenching from the melt in either 100 ps or 3 ns. To this end we used an interatomic potential with a quasi ab-initio accuracy developed in [17] by fitting a huge database of DFT energies by means of the Neural Network (NN) method introduced by Behler and Parrinello [53]. The database consists of the total energies of about 30,000 configurations of 64-, 96-, and 216-atom supercells computed within DFT-PBE and norm conserving pseudopotentials. The NN potential displays an accuracy close to that of the underlying DFT-PBE framework. The reliability of the NN potential in describing the structural and dynamical properties of amorphous and liquid GeTe has been extensively demonstrated in [17]. The structural properties of a-GeTe generated with the NN potential reported in Fig. 15.2 do not show sizeable dependence on the quenching time. The evolution of temperature in the quenching protocol is shown in Fig. 15.3.

A compelling experimental evidence of the presence of tetrahedra in amorphous GeTe and $Ge_2Sb_2Te_5$ comes from Ge L_3-edge XANES measurements [54]. A pronounced differences in the XANES spectra of amorphous and thermally crystallized samples appears as a step-like feature at the absorption edge corresponding to a 2p $\rightarrow$ 5s (4d) electronic transition. Comparison with DFT XANES simulations revealed that the step-like feature is due to the presence of tetrahedrally coordinated Ge atoms in the amorphous phase. Tetrahedra have also a fingerprint in the vibrational properties as they give rise to localized phonons in a-GeTe above the edge of the phonon density of states (DoS) of the crystal at about 170 cm^{-1} [38, 39]. These phonons localized on tetrahedra show up as a distinctive feature in the Raman spectrum of

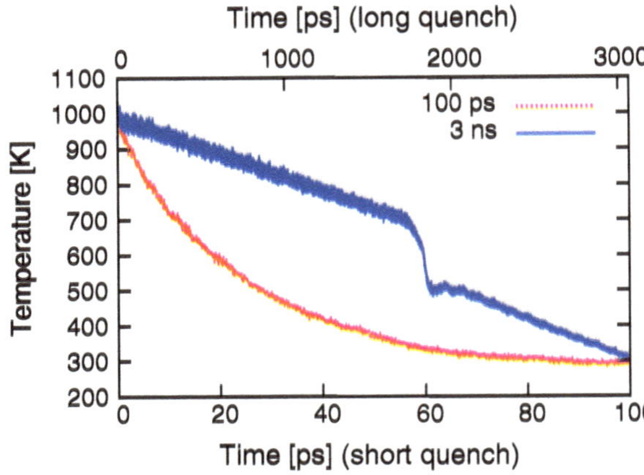

Fig. 15.3 Temperature as a function of time in the slow (*upper curve*) and fast (*lower curve*) quenching protocols used to generate the a-GeTe models whose structural properties are reported in Fig. 15.2. The quenching rate in the temperature range 500–700 K is faster to prevent crystallization

a-GeTe as demonstrated from calculations of the Raman spectrum from DFT phonons and polarizabilities based on the semi-empirical bond polarizability model [39]. The agreement between the theoretical [39] and experimental spectra [55–57] is good and allowed assigning the main peaks to vibrations of defective octahedra and some tiny features above 170 cm^{-1} to vibrations of tetrahedra. Similar calculations have been performed for amorphous and crystalline $Ge_2Sb_2Te_5$ [42]. Inspection on the phonon displacements reveals that the main Raman peaks in c-GST and in a-GST are due to vibrations of defective octahedra. Although in a-GST as well the phonon DoS above 190 cm^{-1} was dominated by vibrations strongly localized on tetrahedra, they do not show up in Raman spectra. This is because of the much lower polarizability of Ge-Te and Ge-Ge/Sb bonds present in tetrahedra with respect to the Sb-Te bonds in defective octahedra. As opposed to a-GeTe, the signatures of tetrahedra in the Raman spectrum of a-GST are hidden by the larger Raman cross section of vibrations of defective octahedra. Moreover, the Raman spectrum of the crystalline phase is broader than the spectrum of the amorphous phase, as opposed to the common behavior of most materials. This is due to the presence of disorder and vacancies in the Ge/Sb sublattice of c-GST, which makes the spread in Ge-Te and Sb-Te bond lengths larger in c-GST than in a-GST. Furthermore, the bonds are stronger in a-GST than in c-GST which is responsible for a blueshift of the Raman peaks in a-GST with respect to c-GST.

The fraction of Ge in tetrahedral configurations also depends on the composition of the GST alloys as shown by the DFT results shown in Table 15.2.

Regarding the medium-range order, the analysis of the ring distribution revealed that the four-membered ring is the most abundant [13–15, 35, 38]. The overwhelming majority of four-membered rings consists of ABAB rings (A=Ge; Sb and B=Te) without wrong bonds corresponding to the square rings typical of the crystalline cubic phase. The similarity of the bonding topology in the amorphous/liquid and crystalline phases with the predominance of square rings has been identified as one of the key features leading to a fast crystallization of phase change materials [13–15]. The alignment of the four-membered rings during crystallization could be also favored by the presence of the nanocavities in the amorphous/liquid phase which is another distinctive feature of phase change alloys [35, 58].

Table 15.2 Fraction of tetrahedra in DFT models of GeSbTe alloys

2 2 5[a]	2 2 5[b]	GeTe[c]	1 2 4[d]	2 4 5[e]	1 1 1[f]	8 2 11[g]	Sb_2Te_3[h]
33 %	58 %	22 %	22 %	28 %	46 %	42 %	No tetra

[a]melt-quenched, [37]
[b]as-dep, [45]
[c][39]
[d][41]
[e][59]
[f][59]
[g][36]
[h][60]

In the search of phase change materials with higher crystallization temperature of interest for automotive applications, other telluride compounds besides GeSbTe alloys are attracting increasing interest in recent years. Models of the amorphous phase of In_3SbTe_2 [61], $InGeTe_2$ [62] and $Ga_4Sb_6Te_3$ [63] have then been generated by quenching from the melt in DFT-MD. In the models of amorphous $InGeTe_2$, In atoms are mostly fourfold coordinated in $InTe_4$ tetrahedra (63 % of tetrahedral In), similarly to the crystalline phase of InTe and In_2Te_5 [62]. In the In_3SbTe_2 compound as well a large fraction of In (47 %) is in a tetrahedral geometry. This ternary system can be actually considered as a pseudobinary alloy of InSb and InTe compounds which have both a crystal structure with tetrahedral coordinations. In fact, very few Sb-Te bond are found in amorphous In_3SbTe_2 [61]. Similarly, the amorphous phase of $Ga_4Sb_6Te_3$ can be seen as a pseudobinary alloy of GaSb and GaTe, still tetrahedral in the crystalline phase, with few Sb-Te bonds [63]. Consequently, also Ga is mostly tetrahedral in $Ga_4Sb_6Te_3$ [63] as well as in the amorphous phase of the binary GaSb compound [64] and of $Ga_{11}Ge_{11}Te_{78}$[65].

As a final remark on the structural properties, we mention that most amorphous models of phase change materials have been generated at fixed volume by choosing the experimental density of the amorphous or of the liquid phase. In some cases, the density of the resulting amorphous models have then been optimized at zero temperature. However, the constraint of fixed volume during quenching might drive the system toward a somehow different structure with respect to that which would have been obtained by quenching at constant pressure. For selenide glasses it has been actually pointed out that the structure generated at the theoretical equilibrium density is in fact in better agreement with experiments, in particular the ratio between corner and edge-sharing tetrahedra, than those generated by fixing the density to the experimental value [66]. In this respect, the optimization of the density in the liquid phase of GeTe, GST and $InGeTe_2$ turned out to be problematic with DFT-GGA because of the presence of nanovoids which tend to coalesce by increasing the cell size. The addition of van der Waals interaction with the semi-empirical prescription given by Grimme [67] hinders the coalescence of voids and brings the theoretical equilibrium density of the liquid in better agreement with experiments [62, 68]. Van der Waals interaction à la Grimme has also been used to model the layered crystalline phase of GeSbTe alloys [69]. Dispersion forces have also been shown to improve the description of structural properties of liquid $Ge_{15}Te_{85}$ which display the presence of Te chains weakly interacting with each other [70].

15.3 Origin of the Electrical Resistivity Contrast Between the Crystal and Amorphous Phases

In spite of the structural differences discussed in Sect. 15.2 the electronic DoS of the cubic and amorphous phases are very similar for GeTe and GST. This is shown experimentally by photoemission spectroscopy [71] and by DFT simulations [35, 38]. The optical gap increases only slightly from 0.5 eV in c-GST to 0.7 eV in a-GST [72]. The contrast in electrical conductivity between the two phases should then come from a different position of the Fermi level and/or a different mobility due for instance to the presence of a mobility gap in the amorphous phase. The trigonal crystalline phase of GeTe and the cubic phase of GST are actually small gap semiconductors which turn into degenerate p-type semiconductors because of the presence of vacancies (non-stoichiometric in the case of $Ge_2Sb_2Te_5$) in the Ge/Sb sublattice. The hole concentration in c-GST is about $8 \times 10^{19}/cm^3$ at 300 K as given by Hall measurements [72]. In the case of trigonal GeTe, it was shown that the formation energy of a neutral Ge vacancy is about 0.25 eV and that the vacancies brings the Fermi level inside the valence band leading to the degenerate p-type character measured experimentally [25, 73]. A Te vacancy and a Ge_{Te} antisite defects cost instead as much as 2.5 and 1.5 eV, respectively [25, 73]. The calculation of the formation energy for the charged vacancies (+/−) actually assigns the position of the Fermi level for the transition between the +/− states just below the valence band edge [25, 73]. A Ge vacancy thus generates two holes in the valence band. The formation energy of a Ge vacancy is low because of the formation of new Ge-Te strong bonds in place of the weak longer bonds of the ideal crystal (see Fig. 15.1b). This process can be seen as the formation of a dative bond from a LP of a Te atom, nearest neighbor to the vacancy, to the Ge atoms of the neighboring bilayer which corresponds to the 3c-4e bond discussed in [33] and mentioned in Sect. 15.2 but for two holes injected in the valence band.

In c-GST, because of the disorder in the Ge/Sb sublattice and the presence of stoichiometric vacancies, the formation energy of defects strongly depends on the defect site as demonstrated by DFT calculations in a 270-atom model [38, 74]. The defect with the lowest average formation energy is actually the Sb vacancy (0.31 ± 0.25 eV) [38, 74]. Deficiency in the Ge or Sb content with respect to the stoichiometric $Ge_2Sb_2Te_5$ composition leads to a shift of the Fermi level close or inside the valence band. On the contrary, a Ge/Sb excess introduced by partially filling the stoichiometric vacancies induces a shift of the Fermi level close to or inside the conduction band. Due to the small size of the simulation cell (270-atom) used in [38, 74], it was not possible to asses whether the Fermi level actually coincides with delocalized states at the edge of the valence band or with the position of shallow localized states above the band edge. The possibility to have a shallow acceptor level close to the valence band, e.g. by localization of a hole on a lone pair of Te pointing to a vacancy, has been addressed by Zhang et al. [75]. By using very large models (1008 atoms) of c-GST, it was shown that the DoS projected on Te atoms has a peak at about E_F whose contribution increases by decreasing the coordination of Te. In fact, spatially localized states at E_F are found around clusters of low-coordinated Te atoms.

These configurations, although higher in energy, might be present as metastable states in crystals generated by recrystallization of the amorphous phase. Ordering of the vacancies and disappearing of strongly undercoordinated Te atoms upon annealing has then proposed to cause the metal-insulator transition observed experimentally in the hexagonal phase of $GeSb_2Te_4$ [75, 76]. Actually, we can understand the possible formation of localized states at the Fermi level in GeTe by pursuing further the idea presented by Kolobov et al. [33]. To this end, we performed DFT calculations with the PBE functional and norm conserving pseudopotentials [77]. The Kohn-Sham orbitals were expanded in a triple-zeta-valence plus polarization (TZVP) Gaussian-type basis set and the charge density was expanded in a plane wave basis set with a cutoff of 240 Ryd to efficiently solve the Poisson equation within the Quickstep scheme [78, 79]. Brillouin zone integration was restricted to the supercell Γ point. An increasing number of vacancies nearest neighbor to a Te atom has been introduced starting from a 360-atoms supercell of trigonal GeTe. To keep the number of electrons constant and to mimic what happens in GST, two Ge atoms are substituted by Sb for each Ge vacancy added to the system. The localization of individual Kohn-Sham (KS) states is quantified by the Inverse Participation Ratio (IPR) defined for the i-th KS state by $\sum_j c_{ij}^4/(\sum_j c_{ij}^2)^2$ where j runs over the Gaussian-Type Orbitals (GTOs) of the basis set and c_{ij} are the expansion coefficients of the i-th KS state in GTOs.

In GeTe the introduction of a single vacancy leads to the formation of the 3c–4e bonds discussed in Sect. 15.2 and no localized states at E_F. By introducing two vacancies in adjacent layers in such a way that Te has no neighbors along one of the three orthogonal bonding directions one should actually induce the formation of a free lone pair. However, the structural relaxation around the undercoordinated Te atoms is so large that only when the number of vacancies is increased to four the resulting two-fold coordinated Te gives rise to a strongly localized state at the edge of the valence band as shown in Fig. 15.4a.

In the case of c-GST, the presence of very many stoichiometric vacancies makes the configurations of 4-, 3- and 2-coordinated Te not so unlikely. However, the degree of localization of the associated KS states is much lower with respect to that of the same complexes in GeTe. In fact, the LP on an even twofold coordinated Te atom in a 270-atom model of c-GST has a rather weak localization as shown in Fig. 15.4b. Localization can, however, be enhanced by further vacancy clustering as discussed in [75].

Regarding the amorphous phase, it was shown in [38] that amorphous models of GST generated from the melt with defects in stoichiometry still show a Fermi level at midgap as in the stoichiometric model. The 270-atom models with composition $Ge_{2.2}Sb_2Te_5$ (Ge excess) and $Ge_2Sb_{1.8}Te_5$ (Sb deficiency), which show a degenerate n-type and p-type character in the cubic phase, are intrinsic semiconductors in the amorphous phase with the Fermi level at midgap. In GST, and presumably also in other materials of this class, the change in number of electrons caused by defects in stoichiometry can be accommodated by changing the number of homopolar/wrong bonds (Ge–Ge, Ge–Sb and Te–Te), the number of Te LP involved in dative bonds and/or by the switch of Ge atoms from octahedral to tetrahedral configurations that brings also the Ge s-electrons to be involved in the chemical bonding. We must

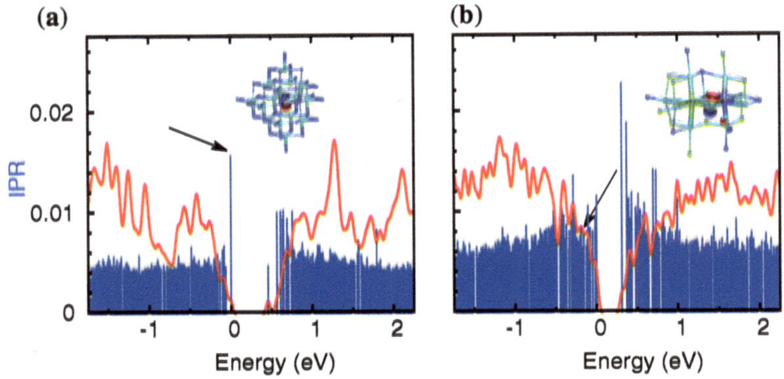

Fig. 15.4 **a** Electronic density of states and Inverse Participation Ratio (*IPR*, see text) of individual KS states in a model of trigonal GeTe with ad-hoc clustered (see text) Ge vacancies. The arrow indicates the Kohn-Sham orbital corresponding to a *p* lone pair of a twofold coordinated Te with four nearest neighbor vacancies. *Inset* Isosurface of the lone pair state under consideration. **b** The same of panel (**a**) for a two-fold coordinated Te atom in a 270-atom model of c-GST

consider, however, that these conclusions are drawn from calculations on relatively small (270 atoms) amorphous models. In a real system we might conceive the occurrence of several other types of defective sites in the amorphous network which might lead to localized states in the band gap. We should therefore consider a mechanism suitable to pin the Fermi level at midgap in the presence of whatever other types of defects that might occur in a real system, but are not seen in our models because of their small size. Three general possibilities leading to the pinning of the Fermi level at midgap in amorphous materials are discussed in literature and summarized in [25] (Fig. 15.5). The first possibility (Fig. 15.5a) is the Cohen-Fritzshe-Ovshinsky model [80] in which E_F is pinned at midgap because of the overlap of Urbach tails from valence and conduction bands. The second is the presence of a dominant defect at midgap, such as the Si dangling bond in a-Si [81]. The third is the presence of defects with an effective negative Hubbard U energy, such as the valence alternation pairs (VAP) present in As_2Se_3 and amorphous Se [2, 82, 83].

In the latter case a pair of neutral defects with energy inside the gap is unstable with respect to charge transfer and formation of a filled and empty states close to the band edges (D^- and D^+ in Fig. 15.5c). This is the case for a couple of undercoordinated atoms in amorphous Se. As shown in Fig. 15.6 two neutral one-coordinated Se atoms (a radical defect, C_1^0) transform into a negatively charged one-coordinated atom (with three filled lone pairs, C_1^-) and a positively charged 3-coordinated atom (C_3^+), i.e. 2 $C_1^0 \rightarrow C_1^- + C_3^+$ because of the negative Hubbard-U character of the C_1^0 defect.

In order to discriminate between these three possible scenarios, we have computed the Bader charges of the somehow larger models (459 atoms) of stoichiometric amorphous GST generated with the BLYP functional as discussed in Sect. 15.2. The Bader charges give a measure of the ionic charges [84]. The Bader ionic charges calculated from the total electronic charge density by using the scheme of [85] are

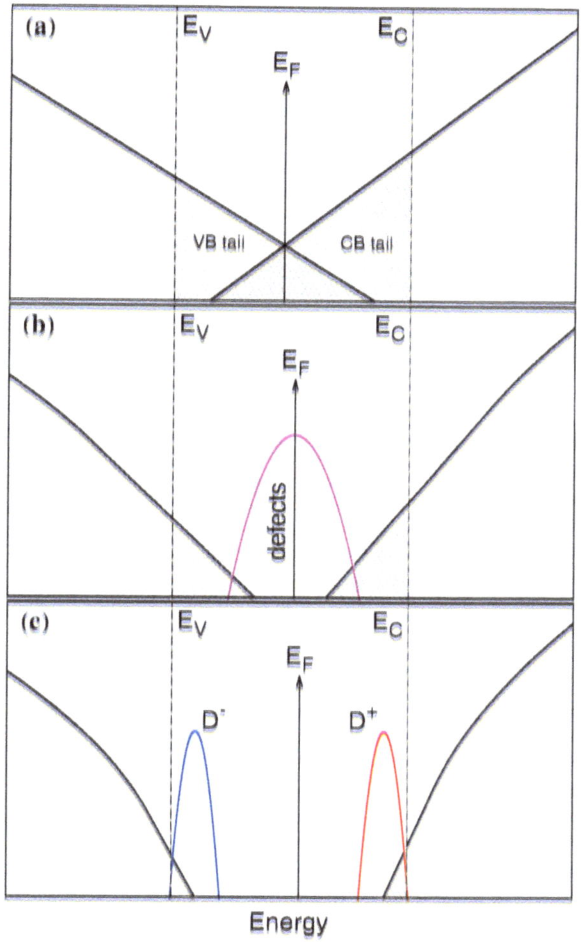

Fig. 15.5 Models of the electronic DoS in amorphous materials leading to a pinning of the Fermi level at midgap. **a** Cohen-Fritzshe-Ovshinsky model with overlapping band tails, **b** dangling bond model, and **c** negative U defect model with charged defects (adapted from [25])

shown for a 270-atom model of c-GST and for four different 459-atom models of a-GST in Fig. 15.7. In a-GST, the distribution of ionic charges tails toward zero due to the presence of homopolar bonds. No highly charged defects such as VAPs are found, confirming the previous outcome from smaller 270-atom models [38, 74]. In fact, the undercoordinated configurations originating from the breaking of a Ge-Te bond can in principle give rise to a charge transfer among the two species as reported in Fig. 15.6 for Se. However, the charged species can then donate/receive dative bonds involving Te lone pairs. Since the dative bond of Te lone pair is part of the bonding network in GST/GeTe, the resulting configurations give rise to electronic states not in the gap but inside the valence/conduction bands [25]. To make one step further we have generated a very large 1728-atom model of amorphous GeTe by quenching from the melt to 300 K in 100 ps with the NN interatomic potential (see Sect. 15.2). The model then optimized at the DFT-PBE level shows a distribution

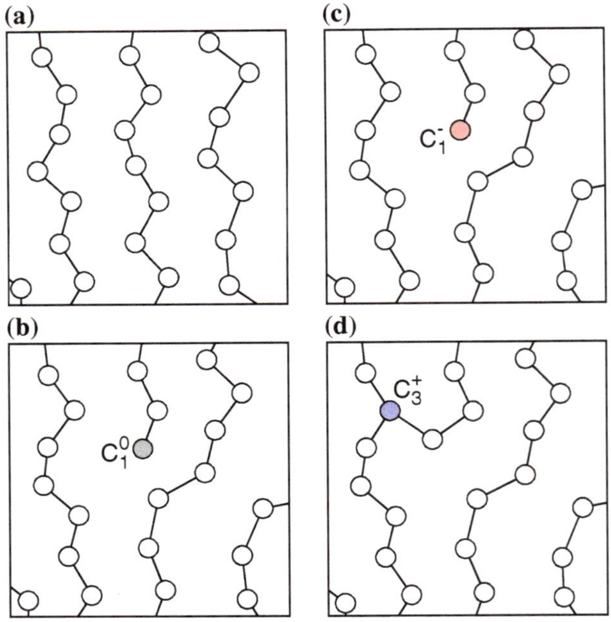

Fig. 15.6 **a** Chain-like structure of amorphous Se. **b** Neutral undercoordinated Se atom (C_1^0). Two C_1^0 defects transform into a valence alternation pair consisting of a one-coordinated negative species C_1^- and a three-coordinated positive species C_3^+ depicted in panel (**c**) and (**d**) respectively (adapted from [83])

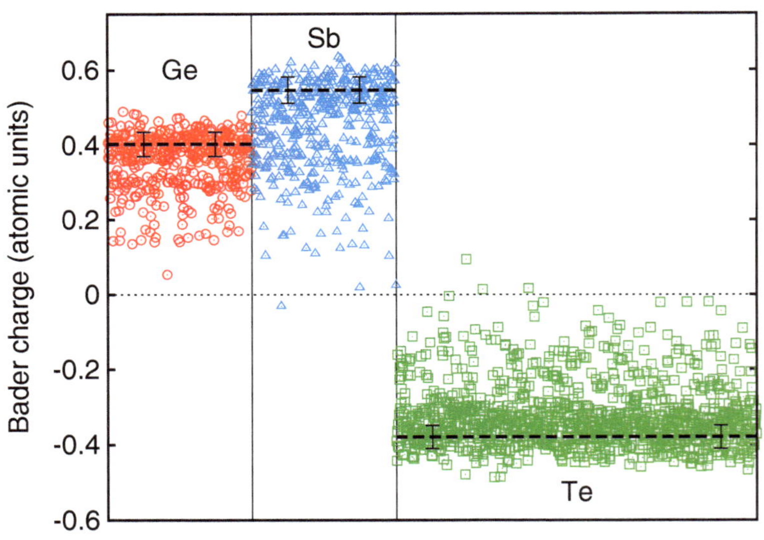

Fig. 15.7 Bader ionic charges (atomic units) of four models of a-GST. Each point corresponds to an individual atom in the supercells. *Thick dashed lines* correspond to the average ionic charges for each species in a 270-atom model of c-GST with standard deviation given by the error bars

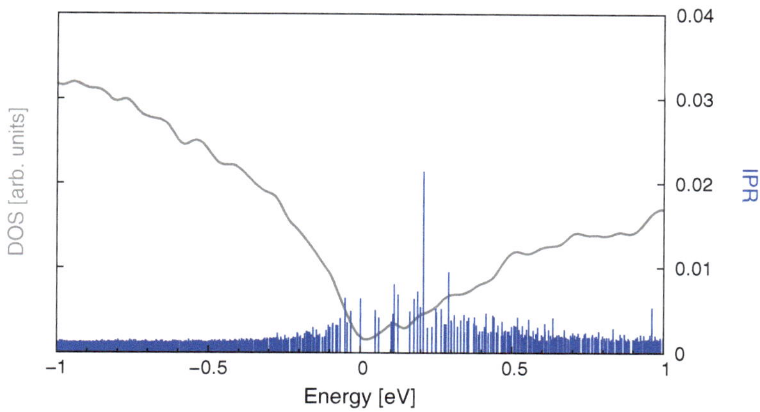

Fig. 15.8 Electronic density of states close to the band gap of a 1728-atom model of a-GeTe. The KS energies are broadened by Gaussian functions of 27 meV width. The zero of energy corresponds to the highest occupied state. The IPR is given on the *left scale* (*blue spikes;* see text for definition) [86]

of Bader charges similar to that reported in Fig. 15.7 with no highly charged defects attributable to VAPs and other negative U centers. The resulting DOS, reported in Fig. 15.8 [86], suggests instead an overlap of the band tails. The states in the gap are mostly associated with homopolar bonds in the form of chains of Ge–Ge bonds.

The most plausible scenario for the pinning of E_F at midgap in the amorphous phase emerging from DFT calculations seems thus to be the Cohen-Fritzshe-Ovshinsky [25]. Actually, the DoS around the band gap in a-GeTe has been measured very recently by modulated photocurrent experiments and photothermal deflection spectroscopy [87]. An overlap of the Urbach tails at midgap is indeed found but it is accompanied by a large density of localized states just above and below the crossing of the Urbach tails at midgap. The microscopic origin of these localized states is at the moment unclear. The identification of the atomistic nature of the Urbach tails and of the midgap defects is particularly important to understand the microscopic origin of the drift in resistance with time which seems a common feature of phase change materials in the amorphous state [88, 89]. The phenomena of resistance drift in the amorphous phase is particularly detrimental for the operation of non-volatile memory, hindering for instance the realization of multibits devices [90].

15.4 Origin of the Optical Contrast Between the Amorphous and Crystalline Phases

Phase change materials based on chalcogenide alloys are of widespread use in rewritable optical media (Digital Versatile Disc, Blu-ray disc) because of the strong optical contrast between the amorphous and crystalline phase (see inset of Fig. 15.9). Such a strong optical contrast is rather unusual since in typical semiconductors (e.g. GaAs in [91]) the amorphous and the crystalline phases have a similar

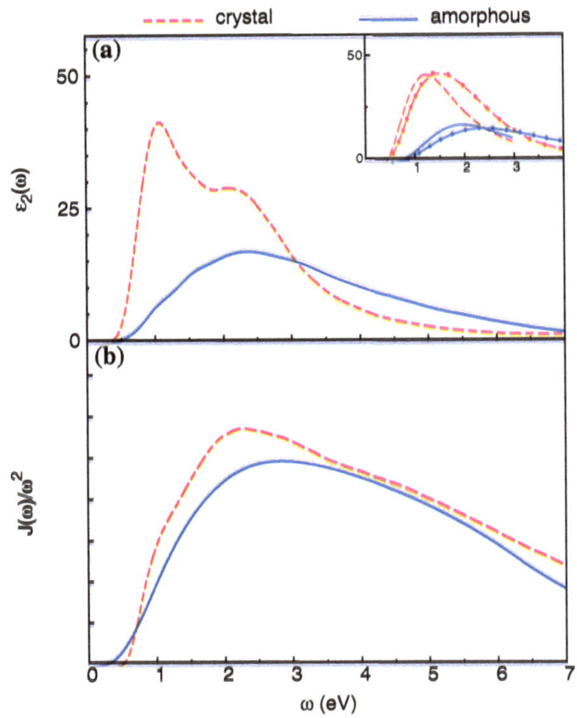

Fig. 15.9 **a** Imaginary part of the dielectric function for the amorphous (*continuous lines*) and cubic crystalline (*dashed lines*) phases of GST computed with the HSE functional [95] to better reproduce the band gap. Two sets of experimental data are given in the insets from [26] (*lines*), and [96] (*lines with circles*). **b** Joint density of states $J(\omega)$ (see 15.2) of crystalline and amorphous GST. To better compare with the ε_2 function (cf. 15.1), $J(\omega)$ is divided by ω^2. Adapted from [94]

optical response. Moreover the electronic DoS of the valence bands has been shown to be very similar in the two phases of GST by photoemission spectroscopy [71]. The change in the optical Tauc gap is also rather small (from 0.5 in c-GST to 0.7 eV in a-GST [72]) and not large enough to account for the large change in optical response across the phase change. It has been thus recognized [92] that the optical contrast must originate from a strong change in the optical matrix elements which enter in the simple expression for the imaginary part of the dielectric function ε_2 given by

$$\varepsilon_2(\omega) = \frac{8\pi^2}{3V_o N_{\mathbf{k}} \omega^2} \sum_{v,c,\mathbf{k}} |\langle c, \mathbf{k}|\mathbf{p}|v, \mathbf{k}\rangle|^2 \delta(\omega - E_{c,\mathbf{k}} + E_{v,\mathbf{k}}), \tag{15.1}$$

where $E_{c,\mathbf{k}}$ and $E_{v,\mathbf{k}}$ refer to the KS energies of conduction and valence bands at the $N_{\mathbf{k}}$ **k**-points in the Brillouin Zone and V_o is the unit cell volume. The optical matrix elements in 15.1 modulate the Joint Density of states $J(\omega)$ (JDOS) given by

$$J(\omega) = \frac{1}{V_o^2} \frac{1}{N_{\mathbf{k}}^2} \sum_{\mathbf{k},\mathbf{k}',c,v} \delta(\omega - E_{c,\mathbf{k}} + E_{v,\mathbf{k}'}). \tag{15.2}$$

The random phase approximation (RPA) expression for ε_2 in 15.1 has been shown to be adequate for GeTe [93]. In fact, local field effects and other many body effects in the GW approximation or from the solution of the Bethe–Salpeter equation have also been shown to give negligible contributions compared to the strong optical contrast between the two phases [93, 94]. The origin of the reduction of the matrix elements in the amorphous phase with respect to the crystal was initially ascribed to the presence of tetrahedra in the amorphous network. A progressive reduction of the optical matrix elements was in fact observed by switching some Ge atoms from the octahedral to tetrahedral sites in small cells of c-GST [93]. However, it was soon recognized that even in the absence of tetrahedral Ge, the loss of medium range order in the form of misalignment of p-orbitals can induce the observed optical contrast in GST and GeTe [26, 28]. In fact, the optical matrix elements are strongly enhanced in the crystal by aligned rows of partially resonantly bonding p orbitals which are absent in the amorphous phase [26, 28]. To prove this, in [28] the optical response has been calculated in trigonal GeTe and in an orthorhombic GeTe structure which was supposed to mimic the amorphous phase as it retains the σ_{pp}-type bonding of trigonal GeTe but lacks the alignment of p orbitals needed to establish a partial resonant bonding. The enhancement of the optical matrix elements in a 3c-2e bonding can actually be deduced from a simple MO-LCAO model. Similarly, a 3c-4e bonding (see Sect. 15.2, [33]) also provides an enhancement of the optical matrix elements with respect to the 2c-2e bonds expected to be present in the amorphous phase. This explanation of the optical contrast provided by the calculations on the crystalline models was then confirmed by the analysis of the optical response in the realistic amorphous models generated by DFT-MD and discussed in Sect. 15.2. The theoretical ε_2 function (15.1) of 270-atom models of crystalline and amorphous phases of GST, reported in Fig. 15.9a reproduces the optical contrast measured experimentally (cf. inset of Fig. 15.9a). The calculation of the JDOS reported in Fig. 15.9b demonstrates that indeed the optical contrast is due to a change in the optical matrix elements. The same features are also found in Sb_2Te_3 which does not have tetrahedra [60, 94], which further confirms that the presence of tetrahedra is not necessary to have a strong optical contrast.

15.5 Atomistic Simulations of Crystal Nucleation and Growth

The key property that makes phase change materials attractive for applications in non-volatile memories is the fast crystallization which allows for a full crystallization in PCM devices on the time scale of 10–100 ns. The fact that both nucleation rate and crystal growth velocity are very large has stimulated direct simulations of the crystallization process by DFT-MD [13–16]. Simulations with up to 180-atom cell and periodic boundary conditions shed light on the atomistic mechanism of formation of the crystalline nucleus and on the role of four-membered rings as seeding structures for nucleation [13–16].

Since the bonding network of both the crystalline and amorphous phases mostly consists of four-membered rings, it has been suggested that the crystallization should require few bonds breaking which would make the transformation particularly fast [13–15]. This has to be contrasted with the behavior of a strong glass former such as silica whose amorphous network is made mostly of five- and seven-membered rings while only six-membered rings are present in crystalline quartz.

Simulations of $Ge_2Sb_2Te_5$ at 600 K [13–15] suggested a critical nucleus size containing about 50 atoms which is, however, not so small compared to the size of the supercell (up to 180-atom) to rule out possible interactions with its periodic images. Indeed simulations of a seeded crystallization with a larger 460-atom cell showed the percolation of crystalline dendrimers connecting the periodic images of the seed before a critical stage is reached [16]. A large atomic mobility was also observed during the seeded crystallization [16]. Quantitative estimate of the growth velocity of supercritical nuclei requires therefore larger simulation cells and longer simulation times than those affordable by DFT-MD simulations. To this end we have undertaken large scale simulations of the binary GeTe compound by using the Neural Network interatomic potential [17] mentioned in Sect. 15.2. We have first studied crystal nucleation and growth from the supercooled liquid [97] at temperatures between the melting temperature T_m and the glass transition temperature T_g. In phase change materials the crystallization temperature and T_g are quite equal as measured from conventional differential scanning calorimetry (DSC) [104]. However, the fast heating rate in the set process of PCM can bring the amorphous phase far above T_g before the onset of crystallization can take place [99]. The fragility of the supercooled liquid has then been proposed as another feature boosting the crystallization speed at the conditions of operation of the devices [98]. In a fragile liquid, the viscosity η is very low down to temperatures close to T_g where a steep, super-Arrhenius increase of η is observed to reach the high value of η expected at T_g. The super-Arrenhius behavior is often described by a Vogel-Tammann-Fulcher (VTF) function $\eta = \eta_o \exp(\frac{E}{k_B(T-T_o)})$ with η_o, E and $T_o < T_g$ [100] as the fitting parameters. The degree of fragility is quantified by the fragility index $m = d(\log_{10}\eta)/d(T_g/T)|_{T=T_g}$ which is up to few hundreds in very fragile liquids and just few tens in strong liquids. In an ideal strong liquid the viscosity follows in fact an Arrhenius behavior with a single activation energy from T_m down to T_g. In a fragile liquid the crystal nucleation rate and the speed of crystal growth can both be very high because the atomic mobility is very high (the viscosity is low) down to temperatures close to T_g where the thermodynamical driving force for the crystallization is also large. In fact, in classical nucleation theory (CNT) the self-diffusion coefficient D controls the kinetic prefactor of both the steady state nucleation rate I^{ss} and the crystal growth velocity u, given by [101, 102]

$$I^{ss} = f_s \frac{6D}{\lambda^2} \Gamma_z \exp\Big(-\frac{\Delta G_c}{kT}\Big) \tag{15.3}$$

and [103]

$$u = \frac{dR}{dt} = \frac{4D}{\lambda}(1 - e^{-\frac{\Delta\mu}{kT}(1-\frac{r_c}{R})}) \tag{15.4a}$$

$$u \xrightarrow[R\gg r_c]{} \frac{4D}{\lambda}(1 - e^{-\frac{\Delta\mu}{kT}}) = u_{kin}(1 - e^{-\frac{\Delta\mu}{kT}}) \tag{15.4b}$$

where ΔG_c and r_c are the formation energy and the radius of the critical nucleus, R is the radius of the growing supercritical nucleus, f_s is the number of adsorption sites at the surface of critical nucleus, λ is a typical atomic jump distance and $\Delta\mu$ is the difference in free energy between the crystal and the liquid/amorphous phase. $\Delta\mu$ is the driving force for crystallization which vanishes at melting and increases upon supercooling. Finally, Γ_z is the Zeldovich factor, typically in the range $0.01 - 0.1$, given by $\Gamma_z = \frac{3}{4\pi r_c^3 \rho}\left(\frac{\Delta G_c}{\pi kT}\right)^{\frac{1}{2}}$ where ρ is the atomic density. Note that in 15.4 the speed of crystal growth becomes a constant for nuclei much larger than the critical one. The expressions 15.3 and 15.4, refers to a diffusion-limited crystallization, which seems to apply to phase change materials as they show a maximum in the crystallization speed for temperatures in between T_m and T_g. In a collision limited crystallization (which does not apply to phase change materials) D/λ has to be replaced by the velocity of sound which is weakly dependent on temperature. The formation energy and the radius of the critical nucleus are given in turn by $\Delta G_c = \frac{16\pi\sigma^3}{3(\rho\Delta\mu)^2}$ and $r_c = \frac{2\sigma}{\rho\Delta\mu}$ where σ is the crystal/liquid(amorphous) interface energy. In the modeling of crystallization kinetics of phase change materials the interface energy is usually taken in the range 0.1–0.033 J/m^3 [105–107]. An interface energy between crystalline and amorphous GeTe of about 0.19 J/m^2 has also been obtained from DFT calculations of slab models [108]. This is possibly an overestimation of the interface energy because a sharp interface model has been used as obtained by joining in a slab geometry two previously optimized crystalline and amorphous models. In this approach the interface energy is computed as $\sigma = (E_{int} - N_{am}E_{am} - N_c E_c)/(2A)$ where E_{int} is the energy of the slab model including the interface, E_{am} and E_c are the bulk energy per atom of N_{am} and N_c atoms, respectively in the amorphous and crystalline phase, and A is the interface area in the slab. A more realistic modeling of the amorphous/crystal is obtained by quenching the liquid in a crystal/liquid interface model by keeping the crystal or part of it fixed during part of the quenching. We followed this procedure by using the NN potential with very large cells which, however, leads to a fuzzy interface as new crystalline layers grow during the quenching from the liquid. This procedure thus introduces an additional error due to the uncertainty in defining which atoms have to be assigned to the crystalline and amorphous regions. A upperbound for this error is the quantity $\frac{\Delta\mu}{2}$ times the interface atomic density. For $\Delta\mu$ of about 0.12 eV corresponding to the crystal and the amorphous phases at low temperature, the upperbound in this error is actually 0.1 J/m^2, i.e. larger than the estimated interface energy itself. This uncertainty does not actually arise for the liquid/crystal interface at equilibrium as the free energy of the liquid and the crystal are equal. We remark, however, that for very small critical nuclei, σ actually corresponds to an effective interface energy which might be sizably different from the interface energy with specific low index surfaces of the crystalline bulk.

The free energy difference between the crystal and the liquid actually depends on temperature as

$$\Delta\mu(T) = \Delta H_f \frac{(T_m - T)}{T_m} + \int_{T_m}^{T} dT' \Delta C_p(T') - T \int_{T_m}^{T} dT' \frac{\Delta C_p(T')}{T'} \tag{15.5}$$

where $\Delta C_p(T)$ is the difference in heat capacity between the two phase and ΔH_f is the latent heat at melting. Thomson and Spaepen [109] proposed the following approximate expression

$$\Delta\mu(T) = \frac{\Delta H_f(T_m - T)}{T_m}\frac{2T}{(T_m + T)} \tag{15.6}$$

which is obtained under the assumption that ΔC_p was independent on temperature and that $\ln\frac{T_m}{T_K} \sim 1$ for the Kauzmann temperature T_K at which the entropy of the crystal equals the entropy of the supercooled liquid. This finally leads to $\Delta C_p \sim \frac{\Delta H_f}{T_m}$. Depending on the relative values of nucleation rate and crystallization speed, phase change materials can exhibit two types of crystallization behavior: (i) nucleation-dominated crystallization in which the nucleation rate is very high and several nuclei appear inside the bulk material which is case of GST, and (ii) growth-dominated crystallization in which crystallization proceeds from the interface with the polycrystalline phase the amorphous bit is embedded in which is the case of doped Sb–Te alloys close to the eutectic composition $Sb_{70}Te_{30}$. However, the crystallization mechanism can change with decreasing size of the amorphous region [110]. A material exhibiting a nucleation-dominated crystallization can crystallize from the crystalline rim once the amorphous bit is very small which might be the case for GST in the 45 nm memory devices.

The self-diffusion coefficient in 15.3 and 15.4 is customarily expressed in terms of the viscosity η by using the Stokes-Einstein relation (SER) $D = \frac{kT}{3\pi\eta\lambda}$. A breakdown of the Stokes-Einstein relation is, however, often observed in fragile liquids [111]. Experimental evidence of the high fragility of the liquid phase and of the breakdown of the SER in GST came from ultrafast DSC, which allowed measuring crystallization speeds of the order of few meters per second in the temperature range of 450–650 K of interest for the operation of PCM devices [98]. The temperature dependence of the crystallization speed measured by ultrafast DSC [98] suggests two regime of crystallization: one at high temperature, much above T_g, with a low activation energy and a second regime near and below T_g with an activation energy for crystal growth of about 2.4 eV as measured by different means in previous works [112, 113]. An activation energy for crystallization of about 0.5 eV at high temperature has been also estimated from electrical measurements on PCM devices [114]. In the latter paper the temperature was, however, not directly accessible during the programming of PCM, but it resulted from a modeling approach. In ultrafast DSC experiments as well, the inferred high fragility of the supercooled liquid phase was the result of several crucial assumptions on the dynamics of the crystallization process that were adopted to deduce the temperature dependence of D and η from the only measurement of the heat released during crystallization. No direct measurements of the diffusivity and viscosity were actually made under the conditions of operation of the device. Molecular dynamics simulations offer instead a route to estimate independently the viscosity, the self-diffusion coefficient and the crystal growth velocity which can allow assessing the breakdown of the SER and the applicability of the CNT relations to describe the crystallization kinetics of phase change materials. To this end we computed in [68] the viscosity of the supercooled liquid phase of GeTe by means of large scale (4096 atoms) NN simulations. The Green-Kubo (GK) expression

$$\eta = \frac{V_o}{k_B T} \int_0^\infty \langle \sigma_{xy}(t)\sigma_{xy}(0)\rangle dt, \tag{15.7}$$

was used, where σ_{xy} is the off diagonal component of the stress tensor and V_o the supercell volume. In the supercooled liquid phase it was possible to reach a converged value of the integral in 15.7 by computing the correlation function up to 100 ps and by averaging over the initial states (< .. >) in simulations up to 1 ns long. Longer simulation times are, however, necessary at lower temperatures as the correlation time of the stress tensor increases by decreasing temperature. Unfortunately, at low temperature the system starts to crystallize on the time scale of few hundreds of ps preventing the integral in 15.7 to converge. For this reason it was possible to reliably compute the viscosity with the GK formula only above 700 K. Still the values of the computed viscosity was so low that it was necessary to assume a fragility index as high as about $m \sim 100$ to fit the computed results with a VTF-like function [115] and a value of T_g of about 400–450 K, equal to the crystallization temperature measured experimentally [116]. On the other hand, the calculated self-diffusion coefficient shows an Arrhenius behavior with an activation energy of about 0.3 eV down to temperatures of about 500 K where a steep rise in viscosity is expected [68]. This result on the self-diffusion coefficient was also confirmed by independent DFT-MD simulations by Liu et al. [108] and it demonstrates the breakdown of SER inferred from DSC [98]. The decoupling between viscosity and atomic mobility has been ascribed [117] to the presence of dynamical heterogeneities in the supercooled liquid, which is another characteristic feature of fragile liquids [111]. Isoconfigurational analysis of atomic trajectories revealed the presence of domains of fast moving particles spatially separated from domains of less mobile particles. The fast moving particles are actually clustered around chains of wrong Ge-Ge homopolar bonds [117].

Simulations with 4096-atom cells lasting up to 4 ns have also allowed estimating the speed of crystal growth as a function of temperatures in the supercooled liquid GeTe [97]. At temperatures below 600 K, several nuclei appeared on the time scale of few hundreds of picoseconds, while a single nucleus forms at temperatures in the range of 625–675 K. Above 675 K, nucleation did not occur in 2 ns due to the decrease of the nucleation rate when approaching T_m. The crystal growth velocity was computed as $u = \frac{dR}{dt}$ in a time interval (up to several hundreds ps) in which the radius of the nuclei increases linearly in time. The resulting kinetic prefactor u_{kin} defined by 15.4 is reported in Fig. 15.10 as function of temperature.

The Thomson-Spaepen approximation (15.6) was used for $\Delta\mu$ with $\Delta H = 0.186$ eV/atom and T_m = 1023 K (exp. T_m = 998 K, [118]) for the NN potential obtained from thermodynamic integration in [68]. The Thomson-Spaepen approximation might appear as a crude approximation in view of DFT estimates of the temperature dependence of C_p for the crystalline and amorphous phases by Liu et al. (see Fig. 15.4 of [108]). However, in the temperature range 500–700 K, the thermodynamic factor has actually a little effect on the crystal growth velocity compared to

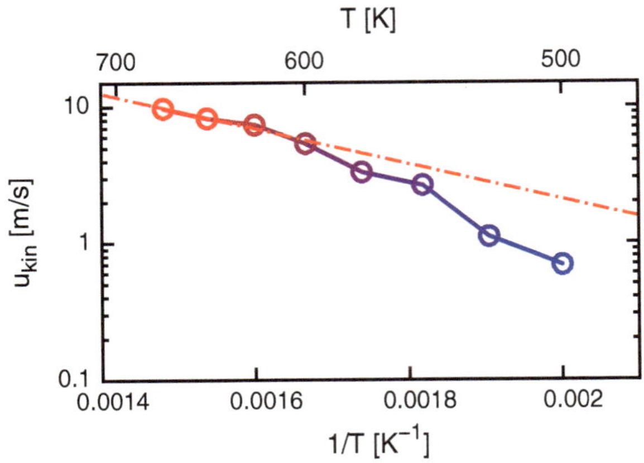

Fig. 15.10 Kinetic prefactor of the crystallization speed u_{kin} as a function of temperature (cf. 15.4) from NN simulations. The *dashed line* is an Arrhenius fit yielding an activation energy of 0.26 ± 0.03 eV [97]

the kinetic prefactor. At higher temperatures u_{kin} clearly follows an Arrhenius behavior with an activation energy of 0.26 eV (cf. Fig. 15.10), which is very close to the computed activation energy of 0.30 eV for the self-diffusion coefficient. At lower temperatures, larger deviations from the Arrhenius behavior were observed because of the uncertainties in the calculation of the growth speed of an individual nucleus due to the presence of several interacting nuclei, which might also affect the application of CNT. The values of u_{kin} extracted from the simulations at higher temperatures are actually well described by CNT. By plugging in the expression $u_{kin} = \gamma 4D/\lambda$ (15.4) the values of D and u_{kin} obtained from the simulations and by setting $\lambda = 3$ Å, one obtains $\gamma = 0.9 \pm 0.1$ in the temperature range of 575–675 K, consistent with the value of γ =1 predicted by CNT [103].

The large scale NN simulations thus allowed assessing the applicability of CNT to the homogeneous crystallization process of GeTe and demonstrating that the high crystallization speed is actually due to the high diffusivity at low temperatures which allows for both a high nucleation rate and high growth velocity of supercritical nuclei. The large self-diffusion coefficient at low temperatures is in turn a manifestation of the fragility of the supercooled liquid and the breakdown of the Stokes-Einstein relation due to the onset of dynamical heterogeneities in the liquid. Quantitative estimates of the crystal growth velocity and activation energies have been provided by NN simulations. Calculation of the formation free energy of the critical nucleus as a function of temperature is also possible in principle from metadynamics simulations by using the techniques described in the Chapter by Mazzarello et al. [119]. The calculation of the free energy of the critical nucleus from the CNT expression is instead very inaccurate due to the ambiguities in the definition and calculation of the interface free energy for small nuclei. A rough estimate of the size of the critical nucleus in the NN simulation was given in [97] by looking at the size over which a constant u is observed according to 15.4. A critical nucleus size containing about 40–50 atoms at 600 K was suggested in good agreement with previous estimates based on DFT-MD simulations on small cells [13–15].

15.6 Concluding Remarks

We have reviewed here the main insights gained in the last few years from atomistic simulations of the structural and functional properties of phase change materials. We have focused mainly on GST and the GeTe compounds. The structure of the amorphous phase of these and other materials in the same class have been elucidated from models generated by quenching from the melt within DFT molecular dynamics simulations. The availability of reliable amorphous models has then allowed to identify the microscopic origin of the electronic and optical contrast between the amorphous and crystalline phases. We have also reviewed the results from large scale simulations based on neural network interatomic potentials that allowed studying the fragility of the supercooled liquid and the kinetics of crystallization process. The fragility of the liquid is responsible for a high atomic mobility down to temperatures close to the glass transition which boosts a large velocity of crystal growth. Large scale simulations also reveal that the size of the critical nuclei is rather small (about 50 atoms) at temperatures of interest for PCM operation which confirms previous outcomes from fully DFT simulations in small cells. Several other issues of interest for the understanding and electrothermal modeling of PCM operation are also being addressed by atomistic simulations. The chapter by Skelton et al. in this book [120] summarizes the results of DFT calculations on doping of the amorphous phase of phase change materials which is under scrutiny to tailor the electronic properties for applications in PCM. Thermal transport in the bulk [121] and at the interface with dielectrics [122] are also being addressed by atomistic simulations and DFT calculation of electron-phonon coupling constant. Thermal transport is indeed another crucial issue for PCM performances as writing and erasing of the memory rest on heat dissipation and transport [123, 124]. The calculation of Schottky barrier at the interface with metallic electrodes and the electrical contact resistance can also be addressed by DFT methods [125, 126]. Finally, we also mention that the availability of large amorphous models with few thousand atoms generated with the neural network potentials would make possible to shed light on the microscopic origin of the Urbach tails and localized states inside the mobility edges which are believed to control two features of great importance for the operation of PCM, namely the resistance drift [88] and the threshold voltage for the Ovonic switching in the amorphous phase.

Acknowledgments We thank all colleagues who worked with us on these problems including S. Angioletti-Uberti, J. Behler, D. Colleoni, D. Donadio, S. Gabardi, M. Krack, T. D. Kühne, J. Los, D. Mandelli, R. Mazzarello, G. Miceli, M. Parrinello, and E. Spreafico.

References

1. M. Wuttig, N. Yamada, Nat. Mater. **6**, 824 (2007)
2. A. Pirovano, A.L. Lacaita, A. Benvenuti, F. Pellizzer, R. Bez, IEEE Trans. Electron. Dev. **51**, 452 (2004)
3. A.L. Lacaita, D.J. Wouters, Phys. Status Solidi A **205**, 2281 (2008)
4. D. Lencer, M. Salinga, M. Wuttig, Adv. Mater. **23**, 2030 (2011)

5. S. Raoux, W. Welnic, D. Ielmini, Chem. Rev. **110**, 240 (2010)
6. G.E. Ghezzi, J.Y. Raty, S. Maitrejean, A. Roule, E. Elkaim, F. Hippert, Appl. Phys. Lett. **99**, 151906 (2011)
7. Y.T. Kim, E.T. Kim, C.S. Kim, J.Y. Lee, Phys. Status Solidi RRL **5**, 98 (2011)
8. T. Morikawa, K. Kurotsuchi, M. Kinoshita, M. Matsuzaki, Y. Matsui, Y. Fujisaki, S. Hanzawa, A. Kotabe, M. Terao, H. Moriya, T. Iwasaki, M. Matsuoka, F. Nitta, M. Moniwa, T. Koga, N. Takaura, *iEEE International Electron Devices Meeting*, (2007) IEDM 2007, pp. 307–310
9. H.-Y. Cheng, S. Raoux, J.L. Jordan-Sweet, Appl. Phys. Lett. **98**, 121911 (2011)
10. S.R. Ovshinsky, Phys. Rev. Lett. **21**, 1450 (1968) (S.R. Ovshinsky and H. Fritzsche, IEEE Trans. Elect. Dev. **20**, 91 (1973))
11. J. Akola, J. Kalikka, R.O. Jones, This volume
12. J.-Y. Raty et al., This volume
13. J. Hegedüs, S.R. Elliott, Nat. Mater. **104**, 399 (2008)
14. T. H. Lee, and S.R. Elliott, Phys. Rev. Lett. 107, 145702 (2011)
15. T. H. Lee, and S.R. Elliott. Phys. Rev. B **84**, 094124 (2011)
16. J. Kalikka, J. Akola, J. Larrucea, R.O. Jones, Phys. Rev. B **86**, 144113 (2012)
17. G.C. Sosso, G. Miceli, S. Caravati, J. Behler, M. Bernasconi, Phys. Rev. B **85**, 174103 (2012)
18. J. Tersoff, Phys. Rev. B **39**, R5566 (1989)
19. F. Zipoli, A. Curioni, New J. Phys. **15**, 123006 (2013)
20. J. Goldak, C.S. Barrett, D. Innes, W. Youdelis, J. Chem. Phys. **44**, 3323 (1966)
21. T. Chattopadhyay, J. Boucherle, H. Von Schnering, J. Phys. C **20**, 1431 (1987)
22. P. Fons, A.V. Kolobov, M. Krbal, 1 J. Tominaga, K.S. Andrikopoulos, S.N. Yannopoulos, G.A. Voyiatzis, T. Uruga. Phys. Rev. B **82**, 155209 (2010)
23. G. Lucovsky, R.M. White, Phys. Rev. B **6**, 660 (1973)
24. X. Yu, J. Robertson, Can. J. Phys. **92**, 671 (2014)
25. B. Huang, J. Robertson, Phys. Rev. B **85**, 125305 (2012)
26. K. Shportko, S. Kremers, M. Woda, D. Lencer, J. Robertson, M. Wuttig, Nat. Mater. **7**, 653 (2008)
27. D. Lencer, M. Salinga, B. Grabowski, T. Hickel, J. Neugebauer, M. Wuttig, Nat. Mater. **7**, 972 (2008)
28. B. Huang, J. Robertson, Phys. Rev. B **81**, 081204(R) (2010)
29. T. Matsunaga, N. Yamada, Y. Kubota, Acta Crystallogr. B **60**, 685 (2004)
30. M. Luo, M. Wuttig, Adv. Mater. **16**, 439 (2004)
31. M.H. Muser, Eur. Phys. J. B **74**, 291 (2010)
32. M. Xu, Y.Q. Cheng, H.W. Sheng, E. Ma, Phys. Rev. Lett. **103**, 195502 (2010)
33. A.V. Kolobov, P. Fons, J. Tominaga, Phys. Rev. B **87**, 155204 (2013)
34. G.C. Sosso, S. Caravati, M. Bernasconi, J. Phys.: Condens. Matter **21**, 095410 (2009)
35. J. Akola, R.O. Jones, Phys. Rev. B **76**, 235201 (2007)
36. J. Akola, R.O. Jones, Phys. Rev. B **79**, 134118 (2009)
37. S. Caravati, M. Bernasconi, T.D. Kühne, M. Krack, M. Parrinello, Appl. Phys. Lett. **91**, 171906 (2007)
38. S. Caravati, M. Bernasconi, T.D. Kühne, M. Krack, M. Parrinello, J. Phys.: Condens. Matter **21**, 255501 (2009) (errata 21, 499803 (2009); errata 22; 399801 (2010))
39. R. Mazzarello, S. Caravati, S. Angioletti-Uberti, M. Bernasconi, M. Parrinello, Phys. Rev. Lett. **104**, 085503 (2010); 107, 039902(E) (2011)
40. Z.M. Sun, J. Zou, H.J. Shin, A. Blomqvist, R. Ahuja, Appl. Phys. Lett. **93**, 241908 (2008)
41. J.Y. Raty, C. Otjacques, J.P. Gaspard, C. Bichara, Solid State Sci. **12**, 193 (2010)
42. G.C. Sosso, S. Caravati, R. Mazzarello, M. Bernasconi, Phys. Rev. B **83**, 134201 (2011)
43. V. L. Deringer, W. Zhang, M. Lumeij, S. Maintz, M. Wuttig, R. Mazzarello, R. Dronskowski, Angew. Chem. Int. Ed. **53**, 10817 (2014)
44. A.V. Kolobov, P. Fons, A.I. Frenkel, A.L. Ankudinov, J. Tominaga, T. Uruga, Nat. Mater. **3**, 703 (2004)
45. J. Akola, J. Larrucea, R.O. Jones, Phys. Rev. B **83**, 094113 (2011)
46. J.P. Perdew, K. Burke, M. Ernzerhof, Phys. Rev. Lett. **77**, 3865 (1996)
47. J. Akola, R.O. Jones, Phys. Status Solidi B **249**, 1851 (2012)

48. J. Tao, J.P. Perdew, V.N. Staroverov, G.E. Scuseria, Phys. Rev. Lett. **91**, 146401 (2003)
49. K.Y. Kim, D.-Y. Cho, B.-Ki Cheong, D. Kim, H. Horii, S. Han. J. Appl. Phys. **113**, 134302 (2013)
50. A.V. Krukau, O.A. Vydrov, A.F. Izmaylov, G.E. Scuseria, J. Chem. Phys. **125**, 224106 (2006)
51. S. Caravati, M. Bernasconi, Phys. Status Solidi B **252**, 260 (2015)
52. A.D. Becke, Phys. Rev. A **38**, 3098 (1988); C. Lee, W. Yang, and R. G. Parr. Phys. Rev. B **37**, 785 (1988)
53. J. Behler, M. Parrinello, Phys. Rev. Lett. **98**, 146401 (2007); J. Behler. J. Chem. Phys. **134**, 074106 (2011)
54. M. Krbal, A.V. Kolobov, P. Fons, K.V. Mitrofanov, Y. Tamenori, J. Hegedüs, S.R. Elliott, J. Tominaga, Appl. Phys. Lett. **102**, 111904 (2013)
55. K.S. Andrikopoulos, S.N. Yannopoulos, A.V. Kolobov, P. Fons, J. Tominaga, J. Phys. Chem. Solids **68**, 1074 (2007)
56. K. S. Andrikopoulos, S. N. Yannopoulos, G. A. Voyiatzis, A. V. Kolobov, M. Ribes, J. Tominaga, J. Phys. Condens. Matter **18**, 965 (2006)
57. R. De Bastiani et al., Phys. Rev. B **80**, 245205 (2009)
58. J. Akola, R.O. Jones, Phys. Rev. Lett. **100**, 205502 (2008)
59. S. Gabardi, S. Caravati, M. Bernasconi, M. Parrinello, J. Phys.: Condens. Matter **24**, 385803 (2012)
60. S. Caravati, M. Bernasconi, M. Parrinello, Phys. Rev. B **81**, 014201 (2010)
61. J. Los, T. Küehne, S. Gabardi, M. Bernasconi, Phys. Rev. B **88**, 174203 (2013)
62. E. Spreafico, S. Caravati, M. Bernasconi, Phys. Rev. B **83**, 144205 (2011)
63. A. Bouzid, C. Massobrio, S. Gabardi, M. Bernasconi, Phys. Rev. B (2015), in press
64. J. Kalikka, J. Akola, R. O. Jones, J. Phys.: Condens. Matter **25**, 115801 (2013)
65. I. Voleska, J. Akola, P. Jovari, J. Gurwirth, T. Wagner, T. Vasileiadis, S.N. Yannopoulos, R.O. Jones, Phys. Rev. B **86**, 094108 (2012)
66. A. Bouzid, C. Massobrio, J. Chem. Phys. **137**, 046101 (2012)
67. S. Grimme, J. Comput. Chem. **27**, 1787 (2006)
68. G.C. Sosso, J. Behler, M. Bernasconi, Phys. Status Solidi B **249**, 1880 (2012) (Erratum: 250, 1453 (2013))
69. B. Sa, N. Miao, J. Zhou, Z. Sun, R. Ahuja, Phys. Chem. Chem. Phys. **12**, 1585 (2010)
70. M. Micoulaut, J. Chem. Phys. **138**, 061103 (2013)
71. J.-J. Kim, K. Kobayashi, E. Ikenaga, M. Kobata, S. Ueda, T. Matsunaga, K. Kifune, R. Kojima, N. Yamada, Phys. Rev. B **76**, 115124 (2007)
72. B.-S. Lee, J. R. Abelson, S. G. Bishop, D.-H. Kang and B.- k. Cheong, K.-B. Kim, J. Appl. Phys. **97**, 093509 (2005)
73. A.H. Edwards, A.C. Pineda, P.A. Schultz, M.G. Martin, A.P. Thompson, H.P. Hjalmarson, C.J. Umrigar, Phys. Rev. B **73**, 045210 (2006)
74. S. Caravati, D. Colleoni, R. Mazzarello, T.D. Kühne, M. Krack, M. Bernasconi, M. Parrinello, J. Phys.: Condens. Matter **23**, 265801 (2011)
75. W. Zhang, A. Thiess, P. Zalden, R. Zeller, P.H. Dederichs, J.-Y. Raty, M. Wuttig, S. Blügel, R. Mazzarello, Nat. Mater. **11**, 952 (2012)
76. T. Siegrist, P. Jost, H. Volker, M. Woda, P. Merkelbach, C. Schlockermann, M. Wuttig, Nat. Mater. **10**, 202 (2011)
77. S. Goedecker, M. Teter, J. Hutter, Phys. Rev. B **54**, 1703 (1996) (M. Krack. Theor. Chem. Acc. **114**, 145 (2005))
78. M. Krack, M. Parrinello, in *High Performance Computing in Chemistry*, NIC Series, vol. 25, ed. by J. Grotendorst (John von Neumann Institute for Computing, 2004), pp. 29–51
79. J. VandeVondele, M. Krack, F. Mohamed, M. Parrinello, T. Chassaing, and J. Hutter, Comput. Phys. Commun. 167, 103 (2005); http://www.cp2k.org
80. M.E. Cohen, H. Fritzsche, S. Ovshinsky, Phys. Rev. Lett. **22**, 1065 (1969)
81. H. Dersch, J. Stuke, J. Beichler, Phys. Status Solidi B **105**, 265 (1981)
82. M. Kastner, D. Adler, H. Fritzsche, Phys. Rev. Lett. **37**, 1504 (1976)
83. D. Vanderbilt, J. Jannopoulos, Phys. Rev. Lett. **49**, 823 (1982) (Phys. Rev. B **23**, 2596 (1981))

84. R. Bader, *Atoms in Molecules: A Quantum Theory* (Oxford University Press, New York, 1990)
85. G. Henkelman, A. Arnaldsson, H. Jonsson, Comput. Mater. Sci. **36**, 354 (2006)
86. G. Gabardi, S. Caravati, G.C. Sosso, M. Bernasconi (submitted)
87. C. Longeaud, J. Luckas, D. Krebs, R. Carius, J. Klomfass, M. Wuttig, J. Appl. Phys. **112**, 113714 (2012)
88. D. Ielmini, A.L. Lacaita, D. Mantegazza, IEEE Trans. Electron Devices **54**, 308 (2007)
89. M. Boniardi, D. Ielmini, Appl. Phys. Lett. **98**, 243506 (2011)
90. N. Papandreou, A. Pantazi, A. Sebastian, E. Eleftheriou, M. Breitwisch, C. Lam, H. Pozidis, Solid State Electron. **54**, 991 (2010)
91. J. Stuke abd G. Zimmerer, Phys. Status Solidi B **49**, 513 (1972)
92. W. Wełnic, S. Botti, L. Reining, M. Wuttig, Phys. Rev. Lett. **98**, 236403 (2007)
93. W. Wełnic, S. Botti, M. Wuttig, L. Reining, C. R. Physique **10**, 514 (2009)
94. S. Caravati, M. Bernasconi, M. Parrinello, J. Phys.: Condens. Matter **22**, 315801 (2010)
95. J. Heyd, G.E. Scuseria, M. Ernzerhof, J. Chem. Phys. **118**, 8207 (2003)
96. J. Orava, T. Wagner, J. Sik, J. Prikyl, M. Frumar, L. Benes, J. Appl. Phys. **104**, 043523 (2008)
97. G.C. Sosso, G. Miceli, S. Caravati, F. Giberti, J. Behler, M. Bernasconi, J. Phys. Chem. Lett. **4**, 4241 (2013)
98. J. Orava, A.L. Greer, B. Gholipour, D.W. Hewak, C.E. Smith, Nat. Mater. **11**, 279 (2012)
99. A.L. Lacaita, A. Redaelli, Microelectron. Eng. **109**, 351 (2013)
100. C.A. Angell, Science **267**, 1924 (1995)
101. D. Turnbull, J.J. Fisher, J. Chem. Phys. **17**, 71 (1949)
102. K.F. Kelton, Solid State Phys. **45**, 75 (1991)
103. K.F. Kelton, A.L. Greer, J. Non-Cryst. Solids **79**, 295 (1986)
104. J. Kalb, F. Spaepen, M. Wuttig, J. Appl. Phys. **93**, 2389 (2003)
105. S. Senkander, C.D. Wright, J. Appl. Phys. **95**, 504 (2004) (K. Kohary, C.D. Wright, Phys. Status Solidi B **250**, 944 (2013)
106. G.W. Burr, P. Tchoulfian, T. Topuria, C. Nyffeler, K. Virwani, A. Padilla, R.M. Shelby, M. Eskandari, B. Jackson, B.-S. Lee, J. Appl. Phys. **111**, 104308 (2012)
107. C. Peng, L. Cheng, M. Mansuripur, J. Appl. Phys. **82**, 4183 (1997)
108. J. Liu, X. Xu, L. Brush, M.P. Anatram, J. Appl. Phys. **115**, 023513 (2014)
109. C.V. Thomson, F. Spaepen, Acta Metall. **27**, 1855 (1979)
110. G. Bruns, P. Merkelbach, C. Schlockermann, M. Salinga, M. Wuttig, T.D. Happ, J.B. Philipp, M. Kund, Appl. Phys. Lett. **95**, 043108 (2009)
111. L. Berthier, G. Biroli, Rev. Mod. Phys. **83**, 587 (2011)
112. J.A. Kalb, C.Y. Wen, F. Spaepen, H. Dieker, M. Wuttig, J. Appl. Phys. **98**, 054902 (2005)
113. S. Privitera, S. Lombardo, C. Bongiorno, E. Rimini, A. Pirovano, J. Appl. Phys. **102**, 013516 (2007)
114. N. Ciocchini, M. Cassinerio, D. Fugazza, D. Ielmini, IEEE Electon. Devices **60**, 3767 (2013)
115. J.C. Mauro, Y. Yueb, A.J. Ellisona, P.K. Guptac, D.C. Allan, Proc. Natl. Acad. Sci. USA **106**, 19780 (2009)
116. S. Raoux, B. Munoz, H.Y. Cheng, J.L. Jordan-Sweet, Appl. Phys. Lett. **95**, 143118 (2009)
117. G.C. Sosso, J. Colombo, J. Behler, E. Del Gado, M. Bernasconi, J. Phys. Chem. B **118**, 13621 (2014)
118. W. Klemm, G. Frischmuth, Z. Anorg, Chemistry **218**, 249 (1934)
119. R. Mazzarello et al., This volume
120. J.M. Skelton, T.H. Lee, S.R. Elliott, This book
121. G.C. Sosso, D. Donadio, S. Caravati, J. Behler, M. Bernasconi, Phys. Rev. B **86**, 104301 (2012)
122. D. Campi, D. Donadio, G. C. Sosso, J. Behler, M. Bernasconi, J. Appl. Phys. **117**, 015304 (2015); D. Campi, E. Baldi, G. Graceffa, M. Bernasconi, J. Phys. Condens. Matter (2015), in press
123. J.P. Reifenberg, K.-W. Chang, M.A. Panzer, S. Kim, J.A. Rowlette, M. Asheghi, H.-S.P. Wong, K. E. Goodson, IEEE Elect. Dev. Lett. **31**, 56 (2010)
124. J.-L. Battaglia, A. Kusiak, V. Schick, A. Cappella, C. Wiemer, M. Longo, E. Varesi. J. Appl. Phys. **107**, 044314 (2010)
125. D. Mandelli, S. Caravati, M. Bernasconi, Phys. Status Solidi B **249**, 2140 (2009)
126. J. Liu, M.P. Anantram, J. Appl. Phys. **113**, 063711 (2013)

Chapter 16
Ab Initio Molecular-Dynamics Simulations of Doped Phase-Change Materials

J.M. Skelton, T.H. Lee and S.R. Elliott

Abstract The physical behaviour and device performance of phase-change, non-volatile memory materials can often be improved by the incorporation of small amounts of dopant atoms. In certain cases, new functionality can also be introduced, for example a contrast in magnetic properties between amorphous and crystalline phases of the host phase-change material when certain transition-metal dopants are included. This Chapter reviews some of the experimental data relating to doped phase-change materials and, in particular, a survey is given of the role played by molecular-dynamics simulations in understanding the atomistic mechanisms involved in the doping process. In addition, some examples are given of the in silico discovery of new phase-change compositions resulting from ab initio molecular-dynamics (AIMD) simulations.

16.1 Introduction

There is much current research being undertaken on developing new non-volatile memory technologies to replace silicon-based flash memory, for which future size down-scaling will become increasingly problematic due to electron-tunnelling leakage between the floating gate and other parts of the memory cell as feature sizes become ever smaller, leading to unavoidable memory volatility. A leading contender for such a flash-memory replacement is resistive random-access memory (RRAM),

J.M. Skelton
Department of Chemistry, University of Bath,
Claverton Down, Bath BA2 7AY, UK
e-mail: j.m.skelton@bath.ac.uk

T.H. Lee · S.R. Elliott (✉)
Department of Chemistry, University of Cambridge,
Lensfield Road, Bath CB2 1EW, UK
e-mail: sre1@cam.ac.uk

T.H. Lee
e-mail: th132@cam.ac.uk

C. Massobrio et al. (eds.), *Molecular Dynamics Simulations of Disordered Materials*, Springer Series in Materials Science 215,
DOI 10.1007/978-3-319-15675-0_16

of which perhaps the most promising variant is 'Phase-change Random-access Memory' (PCRAM). Indeed, PCRAM devices are now being produced and shipped commercially, with Micron's 1 Gb, 45 nm feature-size PCRAM units being installed in Nokia Asha smartphones since the beginning of 2013.

Phase-change memory (PCM) materials encode digital information, e.g. binary bits (0, 1), as different atomic-structural states of the material, e.g. amorphous (glassy) and crystalline phases, which, moreover, have a significant contrast in (interrogatable) physical properties, viz. electrical resistivity in the case of PCRAM (or optical reflectivity in the case of optical memory, such as rewritable digital video discs (DVD-RW)). Switching between the different PC states is accomplished by heat pulses, caused either by the application of voltage pulses (resulting in Joule heating) in the case of PCRAM, or light pulses in the case of optical memory. In both cases, short, intense heat pulses cause local melting of (part of) a PCM memory cell and subsequent vitrification on ultra-rapid cooling (RESET process). Alternatively, longer, less intense heat pulses can cause crystallization of the glassy phase (SET process). Moreover, these structural changes need to be: (i) reversible, so that a large number of write-erase cycles can be performed; (ii) ultra-rapid, so that high data rates can be accommodated; (iii) and yet the material needs to remain in a non-volatile condition in each of the memory states when not subject to rewrite pulses, so that long-term data storage is assured. These are a very unusual and taxing set of conditions, and few classes of materials can satisfy them in their entirety. The primary condition for a PCM, namely that there should be a large contrast in material properties between amorphous (a-) and crystalline (c-) phases, seems to be satisfied only for materials consisting of combinations of 'heavy' metalloid elements, i.e. those lying near the bottom of the Periodic Table (whose metallicity is higher than for lighter elements), as for instance in the canonical PCM composition, $Ge_2Sb_2Te_5$ ('GST-225'). In general, the c-state is more electrically conducting (or, correspondingly, more optically reflective) than is the a-state. In the case of GST materials, the c-state is near-metallic (being a degenerate, narrow-gap p-type semiconductor) whereas the a-state is more electrically insulating (with the Fermi level lying near the middle of a somewhat wider bandgap). These PCMs generally crystallize (the rate-limiting step in the write-erase cycle) in extremely short times, typically 10 s of nanoseconds. Under certain conditions, sub-ns crystallization of GST PCM cells is achievable [13], and such ultra-rapid phase-transformation times open the way for this non-volatile memory technology to supplant the Si-based volatile dynamic random-access memory (DRAM) lying at the heart of current computer architectures, thereby leading to very considerable energy savings by not having continuously to refresh data stored in DRAM.

Current PCRAM cells have dimensions of the order of a few 10 s of nanometers (although further down-size scaling seems to be possible, since PC switching between c- and a-states, and vice versa, has been demonstrated down to dimensions of $\sim$2 nm [5]), with transformation times of the order of nanoseconds. It is extremely challenging to perform experimental studies of PC transformations in situ in actual memory cells, at such very small dimensions and very short times, although synchrotron experiments are now approaching this capability. Thus, computer simulations can play an important and crucial role in establishing an understanding of the atomistic processes underlying PC transformations, particularly the reasons why crystallization can be so rapid in materials such as GST. In this context, ab initio molecular-dynamics (AIMD) simulations can be very informative. They can also be used to aid in the interpretation of time-resolved synchrotron experimental data. Such first-principles calculations, based upon density-functional theory (DFT) and associated approximate schemes, such as the local-density approximation (LDA), generalized-gradient approximation (GGA) etc., provide the most accurate representation of interatomic interactions for molecular-dynamics simulations, but at very considerable computational cost. Nevertheless, recent improvements in computing power mean that AIMD simulations can now be carried out on models containing many hundreds of atoms and for simulation times of a few nanoseconds. Thus, possibly for the first time, AIMD simulations can be performed on models which truly mimic real material systems, in this case PCM cells with real device dimensions switching in realistic times.

In addition to shedding new light on the basic atomistic processes underlying the a-to-c-transformations in PCMs, AIMD simulations can also be used for in silico materials discovery, either by simulating new basic atomic compositions, different from those of established PCM materials (e.g. GST-225), or by simulating existing or other PCM compositions which are 'doped' with low concentrations of other elements in order to improve particular PCM properties or characteristics. This Chapter will include a discussion of some of our recent work in studying doping effects in PCMs using AIMD simulations.

All the AIMD simulations that we have performed have been carried out using the Vienna Ab Initio Simulation Package (VASP) [12], which is a plane-wave electronic-structure code using the projector-augmented wave (PAW) method [3], together with the PBE exchange-correlation functional [15].

16.2 'Doping' of Phase-Change Memory Materials

The term 'doping' in the PCM context does not have the same meaning as in conventional semiconductor physics, where non-isovalent impurity atoms (e.g. P, B) incorporated in a host semiconductor (e.g. Si) at the ppm level cause the Fermi level to move by significant amounts through the bandgap of the semiconductor, rendering it n-type (Si:P) or p-type (Si:B). In fact, GST PC materials (e.g. $Ge_2Sb_2Te_5$) are also p-type semiconductors, but by virtue of the presence of cation (Ge, Sb) vacancies on the cation sub-lattice of the rocksalt (NaCl) structure, which is the metastable crystalline phase of these materials formed on ultra-rapid crystallization, even though the vacancy concentrations are very much higher than for conventional dopants (e.g. 10 at.% in the case of GST-225). However, the aim of PCM 'doping'—perhaps better referred to as 'materials modification'—is to improve physical properties, e.g. decreasing the electrical conductivity of the c-phase so as to reduce the RESET programming current, or changing properties relating to crystallization, e.g. increasing crystallization temperatures or crystallization activation energies so as thermally to stabilize the a-phase, as well as possibly introducing new functionality to the PC process, e.g. magnetism.

Two types of dopants in (GST) PCRAM materials have been investigated experimentally and computationally. The first class of dopants includes light non-metallic elements, such as C, N, O etc., which are incorporated primarily to improve the crystallization, as well as electrical, properties of PC materials. The second class of dopants includes metallic elements, e.g. transition metals, which can introduce magnetic functionality into the PC process.

16.2.1 Carbon Doping

An example of the effect of C doping on the thermal-crystallization and electrical properties of PCRAM materials is in work carried out on the binary material, GeTe. Doped $GeTe_{1-x}C_x$ materials, with $x = 0.04$, 0.1, show an improved amorphous stability (e.g. 10-year data retention at 127 °C for $x = 0.1$) [2], as well as a reduction by 50 % in RESET power for the same composition [16].

AIMD simulations of a similar composition of C-doped GeTe ($(Ge_{0.52}Te_{0.48})_{0.85}C_{0.15}$) have been performed [10]. It was found that in the simulated amorphous models, the C dopant atoms preferentially bond to Ge atoms and also to other C atoms, forming C–C–C chains. In addition, the addition of the C dopants was found to promote the transformation of (defective) octahedrally-coordinated Ge atoms (a structural motif characteristic of the metastable rocksalt c-phase) to tetrahedrally-bonded sites instead. It was speculated that the presence of the C chains, and the difficulty in breaking the strong C–C bonds, is responsible for the experimentally observed increase in the barrier for crystallization with C-doping.

Similar simulations have also been carried out for 5 at.% C-doped GST-225, CGST ($(Ge_2Sb_2Te_5)_{0.95}C_{0.05}$) [6]. Again, it was found that the addition of C dopants caused an increase in the proportion of tetrahedral Ge sites in the a-phase (from 36 % in undoped GST to 50 % in the C-doped material). It was also found that the number of ABAB square rings (where A = Ge, Sb; B = Te), which are another structural motif also characteristic of the rocksalt c-structure, decreased in CGST from the level seen in undoped GST. This could also be a reason for the increased stability of the a-phase in CGST, in that the number of ABAB-ring 'seeds' for nucleation of the rocksalt c-phase is reduced on C-doping.

16.2.2 Nitrogen Doping

The effects of 'doping' nitrogen into PCM materials, particularly into GST (NGST), has been widely studied experimentally, since its introduction imparts a number of favourable PC characteristics [e.g. [8]]: (i) the resistivity of c-GST increases with

N doping, leading to a corresponding reduction of the electrical current needed for RESET; (ii) the crystallization temperature increases on N-doping, leading to improved high-T data retention in the a-phase; (iii) there is a reduction in the crystal grain size with N doping; (iv) and the a-c-a-c PC cycling endurance increases (at least for <4 at.% N).

Not surprisingly, considering the technological usefulness of nitrogen as a dopant, there have been a number of simulational studies of NGST. For instance, a study has been made of the relative energetics of different N-doping crystalline configurations, viz. atomic-substitutional sites (N_{Ge}, N_{Sb}, N_{Te}), interstitial (N_i) and atomic-vacancy (N_{vac}) sites, by placing N atoms 'by hand' in these local configurations in an ideal model of rocksalt c-GST [4]. It was found that the N_{Te} site had the lowest energy, as was also confirmed by [7] in a comparative study of doping-site energetics for the dopants N, O and Si. (Oxygen was also found energetically to prefer the anionic Te substitutional site, O_{Te}, whereas silicon preferred the cationic Ge substitutional site, Si_{Ge}.) Thus, although the energetically-preferred site for *atomic* nitrogen dopants in c-GST appears to be N_{Te}, the lowest-energy N-defect configuration of all seems to be *molecular* nitrogen located on cationic Ge/Sb sites, $N_{2Ge/Sb}$ [4]. There seems to be some experimental support for this finding from nitrogen K-edge XANES (X-ray absorption near-edge structure) spectra [11], where a very sharp absorption peak occurs (a 'white line'), which has been ascribed to the presence of N_2 molecular entities in the c-GST matrix.

AIMD simulations have also been performed of a-NGST. It was found that nitrogen adopts one of two principal bonding configurations, either trigonally bonded, predominantly to Ge (or, less commonly, Sb) atoms, in a near-planar geometry, or to four atoms (again predominantly Ge) in a tetrahedral configuration [4, 7]. The electronic structure of a- and c-NGST has also been calculated for models of $(Ge_2Sb_2Te_5)_{0.96}N_{0.04}$ [4]. It was found that the N_{Te} defect in c-NGST gives rise to a shallow state lying at the top of the valence band, which might be expected to lead to a corresponding increase in the electrical conductivity which, however, is not observed experimentally. It was concluded, therefore, that the N dopants in c-GST must be incorporated either as atomic defects at grain boundaries or as molecular N_2 in order for there not to be any increase in conductivity on doping.

In order to investigate the dynamical behaviour of atoms in a-NGST at elevated temperatures, we have performed AIMD simulations of annealing at 700 K for models of N-doped GST with different N concentrations (i.e. 3 and 10 at.%), and for undoped GST.

Figure 16.1 provides evidence showing the tendency for formation of N-rich clusters during annealing, which consist of N-based tetrahedral, or planar-trigonal, units. Nitrogen atoms have mostly Ge atoms as nearest neighbours, and tend to form fourfold rings by sharing their edges (Fig. 16.1a). This trend is clearly revealed in the pair-correlation function for Ge atoms as origin, averaged over the configurations during annealing, in which the peak marked *ii*, which is a signature of the formation of this type of fourfold ring, increases with N doping (Fig. 16.1b). Such cluster formation leads to a deterioration in the degree of medium-range order in NGST,

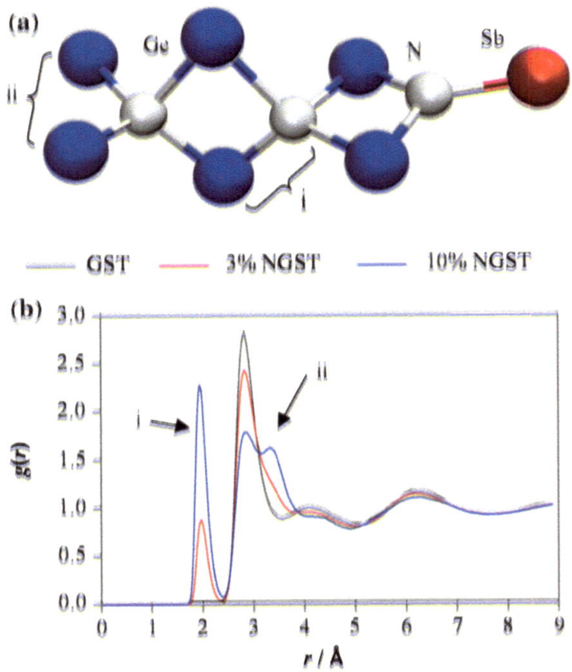

Fig. 16.1 AIMD-simulated annealed nitrogen-doped amorphous $Ge_2Sb_2Te_5$ (NGST). **a** Local atomic configuration taken from the model, showing the formation of annealing-induced clustering of N atoms, in this case consisting of connected N-centred tetrahedral and planar-trigonal units. **b** Ge-centred atomic pair-distribution function for AIMD models of amorphous GST and two concentrations of NGST. The curves show clear evidence for the formation of Ge–N bonds at the expense of Ge-Te bonds, and for the formation of fourfold rings with the –Ge–N–Ge–N– configuration, which is indicated by the increase of the peaks at *i* and *ii*, respectively, corresponding to the interatomic distances so marked in (**a**)

resulting in a diminution of the size of the largest crystalline-like planes found in the models (Fig. 16.2a).

Figure 16.2b shows the temperature dependence of the diffusivity for each type of atom in which, for all temperatures investigated, N atoms showed the lowest diffusivity compared to the other types of atom. Since the medium-range order and diffusivity of atoms are seemingly factors affecting the crystallization speed, as well as amorphous stability, a deeper investigation of these properties could enhance our understanding of the N-doping effect in GST. However, the role of nitrogen doping in controlling such a crystallization process, and hence in determining the resulting crystal-grain size, in GST is still not yet clear, and is the subject of continuing simulational studies.

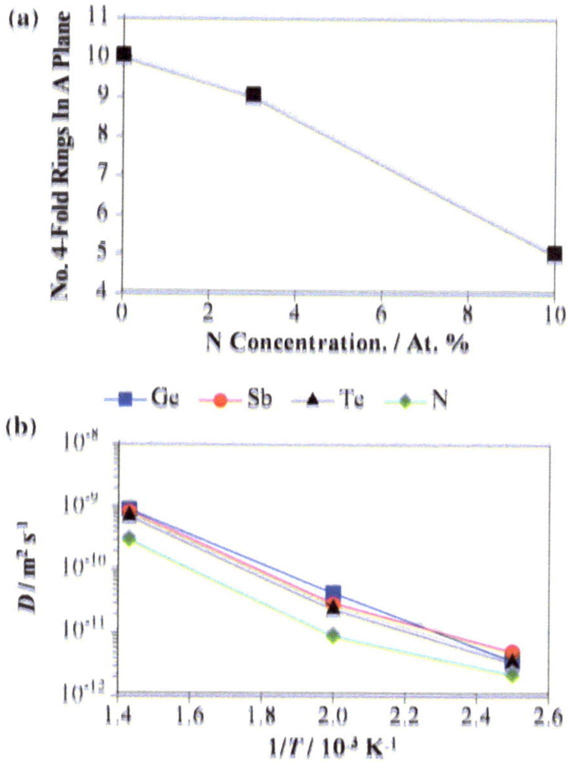

Fig. 16.2 Medium-range order and atomic diffusivity in AIMD-simulated models of NGST at various temperatures. **a** The largest size of planes observed during annealing at 700 K. As the N concentration increases, the size of these planes decreases, which indicates that the N atoms tend to disturb structural ordering during annealing. **b** Dependence of atomic diffusivity on temperature. Each type of atom exhibits an approximately Arrhenius temperature dependence. N atoms show the smallest diffusivity compared to the other types of atoms at all temperatures investigated

16.2.3 Transition-Metal Doping

The motivation behind studies of transition-metal (TM) doping of PCM materials is the possible introduction of new PC functionality (e.g. magnetism), as well as improvement of the normal PCM behaviour, as for other types of dopants. For instance, Song et al. [24] showed that iron incorporated into GST-123 up to the solubility limit of ~19 at.% causes the material to become ferromagnetic. Moreover, on either optically- or electrically-induced PC switching between a- and c-phases, and vice versa, there is a reversible change in magnetization (at 5 K), as well as in the optical reflectivity or the electrical conductivity. In addition to this new magnetic PC functionality introduced by Fe-doping, it has been shown by [25] that Zn-doping of GST-225, up to doping levels of ~20 at.%, can result in more favourable PC characteristics, e.g.: (i) a higher crystallization temperature; (ii) better high-temperature data retention in the a-phase; (iii) and a higher electrical resistivity in the c-phase, leading to a lower RESET current.

A number of simulational studies have been made of TM-doped PCM materials, especially of Fe-doped GST (FeGST) because of its magnetic interest. One such study involved the creation of models of 7 at.% Fe-doped GST-225 in which the TM dopant atoms were introduced 'by hand' into ideal structures of the stable hexagonal (HEX) and metastable rocksalt (RS) crystalline forms of the PCM material in various sites [13]. Amorphous models were simulated by AIMD of quenched liquid configurations. It was found in this way that $Fe_{Ge/Sb}$ substitutional sites (in c-FeGST) appear to be the most energetically stable. The magnetic moments associated with the Fe dopants in $Fe_{Ge/Sb}$ substitutional sites in the models of the two c-phases and the a-phase were calculated, and it was found that the moment was largest in the HEX phase and lowest, on average, in the a-phase, due to a wide range of Fe-dopant configurations in the latter, with the maximum moment for certain sites in a-FeGST being comparable to the value of the moment for the RS c-phase. Thus, the magnetic contrast between a- and c-phases observed experimentally was reproduced. Simulations of GST doped with 7 at.% each of four other TMs, viz. Cr, Mn, Co and Ni, were performed in a similar fashion [26]. It was found that Cr- and Mn-dopants were ferromagnetic, whereas Co- and Ni-dopants were non-magnetic. The latter behaviour was ascribed to the existence of short TM nearest-neighbour distances, implying strong p-d hybridization, leading to quenching of the magnetic moments.

We have performed a series of AIMD simulations of GST-225 doped separately with *all* the elements in the first TM series, viz. Sc to Zn [21, 22], obtaining melt-quenched amorphous models, as well as crystallized models resulting from annealing, as part of a study of the PC cycle for each doped material system. From a structural point of view, the most interesting feature is the nearest-neighbour coordination shell around the TM dopants. It was found from the simulations (see Fig. 16.3) that the most common coordination geometry is (distorted) octahedral for both phases of doped GST for the early TM elements (Sc–Co), but the average coordination number decreases thereafter, reaching a value of 4 for the end-member, Zn, the coordination geometry being tetrahedral for a-ZnGST and defective octahedral for c-ZnGST. There is a marked tendency for the TM dopants to be surrounded with

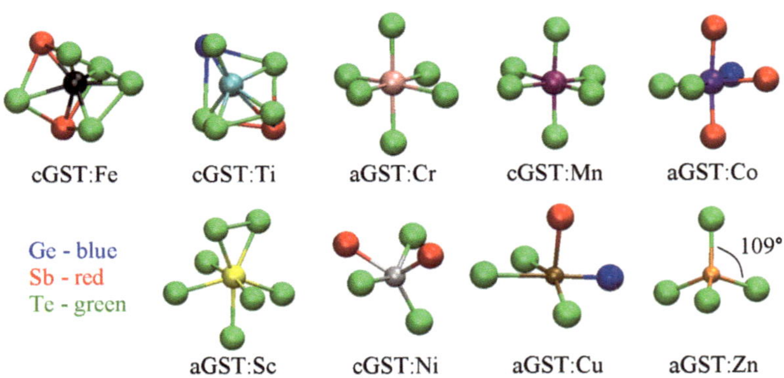

Fig. 16.3 Coordination geometries for some TM dopants in AIMD-simulated models of amorphous (a-) and crystalline (c-) phases of $Ge_2Sb_2Te_5$ (GST)

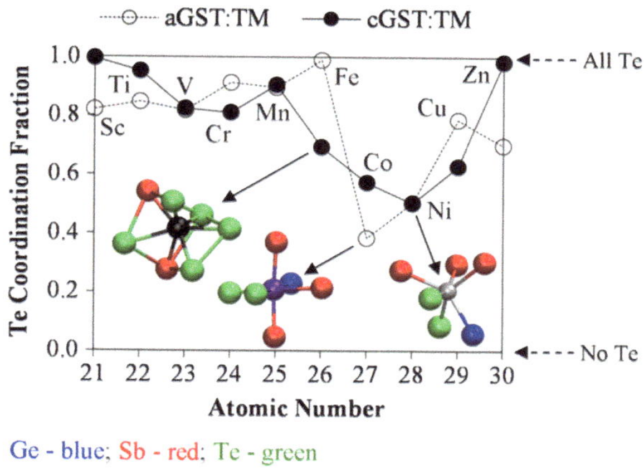

Fig. 16.4 Nearest-neighbour coordination preferences for TM atoms in GST, in terms of the Te coordination fraction. Some TM-centred atomic configurations are illustrated, for Fe, Co and Ni dopants, showing the mixed atomic coordinations characteristic of these dopants

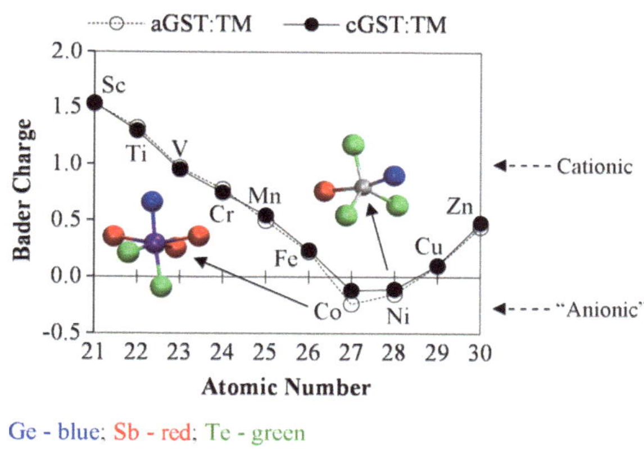

Fig. 16.5 Bader charges of TM dopants in AIMD-simulated models of doped GST. Local TM-centred atomic configurations for Co- and Ni-doped a-GST are illustrated, showing the presence of Ge and Sb atoms in the nearest-neighbour coordination shell, indicative of the 'anionic' nature of these two TM dopants

anionic Te atoms as nearest neighbours (particularly for the c-phase), except for the cases of Ni and Co (and, to a somewhat lesser extent, for Fe and Cu), where the TM dopants are also coordinated to Ge/Sb atoms—see Fig. 16.4. The systematic changes in TM coordination geometry evident in Figs. 16.3 and 16.4 are also manifested in the atomic charges [1] calculated for the TM dopants—see Fig. 16.5. It can be seen that the TM Bader charges are essentially identical for both a- and c-phases for a given dopant, and that there is a clear minimum of charge for the TM elements Ni and Co, the charge actually being *negative*, characteristic of anionic behaviour, for these elements, in contrast to the other first-row TMs which all have positive charges and cationic character. The negative partial charge on Ni and Co atoms accounts for the propensity for these dopants to coordinate also to the cationic matrix materials, Ge and Sb (Figs. 16.3 and 16.4). The TM-centred bond lengths also show a very similar behaviour, with a minimum bond length being exhibited for Ni (Fig. 16.6). Interestingly, the Sc–Te bond length is found to be almost identical to that for

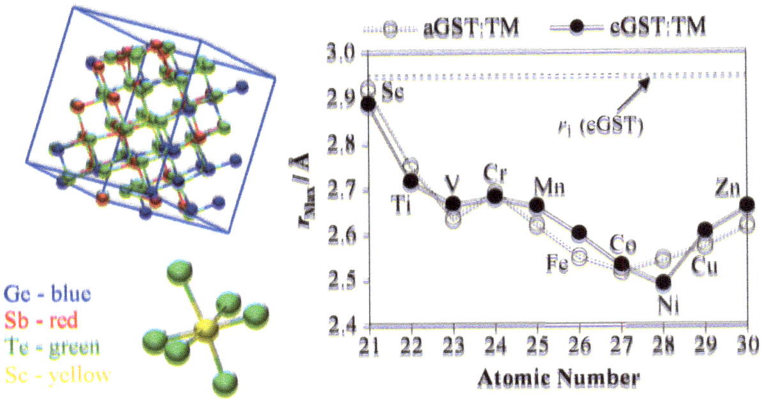

Fig. 16.6 TM-centred bond lengths in AIMD models of amorphous and crystalline doped GST, compared with the Ge-centred average bond length, r_1, in undoped c-GST. A crystallized model of scandium-doped GST is shown in the *left-hand panel*, together with the local bonding environment of the dopant

Ge/Sb–Te bonds, implying that this TM dopant would produce the least structural distortion when incorporated in the host matrix.

It was found that many TM atoms were located near to vacancies, even those that adopt octahedral geometries. This structural configuration has a pronounced influence on the atomic dynamics. For instance, this asymmetric structural arrangement facilitates the occurrence of large-amplitude atomic-vibrational fluctuations of the TM atoms. This behaviour is shown in Fig. 16.7 for the case of ZnGST. It can be seen that the mean-square displacements (MSDs) are small in the a-phase but considerably larger in the c-phase; snapshot atomic configurations corresponding to

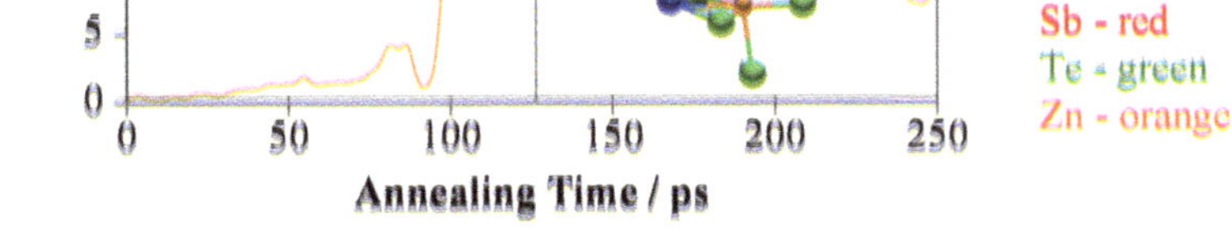

Fig. 16.7 Mean-square-displacement (MSD) fluctuations for Zn-doped GST as a function of annealing time through the crystallization event. Representative instantaneous atomic configurations corresponding to particular extrema in the MSD fluctuations are indicated

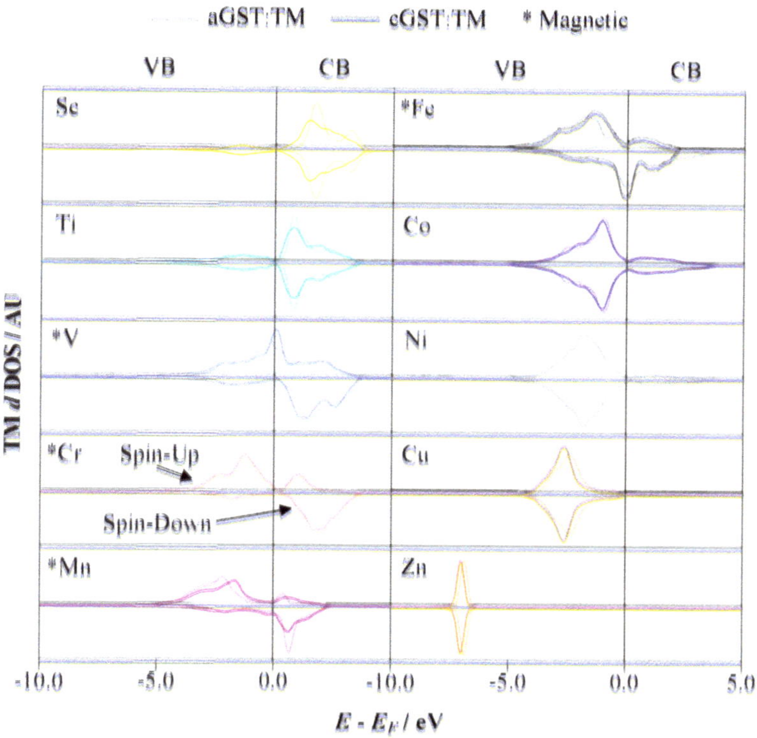

Fig. 16.8 TM-projected partial densities of states (PDOS), relative to the background GST DOS, for a- and c-phases of TM-doped GST. Spin-up and spin-down PDOS are shown; a difference between these indicates the presence of a permanent magnetic moment (as marked with an *asterix*)

extrema in the MSD fluctuations are indicated, which correspond to large-amplitude, 'rattling-motion' atomic-vibrational displacements of the TM dopant in the vicinity of atomic vacancies [22].

Of course, the main motivation for the simulational study of TM-doped PCM materials, such as GST, is to explore the possibility of thereby introducing new functionality, e.g. magnetism, which can also exhibit a reversible contrast in properties when the material undergoes a PC transformation. With this in mind, the electronic density of states (EDOS) has been calculated for all a- and c-models of GST doped with all the first-row TM elements [21]; the results are shown in Fig. 16.8, where the local EDOS projected onto the TM 3d states is shown in each case, referred to the overall DOS of the GST matrix. Several features are apparent. It can be seen that the TM 3d EDOS is very similar for the a- and c-phases of each doped system, reflecting their rather similar local atomic configurations. The most striking feature, however, is the systematic movement of the TM 3d EDOS with respect to the EDOS of the host GST material: in the case of Sc, most of the 3d states lie deep in the energy region of the GST conduction band, whereas for the case of the TM end-member,

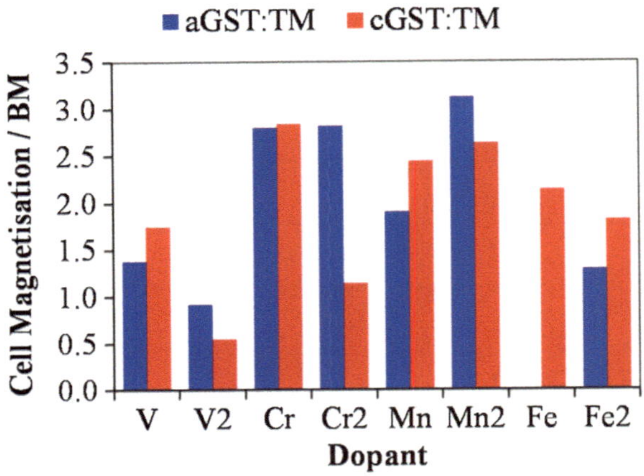

Fig. 16.9 Magnetic moments of TM dopants in a- and c-phases of GST

Zn, its 3d states entirely lie very deep in the valence-band region. Of most interest are those dopants for whose d-states lie in the bandgap region of the GST since these dopants would be expected to have the greatest influence on the electronic behaviour of the PCM; these include the elements V, Mn and Fe.

Of especial interest are those TM dopants which exhibit a permanent magnetic moment when incorporated in GST; these are the elements V, Cr, Mn and Fe, for which the calculated spin-up and spin-down EDOS are different (see Fig. 16.8). The magnetic-moment values of the a- and c-models of V-, Cr-, Mn- and Fe-doped GST are shown in Fig. 16.9.

Thus, from these AIMD simulations, it appears that all four of these TM dopants could potentially be used for magnetic PC applications since, for most model configurations which have been generated, there is a magnetic contrast between a- and c-phases for each of the dopants. However, in some cases, there is an appreciable variation in the magnetic moment between two configurations of each model of a given phase, reflecting variations in the local structural of the dopants in these metastable a- and c-phases. Further simulational work is needed to gain a better idea of these statistical variations and hence what is likely to be the average magnetic moment for each TM dopant in the two phases, and therefore the likely value of the magnetic contrast. In addition, simulations of more highly doped models are required in order to study TM-TM exchange interactions.

Finally, we consider the effect of TM doping of GST on the optical properties of this PCM material for optical-memory applications. Figure 16.10 shows the optical-reflectivity contrast between c- and a-phases for GST, doped separately with each of the first-row TM elements, and calculated at three different wavelengths, viz. 405, 650 and 780 nm [21]. In all cases, the reflectivity of the c-phase is greater than for the a-phase, as for undoped GST. Although it can be seen that TM doping has an effect on the reflectivity contrast, it appears that only V-doping causes an *increase* in the contrast relative to that of undoped GST.

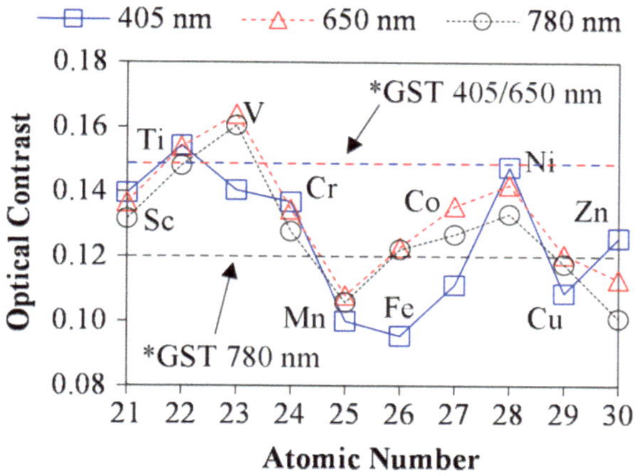

Fig. 16.10 Effect of TM doping of GST on the optical-reflectivity contrast (crystalline reflectivity minus the amorphous reflectivity, normalised by the crystalline value)

16.3 $GeCu_2Te_3$

As can be seen from Figs. 16.10 and 16.11, the normal behaviour for conventional PCM materials, such as (doped) GST, is for the c-phase to be more ‘metallic’ and hence more optically reflective, than is the more semiconducting a-phase. However, there are a very few PCM materials which exhibit the *reverse* optical contrast, where the reflectivity of the a-phase is greater than that of the c-phase; one example of a material showing this behaviour is $GeCu_2Te_3$ (GCT)—see Fig. 16.11 [20]. This

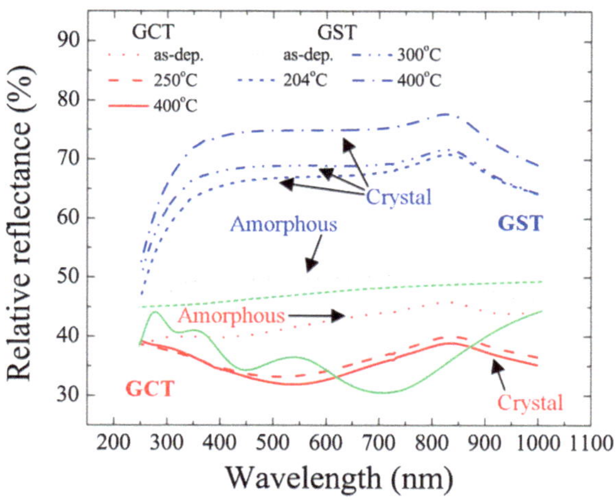

Fig. 16.11 Reflectivity of a- and c-phases of $GeCu_2Te_3$ (GCT) and GST

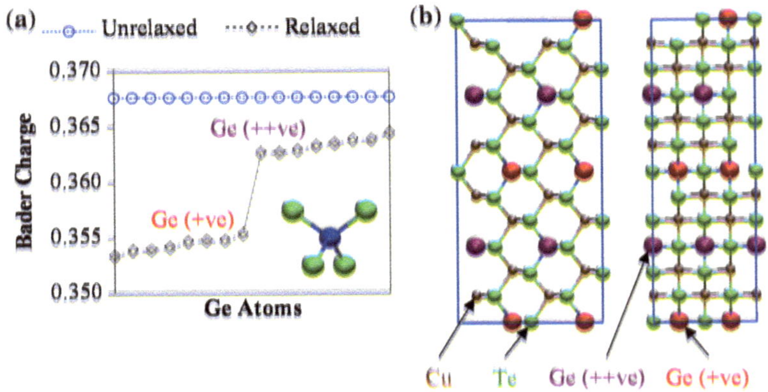

Fig. 16.12 The crystal structure of GCT (**b**), with the Bader charges on Ge atoms before and after energy relaxation of the models, showing the Ge charge disproportionation occurring after relaxation (**a**)

is also a TM-containing PCM material, but with a TM concentration considerably higher than in the doped materials discussed in the previous section. Instead, GCT can be regarded as being a composition along the pseudo-binary GeTe–CuTe tie-line, i.e. the composition GeTe–2CuTe, similar to the canonical PCM material, GST-225 (2GeTe–Sb_2Te_3), being a composition along the GeTe–Sb_2Te_3 tie-line. We have undertaken AIMD simulations of the a- and c-phases of GCT [23]. In contrast to conventional GST-based PCM materials, which have rocksalt-based metastable c-phases with local octahedral coordination, c-GCT has a *tetrahedrally* coordinated structure. This structure has one Cu-centred $CuTe_4$ site and two Te-centred sites ($TeCu_2Ge_2$, $TeCu_3Ge$). Although there also appears to be just a single Ge-centred $GeTe_4$ site in terms of nearest-neighbour coordination, on energy relaxation, the Ge sites seem to undergo charge disproportionation involving subtle structural relaxations—see Fig. 16.12. The Cu- and Te-centred sites do not show such an energy-relaxation-induced charge disproportionation.

Structural characteristics of the simulated a-phase of GCT are shown in Fig. 16.13. The most interesting feature is the prominent peak in the bond-angle distribution, primarily for Cu-centred configurations, at an angle of ~60°. These are indicative of threefold rings in Cu-rich regions, composed of connected Cu_3[Ge,Te] or Cu_4 tetrahedra. These structural configurations characteristic of the a-phase are responsible for the a-phase, unusually, having a *higher* atomic density than the c-phase of GCT. There is evidence for the presence of a substantial fraction of tetrahedral configurations for Cu-, Ge- and Te-centred sites, reminiscent of the c-phase, although Ge-centred sites also exhibit signs of (defective) octahedral configurations with bond angles of 90°.

As mentioned previously, GCT exhibits anomalous PC behaviour compared with 'conventional' PCM materials, such as GST: the optical reflectivity of the a-phase is higher than that of the c-phase (Fig. 16.11), and the mass density of the a-phase is also larger than that of the c-phase. It is expected that, as the atomic density, and

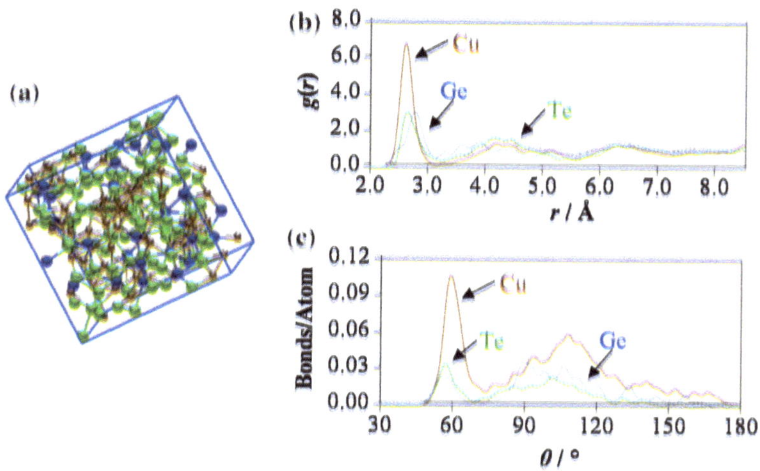

Fig. 16.13 Structural characteristics of a-GCT: **a** view of the model, **b** individual atom-based partial radial-distribution functions (RDFs), **c** partial bond-angle distributions for bond angles subtended at the atom types, as marked

hence the electron density, increases, so does the refractive index, and hence also the reflectivity, which is one explanation for this anomalous behaviour. A very few other PCM materials also exhibit an anomalous negative density change on the a-c phase transition; these include GaSb [17, 18], Fe–Te [9] and Ge-rich Ge–Sb materials [19].

Furthermore, the reflectivity of c-GCT is less than that of c-GST (Fig. 16.11). This behaviour can be understood by the c-phase of GCT being a tetrahedrally-coordinated structure, rather than an octahedrally-coordinated (rocksalt) structure, as in c-GST. The 90° bond angles in the latter structure promote resonant bonding among p-like electron states, leading to metallic-like behaviour, whereas the tetrahedral bonding characteristic of c-GCT does not support such interactions and hence the material remains semiconducting. This effect might also be a second explanation for the reflectivity of the a-phase of GCT being anomalously larger than that of the c-phase (Fig. 16.11)—i.e. the bandgap of the a-phase is correspondingly smaller than that of the c-phase. Simulations of the a-phase of GCT show that there is evidence for the existence of remnants of octahedral coordination of Ge atoms (i.e. 90° bond angles—see Fig. 16.13) which could support partial (local) resonant bonding and hence lead to a reduction of the bandgap from the c-phase value.

16.4 Conclusions

Ab initio molecular-dynamics simulations of phase-change memory materials have proved to be an invaluable method for obtaining an understanding, at an atomistic level, of the processes responsible for the ultrafast crystallization in materials such as Ge–Sb–Te alloys. Such simulational methods can also shed much light on the role played by dopants in improving phase-change behaviour. However, perhaps the most important contribution that such simulations can make in the future is in materials design in silico, wherein the properties of new, hitherto unexplored, material compositions can be investigated computationally, and their behaviour optimized, before they are then fabricated experimentally.

References

1. R.F. Bader, in Atoms in Molecules: A Quantum Theory (Oxford: Oxford University Press, 1990)
2. G. Betti Beneventi et al., Sol. St. Electr. **65–66**, 197 (2011)
3. P.E. Blochl, Phys. Rev. B **50**, 17953 (1994)
4. S. Caravati et al., J. Phys. Cond. Matt. **23**, 265801 (2011)
5. Y.C. Chen et al., IEDM Tech. Dig. **777–780**, S30P3 (2006)
6. E. Cho et al., Appl. Phys. Lett. **99**, 183501 (2011a)
7. E. Cho et al., J. Appl. Phys. **109**, 043705 (2011b)
8. L. Fang, J. Appl. Phys. **107**, 104506 (2010)
9. X.T. Fu et al., Appl. Phys. Lett. **100**, 201906 (2012)
10. G.E. Ghezzi et al., Appl. Phys. Lett. **99**, 151906 (2011)
11. A. Kolobov et al., Appl. Phys. Lett. **100**, 061910 (2012)
12. G. Kresse, J. Hafner, Phys. Rev. B **47**, 558 (1993)
13. Y. Li, R. Mazzarello, Adv. Mater. **24**, 1429 (2012)
14. D. Loke et al., Sci. **336**, 1566 (2012)
15. J.P Perdew et al., Phys. Rev. Lett. **77**, 3865 (1996)
16. L. Perniola et al., Proc. ESSDERC 313 (2010)
17. M. Putero et al., Appl. Phys. Lett. Mater. **1**, 062101 (2013a)
18. M. Putero et al., Appl. Phys. Lett. **103**, 231912 (2013b)
19. S. Raoux et al., J. Appl. Phys. **105**, 064918 (2009)
20. Y. Saito, Y. Sutou, J. Koike, Appl. Phys. Lett. **102**, 051910 (2013)
21. J.M. Skelton, S.R. Elliott, J. Phys. Cond. Matt. **25**, 205801 (2013)
22. J.M. Skelton et al., Appl. Phys. Lett. **101**, 024106 (2012)
23. J.M. Skelton et al., Appl. Phys. Lett. **102**, 224105 (2013)
24. W.-D. Song et al., Adv. Mater. **20**, 2394 (2008)
25. G. Wang et al., Appl. Phys. Lett. **101**, 051906 (2012)
26. W. Zhang et al., Adv. Mat. **24**, 4387 (2012)

Chapter 17
The Prototype Phase Change Material $Ge_2Sb_2Te_5$: Amorphous Structure and Crystallization

Jaakko Akola, Janne Kalikka and Robert O. Jones

Abstract The widespread use of phase change materials in storage media is based on the extremely rapid and reversible switching between the amorphous and crystalline phases of some families of semiconducting alloys. Detailed information about the structure of the amorphous phase and the mechanism of crystallization are essential for the development of new storage media, and we study both aspects here using density functional/molecular dynamics simulations of $Ge_2Sb_2Te_5$, the prototype phase change material of the Ge/Sb/Te semiconductor family.

J. Akola
Department of Physics, Tampere University of Technology,
33101 Tampere, Finland
e-mail: jaakko.akola@tut.fi

J. Akola
COMP Centre of Excellence, Department of Applied Physics,
Aalto University, 00076 Aalto, Finland

J. Kalikka
Singapore University of Technology and Design,
8 Somapah Road, Singapore 487372, Singapore
e-mail: janne_kalikka@sutd.edu.sg

J. Kalikka
Department of Nuclear Science and Engineering,
Massachusetts Institute of Technology, Cambridge, MA 02139, USA

R.O. Jones (✉)
Peter Grünberg Institut PGI-1 and JARA/HPC, Forschungszentrum Jülich,
52425 Jülich, Germany
e-mail: r.jones@fz-juelich.de

R.O. Jones
German Research School for Simulation Sciences,
FZ Jülich and RWTH Aachen University, 52425 Jülich, Germany

C. Massobrio et al. (eds.), *Molecular Dynamics Simulations of Disordered Materials*, Springer Series in Materials Science 215,
DOI 10.1007/978-3-319-15675-0_17

17.1 Introduction

The 1960s saw a range of exciting developments in amorphous materials [1], one of which was the much-cited paper of Ovshinsky [2] on the resistive switching in semiconducting alloys. This was the beginning of the era of phase change (PC) materials. PC materials are based on the extremely rapid and reversible transition between the amorphous and crystalline form of nanosized "bits" in a very thin polycrystalline layer, whose states must show sufficient contrast in resistivity or optical properties to be identified. It is clear that most materials do not satisfy these criteria, and great efforts across more than 30 years were necessary to identify and develop appropriate alloys. The original paper [2] discussed the alloy $As_{30}Ge_{10}Si_{12}Te_{48}$, and Te-based alloys have been well represented ever since.

Alloys of the $Ge_xSb_yTe_{1-x-y}$ family are common in the PC world. Ge_xTe_{1-x} alloys were the first to show real promise as PC storage media [3], and the widely studied $Ge_2Sb_2Te_5$ (GST-225) is used commercially in DVD-RAM (random access memory). There has been particular focus on the pseudobinary alloys along the tie line GeTe–Sb_2Te_3 in Fig. 17.1a, where GST-225 and $Ge_8Sb_2Te_{11}$ (used in Blu-ray Disc memories) are well-known examples. Such alloys are denoted in Fig. 17.1 as Group 1, and all have rock salt structures and numerous vacancies in the ordered phase [4]. Crystallization of amorphous bits in these materials usually begins from small nuclei (Fig. 17.1b). Alloys near the Sb_2Te_3 end of the tie line crystallize more readily, but have lower optical contrast than GeTe-rich alloys. Some compromises are necessary in practice, and these will include the crystallization temperature [5]. Sb-Te based alloys are also well represented in commercial PCM materials, often

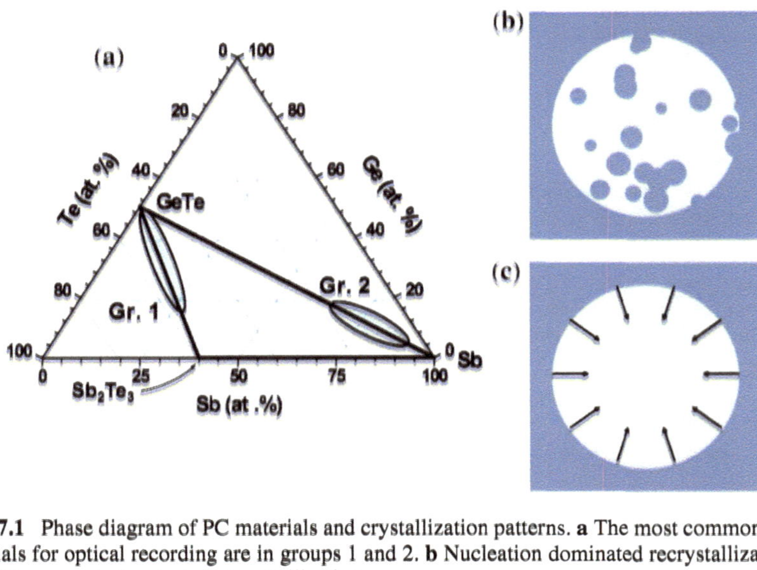

Fig. 17.1 Phase diagram of PC materials and crystallization patterns. **a** The most commonly used materials for optical recording are in groups 1 and 2. **b** Nucleation dominated recrystallization (as in GST), **c** growth-dominated recrystallization

with compositions that are near that of the eutectic mixture $Sb_{70}Te_{30}$, and are denoted as Group 2. Recrystallization in these alloys proceeds from the polycrystalline region surrounding the amorphous bit (Fig. 17.1c).

Central to our understanding of the properties of these materials is a knowledge of the structures of the different phases, but these are difficult to determine in binary or ternary alloys with significant numbers of vacancies. It was noted as recently as 2005 that "the local structural order of glasses in the ternary system $Ge_xSb_yTe_{1-x-y}$ is not well established" [6]. Even the structure of the metastable, ordered phase of GST is still open to question: Yamada [7] proposed that the metastable phase has a rock salt structure with "Na" sites occupied randomly by Ge and Sb atoms and vacancies, and "Cl" sites by Te. Kolobov et al. [8], however, proposed that Ge and Sb atoms are displaced from their ideal positions, enabling the order-disorder transition to occur as an "umbrella flip" of Ge atoms from octahedral to tetrahedral positions. A model of amorphous $Ge_1Sb_2Te_4$ where a spinel structure (tetrahedral Ge) was compared to the octahedral rock salt phase has also been studied [9], and x-ray fluorescence holography of an epitaxial layer of GST indicated a cubic structure with tetrahedral site symmetry about Ge atoms [10]. It is astonishing that PC materials have become the basis of commercially successful products with so much uncertainty about the structures of the phases involved.

Density functional (DF) calculations are free of adjustable parameters, and their combination with molecular dynamics [11] has had a major impact on materials science and chemistry, including chalcogenide materials. Nevertheless, their demands on computational resources restricted simulations on GST systems up to 2007 to relatively small unit cells (up to 56 atoms in [9], 108 atoms and 12 vacancies in [12], 168 atoms in [13]) and time scales that are often much shorter than those measured (e.g. a total of 18 ps in [13]). In such cases, most of the atoms are near the "surface" of the simulation cell, and significant finite size effects are inevitable. Nevertheless, simulations on GST and other PC materials have the advantage that the physical time scale (several nanoseconds) is unusually short. This has encouraged us and other research groups to carry out extensive DF simulations on PC materials, particularly GST alloys, and to take advantage of continuing improvements in computing hardware and software. It is now possible to simulate hundreds of atoms over times that approach the *experimental* time scale of both the quenching and crystallization processes. The following article is restricted to GST (Group 1) materials, but it should reflect the sense of excitement that has accompanied this work.

17.2 Density Functional Calculations

The density functional (DF) formalism [14] is the most widely used method for determining electronic and structural properties of condensed matter that does not require adjustable parameters. It is based on the electron density $n(r)$ and reduces the problem of determining ground state properties of an interacting system of electrons and ions to the solution of single-particle Schrödinger-like equations. An approximation for the exchange-correlation energy E_{xc}, one contribution to the total energy of the system, is unavoidable, but there has been substantial progress in this area in recent years [15]. Crucial to the success of applications to PC materials has been the coupling of DF calculations to molecular dynamics (MD) [11], since the DF energy surfaces and forces can be used to determine the motion of the ions at a known temperature.

Most simulations of amorphous materials are started with a liquid sample at high temperatures (typically above 3000 K), so that all memory of the initial solid configuration is erased. It is important that cooling to the melting point and to room temperature and below be carried out as slowly as possible, as energetically unfavourable configurations may otherwise be frozen in. In the calculations described here, we have used total simulation times of hundreds of picoseconds for samples of up to 648 atoms. DF calculations of this scale are *extremely* demanding of present computer resources, but they do allow definite statements to be made concerning the convergence of the simulation results as a function of sample size.

MD methods allow us to follow the coordinates R_i and velocities v_i of all atoms throughout the simulations, and insight into the local order can be found from the distributions of the bond (θ_{ijk}) and dihedral angles (γ_{ijkl}). The pair distribution function (PDF) $g(r)$ is a spherically averaged distribution of interatomic vectors,

$$g(r) = \frac{1}{\rho^2}\left\langle \sum_i \sum_{i \neq j} \delta(r_i)\delta(r_j - r) \right\rangle , \tag{17.1}$$

where ρ is the density. Partial PDF $g_{\alpha\beta}(r)$ can be calculated by restricting the analysis to the elements α and β. The local structure can also be characterized by the average coordination numbers, which are found by integrating $g_{\alpha\beta}(r)$ to the first minima $R_{\min}$

$$n_{\alpha\beta} = \int_0^{R_{\min}} dr\ 4\pi r^2 \rho_{\alpha\beta} g_{\alpha\beta}(r). \tag{17.2}$$

The partial density $\rho_{\alpha\beta} = \rho\sqrt{c_\alpha c_\beta}$, where c_α and c_β are the concentrations of the elements. We calculate the structure factor $S(Q)$ by Fourier transforming the $g_{\alpha\beta}(r)$ to give the partial structure factors:

$$S_{\alpha\beta}(Q) = \delta_{\alpha\beta} + \rho_{\alpha\beta} \int_0^{\infty} dr\ 4\pi r^2 \left[g_{\alpha\beta}(r) - 1\right] \frac{\sin(Qr)}{Qr}. \tag{17.3}$$

These quantities can then be weighted according to the atomic fractions c_α and form factors $f_\alpha(Q)$ (for x-rays) or the Q-independent coherent scattering lengths b_α (for neutrons) for direct comparison with experiment.

The cavity analysis is performed by introducing the concepts shown in Fig. 17.2 [16]. A *vacancy domain* (I) is a region where the minimum distance to a nearby atom is larger than a given cutoff (here 85 % of the average Ge-Te distance, 2.8 Å), and

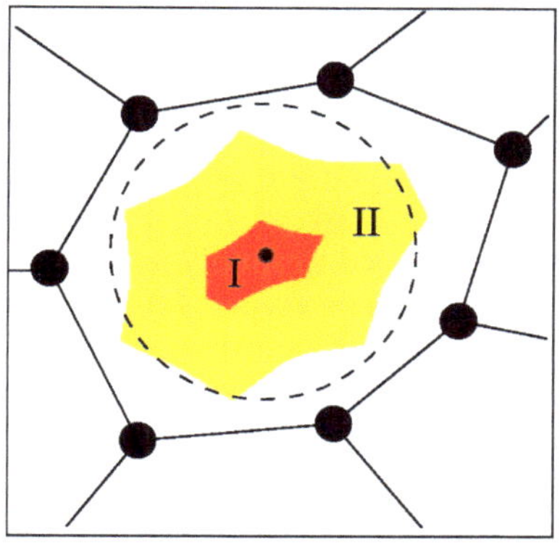

Fig. 17.2 Schematic form for 2D-cavity

each domain is characterized by the point where the distance to all atoms is greatest. The cavity *centre* is found by locating the largest sphere that can be placed inside the cavity, and the centres can be used to calculate RDF, including cavity-cavity correlation functions. Finally, a *vacancy cell* (II) is analogous to the Wigner-Seitz cell in crystals or Voronoi polyhedra in amorphous materials and can be determined on an appropriate mesh with respect to the surface points of the vacancy domains. The importance of cavities and their relatively small number provide additional incentives for choosing large simulation cells.

Dynamical information includes the velocity-velocity autocorrelation function C_v:

$$C_v(t) = \frac{1}{N} \sum_{i=1}^{N} \frac{\langle v_i(0) \cdot v_i(t) \rangle}{\langle v_i(0) \cdot v_i(0) \rangle}, \tag{17.4}$$

where N is the number of particles. The sampling can be improved by averaging the autocorrelation function over several windows where the starting time $(v_i(0))$ is varied. The self-diffusion constant for all atoms or those of species α can be determined from C_v:

$$D_\alpha = \frac{1}{3} \int_0^\infty dt \; C_v(t) \tag{17.5}$$

or directly from the coordinates R_α

$$D_\alpha = \lim_{t \to \infty} \frac{\left\langle |R_\alpha(t) - R_\alpha(0)|^2 \right\rangle}{6t}, \tag{17.6}$$

where the average is over all atoms of species α. Vibration frequency distributions (power spectra) can be obtained by Fourier transforming the velocity-velocity autocorrelation function obtained from long trajectories at the temperature in question.

17.3 Results and Discussion

The structures and their changes with time and temperature provide a rich source of information for comparison with experiment, and we do this now for several Ge/Sb/Te compounds. We provide most details for GST-225, a prototype PC material that is often referred to simply as GST. In these materials it is convenient to separate the components into types *A* (Ge, Sb) and *B*(Te). Details of all calculations can be found in the original articles.

17.3.1 $Ge_2Sb_2Te_5$ (GST-225)

The amorphous (a-), liquid (ℓ-) and crystalline (c-) phases of $Ge_2Sb_2Te_5$ have been studied using a sample of 460 atoms and 52 vacancies in the unit cell over a total of 400 ps [16, 17]. The starting geometry (Fig. 17.3a) was a rock salt crystal structure, with the Na sites occupied randomly by Ge and Sb atoms (20 % each), and the Cl sites by Te atoms [7]. At the end of the simulation (300 K), the amorphous structure in Fig. 17.3b was found. Cavities are shown by light blue isosurfaces, but we note that the multivacancy in Fig. 17.3b is just one of numerous cavities in a-GST.

A striking (and unexpected) feature of the partial PDF (Fig. 17.4) is the medium-range order found among Te atoms up to 10 Å, with peaks at 4.16 and 6.14 Å, a minimum at 5.4 Å, and additional maxima at 7.8 and 9.8 Å. The weak maximum at 2.95 Å and the low coordination number (0.3) show that there are few Te-Te bonds, and there is no significant order at 900 K. Te atoms prefer coordination with Ge/Sb, with Te forming the second-neighbour shell. Thermal fluctuations in c-GST lead to broadening of the Bragg peaks. The "homopolar" Ge-Ge, Ge-Sb, and Sb-Sb bonds in the three phases are shown in Fig. 17.4. While these bonds are absent in c-GST,

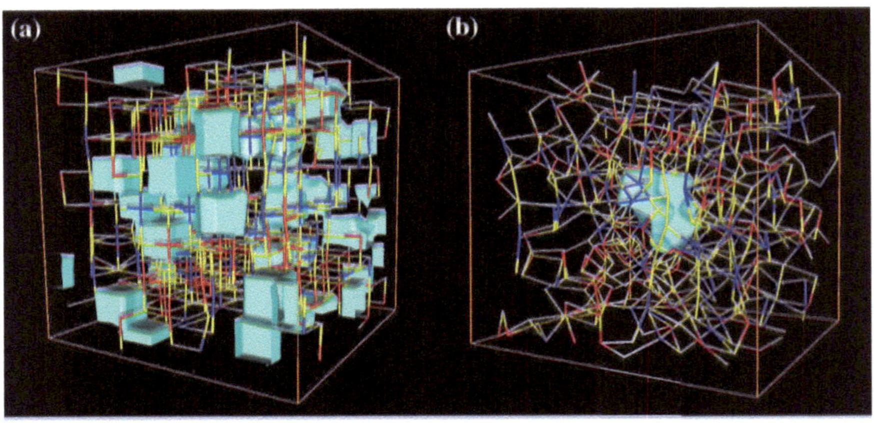

Fig. 17.3 System of 460 atoms and 52 vacancies in **a** c-GST and **b** a-GST with one cavity highlighted. *Red* Ge, *blue* Sb, *yellow* Te. Vacancies/Cavity are shown as *light blue* isosurfaces

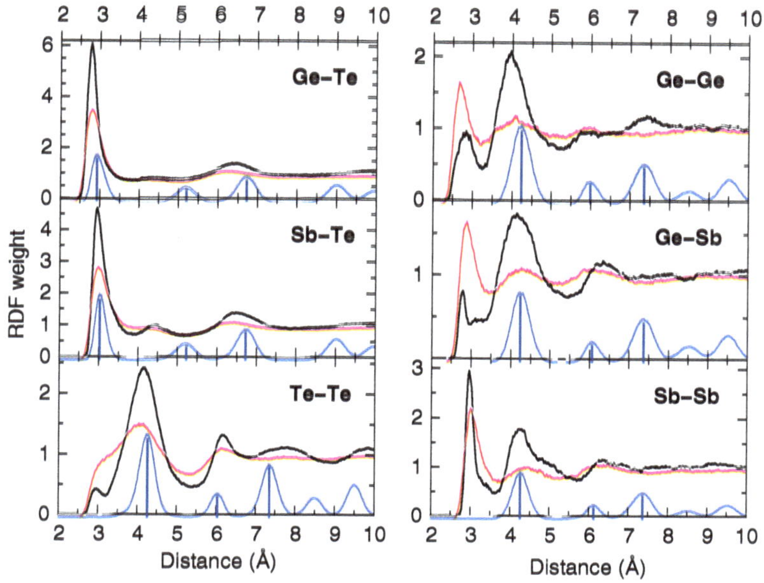

Fig. 17.4 Partial PDF of *a*-(*thick black*) and ℓ-GST (*red*). *Blue curve* and *bars* are for *c*-GST at 300 K (with different scale)

they occur in the amorphous phase ("wrong bonds"). Heating to 900 K results in significant changes in the Ge-Ge and Ge-Sb curves: the first peak now dominates, and the order between 4 and 7 Å moves to shorter range.

All coordination numbers for *a*-GST (Ge: 4.2, Sb: 3.7, Te: 2.9) are lower than in the crystal, but larger than the values (4, 3, 2) suggested by the often-quoted "$8 - N$ rule", where N is the number of valence electrons. Liquid GST (900 K) is less ordered than the amorphous state. The similarities between ring statistics in *a*- and *c*-GST have been proposed as the basis for the structural phase change in GST [18]. The statistics of all irreducible loops (rings) in *a*- and ℓ-GST have been calculated using a bond cutoff of 3.2 Å and periodic boundary conditions (Fig. 17.5). Fourfold rings dominate the statistics at 300 K, and the contribution of "$ABAB$ squares" is 86 % of the total. 52 % of atoms participate in at least one $ABAB$ configuration. In Fig. 17.6 we show that alternating $ABAB$ *cubic* structures also occur.

Cavities play an essential role in Ge/Sb/Te alloys. Since *c*-GST contains 10 % vacancies, *a*-GST, whose density is 7 % lower, must have empty regions (cavities or voids, see Fig. 17.3). Cavities comprise 11.8 % of the total volume of *a*-GST (slightly more than in the crystal), and this increases to 13.8 % at 900 K.

The energy eigenvalues in DF calculations cannot be related directly to the optical excitation spectra, but the electronic density of states (DOS) and its projections often lead to useful insight and are shown in Fig. 17.7. The former are characteristic of materials with average valence five: the two lowest bands can be assigned to s-electrons (σ-band), and the broad band between -5 eV and the Fermi energy

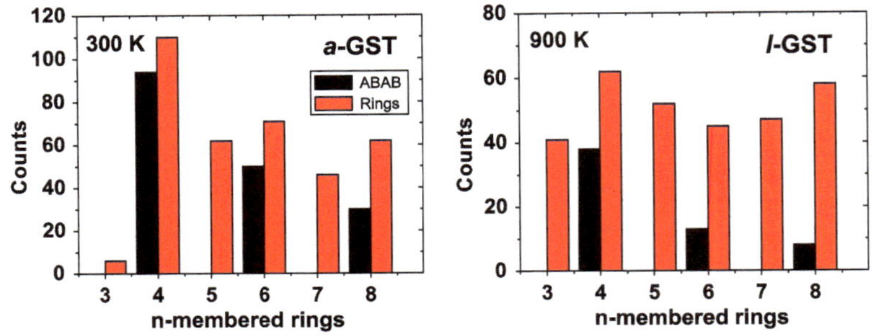

Fig. 17.5 Ring statistics for a-GST (300 K) and ℓ-GST (900 K). $ABAB$ refers to even-membered rings with bond alternation

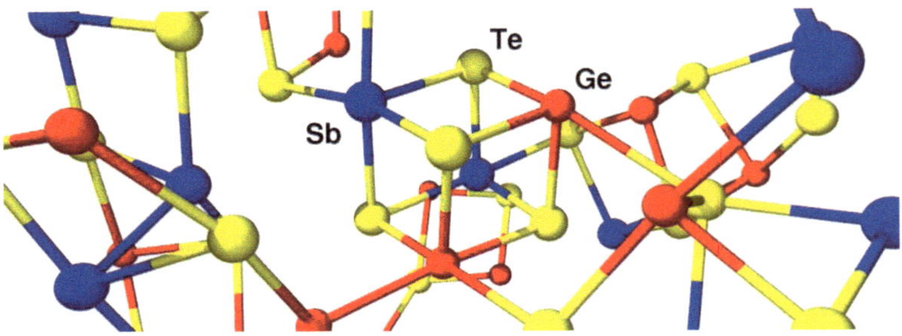

Fig. 17.6 A cubic $ABAB$ subunit in a-GST. A Ge-Ge bond (between the cube and its neighbours) and an Sb-Sb bond are evident. Cavities occur near the cube, and a threefold coordination is typical for Te

has p-character (π-band). The calculations for c- and a-GST show band gaps of around 0.2 eV at the Fermi energy. Photoemission measurements have shown that there are significant differences in the valence band DOS and the core levels of a- and c-GST and other Ge/Sb/Te alloys [19]. The differences measured by hard x-ray photoemission spectroscopy (XPS [19]) are compared with our results in Fig. 17.7. The agreement is remarkably good, particularly for the DOS difference (Fig. 17.7c).

The distribution of vibration frequencies in condensed matter and molecules provides important structural information and can be calculated: (a) from the Fourier transform of the velocity-velocity autocorrelation function, calculated from trajectories of 6000 time steps of 3.025 fs each, and (b) from the dynamical matrix calculated from the second-order energy derivatives of a well-equilibrated structure at 0 K. The results for a-GST (Fig. 17.8) show that the results for the two methods agree very well. The most pronounced features are peaks near 60 and 150 cm^{-1} and a tail for frequencies above 180 cm^{-1}. Projections onto the vibrations of different elements and structural units show that the tail is associated with the vibrations of the lightest element Ge, particularly atoms that are fourfold coordinated.

The original DF calculations [16] agreed reasonably well with measured $S(Q)$ and $g(r)$ for GST, [18] but several details were open to improvement. In particular,

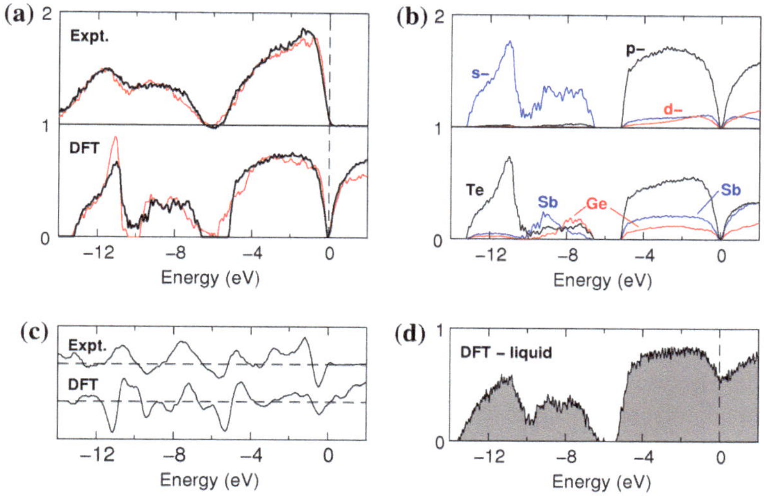

Fig. 17.7 **a** XPS valence band spectrum of *c*-(*thick black*) and *a*-GST (*red*) ([19], *upper panel*) and the calculated electronic DOS (*lower panel*). **b** Theoretical DOS of *a*-GST projected onto atom-centred *s*-, *p*-, and *d*-components and atomic types. **c** DOS difference between *c*-GST and *a*-GST [from (**a**), positive values imply a larger weight on *a*-GST]. **d** Electronic DOS of ℓ-GST. The *vertical dashed lines* mark the Fermi energy

Fig. 17.8 Calculated vibrational density of states for *a*-GST (VDOS, *lowest curve*), and power spectra from simulation trajectories for *a*-GST, for Ge, Sb and Te, and for fourfold coordinated Ge atoms ("4-Ge"). Curves have maxima at 1.3 (*a*-GST) or 0.65 (projections)

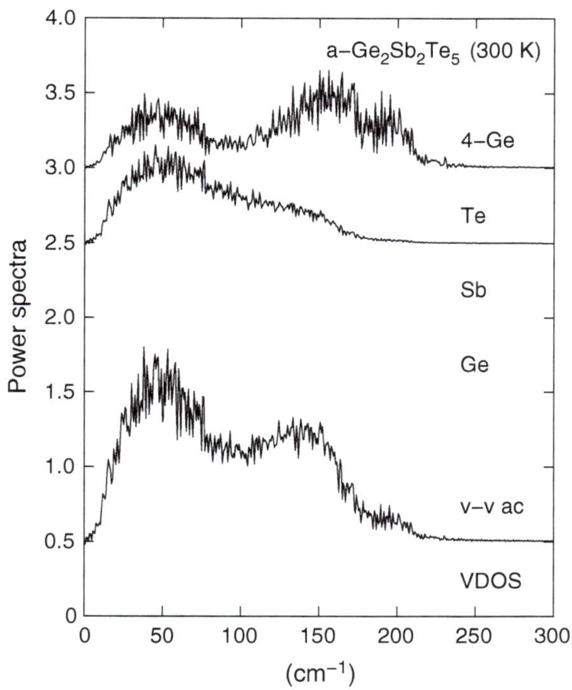

we had used the approximation of Perdew, Burke and Ernzerhof (PBE) [20] for E_{xc}, and the bond lengths that resulted were longer than those measured. A valuable collaboration between theory and experiment began when we discovered that the structure obtained in the original reverse Monte Carlo (RMC) fit to the x-ray diffraction (XRD) measurements [18] gave rise to a *metallic* density of states, i.e. no band gap at the Fermi energy. We combined the strengths of theory and experiment with the goal of finding a structure that:

- reproduces the measured $S(Q)$ and the total pair distribution function $g(r)$ for XRD and/or neutron diffraction,
- has bond lengths that are are consistent with EXAFS measurements and not artificially short,
- has an electronic structure with a band gap at the Fermi energy,
- has a DF total energy close to the minimum.

We do this by combining data from high-energy XRD [18] and XPS measurements on *a*- and *c*-GST [19] with DF simulations of a 460-atom sample over hundreds of picoseconds. This strategy is related to the "experimentally constrained molecular relaxation" (ECMR) of Biswas et al. [21, 22].

We adopted the approximation of Tao et al. (TPSS) [15], which has orbital-dependent terms that increase the computing demands, but the resultant structures and energy differences are generally better for extended systems [15]. The structures of *c*- and *a*-GST [16] were re-optimized using the TPSS functional, and the PDF for *c*-GST agrees very well with XRD measurements (Fig. 17.9). Both EXAFS and XRD results indicate the presence of the double maximum in the PDF, and the calculated Ge-Te and Sb-Te bond distributions have double maxima that were not evident in the PBE results. The shorter Ge-Te and Sb-Te bond lengths (2.84 and 2.93 Å, respectively) agree very well with EXAFS values (2.83 and 2.91 Å) [8]. The longer bonds have maxima above 3.1 Å. TPSS bonds in *a*-GST are 1–2 % shorter than in the calculations using the PBE functional, and the coordination numbers (Ge: 3.9, Sb: 3.5, Te: 2.7) are lower than in [16] (4.2, 3.7, and 2.9). The cohesive energy is 2.62 eV, and *c*-GST is 58 meV/atom more stable than *a*-GST.

The re-optimized *a*-GST geometry was taken as input for RMC refinement using the experimental x-ray total structure factor $S(Q)$. The experimental $S(Q)$ can be reproduced very well with a range of input parameters, but RMC simulations *without* constraints on the bond angles always led to a *metallic* DOS, even when the initial structure was semiconducting. A semiconducting structure that satisfies the x-ray structural information for *a*-GST can be obtained in practice only by using an iterative scheme where the bond angle distributions Te-Ge-Te, Te-Sb-Te, Ge-Te-Ge, Ge-Te-Sb, and Sb-Te-Sb are constrained to have values close to those found in the DF calculations. The TPSS bond lengths are slightly shorter than PBE values, but the use of the PBE approximation instead of TPSS would lead with the present optimization scheme to very similar structures.

The RMC optimization of the TPSS structure yields a structure factor $S(Q)$ in excellent agreement with XRD results (Fig. 17.9a), with minor differences in amplitude near the first two peaks. The excellent description of the high-Q region is

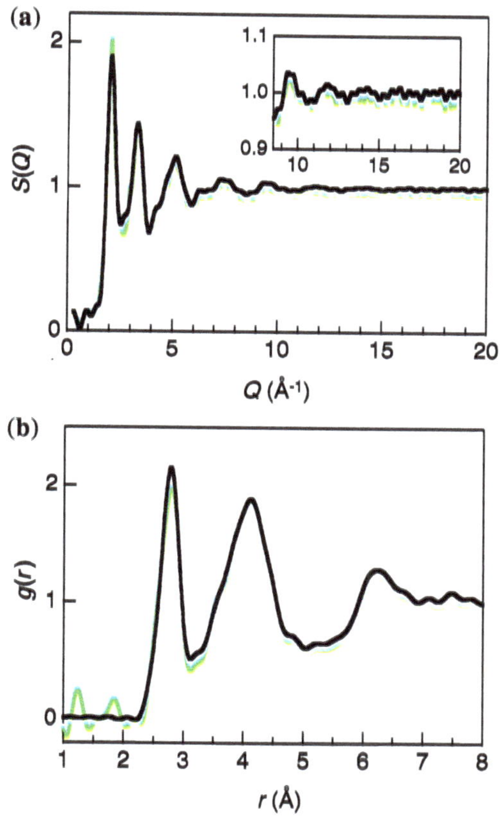

Fig. 17.9 **a** Structure factor $S(Q)$ in a-GST. *Green* XRD [18]; *black* calculated for RMC-optimized geometry. *Inset* data for high Q values. **b** Total PDF (XRD weights)

important, because the short-range atomic correlations in the XRD pair distribution function $g(r)$ (Fig. 17.9b) is obtained by Fourier transforming $Q[S(Q) - 1]$ and is affected strongly by the high-Q portion of $S(Q)$. The calculated $g(r)$ follows the experimental curve almost perfectly with small deviations near the first peak, and the total energy is only 79 meV/atom higher than that of the DF energy minimum. The electronic DOS shows a band gap at the Fermi level.

The ring statistics of amorphous GST-225, $Ge_8Sb_2Te_{11}$, and GeTe are compared in Fig. 17.10. There are many four-membered rings and few triangular configurations in all cases. As in GST-225, most of the four-membered rings in $Ge_8Sb_2Te_{11}$ show AB alternation, confirming that the basic building blocks in the amorphous and crystalline phases are $ABAB$ squares [16, 23, 24]. The odd-even alternation is apparent in GST-8211 and GST-225, and the ring distributions of these two materials are strikingly similar (Fig. 17.10).

For the optimized structure of a-GST-8211, cavities occupy 12.4 % of the total volume, i.e. the small increase in Sb content is accompanied by a dramatic increase in

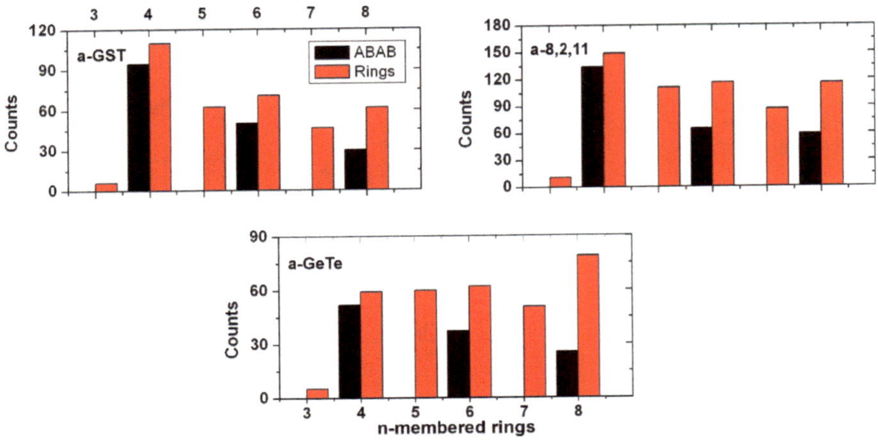

Fig. 17.10 Statistics of irreducible rings (300 K) in $Ge_2Sb_2Te_5$, $Ge_8Sb_2Te_{11}$, and GeTe alloys. *ABAB* configurations refer to alternation of atomic types *A* (Ge, Sb) and *B* (Te). The bond cutoff distance is 3.2 Å. Results for $Ge_2Sb_2Te_5$ and GeTe from [16]

the number of cavities to a value near that in *a*-GST-225. The coordination numbers around the cavity centers are Ge: 1.3, Sb: 0.4, and Te: 3.7, and other cavities: 0.2. Cavities are then surrounded mainly by Te atoms, and Ge is a more common neighbor than Sb. These features reflect the small Sb content and the underlying cubic structure that resembles the crystalline phase.

17.3.2 As-deposited Versus Melt-quenched GST-225

Atomistic simulations on phase-change materials have focused on melt-quenched (MQ) samples, where a liquid sample is cooled rapidly. Experimental samples are often deposited on a substrate, and the structure is likely to differ. We have developed a method to generate an "as deposited" (AD) sample with 648 atoms at 300 K [25] and have compared the results with those for a 460-atom MQ sample, refined as described above with respect to XRD and XPS data [26].

The simulations were carried out at 300 K, beginning with a random template of 36 atoms in an area of 27.61 × 27.61 Å (layer thickness 1.4 Å). These atoms were fixed during deposition to minimize relaxation effects near the base of the system. 17 layers of 36 atoms were then deposited on the sample and allowed to relax for 5–10 ps each. The vertical box dimension was adjusted for each layer to minimize the interaction with the slab replica (vacuum region 10 Å). Deposition took a total of 91 ps, and the result is shown in Fig. 17.11. There is limited mobility at 300 K, but some atoms mix with layers deposited previously, and the lower coordination of Te enhances its concentration at the surface. The final vertical dimension of the 648-atom sample is 15 % larger than in the bulk.

The model AD system was then prepared by releasing the 32 template atoms and decreasing the vertical box size (19 steps at 300 K) to obtain a cubic supercell (size 27.61 Å, density 0.0308 atoms/Å^3). The process lasted 67 ps in steps of 3–4 ps, and the system was equilibrated for 34 ps before data collection (25 ps). The pressure was small (2.9 $\pm$ 1.8 kbar) at the final (experimental) density, and the final structure was optimized at 0 K (Fig. 17.11c–d). The structure of the MQ sample [26] does not correspond to a DF energy minimum, so we performed a simulation at 300 K using the same functional (PBEsol [27]) and atomic density as for AD. Initialization (5 ps, 300 K) was followed by data collection (35 ps). The MQ structure is 11 meV/atom more stable than AD and 58 meV/atom less stable than the rock salt structure.

The structure factors $S(Q)$ for the AD and MQ samples are compared with experiment in Fig. 17.12. The curves for AD and MQ are strikingly similar, and the agreement with the experiment is very good. We note only that the calculated heights for the first two peaks (2.1 and 3.3 Å^{-1}) are somewhat low, and the shape of the third peak differs slightly. The calculated structures are from DF/MD simulations at 300 K, and deviations from experiment are caused in part by the approximations in the E_{xc} functional [25].

The electronic densities of states (DOS) in AD and MQ are very similar, with band gaps of 0.2–0.3 eV at the Fermi energy. The DOS profiles are in satisfactory agreement with XPS data [19], the most significant difference being near the DOS minimum around -10 eV between two σ-bands.

The PDF of both samples [25] show that Ge-Te and Sb-Te bonds dominate (*AB* alternation). EXAFS measurements [8, 28, 30] have shown that Ge-Te distances shrink upon amorphization (2.61–2.63 Å), suggesting that Ge may have tetrahedral coordination. This differs from earlier DF simulations, which found mostly (distorted) octahedral configurations for Ge with Ge-Te bonds of 2.77–2.78 Å. Tetrahedrally bonded Ge atoms also occur. The discrepancies in the Ge-Te bond length appear to have two causes: (a) Most approximations for E_{xc} tend to overestimate bond lengths by 2–3 %, but the use of the PBEsol functional here reduces this error (the Ge-Te PDF maximum is at 2.72 Å in MQ). (b) Sample preparation is crucial:

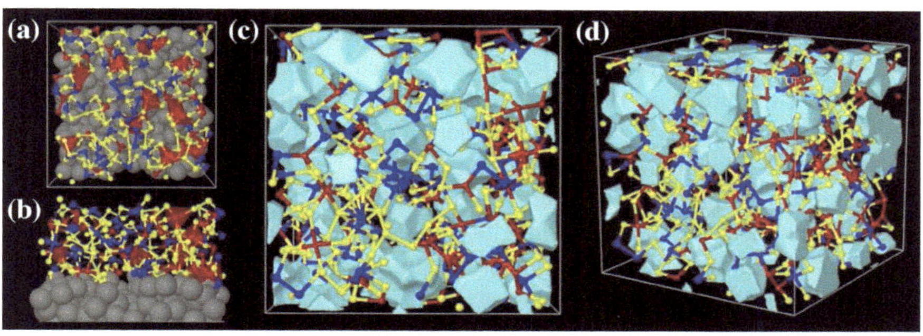

Fig. 17.11 AD sample of *a*-GST at 300 K. **a** Top view after 17 added layers (648 atoms). The uppermost atoms within 1.2 nm are highlighted, and the *red* tetrahedra mark Ge atoms with tetrahedral coordination. Ge, *red*; Sb, *blue*; Te, *yellow*. **b** Side view of the final surface. **c-d** Two perspectives of final sample. Cavities (cyan isosurfaces) comprise 16.3 % of the total volume

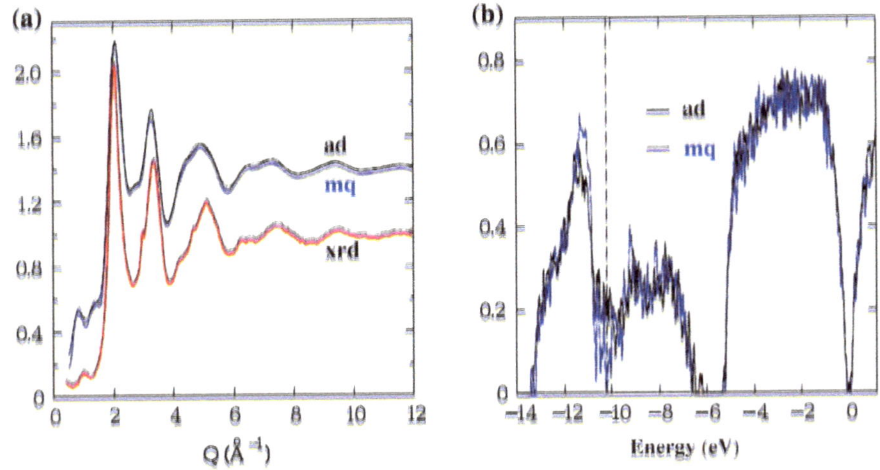

Fig. 17.12 **a** Structure factor $S(Q)$ of AD (*black*) and MQ (*blue*) samples at 300 K, and XRD (*red*). The calculated $S(Q)$ are shifted by 0.4. **b** Electronic DOS of AD and MQ samples. Vertical dashed line marks band gap separating the σ-band with Te-5s character

AD has more tetrahedral Ge atoms, and the Ge-Te maximum shifts to 2.69 Å. Ge-Ge bonds (2.52 Å), which correspond mainly to tetrahedral Ge atoms here, agree well with experiment (2.47 ± 0.03 Å) [28, 29]. Sb-Te bond lengths change little upon amorphization, and the first maximum of AD and MQ is at 2.89 Å.

The other PDF show clear differences between the samples, since the number of "wrong bonds" is significantly larger for AD. This is reflected in the Ge-Te and Sb-Te PDF (AB bonds) as weaker first maxima and emerging second maxima at 4.2 Å. The corresponding partial coordination numbers [25] show that Ge and Sb have twice as many wrong bonds (1.1 on average) as Te (including Te-Te bonds), although more than half of the atoms are Te. The total coordination numbers for AD are Ge: 4.2, Sb: 3.7, and Te: 2.8. Earlier work on MQ samples [16, 23, 26] indicated that the total coordination numbers of Sb and Te in a-GST do not satisfy the "$8-N$ rule" of covalently bonded networks, and additional support for this conclusion is found by calculating the bond orders (number of chemical bonds) for each interatomic connection [25].

The AD structure of a-GST (648 atoms, over 200 ps) differs from MQ (460 atoms, [26]) in essential ways: (1) The environment of Ge atoms is predominantly tetrahedral in AD and disordered octahedral in MQ, and our results for the Ge-Te bond lengths agree with EXAFS measurements [8, 28, 30]. This resolves the contradiction between measured and bond lengths and those found in earlier DF calculations (2) homopolar and Ge-Sb bonds are more common and reduce the number of $ABAB$ squares, which are needed for rapid crystallization. Sb and Te are more highly coordinated than expected from the "$8-N$ rule", so that a-GST cannot be seen as a covalent network glass. The PBEsol functional gives improved results, and both samples tend to form fewer Te-Te bonds, and MQ is 11 meV/atom more stable than AD. Their structure factors and electronic properties are similar, so that distinguishing between them using XRD alone remains a challenge.

17.4 Crystallization of Amorphous $Ge_2Sb_2Te_5$

The essential feature of PC materials is the ultrafast phase transition between amorphous and crystalline structures that occurs on a nanosecond time scale. In the previous sections, we have discussed extensively the amorphous and crystalline structures of GST and their properties. These correspond to the starting and end points for the actual phase transition, which are crucial to understand the function of PC materials. We now present results for the nucleation-driven crystallization process of GST using DF calculations combined with MD [31]. A sample of a-GST with 460 atoms was studied at 500, 600, and 700 K, and a second sample of 648 atoms was simulated at 600 K. In all cases we used a fixed crystalline seed (58 atoms, 6 vacancies) in order to speed up the crystallization process. More recent experience has shown that the time scale for the crystallization is of the order of several nanoseconds for these system sizes in the absence of a fixed seed, while those here are of the order of 0.3–0.6 ns. This means that we cannot discuss the onset of nucleation, but this is also true in the case of smaller systems (<200 atoms) discussed by other groups. In very small systems, periodic boundary conditions bias the process severely. Our larger samples reduce finite-size effects, and we show the effect of choosing different annealing temperatures. Simulations of this scale (up to 648 atoms over 1 ns) are near the limit of present day DF/MD calculations.

17.4.1 Simulation Details

The simulations follow the pattern described in previous sections. The CPMD program is used with Born-Oppenheimer MD and a time step of 3.0236 fs (125 a.u.). We employ an NVT ensemble with cubic simulation cell, periodic boundary conditions, and a single point ($k \equiv 0$) in the Brillouin zone. Simulations have been performed on two samples of a-GST (Fig. 17.1), with 460 and 648 atoms, respectively. We embedded in both a crystalline 'seed' of $4 \times 4 \times 4$ sites (13 Ge, 13 Sb, and 32 Te atoms, 6 vacancies) in a rock salt structure with lattice constant of 3.0 Å. The structure of the crystallite adopted the model of Yamada [7], which assumed that one sublattice of the rock salt structure comprises Te atoms, the other a random arrangement of Ge, Sb, and vacancies. We removed the atoms of the amorphous structure inside this volume, fixed the coordinates of the seed, and optimized the resulting structure.

For the 460-atom model, the initial geometry used was the structure of a-GST that gave excellent agreement between DF simulations and experimental x-ray diffraction (XRD) and x-ray photoemission spectroscopy (XPS) measurements [26]. With fixed coordinates of the seed, we performed DF/MD simulations at 500, 600, and 700 K over 600 ps (500/600 K) and 350 ps (700 K). The path to crystallization at 600 K is shown in Fig. 17.13(a–d). The amorphous and crystalline densities of GST differ, and the size of the cubic simulation cell was changed from 24.629 Å (amorphous density of 0.0308 atoms/Å^3) to 24.060 Å (crystalline density of 0.0330 atoms/Å^3) in five steps of 0.114 Å.

For the 648-atom structure, the initial coordinates of the 648-atom sample were obtained from a previous computer-aided deposition simulation [25]. The size of the cubic box was reduced in two steps from 27.736 Å (amorphous density) to 27.150 Å (density 0.0324 atoms/Å^3), which is incommensurate with the rock salt structure with this number of atoms. Complete crystallization was not expected and was not found at 600 K (see Fig. 17.13f), but the crystallite grew until making contact with its replica at the simulation cell boundaries.

17.4.2 Bond Orientational Order and Percolation

Crystallization is the ordering of a disordered system. In order to follow this process, it is necessary to label atoms which have a "crystalline" local environment. This is done by quantifying the directional order of a system of bonds for each atom. Angular correlations at an atomic scale can be achieved by projecting interatomic vectors $\mathbf{r}_{ij}$ onto a basis of spherical harmonics $Y_{lm}\left(\mathbf{r}_{ij}\right)$, and the order parameter of Steinhardt, Nelson, and Ronchetti has proved to be valuable in several contexts for different symmetries [32].

$$\bar{Q}_l\left(i\right) = \sqrt{\frac{4\pi}{2l+1}\sum_{m=-l}^{l}\left|\bar{Q}_{lm}\left(i\right)\right|^2}, \tag{17.7}$$

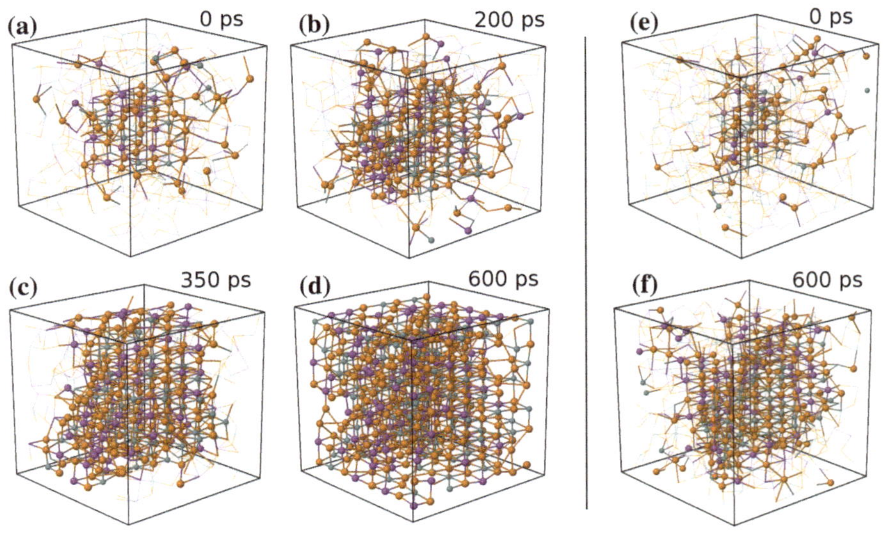

Fig. 17.13 Visualization of **a–d** the 460-atom and **e–f** 648-atom system at different stages of crystallization at 600 K. The atoms with crystalline local environment are highlighted with ball-and-stick. The fixed cubic seed is at the centre of (**a**) and (**e**). *Green* Ge, *purple* Sb, and *orange* Te

where

$$\bar{Q}_{lm}(i) = \frac{1}{N_b(i)} \sum_{k=0}^{N_b(i)} Q_{lm}(k), \tag{17.8}$$

and $N_b(i)$ includes the atom i and its neighbors, thereby averaging over an atom and its first coordination shell and including intermediate range order. The single atom quantity $Q_{lm}(i)$ is

$$Q_{lm}(i) = \frac{1}{N(i)} \sum_{j=1}^{N(i)} Y_{lm}(\mathbf{r}_{ij}) \tag{17.9}$$

where $N(i)$ is the number of neighbors for atom i. For cubic structures, the first nonzero value of $\bar{Q}_\ell$ is for $\ell = 4$, and we use this to define a crystalline atom as one for which $\bar{Q}_\ell \geq 0.6$.

Once we have obtained classification scheme for atoms that have a crystalline local environment, it is useful to analyze their connections between these atoms. In particular, we are interested in continuous paths over these atoms (bonds) which extend over the whole system, i.e. *percolation*. The connection between statistical mechanics, especially phase transitions, and percolation has a vast literature. In the case of GST, simultaneous measurements of electrical resistivity and optical reflectivity showed a significant influence of percolation on electrical properties, but a negligible influence on optical properties [33]. Here, we have analyzed percolation of the simulation trajectories by locating crystalline atoms as defined above and locating continuous paths of such atoms from an atom i to the same atom in a neighboring cell in all three Cartesian directions (for bonds of maximum length 3.20 Å).

17.4.3 Results for Nucleation-Driven Crystallization

We observed complete crystallization for the 460-atom simulations at 600 K and 700 K within the given time scale, and the former simulation is shown in Fig. 17.13a–d. Crystal growth around the seed was evident in the simulation at 500 K and in the 648-atom simulation at 600 K Fig. 17.13e–f, but crystallization was not complete within 600 ps. It should be noted that 500 K is already within 100 K of the experimental crystallization temperature (onset), whereas the process occurs in practice at higher temperatures (laser heating, current).

The amorphous phase is a metastable state, and crystallization results in a lower energy. Figure 17.14 shows the DF energy, the fraction of crystalline atoms (via order parameter $\bar{Q}_4$ as described previously), and the number of $ABAB$ squares in all simulations. Crystallization is faster at 700 K than at other temperatures, with the crystalline fraction increasing most rapidly between 150–205 ps., and it is complete at 600 K within 600 ps. The number of $ABAB$ squares and the strength of the Bragg peaks correlate well with the number of crystalline atoms. At 600 K, the latter increases by 40 % between 330 and 435 ps, and it changes by 50 % between 150 and 205 ps at 700 K. The more rapid crystallization at higher temperatures is consistent with the higher atomic mobility and with the results of recent ultrafast heating calorimetry measurements [34]. For the individual elements, the fraction of crystalline atoms is lowest in Ge until near the end of crystallization at both 600 K and 700 K, which is related to the coexistence of tetrahedral (minority) and octahedral Ge in the amorphous state [16, 23]. The tetrahedral component of tetrahedral Ge becomes negligible during the crystallization process, and the Ge fraction becomes comparable to the other two.

The gradual transition between amorphous and crystalline structures is shown in the evolution of the partial PDF at 600 K in Fig. 17.15. The crystalline seed gives rise to Bragg peaks at 3.00, 4.24, 5.20, 6.00, 6.71 and 7.35 Å, which are enhanced by crystallization. The increase is most rapid between 280–300 and 480–500 ps at 600 K (Fig. 17.15), and the rapid increase at 700 K occurs between 80–100 ps and 180–200 ps. The total coordination numbers were calculated from the PDF for Ge, Sb, and Te atoms at the beginning and end of the 460-atom simulations (500 K, 600 K, 700 K), and they show the expected change from amorphous to crystalline values for the samples that order and smaller changes otherwise.

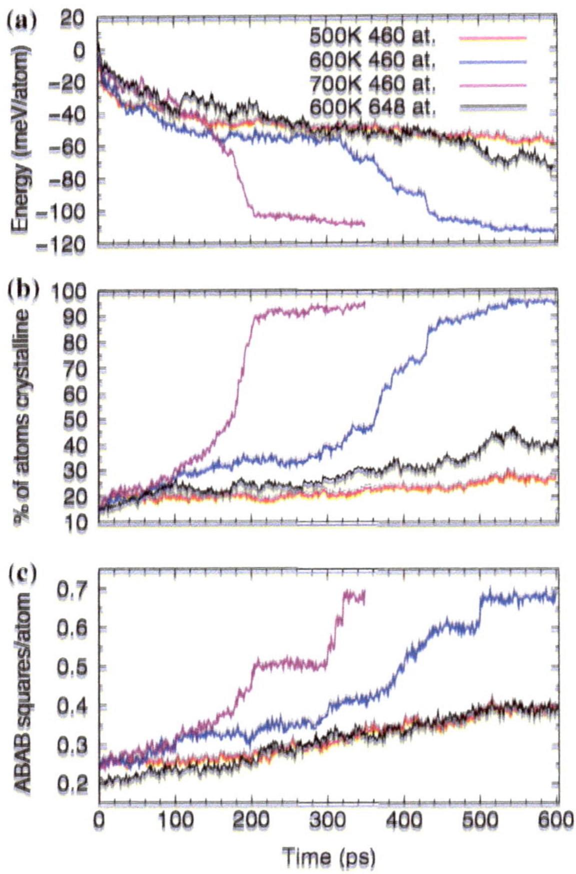

Fig. 17.14 **a** Total energy, **b** fraction of crystalline atoms, and **c** number of *ABAB* rings/atom in each simulation. The discontinuities in slope (steps) in *c* are due to changes in supercell size

PDFs can be constructed for all bond types, and Ge-Te and Sb-Te bonds are by far the most abundant (*AB* alternation). "Wrong bonds" (Ge-Ge, Ge-Sb, Te-Te, and Sb-Sb) are those that do not occur in a widely accepted model of the metastable cubic structure, [7] and are present throughout *all* simulations. The existence of wrong bonds is evident at 600 K in maxima in the partial PDF (Fig. 17.15) of the fully crystallized samples. The number of these bonds decreases during all simulations, most rapidly during fast crystallization, and are largest in the systems that do not crystallize completely (460-atom at 500 K, 648-atom at 600 K). At 700 K, the mobility and the ability to find energetically favorable configurations are higher than at 600 K, resulting in fewer wrong bonds in the fully crystallized structure. There are only four

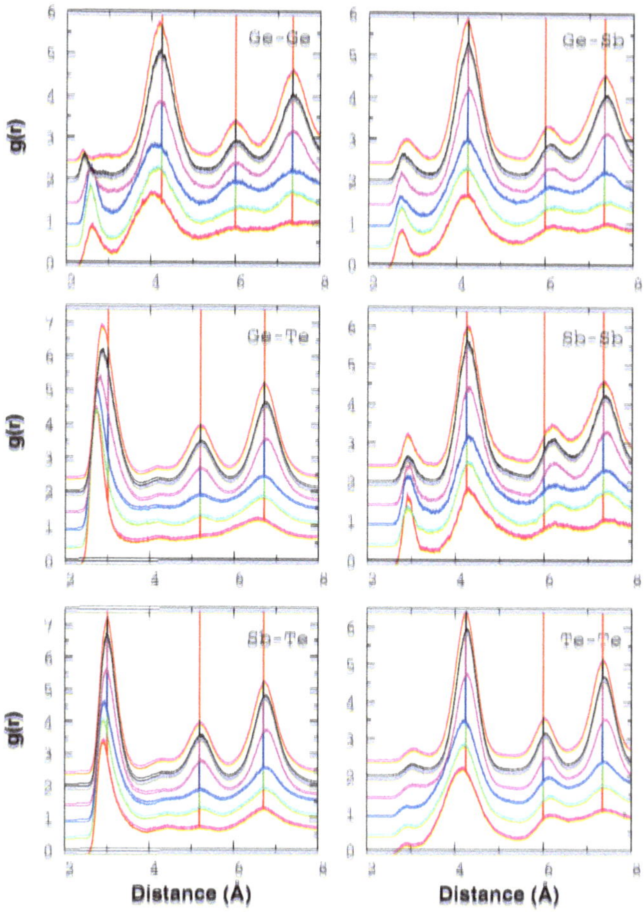

Fig. 17.15 PDF of atom pairs in 600 K simulation. Each plot is the average over 20 ps of a trajectory. *Red* 80–100 ps, *green* 180–200 ps, *blue* 280–300 ps, *magenta* 380–400 ps, *black* 480–500 ps, *orange* 580–600 ps. Successive plots are shifted by 0.5. The vertical lines correspond to the fixed crystalline seed with lattice constant of 3.0 Å

Ge-Ge bonds near the end, but the first peak shows an unusually short bond less than 2.5 Å.

Percolation measures the connectivity of the growing crystallite with its replica, and it can be analyzed from the simulation trajectories in all three directions. The results for three temperatures of the 460-atom model are shown in Fig. 17.16. The simulations began with the same structure, but no preferred direction for crystal growth was found. The order of occurrence is z-, y-, and x-directions at 600 K, and x-, z-, and y-directions for 700 K. Furthermore, percolation is present in the 500 K simulation in two directions, despite the fact that the system did not crystallize fully. Similarly, the 648-atom system is found to percolate at 600 K in all three directions (with some on-off character) although the system does not crystallize and the unit cell does not even match with the dimensions of the crystal structure. As a general feature, we note that percolation can be achieved with as few as ~20 % of the atoms being "crystalline", and its onset occurs well before the critical stage of crystallization. Percolation is evident in Fig. 17.13b–c, where the crystallite has reached the cell boundaries in the 460-atom simulation at 600 K, and this leads to a network of periodic images of the simulation cell prior to a rapid collapse to the crystalline state (~350–435 ps).

The evolution of crystallization is also apparent in the mean-square displacement (MSD) of the atomic coordinates from the starting structure (Fig. 17.17). There should be no atomic diffusion in a crystalline material, and changes in MSD slow down and eventually stop as crystallization occurs. The order of Ge and Sb varies: Sb is slightly more mobile than Ge at 500 K until ~500 ps, where the MSD of Ge increases rapidly. At 600 K, Ge has a higher MSD than Sb, while the reverse is true at 700 K. Furthermore, we have observed that Sb has a higher mobility in the liquid

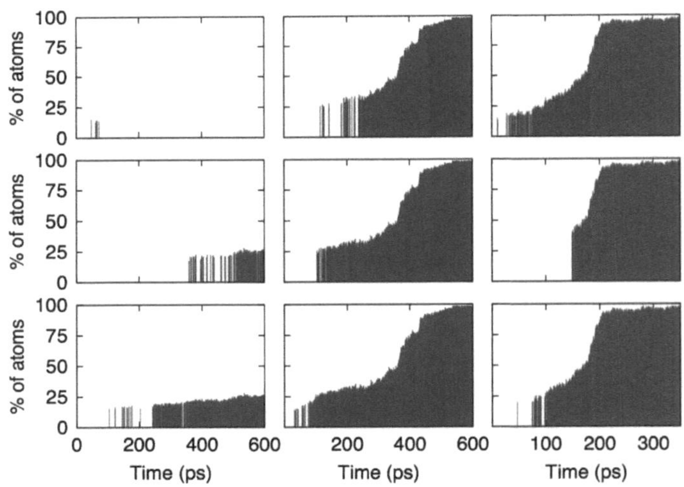

Fig. 17.16 Size of percolating crystalline cluster as a function of time. *Left*, *middle* and *right* columns are 500, 600 and 700 K simulations, respectively. *Top*, *middle* and *bottom rows* are percolation along x, y, and z axes, respectively

(900 K) [16]. The standard deviation of the MSD of individual atoms $\sigma\ (MSD)$ (or the fraction $\sigma\ (MSD)\ /MSD$) displays that there is a wide range of atomic mobilities in individual elements, e.g. for Ge atoms at 600 K and Sb atoms at 700 K. On the other hand, Sb atoms at 600 K and Ge atoms at 700 K have less variance than Te atoms.

Ge and Sb atoms are generally more mobile than Te. Atoms crystallizing in the first shell outside the seed move less than atoms crystallizing in the second shell, and crystal growth is faster in the direction parallel to the crystal lattice alignment of the seed. Some atoms move very little, while others diffuse far during the simulations, as shown for $\sigma\ (MSD)\ /MSD$ (Fig. 17.17). The motion of individual atoms is an interesting aspect of the crystallization process, and examples of mobile Ge/Sb with almost stationary Te, and the reverse, can be found. Atoms can move long distances during crystallization, breaking and forming chemical bonds. There is no simple

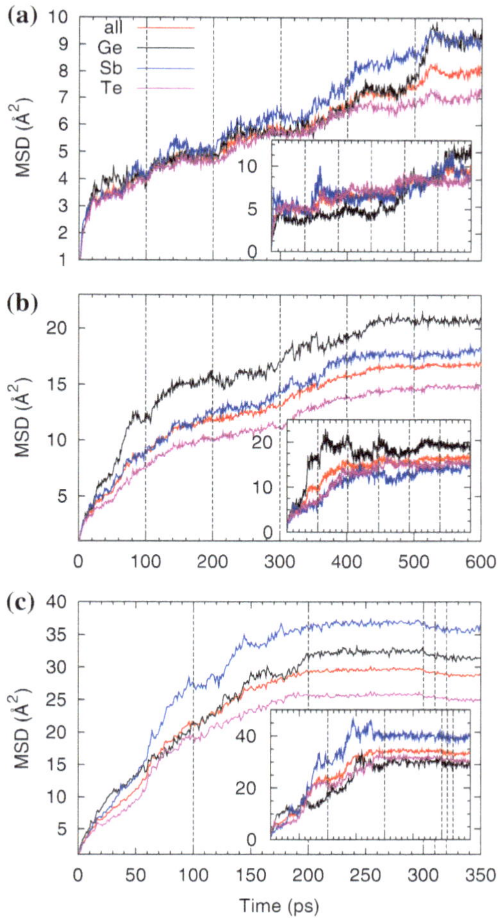

Fig. 17.17 MSD for all atoms and for each element at **a** 500 K, **b** 600 K, and **c** 700 K. Vertical dashed lines denote changes in box size. Insets show $\sigma\ (MSD)\ /MSD$. Fixed atoms were excluded

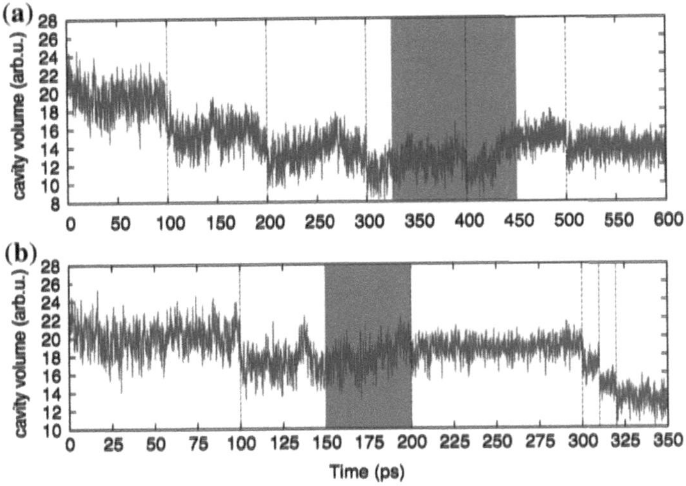

Fig. 17.18 Variation of total cavity volume in 460-atom simulations. **a** 600 K, **b** 700 K. Vertical *dashed lines* show changes in box size, shaded areas correspond to the periods of fastest crystallization [325–450 ps in (**a**) and 150–200 ps in (**b**)]

concerted diffusionless mechanism, such as the *umbrella flip* model, that is consistent with these findings.

The results in Figs. 17.17 show that crystallization is favored by the mobility brought by higher temperatures. Recent ultrafast differential scanning calorimetry measurements by Orava et al. [34] showed the highest crystal growth velocity at 670 K, which is consistent with the faster process rate in our simulations at 700 K. The diffusion constant of Sb is highest above 700 K (as for liquid GST at 900 K [16]), indicating that Sb is important in processes at higher temperatures. The number of *ABAB* squares correlates with the number of crystalline atoms, and they have been considered crucial to the speed of crystallization [16, 24]. However, these structural units can break and re-form during crystallization [31, 35]. Furthermore, the original fixed seed of *AB* motif embedded in an amorphous surrounding raises the question of its stability, and we have performed an unconstrained MD simulation of the initial seeded structure at 600 K. This resulted in a reduction of 20–30 % of the total number of *ABAB* squares within the first 30 ps. The remaining squares were scattered throughout the sample and did not form a crystallite. However, as our more recent simulations have shown, this unconstrained system still bears a *memory effect* that affects its crystallization properties.

The empty regions in space (vacancies, cavities), reflect the arrangements of the surrounding atoms. The existence of cavities in both amorphous and crystalline phases is characteristic of materials in the $(GeTe)_{1-x}(Sb_2Te_3)_x$ family [4], and it is widely assumed that they play an essential role in the rapid phase changes that can occur. The crystalline phase of GST has 10 % vacancies (cavities) and 7 % higher density than the amorphous phase, and it is natural to expect changes in the cavity

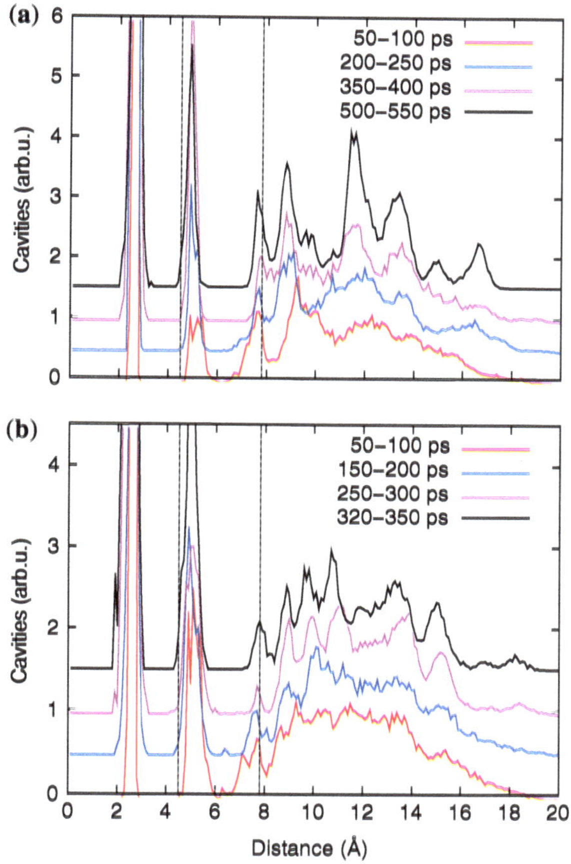

Fig. 17.19 Radial distribution of cavity centres from centre of seed. **a** 600 K and **b** 700 K simulation. *Dashed vertical lines* show the distances to nearest and farthest edges of the seed

volumes upon crystallization. Figure 17.18 shows the evolution of total cavity volume in the 460-atom simulations at 600 K and 700 K. The box size is adjusted smaller during the course of the simulation, and this is associated with the slight downward trend prior to crystallization. The variation for constant cells (e.g. 0–100 ps, 200–300 ps, etc.) is not large at 600 K. The total cavity volume changes little during the rapid stage of crystallization, but there is a small increase towards the end from 400–450 ps and 150–200 ps at 600 and 700 K, respectively. The overall volume reduction from the amorphous to crystalline state is significant, of the order of 1/3, which is consistent with the higher density and smaller fraction of cavities/vacancies in *c*-GST. [26]

It is interesting follow the "motion" of vacancies/cavities during crystallization. Recent simulations of crystallization in GST suggested that there is cavity diffusion to the crystal/glass interface. This was followed by Ge/Sb diffusion to these sites, aiding the formation of cubic, cavity-free crystallites [37]. Here, the fixed crystalline seed comprises, by definition, 6 vacancies as in *c*-GST, and we have followed how the other vacancies rearrange in different shells of the growing crystallite. The radial distribution of vacancies from the centre of the seed (Fig. 17.19) shows that there are cavities from the limit determined by the box size down to the cutoff of the cavity domains (2.8 Å). The ordering of the cavities is less pronounced at 700 K due to thermal fluctuations of atoms which distort vacancies and their positions (centres). However, vacancies are present *throughout* the crystallite during its formation.

At 600 and 700 K, the sizes of individual cavities range between 20-200 Å^3 depending on whether they comprise a single cavity or several joined ones (multi-vacancy). After crystallization, there is a higher abundance with volumes that are multiples of $\sim$35 Å^3, which is approximately the size of a single vacancy in the GST rock salt lattice with the above definition of a cavity. The decreasing size distribution at 700 K before crystallization resembles a liquid, while there is more structure at 600 K pointing to a certain degree of ordering.

Once the cavity centres are known, one can compute partial distribution functions between them (cavity-cavity) as well as with respect to the different atomic types (cavity-atom). This shows that there is a clear tendency for Ge and Sb atoms to move *away* from cavities during crystallization, although the final rock salt structure shows nearest-neighbor cavity-Sb "wrong bonds". This is due to the inevitable anti-site defects, where some atoms (and cavities) are located in the wrong rock salt sublattice (A type atom/vacancy in B sublattice or *vice versa*). Te atoms make up initially more than half of the total number of cavity neighbors, and the number of Te-cavity bonds remains large throughout the simulation. For the crystalline phase, the cavity PDF show maxima ($\sim$4.3 Å) for Ge, Sb, and cavities at distances corresponding to the opposite edges of $ABAB$ squares, whereas for Te atoms and cavities there is a maximum ($\sim$5.2 Å) corresponding to opposite corners of $ABAB$ cubes. Further correlations with the rock salt structure and AB alternation can be found at larger distances. The overall picture of crystallization is motion of Ge and Sb atoms *away* from cavities to occupy sites on one sublattice (A) of the rock salt structure. However, the phase transition occurs so rapidly that "wrong bonds" (anti-site defects) are inevitable.

17.5 Concluding Remarks

We have summarized here results of a series of simulations on $Ge_2Sb_2Te_5$ (GST-225), as well as some work on GeTe, $Ge_8Sb_2Te_{11}$, and $Ge_{15}Te_{85}$. The availability of massively parallel computers and efficient numerical algorithms make density functional calculations possible for hundreds of atoms over time scales of hundreds of picoseconds, i.e. approaching the crystallization time of phase change materials. This is a very encouraging development, and the calculations have enabled us to identify the crucial pattern ("$ABAB$ squares") in the phase transition in GST [16], to study the differences between members of the GST-family, and to identify structural differences between as-deposited and melt-quenched structures of GST-225.

Other calculations have led to important results for GST alloys and other materials. Concurrently with [16], Caravati et al. [23] performed DF/MD simulations on *a*-GST-225 (270 atoms, cooling from 2300 K to 300 K in 58 ps). The simulated $S(Q)$ was very similar to that in [16], and the authors noted the coexistence of tetrahedral and octahedral sites in *a*-GST: One third of Ge atoms were in a tetrahedral environment, while the remaining atoms showed a defective octahedral environment. The same authors studied the effects of stoichiometric defects [38] and pressure-induced amorphization in the same material [39]. Hegedüs and Elliott [24] have performed DF/MD simulations (63 atoms, 1.3 ns) of the crystallization process in *a*-GST-225. Crystallization is the time-limiting process in the write/erase cycle of PC materials and is the focus of much current research.

Our series of crystallization simulations has focused on 460- and 648-atom samples of *a*-GST (GST) using a fixed crystalline nucleus up to 600 ps [31]. Our simulations, the largest performed to date for crystallization in GST, have studied the process at three temperatures (500, 600 and 700 K). The onset of crystallization from a completely disordered sample is being studied in our ongoing work, but the present results go beyond earlier work in several ways. We focus on changes in the sample crystallinity based on a bond orientational order parameter [32], numbers of "$ABAB$ squares" and "wrong bonds", percolation, and the difference in densities between the amorphous and crystalline phases is taken into account during crystallization by adjustment of the cell size. All samples display crystal growth of the nucleus, but the process is complete only in the 460-atom sample at 600 and 700 K. Each process involves a preliminary stage where the nucleus grows gradually and forms percolating connections (narrow necks) with replicas in the neighboring cells. Crystal growth is favored by the increased mobility brought by higher temperatures. The diffusive nature of the atomic motion is an important aspect of crystallization in GST, and atoms can move significant distances during the structural transition. This applies to *all* elements, including Te.

The widely accepted model of the metastable structure of GST by Yamada [7, 40] considers a rock salt (Fm$\bar{3}$m) structure with one sublattice containing Te atoms, the other a random mixture of Ge atoms (40 %), Sb atoms (40 %), and cavities (20 %). However, this model does not include "wrong bonds", and there may be substantial displacements from the ideal crystallographic positions, particularly for Ge [40]. In our crystallized samples, the presence of wrong bonds, especially Te-Te, is particularly striking. Analysis of x-ray diffraction results of GST films have showed that exchanging *all* Sb atoms (atomic number 51) with Te (atomic number 52) in the Yamada model did not diminish the R-factor significantly [41], confirming that XRD measurements alone cannot distinguish between these two elements. Our crystallization simulation show clear indications that total energy favors Te occupancy of one sublattice (B) of the rock salt structure. Nevertheless, the very short crystallization times mean that other contributions to the free energy, particularly configurational entropy, play crucial roles. Vibration frequencies are low in GST (typically 100 cm^{-1} or 3 THz) [17], which means that atoms cannot vibrate more than a few thousand times within 1 ns. The system can easily become trapped in metastable states upon rapid cooling, since configurations with "wrong bonds" (defects) are much more common than structures with a perfect Te sublattice.

We have discussed alloys on the tie line GeTe–Sb_2Te_3 in Fig. 17.1a, which are group 1 materials where crystallization of the amorphous bits proceeds by nucleation from within. We noted above that amorphous bits of Sb-based materials near the $Sb_{70}Te_{30}$ eutectic (group 2), also used as PC materials, crystallize from the polycrystalline surroundings. We have simulated the amorphous structure of $Ag_{3.5}In_{3.8}Sb_{75.0}Te_{17.7}$ and suggested a plausible scheme for its crystallization [42].

Some words of caution are, however, appropriate. The discovery of medium-range order in Te atoms in a-GST and GeTe was unexpected and could not have been found in simulations with fewer atoms. This is also true for studies of cavities, which can occur as large multivacancies in these systems. Earlier simulations with less than 100 atoms certainly aided our understanding of the processes involved, but they cannot be expected to give reliable predictions. It is also crucial that the liquid state be cooled over a time scale that is physically relevant. The use of Born-Oppenheimer MD with an efficient predictor-corrector scheme for converging the orbital eigenfunctions has been decisive in our own work, since the time steps are two orders of magnitude greater than possible with standard techniques. Furthermore, high temperatures favour metallic samples in many of these systems, and the absence of a gap in the eigenvalue spectrum often leads to instabilities in the Car-Parrinello approach. Similar findings were reported in other work [23].

The approximation used for the exchange-correlation is crucial to obtain an accurate description of the structures of several PC materials. Elemental Te provides an interesting material in this context, since experimental and calculated structure factors and pair distribution functions often disagree. The results of DF calculations of structural properties of Te depend significantly on the choice of the exchange-correlation energy functional [43], and the best agreement with experiment is obtained using recently developed functionals that go beyond the gradient of the density(TPSS [15]).

While DF simulations of hundreds of atoms over time scales of nanoseconds are possible today, they make major demands on computational resources and the effort required to analyze the results. This situation is unlikely to change soon, and one should ask whether simulations of "hundreds of atoms" are enough to answer the problems we face. The development of classical force fields that have DF accuracy is important, and the work of Sosso et al. [44] is a very interesting development. These authors used a huge database of DF calculations on Ge/Te systems to develop a force field for GeTe that permits MD simulations with much longer time scales and up to several thousand atoms.

Acknowledgments We thank all colleagues who worked with us on these problems, particularly K. Kobayashi, S. Kohara, J. Larrucea, T. Matsunaga, and N. Yamada. J.A. thanks the Academy of Finland for support. We acknowledge gratefully the computer time provided by the JARA-HPC Vergabegremium on the JARA-HPC partition of the supercomputer JUQUEEN at the Forschungszentrum Jülich, and for time granted on the supercomputer JUROPA at Jülich Supercomputer Centre. The German Research School for Simulation Sciences is a joint venture of the FZ Jülich and the RWTH Aachen University.

References

1. N.F. Mott, Adv. Phys. **16**, 49 (1967)
2. S.R. Ovshinsky, Phys. Rev. Lett. **21**, 1450 (1968)
3. M. Chen, K.A. Rubin, R.W. Barton, Appl. Phys. Lett. **49**, 502 (1986)
4. T. Matsunaga, R. Kojima, N. Yamada, K. Kifune, Y. Kubota, Y. Tabata, M. Takata, Inorg. Chem. **45**, 2235 (2006)
5. N. Yamada, R. Kojima, T. Nishihara, A. Tsuchino, Y. Tomekawa, H. Kusada, in *Proceedings European Symposium on Phase Change and Ovonic Science (E/PCOS)*, Aachen, Germany, September 2009, pp. 23–28
6. J.K. Olson, H. Li, P.C. Taylor, J. Ovonic Res. **1**, 1 (2005) (ISSN 1584–9953)
7. N. Yamada, Mat. Res. Soc. Bull. **21**, 48 (1996)
8. A. Kolobov, P. Fons, A.I. Frenkel, A.I. Ankudinov, J. Tomonaga, T. Uruga, Nat. Mater. **3**, 703 (2004)
9. W. Wełnic, A. Pamungkas, R. Detemple, C. Steimer, S. Blügel, M. Wuttig, Nature Mater. **5**, 56 (2006)
10. S. Hosokawa, T. Ozaki, K. Hayashi, N. Happo, M. Fujiwara, K. Horii, P. Fons, A. V. Kolobov, J. Tominaga, Appl. Phys. Lett. **90**, 131913 (2007)
11. R. Car, M. Parrinello, Phys. Rev. Lett **55**, 2471 (1985)
12. Z. Sun, J. Zhou, R. Ahuja, Phys. Rev. Lett. **98**, 055505 (2007)
13. C. Bichara, M. Johnson, J.P. Gaspard, Phys. Rev. B **75**, 060201(R) (2007)
14. R. O. Jones, O. Gunnarsson, Rev. Mod. Phys. **61**, 689 (1989)
15. J. Tao, J.P. Perdew, V.N. Staroverov, G.E. Scuseria, Phys. Rev. Lett. **91**, 146401 (2003)
16. J. Akola, R.O. Jones, Phys. Rev. B **76**, 235201 (2007)
17. J. Akola, R.O. Jones, J. Phys.: Condens. Matter **20**, 465103 (2008)
18. S. Kohara, K. Kato, S. Kimura, H. Tanaka, T. Usuki, K. Suzuya, H. Tanaka, Y. Moritomo, T. Matsunaga, N. Yamada, Y. Tanaka, H. Suematsu, M. Takata, Appl. Phys. Lett. **89**, 201910 (2006)
19. J.-J. Kim, K. Kobayashi, E. Ikenaga, M. Kobata, S. Ueda, T. Matsunaga, K. Kifune, R. Kojima, N. Yamada, Phys. Rev. B **76**, 115124 (2007)
20. J.P. Perdew, K. Burke, M. Ernzerhof, Phys. Rev. Lett. **77**, 3865 (1996)
21. P. Biswas, R. Atta-Fynn, D. A. Drabold, Phys. Rev. B **76**, 125210 (2007)
22. P. Biswas, D.N. Tafen, F. Inam, B. Cai, D.A. Drabold, J. Phys.: Condens. Matter 21, 084207 (2009)
23. S. Caravati, M. Bernasconi, T.D. Kühne, M. Krack, M. Parrinello, Appl. Phys. Lett. **91**, 171906 (2007)
24. J. Hegedüs, S.R. Elliott, Nat. Mater. **7**, 399 (2008)
25. J. Akola, J. Larrucea, R.O. Jones, Phys. Rev. B **83**, 094113 (2011)
26. J. Akola, R.O. Jones, S. Kohara, S. Kimura, K. Kobayashi, M. Takata, T. Matsunaga, R. Kojima, N. Yamada, Phys. Rev. B **80**, 020201(R) (2009)
27. J.P. Perdew, A. Ruzsinsky, G.I. Csonka, O.A. Vydrov, G.E. Scuseria, L.A. Constantin, X. Zhao, K. Burke, Phys. Rev. Lett. **100**, 136406 (2008)
28. D.A. Baker, M.A. Paesler, G. Lucovsky, S.C. Agarwal, P.C. Taylor, Phys. Rev. Lett. **96**, 2 55501 (2006)
29. M.A. Paesler, D.A. Baker, G. Lucovsky, A.E. Edwards, P.C. Taylor. J. Phys. Chem. Solids **68**, 873 (2007)
30. P. Jovari, I. Kaban, J. Steiner, B. Beuneu, A. Schöps, A. Webb, J. Phys.: Condens. Matter **19**, 335212 (2007)
31. J. Kalikka, J. Akola, J. Larrucea, R. O. Jones, Phys. Rev. B **86**, 144113 (2012)
32. P.J. Steinhardt, D.R. Nelson, M. Ronchetti, Phys. Rev. B **28**, 784 (1983)
33. D.-H. Kim, F. Merget, M. Laurenzis, P.H. Bolivar, H. Kurz, J. Appl. Phys. **97**, 083538 (2005)
34. J. Orava, A.L. Greer, B. Ghoulipour, D.W. Hewak, C.E. Smith, Nature Mater. **11**, 279 (2012)
35. T.H. Lee, S.R. Elliott, Phys. Rev. Lett. **107**, 145702 (2011)

36. T. Matsunaga, R. Kojima, N. Yamada, K. Kifune, Y. Kubota, Y. Tabata, M. Takata, Inorg. Chem. **45**, 2235 (2006)
37. T.H. Lee, S.R. Elliott, Phys. Rev. B **84**, 094124 (2011)
38. S. Caravati, M. Bernasconi, T.D. Kühne, M. Parrinello, J. Phys.: Condens. Matter **21**, 255501 (2009)
39. S. Caravati, M. Bernasconi, T.D. Kühne, M. Parrinello, Phys. Rev. Lett. **102**, 205502 (2009)
40. N. Yamada, T. Matsunaga, J. Appl. Phys. **88**, 7020 (2000)
41. T. Nonaka, G. Ohbayashi, Y. Toriumi, Y. Mori, H. Hashimoto, Thin Solid Films **370**, 258 (2000)
42. T. Matsunaga, J. Akola, S. Kohara, T. Honma, K. Kobayashi, E. Ikenaga, R.O. Jones, N. Yamada, M. Takata, R. Kojima, Nature Mater. **10**, 129 (2011)
43. J. Akola, R.O. Jones, Phys. Rev. B **85**, 134103 (2012)
44. G.C. Sosso, G. Miceli, S. Caravati, J. Behler, M. Bernasconi, Phys. Rev. B **85**, 174103 (2012)

Chapter 18
Amorphous Phase Change Materials: Structure, Stability and Relation with Their Crystalline Phase

Jean-Yves Raty, Céline Otjacques, Rengin Peköz, Vincenzo Lordi and Christophe Bichara

Abstract Phase Change Materials should be stable enough in their amorphous phase to achieve a durable data retention, however they should also be bad glass formers to be able to recrystallise at high speed. To understand these contradicting properties, we construct models of amorphous Ge–Sb–Te systems using Ab Initio Molecular Dynamics and analyse the structures in relation with the relevant crystalline state. We show that structural patterns that are precursors of the crystalline phase exist in the amorphous state and we identify the signature of the various types of local atomic orders in the X-ray absorption spectra that we compute using Density Functional Theory. We first analyse the mechanical properties of the amorphous phase in the framework of the Maxwell rigidity theory, showing that all efficient Phase Change Materials deviate from the perfect glass and are mechanically stressed-rigid. Additionally, we show that the stability of Phase Change Materials is related to the density of low frequency vibrational modes (Boson peak). We describe how an adequate doping can result in an increased stability of the amorphous phase while keeping intact the phase change ability of the material.

J.-Y. Raty (✉) · C. Otjacques
Physics of Solids, Interfaces and Nanostructures, University of Liége B5,
Sart-tilman, 4000 Liége, Belgium
e-mail: jyraty@ulg.ac.be

C. Otjacques
e-mail: cotjac@ulg.ac.be

R. Peköz
Max Planck Institute for Polymer Research,
Ackermannweg 10, P.O. Box 3148, 55128 Mainz, Germany
e-mail: pekoez@mpip-mainz.mpg.de

V. Lordi
Lawrence Livermore National Laboratory,
7000 East Avenue, L-413, Livermore, CA 94550, USA
e-mail: vlordi@llnl.gov

C. Bichara
CINAM - CNRS Université Aix-Marseille,
Campus de Luminy, Case 913, 13288 Marseille Cedex 9, France
e-mail: xtof@cinam.univ-mrs.fr

C. Massobrio et al. (eds.), *Molecular Dynamics Simulations of Disordered Materials*, Springer Series in Materials Science 215,
DOI 10.1007/978-3-319-15675-0_18

18.1 Introduction

Phase Change Materials (PCMs) are nowadays widely used for data recording and increasingly present non-volatile memories [1, 2]. Since Ovshinsky's precursor work [3], PCMs have been studied extensively, both experimentally [4–6] or with ab initio calculations [7–11]. The operating principle of PCMs is based on the easy and reversible switching between the crystalline (C) and amorphous (A) states, each phase possessing very different electrical and optical properties. These materials are nowadays found in commercial applications for optical information storage and dissemination, such as CD-RW, DVD-RW or Blue-Ray disk, and could be used for the achievement of PC-RAM.

Novel compositions for PCMs have been often found by trial and error, after identifying common features such as a small ionicity or a number of p valence electrons that is close to the value of 3 [12]. These materials often belong to the Ge–Sb–Te ternary phase diagram (so-called GST systems), sometimes with additional doping, such as in the Ag–In–Sb–Te compounds.

To be suitable for applications, PCMs should possess an ensemble of properties besides the fact that they could be switched reversibly between an amorphous and a crystalline phase.

- Both phases should be sufficiently contrasted in properties (either optical, electrical, or both) to allow for the reading of data.
- The recrystallisation of the amorphous phase should be fast.
- The melting temperature should be low enough to allow for amorphisation at minimal energy cost.
- The amorphous phase should be stable enough to maintain the data over long times under operating conditions (i.e. moderate heating).

The synthesis of all these properties in a single compound appears at first to lead to contradictions. The fast crystallisation implies that the structure of the amorphous and crystalline phases should not be too dissimilar, but the contrast in properties and the stability of the amorphous over long times implies strong differences in the nature of the crystal and amoprhous phase, as well as a significant activation barrier to recrystallisation.

To understand the relation between structure and phase change properties, Ab Initio Molecular Dynamics has proven to be a powerful tool as it can produce structural models of the amorphous phase that can help understanding the results of many experiments.

In the following, we will focus on the description of the atomistic structure of two classes of PCMs, respectively based on Sb–Te and on Ge–Te and compare the local atomic ordering of the amorphous with that of the crystal.

In the second part, we will show in particular how structural features can be related to near-edge X-ray absorption spectroscopy (XANES), giving further validation of the ab initio generated models.

In the third part of this work, we will use the Maxwell rigidity theory framework to show that, despite the apparent proximity between the structure of the A and C phases, all GST-based PCMs deviate from the isostatic ideal glasses and are mechanically stressed rigid.

To obtain further information about the stability of these stressed glasses, we will describe their dynamical properties in the last section. We will show that low frequency vibrational modes dominate the Sb–Te vibrational spectrum in agreement with its extremely easy recrystallisation under ambient conditions. This observation will be used to provide a way to improve the stability of the amorphous Ge–Te system by introducing high frequency modes and lowering the density of low frequency, 'floppy', modes thanks to proper doping.

18.2 Structure of Ge–Sb–Te Amorphous Phase Change Materials

The building compounds of a wide number of PCM are the Sb_2Te, Sb_2Te_3 and GeTe alloys. The crystalline structure of these compounds results from the Peierls instability, its nature depending on the number of *p*-electrons [13, 14]. It is remarkable that all these compounds crystallize in layered structures with rhombohedral symmetries. In these three structures, the atomic environment is octahedral with some distortions depending on the site (see Fig. 18.1).

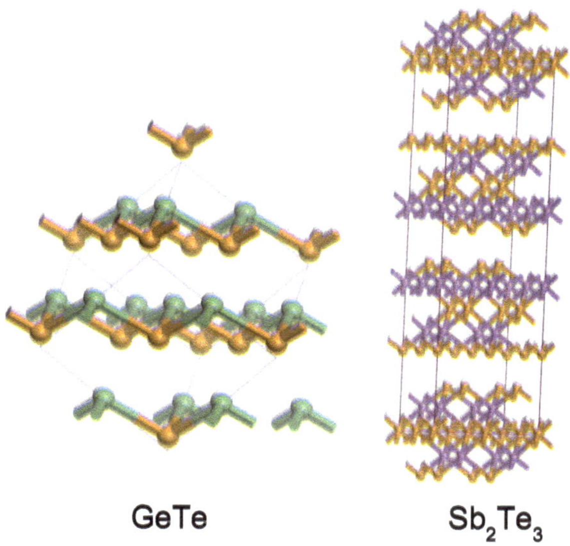

Fig. 18.1 Crystalline structure of GeTe and Sb_2Te_3 (R-3m). For GeTe, a 2 × 2 × 2 supercell is represented

The PCM lying along the pseudo-binary $(GeTe)_xSb_2Te_3$ line are also observed in a stable hexagonal form that consists of stacked layers, and in a metastable cubic phase, the latter being the one used in phase change applications. Both structures are stabilized by Peierls distortions and can be viewed as stacks of distorted cubes, that are further stabilized by vacancies [15–17]. In the liquid state, neutron diffraction study on a wide variety of PCM indicates that they all possess an octahedral local order as well [18].

The situation in the amorphous phase is less clear, as no explanation is unanimously accepted. Many of the proposed models stand for a transition from a sixfold (crystal) to a fourfold (amorphous) coordination for the Ge atoms in the structure. While a previous model supposed a switching from an octahedral to a tetrahedral local order around the Ge atoms [19], Huang and Robertson suggested that the Ge environment could remain octahedral in the amorphous [20] state, the optical contrast being attributed to the breakdown of the resonant bonding that exists in the crystalline state [21].

In this work, we used Ab Initio Molecular Dynamics (AIMD) to produce amorphous models by the melt-and-quench method. The PWSCF and VASP codes have been used together with pseudopotentials (PAW, Ultrasoft) and GGA exchange-correlation functionals (PBE). In all cases, the systems were initially equilibrated in the liquid phase for several tens of picoseconds, before being gradually cooled down to ambient temperature with a typical cooling rate of 3–5 ps/100 K. In all calculations, thermostats were used to control the temperature, however their influence was kept as minimal as possible at the lowest temperature (large fictitious thermostat mass, slow thermostat response to temperature fluctuations). The amorphous systems were finally annealed for several tens of picoseconds before gathering average structural and dynamical data.

18.2.1 Structure of Sb–Te Compounds

The structure factors, $S(q)$, and pair correlation functions, $g(r)$ of Sb_2Te_3 and Sb_2Te in the liquid and amorphous phases are shown in Fig. 18.2. If the overall agreement between experiment and simulation is reasonable, it is not as good as in the case of other chalcogenide glasses, such as $GeSe_2$ [22]. It was established [23] that the discrepancy is mostly due to an over coordination of Te atoms in DFT calculations. However, AIMD simulation reproduce all trends observed experimentally.

The first $g(r)$ peaks positions are at slightly larger distances (by $\sim$0.1 Å) in the AIMD curve than in the experiment, as discussed in [23], otherwise, the overall agreement is reasonable.

The first peak position, r_1, first minimum, $r_{\min}$, and total coordination numbers N_c from both the experiment and the AIMD are indicated in Table 18.1. Coordination numbers were calculated by integration of $4\pi r^2 \rho g(r)$ from a fixed distance equal to 2.4 Å to a cutoff distance equal to $r_{\min}^{exp}$ for experiment and $r_{\min}^{FPMD}$ for calculations.

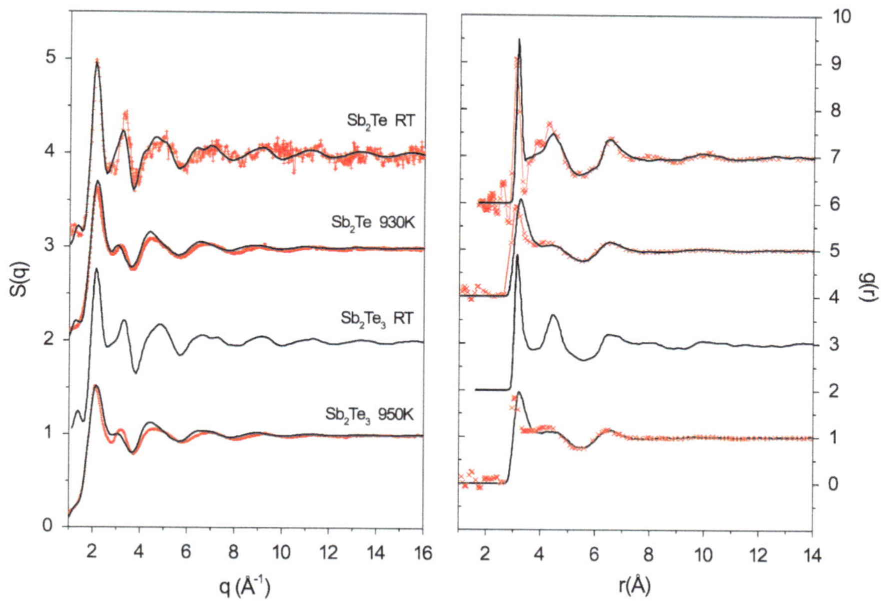

Fig. 18.2 Structure factors (*left*) and pair correlation functions (*right*) of Sb_2Te and Sb_2Te_3 in their liquid and amorphous state. The experimental data (to be published) were recorded by neutron scattering and are shown with *symbols*. The *thin black line* corresponds to the AIMD simulation data, for which the total g(r) are computed with proper weighting of the partials by the elements neutron scattering lengths

Table 18.1 Compounds and temperatures studied by neutron diffraction experiment and AIMD simulations

Compound	$T(K)$	$\rho(at/\text{Å}^{-3})$	r_1^{exp}	r_{min}^{exp}	N_c^{exp}	r_1^{fpmd}	r_{min}^{fpmd}	N_c^{fpmd}
Sb_2Te	930	0.0301	2.9(6)	3.5(4)	4.5(9)	3.0(6)	3.6(3)	4.8(6)
Sb_2Te_3	950	0.0283	2.9(2)	3.5(1)	3.9(4)	3.0(1)	3.5(9)	4.3(8)
Sb_2Te	300	0.0289	2.8(7)	3.1(5)	2.5(2)	2.9(5)	3.1(8)	2.7(1)
Sb_2Te_3	300	0.0283				2.9(3)	3.4(7)	3.7(5)

Densities [25], positions of the first maximum r_1, first minimum r_{min} and coordination number N_c from experimental and calculated pair correlation functions $g(r)$. All distances are given in Å

Because of the interatomic distances overestimate mentioned before, we observe coordination numbers from AIMD that are slightly larger (by 0.4 in average) than the experimental ones. The values for Sb_2Te_3 are in line with those of Caravati et al. [24], given that the cutoff distance was smaller in that study.

18.3 Structural Properties

The interatomic distances computed for the 6 first neighbors of Sb and Te atoms in the simulation box and averaged over the FPMD trajectories are shown in Fig. 18.3. In the liquid phase, the situation is similar for both compounds and around each element: the distances are rather equally distributed between ~2.9 and ~3.8 Å. Upon quenching, this shell of six neighbors splits progressively into two sub-shells made of 3 'short' +3 'long' distances around Sb, and 2 'short' +4 'long' distances around Te.

To interpret this evolution of distances, it is necessary to analyze quantities that can only be obtained by the AIMD simulations. First, we plot the angle distributions around Sb and Te atoms in Fig. 18.4. These are computed by summing the angles $\widehat{i0j}$ between one atom 0 in the structure and its neighbors i and j, including the first 3 neighbors for Sb and 2 for Te. For both compounds, the total angle distributions are peaked around 90° (and a small peak at 180°). This indicates that the local atomic order remains mostly octahedral (the peaks become higher and thinner upon quenching but keep the same position as in the liquid).

The local order around Sb atoms is similar to that found in liquid As [26]. In that case, it was shown that a correlation existed between the first 3 neighbors and 3 more distant neighbors. That correlation is similar to that of the crystal phase, so that the local environment of an atom inside the liquid can be seen as a distorted octahedron, with 3 shorter bonds are facing 3 longer bonds, forming an angle that is close to

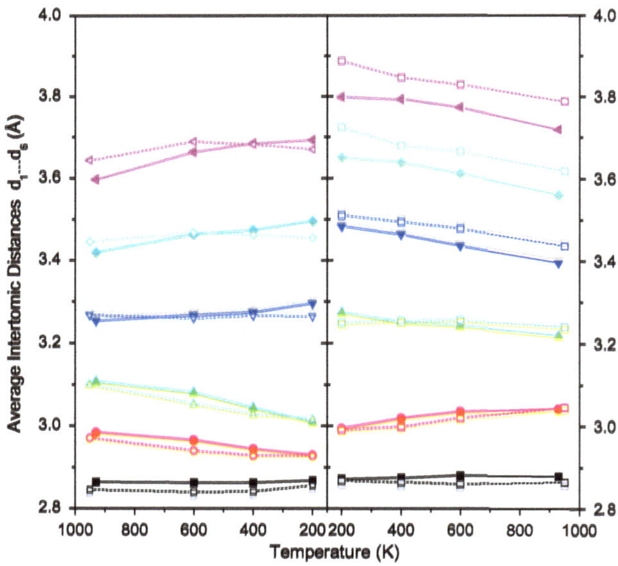

Fig. 18.3 Evolution of the average six first interatomic distances around Sb (*left panel*) and Te atoms (*right panel*) upon quenching from the liquid phase to 200 K (Sb_2Te: *lines* with *full symbols*; Sb_2Te_3: *dotted lines* with *empty symbols*)

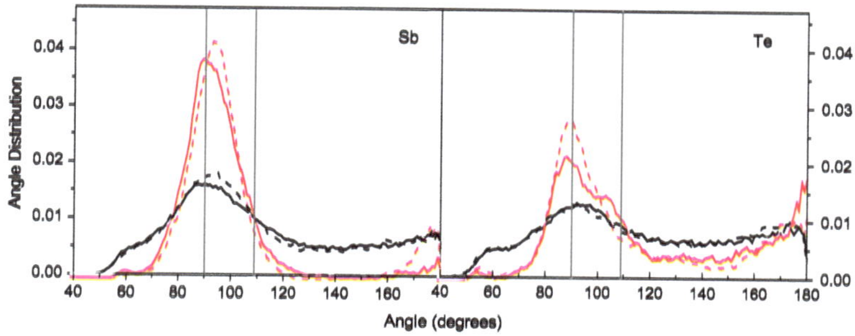

Fig. 18.4 Angle distributions of liquid and amorphous Sb_2Te (*plain lines*) and Sb_2Te_3 (*dotted lines*) in the liquid (*dark line*) and amorphous state (*red line*)

180°. Gaspard et al. showed that this local alternation of a short and long bond is sufficient to stabilize a covalent structure for a group V element (3p electrons fill one half of the valence p-DOS), so that the Peierls distortion mechanism can hold in the case of a disordered structure under some conditions [13].

In order to verify that such a Peierls distorted structure is also present in amorphous Sb_2Te and Sb_2Te_3, we compute Three Body Correlations (TBC) in the following way: the TBC is a function of two distances r1 and r2 and is the probability to find two neighbors of a given atom at distances r1 and r2, the angle between the two bonds deviating from 180° by less than 15°.

In the liquid phase, the TBC correlation value around Sb and Te is very low, the maximal correlation between being broadly located around (r1, r2) = (3.2 Å, 3.2 Å) the diagonal in the TBC map (see Fig. 18.5). In the amorphous, however, this broad peak around r1 = r2 ≡ 3.2 Å splits into two (symmetric, by construction) peaks located off the diagonal. The maximal correlation value is increased, especially in the Sb case), and clear maxima are observed around (r1, r2) = (3.05 Å, 3.2 Å) and (2.95 Å, 3.35 Å) for Sb in Sb_2Te_3 and Sb_2Te, respectively, and (r1, r2) = (2.9 Å, 3.45 Å) for Te in both compounds. This correlation expresses the existence of a local distortion, particularly in the case of Sb, such that the 3 shorter bonds are almost aligned with the 3 longer ones.

These three analyses of interatomic distances in our amorphous SbTe structures lead to the picture of a distorted octahedral local environment around atoms, in which the type of sub-shells of first neighbors depends on the element and the nature of the alloys. Altogether, this local environment is similar to what is found in the crystalline phase.

In the case of glasses, it is well known that the atomic ordering can extend farther than the first shell of neighbors, as sometimes reflected in the presence of a First Sharp Diffraction Peak in the diffraction spectrum. To investigate this possible extended range order, we develop a specific tool of analysis by extending the TBC to include higher order correlations (see Fig. 18.6). The resulting maps obtained for Sb_2Te_3 and Sb_2Te are shown in Fig. 18.7. In the case of the liquid (not shown here), the maps

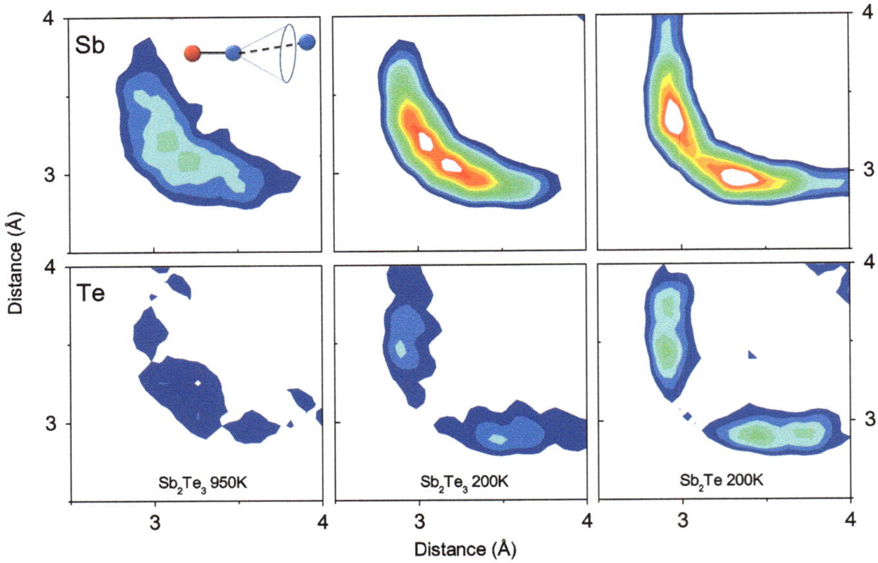

Fig. 18.5 TBC correlation maps TBC (r1, r2) for central Sb (*top panels*) and Te atoms (*lower panels*) in liquid Sb_2Te_3 (Sb_2Te analysis produces similar maps, not shown here), and amoprhous Sb_2Te_3 and Sb_2Te. The maps are plotted between 30 and 90 % of the maximal correlation (found in Sb_2Te_3)

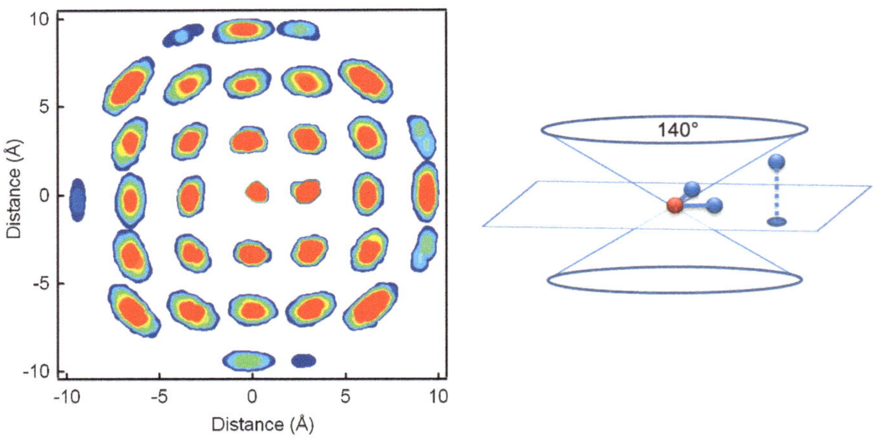

Fig. 18.6 Example of a correlation map produced in the case of a simple cubic structure simulated at finite temperature (*left*). This map is obtained by averaging over the atoms and configurations of the MD trajectory. For each atom, the two first neighbors are selected, which define a projection plane. Other atoms from the simulation box are projected perpendicularly onto that plane if they belong to the space complementary to two cones centered on the defining atom and perpendicular to the plane (*right panel*). By construction, the map is not symmetric (the shortest interatomic distance is located on the positive x axis)

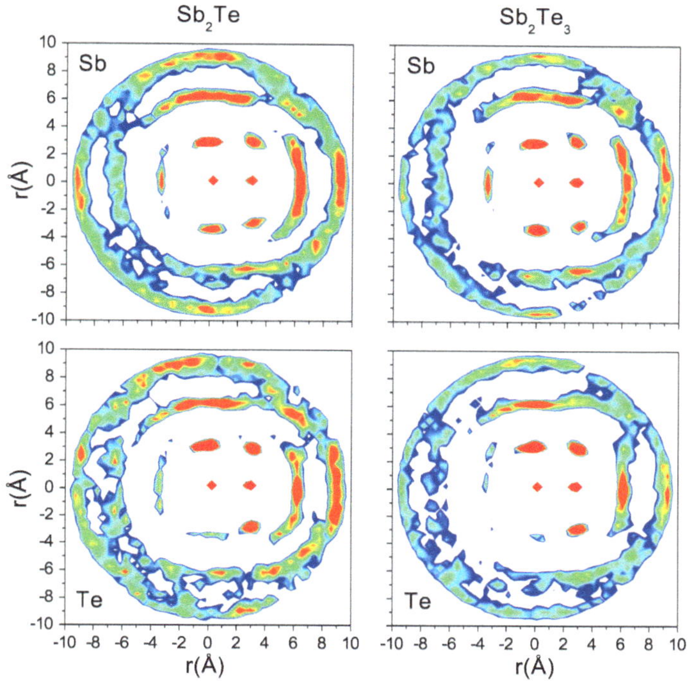

Fig. 18.7 2D correlation maps for amorphous Sb_2Te (*left panel*) and Sb_2Te_3 (*right panel*)

only shows weak correlation between the central atom and its first shell of neighbors, as expected from the weak correlation observed in the TBC (see Fig. 18.5).

Upon quenching, more correlations appear on the maps, correlations that extend up to the second and third neighbors shell. These spots are not random and are ordered in a way that suggests the existence of an underlying distorted cubic structure in the amorphous. The correlations extend up to 9 Å for some neighbors (i.e. up to the simulation box limit). This analysis indicates that the Peierls distortion extends somehow farther than the first shell of neighbors. The fact that the bonding angles are strongly peaked around 90° helps to generate this extended order, which does not differ much from that of the crystal phase.

18.3.1 Ge Containing Compounds

As seen in Sect. 18.2.1, the structure of Sb_2Te_3 and Sb_2Te , which are building blocks for most of current PCMs, is intrinsically close to that of their crystalline phase. The other building block for GST PCMs is GeTe, and its amorphous structure, as well as that of GST amorphous materials has been shown to be more complex.

Our understanding of the detailed structural differences between crystalline (C) and amorphous (A) phases, as well as of the mechanisms underlying the fast switching between these phases, still is not complete. An important step forward was achieved in [19], with the assignment of the most important structural difference between C and A phases of the typical PC alloy $Ge_2Sb_2Te_5$ (GST225) to a change of the local environment of Ge atoms, based on measurements of the near-edge (XANES) and extended range (EXAFS) X-ray absorption spectra. It was suggested that Ge atoms that lie at the center of slightly distorted octahedral cages in the (metastable) crystal form (having a rocksalt structure, RS), shift to a tetrahedral position in the amorphous structure. This mechanism, known as the umbrella-flipping model, has been further supported by later theoretical [17, 27] and experimental works [28]. The somewhat oversimplified picture of the umbrella-flipping model has since been under debate. Other EXAFS experiments [29], analyzed with a reverse Monte Carlo fitting method, indicated a predominance of fourfold rings. Photoelectron spectroscopy [4] clearly indicated the presence of a mixed environment around Ge atoms in the A phase. This mixing of local environments around Ge atoms in the amorphous phase is also a common feature of all density functional theory (DFT)-based molecular dynamics simulations of GST [7, 30–33], which find 20–40 % of Ge atoms to be tetrahedrally bonded. Some of these works pointed out the role of large voids [34] or that of square planar rings in the recrystallization [32], while others indicated that tetrahedrally bonded Ge atoms account for the high energy excitations observed in Raman spectra [35] in addition to providing the correct optical contrast between the A and C phases [27]. However, the AIMD models also have in common that all of them are rather at odds with the initial XANES data interpretation and modeling in [19]. It has been show that both type of environments have very different electronic nature, as it can be seen in Fig. 18.8.

The precise structure of the metastable crystalline phase involved in the phase change mechanism of GST alloys is also not fully understood. In particular, its metastable character leaves a whole range of possibilities for its atomic structure. Indeed, X-ray diffraction has determined only that the structure is globally RS with a regular fcc lattice occupied by Te atoms and some kind of disorder on the second fcc lattice shared by Ge atoms, Sb atoms and vacancies [15, 36]. Recent works have shown that the structure of the metastable cubic phase is strongly varying with the annealing conditions [37], and total energy calculations have shown that this was related to the varying degree of ordering of vacancies [38].

To compare model structures against experimental X-ray absorption data, a widely used method is the multiple scattering (MS) theoretical approach [39], which has been shown to reproduce accurately the experimental spectra of many crystals. When extended to the study of liquid and amorphous systems, it has been demonstrated [40] that a proper averaging over inequivalent sites is required, since the fluctuations of the local atomic ordering produce strong effects, at least on the EXAFS part of the signal, that cannot be reproduced by using an ad hoc Debye-Waller factor. On the other hand, for the computation of the XANES part of the spectrum, full potential (FP) ab initio calculations are needed. However, these calculations are so demanding that

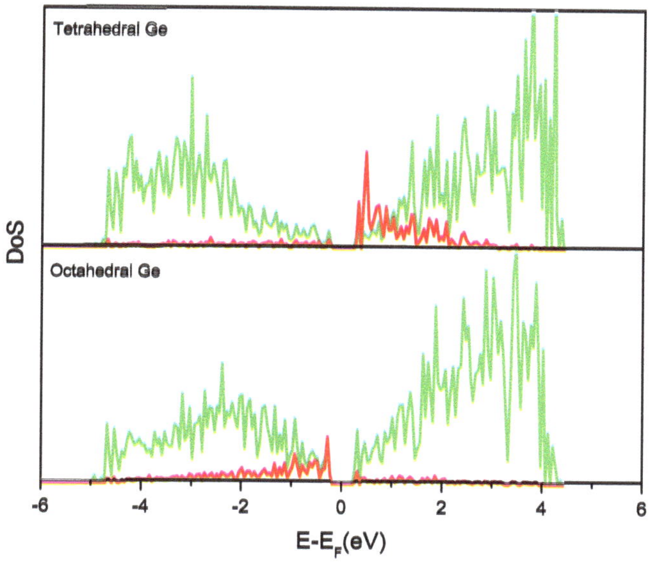

Fig. 18.8 Density of states, DOS, of amorphous $Ge_1Sb_2Te_4$. The densities are projected onto tetrahedral and octahedral Ge atoms [30]. p- and s-states are plotted in *green* and *red*, respectively. p- and sp3-bonded Ge have different partial DOS, tetrahedral Ge having large contributions of s states at the bottom of the conduction band

current computing resources effectively limit the application of AIMD calculations only to crystals and limited-size cluster models [41]. A recent detailed analysis of an AIMD model using this approach has shown the lack of sensitivity of the computed XANES spectra with respect to the local Ge environments, even if these exhibit very strong distortions from ideal tetrahedral or octahedral geometries [42]. It is actually well known however that AIMD calculations perform very poorly in the cases of the lighter elements, and germanium is known for being problematic [43], as evidenced in the case of $GeCl_4$ molecules. These difficulties have been attributed to the spherical muffin-tin orbitals approximation used to construct the total potential. This spherical description of the core states is clearly insufficient in the case of the Ge near-edge calculations. Thus, for the amorphous phases and highly defective crystal structures encountered in this work, there is a need for a method that still accurately describes the fine details of the effective potential, but at a more affordable computational cost than all electron calculations.

It is actually possible to use DFT calculations to compute XANES for amorphous and crystalline PCMs. To compute the XANES spectra, we use a new implementation by one of the present authors (V.L.), employing the PAW formalism [44] (as implemented in the VASP code [45, 46]) and including a self-consistent treatment of the core hole effects on the unoccupied states.

The XANES spectrum is calculated using Fermi's golden rule as

$$\sigma(\omega) = \frac{S_0}{\omega} \sum_f (1 - w_f) \left|\langle \psi_f | \mathbf{e} \cdot \nabla | \psi_i \rangle\right|^2 \delta(E_f - E_i - \hbar\omega),$$

where S_0 involves only fundamental constants, ω is the angular frequency of the incident X-ray, $|\psi_i\rangle$ and $|\psi_f\rangle$ are the initial and final state wavefunctions of the excitation, $\mathbf{e}$ is the electric field polarization vector of the incident X-ray, E_i and E_f are the eigenenergies of the initial and final states, and w_f is the Fermi-Dirac occupation factor for the final state. The initial core state $|\psi_i\rangle$ is found by solving the radial Schrödinger equation within the PAW augmentation sphere of the excited atom in the presence of the core hole. The final states $|\psi_f\rangle$ are computed in the presence of the core hole by modifying the excited pseudopotential. The PAW formalism makes the calculation of the required matrix elements especially simple via

$$\langle \psi_f | \mathbf{e} \cdot \nabla | \psi_i \rangle = \sum_j \langle \tilde{\psi}_f | \hat{p}_j \rangle \langle \phi_j | \mathbf{e} \cdot \nabla | \psi_i \rangle ,$$

where $|\tilde{\psi}_f\rangle$ is the plane-wave part of the wavefunction, $|\hat{p}_j\rangle$ are the projector functions, $|\phi_j\rangle$ are the all-electron partial waves within the augmentation sphere, and j runs over all projector channels of the excited atom. A large number of unoccupied bands and 80 independent k-points were included in the final state sum to obtain convergence of the spectra up to 10 eV above the absorption edge. Spectra were further averaged over all similar Ge core excitations in the supercell.

For the C phase, we considered a series of possible crystal structures for GST124, including both ordered and disordered RS structures. In Fig. 18.9 we show the computed Ge-K edge XANES signatures for these structures, compared to those for α-GeTe and a spinel structure with the GST124 composition.

The absolute energy scales of the computed and experimental spectra differ slightly, which is a well-known phenomenon when using DFT [47]. The white line (WL) in the simulated spectra is clearly identified, however its width is smaller than in the experiment. This has been related to the DFT underestimate of the eigenvalue spectral bandwidth, which, as shown in [47], may be corrected by applying a scaling factor proportional to the ratio between the experimental band gap and the computed one. In any case, relative comparison of the spectra is possible and several points are noted. First, the shape of the WL of the spinel structure in which all Ge atoms are tetrahedrally bonded is clearly trimodal if a low (0.06 eV) energy smearing of states (spectral resolution) is used. For the other crystal structures, with Ge located inside distorted octahedral environments, the spectra consist of a main peak followed by a variable shoulder. The second characteristic feature is the width of the WL, which is narrower for the model spinel structure compared to the other models. Finally, the height of the white line varies among the models and is largest for the most disordered crystal structure, which is rather non-intuitive. The lack of structure in the computed

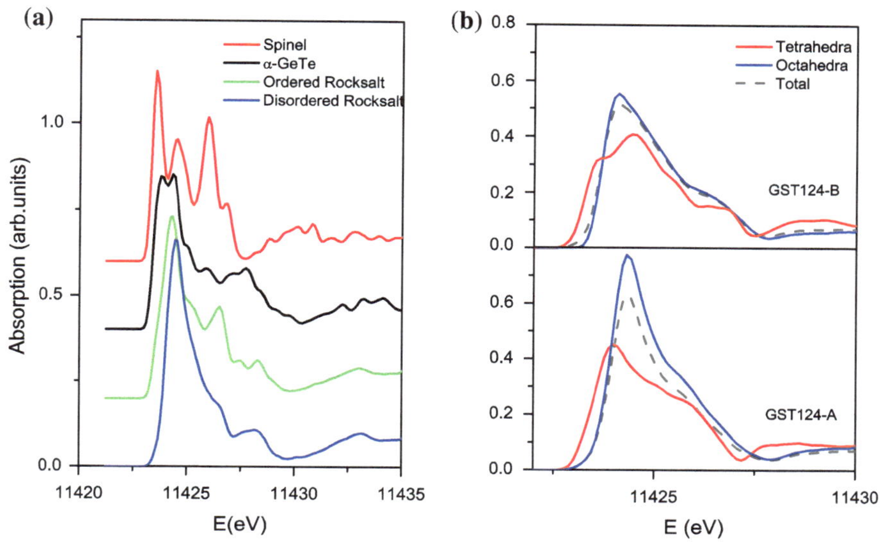

Fig. 18.9 Ge-*K* edge XANES spectra computed for GST124. *Left panel* crystalline models with 0.2 offsets. The difference in energy between the ordered and disordered RS crystals is 0.048 eV/atom. *Right panel* equilibrium (A) and compressed (B) amorphous GST124 (*black* total; *red* tetrahedral Ge; *blue* distorted octahedral Ge)

spectra beyond the WL region can be related to the neglect of shake-off excitations, reported by Lucovsky et al. [48], that are not included in our DFT single excitation calculations.

To understand these features, the computed spectra can be usefully compared to those measured for β-Ge_3N_4 and γ-Ge_3N_4 by Bull et al. [49]. In that case, a shoulder on the first peak is observed to be shifted to higher energy for the γ-Ge_3N_4 phase. The spectral shift stems from the fact that all Ge atoms are tetrahedrally coordinated in β-Ge_3N_4, while some sixfold coordinated Ge atoms are also present in γ-Ge_3N_4. For crystalline GST124, our calculations indicate that the degree of order of the Ge/Sb/vacancy sublattice also has an influence on the shoulder of the WL, which is less pronounced for a random distribution of the atomic species. The random distribution gives the best match to the shape of the published experimental XANES data (Fig. 18.10).

A model for the amorphous GST124 structure was generated by FPMD as described in [30, 50]. In Fig. 18.9 we compare the computed Ge-*K* edge XANES of amorphous GST124 at two densities: the experimental one (0.027 Å^{-3}; denoted GST124-A) and a more dense one (0.0296 Å^{-3}; denoted GST124-B) whose atomic structure is slightly different. The equilibrium structure GST124-A contains a fraction $\eta = 0.26$ of tetrahedral Ge and $(1 - \eta)$ (distorted) octahedral Ge, both types being globally fourfold coordinated. The denser model GST124-B contains more tetrahedral Ge ($\eta = 0.41$) and these include more "wrong bonds" (Ge–Sb, Ge–Ge)

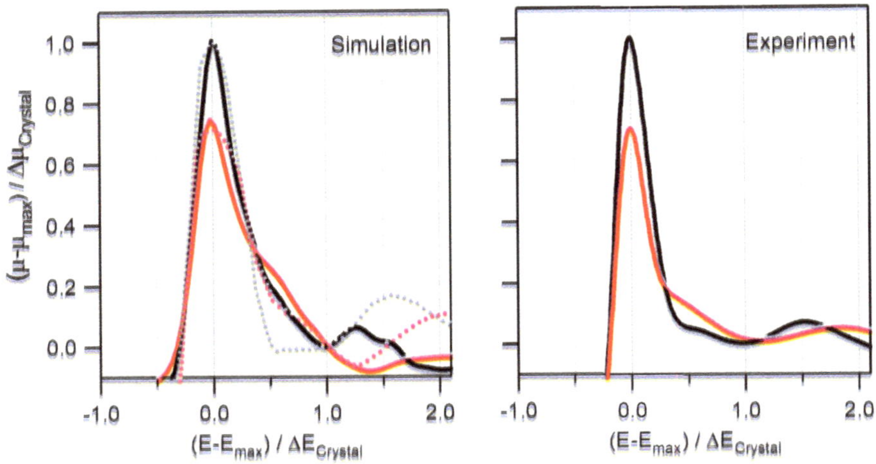

Fig. 18.10 Comparison between experimental and simulated GST124 crystal (*black curves*) and amorphous (*red curves*) Ge-*K* XANES. μ_{max} is the maximal peak value, $\Delta\mu_{crystal}$ is the difference between μ_{max} and μ_{min} (the value at the first minimum) for the crystal. Similar rescaling is applied to the energy axis. All spectra are normalized per atom. *Dotted lines* LMTO MS crystal model calculations from [19]

than in GST124-A. The main peak is about 30 % higher for the octahedral Ge and also exhibits a larger bandwidth and a smaller high energy shoulder (around 11,426 eV) than for tetrahedral Ge. The main peak of tetrahedral Ge in GST124-B is trimodal, reminiscent of the three peak structure observed for the spinel structure in Fig. 18.9. Several general features emerge from our set of calculations. Perfect tetrahedra (no wrong bonds) in the A structure very strongly diminish the amplitude of the WL peak, reduce its width, and shift it towards lower energies; in the experiment, the trends are opposite as the height decreases slightly but the width increases along with a slight shift towards higher energy. A larger fraction of tetrahedra compared to octahedra produces a rounder peak shape (see Fig. 18.9, GST124-B, as already noticed in the spinel model MS calculations of [19]). Altogether, the disordered structures containing octahedrally bonded Ge atoms bear more characteristics of the experimental A phase spectra than those including tetrahedrally bonded Ge atoms.

Since XANES is directly sensitive to the local atomic arrangement, it is natural to visualize separately the two kinds of Ge atomic environments that contribute distinctly to the total signal. Figure 18.11 shows atomic density maps averaging the positions of the atoms surrounding a central Ge in tetrahedral and distorted octahedral geometry in model GST124-A. For comparison, maps obtained from the crystalline structure are superimposed. In the right panel, the tetrahedral symmetry around specified Ge atoms is clearly seen with a bond angle at 109° and distances around 2.66 Å. The two other neighbors are out of the sampled volume. In the left panel, four nearly in-plane neighbors are seen around the octahedral Ge. Quite interestingly, a short/long bond alternation is present in both C and A structures, with a

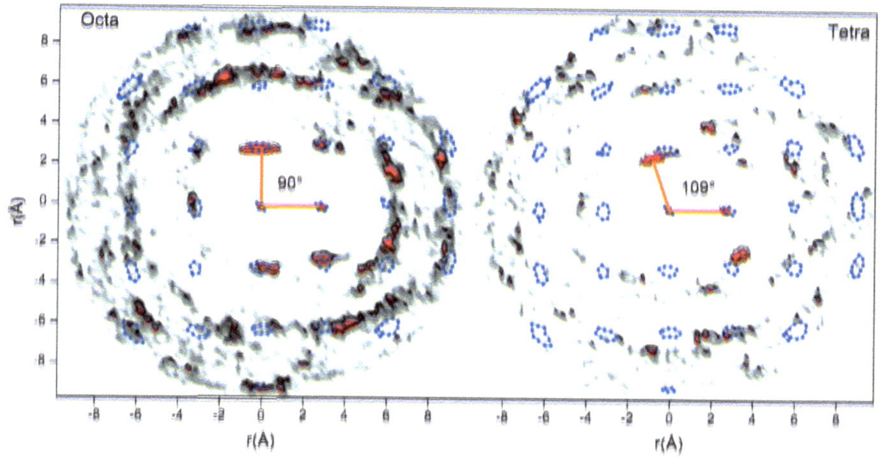

Fig. 18.11 2D correlation map for amorphous GST124-A; *dotted black isocurves* correspond to a 168 atoms disordered RS crystal. *Left* octahedral Ge atoms; *right* tetrahedral Ge atoms

slightly larger distortion in the A phase (~2.70/3.40 Å, compared to 2.90/3.10 Å in the C phase). In this A structure, the distorted cubic organization of atomic shells survives somewhat up to the second neighbors (around 3.9–4.0 Å) and then is lost, in agreement with the finding of squares and cubes in A-GST by Akola et al. [7] and different from the local order in GeTe amorphous compounds [34]. Therefore, we conclude that the A phase local order is globally similar to that of the C phase with a slightly larger degree of octahedral distortion in the amorphous phase, an additional disorder (such as more tetrahedral units), and without any quantifiable vacancies organization. In contradiction with the initial umbrella-flipping model, these findings are in direct line with the recent model proposed in [51]. We can also correlate the high energy shoulder observed in the amorphous spectrum with an increased distortion of the local octahedral coordination (see Fig. 18.9, GST124-A). This provides support to the resonant bonding picture [21] that explains the optical properties of the C phase, in the sense that increasing the distortion inside a globally *p*-bonded structure lifts the resonance between excitations which in turn creates the optical contrast [52].

This approach shows that DFT calculations can be advantageously used to compute the XANES WL spectra of A and C GST materials, allowing for a consistent reinterpretation of the experimental data. The fine features of the XANES spectra can be correlated with the presence of tetrahedrally and octahedrally coordinated Ge atoms, thus reconciling the experimental data with the more complex picture of the amorphous phase that has emerged in recent years based on AIMD simulations.

18.4 Stability of GST Phase Change Materials

As we have shown in the previous sections, the structure of both the Sb–Te and the Ge–Sb–Te class of compounds is locally close to that of the crystalline phase, of course with large fluctuations, much disorder (chemical and structural) and 'defects' such as tetrahedral Ge environments in GST. Instead of performing a direct simulation of the crystallization process, that would imply simulations over hundreds of picoseconds on large systems [53], we address the stability of the crystalline phase using two different approaches. The Maxwell theory of rigidity is used to compute the degree of 'locking' of the amorphous phase by denumbering the mechanical constraints that prevent each atom from diffusing. This approach however does not tell anything about the energy barrier that has to be overcome by the atoms to evolve away from their (metastable) equilibrium position and induce the crystallization of the system. To get information about the activation energy for crystallization, it would be necessary to perform a statistical study of the nucleation and growth process, which is definitively out of reach of AIMD simulations. However, AIMD simulations with limited simulations time give access to dynamical properties, such as the vibrational densities of states, that can be correlated with the activation energy for crystallization, as we will show at the end of this section.

18.4.1 Static Approach to the Mechanical Stability

The rigidity theory offers a practical computational scheme using topology, namely the Maxwell counting procedure, and has been central to many contemporary investigations on non-crystalline solids, such as sulphur and selenium based amorphous networks [54, 55]. It has led to the recognition of a rigidity transition [56, 57] which separates flexible glasses, that have internal degrees of freedom that allow for local deformations, from stressed rigid glasses, which are locked by their high bond connectivity. Mathematically, this transition, i.e. the isostatic case, is reached when the number of mechanical constraints per atom equals the number of degrees of freedom, that is 3 in three dimensions. Simply speaking, an atom that is supported by less than 3 constraints has the possibility to move away from its position, while an atom with more than 3 constraints is strongly blocked into his position. The telluride glasses considered here have been shown to exhibit complex local atomistic structures for which a simple counting procedure of the mechanical constraints may not be adequate. An attempt was made on the basis of EXAFS measurements by Baker et al. [28], that concluded that GST glasses are ideal in the sense that they are isostatic. It is however possible to enumerate unambiguously the number of constraints that arise from the bond-stretching (BS) and bond-bending (BB) interactions inside of a system that is simulated with AIMD by determination of the fluctuations of the bond lengths and bond angles, respectively, around each individual atom.

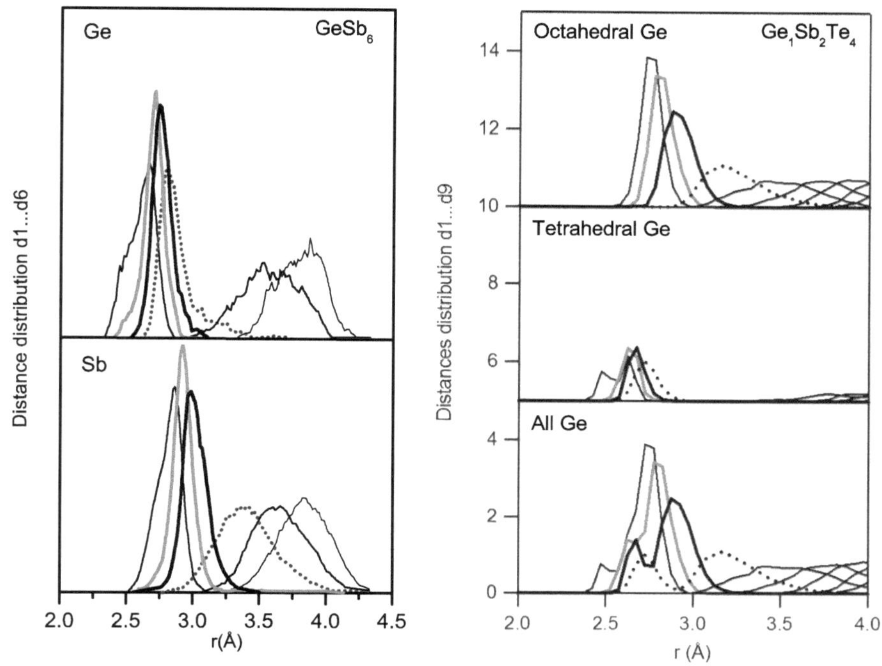

Fig. 18.12 Distance distributions $d_i(r)$ in amorphous $GeSb_6$ (*left*, i = 1 ... 6) and GST124 (*right*) (i = 2 in *grey*, i = 3 in *bold*, i = 4 with *dots*)

The total number of constraint on an atom is given by $n = C_{BS} + C_{BB}$, with C_{BS} the number of BS constraints, that is equal to $Z/2$, with Z the coordination number, and C_{BB} the number of independent rigid angles around the considered atom.

The actual value of Z can be easily obtained on average by integrating the g(r). However locally, this requires a detailed analysis of the individual bond length distributions. To do so, we first sort the neighbors i of an atom according to the bond length, $d_i(r)$, and we compute the standard deviation of $d_i(r)$ by averaging over the trajectory. Very often, atoms of the same nature share the same distribution $d_i(r)$, such as in the $GeSb_6$ case shown in Fig. 18.12. Sb atoms have distance distributions $d_i(r)$ with $i = 1, 2, 3$ which have much smaller width than for $i = 4, 5, 6$. This corresponds to the fact that the partial pair correlation around Sb atoms, $g_{Sb}(r) = g_{Sb-Ge}(r) + g_{Sb-Sb}(r) = \sum_i d_i(r)/\left(4\pi\rho r^2\right)$, has a minimum located somewhere between the position of the maxima of $d_3(r)$ and $d_4(r)$. In the case of the Ge atoms, C_{BS} is clearly equal to 4, but the specific shapes of $d_i(r)$, $i = 1...4$ are slightly different. The low-r tail of $d_1(r)$ reflects the presence of some shorter Ge–Ge bonds and the high-r tail of $d_4(r)$ is due to the fact that a small proportion of atoms have only three first neighbors that are actually chemically bonded.

The situation is often more complex, as for the Ge environments in GST124 (see Fig. 18.12, right panel). The distance distributions averaged over all Ge atoms are bimodal, which reflects the existence of two distinct atomic environments. As shown previously, these are those of tetrahedrally bonded Ge (Ge4) and octahedrally bonded Ge (Ge3). Indeed, the distance distributions for Ge4 and Ge3 atoms plotted in Fig. 18.12 are totally different, which corresponds to $C_{BS} = 4$ for Ge4 and $C_{BS} = 3$ for Ge3, as expected.

The counting of the BB constraints is performed in a similar way, with the additional difficulty that if we consider N neighbors of a given atom, this defines $N(N-1)/2$ angles. We identify the angles in the following way. After sorting the neighbors with increasing interatomic distances, $i = 1 \ldots N$, we compute the distributions $A_{i0j}(\alpha)$, *0* being the central atom and i, j being the ith and jth neighbor. Distributions A that have a small standard deviation correspond to a rigid angle with a limited angular excursion over time, and distributions with a large SD mean a free angle. The only parameter is the critical SD value under which an angle is considered as rigid. However, in practice, the distinction between rigid and free angles is unambiguous (see Fig. 18.13). Free angles have typical SD that are larger than 30° and rigid angles have a typical SD that is smaller than 20°. For $GeSb_6$, we find 6 constrained angles around Ge and 3 around Sb. However in three dimensions, not all

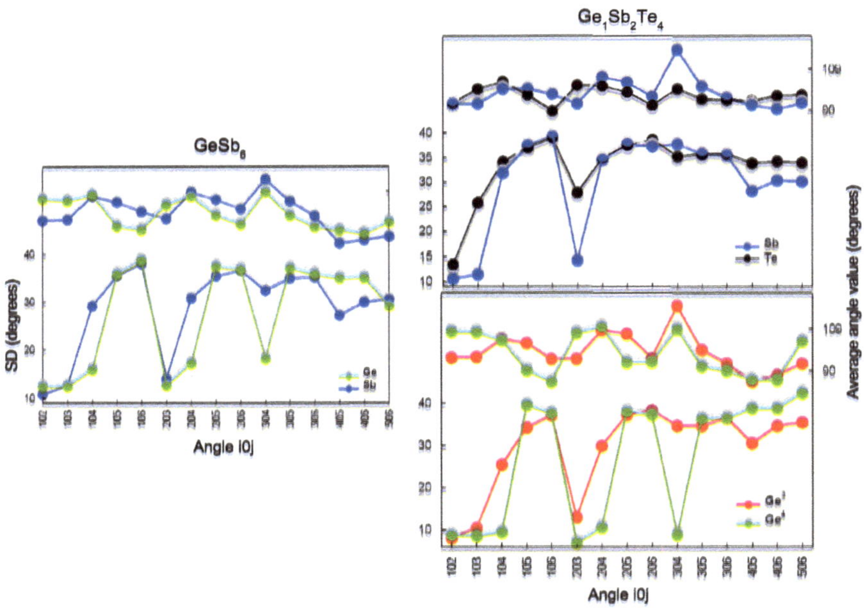

Fig. 18.13 Angle distributions $A_{i0j}(\alpha)$ defined by the first 6 neighbors in $GeSb_6$ (*left*) and GST124 (*right*). In each panel, the *top curves* show the angle of the maximum of the distribution and the *lower curves* correspond to the standard deviation of each distribution. For Ge atoms in GST124 (*bottom right panel*) Ge3 and Ge4-centred angles are analyzed separately

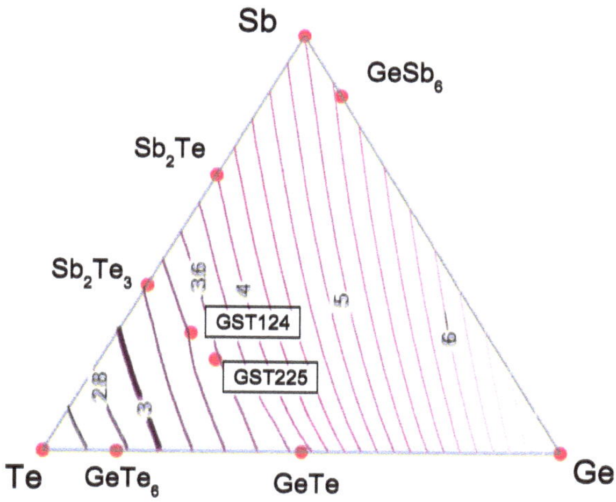

Fig. 18.14 Map of constraints in the Ge–Sb–Te diagram as obtained by analysis of AIMD simulations. Note that the value for Sb has not been computed. Values for Te and Ge are deduced from the experimental coordination number of 2 and 4, respectively [58]

angles are independent so that the actual number of angular constraints is 5 for each Ge atom and 3 for Sb atoms (See [58]).

The GST124 case shown in Fig. 18.13 is more complex as it is necessary to consider Ge3 and Ge4 atoms separately (Some angular distributions averaged over all Ge atoms have larger SD than for Ge3 and Ge4 taken separately, leading to the wrong conclusion that these angles are free). We obtain that Ge3 have 3 angular constraints (so have Sb atoms) and Ge4 have 5 angular constraints. The situation around Te atoms is less clear, however it is reasonable to estimate that there is on average only one angular constraint.

We computed the number of constraints for a series of compounds in the Ge-Sb-Te ternary diagram (See [58]) and plotted them on a map (See Fig. 18.14). It is remarkable that the isoconstraints curves are regular in the diagram although the atomistic structure of the amorphous are very different. To the contrary of [28], all phase change materials are located in the stressed-rigid part of the diagram, the isostatic line being located somewhere between Sb_2Te_3 and $GeTe_6$. It should be noted that the $GeSb_6$ compound is the most stressed-rigid, and it has been shown that this compound was prone to undergo phase separation upon cycling [59]. On the other hand, Sb_2Te_3 is the closest to the isostatic line and is undergoing spontaneous recrystallization at ambient temperature, as does Te which is in the flexible part of the diagram.

18.4.2 Dynamical Approach to Stability

If the AIMD simulation is performed over a sufficiently long time, it is possible to compute the dynamical structure factor $S(q, w)$ for frequencies ranging from $1/T$, T being the total simulation time, and $1/\tau$, τ being the simulation time step, and *for q*

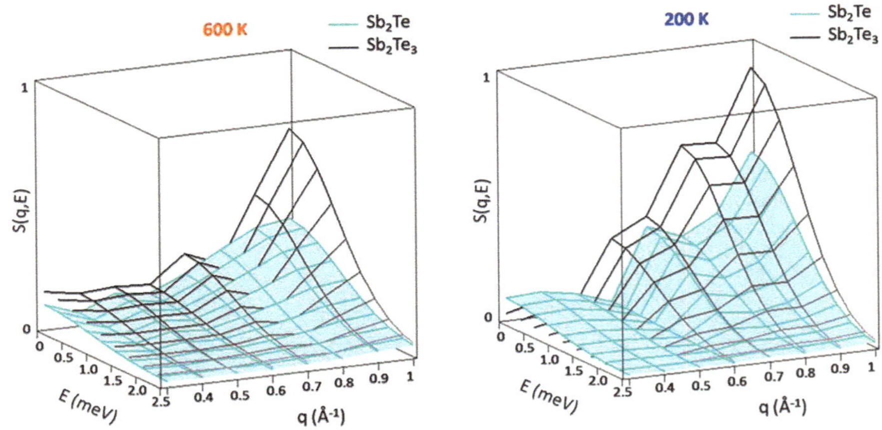

Fig. 18.15 Dynamical structure factor, $S(q, w)$ for amorphous Sb_2Te_3 and Sb_2Te at high and low temperatures

values down to $2\pi/a$ with a equal to half of the simulation box size. The small angle and small frequency part of $S(q, w)$ computed for Sb_2Te_3 and Sb_2Te are plotted in Fig. 18.15. The simulation parameters give access to E higher than 10 meV and q larger than 0.3 Å^{-1}. As shown in Fig. 18.7, correlations in the amorphous extend up to $r \sim 8$–10 Å, so that the corresponding part of $S(q, w)$ is in the range $q \sim 0.6$–0.8 Å^{-1}. It can be seen in Fig. 18.15 that $S(q, w)$ has little contribution in that range at high temperature, but it increases at low T. Sb_2Te_3 has an especially high contribution to $S(q, w)$ in that q range, for low frequencies. This indicates an excess of 'floppy' modes associated to medium range interactions. These low frequency express the fact that the amorphous system can be easily destabilized by interactions over an extended range. In the present case, we extrapolate that the excess of these modes in Sb_2Te_3 as compared to Sb_2Te is linked to the fact that amorphous Sb_2Te_3 is unstable against recrystallization at ambient condition while amorphous Sb_2Te is not.

This study of the dynamical properties of Sb–Te systems actually suggests a way to improve the stability of amorphous PCMs so as to reach longer retention times, especially at high temperatures. This is particularly important for embarked applications, such as in automobiles. It was shown experimentally that the crystallization temperature of GeTe is increasing upon doping (alloying) with some light elements, such as C and N [60], as shown in Table 18.2. This increase is linked to an increase in activation energy. On the other hand, this increased stability is correlated with an increase in the average number of constraints, as described previously.

The fact that GeTe is getting more stressed-rigid upon doping is not the only parameter that influences the stability of the compound. If we compute the vibrational density of states by performing a Fourier transform of the velocity autocorrelation function computed for the AIMD trajectory, we obtain curves plotted in Fig. 18.16. As already found experimentally [62], amorphous GeTe is at the same time elastically

Table 18.2 Activation energy, crystallization temperature and average number of constraints η for undoped and doped GeTe [60, 61]

	E_a(eV)	T_C (C)	η
$Ge_{52}Te_{48}$	2 ± 0.1	180	4.8
$Ge_{52}Te_{48}$ + 10 at.%N	3.9 ± 0.2	264	5
$Ge_{52}Te_{48}$ + 16 at.%C	4.52 ± 0.2	317	5.2

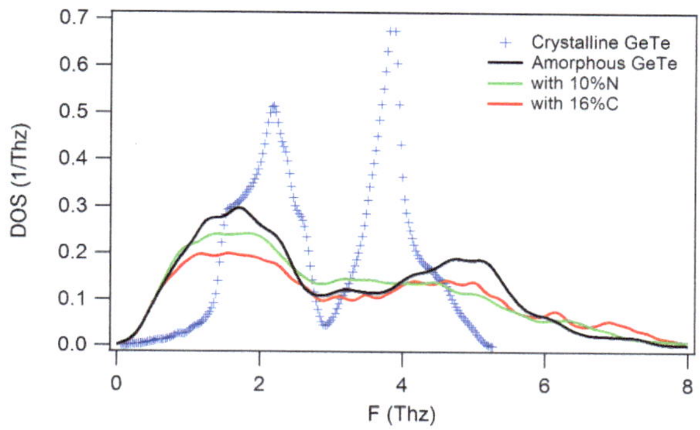

Fig. 18.16 Power spectrum (equal to the vibrational density of states, VDOS, as there is no diffusion) of amorphous GeTe, doped and undoped [61]. The VDOS of the crystal [65] is shown for comparison (*symbols*)

softer and vibrationally harder than the crystal. The low frequency modes observed below 2.5 THz correspond to the so-called Boson peak (BP) that is present in glasses. The properties of the BP have been investigated a lot in the literature and are still under debate. However, it was safely determined that the height of this peak is related to the stability of the glass [63] and linked to the number of mechanical constraints [64]. In the present case, doping GeTe with N, and even more with C, is decreasing significantly the height of the BP, therefore explaining the improved stability at high T.

The vibrational density of states is a normalized quantity, so that the lowering of the BP should be compensated at some frequencies. Figure 18.17 shows that doping with C (the situation is similar for in the nitrogen case) introduces (localized) high frequency modes in the system, modes that indirectly stabilize GeTe by removing floppy modes in the frequency range of the BP. This suggests that the stability of a PCM in general can be improved by doping with a well chosen element. This dopant should be light enough to create high frequency modes and bind covalently to the highest number of species in the system. Actually it was shown that the difference between doping GeTe with C and N comes from the fact that N does not bind to Te atoms whereas C binds to both Ge and Te [61]. Nitrogen atoms also tend to form N_2 molecules that are useless to stabilize the global amorphous structure.

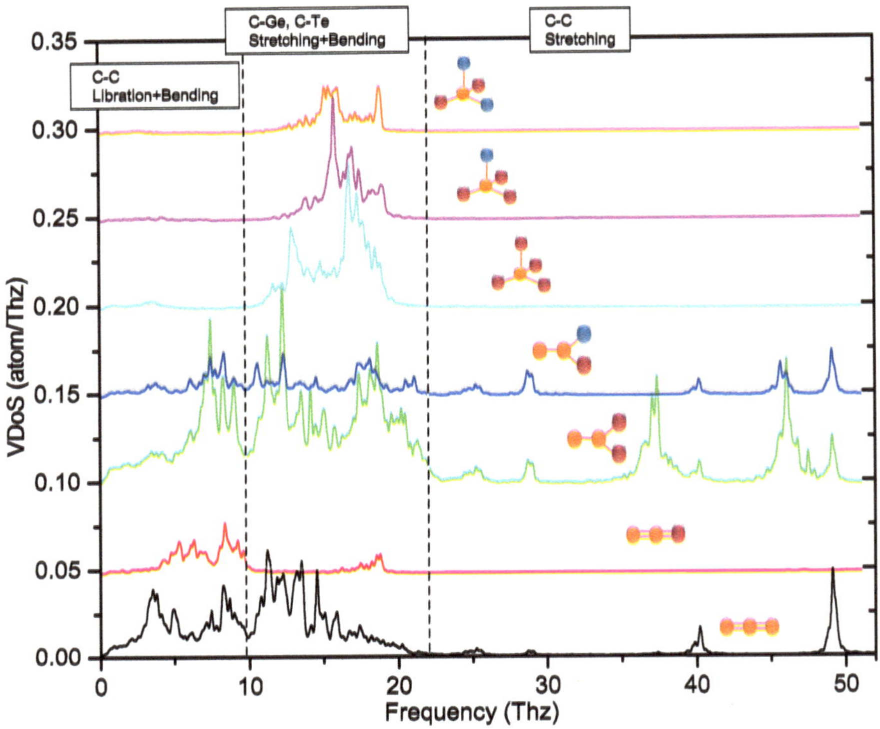

Fig. 18.17 Vibrational density of states of GeTe doped with 16 at.% C. The contribution of the various C atoms, sorted according to their local environment, are plotted separately (shifted curves for clarity). The *dotted vertical lines* approximately separate the frequency ranges of the different types of modes. The local environments are shown with C atoms in *orange*, Ge in *red* and Te atoms in *blue*

18.5 Conclusion

Ab Initio Molecular Dynamics is a very powerful tool to study amorphous phase change materials. Not only does it produce atomistic structures that help interpreting the experiment, such as XANES spectra, but it also allows to extract important trends across the phase diagram. A detailed analysis of the various correlations shows that structural motifs that resemble the crystal phase are present in many GST amorphous materials. In some cases, such as the Sb–Te systems, the correlation extend up to the third shell of neighbors. Thanks to the analysis of the distance and angle fluctuations, it is possible to use Maxwell rigidity theory to establish that PCMs are stressed-rigid glasses. The less stressed rigid glasses possess an excess of low frequency vibrational modes and AIMD simulations have shown that it is possible to remove these floppy vibrational modes from the system (GeTe) by performing an adequate doping, therefore improving the stability.

Acknowledgments The authors wish to than Profs. Jean-Pierre Gaspard, Pierre Noé and Françoise Hippert for fruitful discussions. J.Y.R. acknowledges support the FNRS (FRFC.2405.09), BelSpo (PAI 6/42), the University of Liége (ARC "Themoterm"), and computing support from the Lawrence Livermore National Laboratory and the Julich Supercomputing Center. Support from the "Agence Nationale de la Recherche" (project ANR-11-BS08-0012) is gratefully acknowledged. Part of this work was performed under the auspices of the U.S. Department of Energy by Lawrence Livermore National Laboratory under Contract No. DE-AC52-07NA27344.

References

1. M. Wuttig, N. Yamada, Nat. Mater. **6**(11), 824 (2007). doi:10.1038/nmat2009. http://www.ncbi.nlm.nih.gov/pubmed/17972937
2. M.H.R. Lankhorst, B.W.S.M.M. Ketelaars, R.A.M. Wolters, Nat. Mater. **4**, 347 (2005)
3. S.R. Ovshinsky, Phys. Rev. Lett. **21**(20) (1968)
4. A. Klein, H. Dieker, B. Späth, P. Fons, A. Kolobov, C. Steimer, M. Wuttig, Phys. Rev. Lett. **016402**, 1 (2008). doi:10.1103/PhysRevLett.100.016402
5. N. Yamada, E. Ohno, K. Nishiuchi, N. Akahira, Glass 2849–2856 (1991)
6. J.A. Kalb, F. Spaepen, M. Wuttig, J. Appl. Phys. **98**(5), 054910 (2005). doi:10.1063/1.2037870. http://link.aip.org/link/JAPIAU/v98/i5/p054910/s1&Agg=doi
7. J. Akola, R. Jones, Phys. Rev. B **76**(23), 1 (2007). doi:10.1103/PhysRevB.76.235201. http://link.aps.org/doi/10.1103/PhysRevB.76.235201
8. J. Akola, R. Jones, S. Kohara, S. Kimura, K. Kobayashi, M. Takata, T. Matsunaga, R. Kojima, N. Yamada, Phys. Rev. B **80**(2), 2 (2009). doi:10.1103/PhysRevB.80.020201. http://link.aps.org/doi/10.1103/PhysRevB.80.020201
9. C. Bichara, M. Johnson, J.P. Gaspard, 4–7 (2007). doi:10.1103/PhysRevB.75.060201
10. S. Caravati, M. Bernasconi, T. Kuhne, M. Krack, M. Parrinello, Phys. Rev. Lett. **102**(20), 1 (2009). doi:10.1103/PhysRevLett.102.205502. http://link.aps.org/doi/10.1103/PhysRevLett.102.205502
11. S.R. Elliott, J. Hegedus, Nat. Mater. **56**, 399 (2008). doi:10.1038/nmat2157
12. D. Lencer, M. Salinga, B. Grabowski, T. Hickel, J. Neugebauer, M. Wuttig, Nat. Mater. **7**(12), 972 (2008). doi:10.1038/nmat2330. http://www.ncbi.nlm.nih.gov/pubmed/19011618
13. J.P. Gaspard, F. Marinelli, A. Pellegatti, Europhys. Lett. (EPL) **3**(10), 1095 (1987). doi:10.1209/0295-5075/3/10/007. http://stacks.iop.org/0295-5075/3/i=10/a=007?key=crossref.4f75dfc829590834357e1699c95b8d27
14. F.M.J.P. Gaspard, A. Pellegatti, Philos. Mag. B **77**(3), 727 (1998). doi:10.1080/014186398259149. http://www.informaworld.com/openurl?genre=article&doi=10.1080/014186398259149&magic=crossref||D404A21C5BB053405B1A640AFFD44AE3
15. S. Shamoto, N. Yamada, T. Matsunaga, T. Proffen, J.W. Richardson, J.H. Chung, T. Egami, Appl. Phys. Lett. **86**(8), 081904 (2005). doi:10.1063/1.1861976. http://link.aip.org/link/APPLAB/v86/i8/p081904/s1&Agg=doi
16. T. Matsunaga, N. Yamada, 1–8 (2004). doi:10.1103/PhysRevB.69.104111
17. W. Welnic, A. Pamungkas, R. Detemple, C. Steimer, S. Blügel, M. Wuttig, Nat. Mater. **5**(1), 56 (2005). doi:10.1038/nmat1539. http://www.nature.com/doifinder/10.1038/nmat1539
18. C. Steimer, V. Coulet, W. Welnic, H. Dieker, R. Detemple, C. Bichara, B. Beuneu, J.P. Gaspard, M. Wuttig, Adv. Mater. **20**(23), 4535 (2008). doi:10.1002/adma.200700016. http://doi.wiley.com/10.1002/adma.200700016
19. A.V. Kolobov, P. Fons, A.I. Frenkel, A.L. Ankudinov, J. Tominaga, T. Uruga, Nat. Mater. **3**, 703 (2004). doi:10.1038/nmat1215
20. B. Huang, J. Robertson, Phys. Rev. B **81**(8), 1 (2010). doi:10.1103/PhysRevB.81.081204. http://link.aps.org/doi/10.1103/PhysRevB.81.081204

21. K. Shportko, S. Kremers, M. Woda, D. Lencer, J. Robertson, M. Wuttig, Nat. Mater. **7**(8), 653 (2008). doi:10.1038/nmat2226. http://www.ncbi.nlm.nih.gov/pubmed/18622406
22. C. Massobrio, M. Micoulaut, P.S. Salmon, Solid State Sci. **12**(2), 199 (2010). doi:10.1016/j.solidstatesciences.2009.11.016. http://dx.doi.org/10.1016/j.solidstatesciences.2009.11.016
23. C. Bichara, M. Johnson, J.Y. Raty, Phys. Rev. Lett. **267801**, 1 (2005). doi:10.1103/PhysRevLett.95.267801
24. S. Caravati, M. Bernasconi, M. Parrinello, Phys. Rev. B **81**(1), 1 (2010). doi:10.1103/PhysRevB.81.014201. http://link.aps.org/doi/10.1103/PhysRevB.81.014201
25. K. Singh, R. Satoh, K. Tsuchiya, J. Phys. Soc. Jpn. **72**, 2546 (2003)
26. C. Bichara, A. Pellegatti, J.P. Gaspard, Phys. Rev. **140**, A1133–1138; **47**(9), 5002 (1993)
27. W. Welnic, S. Botti, L. Reining, M. Wuttig, Phys. Rev. Lett. **98236403** (2007)
28. D. Baker, M. Paesler, G. Lucovsky, S. Agarwal, P. Taylor, Phys. Rev. Lett. **96**(25), 5 (2006). doi:10.1103/PhysRevLett.96.255501. http://link.aps.org/doi/10.1103/PhysRevLett.96.255501
29. S. Kohara, K. Kato, S. Kimura, H. Tanaka, T. Usuki, K. Suzuya, H. Tanaka, Y. Moritomo, T. Matsunaga, N. Yamada, Y. Tanaka, H. Suematsu, M. Takata, **3** (2006). doi:10.1063/1.2387870
30. J.Y. Raty, C. Otjacques, J.P. Gaspard, C. Bichara, Solid State Sci. **12**(2), 193 (2010). doi:10.1016/j.solidstatesciences.2009.06.018. http://linkinghub.elsevier.com/retrieve/pii/S1293255809002295
31. S. Caravati, M. Bernasconi, T.D. Kuhne, M. Krack, M. Parrinello, Appl. Phys. Lett. **91**(17), 171906 (2007). doi:10.1063/1.2801626. http://link.aip.org/link/APPLAB/v91/i17/p171906/s1&Agg=doi
32. J. Hegedüs, S.R. Elliott, Nat. Mater. **7**(5), 399 (2008). doi:10.1038/nmat2157. http://www.ncbi.nlm.nih.gov/pubmed/18362909
33. Z. Sun, J. Zhou, R. Ahuja, Phys. Rev. Lett. **96055507** (2006)
34. J. Akola, R. Jones, Phys. Rev. Lett. **100**(20), 21 (2008). doi:10.1103/PhysRevLett.100.205502. http://link.aps.org/doi/10.1103/PhysRevLett.100.205502
35. R. Mazzarello, S. Caravati, S. Angioletti-Uberti, M. Bernasconi, M. Parrinello, Phys. Rev. Lett. **104**(8), 1 (2010). doi:10.1103/PhysRevLett.104.085503. http://link.aps.org/doi/10.1103/PhysRevLett.104.085503
36. T. Matsunaga, N. Yamada, Phys. Rev. B **69**(10), 1 (2004). doi:10.1103/PhysRevB.69.104111. http://link.aps.org/doi/10.1103/PhysRevB.69.104111
37. T. Siegrist, P. Jost, H. Volker, M. Woda, P. Merkelbach, C. Schlockermann, M. Wuttig, Nat. Mater. **10**(3), 202 (2011). doi:10.1038/nmat2934. http://www.ncbi.nlm.nih.gov/pubmed/21217692
38. W. Zhang, A. Thiess, P. Zalden, R. Zeller, P.H. Dederichs, J.Y. Raty, M. Wuttig, S. Blügel, R. Mazzarello, Nat. Mater. **11**(11), 952 (2012). doi:10.1038/nmat3456. http://www.ncbi.nlm.nih.gov/pubmed/23064498
39. J.J. Rehr, R.C. Albers, Phys. Rev. B **41**, 8139 (1990)
40. G. Ferlat, J.C. Soetens, A.S. Miguel, P.A. Bopp, J. Phys. Condens. Matter **17**(5), S145 (2005). doi:10.1088/0953-8984/17/5/015. http://stacks.iop.org/0953-8984/17/i=5/a=015?key=crossref.820fb7775c2250b70fa4ce9c7beb63c7
41. G. Pfeiffer, J. Rehr, D. Sayers, Phys. Rev. B **51**(2), 804 (1995)
42. M. Krbal, A. Kolobov, P. Fons, J. Tominaga, S. Elliott, J. Hegedus, T. Uruga, Phys. Rev. B **83**(5), 1 (2011). doi:10.1103/PhysRevB.83.054203. http://link.aps.org/doi/10.1103/PhysRevB.83.054203
43. J. Kas, Toward Quantitative Calculation and Analysis of X-Ray Absorption Near Edge Spectra, Ph.D. Thesis, University of Washington, 2009
44. P.E. Blöchl, Phys. Rev. B **50**(24), 17953 (1994). http://link.aps.org/doi/10.1103/PhysRevB.50.17953
45. G. Kresse, J. Hafner, Phys. Rev. B **49**(20) (1994)
46. G. Kresse, J. Furthmüller, Comput. Mater. Sci. **6**, 15 (1996)
47. D. Prendergast, G. Galli, Phys. Rev. Lett. **96**(21), 215502 (2006). doi:10.1103/PhysRevLett.96.215502. http://link.aps.org/doi/10.1103/PhysRevLett.96.215502

48. G. Lucovsky, J.P. Washington, L. Miotti, M. Paesler, Phys. Status Solidi (C) **847**(3) (2010). doi:10.1002/pssc.200982887. http://doi.wiley.com/10.1002/pssc.200982887
49. C.L. Bull, P.F. McMillan, J.P. Itiïœ, A. Polian, Phys. Status Solidi (A) **201**(5), 909 (2004). doi:10.1002/pssa.200306786. http://doi.wiley.com/10.1002/pssa.200306786
50. M. Micoulaut, J.Y. Raty, C. Otjacques, C. Bichara, Phys. Rev. B **81**(17), 1 (2010). doi:10.1103/PhysRevB.81.174206. http://link.aps.org/doi/10.1103/PhysRevB.81.174206
51. A.V. Kolobov, M. Krbal, P. Fons, J. Tominaga, T. Uruga, Nat. Chem. **3**(4), 311 (2011). doi:10.1038/nchem.1007. http://www.ncbi.nlm.nih.gov/pubmed/21430691
52. B. Huang, J. Robertson, Phys. Rev. B **81**, 081204R (2010)
53. J.S. Van Duijneveldt, D. Frenkel, J. Chem. Phys. **96**, 4655 (1992). doi:10.1063/1.462802
54. J. Chung, M. Thorpe, Phys. Rev. B **59**(7), 4807 (1999). doi:10.1103/PhysRevB.59.4807. http://link.aps.org/doi/10.1103/PhysRevB.59.4807
55. S. Chakravarty, D.G. Georgiev, P. Boolchand, M. Micoulaut, J. Phys. Condens. Matter **17**, 1 (2005). doi:10.1088/0953-8984/17/1/L01
56. J.C. Phillips, J. Non-Cryst. Solids **34**, 153 (1979)
57. M.F. Thorpe, J. Non-Cryst. Solids **57**(3), 355 (1983). http://www.scopus.com/inward/record.url?eid=2-s2.0-0021093473&partnerID=40&md5=4854ab8fab097a30aa0808d48fef3517
58. M. Micoulaut, J. Raty, C. Otjacques, C. Bichara, Phys. Rev. B **81**, 174206 (2010). doi:10.1103/PhysRevB.81.174206
59. C. Cabral, J. Bruley, S. Raoux, V. Deline, A. Madan, T. Pinto, Appl. Phys. Lett. **93**, 071906 (2008). doi:10.1063/1.2970106
60. G.E. Ghezzi, J.Y. Raty, S. Maitrejean, A. Roule, E. Elkaim, F. Hippert, Appl. Phys. Lett. **99**, 151906 (2011). doi:10.1063/1.3651321
61. J.Y. Raty, P. Noé, G. Ghezzi, S. Maîtrejean, C. Bichara, F. Hippert, Phys. Rev. B **88**(1), 014203 (2013). doi:10.1103/PhysRevB.88.014203. http://link.aps.org/doi/10.1103/PhysRevB.88.014203
62. T. Matsunaga, N. Yamada, R. Kojima, S. Shamoto, M. Sato, H. Tanida, T. Uruga, S. Kohara, M. Takata, P. Zalden, G. Bruns, I. Sergueev, H.C. Wille, R.P. Hermann, M. Wuttig, Adv. Funct. Mater. **21**(12), 2232 (2011). doi:10.1002/adfm.201002274. http://doi.wiley.com/10.1002/adfm.201002274
63. G. Naumis, H. Flores-Ruiz, Phys. Rev. B **78**(9) (2008). doi:10.1103/PhysRevB.78.094203. http://link.aps.org/doi/10.1103/PhysRevB.78.094203
64. H.M. Flores-ruiz, G.G. Naumis, 1–5 (2010). doi:10.1103/PhysRevB.82.214201
65. R. Shaltaf, E. Durgun, J.Y. Raty, P. Ghosez, X. Gonze, Phys. Rev. B **78**(20), 1 (2008). doi:10.1103/PhysRevB.78.205203. http://link.aps.org/doi/10.1103/PhysRevB.78.205203

Chapter 19
Transition Metals in Phase-Change Memory Materials: Impact upon Crystallization

Binay Prasai and D.A. Drabold

Abstract We employed plane wave density functional molecular dynamics to simulate the crystallization of $Ge_2Sb_2Te_5$ materials alloyed with ~2 % transition metals (Cu, Ag, and Au) and studied the resulting structural modifications. Despite having very different chemistry, we show that under many circumstances, transition metals join the crystalline structure essentially substitutionally. Bader charge analysis revealed similar positive atomic charges for Cu and Ag whereas negative charge for Au was observed. The optical contrast is preserved in Ag or Au doped $Ge_2Sb_2Te_5$, but not in Cu doped $Ge_2Sb_2Te_5$. The estimation of the crystallization time for the transition metal doped $Ge_2Sb_2Te_5$ showed large variation which were attributed to the presence of different fractions of wrong bonds in the alloys.

19.1 Introduction

Phase change memory materials offer a remarkable proving ground for ab-initio simulation since direct simulation of the key process of phase change is possible, as discovered in pioneering work of Hegedus and Elliott [1]. In this paper, we extend this research by investigating the role of transition metal (TM) impurities in the phase change process. Remarkably, we find that moderate concentration of certain TMs (Ag and Au) readily join the octahedral crystalline phase, while Cu does not. We report the atomic trajectories leading to full crystallization in detail. We confirm, as in our earlier report, that Ag enhances the crystallization speed [2].

B. Prasai (✉) · D.A. Drabold (✉)
Department of Physics and Astronomy, Ohio University, Athens, OH 45701, USA
e-mail: bp249507@ohio.edu

D.A. Drabold
e-mail: drabold@ohio.edu

C. Massobrio et al. (eds.), *Molecular Dynamics Simulations of Disordered Materials*, Springer Series in Materials Science 215,
DOI 10.1007/978-3-319-15675-0_19

19.2 Methods

Ab-initio molecular dynamic (AIMD) calculations were performed using the Vienna Ab-initio Simulation Package (VASP) [3] to generate models of $Ge_2Sb_2Te_5$ (GST) and $X_{0.17}Ge_2Sb_2Te_5$ (X = Cu, Ag, Au). Each of the models was created in a cubic supercell with 108 (24 Ge atoms, 24 Sb atoms and 60 Te atoms) host atoms and two dopant atoms.

The models were prepared using the "Melt and Quench" method [4], starting with a random configuration at 3000 K. After annealing the random configuration at 3000 K for 25 ps, each model was cooled to 1200 K in 10 ps and equilibrated for 60 ps in the liquid state at 1200 K. A cooling rate of 12 K/ps was adopted to obtain the amorphous models from the melt at 1200–300 K and followed by equilibration at 300 K for another 50 ps. Each of these models was subsequently annealed at 650 K for up to 450 ps to simulate crystallization. The density chosen, 6.05 g/cm^3, for undoped $Ge_2Sb_2Te_5$ is intermediate between the amorphous and crystalline densities. Due to lack of experimental values, same lattice spacing was chosen for doped $Ge_2Sb_2Te_5$ with the hope that dopants would take the vacancy sites. The density of each system is shown in Table 19.1, and was fixed during the MD runs, and a time step of 5 ps was used.

To compute ground state properties, the models were fully relaxed to a local minimum energy configuration at zero pressure (allowing the cell shape and volume to vary). The final relaxed densities are illustrated in Table 19.1. Three independent models were generated for each of the structures to check the consistency of the results.

The calculations were implemented with the projector augmented-wave (PAW) [5, 6] method to describe electron-ion interactions. The Perdew-Burke-Ernzerhof (PBE) [6, 7] exchange correlation functional was used throughout. Molecular-dynamics (MD) simulations were performed in a cubic supercell at constant volume for annealing, equilibrating and cooling, whereas, zero pressure conjugate gradient (CG) simulations were performed for relaxation. Both the MD and CG simulations were performed by using only the Γ point to sample the supercell Brillouin zone for computing the total energies and forces.

Table 19.1 Density (in gm cm^{-3}) chosen in different unrelaxed models. The densities employed are compared to the amorphous and crystalline densities of fully relaxed models

Systems	Unrelaxed	Relaxed	
		Amorphous	Crystalline
$Ge_2Sb_2Te_5$	6.05	5.70	6.12
Cu-$Ge_2Sb_2Te_5$	6.11	5.82	6.20
Ag-$Ge_2Sb_2Te_5$	6.20	5.94	6.21
Au-$Ge_2Sb_2Te_5$	6.24	5.88	6.23

19.3 Structural Properties

19.3.1 Correlation Functions

There have been a number of reports discussing where a dopant sits in the $Ge_2Sb_2Te_5$ host network [8]. It is therefore interesting to see how these dopants (Cu, Ag, Au) adapt to the local structure. The atomic structure of our models is studied through a set of pair correlation functions. A pair correlation function is a position distribution function based on the probability of finding atoms at some distance **r** from a central atom. Following [9], we tersely develop the expressions for correlation functions and present this below. A general expression for the pair distribution function [9] is:

$$g(\mathbf{r}) = \frac{1}{\rho^2 V} N(N-1)\langle \delta(\mathbf{r} - \mathbf{r_{ij}}) \rangle \tag{19.1}$$

Here, ρ and V are the number density and volume respectively of the model, N the number of neighboring atoms of the central atom and r_{ij} is the distance of any atom from the central atom. The symbol $\langle .. \rangle$ represents a configuration average. The radial pair correlation function can be obtained as

$$g(r) = \int \frac{d\Omega}{4\pi} g(\mathbf{r}) \tag{19.2}$$

or,

$$g(r) = \frac{1}{\rho^2 V} \sum_{i,i \neq j} \int \int \frac{\sin\theta}{4\pi} d\theta d\phi \frac{1}{r^2 \sin\theta} \delta(r - r_{ij}) \delta(\theta - \theta_{ij}) \delta(\phi - \phi_{ij}) \tag{19.3}$$

or,

$$g(r) = \frac{1}{4\pi \rho^2 V r^2} \sum_{i,i \neq j} \delta(r - r_{ij}) \tag{19.4}$$

The pair correlation functions (PCFs) provide local structural information of essential interest for amorphous materials. The peaks in these functions describe the average distance of neighboring atoms from a central atom. Since amorphous materials do not possess long range order, $g(r) \rightarrow 1$ as $r \rightarrow \infty$. For crystalline structures, $g(r)$ is a sum of delta functions, with each term coming from a coordination shell.

For systems with more than one species, structural correlations are usually investigated through partial pair correlation functions $g_{\alpha\beta}(r)$, which are expressed as

$$\langle n_{\alpha\beta}(r) \rangle \Delta r = 4\pi r^2 \Delta r \rho c_\alpha g_{\alpha\beta}(r) \tag{19.5}$$

where $n_{\alpha\beta}(r)\Delta r$ is number of particles of species β in a shell between r and r+Δr around a central atom α, $c_\alpha = N_\alpha / N$ is the concentration of species α. The total pair correlation function is then defined as the sum of all partial contributions as

$$g(r) = \sum_{\alpha\beta} c_\alpha c_\beta g_{\alpha\beta}(r) \tag{19.6}$$

In addition to local bonding, integration of the PCF up to the first minimum provides information on mean coordination numbers. Coordination analysis is particularly interesting in these materials since the coordination changes from six in the crystalline phase (rocksalt structure) to about four in the amorphous phase. It should also be pointed out that the information obtained from the PCF alone may not be sufficient to describe the local structure and hence require introduction of simple three-body distribution functions like bond angle distributions (BADs). BAD would lend insight into the local environment of each impurity atom and the network structure of doped phase change materials.

Since crystalline $Ge_2Sb_2Te_5$ contains of order 20 % vacant octahedral sites, these vacant sites would seem to be an ideal place for the dopant atoms to sit. However, as there is no certain definition of vacancy in the amorphous phase, the local structure of the dopant atoms in the amorphous phase is all the more interesting to investigate. The local structures of the dopants are presented in Fig. 19.1. Figure 19.1a–f represent the local structure of the dopants in the amorphous phase, while the bottom half (Fig. 19.1g–l) depict the crystalline phase. Figure 19.2 presents a set of PPCFs corresponding to different dopant atoms. At a dopant concentration near 2 %, there is no major change in pair correlation functions for either amorphous or crystalline phases. As the position of the first maximum in the PPCF reflects the average bond lengths, we were able to estimate the average bond lengths and present them in Table 19.2. We observed no dopant-induced change in Ge–Te and Sb–Te average

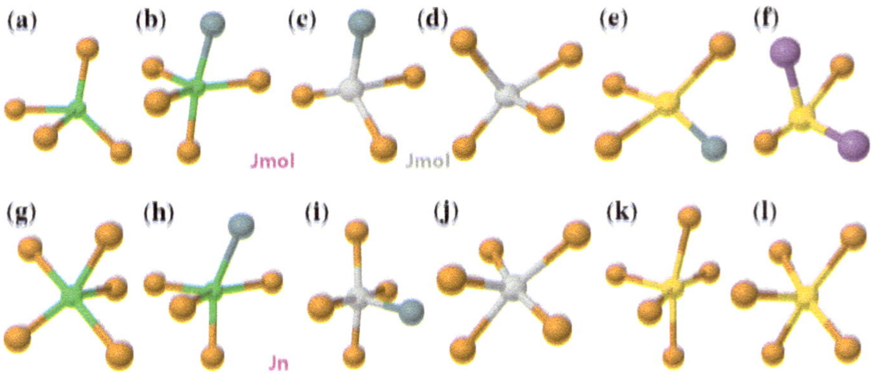

Fig. 19.1 Local atomic structures surrounding the dopants. **a, b** Cu, **c, d** Ag, and **e, f** Au local geometries in amorphous phases (*top* images). The *bottom* images correspond to structures around the dopants in crystalline phases. Color code: *Orange* Te, *Blue* Ge, *Purple* Sb, *Green* Cu, *Silver* Ag, *Yellow* Au

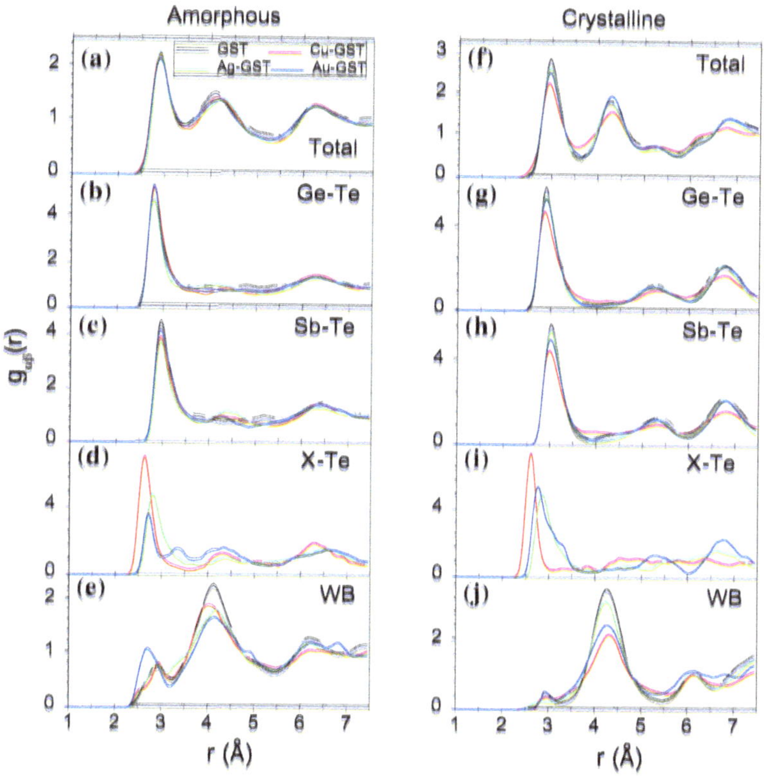

Fig. 19.2 Pair correlation functions at 300 K for pure and doped $Ge_2Sb_2Te_5$. Partial pair correlation functions for amorphous structures are on the *left* (**a**–**e**) and for crystalline structures are on the *right* (**f**–**j**). WB stands for wrong bonds

Table 19.2 Average bond lengths (Å) in amorphous and crystalline phases of pure and doped $Ge_2Sb_2Te_5$

Bonds	Dopant	Amorphous	Crystalline
Ge-Te	–	2.78	2.92
	Cu	2.78	2.91
	Ag	2.78	2.92
	Au	2.78	2.92
Sb-Te	–	2.96	3.02
	Cu	2.96	3.00
	Ag	2.96	3.02
	Au	2.96	3.02
Cu-Te	Cu	2.59	2.62
Ag-Te	Ag	2.82	2.88
Au-Te	Au	2.72	2.78

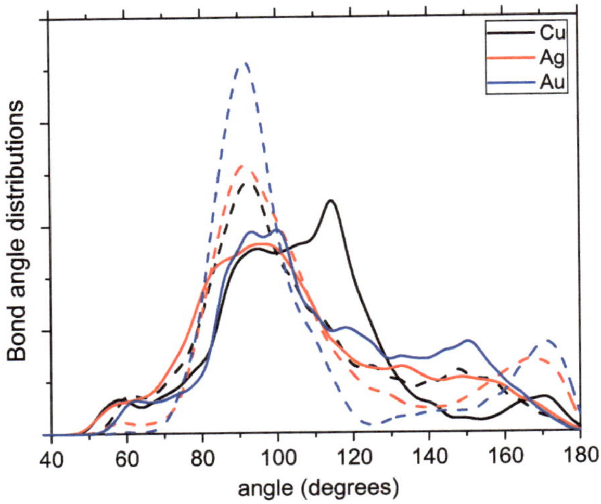

Fig. 19.3 Bond angle distributions around the dopants at 300 K for doped $Ge_2Sb_2Te_5$. Dashed lines are bond angle distributions for crystalline $Ge_2Sb_2Te_5$

bond lengths in the amorphous phase, whereas the change in Ge–Te and Sb–Te average bond lengths in crystalline phase is also negligible being within the uncertainty of ∼0.01 Å. On the other hand, the Cu–Te average bond length is observed to be the shortest while Ag–Te average bond length is the longest among the three dopants. To further detail the local topology of the dopants, we computed the bond angle distribution (BAD) centered on the dopants and illustrate the results in Fig. 19.3, which reveals a substantial distortion of the BADs in the amorphous phase while the BADs in the crystalline phase illustrate the remarkable tendency of the dopants to adopt an octahedral geometry, just as the host atoms. Among the three dopants, Cu is least likely to adopt octahedral geometry as compared to Ag and Au since no peak at around 180° is observed for Cu BAD. The reason could be the shorter Cu–Te bond lengths as compared to Ag–Te and Au–Te bond lengths or the wide variation of acceptable valency of Cu.

To investigate the dopant impact on coordination, we computed coordination numbers by integrating the PPCFs up to the first minimum in the RDF, and present results in Table 19.3. Among the three dopants, Cu has the highest coordination number. Furthermore, the coordination number of Cu does not change significantly from the amorphous to crystalline phase. Beside this, the coordination numbers of the host species (Ge/Sb/Te) do not change significantly as in pure GST or doped with Ag or Au, suggesting that the presence of Cu in the GST might suppress the phase transition, thereby reducing the crystallization speed. Au on the other hand, tends to exhibit a significant fraction of Ge/Sb atoms as neighbors although Te is the most anionic species (no Ge/Sb neighbors are observed in the crystalline phase). Unlike Cu, the coordination numbers of Au as well as the host species in Au-doped GST increase significantly. Ag also shows some Ag–Ge and Ag–Sb ($<10\,\%$) bonds, which persist even to the crystalline phase of Ag–doped GST however, Ag does not suppress the increase in the coordination numbers like Cu.

Table 19.3 Comparison of coordination numbers in amorphous and crystalline phases of pure and doped $Ge_2Sb_2Te_5$. A cutoff distance of 3.2 Å was chosen for integration of the first peak of PPCF

Species	Dopant	Amorphous					Crystalline				
		Ge	Sb	Te	X	Total	Ge	Sb	Te	X	Total
Ge	–	0.4	0.3	3.5	–	4.2	–	0.1	5.4	–	5.5
	Cu	0.2	0.3	3.8	–	4.3	–	0.1	4.6	–	4.7
	Ag	0.1	0.5	3.2	–	3.8	0.1	0.2	5.1	–	5.4
	Au	0.4	0.2	3.6	–	4.2	0.1	–	5.0	–	5.1
Sb	–	0.3	0.5	2.8	–	3.6	0.1	0.3	5.2	–	5.6
	Cu	0.3	0.7	2.7	–	3.7	0.1	0.5	4.2	–	4.8
	Ag	0.5	0.6	2.6	–	3.7	0.2	0.1	4.9	–	5.2
	Au	0.2	0.5	2.9	0.1	3.7	–	0.3	4.7	–	5.0
Te	–	1.4	1.1	0.2	–	2.7	2.2	2.1	0.2	–	4.5
	Cu	1.5	1.1	0.4	0.1	3.1	1.8	1.7	0.2	0.1	3.8
	Ag	1.3	1.0	0.4	0.1	2.8	2.0	2.0	0.2	0.2	4.4
	Au	1.5	1.1	0.3	0.1	3.0	1.8	1.7	0.2	0.1	3.8
Cu	Cu	0.6	–	4.2	–	4.8	0.5	–	4.5	–	5.0
Ag	Ag	0.3	0.1	3.5	–	3.9	0.3	0.2	4.5	–	5.0
Au	Au	0.5	1.0	1.9	–	3.4	–	–	5.2	–	5.2

Table 19.4 Computed average atomic charges on the dopant atoms in relaxed amorphous and crystalline phases of Cu, Ag, and Au doped $Ge_2Sb_2Te_5$. The charges are in the unit of e

Dopants	Amorphous	Crystalline	Change (%)
Cu	0.16	0.12	25
Ag	0.14	0.08	43
Au	–0.44	–0.28	36

Coordination analysis reveals that the majority of the dopant atoms bond to Te rather than Ge and Sb (see Table 19.3). Cu, Ag, and Au have 88, 90, and 56 % Te nearest neighbors, respectively. Au seems noticeably different than Cu or Ag with only 56 % Te as first neighbor. This contrast is interesting, since the three elements lie on the same column of the periodic table and might be supposed to have similar coordination. To further explore the coordination, we computed the average atomic charge on the dopant atoms using a Bader analysis [10–12] and present the results in Table 19.4. We observe similar atomic charges for Cu and Ag whereas we obtain a negative charge for Au. The negative charge on Au makes it anionic, and a consequence is that a large fraction of Ge and Sb are bonded to Au. The atomic charge on Au becomes significantly less negative in the crystalline phase and the fraction of Ge and Sb neighbors drops.

Beside Ge–Te, Sb–Te and X–Te (X = Cu, Ag, and Au) bonds, there are a number of other bond pairs such as, Ge–Sb, Ge–X, and Sb–X bonds which are often termed "wrong bonds" (Fig. 19.2e, j). About 14 % of the total bond pairs are wrong bonds (WB) in the amorphous phase, whereas about 5 % of wrong bonds persist in the crystalline phase.

19.4 Electronic and Optical Properties

Practical utilization of GST materials depends upon a substantial optical or electrical contrast between the amorphous and crystalline phases. As both the resistivity and optical absorption depend upon the electronic structures of materials, it is necessary to investigate the influence of dopants on the electronic structure of the host network.

The electronic structure is described by the electronic density of states (EDOS), projected density of states (PDOS), and inverse participation ratio (IPR) of each individual state. The EDOS is defined as:

$$g(E) = \frac{1}{N_b} \sum_{l=1} \delta(E - E_l), \qquad (19.7)$$

in which N_b is the number of basis orbitals, and in this work, E_l are Kohn-Sham eigenvalues. The EDOS provides information about the electronic gaps, the PDOS provides information on the defects or irregularities in the topology. A common expression for PDOS is:

$$g_n(E) = \frac{1}{N_b} \sum_{l=1} \delta(E - E_l)|\langle\phi_n|\Psi_l\rangle|^2 \qquad (19.8)$$

where $g_n(E)$ is site projected DOS for the site n, ϕ_n is the local orbital and Ψ_l is the lth Kohn-Sham eigenvector.

To investigate the effect of dopants on the electronic properties we computed both the species-projected and orbital-projected (PDOS) and present these in Fig. 19.4a–e. At ~2 % dopant level, we were not able to observe a significant difference in the total density of states except for slight variation in the energy range of –5 to –1 eV below the Fermi level. This change mainly corresponds to the d states of the dopants (Fig. 19.4d, e). Among the dopants, Cu d states lie closer to the band gap and therefore the presence of Cu may significantly affect the electrical and optical properties. This is presumably the ultimate origin of the difference in the local topology of Cu impurities. Of course, the present calculations underestimate the experimental band gaps as expected for DFT [13].

To observe the influence of dopants on the optical properties of $Ge_2Sb_2Te_5$, we computed the dielectric functions of the pure and doped $Ge_2Sb_2Te_5$. Figure 19.5 presents the imaginary and real part of the dielectric functions of pure and doped $Ge_2Sb_2Te_5$ in both the amorphous and the crystalline phases. Figure 19.5 confirms that the addition of Ag or Au does not affect the optical properties strongly,

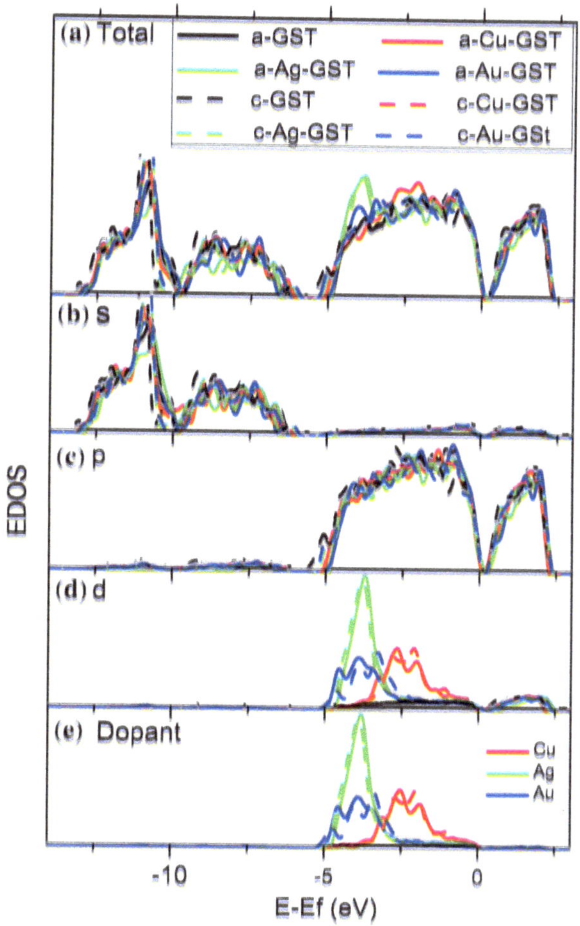

Fig. 19.4 Comparison of partial density of states (PDOS) for different dopants in $Ge_2Sb_2Te_5$. The Fermi level is set to 0 eV

preserving the optical contrast in the visible regions (1–3 eV) whereas the addition of Cu does not look promising in terms of optical contrast. Table 19.5 presents the optical dielectric constant, i.e., the lower energy-limit of the real part of the dielectric function ($\omega \rightarrow 0$). The reduction in optical contrast on adding Cu can be attributed the structure of Cu in the host network. We observe almost no change in the Cu–Te bond lengths as well as the Cu coordination numbers during the phase transition suggesting Cu very immobile in the host network. As discussed by Shportko et al. [14], the contrast in the optical properties in PCMM is attributed to a large electronic polarizability in the crystalline phase due to resonant (electron deficient) p-bonding. Unlike Ag and Au, Cu is unable to integrate itself fully into the crystalline structure with octahedral structure (see Fig. 19.1) introducing a significant structural disorder and hence reducing the contribution to the resonant bonding. This reduction would cause the decrease in the optical contrast.

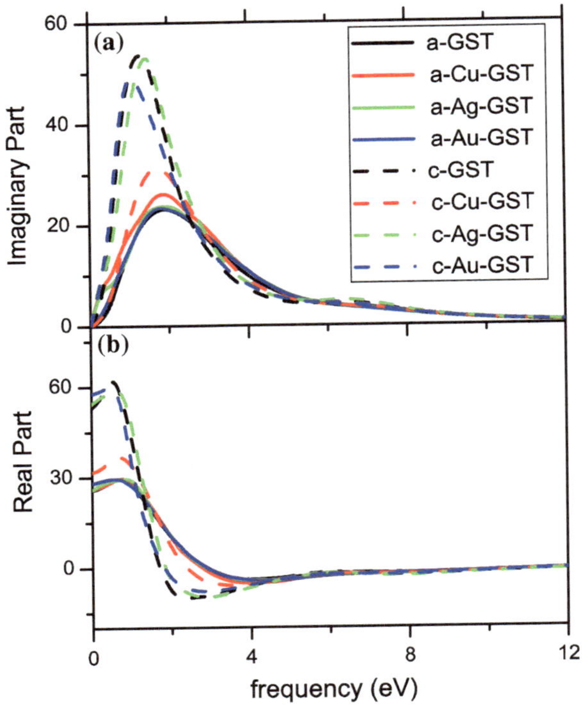

Fig. 19.5 Comparison of optical contrast in Cu, Ag, and Au doped $Ge_2Sb_2Te_5$

Table 19.5 Comparison of the optical dielectric constant between the two phases of pure and doped $Ge_2Sb_2Te_5$

Dopants	Amorphous	Crystalline	Increase (%)
–	25.9	53.0	105
Cu	26.0	31.7	22
Ag	26.1	54.3	108
Au	28.1	57.7	105

19.5 Crystallization Dynamics

The dopant centered bond angle distributions (Fig. 19.3) illustrate the modification of local geometry of dopants from wide distortion to more ordered octahedral structures. This makes it possible to observe the atomic motion of dopants during and after the crystallization of the host network is completed. There have been few studies on the dynamics of dopants in $Ge_2Sb_2Te_5$ during crystallization [8, 15]. In the simulations with the first row dopants (Sc-Zn), Skelton and Elliott have reported a range of atomic motion even after crystallization [8]; however only one dopant was investigated. In their recent work, Prasai et al. investigated the dynamics of Ag in $Ge_2Sb_2Te_5$ with up to 12 % Ag, and reported mixed diffusion of Ag. We observe atomic motion during and after crystallization of host network in the present work as presented in Fig. 19.6. A thorough investigation of atomic diffusion reveals that the dopant atoms

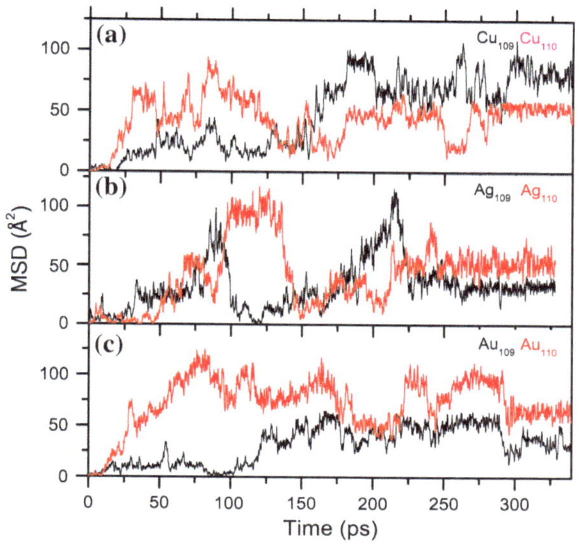

Fig. 19.6 Mean squared displacements (MSD) of dopants during the crystallization process

become less diffusive whenever the atom is close to octahedral geometry. Figure 19.6 shows the mean-squared displacement of each of the dopant atoms through the entire crystallization process. Both Ag or Au atoms show no hopping after crystallization, whereas a particular Cu atom, Cu_{109} shows hopping even after the crystallization has occurred. The local geometries of the dopant atoms are shown in the Fig. 19.1a–l. The dopant atom that achieve octahedral geometry exhibit no hopping after the crystallization. Cu_{109} (one of two Cu atoms) showed hopping after the crystallization, and did not achieve the octahedral geometry over the time of the simulation. Cu_{109} is represented by Fig. 19.1a, g in amorphous and crystalline phases respectively.

We also analyze the average atomic charge on the dopants during the crystallization of $Ge_2Sb_2Te_5$. Figure 19.7 presents the time evolution of the average atomic charge on Au along with the nearest neighbors of Au within the first coordination shell. The fraction of Ge/Sb neighbors decreases while increasing the fraction of Te neighbors as the charge on Au decreases.

The speed of crystallization was investigated by observing the time evolution of the total energy of the system (Fig. 19.8). Each of the models was annealed at 650 K until crystallized. Based on the evolution of energy, the total crystallization process can be divided into three regions as also explained by Lee et al. [16]. During Period I (incubation period), although the total energy does not change significantly, the number of four-fold (square) rings kept increasing forming cubes or planes with random orientations driven by thermal fluctuations [2, 16]. Period II, in which the total energy shows a monotonous decrease, is the time in which the cubes or planes start arranging themselves into more globally ordered structures. By the end of this

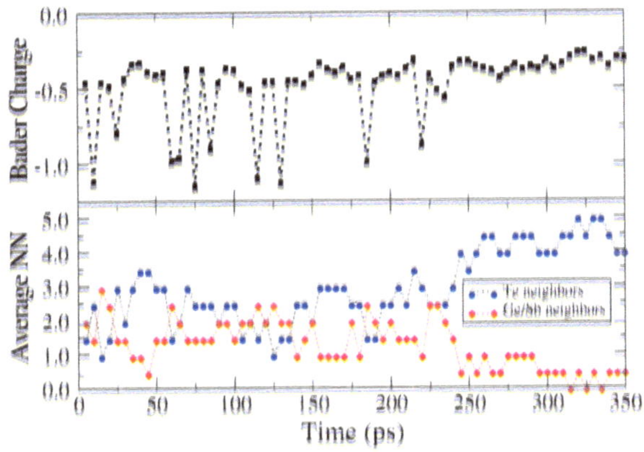

Fig. 19.7 Evolution of the average charge on Au during crystallization of Au–doped $Ge_2Sb_2Te_5$. The number of nearest neighbors in the first coordination shell is presented for comparison. Only Au charge evolution is presented for comparison purpose

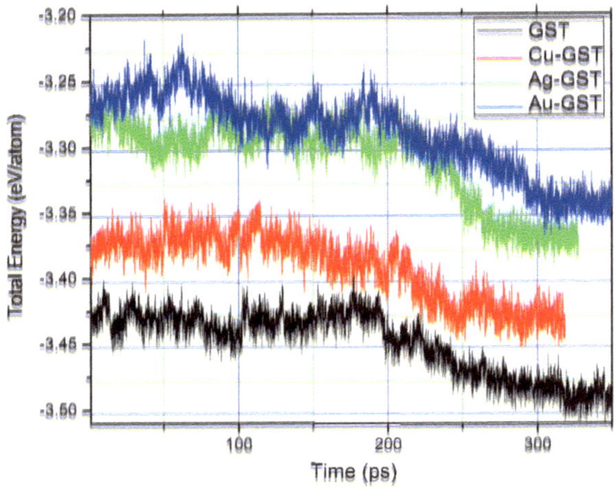

Fig. 19.8 Evolution of total energy in pure and doped $Ge_2Sb_2Te_5$ annealed at 650 K. The drop in the total energy represents the crystallization

period the energy goes to a minimum and remains more or less constant thereafter reaching Period III where the crystallization is complete. The duration of Periods I and II shows large fluctuation for three independent models for same structure resulting in significant uncertainty for estimation of crystallization time. These fluctuations are mainly associated with the fraction of wrong bonds present in the amorphous phase as well as the random orientations of the square rings due to thermal fluctuations. Table 19.6 presents the average fraction of wrong bonds and the crystallization time from three different models of varying composition. Although the crystallization periods show large variation, the time of about 400 ps was enough for each of the models to complete crystallization. While we may not be able to predict the exact modification on the crystallization speed at this level but it is confirmed that the crystallization speed of the host network is preserved.

Table 19.6 Computation of wrong bonds and the estimation of the crystallization time in doped and undoped $Ge_2Sb_2Te_5$

Dopant	Wrong bonds in fraction	Incubation period (ps)	Crystallization period (ps)	Total (ps)
–	0.13±0.03	150±50	120±40	270±90
Cu	0.14±0.03	180±30	120±45	300±75
Ag	0.15±0.04	130±40	100±20	230±60
Au	0.14±0.04	170±50	120±30	290±80

19.6 Conclusion

The phase change memory materials present an opportunity unique in materials theory: a chemically complex system of huge technical importance, and a rapidity of phase transformations that enables the direct simulation with the most accurate methods.

We used ab initio molecular dynamic simulations to study crystallization of Ge–Sb–Te phase change alloys in the presence of transitions metals (Cu, Ag, and Au) impurities. We were able to simulate the ultrafast phase transitions from amorphous to crystalline phase through the MD simulations. At ~2 % of impurities, we did not observe any significant dopant-induced structural change in Ge–Te or Sb–Te average bond lengths. Meanwhile the Bader charge analysis confirmed similar positive charge for Cu and Ag whereas negative charge for Au as a consequence Au was observed to have significant amount of Ge/Sb as neighbors. The estimation of dielectric constant in amorphous and crystalline phase implies that the optical contrast is preserved in Ag or Au doped $Ge_2Sb_2Te_5$ while Cu doped $Ge_2Sb_2Te_5$ did not look too promising in terms of optical contrast. We were also able to estimate the crystallization time for the transition metal doped $Ge_2Sb_2Te_5$ however with large variations which may be attributed to the presence of WB in the system.

Acknowledgments The authors would like to thank Professor S. R. Elliott for many helpful suggestions. This work was supported by the Army Research Office under Grant No W911NF1110358. This work was also supported in part by an allocation of computing time from the Ohio Supercomputer Center.

References

1. J. Hegedus, S.R. Elliott, Microscopic origin of the fast crystallization ability of Ge–Sb–Te phase-change memory materials. Nat. Mater **7**, 399–405 (2008)
2. B. Prasai, G. Chen, D.A. Drabold, Direct ab-initio molecular dynamic study of ultrafast phase change in Ag-alloyed $Ge_2Sb_2Te_5$. Appl. Phys. Lett. 102 (2013)
3. G. Kresse, J. Furthmüller, Efficiency of ab-initio total energy calculations for metals and semiconductors using a plane-wave basis set. Comput. Mater. Sci. **6**(1), 15–50 (1996)
4. D.A. Drabold, Topics in the theory of amorphous materials. Eur. Phys. J. B **68**, 1–21 (2009)
5. P.E. Blöchl, Projector augmented-wave method. Phys. Rev. B **50**, 17953–17979 (1994)
6. G. Kresse, D. Joubert, From ultrasoft pseudopotentials to the projector augmented-wave method. Phys. Rev. B **59**, 1758–1775 (1999)

7. J.P. Perdew, K. Burke, M. Ernzerhof, Generalized gradient approximation made simple. Phys. Rev. Lett. **77**, 3865–3868 (1996)
8. J.M. Skelton, S.R. Elliott, In silico optimisation of phase-change materials for digital memories: a survey of first-row transition-metal dopants for $Ge_2Sb_2Te_5$. J. Phys. Condens. Matter **25**, 205801 (2013)
9. N.E. Cusack, *in The Physics of Structurally Disordered Matter: An Introduction*, ed. by D.F. Brewer, D. Phil, pp. 30–36 (1987)
10. W. Tang, E. Sanville, G. Henkelman, A grid-based bader analysis algorithm without lattice bias. J. Phys. Condens. Matter **21**, 084204 (2009)
11. E. Sanville, S.D. Kenny, R. Smith, G. Henkelman, An improved grid-based algorithm for bader charge allocation. J. Comp. Chem. **28**, 899–908 (2007)
12. G. Henkelman, A. Arnaldsson, H. Jonsson, A fast and robust algorithm for bader decomposition of charge density. Comput. Mater. Sci. **36**, 254–360 (2006)
13. A. Seidl, A. Gorling, P. Vogl, J.A. Majewski, M. Levy, Generalized kohn-sham schemes and the band-gap problem. Phy. Rev. B. **53**, 3764–74 (1996)
14. K. Shportko, S. Kremers, M. Woda, D. Lencer, J. Robertson, M. Wuttig, Resonant bonding in crystalline phase-change materials. Nat. Mater. **7**, 653–658 (2008)
15. B. Prasai, M.E. Kordesch, D.A. Drabold, G. Chen, Atomistic origin of rapid crystallization of Ag-doped Ge-Sb-Te alloys: a joint experimental and theoretical study. Physica Status Solidi B **102**, 1–6 (2013)
16. T.H. Lee, S.R. Elliott, Ab-Initio computer simulation of the early stages of crystallization: application to $Ge_2Sb_2Te_5$ phase-change materials. Phys. Rev. Lett. **107**, 145702 (2011)

Index

C. Massobrio et al. (eds.), *Molecular Dynamics Simulations of Disordered Materials*, Springer Series in Materials Science 215,
DOI 10.1007/978-3-319-15675-0

S

T

U

V

W

X

Z

MIX
Papier aus verantwortungsvollen Quellen
Paper from responsible sources
FSC® C105338

If you have any concerns about our products,
you can contact us on
ProductSafety@springernature.com

In case Publisher is established outside the EU,
the EU authorized representative is:
Springer Nature Customer Service Center GmbH
Europaplatz 3, 69115 Heidelberg, Germany

Printed by Libri Plureos GmbH
in Hamburg, Germany